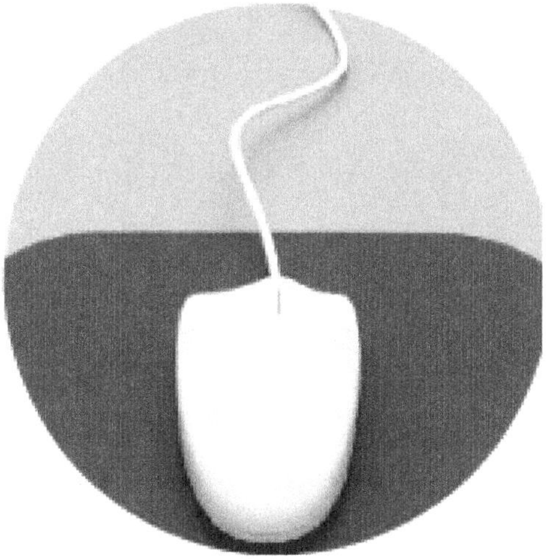

Free Online A

MW01709288

Plunkett's Apparel & Textiles Industry Almanac 2013

Your purchase includes access to Book Data and Exports online

As a book purchaser, you can register for free, 1-year, 1-seat online access to the latest data for your book's industry trends, statistics and company profiles. This includes tools to export company data. Simply send us this registration form, and we will send you a user name and password. In this manner, you will have access to our continual updates during the year. Certain restrictions apply.

____ YES, please register me for free online access. I am the actual, original purchaser. (Proof of purchase may be required.)

Customer Name ______________________________

Title______________________________

Organization ______________________________

Address ______________________________

City______________________State________Zip______________

Country (if other than USA) ______________________________

Phone__________________Fax ______________________

E-mail ______________________________

Return to: Plunkett Research®, Ltd.

Attn: Registration
P.O. Drawer 541737, Houston, TX 77254-1737 USA
713.932.0000 · Fax 713.932.7080 · www.plunkettresearch.com
customersupport@plunkettresearch.com

* Purchasers of used books are not eligible to register. Use of online access is subject to the terms of the end user license agreement.

PLUNKETT'S APPAREL & TEXTILES INDUSTRY ALMANAC 2013

The only comprehensive guide to apparel companies & trends

Jack W. Plunkett

Published by:
Plunkett Research®, Ltd., Houston, Texas
www.plunkettresearch.com

PLUNKETT'S APPAREL & TEXTILES INDUSTRY ALMANAC 2013

Editor and Publisher:
Jack W. Plunkett

Executive Editor and Database Manager:
Martha Burgher Plunkett

Senior Editors and Researchers:
Jeremy Faulk
Jill Steinberg

Editors, Researchers and Assistants:
Alex Grandstaff
Trung Le
Jillian Claire Lim
Erika Lyles
Isaac Snider
Suzanne Zarosky

Enterprise Accounts Manager:
Meghan Zarate

Information Technology Manager:
Seifelnaser Hamed

Special Thanks to:
International Services for the Acquisition of Agri-Biotech Applications
U.S. Bureau of the Census
U.S. Bureau of Labor Statistics
U.S. Department of Agriculture
U.S. Department of Commerce
U.S. International Trade Administration
World Trade Organization

Plunkett Research®, Ltd.
P. O. Drawer 541737, Houston, Texas 77254 USA
Phone: 713.932.0000 Fax: 713.932.7080
www.plunkettresearch.com

Plunkett Research®, Ltd.
P. O. Drawer 541737
Houston, Texas 77254-1737
Phone: 713.932.0000, Fax: 713.932.7080 www.plunkettresearch.com

ISBN13 # 978-1-60879-702-8 (eBook Edition # 978-1-60879-962-6)

Limited Warranty and Terms of Use:

Users' publications in static electronic format containing any portion of the content of this book (and/or the content of any related Plunkett Research, Ltd. online service to which you are granted access, hereinafter collectively referred to as the "Data") or Derived Data (that is, a set of data that is a derivation made by a User from the Data, resulting from the applications of formulas, analytics or any other method) may be resold by the User only for the purpose of providing third-party analysis within an established research platform under the following conditions: (However, Users may not extract or integrate any portion of the Data or Derived Data for any other purpose.)

a) Users may utilize the Data only as described herein. b) User may not export more than an insubstantial portion of the Data or Derived Data, c) Any Data exported by the User may only be distributed if the following conditions are met:

i) Data must be incorporated in added-value reports or presentations, either of which are part of the regular services offered by the User and not as stand-alone products.
ii) Data may not be used as part of a general mailing or included in external websites or other mass communication vehicles or formats, including, but not limited to, advertisements.
iii) Except as provided herein, Data may not be resold by User.

"Insubstantial Portions" shall mean an amount of the Data that (1) has no independent commercial value, (2) could not be used by User, its clients, Authorized Users and/or its agents as a substitute for the Data or any part of it, (3) is not separately marketed by the User, an affiliate of the User or any third-party source (either alone or with other data), and (4) is not retrieved by User, its clients, Authorized Users and/or its Agents via regularly scheduled, systematic batch jobs.

LIMITED WARRANTY; DISCLAIMER OF LIABILITY: **While Plunkett Research, Ltd. ("PRL") has made an effort to obtain the Data from sources deemed reliable, PRL makes no warranties, expressed or implied, regarding the Data contained herein. This book and its Data are provided to the End-User "AS IS" without warranty of any kind. No oral or written information or advice given by PRL, its employees, distributors or representatives will create a warranty or in any way increase the scope of this Limited Warranty, and the Customer or End-User may not rely on any such information or advice.** Customer Remedies: PRL's entire liability and your exclusive remedy shall be, at PRL's sole discretion, either (a) return of the price paid, if any, or (b) repair or replacement of a book that does not meet PRL's Limited Warranty and that is returned to PRL with sufficient evidence of or receipt for your original purchase.

NO OTHER WARRANTIES: TO THE MAXIMUM EXTENT PERMITTED BY APPLICABLE LAW, PRL AND ITS DISTRIBUTORS DISCLAIM ALL OTHER WARRANTIES AND CONDITIONS, EITHER EXPRESSED OR IMPLIED, INCLUDING, BUT NOT LIMITED TO, IMPLIED WARRANTIES OR CONDITIONS OF MERCHANTABILITY, FITNESS FOR A PARTICULAR PURPOSE, TITLE AND NON-INFRINGEMENT WITH REGARD TO THE BOOK AND ITS DATA, AND THE PROVISION OF OR FAILURE TO PROVIDE SUPPORT SERVICES. LIMITATION OF LIABILITY: TO THE MAXIMUM EXTENT PERMITTED BY APPLICABLE LAW, IN NO EVENT SHALL PRL BE LIABLE FOR ANY SPECIAL, INCIDENTAL OR CONSEQUENTIAL DAMAGES WHATSOEVER (INCLUDING, WITHOUT LIMITATION, DAMAGES FOR LOSS OF BUSINESS PROFITS, BUSINESS INTERRUPTION, ABILITY TO OBTAIN OR RETAIN EMPLOYMENT OR REMUNERATION, ABILITY TO PROFITABLY MAKE AN INVESTMENT, OR ANY OTHER PECUNIARY LOSS) ARISING OUT OF THE USE OF, OR RELIANCE UPON, THE BOOK OR DATA, OR THE INABILITY TO USE THIS DATA OR THE FAILURE OF PRL TO PROVIDE SUPPORT SERVICES, EVEN IF PRL HAS BEEN ADVISED OF THE POSSIBILITY OF SUCH DAMAGES. IN ANY CASE, PRL'S ENTIRE LIABILITY SHALL BE LIMITED TO THE AMOUNT ACTUALLY PAID BY YOU FOR THE BOOK.

PLUNKETT'S APPAREL & TEXTILES INDUSTRY ALMANAC 2013

CONTENTS

Continued on next page

Continued from previous page

INTRODUCTION

PLUNKETT'S APPAREL & TEXTILES INDUSTRY ALMANAC, the ninth edition of our guide to the apparel and textiles field, is designed as a general source for researchers of all types.

The data and areas of interest covered are intentionally broad, ranging from design trends to developments in logistics and technologies, to an in-depth look at the major firms (which we call "THE APPAREL & TEXTILES 350") within the many segments that make up the apparel and textiles industry.

This reference book is designed to be a general source for researchers. It is especially intended to assist with market research, strategic planning, employment searches, contact or prospect list creation and financial research, and as a data resource for executives and students of all types.

PLUNKETT'S APPAREL & TEXTILES INDUSTRY ALMANAC takes a rounded approach for the general reader. This book presents a complete overview of the apparel & textiles field (see "How To Use This Book"). For example, the globalization of apparel manufacturing is covered in exacting detail, as are the changes in clothing sizes and the impact of nanotextiles and smart fabrics.

THE APPAREL & TEXTILES 350 is our grouping of the biggest, most successful corporations in all segments of the apparel and textiles industry. Tens of thousands of pieces of information, gathered from a wide variety of sources, have been researched and are presented in a unique form that can be easily understood. This section includes thorough indexes to THE APPAREL & TEXTILES 350, by geography, industry, sales, brand names, subsidiary names and many other topics. (See Chapter 4.)

Especially helpful is the way in which PLUNKETT'S APPAREL & TEXTILES INDUSTRY ALMANAC enables readers who have no apparel or financial background to readily compare the strategies, financial records and growth plans of apparel and textiles companies and major industry groups. You'll see the mid-term financial record of each firm, along with the impact of earnings, sales and strategic plans on each company's potential to fuel growth and provide investment and employment opportunities.

No other source provides this book's easy-to-understand comparisons of growth, expenditures, technologies, corporations, research and many other items of great importance. The information within is crucial to people of all types who may be studying this, one of the most exciting industries in the world today.

By scanning the data groups and the unique indexes, you can find the best information to fit your personal research needs. The major companies are profiled using several different groups of specific criteria. Which firms are the biggest employers? Which companies earn the most profits? These things and much more are easy to find.

In addition to individual company profiles, an analysis of apparel and textiles technologies and trends is provided. This book's job is to help you sort through clear summaries of today's technologies and trends in a quick and effective manner.

Whatever your purpose for researching the apparel and textiles field, you'll find this book to be a valuable guide. Nonetheless, as is true with all resources, this volume has limitations that the reader should be aware of:

- Financial data and other corporate information can change quickly. A book of this type can be no more current than the data that was available as of the time of editing. Consequently, the financial picture, management and ownership of the firm(s) you are studying may have changed since the date of this book. For example, this almanac includes the most up-to-date sales figures and profits available to the editors as of early 2013. That means that we have typically used corporate financial data as of mid-2012.

- Corporate mergers, acquisitions and downsizing are occurring at a very rapid rate. Such events may have created significant change, subsequent to the publishing of this book, within a company you are studying.

- Some of the companies in THE APPAREL & TEXTILES 350 are so large in scope and in variety of business endeavors conducted within a parent organization, that we have been unable to completely list all subsidiaries, affiliations, divisions and activities within a firm's corporate structure.

- This volume is intended to be a general guide to a rapidly changing industry. That means that researchers should look to this book for an overview and, when conducting in-depth research, should contact the specific corporations or industry associations in question for the very latest changes and data. Where possible, we have listed contact names, toll-free telephone numbers and Internet site addresses for the companies, government agencies and industry associations involved so that the reader may get further details without unnecessary delay.

- Tables of industry data and statistics used in this book include the latest numbers available at the time of printing, generally through mid-2012. In a few cases, the only complete data available was for earlier years.

- We have used exhaustive efforts to locate and fairly present accurate and complete data. However, when using this book or any other source for business and industry information, the reader should use caution and diligence by conducting further research where it seems appropriate. We wish you success in your endeavors, and we trust that your experience with this book will be both satisfactory and productive.

Jack W. Plunkett
Houston, Texas
May 2013

HOW TO USE THIS BOOK

The two primary sections of this book are devoted first to the apparel and textiles industry as a whole and then to the "Individual Data Listings" for THE APPAREL & TEXTILES 350. If time permits, you should begin your research in the front chapters of this book. Also, you will find lengthy indexes in Chapter 4 and in the back of the book.

Video Tip
For our brief video introduction to the apparel and textiles industry, see www.plunkettresearch.com/video/apparel.

THE APPAREL & TEXTILES INDUSTRY

Glossary: A short list of apparel and textiles industry terms.

Chapter 1: Major Trends Affecting the Apparel & Textiles Industry. This chapter presents an encapsulated view of the major trends that are creating rapid changes in the apparel industry today.

Chapter 2: Apparel & Textiles Industry Statistics. This chapter contains an extensive set of industry statistics.

Chapter 3: Important Apparel & Textiles Industry Contacts – Addresses, Telephone Numbers and Internet Sites. This chapter covers contacts for important government agencies and trade groups. Included are numerous important Internet sites.

THE APPAREL & TEXTILES 350

Chapter 4: THE APPAREL & TEXTILES 350: Who They Are and How They Were Chosen. The companies compared in this book (the actual count is 359) were carefully selected from the apparel and textiles industry, largely in the United States. 121 of the firms are based outside the U.S. For a complete description, see THE APPAREL & TEXTILES 350 indexes in this chapter.

Individual Data Listings:

Look at one of the companies in THE APPAREL & TEXTILES 350's Individual Data Listings. You'll find the following information fields:

Company Name:

The company profiles are in alphabetical order by company name. If you don't find the company you are seeking, it may be a subsidiary or division of one of the firms covered in this book. Try looking it up in the Index by Subsidiaries, Brand Names and Selected Affiliations in the back of the book.

Industry Code:

Industry Group Code: An NAIC 2012 code used to group companies within like segments. (See Chapter 4 for a list of codes.)

Types of Business:

A listing of the primary types of business specialties conducted by the firm.

Brands/Divisions/Affiliations:

Major brand names, operating divisions or subsidiaries of the firm, as well as major corporate affiliations—such as another firm that owns a significant portion of the company's stock. A complete Index by Subsidiaries, Brand Names and Selected Affiliations is in the back of the book.

Contacts:

The names and titles up to 27 top officers of the company are listed, including human resources contacts.

Growth Plans/ Special Features:

Listed here are observations regarding the firm's strategy, hiring plans, plans for growth and product development, along with general information regarding a company's business and prospects.

Financial Data:

Revenue (2012 or the latest fiscal year available to the editors, plus up to five previous years): This figure represents consolidated worldwide sales from all operations. These numbers may be estimates.

R&D Expense (2012 or the latest fiscal year available to the editors, plus up to five previous years): This figure represents expenses associated with the research and development of a company's goods or services. These numbers may be estimates.

Operating Income (2012 or the latest fiscal year available to the editors, plus up to five previous years): This figure represents the amount of profit realized from annual operations after deducting operating expenses including costs of goods sold, wages and depreciation. These numbers may be estimates.

Operating Margin % (2012 or the latest fiscal year available to the editors, plus up to five previous years): This figure is a ratio derived by dividing operating income by net revenues. It is a measurement of a firm's pricing strategy and operating efficiency. These numbers may be estimates.

SGA Expense (2012 or the latest fiscal year available to the editors, plus up to five previous years): This figure represents the sum of selling, general and administrative expenses of a company, including costs such as warranty, advertising, interest, personnel, utilities, office space rent, etc. These numbers may be estimates.

Net Income (2012 or the latest fiscal year available to the editors, plus up to five previous years): This figure represents consolidated, after-tax net profit from all operations. These numbers may be estimates.

Operating Cash Flow (2012 or the latest fiscal year available to the editors, plus up to five previous years): This figure is a measure of the amount of cash generated by a firm's normal business operations. It is calculated as net income before depreciation and after income taxes, adjusted for working capital. It is a prime indicator of a company's ability to generate enough cash to pay its bills. These numbers may be estimates.

Capital Expenditure (2012 or the latest fiscal year available to the editors, plus up to five previous years): This figure represents funds used for investment in or improvement of physical assets such as offices, equipment or factories and the purchase or creation of new facilities and/or equipment. These numbers may be estimates.

EBITDA (2012 or the latest fiscal year available to the editors, plus up to five previous years): EBITDA is an acronym for earnings before interest, taxes, depreciation and amortization. It represents a company's financial performance calculated as revenue minus expenses (excluding taxes, amortization, depreciation and interest), and is a prime indicator of profitability. These numbers may be estimates.

Return on Assets % (2012 or the latest fiscal year available to the editors, plus up to five previous years): This figure is an indicator of the profitability of a company relative to its total assets. It is calculated by dividing annual net earnings by total assets. These numbers may be estimates.

Return on Equity % (2012 or the latest fiscal year available to the editors, plus up to five previous years): This figure is a measurement of net income as a percentage of shareholders' equity. It is also called the rate of return on the ownership interest. It is a vital indicator of the quality of a company's operations. These numbers may be estimates.

Debt to Equity (2012 or the latest fiscal year available to the editors, plus up to five previous years): A ratio of the company's long-term debt to its shareholders' equity. This is an indicator of the overall financial leverage of the firm. These numbers may be estimates.

Address:
The firm's full headquarters address, the headquarters telephone, plus toll-free and fax numbers where available. Also provided is the Internet address.

Stock Ticker, Exchange: When available, the unique stock market symbol used to identify this firm's common stock for trading and tracking purposes is indicated. Where appropriate, this field may contain "private" or "subsidiary" rather than a ticker symbol.

Total Number of Employees: The approximate total number of employees, worldwide, as of the end of 2012 (or the latest data available to the editors).

Parent Company: If the firm is a subsidiary, its parent company is listed.

Salaries/Bonuses:

(The following descriptions generally apply to U.S. employers only.)

Highest Executive Salary: The highest executive salary paid, typically a 2012 amount (or the latest year available to the editors) and typically paid to the Chief Executive Officer.

Highest Executive Bonus: The apparent bonus, if any, paid to the above person.

Second Highest Executive Salary: The next-highest executive salary paid, typically a 2012 amount (or the latest year available to the editors) and typically paid to the President or Chief Operating Officer.

Second Highest Executive Bonus: The apparent bonus, if any, paid to the above person.

Other Thoughts:

Estimated Female Officers or Directors: It is difficult to obtain this information on an exact basis, and employers generally do not disclose the data in a public way. However, we have indicated what our best efforts reveal to be the apparent number of women who either are in the posts of corporate officers or sit on the board of directors. There is a wide variance from company to company.

Hot Spot for Advancement for Women/Minorities: A "Y" in appropriate fields indicates "Yes." These are firms that appear either to have posted a substantial number of women and/or minorities to high posts or that appear to have a good record of going out of their way to recruit, train, promote and retain women or minorities. (See the Index of Hot Spots For Women and Minorities in the back of the book.) This information may change frequently and can be difficult to obtain and verify. Consequently, the reader should use caution and conduct further investigation where appropriate.

Glossary: A short list of apparel and textile industry terms.

Chapter 1

MAJOR TRENDS AFFECTING THE APPAREL & TEXTILES INDUSTRY

Major Trends Affecting the Apparel & Textiles Industry:

1) Introduction to the Apparel & Textiles Industry
2) Globalization: Wages Rise in China/Vietnam, Cambodia and Other Countries Gain Market Share
3) Western Apparel Firms Enter the Chinese Market
4) Working Conditions and Workers' Rights Become Major Considerations as Work Moves from China to Lower-Cost Nations
5) High-Tech, Nanotech and Smart Fabrics Proliferate
6) Wearable Electronics and Electronics Incorporated into Apparel Begin to Enter the Marketplace
7) Exercise Apparel Sales Rise
8) Athletic Footwear Sales Soar, Draws Big Names from Athletes to Designers
9) 3PL Supply Chain Management Evolves to Serve the Global Market
10) The Vast Majority of Shoes Sold in the U.S. Are Made Elsewhere
11) Bricks, Clicks and Catalogs Create Synergies While Online Sales Growth Surges
12) Alternative Sizing Is Big, But Plus-Size Sales Are Slowing
13) Discount Clothing Retailers See Promise in Designer Lines
14) Luxury Lines Bounce Back
15) Luxury Goes Online
16) Self Service Apparel Fitting Technologies Grow in Stores and Online
17) Fast Fashion: Designers and Retailers Speed Up
18) Apparel Manufacturing Gains Ground in the U.S.
19) Apparel Goes Green

1) Introduction to the Apparel & Textiles Industry

Video Tip
For our brief video introduction to the Apparel and Textiles industry, see www.plunkettresearch.com/video/apparel.

While Asia, particularly China, has enjoyed a dominant position in shoes, apparel and household textiles manufacturing for several years, makers of these items located in developed nations such as the U.S. and Canada have suffered a long period of decline. For example, over 98% of the shoes sold in

America each year are imports, and the majority of these imports come from Asia. To consumers in Europe and North America, this growing reliance on Asia as a low-cost producer has meant very low retail prices for goods of reasonable quality.

However, recent increases in the value of the Chinese currency, combined with rapidly rising labor costs, have put Chinese manufacturers in a much less competitive position. Competition from very low-cost nations in Africa as well as Vietnam, Sri Lanka, Mauritius, Malaysia, Cambodia, Bangladesh, Pakistan, the Philippines and elsewhere is intense, and a large portion of apparel manufacturing formerly done in China is moving to these areas at a rapid pace. For example, Vietnam's apparel exports to the U.S. jumped to $7.1 billion in 2012 from $6.6 billion a year earlier. China's apparel exports to the U.S. were $29.0 billion in 2012, down slightly from 2011's $29.3 billion.

While China continues to have a robust apparel manufacturing industry, it is moving up the industrial chain by fostering manufacturing that requires greater skills, better technology and more investment in advanced equipment. Such segments that are rapidly evolving in China include InfoTech, automobiles, trains, aerospace, medical equipment and telecommunications gear.

China's textile and clothing exports soared from $16.89 billion in 1990 to $206.74 billion in 2010, according to the World Trade Organization. For 2011, China's total was $253.2 billion, nearly a 23% increase. India is a distant second in this category, at $29.4 billion in 2011 (up from $4.71 billion in 1990). Other nations in the top ten for global apparel and textiles exports in 2011 included Turkey ($24.7 billion), Bangladesh ($21.5 billion), the United States ($20.0 billion by WTO estimates), Vietnam ($17.0 billion), the Republic of Korea ($14.2 billion), Pakistan ($13.7 billion) and Indonesia ($12.8 billion). The EU, in broad terms counting 27 nations, enjoyed exports of apparel and textiles totaling $193.0 billion in 2011.

As wages rise in China, workers are also calling for better working conditions. Workers in many parts of the world are beginning to demand shorter hours and safer working environments. The collapse of a multi-story apparel factory in Bangladesh in 2013, killing more than 800 people, will bring a new focus on workers' rights and working conditions in the apparel manufacturing sector.

The changes that will result from China's rapidly increasing wage rates should not be overlooked. The apparel and textile manufacturing industry has historically been one that has few barriers to entry, little capital investment needed and a high level of low-skilled labor involved. When China's wages were extremely low, the nation had a clear advantage in this industry, particularly in light of its tendency to cluster factories near shipping ports and logistics centers so that goods can be sent to customers with minimal delay. Today however, wages are climbing very rapidly as China's urban workers have many options in terms of places to work. Demand for workers is high, and workers are able to demand much higher pay than those in less-developed nations.

Many Chinese firms are massive in size, with tightly integrated units providing rapid design, manufacturing and logistics. One of the most interesting organizations is the Esquel Group of Companies. Esquel (www.esquel.com) is one of the world's largest producers of cotton shirts, with an output of more than 60 million garments each year. The firm's vertically integrated operation starts in China where it oversees nearly 4,700 acres of cotton farms, which supply Esquel's spinning, dyeing and knitting facilities that produce 90 million yards of high-quality cotton fabric annually.

Esquel has manufactured clothing on behalf of brands that include Banana Republic, Tommy Hilfiger, Hugo Boss, Brooks Brothers, Abercrombie & Fitch, Nike, Nordstrom and Lands' End, as well as private-label items for retailers such as Marks & Spencer. The Esquel companies transform their fabric into premium men's and women's wear at plants in China, Hong Kong, Malaysia, Mauritius, Sri Lanka and Vietnam. The company also sells its products through its proprietary brand, PYE, which markets high-end cotton apparel in China with a flagship luxury retail store in Beijing. Esquel maintains group sales offices in select locations around the world, including U.S. offices in New York City and regional locations convenient to key accounts. Unlike many textile companies, the firm has an expressed strategy of in-company resource development, seeking to ensure high standards of quality and consistency across all of its product lines.

Meanwhile, some manufacturers outside of Asia have become very efficient and quick to react to consumer trends. The most notable company in this category is Spain's Inditex. The firm is famous for its lean inventory and fast-fashion strategy. It can get a new item of clothing from its 300-person design team, through its manufacturing plants and into its more than 4,600 stores in as little as two weeks. The company operates several manufacturing facilities in

Spain for its clothing. Many of its outside suppliers are located in Spain and elsewhere in Europe. The firm has been opening large numbers of new retail stores under the Zara, Pull & Bear, Massimo Dutti and Oyosho brands.

One of the more interesting recent developments in apparel manufacturing has been to reshore, which is the return of some of the business to plants based in the country where most sales are made. Some designers and retailers find that their orders are not of sufficient size to interest major offshore plants, or their need for fast delivery makes it impossible to use overseas manufacturers. There is promising growth among many American manufacturers. Likewise, the need for extremely high quality is keeping some manufacturing of luxury apparel and accessories at home by Italian designers. As wages continue to rise in emerging nations, and shipping costs continue to be subject to spikes due to high fuel prices, some portion of clothing manufacturing will continue to reshore. Growing productivity from better manufacturing technology will add to this trend.

Apparel and accessory stores have generally been enjoying a rebound in sales around the world. Meanwhile, the luxury apparel, shoe and accessories market has enjoyed exceptional growth in hot economies such as Brazil and China. Leading brands have been opening large numbers of new stores in these markets. For example, LVMH is focusing heavily on new stores in China for its luxury brands. However, as of mid-2013, China's growth was slowing, and the entire nation was going through a new focus on reducing corruption among public officials as well as conspicuous consumption in general. This has been hard on the luxury goods market, to the extent that Armani closed a flagship store on Shanghai's famous Bund.

During 2012, America exported $22.6 billion in textiles and apparel (up slightly from $22.3 billion) and imported $100.93 billion (down from $102.0 billion). These numbers are from the U.S. International Trade Administration.

In the European Union (EU), the textile and apparel sector is quite large, particularly in nations that enjoy lower operating costs, such as the Baltic States and Eastern European States. Nearby, the textile and apparel industry remains a major part of the economy of Turkey. South America, Central America and Africa also play minor roles in world apparel trade.

Globally, the World Trade Organization (WTO) reports that apparel and textile exports grew 17% in 2011 (compared to 14.6% in 2010). Textile exports totaled $294 billion and apparel exports totaled $412 billion. For the decade ending in 2010, compound annual growth averaged 5.5% worldwide. Among the fastest growing nations in that period were Vietnam, China, Bangladesh, Turkey and India.

Over the past several years, manufacturers of basic synthetic textiles, such as polyester fabrics, have been dealing with a global manufacturing glut. Synthetic textile manufacturing has been dominated by the largest global chemical firms, but many of them have exited the business by spinning off or selling their holdings.

Trade agreements among the U.S. and its trading partners attempt to foster employment in certain parts of the world (such as low-income areas in the Caribbean) and allow U.S. consumers fair access to reasonably priced goods while providing some sort of relief to U.S.-based businesses. Because trade agreements will never satisfy all parties concerned, they tend to lead to controversy and much critical discussion. On the retail end, consumers of apparel have long enjoyed wide selections and moderate prices in North America, Asia, Europe and elsewhere.

During 2012, retail sales at clothing and accessories stores in the U.S. totaled $239.2 billion (up from $226.5 billion the previous year), according to the Bureau of the Census. In addition, Plunkett Research estimates that 80% ($148.5 billion) of sales at department stores, 60% ($25.5 billion) of sales at sporting goods stores and 35% ($41.9 billion) of sales at discount department stores were for clothing and accessories in 2012. Assuming that about 30% ($67.2 billion) of the goods sold via e-commerce in America are apparel, shoes and accessories, this would put the total retail and online market in America at $522.3 billion.

Apparel retailing has always been a tough, highly competitive business, and many chains rise dramatically and then fail. Retail fashion merchandising is a vast challenge (witness the recent ups and downs of The Gap). Just-in-time inventory, driven by highly computerized supply chain management systems, is now an immense asset to major retailers. Nonetheless, price pressure from major discounters like Wal-Mart, Target and Kohl's can keep profit margins thin at stores that sell moderately priced apparel. Some of the most successful retail chains are those that focus on niche markets with special tastes and needs, such as Chico's FAS, which caters to 35- to 60-year-old women who want flattering fashions that suit their figures, and Abercrombie & Fitch, which focuses on the 18- to 22-year old market with great success.

So-called "fast fashion" stores that sell the latest designs at very low prices have been enjoying soaring growth in the world's largest markets. These stores include H&M, headquartered in Europe, and Uniqlo, from Japan.

Speaking of figures, the well-documented expanding girth of consumers in many nations is placing new challenges upon fashion merchandisers as overweight people of all ages, tastes and income brackets require clothes in larger sizes. Designers and merchandisers face the task of developing and presenting larger clothes in a flattering light.

While Americans (as well as residents of many other countries from Mexico to China) have been putting on weight, they have also developed a keen interest in sports apparel and workout gear to wear at the gym and in other leisure activities. This is one of the fastest-growing product categories in the apparel and shoe sector. Over 50 million Americans have some sort of gym membership, and they need appropriate clothing to wear while they work out. Plunkett Research estimates the active sports apparel segment of the U.S. retail clothing market at approximately $58 billion for 2012.

Demographic changes will offer immense opportunities to U.S. fashion merchandisers. To begin with, many of the nation's 75 million surviving Baby Boomers (post World War II babies born from 1946 through 1964) are beginning to enter the 65+ age category. As more and more of these people become seniors, their tastes and needs will bring great revenues to savvy apparel sellers who learn how to cater to this market. Meanwhile, the rapid growth of ethnic consumer groups in America, Hispanics in particular, will offer superb marketing and product development opportunities.

Department stores have changed their business models drastically. While they were historically sellers of virtually every type of product, arranged by category in well-defined spaces within giant buildings (thus the use of the word "department" to describe them), most department stores in America today are primarily apparel and accessories stores. When consumers shop at stores like Nordstrom, Neiman Marcus or Dillard's, they find floor after floor of shoes, clothing, accessories and cosmetics. This change has created problems within the department store industry, as managers, faced with intense competition, developed the habit of continuously discounting apparel in sale events, consequently putting pressure on profitability. Consumers have been trained to wait for items to go on sale before they make purchases, thus lowering profit margins at stores. Nonetheless, department stores remain major forces in apparel retailing today.

In many Asian nations including China, however, a department store of today typically looks more like one did in America 100 years ago. Department stores in China are known to sell everything from medicinal herbs to clothing to furniture and even automobiles. The world's largest department stores by far are found in Asia.

Another sweeping change in apparel retailing is the rising success of e-commerce. Retail apparel chains are employing bricks and clicks together successfully. That is, they create synergies between very active web sites and their retail stores. Other firms, such as Bluefly.com, sell apparel through the Internet only, often at everyday discount prices. Catalog retailers continue to do reasonably well, particularly if they operate well-designed web sites to supplement their printed catalogs. Meanwhile, a growing number of fashion companies, such as Worth and The Carlisle Collection, are enjoying success selling women's fashions in the home via independent reps—somewhat like the success of similar companies that sell cosmetics.

2) Globalization: Wages Rise in China/Vietnam, Cambodia and Other Countries Gain Market Share

The apparel and textiles industry has long been ruled by complex import and export agreements that limit the amounts of particular garments (such as t-shirts, pants and sweaters) and textiles (such as yarns and fabrics) that may be exported to specific markets around the world. In an effort to safeguard domestic production, the United States, Canada and several additional countries now part of the European Union established the Multifiber Agreement (MFA) in 1973. Under the provisions of the agreement, a quota system was put in place that established the maximum numbers of products produced in developing countries that could be legally exported to MFA member countries. These amounts differed from country to country and were based on historic purchasing patterns.

Internet Research Tip:

For in-depth explanation of the Multifiber Agreement and subsequent treaties, see the World Trade Organization's "Textiles Monitoring Body" page at www.wto.org/english/tratop_e/texti_e/texintro_e.htm

By the mid 1990s, many factors, including global demand for cheaper goods and political pressures for free trade, brought about the World Trade Organization's (WTO) Agreement on Textiles and Clothing. This agreement defined a 10-year phase-out of the MFA quotas, finalizing on January 1, 2005, after which all WTO member countries were to have virtually unrestricted access to U.S., Canadian and European markets. The abolition of these quotas had a marked effect, particularly in apparel and textiles exported from China. Chinese apparel manufacturing grew at a very rapid rate for years. This apparel is eventually sold under thousands of leading brand names and store names worldwide, although it was manufactured by firms in China that are completely unknown to consumers. For many years, there was remarkable growth in Chinese exports. For example, the U.S. imported $40.6 billion in apparel and textiles alone from China in 2012, compared to $40.7 billion in 2011 and $39.0 billion in 2010, according to the U.S. Department of Commerce.

As of 2013, a new agreement was under consideration that would significantly impact global trade. The Trans-Pacific Partnership (TPP) is a free trade agreement under negotiation between 11 countries: Australia, Brunei Darussalam, Canada, Chile, Malaysia, Mexico, New Zealand, Peru, Singapore, Vietnam and the United States. If agreed upon, the pact could greatly reduce tariffs between member countries (or eliminate them altogether in some cases).

Some Chinese companies have been laying off workers and moving labor-intensive manufacturing to countries such as Bangladesh, Nigeria and Cambodia, the very countries that world economists predicted would suffer when some global quotas were lifted in early 2005. Chinese firms were finding lower operating costs, particularly for labor, in those nations than they enjoyed at home in China.

The Federation of Hong Kong Industries reported that Chinese factory wages rose 18% to 20% annually starting in 2009. The Economist Intelligence Unit reported Chinese average nominal wage per hour of just over $3 in 2012, and projected hourly labor costs to reach $4 by 2014. This means that many workers will be earning $10,000 to $12,000 yearly with overtime, in addition to the factory-provided housing and meals that many of them enjoy.

As wages increase in China, manufacturing in lower cost countries is on the rise. Cambodia, especially, is reaping the rewards of cheap labor. The *Phnom Penh Post* reported that exports of garments and textiles in Cambodia rose from $4.24 billion in 2011 to $4.61 billion in 2012 (although this growth was lower than the country's overall 10.2% increase in exports, due to the European economic crisis and numerous strikes by Cambodian garment workers in 2012). Meanwhile, Vietnam's textiles and apparel exports to the U.S. rose from $7.2 billion in 2011 to $7.7 billion in 2012.

China has been forced to improve the quality of its goods in order to stay competitive. Since many developing nations are capable of providing low-cost, low-quality apparel manufacturing, China needed to move ahead in terms of quality in order to maintain a competitive advantage. Manufacturing trends in China evolved somewhat like they did in post-World War II Japan. In the 1950s and 1960s, Japan was seen as a very low cost producer of goods that were generally poor in quality. However, as Japan's post-war industrial giants gained engineering prowess and a greater understanding of global consumer markets, Japanese quality (and consumers' awareness of that quality) began to rise steadily. In fact, in many respects Japan is now a world leader in producing goods of the highest quality. Watch for the Chinese manufacture of all manner of goods to evolve from among the world's cheapest to become eventually some of the world's best. This is already happening in telecommunications equipment, aerospace, biotech, automobiles and certain other goods.

India has been reaping the expected rewards of free markets in apparel and textiles. After the expiration of the MFA agreement, Indian apparel and textile exports reached $26.8 billion in the fiscal year ending March 31, 2011. Indian exports reached $33.31 billion the following year. However, due to the economic crisis in Europe and a slower-than-expected rebound of the U.S. economy, figures for January through April 2013 were down by 7.5%.

The impact of globalization on the apparel and textiles industry in the U.S. has been devastating. Especially hard hit are the southern states of North and South Carolina, Georgia, Virginia and Alabama, where apparel and textiles were historically a major economic mainstay. According to the Department of Labor, the total number of workers in U.S. textile mills fell from 332,900 in 2001 to 118,000 in 2012.

(A large number of these displaced workers became employed at automobile manufacturing plants recently constructed by foreign firms such as Toyota and BMW.)

3) Western Apparel Firms Enter the Chinese Market

While China's exports grew significantly for years, many Western apparel firms are opening retail locations in China, hoping to tap into the country's growing upper and middle class. Highly regarded foreign brands, luxury items and logos are very much in demand from newly affluent Chinese consumers.

Hennes & Mauritz AB, which owns the H&M chain of clothing stores, planned 350 new stores worldwide for 2013, with its primary focus being China and the U.S. Coach, Inc. has added large numbers of stores in China. Nike is a major presence in the Chinese retail market. Towering multi-story ads featuring American NBA athletes are prominently displayed in downtown Shanghai and Beijing. Gap, Inc. opened its first Chinese store in late 2010. Many major apparel retailers look upon China as an important long-term market. In particular, luxury fashion and accessories firms have been enjoying brisk sales in both mainland China and Hong Kong, and many chains have been expanding there rapidly. Vast new malls have been opening in China on a regular basis, giving apparel store operators plenty of opportunities to create new stores. Many of these malls are within walking distance of new luxury hotels and major office complexes.

However, as of mid-2013, China's growth was slowing, and the entire nation was going through a new focus on reducing corruption among public officials as well as conspicuous consumption in general. This has been hard on the luxury goods market, to the extent that Armani closed a flagship store on Shanghai's famous Bund.

One issue that Western retailers must face is a general difference in Chinese body shapes and sizes. When compared to Westerners, most Chinese people have narrower hips which pose a problem when it comes to fit. Retailers could, of course, design separate lines for the Chinese market. However, doing so would incur great expense and complicate the production process. The booming market in China for goods of all kinds poses an economic potential of huge proportions. Global retailers cannot afford to overlook it.

4) Working Conditions and Workers' Rights Become Major Considerations as Work Moves from China to Lower-Cost Nations

A heart-breaking tragedy in very poor nation of Bangladesh may prove to be a game-changer for the apparel manufacturing industry. In April 2013, a multi-story apparel factory building collapsed suddenly, trapping workers and resulting in a death toll that exceeded 800. This was only one of multiple accidents, tragic fires and other mishaps in apparel factories in recent months, in nations including China and India. In fact, the industry suffered a devastating accident only a few months prior to the April 2013 debacle. In September 2012, more than 300 workers died in two factory fires in Pakistan. In Cambodia, as many as 2,000 factory workers passed out in a short period of time as a result of poor nutrition, heat and/or poor ventilation.

At the same time, workers in many locations are demonstrating, demanding higher wages and better working conditions. The bottom line is that apparel brands simply cannot afford to be associated with unsafe facilities and factory deaths, despite their unending need to keep costs low so that they can be competitive in the challenging global marketplace. We are no doubt about to see a significant tightening of demands that customers place on factories in terms of factory safety, child labor, forced overtime and other workplace issues. The end result will be the closing of some poorly run firms, and higher costs at others. These costs will eventually be passed along to consumers around the world.

5) High-Tech, Nanotech and Smart Fabrics Proliferate

With the advent of Rayon in the 1920s, high-tech fabrics were born. Rayon, introduced by DuPont as a "continuous filament viscose fiber," is durable and silky-soft, and it costs much less than organic fabrics made of silk. Today's high-tech fabrics are either treated with chemicals, polymers or combinations thereof, or specially engineered for durability, stain-proofing, wrinkle-resistance or weather protection. In addition to these wonders, scientists are producing smart fabrics, which are engineered on a molecular level for wearability advantages as well as an astounding array of potential protective and clinical abilities. Through the use of special fabrics capable of conducting data flow, telecommunications and computing abilities may also be enabled in apparel.

The abilities of high-tech materials to keep athletes dry and comfortable have sparked a multi-billion dollar market for high-performance outdoor apparel. An excellent example of the power of this niche is the wildly successful Gore-Tex fabric produced by W.L. Gore & Associates, Inc. The fabric is a fluoropolymer, a manufactured fiber that forms a barrier to wind and water but still allows air to pass through it. Body heat and moisture can evaporate through Gore-Tex even while it protects from wind, rain, sleet or snow. It's used in everything from jackets to shoes to backpacks, and it generates impressive sales since apparel treated with Gore-Tex often costs three times as much as non-treated pieces. W.L. Gore's annual revenues are about $3 billion, and it employs more than 9,500 people.

A surprising material found in some apparel is silver. Long known to fight bacteria and odors, silver has been used historically to store foodstuffs. However, many manufacturers are finding new ways to use the metal. In 2006, Samsung Electronics launched a household clothes washer that utilizes silver ions to sanitize laundry without using hot water. Today, Adidas, Brooks Sports, Victoria's Secret and Polartec are using a silver-coated nylon fiber called X-Static (which is produced by Noble Biomaterials, Inc.). The fibers, which are used in shirts, pants, caps and socks, protect against odor and promote thermal regulation, since silver is a natural energy conductor. A company called NanoHorizons, Inc., www.nanohorizons.com, furthered the cause of silver for use in apparel by developing an engineering process that disperses silver pellets uniformly through the materials used for textile applications. The process, which is marketed under the SmartSilver brand, prolongs the silver's presence in fabric even through repeated washings.

Yet another high-tech sports apparel example is Textronics, Inc.'s (www.textronicsinc.com) use of electro textiles. The fabric has circuitry, sensors and a functional component woven in that render the material capable of enabling electronic devices such as cell phones or music players, providing heat for the wearer or sensing external changes such as shifts in temperature. Textronics makes a sports bra under the NuMetrex brand which senses the wearer's heart rate and communicates that data to a wrist monitor. Retailing for about $99 (including a combination wrist monitor/watch), the bras free female athletes from wearing separate heart rate monitor straps and sensors.

Swimwear maker Speedo International launched its LZR Racer suit in 2008. The $550 LZR Racer is a seamless suit made of fabric tested in NASA wind tunnels for its ability to reduce surface friction. In addition, the suit covers the torso, gluteal and upper hamstring muscles with panels that reduce muscle and skin vibration. The result was a staggering number of new world records set by swimmers wearing the new suits. Speedo was inundated with orders, despite the suit's high price tag. NCAA swim teams were spending as much as $20,000 for a season's supply. Rival manufacturers are producing their own versions of the suit, including Tyr's Racer Light, which retails for $320 and BlueSeventy's Nero, which retails for $395. However, in mid-2009, FINA, the international governing body of swimming, banned the suits from official competition. The ban left Speedo with approximately 18,000 units with no market among competitive swimmers. The company donated them to fashion design students at three schools in the U.K., where they were recycled into fashions for men and women.

Manufacturers are hoping to take high-tech fabrics to everyday clothing by applying new technology to fabrics used in suits, sweaters, ties and more. They believe that consumers will buy into clothing that can't be crushed when packed and is stain-proof as well as waterproof. Retailers such as Eddie Bauer are already stocking casual cotton pants made of nanotextiles—materials constructed at the molecular level of particles so tiny that moisture cannot penetrate them. Even spills of coffee or red wine will not stain these fabrics.

Another nanotech company, Schoeller Textil of Switzerland, produces a nanofinish called NanoSphere which repels liquids without affecting the feel and drape of the underlying fabric. Clothing manufacturers such as Beyond of Eugene, Oregon, and Contourwear of Alameda, California, use NanoSphere to coat everything from pants to jackets. The coating is also used for a line of luggage under the Victorinox brand, made by a division of TRG Group.

High-tech and nanotech fabrics are increasingly used in uniforms where durability, comfort and stain resistance are key selling points. Demand is on the rise, according to the North American Association of Uniform Manufacturers & Distributors, which reports that as of 2010, one in five workers are required to wear a uniform (about double the number in 2000 due to growth among service jobs). Cintas Corp., a uniform manufacturer, is using a process in which water bottles are recycled into tiny flakes woven into yarn. The yarn is used in tailored suits that are washable for hotel, transportation and food service workers. Cintas also manufactures sport shirts made of recycled polyester and charcoal to eliminate odors.

Nike launched a revolutionary new running shoe called the Flyknit in February 2013. The 5.6-ounce high tech shoe is made from synthetic yarn, using a machine that knits together the upper part of the shoe, which is then attached to the sole. This process not only produces a lighter-weight shoe (which running enthusiasts have long awaited) with less manufacturing waste, the Flyknit requires far less labor since it has 35 fewer pieces to assemble than comparable shoes.

Specialized fabrics can be sensitive to a wide variety of external substances, including harmful toxins or chemical agents. The U.S. military has already incorporated patches worn on uniform cuffs that change color when harmful agents are detected in the surrounding air. Scientists at MIT have created smart fabrics that filter or shield wearers from radiation by combining synthetic fibers with an optical device called a dielectric mirror.

In 2012, Wake Forest University's Center for Nanotechnology and Molecular Materials announced a fabric, called Power Felt, that can generate electricity from body heat. For example, the fabric can be utilized to recharge a cell phone. It is manufactured by imprinting carbon nanotubes onto a woven textile made of plastic fibers. Placing such a fabric under a roof, where it could capture the power of the very high heat that resides in most attics, could conceivable provide part of a household's electricity needs. Worn on a person's body, it might power personal health monitoring systems. What remains to be seen is how consumers respond to such innovations and how much they will be willing to pay for them.

6) Wearable Electronics and Electronics Incorporated into Apparel Begin to Enter the Marketplace

While Google Glass will be the biggest newsmaker for some time in wearable electronics, many other interesting products are entering the market that may have strong appeal to consumers. This trend will build rapidly as the price of chips and components, such as highly sensitive motion sensors, has plummeted, and some devices will offer true convenience as well as a cool factor. Researchers at major labs including PARC and Intel are deeply involved in exploring the potential of electronics integrated into shoes, jackets, shirts, belts, gloves, wristbands or watches, eye glasses, hats and more. The integration of custom designed mobile apps, created specifically for wearable electronics, will become widespread. The sweeping and inexpensive availability of Bluetooth, cellular networks, Wi-Fi and cloud-based databases will enhance this trend.

Hoodies (hooded sweatshirts) with earphones built in to the hoods are offered by Hoodie Buddie and sold on Amazon and other outlets. The Hoodie Buddie has a standard headphone jack that plugs into any headphone-compatible device, like an MP3 music player, iPod or iPhone. The sound travels through the Hoodie Buddie and into the built in headphones at the end of the hood's drawstrings. It is machine washable in cold water, including the headphones.

Japan-based apparel retailer Uniqlo offers Heattech clothing that utilizes a special fabric with air pockets to encourage body heat. The air pockets utilize moisture evaporating naturally from the body in order to heat items like knee-high socks for use in cold weather. WarmX clothing, made in Germany, sells clothing that has silver threads woven into the fabric. When the clothing is hooked up to a battery pack, it bathes the body in heat.

Google's Glass is an eyeglasses-like computer interface that will retail for approximately $1,500 when it hits the market. The device will enable wearers to access the Internet, take photos and shoot short videos. The interface is two-fold, as some voice commands are available, along with the ability for the user to tap on the side of the unit to bring it to life. A tiny display resides on a top corner of the unit where the eye can easily watch it. Even though the unit was not on the market as of mid-2013, many groups were already banning its use and complaining about privacy invasion. The fact that Glass can take photos without an obvious camera in the wearer's hands is a cause of great concern. Casinos have stated they will not allow the device, and legislators in at least one U.S. state were considering banning use while driving.

Oakley Airwave googles for snowboarders and other sports enthusiasts retail for $599 and include built-in GPS and streaming features. Oakley includes a built-in heads-up display that integrates GPS, Bluetooth and more. Onboard sensors give the wearer instant access to snowboard jump analytics that show distance, height and airtime. Additional features include preloaded maps, music playlist control and buddy tracking. Oakley describes it as a "mobile dashboard."

Industrial applications for wearable electronics will also evolve. For example, police officers, soldiers and emergency medical technicians may benefit from the ability to see up-to-date data on a wearable dashboard such as Google Glass, while built-in cameras and microphones may exchange vital real-time information with experts and analysts in remote locations. This could move remote collaboration in highly-critical situations to a new level.

SPOTLIGHT: CuteCircuit

London-based CuteCircuit is a fashion house that produces women's and men's clothing made of fabric embedded with mico-electronics that create changing, lighted patterns, shapes and texts. CuteCircuit projects include the Kinetic Dress, which lights up and changes patterns in response to the wearer's movements; the M Dress, which accepts an SIM card and allows the wearer to make and receive calls without the use of a cellular phone; the Hug Shirt, a garment that generates the warmth, pressure and duration of a hug from a friend of loved one (it is activated by a message sent via cellphone); and the Galaxy Dress, which is a wearable LED display with 24,000 full color LEDs. CuteCircuit produces custom clothing through its Haute Couture line, and offers ready-to-wear through its web site, cutecircuit.com.

7) Exercise Apparel Sales Rise

While Americans (as well as residents of many other countries from Mexico to China) have been putting on weight, they have also developed a keen interest in sports apparel and workout gear to wear at the gym and in other leisure activities. This is one of the fastest-growing product categories in the apparel and shoe sector. Over 50 million Americans have some sort of gym membership, and they need appropriate clothing to wear while they work out.

Sports apparel is divided into several categories including branded athletic apparel, performance apparel, fitness apparel and branded activewear. The U.S. Census Bureau reported an increase in U.S. retail clothing sales at sporting goods stores from $37.7 billion in 2009 to $40.2 billion in 2010. For 2011, sales rose again to $40.0 billion. Plunkett Research estimates the American sports apparel market at $45 billion for 2013 and the global market at $150 billion.

Retailers such as Sports Authority, Inc. and Dick's Sporting Goods, Inc. are also promoting women's merchandise. They have been expanding their lines of women's apparel and devoting more floor space to it.

Sports and fitness apparel are big hits with consumers in Asia. For example, in China's major cities, including Beijing and Shanghai, major brands have very high visibility and consumer acceptance. In the malls and biggest shopping districts, large stores are devoted to Nike, Adidas and other global brands. Chinese consumers are eager to be attached to Western sports and sports figures. NBA (the U.S.-based National Basketball Association) has large stores in China dedicated to basketball fashions and accessories, and immense graphics of American basketball stars are visible from streets and sidewalks.

Technology is playing an increasing role in fitness apparel. Nike, Inc. and Apple Computer, Inc. sell a wireless system that allows a sensor inserted into specially designed Nike shoes (as well as other kinds of apparel including t-shirts, shorts and running pants) to transmit performance data wirelessly to Apple's iPod nano digital music player. Data recorded includes distance, pace, time and calories burned. The data can be downloaded to a computer using Apple's iTunes software and permanently stored on nikeplus.com. The system can be used by runners outdoors, and by people using cardio equipment indoors. Compatible cardio machines such as treadmills, elliptical machines and stationary bikes made by firms such as LifeFitness, Precor and Cybex have ports to which iPods can be connected to record pace, time and distance. In addition, iTunes is featuring upbeat music for workouts, favorite playlists from famous athletes and even real-time voice feedback during exercise. The sensor and transmitter, dubbed the Nike+iPod Sport Kit, retails for about $24 (plus the cost of the shoes or other apparel and the iPod nano). The result has been the world's largest online running community, with more than 7 million members of nikeplus by mid-2012. Members can post comments about their runs, share tips and engage in friendly competition to see who makes the best time for a given distance or track.

Adidas has attempted to provide similar sensor technology. Its latest effort is a partnership with Samsung called miCoach in which a sensor attached to Adidas running shoes links to the Samsung F110 mobile phone and a heart rate monitor as well as an online workout journal. The Adidas-Samsung system retails for about $140 for the miCoach Pacer unit (a combination pedometer, heart rate monitor, calorie counter and coaching advisor).

SPOTLIGHT: lululemon athletica, inc.

One of the biggest growth stories in exercise apparel is Canada-based retailer lululemon. Known for high quality, and high priced, yoga and exercise apparel, the company has been growing at a blazing speed while maintaining impressive profit margins. The firm had 186 corporate-owned stores in North America as of early 2013. Even more impressive were its financial results. The stores enjoy one of the highest retail sales per square foot rates in the apparel industry, at $2,000. Revenues grew from $711.7 million in 2011 to $1.0 billion in 2012. The firm has done a remarkable job of building a community around its stores, with free yoga lessons and a strategy that sees sales personnel as educators about lifestyle and apparel, while it has attracted legions of professional fitness instructors to act as ambassadors in gyms for lululemon apparel.

8) Athletic Footwear Sales Soar, Draws Big Names from Athletes to Designers

Athletic footwear weathered the global economic recession of 2008-09 surprisingly well. Sales in the U.S. fell less than 1% to $12.3 billion in 2009 from $12.4 billion in 2008, according to the Sports & Fitness Industry Association (SFIA, formerly the Sporting Goods Manufacturers Association). For 2010, SFIA reported sports footwear sales of $12.6 billion and $13.18 billion in 2011. Athletic shoes have long been sporting the names and endorsements of famous professional athletes, particularly basketball legend Michael Jordan. Nike's line of Air Jordan shoes has been the top selling shoe brand of all time. Other athletes in the Nike stable at one time include Kobe Bryant, Dwayne Wade, LeBron James and Carmelo Anthony. Adidas-Salomon AG offers shoes from Tracy McGrady and Kevin Garnett. This trend to sell a shoe under an athlete's name is called "marquee footwear."

A new twist to marquee footwear is that which sports a famous designer's name instead of an athlete's. Stella McCartney, a high-fashion designer and daughter of former Beatle Paul McCartney, began collaborating with Adidas in 2005 for hip athletic shoes and clothing. Another designer/athletic shoe partnership is that between traditional shoe maker Cole Haan and its parent Nike. The combination resulted in a line of comfortable street shoes called Cole Haan Nike Air.

All is not golden in the land of big name shoes, however, since many consumers balked at high prices (especially since the advent of the global economic crisis). Most marquee athletic shoes in the boom times during 2007 ran in the $125 to $175 range, but prices dropped during the global recession. In 2009 for example, Nike Zoom LeBron V men's basketball shoes retailed for $79.99 at Foot Locker stores. Prices rebounded during the economic recovery starting in 2010. By early 2013, Foot Locker's prices for top basketball shoes were back in the $125 to $175 range, with the most expensive styles costing in excess of $275.

There are cheaper, more fashionable alternatives on the market which are attracting teen consumers especially. One big leader in the trend toward chic athletic footwear is Germany-based Puma. Puma, which experienced dismal sales in the 1980s while Nike leaped to the top of the market, is enjoying a renaissance. Puma's 2009 revenues fell to $2.9 billion due to the global recession, but rebounded in 2010 to $3.8 billion, surging again in 2011 to almost $4.0 billion. Revenue in 2012 jumped to $4.3 billion. What helped to turn the company around was a focus on fashion, aided by designers such as Germany's Jil Sander and American supermodel-turned-yoga guru Christy Turlington. Although Puma's sales figures are still dwarfed by heavy hitters like Nike and Reebok, it continues to do well.

Nike, which garnered $24.1 billion in 2012 revenues, up from $20.9 billion in 2011 and $19.0 billion in sales in 2010, overhauled its computerized supply systems to increase efficiency and focused on international markets for soccer and fashion shoes and apparel. (Nike gets more than one-half of its revenues from markets outside the U.S.) Nike's marketing budget is legendary. Basketball great LeBron James signed a $90 million Nike endorsement contract in 2003, while still a high school senior. He signed another contract with Nike in 2010, again for $90 million.

9) 3PL Supply Chain Management Evolves to Serve the Global Market

The challenges faced by supply chains are multifaceted: coordinating the arrival of supplies in factories; bringing together all the necessary parts and assembling them into consumer-ready products; as well as distributing finished products across oceans, highways and airways to arrive in the correct locations in the right quantities, colors and styles to satisfy consumer demand. All of this must be done at the lowest possible cost. Compounded by delays and mistakes that can be made along the way due to bad weather, communication breakdowns, accidents, inspections or simple human error, challenges can quickly become catastrophes. In order to prevent mishaps and manage day-to-day supply issues, companies hire supply chain managers and utilize advanced data systems. In many cases, supply chain services are outsourced altogether.

Third-party logistics companies (known as 3PLs) are quickly assuming a vital role in the supply chain. The idea here is to provide, on an outsourced basis, a seamless link between freight services, inventory management, warehousing and distribution. Logistics services are generally defined as services added onto regular transportation activities, including freight forwarding, which is the handling of freight from one form of transport to another. Transportation managers determine the most viable mode of transport (by train, truck, boat, plane or a combination thereof). Value-added warehouses store the stock of client companies and ship the stock out as needed. Supply chain management (SCM) software makers specialize in software that can track and/or allow communication between the different parts of a supply chain.

SPOTLIGHT: CH Robinson Worldwide, Inc.

Sector:	3PL
2012 Revenues:	$11.4 bil., up 9.0%
2012 Profits:	$593 mil., up 27.3%
2012 Employees:	10,500

C.H. Robinson Worldwide, Inc. (CHRW) is one of the world's largest third-party logistics (3PL) providers and a global provider of multimodal transportation services. It maintains over 276 offices around the world. CHRW operates in two sectors: multimodal transportation services and fresh produce sourcing. In the multimodal transportation services sector, the company (which does not own any of its own equipment) maintains one of the largest global networks of motor carrier capacity through contracts with approximately 56,000 carriers. Annually, CHRW serves more than 42,000 customers and handles over 11.5 million shipments. The group also contracts air carriers, oceans carriers and specialty motor carriers. In addition, it provides value-added logistics services, supply chain analysis, freight consolidation, core carrier program management and information reporting. The fresh sourcing sector focuses on procuring fresh produce for retailers, wholesalers and foodservice operators nationwide. CHRW markets its own brands of produce under The Fresh 1 and Pink Ribbon Watermelon brands, which are sourced through various growers and packed through contract agreements with other packaging firms.

Additionally, the company maintains exclusive licensing agreements to distribute fresh produce under various third-party brand names, such as Tropicana, Welch's and Mott's. In 2012, the company sold its T-Chek Systems, Inc. subsidiary, which offered funds transfer, fuel purchasing and online expense management services for motor carriers to Electronic Funds Source, LLC. CHRW introduced its ChemSolutions division in 2012 which provides logistics services to chemical manufacturing and distribution customers globally. In September 2012, the firm acquired privately-held freight forwarder Phoenix International, Inc. for $571.5 million in cash and $63.5 million in C.H. Robinson stock. That same month, CHRW agreed to acquire Apreo Logistics, based in Warsaw, Poland.

Many freight and parcel shipping companies have created businesses in the 3PL sector to provide their customers with turnkey shipping services. Deutsche Post AG (the owner of DHL), UPS and FedEx have all made logistics acquisitions as they battle for market share. DHL Worldwide Network, one of the largest providers of logistics services in the world, made several strategic acquisitions in order to offer domestic and international supply chain management from beginning to end, with services including freight forwarding, warehouse management, multimodal planning and powerful information technology. However, the industry has not consolidated to the point where there is no longer room for small or start-up companies. Many regional or specific service specialists have found a great deal of success in their own niche markets.

The rapid growth of 3PL is a global phenomenon. In China, for example, UPS, FedEx and DHL all maintain extensive 3PL networks throughout the country. In fact, as the world's reliance on Chinese manufacturers expanded, 3PL service operators enjoyed booming business. UPS, FedEx and DHL are offering supply chain management services for foreign companies that purchase large quantities of Chinese goods. Competition among the carriers in the Chinese market is fierce.

SPOTLIGHT: Supply Chain Evolution in the Global Apparel Industry

Many industries have benefited greatly from the leaps and bounds in supply chain intelligence that has come about in the recent past. However, one of the last to see a return on investment was the apparel and textiles industry.

The failure of logistics to meet the demands of this industry stemmed from several factors. For one thing, most data systems for purchasing and invoicing, such as EDI (electronic data interchange), were too inflexible to account for the incredible variety that is found in textile goods, such as weave, fiber, color, texture, quality, style and size. Many efforts to implement data systems met with failure. Another factor that compounded the problem was the mercilessly fickle demand found in apparel consumers worldwide, who create demand for a particular shirt one week, for example, and then just as quickly abandon that fashion. This makes it difficult for manufacturers and retailers to maintain the correct amount of stock.

Some trendy garments in lower price categories might be sold for only a few weeks at full price, then at 30% to 40% discounts, and after that the price can plummet. To top it off, new manufacturers and textile mills pop up daily in China, India and elsewhere, each offering its services at a lower price than the one before, slicing profit margins to a razor edge and leaving little room for error.

Advances in information technology and entirely new models for manufacturing and supply have recently been coming to the fore (and to the rescue) in the apparel industry. Major improvements have included cutting time-to-market so as to turn over new fashions as quickly as possible and working more closely with retailers in order to monitor and react to demand with greater speed and efficiency. It is now common to see garments move from the factory floor in China to the retail rack in America in 30 to 45 days.

A leader in quick apparel turnaround is Zara, an international retailer owned by Spanish textile conglomerate Inditex Group. Zara takes no more than five weeks to get new designs into stores from its manufacturing facilities. While it buys some clothing and materials from Asia (usually unfinished), about one-half its merchandise is manufactured in-house at its central facilities in Spain. Finished pieces are delivered by truck to European outlets and by air to stores in other parts of the world.

Zara relies on superb communication between its stores and headquarters as to what is selling (and what is not) supported by real-time sales data. The firm has little need for costly warehousing since its designers and manufacturers work so closely with the store managers to deliver goods on an as-needed basis to the stores.

10) The Vast Majority of Shoes Sold in the U.S. Are Made Elsewhere

U.S. retail stores specializing in shoes sold $28.7 billion in goods during 2012, up from $27.5 billion in 2011, per U.S. Census Bureau reports. (This number does not include the massive number of shoes sold in department stores, in sporting goods stores and on Internet sites like Zappos, which sells about $1 billion yearly by itself.)

According to the American Apparel & Footwear Association (www.apparelandfootwear.org), Americans purchase more than 2.2 billion pairs of shoes yearly. More than 98% of those shoes are imported, and China is one of the most important sources. (However, as wages and other costs are rising in China, manufacturing is starting to migrate to lower cost nations such as Vietnam and Cambodia.) Major Chinese shoe manufacturers include Yue Yuen Industrial Holdings Ltd. (which had about $7.3 billion in 2012 sales, up from $7.0 billion in 2011 and $5.8 billion in 2010), Li Ning Co. Ltd. ($1.09 billion in 2012 sales, down from $1.3 billion in 2011 and $1.5 billion in 2010 sales) and Belle International Holdings Ltd. ($5.3 billion in 2012, up from $4.7 billion in 2011 and $3.8 billion in 2010).

While a handful of manufacturers, such as Allen-Edmonds (a high-end maker of men's shoes with about $100 million in annual sales), are able to maintain factories in the U.S., domestic manufacturing is all but dead. Another exception is New Balance. This maker of high-end running and athletic shoes operates five factories in the U.S. While New Balance gets much of its inventory from overseas factories, the firm has taken an interesting position with its manufacturing philosophy. As of 2012, New Balance maintained 1,300 manufacturing employees in America. It is the only major running shoe brand that continues to operate plants in the U.S., while its major competitors, such as Nike and Reebok, get most or all of their shoes from Asia. New Balance may have to rethink its American-made strategy, however, if shoe tariffs (generally between $3 and $5 per pair) on imported products are eliminated as part of a larger trade deal such as the

Trans-Pacific Partnership, which was in negotiations as of 2013.

Nike, which as of late 2011 produced 96% of all its shoes in Vietnam, China and Indonesia, may revolutionize footwear manufacturing with the launch of its Flyknit running shoe in February 2013. The shoe is made using a knitting machine that weaves the upper part using synthetic yarn. This upper has 35 fewer pieces to assemble, making the need for cheap labor far less necessary. The savings may make it possible for Nike to have the Flyknit manufactured in the U.S. While still more expensive production overseas, the company will spend less on shipping and fulfill more orders quickly to meet demand.

11) Bricks, Clicks and Catalogs Create Synergies While Online Sales Growth Surges

Analysts at eMarketer reported American e-commerce sales in 2012 of an estimated $343 billion. This figure includes online retail sales, travel sales ($119 billion) and digital downloads, but not sales of tickets to events or online gambling. (For 2012 in all of North America, the firm reported sales of $364 billion, up 13.9% from 2011.)

Global Internet retail sales exceeded $1 trillion for the first time in 2012, up 21.1% from 2011. For 2013, the Asia-Pacific region is expected to see 30% growth in online sales, reaching $433 billion. China alone is expected to be the second largest e-commerce market in the world with $181 billion (up 65% from 2012) behind the U.S., which should hit $384 billion in 2013. Online sales have been supported by very strong results at Amazon.com, largely due to its competitive pricing, expansion of categories, free shipping for members of its "Prime" service and convenience of use.

By the end of 2012, there were about 90 million homes and businesses in the U.S. with broadband connections (not including 180 million of wireless Internet access subscriptions for smartphones and tablets). Several factors will encourage consumers to do more of their shopping online, including today's wildly fluctuating gasoline costs, the fact that consumers feel pressed for time, the widespread adoption of high-speed Internet access, and the fact that the lowest prices can often be found online.

Online shopping often goes hand-in-hand with in-store shopping. A large number of shoppers browse web sites to gain information, later visiting a physical store to make a purchase. The reverse is also true—stores like Best Buy worry that shoppers come into be able to touch and feel merchandise, and then look for the best possible prices at online sites.

Many of the most successful companies among retail firms of all types will be those that take full advantage of the stability of traditional, store-based retailing and combine it with the growing popularity of catalog and Internet-based retailing. Such a strategy would entail:

- Seamless integration of store, catalog and Internet-based offerings to consumers, providing choices of 1) place and method of purchase, 2) method of pickup or shipment and 3) place or method of returns, repairs and additional services as needed.
- Communication of a seamless brand identity and level of service throughout catalogs, retail stores and web sites.
- Opportunities to offer personalized marketing and service through the use of customer profile data.

Few companies have reached this level of integration of traditional and non-traditional retailing. However, for good examples of companies that are evolving toward such "seamless" strategies, study Costco, Wal-Mart, REI, The Gap, Staples, Nordstrom and Victoria's Secret.

At lingerie giant Victoria's Secret, customers find enhanced flexibility and customer service thanks to the opportunity to shop via the web, the Victoria's Secret catalog or Victoria's Secret stores. In addition to the millions of catalogs that are mailed every week, stores hand out copies of the catalogs—which feature the web address of VictoriasSecret.com, as well as phone and fax ordering options. The point is to create loyalty-inducing convenience for customers, giving them options for purchasing when, where and how they please. Online apparel sales are soaring.

Nordstrom began offering customers the option to browse and buy online and then pick up merchandise in stores as early as 2008. A year later, Nordstrom offered merchandise from all of its stores on its web site, effectively positioning each store as a warehouse for online shoppers. The response was immediately positive.

In 2011, Wal-Mart launched "Pick Up Today," a program in which registered Wal-Mart shoppers order items online and then pick them up at their local store the same day. Best Buy offers similar services.

These strategies were born because retailers were fighting a practice called "showrooming," in which shoppers browse for merchandise in stores but buy on rival retailers' web sites, especially Amazon.com. (Electronics retailer Best Buy, which closed 50 of its big-box stores in 2012, may be the poster child of showrooming problems.) Not only does Amazon have a vast selection of items, its prices are highly competitive. William Blair & Co. reported in a recent study that Amazon's prices are 9% lower than those found on Walmart.com (when sales taxes are excluded for Amazon), and 14% lower than prices at Target.com.

One caveat in favor of brick and mortar stores is their ability to accept payments in cash. A sizeable number of Americans remain wary of online security and the possibility of identity theft and refuse to purchase items online, even though they may comparison shop for prices. There are also considerable numbers of people who are without bank accounts or credit cards. Wal-Mart, as of April 2012, offers customers the ability to order items online and then pick them up in stores using cash for payment. By year end 2012, Wal-Mart reported that 2% of Walmart.com's sales were picked up in stores and paid for in cash.

Department stores are making a major shift in operations to support online orders. In yet another effort to compete with Amazon.com's success, Macy's began, in 2011, to convert 292 of its more than 800 stores to include expanded storerooms and cutting-edge technology to track inventory and generate shipping labels. Excess store inventory is shifted to highlighted positions on the company's web site, and merchandise that has sold out at the online distribution center may be found in stores and thereby remain on the web site for sale and delivery by UPS. Online orders are being filled by stores closest to consumers, increasing efficiency and lowering costs.

12) Alternative Sizing Is Big, But Plus-Size Sales Are Slowing

The results of a study of American body shapes and sizes reveal trends that will impact a number of industries, including apparel. While a well-known 2004 study by the Centers for Disease Control and Prevention (CDC) has already shown that Americans are getting a lot heavier, a more recent study, which was sponsored by the U.S. military as well as clothing companies such as Lands' End, J.C. Penney, Target and Dillard's, shows how factors such as age, income and occupation affect the size and shape of the human body. The study utilized an advanced 3D body scanner developed by $[TC]^2$ of Cary, North Carolina. Researchers scanned over 10,800 men and women in several cities around the U.S. For example, those who earn up to $25,000 per year weigh 152 pounds on average, while those in the $50,000-to-$75,000 category average 160 pounds. Those earning above $75,000 weighed nearly the same as those in the lowest income category. Regardless of income, the average adult American woman is 5'4" tall, weighs 155 pounds and wears a size 14. (American children are growing in size as well, and childhood obesity is now a very difficult health problem.)

While the study, known as SizeUSA, affirms the CDC's findings, it also shows that as the American population has grown more ethnically diverse, its average body size has changed. What was considered "average" 40 years ago is no longer average today. These developments may mean important changes in the way that apparel designers, retailers and manufacturers anticipate consumers' needs and desires.

The plus-size market usually refers to women's sizes 14 and up. However, there is also a larger set of sizes which range from 14W to 24W (the W stands for "woman"), and an extended range that goes from 24W to 34W. As of late 2010, an estimated 67% of American women are plus-size or larger, while only 17% of women's apparel is produced for them.

Plus sizes also have more difficulty when it comes to fit than regular sizes, since heavier people tend to assume a larger variety of shapes and measurements. For manufacturers, challenges exist both in terms of labor and material costs. The costs of additional fabric and thread can be passed on to the customer fairly easily. However, passing on the costs associated with changing equipment, creating new patterns and training employees on new designs may be more difficult. Some retailers have cut back on plus sizes due to the costs, including Ann Taylor, which no longer carries size 16s from its stores (the store continues to sell 16s online). Bloomingdale's reduced its plus-size offerings and manufacturer Ellen Tracy no longer makes plus sizes.

For retailers, part of the challenge lies in finding the best way to effectively target these consumers. Factors that retailers must consider are product placement in stores and store layout. Having clearly marked sections devoted exclusively to plus-sized garments, rather than combining these garments on racks with non-plus-sizes, seems to help, as do larger fitting rooms with flattering lighting and fixtures.

There are also design issues. Many plus-size consumers, especially women, do not want to sacrifice style for proper fit, and they enjoy wearing youthful, chic clothing instead of the tent-like tops and pants that have been their traditional choices. However, not all designs and fabrics fit all bodies in the same way, and designers must give careful consideration to creating clothes that are both flattering and comfortable for plus-size consumers.

Charming Shoppes, Inc., which owns plus-size women's clothing stores Lane Bryant, Fashion Bug, Sonsi and Catherines, is betting on an online system called Fashion Genius. Shoppers fill out an online survey as to their body shape, clothing and style preferences. Fashion Genius assigns a profile (such as "Fearlessly Confident") and suggests clothing matches. The company is taking a gamble on the system, since Forrester Research reports that plus-size sales are only 11% of total online apparel sales. On the other hand, the firm reports that plus-size customers are likely to spend twice as much as smaller shoppers when they find apparel they like.

Over the years, as more and more Americans have approached the "overweight" category (women are generally moving from size 12 to 16, and men from size 40 to 46), manufacturers have adjusted their sizes to stroke consumer egos and generate more sales. The trend is called "vanity sizing." What used to be labeled a women's size 10 is now a size six. Some manufacturers are even making their sizes slightly looser—a pair of men's slacks with a 32-inch labeled waist size may actually measure 33 inches. Others, such as The Gap and Victoria's Secret, are designing clothes in varying styles to conform to different body shapes.

However, the sponsors of the SizeUSA study hoped that their findings would bring more accuracy to manufacturers' sizing conventions, as well as help them to target certain age and income markets more effectively. These changes should pay off for consumers, who will then be able to find properly fitting clothes more quickly. Several clothing companies have adjusted the fit of their garments based on the study's findings. Jockey, for example, updated the fit of its line of bras while Liz Claiborne, Inc. altered the sizing of all 42 of its brands. Victoria's Secret and The Gap offer pants in a variety of different fits. Victoria's Secret offers the Christie, Marisa, Kate, Eva and Bridget fits for different body types while the Gap is selling Curvy, Classic and Straight fits.

Although Wal-Mart and Target have struggled in this market, some retailers and manufacturers are clearly profiting from these sizing trends. Torrid, a retailer that specializes in plus-size clothing for young women, has stores nationwide.

13) Discount Clothing Retailers See Promise in Designer Lines

The most successful discount stores that offer apparel focus on fashion as well as low prices. Among the key players in this business are Wal-Mart, followed by T.J. Maxx/Marshall's, Target, Old Navy, Kmart and Kohl's. In terms of apparel, discount stores such as Target, Wal-Mart and Kmart have typically been the places where consumers shop for inexpensive clothing basics such as socks, hosiery, underwear and sweatpants. But all of these retailers have tried, at times, to gain a firmer hold on casual and career wear as well, by offering trendier, more upscale lines of clothing.

In 1999, Wal-Mart acquired Asda, a UK-based retailer specializing in "cheap chic." Asda operates several stand-alone stores called "George," which sell its upscale but inexpensive apparel goods under a label bearing the same name. Designed in the UK, the clothes feature sumptuous fabrics such as leather and cashmere at competitive prices. In the U.S., slightly more conservative versions of these clothes are available for men, women and children at Wal-Mart stores.

Meanwhile, Target paved the way in offering designer-name products. In recent years, Target has entered agreements with a number of hip designers, such as Michael Graves (for housewares) and Isaac Mizrahi and Mossimo (for apparel and accessories) to develop products for its stores. In early 2010, Target acquired the upscale Smith & Hawken brand of outdoor furniture and home décor. In 2011 and 2012, the brand made headlines when highly advertised promotions for designers Missoni and Jason Wu sold out in a matter of hours.

Wal-Mart and Target spent millions to advertise their fashion-forward apparel in upscale media including *Vogue*. Despite a sizeable advertising budget, Wal-Mart has been frustrated in the trendy designer-name clothing market. By 2010, Wal-Mart curtailed its pricey ads and refocused its apparel positioning on high-value/low prices rather than high fashion. Target, on the other hand, has found notable success with major designers and continues to advertise in upscale fashion magazines while pushing its positioning as a trendy place for fashion-conscious/cost-conscious consumers.

Designer Vera Wang, who is famous for her couture bridal gowns, launched a line of dresses and handbags at Kohl's department stores. The Simply

Very Vera Wang line was first available in the fall of 2007, and features dresses for around $69 and handbags for about $89. Since then, Kohl's has expanded its list of exclusive discount lines by noted designer brands to include Dana Buchman, ELLE and Chaps. Jennifer Lopez has an exclusive agreement for a discount line of clothing at Kohl's also, as does Derek Lam who had a line of women's clothes available for a limited time in 2013.

While these developments point to the increasing competition among discount retailers, they also reflect the general public's greater ease in shopping for discount apparel. Wal-Mart claims that 93% of all U.S. households include at least one shopper who buys at its stores.

14) Luxury Lines Bounce Back

The economic crisis hit the global luxury clothing and accessories industry hard. Bain & Co. projected worldwide sales of luxury goods in 2009 to fall by 10% from 2008 to about $225 billion. During the worst of the recession in 2008 and 2009, consumers in most markets abruptly cut their purchases of discretionary items and started paying down debt. Luxury goods sales fell sharply.

The impact of the recession brought about a number of changes in the luxury apparel and accessories industry. Retail prices dropped, as designers responded to consumer cutbacks. Fabric makers and manufacturers who supply the designers also had to lower prices to remain competitive. This means that special luxury touches, such as high-end trim, may have been altered.

During 2008, luxury department and clothing stores were desperately trying to cut inventories, and many famous luxury brands were marked down as much as 70%. This disastrous year ended in a very poor Christmas holiday selling season. By 2009, retailers had learned their lesson, and were keeping inventories lean while requiring lower price points and design innovation from their suppliers.

A notable exception to the luxury goods retail meltdown during the global economic recession was China, where many high-end stores enjoyed brisk sales as China's economy continued to expand. China's sales of luxury goods have been nothing short of astronomical in recent years. At one time, McKinsey forecast that the retail value of China's luxury market would more than double by from 2011 to 2016, reaching $27 billion. Many analysts have forecast that China will surpass Japan as the world's largest luxury goods market by 2015 or so.

However, a sea change occurred in China in early 2013, along with the nation's newly installed president, Xi Jinping. Conspicuous consumption, ostentation and the giving of expensive gifts to officials in return for favors are suddenly out of fashion, amid a well publicized crackdown on graft and corruption. The result may be devastating, at least temporarily, to the Chinese luxury goods market.

In the West, discount luxury web sites including Rue La La and Gilt Group proliferated, offering members daily sales of a wide variety of luxury apparel, accessories and housewares.

The year 2010 brought a welcome end to the luxury drought. One of the world's largest luxury groups, LVMH Moet Hennessy Louis Vuitton, reported a 19% increase in revenue in 2010 over 2009. Its revenues jumped from $23.3 billion in 2009 to $31.2 billion in 2011, and jumped even more significantly in 2012 to $36.7 billion. Much of its growth was achieved via its large numbers of new stores in emerging markets, particularly China. While seasonal influences had a positive effect on store sales, some gains were also made due to pent-up demand from shoppers who had been deferring purchases, and from a "wealth effect" felt by some consumers who had seen the values of their investment portfolios rebound substantially from dismal lows.

Some European designers will shift production from France and Italy to other countries where labor is cheaper, such as China. In the U.S., Lafayette 148, a New York-based womenswear label, sent a production veteran to China to teach 120 workers there how to produce and finish fine-gauge knits for sweaters using cashmere or merino wool. Lafayette 148 now handles all production in a new factory in China.

Designers also mix their haute couture with lower-priced ready-to-wear pieces. In the past, the finest custom lines from the runway couture rarely appeared in the same showrooms with less expensive lines, but no more. So-called "diffusion" lines, or lower-priced secondary labels that more consumers can afford, are shown side-by-side with this season's couture.

One example of a diffusion line is Giorgio Armani's Emporio Armani. With more than 170 stores located around the world, Emporio Armani offers ready-to-wear apparel, sunglasses, perfume, accessories and watches for far less that the exclusive Giorgio Armani or Armani Collezioni brands. Merchandise is mass produced, and designed to appeal to younger buyers. Likewise, Dolce & Gabbana's D&G line (which was begun in 1993 but licensed off to Italian fashion conglomerate IT Holding SpA) is now back in the hands of Domenico Dolce and Stefano Gabbana who spent $100 million to design and market the mid-priced line. Other diffusion lines include Prada's Miu Miu and Versace's Versus.

Diffusion lines can be up to 50% cheaper in retail pricing than top-of-the-line designer goods. However, their profit margins are approximately 20% compared to up to 45% for high-end labels.

The market for luxury goods for the near term appears mixed. A record high stock market and a considerable improvement in home values will help the U.S. market, although many consumers remain low on cash and unemployment remains a serious challenge. China, Brazil and other emerging markets that once offered booming markets for luxury items have slowed. Meanwhile, the economy in Europe was very subdued as of early 2013, with no immediate relief in sight.

15) Luxury Goes Online

After a few false starts, luxury apparel makers are marketing more and more goods over the Internet. High-fashion sites featuring many different brands of apparel and accessories are proliferating. In the face of soft economies in Europe and the U.S., online channels are growing in importance as apparel firms hunt for every possible sale.

One example is YOOX SpA (www.yoox.com). The Italian site offers more than 250,000 items from designers such as Chloe, Dolce & Gabbana, Jean Paul Gaultier and Prada. Another haute site is Net-A-Porter.com which was launched in 2000. Having grown to the level of millions of visitors each month, Net-A-Porter became a valuable company. In April 2010, luxury goods giant Financiere Richemont agreed to acquire the firm. In 2012, Net-A-Porter launched a Chinese web site, TheOutnet.com, after its acquisition of Hong Kong-based Shouke.com.

Haute couture companies such as Dior, Louis Vuitton and Bottega Veneta are also operating their own web sites. Although once seen as unnecessary (or even unwanted) competition to luxe boutiques in glamorous world capitals, company web sites are proving to be absolutely necessary. At Dior, www.dior.com is becoming one of the firm's fastest-growing points of sale. Unlike the company's brick and mortar stores, the web site does not require exorbitant rent payments, world-class architecture and staggering overhead in the form of staff and customer perks. The company garnered attention in 2009 and 2010 with a series of short videos that attract viewers back regularly to view the latest offering. The videos continued as of early 2013.

Another trend in online luxury is the broadcasting of designer trunk shows. Once held only in stores and open to favored customers only, online trunk show videos are open to the public and can be accessed for up to two weeks instead of limited to a single day in stores. Online shows can also be augmented by footage from runway shows and interviews with designers.

SPOTLIGHT: Amazon Fashion

Amazon.com is attempting to dominate online fashion shopping in the same way it has revolutionized other retail categories including books, electronics and groceries. It first jumped into the fashion pool with the acquisition of Shopbop in 2006, followed by the purchase of online footwear site Zappos in 2009 and discount luxury clothing and accessories site MyHabit in 2011. It also launched its own shoes and accessories venture, Endless.com, in late 2006, but shut it down in late 2012. As of 2013, Amazon was taking on high fashion, working out deals with designers such as Michael Kors, Calvin Klein, Catherine Malandrino and Vivienne Westwood. The Amazon Fashion division produced and aired its first TV commercial in 2013, shot at a rented 40,000-foot Brooklyn studio devoted to promoting the new business. Look for further ads on TV, in print, on billboards and online.

16) Self Service Apparel Fitting Technologies Grow in Stores and Online

Clothing Size Scanners

For those looking for better fit in a brick and mortar setting, there's the Virtual Fitting Room (VFR) by Intellifit, a private company outside of Philadelphia, Pennsylvania which was acquired by Unique Solutions in 2009. The VFR is a high-tech scanning booth that promises to scan customers for their exact measurements for use in making custom jeans. The process takes about 20 seconds, and is the result of technology developed by a U.S. Department of Energy research facility. Unique Solutions installed a VFR in a mall in King of Prussia, Pennsylvania in 2010. In 2011, the company received $30 million in private equity funding. It planned to install more than 300 scanning kiosks across the U.S. over the mid-term.

Another initiative, True Fit, makes purchase suggestions from about 350 different brands based on favorite items already in shoppers' closets. True Fit was founded in Boston by statisticians interested in decreasing the 20% to 40% return rate for online clothing purchases. Rather than relying on measurements supplied by users, True Fit analyzes exact pattern measurements submitted by retailers such as Nordstrom, Macy's and Joe's Jeans. Shoppers enter information about their favorite clothes, including brand style and size. The system then matches those measurements with items available in participating stores and makes suggestions as to sizing. True Fit was rolled out on Macys.com and Nordstrom.com, among others, in early 2012.

Interactive Mirrors & Windows

Mirrors that assist shoppers in matching styles or warn shoppers when items do not match were being tested by specialty retailer The Limited in 2011. The mirrors use touch screens that allow shoppers to access information about apparel items such as available sizes or alternative item suggestions. Some mirrors utilize rear-projection that allows a shopper to stand in front of it while different outfits are projected onto the image (the shopper doesn't have to actually try an item on). Mirrus, a Winston Salem, North Carolina-based company, had about 3,000 of its interactive mirrors installed in clothing stores, airports, beauty salons and entertainment and sports venues around the world in 2012. A similar product, called Swivel, is made by FaceCake Marketing Technologies of Calabasas, California. Swivel uses Microsoft's Kinect Sensor to scan a shopper and display the shopper's image on a nearby monitor. The image can "model" selected apparel and accessories from every angle. Bloomingdale's New York City store and 19 other locations had installed Swivel as of mid-2012.

In a related technology, store windows have the ability to display selected items on an avatar via a touch screen. Shoppers can touch the window and use a menu to view a wide variety of items and combinations.

Virtual Sales Associate Kiosks

Intel, in partnership with a creative design firm in San Francisco, California called Frog Design, Inc. (www.frogdesign.com), is working on a prototype of an apparel kiosk that will serve as a virtual sales associate. To be used in department and other clothing stores, this kiosk enables shoppers to stand before a counter with two touch-screens, one at eye level and the other embedded in a waist-high counter. Wave a personal card (similar to shopping cards issued by grocery stores) before one of the screens and the system recognizes the shopper and has access to data regarding previous purchases. Wave the tag of a shirt from the store and the system displays information about coordinating pants, skirts, shorts or accessories for sale in the store; data regarding previous purchases that will complement the new shirt; and available discounts or sales prices that may apply. Intel hopes to outsource the building of these kiosks. Similar systems are being developed by other firms, including IBM's Virtual Mirror for assistance with cosmetics purchases (IBM is working with EZface on the project); and Microsoft, which is also working on an apparel concept.

17) Fast Fashion: Designers and Retailers Speed Up

Several European mass retailers are experimenting with a new business model known as "fast fashion." The key players in this business are Sweden's Hennes & Mauritz, also known as H&M, as well as Zara and Mango, two Spanish retailers, and the U.K. chain New Look. For these retailers, success comes from designing trendy, inexpensive clothes that mimic high-end fashion and that are delivered to consumers at lightning speed—two to four weeks after conception. This turnaround time is a far cry from the four to nine months that U.S. mass retailers typically require.

Fast fashion wasn't always part of New Look's business strategy. As the phenomenon began to heat up, New Look changed its business model to get a piece of the action. Originally a discount retailer, New Look's more than 1,000 stores in the UK, France, Belgium and the Middle East now stock new items every week in order to catch the attention of fashion-conscious shoppers. In 2011, New Look announced a partnership with Fourlis Holdings S.A., with Fourlis attaining the franchise rights to develop New Look stores throughout Romania and Bulgaria. As of 2012, the company had more than 1,100 stores in 16 countries.

Mango (headquartered in Spain), which operated more than 2,415 stores in 140 countries as of early 2013, has developed a clever distribution system that keeps its products and its inventory current. Clothes are categorized according to basic style traits, such as dressy or casual, and then shipped to stores that are most successful at selling specific categories of merchandise. This system has given it a competitive edge in the world of fast fashion and its products can hit the stores in just four weeks. Inditex's Zara, which has more than 1,830 stores in 82 markets, boasts the quickest turnaround time of just two weeks.

H&M is the largest player in the fast fashion market, both in terms of sales and stores. It operates nearly 2,700 stores in the U.S. and Europe and posted remarkable recession-proof sales of $16.3 billion in 2009, up from $12.6 billion in 2008. For 2010, H&M sales topped $17.49 billion, followed by a drop in 2011 to $16.3 billion. Revenues for 2012 soared to $18.5 billion. From design table to store rack, its lead time is three weeks. In keeping with the fast fashion model, H&M typically does not restock items—once they're gone, customers won't see them again. This "now or never" policy helps to promote H&M's image as a purveyor of up-to-the-second style. Mango uses a slightly different strategy to cultivate a sense of fashion immediacy—it displays only one item per size in its stores. Though others may be in the stockroom, the customer is left wondering if that size-10 shirt will be around tomorrow. With the advent of retailers such as Neiman-Marcus stocking less expensive items to boost sales, some analysts believe H&M will suffer in the near- to mid-term.

Although The Gap has yet to meaningfully compete in the world of fast fashion (though it was reported that it had begun to decrease its lead time on some apparel pieces to just a few weeks), it is making some changes in order to keep its styles current. Meanwhile, in recognition of this market, its Old Navy stores offered three different t-shirt size types, some of which are geared to teenagers, who like a snugger fit and bare midriffs. Young adults and baby boomers generally prefer a looser, more comfortable fit.

While The Gap sees lots of potential overseas, it is still trying to find the right business model for non-U.S. stores. It is learning that German and Japanese tastes are far different from Americans', both in terms of style and fit. Rather than try to aggressively compete in the European market, however, The Gap closed its stores in Germany, a country where H&M dominates. In late 2011, The Gap announced a major cutback. By the end of 2013, it plans to shrink its North American store base by nearly 200 units. Meanwhile, it hopes to expand in emerging nations such as China where it already has a minor presence.

A related practice to fast fashion is "chasing," in which retailers order a small number of units for fast delivery to test the waters. Should the item(s) sell, retailers place a quick re-order. Teen-targeted retailer Aeropostale, Inc. is known for its chasing technique. The practice can be difficult for manufacturers, many of which were hard hit during the global recession, finding themselves with scaled-back orders or cancellations. Leery of ramping up production too quickly, some manufacturers are refusing late orders, leaving retailers short on merchandise to sell.

A number of traditional department stores are trying to get into the fast fashion act. Nordstrom signed a deal with British fast fashion firm Topshop in 2012, in which Topshop apparel is featured in Nordstrom stores.

18) Apparel Manufacturing Gains Ground in the U.S.

Up until recently, many American and European clothing manufacturers, unable to resist the less expensive labor costs available in such places as Africa, Latin America and Asia, have given in to outsourcing. H&M, a European mega-retailer known for its aggressive pricing, manufactures most of its clothing in China and Southeast Asia, as does The Gap. Some, however, prefer to manufacture in their home countries, even if it costs far more. This is often the case with haute couture houses, such as Chanel, where quality, both in fabric and workmanship, is of critical importance. All of Chanel's haute couture designs are made in France, one of the most costly countries in the world in terms of labor.

However, this is also the case for Allen-Edmonds, a men's shoe manufacturer based in Wisconsin. Its shoes, which are aimed at traditional, affluent men, have been worn by both President Bushes. How it has been able to sustain a successful American shoe manufacturing company is a lesson in Business 101: train your employees well, reward them with good pay and benefits and make a top-quality product. With products priced at $200 and up and available in a highly diverse array of sizes and widths, Allen-Edmonds is an unusual player in this business. In a market where 98.4% of U.S. shoes are manufactured abroad by laborers earning between $0.60 and $1.00 per hour, Allen-Edmonds' average employee earns $15 per hour. Meanwhile, the firm's high prices mean far fewer customers than a company like Nike. However, its annual sales run an estimated $100 million, a very impressive amount.

Allen-Edmonds' employees are also loyal, which has actually begun to pose a problem. Some of its most highly skilled employees are nearing retirement, and the company must now find and train new, younger employees to fill their shoes. While Allen-Edmonds has thus far resisted the temptation to outsource, it may be forced to look overseas for labor.

Starting in 2011 and on into 2013, labor costs were rising in China and transportation costs were high, aiding Allen-Edmonds and other U.S. apparel companies who choose to manufacture at home. Another "make it at home" firm is Boathouse Sports, an athletic apparel company based in Philadelphia, Pennsylvania, which is seeing record orders due to its ability to deliver goods quickly. The trend is also a boon for small, general manufacturers based in the U.S., which are picking up orders formerly outsourced to Asia. One such company, Contract Sew & Repair of Kent, Washington, provides sewing services to lululemon athletica, Nordstrom and Tommy Bahama.

19) Apparel Goes Green

As industries from packaged foods to automobiles focus on eco-friendly products and processes, apparel and textiles are going green as well. Organic cotton is a major trend in apparel retailing with firms such as Nike, Timberland, Patagonia, Eileen Fisher and Levi Strauss. Textile Exchange (formerly Organic Exchange), a nonprofit organization that provides information and advises farmers and retailers, reported that the organic cotton market reached $5.2 billion in 2010. Textile Exchange projections for 2011 and 2012 were $6.2 billion and $7.4 billion respectively.

Patagonia embraced the organic cotton trend in 1996 when it announced that its cotton items would be made exclusively of organic fibers. The switch doubled the firm's raw material costs. Today, Patagonia's sportswear line is still made of 100% organic cotton, while it is studying ways to use dyes that are organic without suffering yet another significant hike in costs.

In 2011, a group of apparel retailers, manufacturers and government and non-government agencies from around the world launched the Sustainable Apparel Coalition. The coalition promotes environmental sustainability by measuring and evaluating apparel and footwear, spotlighting promising new technologies and watching for chances to improve social and environmental practices. Members include Adidas, Duke Center for Sustainability and Commerce, Gap, Inc., H&M, Nike, the U.S. Environmental Protection Agency and Wal-Mart. Initial focus for the group is on the global supply chain. The Coalition created The Higg Index, which catalogs apparel based on fuel usage, pollutant production and other issues seen as threats to the environment as of early 2013. Future versions of The Higg Index are planned to include footwear products.

There is a wrinkle to the organic fiber trend, and that is that apparel made with synthetic fibers such as polyester is easier to care for and therefore may require less energy over the long run. Although polyester is made of materials that take more energy to make (such as wood and oil), it does not require hot water to wash and can air-dry. Many cotton garments suggest hot or warm water washing and tumble drying followed by ironing.

Some green apparel causes other problems such as durability. Apparel made with organic fibers such as corn melts when ironed. Bamboo fiber quickly wears out when abraded and banana fiber tends to scratch. Manufacturers are experimenting with ways to solve these problems. For example, corn fiber shirts made with enhanced seams may not need ironing.

Levi Strauss conducted a life-cycle assessment in 2007 of some of its major products, finding that 49% of the water use during the lifetime of its 501 brand jeans occurred during the beginning with cotton farmers. Another 45% of the water was used by consumers to wash the jeans (on average 100 times). As a result, Levi's joined the Better Cotton Initiative (http://bettercotton.org), which focuses on teaching cotton farmers in Pakistan, India, Brazil and Mali how to grow the crop with less water. Levi's hopes to use a 20% blend of the new cotton in its products by 2015. Meanwhile, the company launched the Waste<Less jeans brand in late 2012, which is made of at least 20% recycled plastic.

Another ecologic concern is that materials needed for "green" clothing often come from a variety of countries around the world. The carbon footprint necessary to transport the goods and then the finished items may be more harmful to the environment than the energy saved by wearing and caring for the garment.

Regardless of these wrinkles, eco-friendly apparel is a hot trend that is likely to grow over the near- to mid-term. Watch for further innovations in fabrics and their care as more manufacturers embrace the idea.

Chapter 2

APPAREL & TEXTILES INDUSTRY STATISTICS

Contents:

Apparel & Textiles Industry Overview

Category	Quantity	Unit	Date	Source
Global Trade				
Value of Exports of Textiles & Apparel: Global	706.0	Bil. US$	2011	WTO
China	253.2	Bil. US$	2011	WTO
EU 27	193.0	Bil. US$	2011	WTO
India	29.4	Bil. US$	2011	WTO
Turkey	24.7	Bil. US$	2011	WTO
Bangladesh	21.5	Bil. US$	2011	WTO
U.S. Retail Sales[1]				
Clothing & Accessories Stores	239.2	Bil. US$	2012	Census
Clothing Stores, General	177.2	Bil. US$	2012	Census
Men's Clothing Stores	8.7	Bil. US$	2012	Census
Women's Clothing Stores	41.6	Bil. US$	2012	Census
Family Clothing Stores	93.2	Bil. US$	2012	Census
Other Clothing Stores	13.8	Bil. US$	2012	Census
Shoe Stores	28.7	Bil. US$	2012	Census
Sporting Goods Stores	42.6	Bil. US$	2012	Census
Department Stores	185.7	Bil. US$	2012	Census
Discount Department Stores	119.9	Bil. US$	2012	Census
U.S. Textile & Apparel Production				
Value of Manufacturers' Shipments in the Textiles Industry, U.S.	23.5	Bil. US$	2012	Census
Value of Manufacturers' Shipments in the Apparel Industry, U.S.	1.3	Bil. US$	2012	Census
Value of U.S. Exports				
Total Apparel & Textiles	22.62	Bil. US$	2012	ITA
Apparel	5.51	Bil. US$	2012	ITA
Textile Mill Products	17.11	Bil. US$	2012	ITA
Fabric	8.49	Bil. US$	2012	ITA
Value of U.S. Imports				
Total Apparel & Textiles	100.93	Bil. US$	2012	ITA
Apparel	76.81	Bil. US$	2012	ITA
Non-Apparel	24.12	Bil. US$	2012	ITA
Fabric	5.73	Bil. US$	2012	ITA
Man-Made Fiber Products	43.89	Bil. US$	2012	ITA
Cotton Products	50.50	Bil. US$	2012	ITA
Wool Products	4.75	Bil. US$	2012	ITA
Cotton Statistics				
Worldwide Cotton Production	119.87	Mil. Bales	2012/13[2]	USDA
Total Foreign Cotton Production	102.86	Mil. Bales	2012/13[2]	USDA
Cotton Production, U.S.	17.01	Mil. Bales	2012/13[2]	USDA
Worldwide Biotech Cotton Planting	24.7	Mil. Hectares	2011	ISAAA
Percent of Total Cotton Crop	68	%	2011	ISAAA

[1] Includes sales of non-apparel and all other merchandise carried by these stores; clothing represents a significant portion of these sales.

[2] Marketing year beginning August 1, projected.

WTO = World Trade Organization

Census = U.S. Census Bureau

USDA = U.S. Department of Agriculture

ITA = U.S. International Trade Administration

ISAAA = International Service for the Acquisition of Agri-Biotech Applications

U.S. Apparel Manufacturing Producer Price Index (PPI): 2003-2013

(Index Base 12/03 = 100; Latest Data Available)

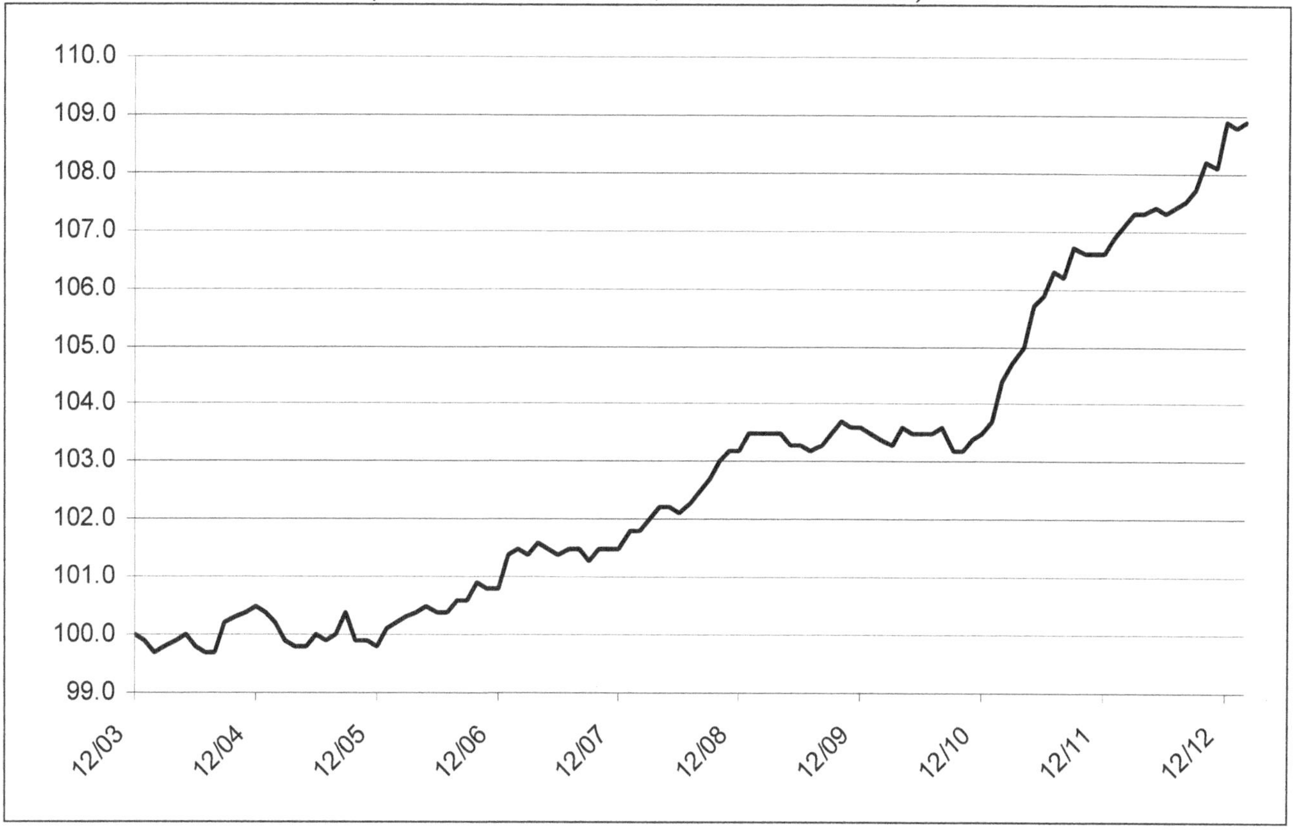

Notes: The Producer Price Index (PPI) program measures the average change over time in the selling prices received by domestic producers for their output. The prices included in the PPI are from the first commercial transaction. Nov. 2012-Feb. 2013 data are preliminary.

Source: U.S. Bureau of Labor Statistics
Plunkett Research, Ltd.
www.plunkettresearch.com

Percent Change in Annual U.S. Apparel Manufacturing Producer Prices: 2006-2012

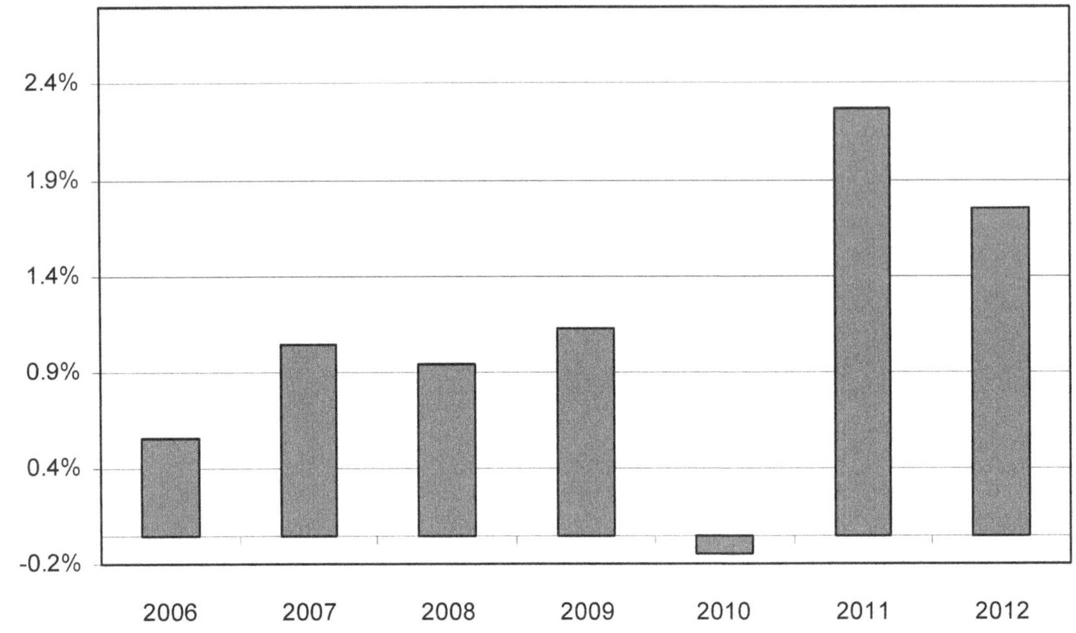

Producer Prices	2011	2012
Producer Price Index	105.7	107.5
Percent change from previous year	2.2%	1.7%

Source: U.S. Bureau of Labor Statistics
Plunkett Research, Ltd.
www.plunkettresearch.com

Top 50 Countries Exporting Textiles & Apparel to the U.S., All MFA Fibers: 2011-2012

(In Millions of US$)

	Country	2011	2012		Country	2011	2012
	World	**101,324.4**	**100,931.5**	26	Malaysia	534.7	495.9
1	China	40,657.8	40,582.7	27	Japan	387.3	444.2
2	Vietnam	7,196.0	7,654.6	28	France	305.8	326.8
3	India	5,934.0	5,895.7	29	Lesotho	315.3	300.9
4	Indonesia	5,318.9	5,197.4	30	Germany	286.4	294.4
5	Mexico	4,761.9	4,628.7	31	Israel	291.5	289.7
6	Bangladesh	4,652.9	4,621.6	32	Portugal	244.2	265.7
7	Pakistan	3,357.2	3,009.9	33	Kenya	261.0	254.8
8	Honduras	2,626.5	2,574.2	34	U.K.	236.0	252.2
9	Cambodia	2,622.0	2,569.0	35	Colombia	263.6	248.5
10	El Salvador	1,780.7	1,873.7	36	Hong Kong	201.5	183.9
11	Italy	1,640.9	1,702.1	37	Bahrain	171.5	166.4
12	Sri Lanka	1,403.6	1,482.0	38	Costa Rica	172.7	164.4
13	Canada	1,446.9	1,411.6	39	Mauritius	157.1	163.1
14	Nicaragua	1,357.8	1,349.0	40	Belgium	152.8	152.6
15	Thailand	1,411.7	1,270.8	41	Romania	116.7	146.5
16	Guatemala	1,331.5	1,251.1	42	Spain	89.9	106.3
17	The Philippines	1,249.1	1,226.6	43	Morocco	84.6	105.2
18	Egypt	1,080.3	1,038.1	44	Tunisia	83.9	96.2
19	Jordan	901.7	1,005.4	45	The Netherlands	96.3	93.3
20	Turkey	901.6	999.8	46	Switzerland	125.1	82.3
21	South Korea	928.1	925.1	47	United Arab Emirates	81.1	74.2
22	Taiwan	878.9	866.5	48	Bulgaria	51.2	66.9
23	Haiti	701.5	730.1	49	Poland	61.2	61.8
24	Dominican Republic	664.4	657.6	50	Swaziland	76.9	60.0
25	Peru	743.4	642.6				

MFA = Multifiber Agreement. The Multifiber Agreement was established in 1973 between the U.S., Canada and several countries now part of the European Union. It set a quota system for the maximum number of products produced in developing countries that could be legally exported to MFA member countries. These amounts were based on historic purchasing patterns and differed from country to country.

Source: U.S. Department of Commerce, International Trade Administration

Plunkett Research, Ltd.

www.plunkettresearch.com

Top 50 Destinations of U.S. Textiles & Apparel Exports: 2011-2012

(In Millions of US$)

	Country	2011	2012		Country	2011	2012
	World	**22,405.3**	**22,619.4**	26	France	**122.2**	**121.0**
1	Canada	4,991.5	5,253.7	27	Thailand	96.0	117.1
2	Mexico	4,898.7	5,210.5	28	Costa Rica	100.6	114.9
3	Honduras	1,839.3	1,462.2	29	Venezuela	93.8	104.0
4	China	1,284.3	1,298.8	30	Russia	114.6	101.3
5	Japan	641.8	720.1	31	Turkey	90.6	89.9
6	Dominican Republic	818.0	680.4	32	Saudi Arabia	88.9	89.8
7	El Salvador	632.0	595.3	33	Spain	84.3	83.9
8	U.K.	556.6	595.2	34	Peru	69.9	83.3
9	Belgium	428.3	425.6	35	Malaysia	64.5	74.6
10	Hong Kong	426.7	408.1	36	Vietnam	45.8	70.8
11	Germany	408.0	382.7	37	Panama	63.5	69.1
12	South Korea	417.4	365.4	38	Israel	62.3	67.5
13	Australia	282.3	299.7	39	Argentina	67.9	63.9
14	The Netherlands	312.0	292.5	40	The Philippines	54.6	59.6
15	Brazil	271.0	272.7	41	New Zealand	49.1	57.8
16	Guatemala	287.7	249.0	42	South Africa	54.8	55.5
17	Chile	190.2	189.3	43	Poland	32.5	52.6
18	United Arab Emirates	165.5	182.1	44	Ecuador	34.7	46.5
19	India	165.4	169.4	45	Ireland	46.1	40.6
20	Colombia	165.2	161.0	46	Haiti	38.0	38.0
21	Italy	158.4	153.4	47	Egypt	27.9	37.7
22	Singapore	147.4	150.2	48	Switzerland	42.9	37.4
23	Indonesia	131.6	147.9	49	Ukraine	32.2	37.2
24	Taiwan	126.1	135.5	50	Finland	37.0	37.1
25	Nicaragua	137.3	132.7				

Source: U.S. Department of Commerce, International Trade Administration
Plunkett Research, Ltd.
www.plunkettresearch.com

Top 25 Countries Exporting Apparel & Non-Apparel Textiles to the U.S., All MFA Fibers: 2011-2012

(In Millions of US$)

Apparel				Non-Apparel			
Country		2011	2012	Country		2011	2012
	World	**77,659.1**	**76,810.6**		**World**	**23,665.3**	**24,120.9**
1	China	29,392.0	29,059.6	1	China	11,265.8	11,523.0
2	Vietnam	6,644.0	7,101.1	2	India	2,618.0	2,854.5
3	Indonesia	5,051.7	4,935.4	3	Pakistan	1,702.3	1,538.1
4	Bangladesh	4,509.9	4,470.0	4	Mexico	958.3	932.9
5	Mexico	3,803.5	3,695.8	5	Canada	940.8	896.7
6	India	3,316.0	3,041.2	6	South Korea	684.8	706.3
7	Honduras	2,615.4	2,559.4	7	Turkey	568.0	612.3
8	Cambodia	2,592.3	2,533.9	8	Italy	539.7	565.0
9	El Salvador	1,738.5	1,840.6	9	Vietnam	552.0	553.5
10	Sri Lanka	1,399.6	1,476.9	10	Taiwan	442.9	472.4
11	Pakistan	1,654.9	1,471.8	11	Japan	354.3	408.5
12	Nicaragua	1,356.9	1,348.1	12	Germany	259.4	270.8
13	Guatemala	1,321.4	1,240.0	13	Indonesia	267.2	262.0
14	The Philippines	1,180.0	1,148.7	14	Israel	195.9	202.7
15	Italy	1,101.1	1,137.2	15	Thailand	224.7	197.5
16	Thailand	1,186.9	1,073.3	16	France	160.5	174.8
17	Jordan	895.5	980.6	17	Egypt	177.5	166.6
18	Egypt	902.8	871.5	18	U.K.	151.7	154.2
19	Haiti	701.4	730.0	19	Bangladesh	143.0	151.6
20	Dominican Republic	653.7	649.3	20	Portugal	149.0	146.6
21	Peru	713.8	610.6	21	Belgium	145.0	145.5
22	Canada	506.1	514.8	22	The Netherlands	93.6	90.7
23	Malaysia	505.4	464.2	23	The Philippines	69.1	77.9
24	Taiwan	436.0	394.1	24	Bahrain	88.7	73.7
25	Turkey	333.6	387.6	25	Spain	49.1	60.7

MFA = Multifiber Agreement. The Multifiber Agreement was established in 1973 between the U.S., Canada and several countries now part of the European Union. It set a quota system for the maximum number of products produced in developing countries that could be legally exported to MFA member countries. These amounts were based on historic purchasing patterns and differed from country to country.

Source: U.S. Department of Commerce, International Trade Administration

Plunkett Research, Ltd.

www.plunkettresearch.com

Top 15 Countries Exporting Fabric, Man-Made Fiber Products, Cotton Products & Wool Products to the U.S., All MFA Fibers: 2011-2012

(In Millions of US$)

Fabric				Man-Made Fiber Products			
	Country	2011	2012		Country	2011	2012
	World	**5,490.4**	**5,732.7**		**World**	**41,151.8**	**43,885.7**
1	China	1,433.2	1,607.4	1	China	19,509.3	20,665.4
2	South Korea	536.8	547.9	2	Vietnam	3,251.0	3,726.7
3	Canada	429.4	428.3	3	Indonesia	2,111.7	2,225.9
4	Japan	308.2	360.8	4	Mexico	1,871.3	1,852.7
5	Taiwan	285.9	313.7	5	Canada	1,157.2	1,110.8
6	Italy	295.8	303.2	6	Honduras	925.6	1,087.7
7	India	312.2	300.1	7	India	931.2	1,070.5
8	Mexico	270.3	278.2	8	Bangladesh	962.7	911.5
9	Germany	176.4	186.6	9	Cambodia	776.3	862.8
10	Vietnam	219.1	179.6	10	Taiwan	722.5	727.5
11	Turkey	137.0	145.8	11	El Salvador	576.9	689.8
12	Pakistan	153.4	133.5	12	Thailand	656.1	656.8
13	Israel	110.6	101.9	13	Jordan	550.9	646.8
14	U.K.	100.2	100.2	14	South Korea	629.3	645.1
15	France	89.9	98.1	15	The Philippines	621.5	638.7
Cotton Products				**Wool Products**			
	Country	2011	2012		Country	2011	2012
	World	**53,491.9**	**50,502.8**		**World**	**4,819.2**	**4,747.7**
1	China	17,840.3	16,914.8	1	China	2,034.5	1,863.3
2	India	4,530.8	4,311.8	2	Italy	604.4	634.5
3	Vietnam	3,753.2	3,729.0	3	India	395.6	424.6
4	Bangladesh	3,593.2	3,610.6	4	Mexico	243.9	270.7
5	Indonesia	3,046.9	2,828.2	5	Canada	151.8	158.8
6	Pakistan	3,101.1	2,712.9	6	Vietnam	145.5	138.6
7	Mexico	2,631.2	2,489.5	7	Turkey	87.4	100.7
8	Cambodia	1,822.1	1,685.4	8	Indonesia	101.4	97.1
9	Honduras	1,700.5	1,485.0	9	U.K.	81.5	88.3
10	El Salvador	1,198.1	1,174.7	10	Pakistan	55.2	66.9
11	Nicaragua	1,032.8	957.7	11	The Philippines	48.0	48.0
12	Sri Lanka	867.2	847.3	12	Bangladesh	50.8	45.8
13	Guatemala	888.9	770.2	13	France	50.0	45.6
14	Egypt	658.6	592.9	14	Dominican Republic	46.7	43.3
15	Thailand	702.6	568.3	15	Romania	38.6	42.7

MFA = Multifiber Agreement. The Multifiber Agreement was established in 1973 between the U.S., Canada and several countries now part of the European Union. It set a quota system for the maximum number of products produced in developing countries that could be legally exported to MFA member countries. These amounts were based on historic purchasing patterns and differed from country to country.

Source: U.S. Department of Commerce, International Trade Administration
Plunkett Research, Ltd.
www.plunkettresearch.com

Gross Output in the Textile & Apparel Industries, U.S.: Selected Years, 2000-2011

(In Millions of US$; Latest Year Available)

IO Code	Description	2000	2005	2008	2009	2010	2011
313100	Fiber, yarn, & thread mills	11,259	10,945	8,959	6,607	7,730	7,981
313210	Broadwoven fabric mills	15,345	8,507	5,055	3,815	4,000	4,186
313220	Narrow fabric mills & schiffli machine embroidery	1,622	1,332	1,142	1,000	1,070	1,134
313230	Nonwoven fabric mills	4,885	6,149	6,316	6,100	6,874	7,158
313240	Knit fabric mills	4,188	2,898	1,742	1,293	1,543	1,602
313310	Textile & fabric finishing mills	11,849	8,523	4,926	4,087	4,498	4,712
313320	Fabric coating mills	2,323	2,401	2,232	1,852	2,315	2,377
314110	Carpet & rug mills	13,193	17,050	12,944	9,614	9,841	10,011
314120	Curtain & linen mills	9,706	7,561	4,323	3,540	3,495	3,594
314910	Textile bag & canvas mills	2,528	3,395	2,899	2,417	2,543	2,650
314990	All other textile product mills	8,206	7,505	6,836	5,610	6,227	6,418
315100	Apparel knitting mills	7,453	4,241	2,222	1,690	1,845	2,006
315210	Cut & sew apparel contractors	4,085	4,198	2,620	2,645	2,583	2,823
315220	Men's & boys' cut & sew apparel mfg.	16,400	6,478	2,461	1,584	1,627	1,742
315230	Women's & girls' cut & sew apparel mfg.	24,831	12,001	6,567	4,689	4,768	4,930
315290	Other cut & sew apparel mfg.	1,942	1,425	1,280	854	966	1,070
315900	Apparel accessories & other apparel mfg.	3,973	2,528	1,337	1,181	1,179	1,270
316100	Leather & hide tanning & finishing	2,922	2,151	1,443	1,078	1,895	2,080
316200	Footwear mfg.	3,357	1,992	1,959	1,794	1,861	2,120
316900	Other leather & allied product mfg.	2,339	2,106	1,713	3,546	4,056	3,891

Source: U.S. Bureau of Economic Analysis
Plunkett Research, Ltd.
www.plunkettresearch.com

Shipments in the Textile & Apparel Industries: Selected Years, 2000-2011

(In Millions of Dollars; Latest Year Available)

NAICS	Description	2000	2005	2008	2009	2010	2011
313111	Yarn Spinning Mills	6,374	6,239	5,136	3,820	4,269	4,591
313112	Yarn Texturizing, Throwing, & Twisting Mills	4,586	4,363	3,591	2,671	2,985	3,210
313113	Thread Mills	374	284	234	174	194	209
313210	Broadwoven Fabric Mills	15,562	8,572	5,228	3,990	4,458	4,226
313221	Narrow Fabric Mills	1,448	1,361	1,133	954	1,066	1,034
313222	Schiffli Machine Embroidery	311	154	128	108	121	117
313230	Nonwoven Fabric Mills	4,873	6,078	6,350	6,306	7,047	7,045
313241	Weft Knit Fabric Mills	2,193	1,666	994	667	745	780
313249	Other Knit Fabric & Lace Mills	2,023	1,199	715	480	536	561
313311	Broadwoven Fabric Finishing Mills	8,420	5,636	3,362	2,788	3,116	3,137
313312	Textile & Fabric Finishing (except Broadwoven Fabric) Mills	3,605	2,996	1,787	1,482	1,657	1,668
313320	Fabric Coating Mills	2,343	2,540	2,298	1,931	2,158	2,427
314110	Carpet & Rug Mills	12,748	16,620	12,620	9,395	9,511	9,638
314121	Curtain & Drapery Mills	2,182	1,416	825	668	677	682
314129	Other Household Textile Product Mills	7,506	6,713	3,911	3,168	3,207	3,234
314911	Textile Bag Mills	877	961	842	723	732	781
314912	Canvas & Related Product Mills	1,720	2,484	2,176	1,868	1,891	2,019
314991	Rope, Cordage, & Twine Mills	821	854	787	833	843	866
314992	Tire Cord & Tire Fabric Mills	1,479	1,486	1,383	1,052	1,065	1,264
314999	All Other Miscellaneous Textile Product Mills	6,321	5,750	5,245	4,260	4,312	4,797
315111	Sheer Hosiery Mills	1,185	969	553	469	471	515
315119	Other Hosiery & Sock Mills	2,952	2,278	1,300	1,102	1,108	1,211
315191	Outerwear Knitting Mills	2,917	1,029	444	289	290	382
315192	Underwear & Nightwear Knitting Mills	515	527	227	148	149	195
315211	Men's & Boys' Cut & Sew Apparel Contractors	1,344	1,018	686	668	671	711
315212	Women's, Girls', & Infants' Cut & Sew Apparel Contractors	2,746	3,101	2,089	2,034	2,044	2,166
315221	Men's & Boys' Cut & Sew Underwear & Nightwear Mfg.	1,858	161	74	54	54	48
315222	Men's & Boys' Cut & Sew Suit, Coat, & Overcoat Mfg.	1,842	656	301	220	221	194
315223	Men's & Boys' Cut & Sew Shirt (except Work Shirt) Mfg.	4,128	1,187	544	398	400	352
315224	Men's & Boys' Cut & Sew Trouser, Slack, & Jean Mfg.	6,059	3,193	1,462	1,069	1,075	946
315225	Men's & Boys' Cut & Sew Work Clothing Mfg.	1,909	1,193	547	400	402	354
315228	Men's & Boys' Cut & Sew Other Outerwear Mfg.	1,400	673	308	225	227	199
315231	Women's & Girls' Cut & Sew Lingerie, Loungewear, & Nightwear Mfg.	3,687	1,852	1,259	863	867	809
315232	Women's & Girls' Cut & Sew Blouse & Shirt Mfg.	4,462	3,260	2,216	1,519	1,527	1,423
315233	Women's & Girls' Cut & Sew Dress Mfg.	5,262	2,591	1,761	1,207	1,214	1,131
315234	Women's & Girls' Cut & Sew Suit, Coat, Tailored Jacket, & Skirt Mfg.	1,789	897	610	418	420	392
315239	Women's & Girls' Cut & Sew Other Outerwear Mfg.	9,961	5,002	3,401	2,331	2,343	2,184
315291	Infants' Cut & Sew Apparel Mfg.	719	285	256	170	171	180

(continued on next page)

Shipments in the Textile & Apparel Industries: Selected Years, 2000-2011 (cont.)

NAICS	Description	2000	2005	2008	2009	2010	2011
315292	Fur & Leather Apparel Mfg.	308	171	153	102	103	108
315299	All Other Cut & Sew Apparel Mfg.	941	982	882	587	590	622
315991	Hat, Cap, & Millinery Mfg.	1,018	634	350	326	328	338
315992	Glove & Mitten Mfg.	431	256	142	132	132	136
315993	Men's & Boys' Neckwear Mfg.	457	341	188	175	176	182
315999	Other Apparel Accessories & Other Apparel Mfg.	2,450	1,337	739	688	691	712
316110	Leather & Hide Tanning & Finishing	3,161	2,434	1,663	1,100	980	2,315
316211	Rubber & Plastics Footwear Mfg.	977	374	362	322	287	360
316212	House Slipper Mfg.	254	127	123	109	97	122
316213	Men's Footwear (except Athletic) Mfg.	2,022	1,123	1,087	967	862	1,082
316214	Women's Footwear (except Athletic) Mfg.	369	214	207	184	164	206
316219	Other Footwear Mfg.	139	119	115	102	91	115
316991	Luggage Mfg.	1,288	859	738	596	531	648
316992	Women's Handbag & Purse Mfg.	247	158	136	110	98	119
316993	Personal Leather Good (except Women's Handbag & Purse) Mfg.	355	466	401	324	289	352
316999	All Other Leather Good Mfg.	836	934	803	648	578	705

Note: The industry shipments in this table are based on industry shipments published by the US Census Bureau and reported in the Census of Manufactures, the Annual Survey of Manufactures, and the Manufacturers' Shipments, Inventories, and Orders (M3) survey. These statistics are not comparable to BEA's gross output measures. Gross output represents the market value of an industry's production of which shipments is only one component. The gross output statistics are part of the integrated industry accounts that provide consistent measures of industry output, intermediate inputs, and value added within a balanced input-output framework.

Source: U.S. Bureau of Economic Analysis

Plunkett Research, Ltd.

www.plunkettresearch.com

U.S. Retail Sales, Selected Kinds of Businesses: 2006-2012

(Unadjusted Estimates in Millions of US$)

NAICS	Kind of Business	2006	2007	2008	2009	2010	2011	2012
448	Clothing & accessories stores	213,189	221,205	215,776	204,626	213,735	226,748	239,224
4481	Clothing stores	154,546	161,622	157,937	151,675	158,849	167,976	177,184
44811	Men's clothing stores	8,844	8,772	8,538	7,707	7,803	8,530	8,656
44812	Women's clothing stores	38,809	40,294	38,315	35,860	37,431	38,609	41,589
44814	Family clothing stores	81,712	84,592	83,059	81,420	85,282	88,723	93,207
44819	Other clothing stores	11,042	12,003	11,761	11,309	11,675	13,365	13,846
4482	Shoe stores	26,682	26,811	26,688	25,060	26,359	27,412	28,667
45111	Sporting goods stores	33,869	35,804	37,052	37,071	38,297	39,878	42,611
4521	Department stores*	218,132	213,941	202,915	190,788	189,265	188,065	185,712
452112	Discount department stores	135,470	134,926	130,113	126,426	123,697	121,718	119,896

Note: Estimates are based on data from the Monthly Retail Trade Survey, Annual Retail Trade Survey and administrative records. Total sales may include data for kinds of business not shown.

* Includes data for leased departments operated within department stores. Includes sales of apparel and all other types of merchandise carried by these stores; clothing represents a significant portion of total sales.

Source: U.S. Census Bureau

Plunkett Research, Ltd.

www.plunkettresearch.com

Worldwide Cotton Area, Yield & Production: 2010-2013

Country/Region	Area (Million hectares)				Yield (Kilograms per hectare)				Production (Million 480 lb. bales)			
		Prel.	2012/13 Proj.			Prel.	2012/13 Proj.			Prel.	2012/13 Proj.	
	2010/11	2011/12	Feb	Mar	2010/11	2011/12	Feb	Mar	2010/11	2011/12	Feb	Mar
World	33.45	35.52	34.13	34.17	757	761	759	764	116.33	124.13	118.95	119.87
United States	4.33	3.83	3.82	3.82	910	886	971	971	18.10	15.57	17.01	17.01
Total Foreign	29.12	31.69	30.31	30.36	734	746	732	738	98.23	108.56	101.94	102.86
China	5.25	5.40	5.28	5.28	1,265	1,335	1,403	1,445	30.50	33.10	34.00	35.00
South Asia	14.03	15.29	14.8	14.8	546	545	519	515	35.19	38.27	35.28	34.98
India	11.14	12.20	11.70	11.70	516	491	475	475	26.40	27.50	25.50	25.50
Pakistan	2.80	3.00	3.00	3.00	672	769	697	675	8.64	10.60	9.60	9.30
Former Soviet Union	2.32	2.33	2.29	2.36	639	641	667	674	6.82	6.86	7.03	7.3
Uzbekistan	1.33	1.31	1.29	1.35	671	698	729	726	4.10	4.20	4.30	4.50
Turkmenistan	0.64	0.58	0.60	0.60	595	530	544	581	1.75	1.40	1.50	1.60
Tajikistan	0.16	0.20	0.20	0.20	550	619	599	599	0.41	0.58	0.55	0.55
Kazakhstan	0.14	0.16	0.15	0.15	664	590	736	662	0.42	0.44	0.50	0.45
Sub-Saharan Africa	3.29	4.02	4.29	4.29	296	313	332	334	4.47	5.77	6.53	6.57
Burkina Faso	0.40	0.40	0.53	0.53	354	381	435	456	0.65	0.70	1.05	1.10
Mali	0.26	0.48	0.50	0.50	398	386	435	435	0.48	0.85	1.00	1.00
Zimbabwe	0.40	0.47	0.43	0.43	272	278	243	243	0.50	0.60	0.48	0.48
Benin	0.18	0.20	0.25	0.25	363	381	479	479	0.30	0.35	0.55	0.55
Cote d'Ivoire	0.20	0.25	0.33	0.33	370	392	402	402	0.34	0.45	0.60	0.60
Cameroon	0.14	0.18	0.23	0.23	428	404	445	445	0.28	0.33	0.46	0.46
Nigeria	0.41	0.33	0.30	0.30	252	218	236	236	0.48	0.33	0.33	0.33
Sudan	0.04	0.10	0.06	0.06	518	435	454	454	0.10	0.20	0.13	0.13
South America	2.15	2.14	1.55	1.52	1,102	1,051	1,110	1,100	10.89	10.31	7.88	7.68
Brazil	1.40	1.40	1.00	0.98	1,400	1,353	1,415	1,407	9.00	8.70	6.50	6.30
Argentina	0.61	0.53	0.36	0.36	486	411	484	484	1.35	1.00	0.80	0.80
Paraguay	0.03	0.08	0.07	0.07	399	340	373	373	0.06	0.13	0.12	0.12
Middle East	0.61	0.8	0.68	0.68	1,180	1,298	1,202	1,202	3.29	4.79	3.73	3.73
Iran	0.12	0.12	0.11	0.11	635	587	591	591	0.35	0.31	0.29	0.29
Syria	0.13	0.15	0.13	0.13	1,280	1,306	1,263	1,263	0.74	0.90	0.73	0.73
Turkey	0.32	0.49	0.40	0.40	1,436	1,529	1,415	1,415	2.11	3.44	2.60	2.60
Australia	0.54	0.58	0.45	0.45	1,681	2,065	2,055	2,055	4.20	5.50	4.20	4.20
European Union	0.30	0.35	0.33	0.33	854	994	930	930	1.16	1.62	1.41	1.41
Greece	0.23	0.29	0.26	0.26	890	1,016	963	963	0.94	1.33	1.15	1.15
Spain	0.06	0.07	0.07	0.07	715	897	800	800	0.21	0.28	0.25	0.25
Egypt	0.16	0.22	0.15	0.15	763	737	773	773	0.55	0.75	0.52	0.52
Mexico	0.11	0.19	0.15	0.15	1,436	1,338	1,343	1,494	0.73	1.18	0.94	1.05
Others	0.37	0.37	0.37	0.37	248	251	250	250	0.42	0.43	0.43	0.43

Source: U.S. Department of Agriculture
Plunkett Research, Ltd.
www.plunkettresearch.com

Apparel & Textiles Industry Employment by Business Type, U.S.: 2008-2013

(Annual Estimates in Thousands of Employed Workers; Not Seasonally Adjusted)

NAICS[1]	Industry Sector	2008	2009	2010	2011	2012	2013*
Nondurable Goods							
313	**Textile mills**	**151.2**	**124.4**	**119.0**	**120.1**	**118.0**	**114.3**
3131	Fiber, yarn & thread mills	37.1	30.1	29.0	29.4	28.0	27.8
3132	Fabric mills	65.5	55.3	53.4	55.5	55.2	53.5
31321	Broadwoven fabric mills	34.1	29.0	27.5	29.2	28.8	28.0
3133	Textile & fabric finishing mills	48.7	39.0	36.6	35.2	34.8	33.0
314	**Textile product mills**	**147.2**	**125.7**	**119.0**	**117.6**	**116.6**	**115.5**
3141	Textile furnishings mills	75.1	62.2	56.9	54.6	52.1	51.5
3149	Other textile product mills	72.2	63.5	62.1	63.0	64.4	64.0
31491	Textile bag & canvas mills	28.2	25.2	24.3	24.9	24.7	23.8
31499	All other textile product mills	43.9	38.3	37.9	38.2	39.7	40.2
315	**Apparel**	**199.0**	**167.5**	**156.6**	**151.7**	**148.1**	**148.4**
3152	Cut & sew apparel	155.4	132.0	124.0	123.8	122.1	123.3
31521	Cut & sew apparel contractors	60.7	49.9	46.2	46.0	46.1	47.8
31522	Men's & boys' cut & sew apparel	37.4	31.3	29.0	29.5	29.4	28.9
31524,8	Women's & all other cut & sew apparel	57.3	50.8	48.7	48.3	46.5	46.6
3151,9	All other apparel manufacturing	43.6	35.5	32.6	27.9	26.1	25.1
316	**Leather & allied products**	**33.1**	**29.0**	**27.8**	**29.2**	**29.4**	**27.7**
Wholesale Trade							
4243	**Apparel & piece goods**	**148.6**	**138.0**	**136.7**	**138.4**	**142.9**	**147.0**
42432	Men's & boys' clothing	30.7	27.5	26.2	25.5	25.9	27.6
42433	Women's & children's clothing	65.4	61.1	63.1	64.6	67.3	69.0
Retail Trade							
448	**Clothing & clothing accessories stores**	**1,468.0**	**1,363.9**	**1,352.5**	**1,360.9**	**1,407.9**	**1,392.1**
4481	Clothing stores	1,121.4	1,047.6	1,039.8	1,043.1	1,080.9	1,064.3
44811	Men's clothing stores	67.8	57.1	55.1	53.7	52.6	52.0
44812	Women's clothing stores	288.4	275.9	279.1	285.5	291.7	285.1
44813	Children's & infants' clothing stores	73.5	72.2	73.4	76.7	82.0	79.9
44814	Family clothing stores	525.9	485.8	470.1	453.4	469.2	451.3
44815	Clothing accessories stores	52.3	51.8	53.6	58.9	59.5	61.5
44819	Other clothing stores	113.5	104.8	108.5	114.9	125.9	134.5
4482	Shoe stores	188.3	178.8	181.5	184.9	189.4	188.9
4483	Jewelry, luggage, & leather goods stores	158.4	137.4	131.2	133.0	137.7	138.9
Dry Cleaning & Laundry Services							
8123	**Dry-cleaning & laundry services**	**332.4**	**311.1**	**301.6**	**298.8**	**297.7**	**296.7**
81231	Coin-operated laundries & drycleaners	34.3	32.9	33.4	33.3	34.5	35.1
81232	Dry-cleaning & laundry services, except coin operated	163.5	151.3	145.9	143.5	141.6	139.5
81233	Linen & uniform supply	134.7	126.8	122.3	122.1	121.6	122.1
812331	Linen supply	75.8	72.3	69.7	70.3	70.2	70.0
812332	Industrial launderers	58.9	546.0	52.6	51.7	51.4	52.1

[1] For a full description of the NAICS codes used in this table, see www.census.gov/eos/www/naics/

* Preliminary estimate as of February 2013.

Source: U.S. Bureau of Labor Statistics
Plunkett Research, Ltd.
www.plunkettresearch.com

Chapter 3

IMPORTANT APPAREL & TEXTILES INDUSTRY CONTACTS

Addresses, Telephone Numbers and Internet Sites

Contents:

I. Apparel Associations

American Apparel and Footwear Manufacturing Association (AAFA)
1601 N. Kent St., Ste. 12
Arlington, VA 22209 US
Phone: 703-524-1864
Fax: 703-522-6741
Toll Free: 800-520-2262
Web Address: www.wewear.org
The American Apparel and Footwear Manufacturing Association (AAFA) is the national trade association for the apparel, footwear and fashion industries and their suppliers.

American Apparel Producers' Network (AAPN)
P.O. Box 720693
Atlanta, GA 30358 US
Phone: 404-843-3171
Fax: 404-671-9456
E-mail Address: *sue@aapnetwork.net*
Web Address: www.aapnetwork.net
American Apparel Producers' Network (AAPN) is a global network of over 600 contractors, suppliers and manufacturers for the apparel industry. It recently launched AAPN apparelpedia, at Aapnetwork.pbwiki.com, a resource written by AAPN members for others in the apparel industry.

Association of Sewing and Design Profesionals (ASDP)
2885 Sanford Ave. SW, Ste. 19588
Grandville, MI 49418 US
Toll Free: 877-755-0303
E-mail Address: *admin@sewingprofessionals.org*
Web Address: www.paccprofessionals.org
The Association of Sewing and Design Profesionals (ASDP), formerly the Professional Association of Custom Clothiers, is a professional organization created to meet the special needs of those engaged in the businesses of sewing and design, in both home-based and professional settings.

British Footwear Association
3 Burystead Pl.
Wellingborough
Northants, NN8 1AH UK
Phone: 44-1933-229005
Fax: 44-1933-225009
E-mail Address:
info@britishfootwearassociation.co.uk
Web Address: www.britishfootwearassociation.co.uk
The British Footwear Association represents those in all aspects of the footwear industry. The site serves as a portal for the industry with links to member companies, partners, trends and information.

California Fashion Association (CFA)
444 S. Flower St., Fl. 34
Los Angeles, CA 90071 US
Phone: 213-688-6288
Fax: 213-688-6290
E-mail Address: *info@calfashion.org*
Web Address: www.calfashion.org
The California Fashion Association (CFA) is an umbrella association of 16 organizations representing the apparel industry in California.

Canadian Apparel Federation (CAF)
151 Slater St., Ste. 708
Ottawa, ON K1P 5H3 Canada
Phone: 613-231-3220
Fax: 613-231-2305
E-mail Address: *info@apparel.ca*
Web Address: www.apparel.ca
The Canadian Apparel Federation (CAF) is a national trade association for the apparel industry in Canada. The CAF also publishes Canadian Apparel Magazine.

Canton Footwear Association (CFA)
No. 103 Huan Shi Xi Lu
Huan Qiu Guo Ji Business Ctr.
Canton, China
Phone: 86-20-3616-8888
Fax: 86-20-3616-8666
E-mail Address: *secretariat@cantonfootwear.org*
Web Address: cantonfootwear.cfa020.org
The Canton Footwear Association (CFA) is a membership organization representing manufacturers in Mainland China related to the footwear industry.

China Chamber of Commerce for Import & Export of Textiles (CCCT)
No. 12 Panjiayuan Nanli
Chaoyang District
Beijing, 100021 China
Phone: 86-10-6773-9246
Fax: 86-10-6771-9235
E-mail Address: *info@ccct.org.cn*
Web Address:
english.chinaassn.com/show.aspx?id=34949
The China Chamber of Commerce for Import and Export of Textiles (CCCT), established in October 1988, is a leading national trade organization representing exporters and importers of textiles and clothing in China, with a membership of more than 6,300 companies.

China National Garment Association (CNGA)
Phone: 86-10-8522-9073
Fax: 86-10-8522-9461
E-mail Address: *contact@cnga.org.cn*
Web Address: www.cnga.org.cn
The China National Garment Association (CNGA) promotes the development of the garment related industry in China.

Clothing Manufacturers' Association of India (The, CMAI)
902 Mahalaxmi Chambers
22 Bhulabhai Desai Rd.
Mumbai, 400 026 India
Phone: 022-2353-82-45
Fax: 022-2351-59-08
E-mail Address: *info@cmai.in*
Web Address: www.cmai.in
The Clothing Manufacturers Association Of India (CMAI) is the representative association of the Indian apparel industry for over four decades. It has a membership base of over 20,000 companies.

Council of Fashion Designers of America (CFDA)
65 Bleeker, Fl. 11
New York, NY 10012
E-mail Address: *info@cfda.com*

Web Address: ww.cfda.com
The Council of Fashion Designers of America, Inc. (CFDA) is a not-for-profit trade association of over 300 of America's foremost fashion and accessory designers. Founded in 1962, the CFDA continues to advance the status of fashion design as a branch of American art and culture; to raise its artistic and professional standards; to define a code of ethical practices of mutual benefit in public and trade relations; and, to promote appreciation of the fashion arts through leadership in quality and aesthetic discernment.

Council of Fashion Designers, Tokyo
Phone: 03-5413-3188
Web Address: www.cfd.or.jp
Reorganized in September 2008, the group is focused on enhancing creativity and business opportunities for its fashion designer members.

Custom Tailors & Designers Association (CTDA)
42732 Ridgeway Dr.
Broadlands, VA 20148 US
Fax: 866-661-1240
Toll Free: 888-248-2832
Web Address: www.ctda.com
Custom Tailors & Designers Association (CTDA), one of the oldest trade organizations in the United States, represents the interests of the U.S. tailoring industry.

Fashion Accessories Shippers' Association, Inc. (FASA/Gemini)
137 W. 25th St., Fl. 3
New York, NY 10001 US
Phone: 212-947-3424
Fax: 212-629-0361
E-mail Address: *info@accessoryweb.com*
Web Address: www.accessoryweb.com
The Fashion Accessories Shippers' Association, Inc. (FASA/Gemini), formerly known as the National Handbag Association, represents the handbag and fashion accessories industry.

Fashion Footwear Association of New York (FFANY)
274 Madison Ave., Ste. 1701
New York, NY 10016 US
Phone: 212-751-6422
Fax: 212-751-6404
E-mail Address: *info@ffany.org*
Web Address: www.ffany.org
The Fashion Footwear Association of New York (FFANY) includes 300 footwear manufacturers representing 800 of the most well-known footwear brands in the world.

Fashion Group International, Inc. (FGI)
8 W. 40th St., Fl. 7
New York, NY 10018 US
Phone: 212-302-5511
Fax: 212-302-5533
E-mail Address: *info@fgi.org*
Web Address: www.fgi.org
The Fashion Group International, Inc. (FGI) is a nonprofit association of professionals in all areas of the fashion, apparel, accessories, beauty and home industries.

Garment Contractors Association of Southern California, Inc. (GCA)
110 E. 9th St., Ste. A701
Los Angeles, CA 90079 US
Phone: 213-629-4422
Fax: 213-629-4517
E-mail Address:
information@garmentcontractors.org
Web Address: www.garmentcontractors.org
The Garment Contractors Association of Southern California, Inc. (GCA) is a nonprofit association representing the retail garment contracting businesses in Southern California.

Hong Kong Footwear Association
185 Prince Edward Rd.
Block A 2/F, Kar Tseuk Bldg.
Kowloon, Hong Kong, China
Phone: 852-2381-2297
Fax: 852-2397-6927
E-mail Address: *hkfootwear@netvigator.com*
Web Address: www.hkfootwear.org
Hong Kong Footwear Association represents footwear manufacturing in Hong Kong.

Hong Kong Garment Manufacturers Association
25 Kimberley Rd.
TST, Rm 401-3 4/F Cheung Lee Comm. Bldg.
Kowloon, New Territories, China
Phone: 852-2-305-2893
Fax: 852-2-305-2493
E-mail Address: *sec@textilecouncil.com*
Web Address:
www.textilecouncil.com/member_hkgm.html
The Hong Kong Garment Manufacturers Association is a membership organization open to any

incorporated firm in Hong Kong which engages in the garment trade.

Hong Kong Knitwear Exporters & Manufacturers Association
25 Kimberley Rd.
TST, Rm 401-3 4/F Cheung Lee Comm. Bldg.
Kowloon, New Territories, China
Phone: 852-2-2755-2621
Fax: 852-2-2756-5672
Web Address:
www.textilecouncil.com/member_hkkema.html
Hong Kong Knitwear Exporters & Manufacturers Association promotes manufacturing related to knitwear products in Hong Kong.

Hosiery Association (The, THA)
7421 Carmel Executive Park, Ste. 200
Charlotte, NC 28226 US
Phone: 704-365-0913
Fax: 704-362-2056
E-mail Address: *THAinfo@hosieryassociation.com*
Web Address: www.hosieryassociation.com
The Hosiery Association (THA) provides standards, job opportunities and advocates for the sock and hosiery industry.

International Apparel Federation (IAF)
P.O. Box 428
Zeist, 3700 AK The Netherlands
Phone: 31-30-232-09-08
E-mail Address: *vanstaden@iafnet.com*
Web Address: www.iafnet.com
The International Apparel Federation (IAF) is the world's leading association of apparel manufacturers, representing 150,000 companies in 30 countries. It recently moved its headquarters from London to Zeist (Amsterdam).

International Swimwear and Activewear Market (ISAM)
13351-D Riverside Dr., Ste. 658
Sherman Oaks, CA 91423 US
Phone: 818-986-2152
Fax: 818-986-2637
E-mail Address: *barbara@isamla.com*
Web Address: www.isamla.com
The International Swimwear and Activewear Market (ISAM) is an association of retail swimwear and activewear manufacturers, related businesses and industries.

Los Angeles Fashion District
110 E. 9th St., Ste. A-1175
Los Angeles, CA 90079 US
Phone: 213-488-1153
Fax: 213-488-5159
E-mail Address: *info@fashiondistrict.org*
Web Address: www.fashiondistrict.org
The Los Angeles Fashion District is a business improvement district that promotes the L.A. fashion industry; it is run by the Downtown Property Owners Association.

Organization of Black Designers
300 M St. SW, Ste. N110
Washington, DC 20024-4019 US
Phone: 202-659-3918
E-mail Address: *OBDesign@aol.com*
Web Address: www.core77.com/OBD/welcome.html
DesigNation, the Organization of Black Designers' Multidiscipline & Multicultural Design Conference, promotes design opportunities for minorities.

SEAMS
4921-C Broad River Rd.
Columbia, SC 29212 US
Phone: 803-772-5861
Fax: 803-731-7709
E-mail Address: *info@seams.org*
Web Address: www.seams.org
SEAMS is a professional organization representing the sewn products industry.

Sewn Products Equipment and Suppliers of the Americas (SPESA)
9650 Strickland Rd., Ste. 103-234
Raleigh, NC 27615 US
Phone: 919-872-8909
Fax: 919-872-1915
E-mail Address: *info@spesa.org*
Web Address: www.spesa.org
Sewn Products Equipment and Suppliers of the Americas (SPESA) is dedicated to the promotion of the global sewn products industry.

Sub-Council of Textile Industry, China Council (CCPIT TEX)
12 E. Chang An St.
Beijing, China
Phone: 86-010-85229450
Fax: 8-010-65121732
E-mail Address: *hwb@ccpittex.com*
Web Address: www.ccpittex.com

Sub-Council of Textile Industry, China Council for the Promotion of Int'l Trade (CCPIT TEX) promotes the international textile and garment trade.

Sustainable Apparel Coalition
901 Mission St., Ste. 105
San Francisco, CA 94109 US
Web Address: www.apparelcoalition.org
The Sustainable Apparel Coalition is an industry-wide group of over 60 leading apparel and footwear brands, retailers, suppliers, nonprofits, and NGOs working to reduce the environmental and social impacts of apparel and footwear products around the world. The focus of the Sustainable Apparel Coalition is The Higg Index that measures the environmental performance of apparel products. Future versions will include footwear products and measure social performance.

Textile & Fashion Federation, Singapore (TaFF)
No. 2 Leng Kee Rd.
02-09 Thye Hong Ctr.
159086 Singapore
Phone: 65-6735-8390
Fax: 65-6735-8409
E-mail Address: *admin@taff.org.sg*
Web Address: www.taff.org.sg
The Textile & Fashion Federation, Singapore (TaFF) represents the textile and apparel industry in Singapore, which is one of the largest textile and apparel sourcing centers in the Asia-Pacific region.

UK Fashion & Textile Association (UKFT)
3 Queen Sq.
Bloomsbury
London, WC1N #AR UK
Phone: 44-20-7843-9460
Fax: 44-20-7843-9478
E-mail Address: *info@ukft.org*
Web Address: www.ukft.org
The UK Fashion & Textile Association (UKFT), formerly the British Clothing Industry Association, works to promote the fashion, clothing and knitting businesses across the UK.

II. Apparel Industry Resources

Apparel Magazine
801 Gervais St., Ste. 101
Columbia, SC 29201 US
Phone: 803-771-7500
Fax: 803-799-1461
E-mail Address: *jspeer@apparelmag.com*
Web Address: www.apparelmag.com
Apparel Magazine provides news and information for the apparel industry.

Apparel Technology & Research Center (ATRC)
3801 W. Temple Ave.
Cal Poly Pomona
Pomona, CA 91768 US
Phone: 909-869-3377
E-mail Address: *atrc@csupomona.edu*
Web Address: www.csupomona.edu/~atrc
The Apparel Technology & Research Center (ATRC) at California State Polytechnic University in Pomona is dedicated to the advancement of apparel manufacturing.

California Apparel News
110 E. 9th St., Ste. A-777
Los Angeles, CA 90079-1777 US
Phone: 213-627-3737
E-mail Address: *info@apparelnews.net*
Web Address: www.apparelnews.net
Apparel News is a newspaper that covers the fashion industry in California.

China Textile Network Company (CNTEX)
12 Dong Chang'an St.
Beijing, 100742 China
Phone: 86-10-65954825
Fax: 86-10-65135947
E-mail Address: *b2b@cntextile.com*
Web Address: www.cntextile.com
The China Textile Network Company (CNTEX), sponsored by the China National Textile & Apparel Council and the China Textile Information Center, is an e-Market database for buyers to find and purchase from producers and suppliers in the fabric, apparel and home furnishings market.

Costume Society of America (CSA)
390 Amwell Rd., Ste. 402
Hillsborough, NJ 08844 US
Phone: 908-359-1471
Fax: 908-450-1118
Toll Free: 800-272-9447
E-mail Address:
national.office@costumesocietyamerica.com
Web Address: www.costumesocietyamerica.com
The Costume Society of America (CSA) advances the preservation, research and education of all aspects of dress and appearance. The Society tends to view costume more in the sense of what people wear, that is, what clothing defines a time and a place, rather

than the word's more common interpretation as something extraordinary worn on rare occasions, as a Halloween costume. Members often work in museums, the theater and libraries. CSA publishes the annual journal, Dress; the CSA News newsletter; and CSA E-News electronic newsletter.

Material Connexion
60 Madison Ave., Fl. 2
New York, NY 10010 US
Phone: 212-842-2050
Fax: 212-842-1090
E-mail Address: *info@materialconnexion.com*
Web Address: www.materialconnexion.com
Material Connexion maintains a carefully curated library of approximately 7,000 textiles and materials for use in apparel, interior design and other applications. The materials are often very innovative and may include polymers, ceramics, glass, metals and sustainable products. Access can be through their physical library or their online database. They have offices and libraries in major cities in North America, Europe and Asia.

Zhejiang China Apparel Network Technology Co., Ltd
276 Shuanglong Rd. S.
Fl. 15, Jinhua Daily Bldg.
Jinhua City, Zhejiang 321017 China
Phone: 86-579-82057999
Fax: 866-579-82057998
Toll Free: 800-857-9699
E-mail Address: *info@efu.com.cn*
Web Address: www.chinaapparel.net
Zhejiang China Apparel Network Technology Company is one of the biggest e-commerce marketplaces in China for international and domestic garment trade. The company's web site, ChinaApparel.net provides users with trade leads, news, network broadcast ads, a company directory and overview, electronic commerce, website creation and special member services.

III. Careers-Apparel

24 Seven Fashion Recruitment
120 Wooster St., Fl. 4
New York, NY 10012 US
Phone: 212-966-4426
Fax: 212-966-2313
E-mail Address: *newyork@24seventalent.com*
Web Address: www.24seveninc.com
24 Seven Fashion Recruitment is an employment agency serving the fashion, beauty, entertainment, advertising, marketing and retail industries.

Fashion Career Center
950 Towe Ln., Fl. 6
Foster City, CA 94404
E-mail Address: *info@quinstreet.com*
Web Address: www.fashioncareercenter.com
The Fashion Career Center site provides employees and employers with a place to meet and access information about employment in the fashion industry. The FashionCareerCenter.com web site offers links to fashion jobs and fashion schools, as well as offering fashion career advice.

IV. Careers-First Time Jobs/New Grads

Black Collegian Online (The)
140 Carondelet St.
New Orleans, LA 70130 US
Phone: 504-523-0154
Web Address: www.blackcollegian.com
The Black Collegian Online features listings for job and internship opportunities, as well as other tools for students of color; it is the web site of The Black Collegian Magazine, published by IMDiversity, Inc. The site includes a list of the top 100 minority corporate employers and an assessment of job opportunities.

CollegeGrad.com, Inc.
950 Tower Ln., Fl. 6
Foster City, CA 94404 US
E-mail Address: *info@quinstreet.com*
Web Address: www.collegegrad.com
CollegeGrad.com, Inc. offers in-depth resources for college students and recent grads seeking entry-level jobs.

MBAjobs.net
Web Address: www.mbajobs.net
MBAjobs.net is an international service for MBA students and graduates, employers, recruiters and business schools. It aims to match MBA graduates with employers and provides links to online MBA programs.

MonsterCollege
799 Market St., Ste. 500
San Francisco, CA 94103 US
E-mail Address: *info@college.monster.com*
Web Address: www.college.monster.com

MonsterCollege provides information about internships and entry-level jobs, as well as career advice and resume tips, to recent college graduates.

National Association of Colleges and Employers (NACE)
62 Highland Ave.
Bethlehem, PA 18017-9085 US
Phone: 610-868-1421
Fax: 610-868-0208
Toll Free: 800-544-5272
E-mail Address: *customer_service@naceweb.org*
Web Address: www.naceweb.org
The National Association of Colleges and Employers (NACE) is a premier U.S. organization representing college placement offices and corporate recruiters who focus on hiring new grads.

V. Careers-General Job Listings

Career Exposure, Inc.
3934 SW Corbett Ave.
Portland, OR 97239 US
Phone: 503-221-7779
E-mail Address: *pr@careerexposure.com*
Web Address: www.careerexposure.com
Career Exposure, Inc. is an online career center and job placement service, with resources for employers, recruiters and job seekers.

CareerBuilder, Inc.
200 N. LaSalle St., Ste. 1100
Chicago, IL 60601 US
Phone: 773-527-3600
Fax: 773-353-2452
Toll Free: 866-438-1485
E-mail Address: *jennifer.grasz@careerbuilder.com*
Web Address: www.careerbuilder.com
CareerBuilder, Inc. focuses on the needs of companies and also provides a database of job openings. The site has over 1 million jobs posted by 300,000 employers, and receives an average 23 million unique visitors monthly. The company also operates online career centers for 140 newspapers and 9,000 online partners. Resumes are sent directly to the company, and applicants can set up a special e-mail account for job-seeking purposes. CareerBuilder is primarily a joint venture between three newspaper giants: The McClatchy Company, Gannett Co., Inc. and Tribune Company.

CareerOneStop
Toll Free: 877-348-0502
E-mail Address: *info@careeronestop.org*
Web Address: www.careeronestop.org
CareerOneStop is operated by the employment commissions of various state agencies. It contains job listings in both the private and government sectors, as well as a wide variety of useful career resources and workforce information. CareerOneStop is sponsored by the U.S. Department of Labor.

LaborMarketInfo (LMI)
Employment Development Dept.
800 Capitol Mall, MIC 83
Sacramento, CA 95814 US
Toll Free: 800-480-3287
Web Address: www.labormarketinfo.edd.ca.gov
LaborMarketInfo (LMI) provides job seekers and employers a wide range of resources, namely the ability to find, access and use labor market information and services. It provides statistics for employment demographics on both a local and regional level, as well as career searching tools for California residents. The web site is sponsored by California's Employment Development Office.

Recruiters Online Network
E-mail Address: *info@bizjournals.com*
Web Address: www.recruitersonline.com
The Recruiters Online Network provides job postings from thousands of recruiters, Careers Online Magazine, a resume database, as well as other career resources.

USAJOBS
1900 E St. NW
Washington, DC 20415 US
Phone: 202-606-1800
Web Address: www.usajobs.gov
USAJOBS, a program of the U.S. Office of Personnel Management, is the official job site for the U.S. Federal Government. It provides a comprehensive list of U.S. government jobs, allowing users to search for employment by location; agency; type of work; or by senior executive positions. It also has special employment sections for individuals with disabilities, veterans and recent college graduates; an information center, offering resume and interview tips and other information; and allows users to create a profile and post a resume.

VI. Careers-Job Reference Tools

Vault.com, Inc.
132 W. 31st St., Fl. 17
New York, NY 10001 US
Phone: 212-366-4212
Fax: 212-366-6117
E-mail Address: *publicity@vault.com*
Web Address: www.vault.com
Vault.com, Inc. is a comprehensive career web site for employers and employees, with job postings and valuable information on a wide variety of industries. Its features and content are largely geared toward MBA degree holders.

VII. Corporate Information Resources

bizjournals.com
120 W. Morehead St., Ste. 400
Charlotte, NC 28202 US
Phone: 855-875-4288
Web Address: www.bizjournals.com
Bizjournals.com is the online media division of American City Business Journals, the publisher of dozens of leading city business journals nationwide. It provides access to research into the latest news regarding companies both small and large. The organization maintains 42 websites and 64 print publications and sponsors over 700 annual industry events.

Business Wire
44 Montgomery St., Fl. 39
San Francisco, CA 94104 US
Phone: 415-986-4422
Fax: 415-788-5335
Toll Free: 800-227-0845
E-mail Address: *info@businesswire.com*
Web Address: www.businesswire.com
Business Wire offers news releases, industry- and company-specific news, top headlines, conference calls, IPOs on the Internet, media services and access to tradeshownews.com and BW Connect On-line through its informative and continuously updated web site.

Edgar Online, Inc.
11200 Rockville Pike, Ste. 310
Rockville, MD 20852 US
Phone: 301-287-0300
Fax: 301-287-0390
Toll Free: 888-870-2316
Web Address: www.edgar-online.com
Edgar Online, Inc. is a gateway and search tool for viewing corporate documents, such as annual reports on Form 10-K, filed with the U.S. Securities and Exchange Commission.

PR Newswire Association LLC
350 Hudson St., Ste. 300
New York, NY 10014-4504 US
Fax: 800-793-9313
Toll Free: 800-776-8090
E-mail Address: *meryl.serouya@prnewswire.com*
Web Address: www.prnewswire.com
PR Newswire Association LLC provides comprehensive communications services for public relations and investor relations professionals, ranging from information distribution and market intelligence to the creation of online multimedia content and investor relations web sites. Users can also view recent corporate press releases from companies across the globe. The Association is owned by United Business Media plc.

VIII. Economic Data & Research

Centre for European Economic Research (The, ZEW)
L 7, 1
Mannheim, 68161 Germany
Phone: 49-621-1235-01
Fax: 49-621-1235-224
E-mail Address: *info@zew.de*
Web Address: www.zew.de/en
Zentrum fur Europaische Wirtschaftsforschung, The Centre for European Economic Research (ZEW), distinguishes itself in the analysis of internationally comparative data in a European context and in the creation of databases that serve as a basis for scientific research. The institute maintains a special library relevant to economic research and provides external parties with selected data for the purpose of scientific research. ZEW also offers public events and seminars concentrating on banking, business and other economic-political topics.

Economic and Social Research Council (ESRC)
Polaris House
North Star Ave.
Swindon, SN2 1UJ UK
Phone: 44-01793 413000
Fax: 44-01793 413001
E-mail Address: *comms@esrc.ac.uk*
Web Address: www.esrc.ac.uk

The Economic and Social Research Council (ESRC) funds research and training in social and economic issues. It is an independent organization, established by Royal Charter. Current research areas include the global economy; social diversity; environment and energy; human behavior; and health and well-being.

Eurostat
5 Rue Alphonse Weicker
Joseph Bech Bldg.
Luxembourg, L-2721 Luxembourg
Phone: 352-4301-33444
Fax: 352-4301-35349
E-mail Address: *eurostat-pressoffice@ec.europa.eu*
Web Address: www.epp.eurostat.ec.europa.eu
Eurostat is the European Union's service that publishes a wide variety of comprehensive statistics on European industries, populations, trade, agriculture, technology, environment and other matters.

Federal Statistical Office of Germany
Gustav-Stresemann-Ring 11
Wiesbaden, D-65189 Germany
Phone: 49-0-611-75-3444
Fax: 49-0-611-75-3976
Web Address: www.destatis.de
Federal Statistical Office of Germany publishes a wide variety of nation and regional economic data of interest to anyone who is studying Germany, one of the world's leading economies. Data available includes population, consumer prices, labor markets, health care, industries and output.

India Brand Equity Foundation (IBEF)
Apparel House, Fl. 5
#519-22, Sector 44
Gurgaon, Haryana 122003 India
Phone: 91-124-4499600
Fax: 91-124-4499615
E-mail Address: *ceo@ibef.org*
Web Address: www.ibef.org
India Brand Equity Foundation (IBEF) is a public-private partnership between the Ministry of Commerce and Industry, the Government of India and the Confederation of Indian Industry. The foundation's primary objective is to build positive economic perceptions of India globally. It aims to effectively present the India business perspective and leverage business partnerships in a globalizing marketplace.

National Bureau of Statistics (China)
57, Yuetan Nanjie, Sanlihe
Xicheng District
Beijing, 100826 China
Fax: 86-10-6878-2000
E-mail Address: *info@gj.stats.cn*
Web Address: www.stats.gov.cn/english
The National Bureau of Statistics (China) provides statistics and economic data regarding China's economy and society.

Organization for Economic Co-operation and Development (OECD)
2 rue André Pascal
Cedex 16
Paris, F-75775 France
Phone: 33-1-45-24-82-00
Fax: 33-1-45-24-85-00
E-mail Address: *webmaster@oecd.org*
Web Address: www.oecd.org
The Organization for Economic Co-operation and Development (OECD) publishes detailed economic, government, population, social and trade statistics on a country-by-country basis for over 30 nations representing the world's largest economies. Sectors covered range from industry, labor, technology and patents, to health care, environment and globalization.

Statistics Bureau, Director-General for Policy Planning (Japan)
19-1 Wakamatsu-cho
Shinjuku-ku
Tokyo, 162-8668 Japan
Phone: 81-3-5273-2020
E-mail Address: *stat_webmaster@soumu.go.jp*
Web Address: www.stat.go.jp/english
The Statistics Bureau, Director-General for Policy Planning (Japan) and Statistical Research and Training Institute, a part of the Japanese Ministry of Internal Affairs and Communications, plays the central role of producing and disseminating basic official statistics and coordinating statistical work under the Statistics Act and other legislation.

Statistics Canada
150 Tunney's Pasture Driveway
Ottawa, ON K1A 0T6 Canada
Phone: 613-951-8116
Fax: 877-287-4369
Toll Free: 800-263-1136
E-mail Address: *infostats@statcan.gc.ca*
Web Address: www.statcan.gc.ca

Statistics Canada provides a complete portal to Canadian economic data and statistics. Its conducts Canada's official census every five years, as well as hundreds of surveys covering numerous aspects of Canadian life.

IX. Engineering, Research & Scientific Associations

American National Standards Institute (ANSI)
1899 L St. NW, Fl. 11
Washington, DC 20036 US
Phone: 202-293-8020
Fax: 202-293-9287
E-mail Address: *info@ansi.org*
Web Address: www.ansi.org
The American National Standards Institute (ANSI) is a private, nonprofit organization that administers and coordinates the U.S. voluntary standardization and conformity assessment system. Its mission is to enhance both the global competitiveness of U.S. business and the quality of life by promoting and facilitating voluntary consensus standards and conformity assessment systems and safeguarding their integrity.

American Society of Mechanical Engineers (ASME), Textile Engineering Division
Two Park Ave.
New York, NY 10016-5990 US
Phone: 973-882-1170
Toll Free: 800-843-2763
E-mail Address: *CustomerCare@asme.org*
Web Address: www.asme.org
The textile engineering division of the American Society of Mechanical Engineers (ASME) promotes product and process technology improvement in the retail fiber industry.

X. Fur Industry Associations

American Legend Cooperative
P.O. Box 58308
Seattle, WA 98138
Phone: 425-251-3200
Fax: 425-251-3222
E-mail Address: *info@americanlegend.com*
Web Address: www.americanlegend.com
The American Legend Cooperative is an association of mink producers from Canada and the U.S. It operates as a marketing company with representatives in countries throughout the world.

International Fur Trade Federation (IFTF)
P.O. Box 495
Weybridge, Surrey KT13 8WD UK
Phone: 44-1932-85-00-20
E-mail Address: *info@iftf.com*
Web Address: www.iftf.com
The International Fur Trade Federation (IFTF) promotes innovation, high standards and a positive image or fur and the fur industry worldwide.

XI. Industry Research/Market Research

Forrester Research
60 Acorn Park Dr.
Cambridge, MA 02140 US
Phone: 617-613-5730
E-mail Address: *press@forrester.com*
Web Address: www.forrester.com
Forrester Research is a publicly traded company that identifies and analyzes emerging trends in technology and their impact on business. Among the firm's specialties are the financial services, retail, health care, entertainment, automotive and information technology industries.

Kantar Retail
6 More London Place
Tooley Street
London, SE1 2QY UK
Phone: 44-1932-833-600
Fax: 44-1932-833-601
E-mail Address: *info@kantarretail.com*
Web Address: www.kantarretail.com
Kantar Retail, formed by the amalgamation of Cannondale Associates, Glendinning Management Consultants, Management Ventures Inc. and Retail Forward, is a consulting business that provides in-depth market research on retail. This firm's web site contains press releases with forecasts and statistics regarding retail activity and trends.

MarketResearch.com
11200 Rockville Pike, Ste. 504
Rockville, MD 20852 US
Phone: 240-747-3000
Fax: 240-747-3004
Toll Free: 800-298-5699
E-mail Address:
customerservice@marketresearch.com
Web Address: www.marketresearch.com
MarketResearch.com is a leading broker for professional market research and industry analysis. Users are able to search the company's database of

research publications including data on global industries, companies, products and trends.

NPD Group (The)
900 W. Shore Rd.
Port Washington, NY 11050 US
Phone: 516-625-0700
Toll Free: 866-444-1411
Web Address: www.npd.com
The NPD Group is one of the world's leading market research firms covering the retailing and related sectors. NPD covers industries including automotive, beauty, technology, entertainment, fashion, food & beverage, home, software, toys and wireless.

Plunkett Research, Ltd.
P.O. Drawer 541737
Houston, TX 77254-1737 US
Phone: 713-932-0000
Fax: 713-932-7080
E-mail Address:
customersupport@plunkettresearch.com
Web Address: www.plunkettresearch.com
Plunkett Research, Ltd. is a leading provider of market research, industry trends analysis and business statistics. Since 1985, it has served clients worldwide, including corporations, universities, libraries, consultants and government agencies. At the firm's web site, visitors can view product information and pricing and access a large amount of basic market information on industries such as financial services, InfoTech, e-commerce, health care and biotech.

XII. Labor Unions

UNITE HERE
275 7th Ave.
New York, NY 10001-6708 US
Phone: 212-265-7000
E-mail Address: *tsnyder@unitehere.org*
Web Address: www.unitehere.org
UNITE HERE is a union representing hotel, gaming, food service, manufacturing, distribution, laundry, airport, apparel and textile industries.

XIII. Logistics & Supply Chain Associations

GS1 Hong Kong
160 Gloucester Rd.
22/F, OTB Bldg.
Wanchai, Hong Kong, China
Phone: 852-2861-2819
Fax: 852-2861-2423
E-mail Address: *info@gs1hk.org*
Web Address: www.gs1hk.org
GS1 Hong Kong, a new name of Hong Kong Article Numbering Association, is a not-for-profit, industry led organization to promote global standards, best practices and enabling technologies in the arena of global value and supply chain management.

XIV. Manufacturing Associations-General

China National Light Industry Council (CNLIC)
22B Fuwai St.
Beijing, 100833 China
Phone: 86-10-6839-6613
Fax: 86-10-6839-6264
E-mail Address: *webmaster@clii.com.cn*
Web Address: www.clii.com.cn
The China National Light Industry Council (CNLIC) serves as an intermediary organization representing industry and services and to promote light industry development in China.

XV. MBA Resources

MBA Depot
Web Address: www.mbadepot.com
MBA Depot is an online community and information portal for MBAs, potential MBA program applicants and business professionals.

XVI. Printers & Publishers Associations

Specialty Graphic Imaging Association (SGIA)
10015 Main St.
Fairfax, VA 22031-3489 US
Phone: 703-385-1335
Fax: 703-273-0456
Toll Free: 888-385-3588
E-mail Address: *sgia@sgia.org*
Web Address: www.sgia.org
The Specialty Graphic Imaging Association (SGIA) is devoted to specialty imaging, digital printing and screen printing providers in markets ranging from from advertising to apparel.

XVII. Retail Industry Associations

Alabama Retail Association (ARA)
7265 Halcyon Summit Dr.

Montgomery, AL 36117-3502 US
Phone: 334-263-5757
Fax: 334-262-3991
Toll Free: 800-239-5423
E-mail Address:
alabamaretailassociation@alabamaretail.org
Web Address: www.alabamaretail.org
The Alabama Retail Association (ARA) represents the retail industry in Alabama, with independent merchants and national companies as members.

Illinois Retail Merchants Association (IRMA)
19 S. LaSalle St., Ste. 300
Chicago, IL 60603 US
Phone: 312-726-4600
Fax: 312-726-9570
E-mail Address: *info@irma.org*
Web Address: www.irma.org
The Illinois Retail Merchants Association (IRMA) is one of the largest state retail organizations and it represents more than 23,000 stores of all sizes and merchandise lines.

Kentucky Retail Federation (KRF)
512 Capitol Ave.
Frankfort, KY 40601 US
Phone: 502-875-1444
Fax: 502-875-1595
E-mail Address: *info@kyretail.com*
Web Address: www.kyretail.com
The Kentucky Retail Federation (KRF) is Kentucky's retail association, providing services such as public relations information and business relations screening as well as lists of important numbers.

Maryland Retailers Association (MRA)
171 Conduit St.
Annapolis, MD 21401 US
Phone: 410-269-1440
Fax: 410-269-0325
Web Address: www.mdra.org
The Maryland Retailers Association (MRA) acts as the retail industry's major statewide trade association in Maryland, and consists of over 600 members at approximately 1,400 locations.

Michigan Retailers Association (MRA)
603 S. Washington Ave.
Lansing, MI 48933 US
Phone: 517-372-5656
Fax: 517-372-1303
Toll Free: 800-366-3699
E-mail Address: *mra@retailers.com*
Web Address: www.retailers.com
The Michigan Retailers Association (MRA) is the state retail association in Michigan, providing legislative representation, daily retail news, event information and membership news.

Minnesota Retailers Association (MnRA)
400 Robert St. N., Ste. 1540
St. Paul, MN 55101 US
Phone: 651-227-6631
Fax: 651-297-6260
Toll Free: 800-227-6762
E-mail Address: *mnra@mnretail.org*
Web Address: www.mnretail.org
The Minnesota Retailers Association (MnRA) is the voice of retailing for the merchant community in Minnesota, consisting of small independent retail establishments as well as large national chains. The association provides legislative services, money saving programs and information services.

National Retail Federation (NRF)
325 7th St. NW, Ste. 1100
Washington, DC 20004 US
Phone: 202-783-7971
Fax: 202-737-2849
Toll Free: 800-673-4692
E-mail Address: *cantrellv@nrf.com*
Web Address: www.nrf.com
The National Retail Federation (NRF) is one of the world's largest retail trade organizations. Its membership includes the leading department, specialty, independent, discount and mass merchandise stores in the United States and 50 nations worldwide.

National Shoe Retailers Association (NSRA)
7386 N. La Cholla Blvd.
Tucson, AZ 85741 US
Phone: 520-209-1710
Toll Free: 800-673-8446
E-mail Address: *info@nsra.org*
Web Address: www.nsra.org
The National Shoe Retailers Association (NSRA) is a trade organization that provides practical information and business services. In addition, it serves as an advocate for the independent shoe retailing community, representing storeowners and operators within the broader retail industry.

North Carolina Retail Merchants Association (NCRMA)
601 St. Mary's St.

Raleigh, NC 27605 US
Phone: 919-832-0811
Fax: 919-832-0812
Toll Free: 800-662-7211
E-mail Address: *info@ncrma.org*
Web Address: www.ncrma.org
The North Carolina Retail Merchants Association (NCRMA) is a nonprofit trade association for retailers in the state of North Carolina. On its web site, business owners and merchants can find links to the state government, as well as news and information on area retail trends.

North Dakota Retail Association (NDRA)
1025 N. 3rd St.
P.O. Box 1956
Bismarck, ND 58502 US
Phone: 701-223-3370
Fax: 701-223-5004
Web Address: www.ndretail.org
The North Dakota Retail Association (NDRA) is a trade association composed of large and small retail businesses across North Dakota. Its site offers organization information and news affecting North Dakota merchants.

Retail Council of Canada (RCC)
1255 Bay St., Ste. 800
Toronto, ON M5R 2A9 Canada
Phone: 416-922-6678
Fax: 877-790-4271
Toll Free: 888-373-8245
E-mail Address: *info@retailcouncil.org*
Web Address: www.retailcouncil.org
The Retail Council of Canada (RCC) is a trade association for Canadian retailers, representing more than 40,000 stores.

Retail Industry Leaders Association (RILA)
1700 N. Moore St., Ste. 2250
Arlington, VA 22209 US
Phone: 703-841-2300
Fax: 703-841-1184
E-mail Address: *chrissa.benson@rila.org*
Web Address: www.rila.org
The Retail Industry Leaders Association (RILA) is a world leading alliance of retailers and their product and service suppliers.

Retail Merchants Association of New Hampshire (RMANH)
35A S. Main St.
Concord, NH 03301 US
Phone: 603-225-9748
Fax: 603-229-0060
E-mail Address: *rmanh@rmanh.org*
Web Address: www.rmanh.org
The Retail Merchants Association of New Hampshire (RMANH) is a statewide trade association representing over 800 businesses in New Hampshire. Its members range from small independent business owners to major corporations.

Retailers Association of India
111/112 Ascot Ctr., Next to Hotel ITC Maratha
Sahar Rd., Sahar, Andheri (E)
Mumbai, 400 099 India
Phone: 91-22-28269527
Fax: 91-22-28269536
E-mail Address: *heena@rai.net.in*
Web Address: www.rai.net.in
The Retailers Association of India promotes and supports retailers and the retail industry in India.

Singapore Retailers Association (SRA)
371 Beach Rd.
02-04/05 Keypoint
199597 Singapore
Phone: 65-6295-2622
Fax: 65-6295-2722
E-mail Address: *info@sra.org.sg*
Web Address: www.retail.org.sg
Singapore Retailers Association (SRA) is a nonprofit, independent, non-governmental retail trade body funded entirely by the private sector. SRA seeks to develop and promote the retail sector in Singapore.

Vermont Retail Association (VRA)
148 State St.
Montpelier, VT 05602 US
Phone: 802-839-1931
Fax: 802-839-1927
Toll Free: 800-839-1927
E-mail Address: *mail@vtretailer.com*
Web Address: www.vtretailers.com
The Vermont Retail Association (VRA) is a member organization that represents Vermont retailers. It offers various publications and advocacy services to Vermont businesses.

Wyoming Retail Merchants Association (WRMA)
1825 Carey Ave.
P.O. Box 1003
Cheyenne, WY 82003 US
Phone: 307-634-8816
Fax: 307-632-0249

E-mail Address: *info@wyoretail.org*
Web Address: www.wyoretail.org/index2.cfm
The Wyoming Retail Merchants Association (WRMA) represents large and small retailers across Wyoming.

XVIII. RFID Industry Associations

International RFID Business Association (RFIDba)
Harry P. Pappas
900 N. Orange St., Ste. 195
Media, PA 19063 USA
E-mail Address: *info@rfidba.org*
Web Address: www.rficba.org
The International RFID Business Association (RFIDba) was founded in April 2004 as a not-for-profit, educational, technology and frequency agnostic, trade association dedicated to serving the business needs of the end user community with vendor neutral information on RFID and RTLS technologies along with information on other associated, complimentary technologies. It has branches that are focused on construction, apparel and footware, and healthcare.

XIX. Shopping Center Directories

Chain Store Guide (CSG)
10117 Princess Palm Dr.
Tampa, FL 33610 US
Fax: 813-627-6888
Toll Free: 800-927-9292
E-mail Address: *webmaster@chainstoreguide.com*
Web Address: www.chainstoreguide.com
The Chain Store Guide (CSG) is a provider of comprehensive retail and foodservice intelligence. The CSG database contains over 700,000 retailers, foodservice operators, distributors and wholesalers in the U.S. and Canada.

XX. Sporting Goods Industry Associations

National Sporting Goods Association (NSGA)
1601 Feehanville Dr., Ste. 300
Mt. Prospect, IL 60056 US
Fax: 847-391-9827
Toll Free: 800-815-5422
E-mail Address: *info@nsga.org*
Web Address: www.nsga.org
The National Sporting Goods Association (NSGA) is a trade association designed to help its members profit in a competitive marketplace. The group publishes excellent market research in addition to hosting a major annual conference.

Sports & Fitness Industry Association, The (SFIA)
8505 Fenton St., Ste. 211
Silver Spring, MD 20910 US
Phone: 301-495-6321
Fax: 301-495-6322
E-mail Address: *info@sfia.org*
Web Address: www.sfia.org
The The Sports & Fitness Industry Association (SFIA), formerly the Sporting Goods Manufacturers Association, is a trade organization representing manufacturers of sporting goods equipment.

XXI. Technology Transfer Associations

Licensing Executives Society (U.S.A. and Canada), Inc.
1800 Diagonal Rd., Ste. 280
Alexandria, VA 22314 US
Phone: 703-836-3106
Fax: 703-836-3107
E-mail Address: *info@les.org*
Web Address: www.lesusacanada.org
Licensing Executives Society (U.S.A. and Canada), Inc., established in 1965, is a professional association composed of about 5,000 members who work in fields related to the development, use, transfer, manufacture and marketing of intellectual property. Members include executives, lawyers, licensing consultants, engineers, academic researchers, scientists and government officials. The society is part of the larger Licensing Executives Society International, Inc. (same headquarters address), with a worldwide membership of some 12,000 members from approximately 80 countries.

XXII. Textile & Fabric Associations

American Association of Textile Chemists and Colorists (AATCC)
1 Davis Dr
Research Triangle Park, NC 27709-2215 US
Phone: 919-549-8141
Fax: 919-549-8933
E-mail Address: *colemanb@aatcc.org*
Web Address: www.aatcc.org

The American Association of Textile Chemists and Colorists (AATCC) is one of the world's largest technical and scientific societies devoted to the advancement of textile chemistry.

American Fiber Manufacturers Association (AFMA)
1530 Wilson Blvd., Ste. 690
Arlington, VA 22209 US
Phone: 703-875-0432
Fax: 703-875-0907
E-mail Address: *oday@afma.org*
Web Address: www.fibersource.com/afma/afma.htm
The American Fiber Manufacturers Association (AFMA) is the trade association representing United States producers of manufactured fibers, filaments and yarns. The organization also offers educational services.

American Flocking Association (AFA)
469 Hospital Dr., Ste. C
Gastonia, NC 28054 US
Phone: 617-303-6288
Fax: 617-542-2199
E-mail Address: *info@flocking.org*
Web Address: www.flocking.org
The American Flocking Association (AFA) promotes the flocking process in the textile industry in Mexico, Canada and the U.S.; it is affiliated with the National Textile Association.

American Textile Machinery Association (ATMA)
201 Park Washington Ct.
Falls Church, VA 22046 US
Phone: 703-538-1789
E-mail Address: *info@atmanet.org*
Web Address: www.atmanet.org
The American Textile Machinery Association (ATMA) is one of the only associations devoted to the advancement of manufacturing textile machinery, parts and accessories.

American Textile Manufacturers Institute (ATMI)
1130 Connecticut Ave. NW
Washington, DC 20036-3954 US
Phone: 202-862-0566
Fax: 202-862-0570
Web Address: www.textileweb.com/companyprofile/d67a4c38-c94f-47b3-aa2f-c7d7bd5e164d
The American Textile Manufacturers Institute (ATMI) is a national trade association for the textile industry.

Association of Suppliers to the British Clothing Industry (ASBCI)
25 Square Rd.
Halifax, Unit 5
West Yorkshire, HX1 1QG UK
Phone: 01-42-235-4666
Fax: 01-42-238-1184
E-mail Address: *info@asbci.co.uk*
Web Address: www.asbci.co.uk
The Association of Suppliers to the British Clothing Industry (ASBCI) represents members in all the key sectors connected with the industry. Examples of sectors are: fibers, fabrics, adhesives, dyers, threads, machinery, presses, garment processors, manufacturers, retailers, drycleaners, launderers, domestic detergent manufacturers, chemical suppliers, computer suppliers to the clothing industry, textile testing houses, research organizations.

Brazilian Textile and Apparel Industry Association
Rua Marques de Itu, 968
Sao Paulo, SP 01223-000 Brazil
Phone: 55-11-3823-6100
Fax: 55-11-3823-6122
E-mail Address: *abit@abit.org.br*
Web Address: www.abit.org.br
The Brazilian Textile and Apparel Industry Association (Associacao Brasileira da Industria Textil e de Confeccao) works towards the integration of the textile supply chain in Brazil. Member companies include cotton growers, synthetic raw materials manufacturers, spinning, weaving, dying and printing firms, and clothing producers. The organization supports the sustainable growth of the industry at the local, federal and international level through its activities.

Carpet and Rug Institute (CRI)
100 S. Hamilton St.
Dalton, GA 30720 US
Phone: 706-278-3176
Fax: 706-278-8835
Web Address: www.carpet-rug.org
Carpet and Rug Institute (CRI) is a national trade association representing 90% of carpet and rug manufactures in the US.

Cashmere & Camel Hair Manufacturers Institute (CCMI)
3 Post Office Sq., Fl. 8
Boston, MA 02110 US
Phone: 617-542-7481
E-mail Address: *jlomas@cashmere.org*
Web Address: www.cashmere.org
The Cashmere & Camel Hair Manufacturers Institute (CCMI) is a membership organization that promotes the use of genuine cashmere and camel hair in the textile industry.

China National Textile and Apparel Council (CNTAC)
#39 Dongsihuan Zhonglu, Chaoyang
Rm. 207, Tower B, Huaye Int'l Ctr.
Beijing, 100025 China
Phone: 86-10-85872528
Fax: 86-10-85872555
E-mail Address: *interco@ctei.gov.cn*
Web Address: english.ctei.gov.cn
The China National Textile and Apparel Council (CNTAC) is the nonprofit, national federation of all textile-related industries in China. The company web site offers market research, directories, a consultancy, a calendar of relevant events, textile base locations and daily news reports on all aspects of the Chinese textile industry.

China Textile Machinery Association (CTMA)
No. 12 E. Chang'an St., Rm. 406
Beijing, 100742 China
Phone: 86-10-5822-1177
Fax: 86-10-5822-1076
E-mail Address: *ctma@ctma.net*
Web Address: www.ctma.net
The China Textile Machinery Association (CTMA) serves the interests of the textile machinery and manufacturing industry and participates in the related trade exhibitions in China and overseas.

Color Association of the United States
33 Whitehall St., Ste. M3
New York, NY 10004 US
Phone: 212-947-7774
E-mail Address: *info@colorassociation.com*
Web Address: www.colorassociation.com
The Color Association provides color standards for the textile industry.

Cotton, Inc.
6399 Weston Pkwy.
Cary, NC 27513 US
Phone: 919-678-2220
Fax: 919-678-2230
Web Address: www.cottoninc.com
Cotton, Inc. is a research and promotion company founded by United States cotton producers and importers.

Embroidery Trade Association (ETA)
P.O. Box 794534
Dallas, TX 75379-3967 US
Fax: 972-755-2561
Toll Free: 888-628-2545
E-mail Address: *info@embroiderytrade.org*
Web Address: www.embroiderytrade.org
The Embroidery Trade Association (ETA) focuses on developing retail embroidery industry solutions by providing education, business support, networking, research and consumer outreach.

European Apparel and Textile Association (EURATEX)
24 rue Montoyer, Ste. 10
Brussels, B-1000 Belgium
Phone: 32-2-285-48-83
Fax: 32-2-230-60-54
E-mail Address: *info@euratex.org*
Web Address: www.euratex.org
The European Apparel and Textile Association (EURATEX) is an international trade association representing the interests of textile manufacturers in the European Union.

European Textile Network (ETN)
ETN Secretariat
Friedenstr. 5
Hannover, D-30175 Germany
Phone: 49-511-817006
Fax: 49-511-813108
E-mail Address: *ETN@etn-net.org*
Web Address: www.etn-net.org
The European Textile Network (ETN) is a group of independent organizations that work together to promote common interests and cooperation in the European textile industry.

INDA, Association of the Nonwoven Fabrics Industry
1100 Crescent Green, Ste. 115
Cary, NC 27518 US
Phone: 919-233-1210
Fax: 919-233-1282
E-mail Address: *info@inda.org*
Web Address: www.inda.org

INDA, the Association of the Nonwoven Fabrics Industry, has been representing this sector since 1968.

International Rayon & Synthetic Fibres Committee (CIRFS)
Ave. E. Van Nieuwenhuyse 6
Bruxelles, 1160 Belgium
Phone: 32-2-676-74-55
Fax: 32-2-676-74-54
E-mail Address: *secretariat@cirfs.org*
Web Address: www.cirfs.org
The International Rayon & Synthetic Fibres Committee (CIRFS) represents the man-made fibers industry in Europe.

International Textile and Apparel Association (ITAA)
P.O. Box 70687
Knoxville, TN 37938-0687 US
Phone: 865-992-1535
E-mail Address: *nruth@utk.edu*
Web Address: www.itaaonline.org
The International Textile and Apparel Association (ITAA) is a nonprofit educational and scientific corporation dedicated to providing opportunities to scholars in the retail, textile and apparel industries.

International Textile Manufacturers Federation (ITMF)
Wiedingstrasse 9
Zurich, CH-8055 Switzerland
Phone: 41-44-283-63-80
Fax: 41-44-283-63-89
E-mail Address: *secretariat@itmf.org*
Web Address: www.itmf.org
The International Textile Manufacturers Federation (ITMF) is dedicated to providing news and information to the world's textile manufacturers. It is also a government liaison between the textile industries and governments.

International Wool Textile Organization (IWTO)
Rue de l'Industrie 4
Brussels, 1000 Belgium
Phone: 32-2-505-40-10
Fax: 32-2-503-47-85
E-mail Address: *info@iwto.org*
Web Address: www.iwto.org
International Wool Textile Organization (IWTO) is an international body representing the interests of the world's wool-textile trade and industry.

Man-Made Textiles Research Association (MANTRA)
Near Textile Market,Telephone Exchange
Ring Rd., Surat
Gujarat, 395 002 India
Phone: 91-0261-2323211
Fax: 91-0261-2322500
E-mail Address: *director@mantrasurat.org*
Web Address: www.mantrasurat.org
Man-Made Textiles Research Association of India (MANTRA) seeks to serve the local indian textile industry more effectively. MANTRA keeps close liaison with the industry.

Mohair Council of America (MCA)
233 W. Twohig Ave.
San Angelo, TX 76903 US
Phone: 325-655-3161
Toll Free: 800-583-3161
E-mail Address: *mohair@mohairusa.com*
Web Address: www.mohairusa.com
The Mohair Council of America (MCA) is dedicated to the promotion of U.S. mohair in the worldwide market.

National Cotton Council of America
7193 Goodlett Farms Pkwy.
Cordova, TN 38016 US
Phone: 901-274-9030
Fax: 901-725-0510
Web Address: www.cotton.org
The National Cotton Council of America represents the interests of producers, ginners, warehouses, merchants, cottonseed crushers, cooperatives and textile manufacturers.

South Carolina Manufacturers Alliance (SCMA)
1340 Bull St.
Columbia, SC 29201 US
Phone: 803-799-9695
Fax: 803-771-8738
Toll Free: 800-889-9695
E-mail Address: *admin@myscma.com*
Web Address: www.myscma.com
The South Carolina Manufacturers Alliance (SCMA) is an organization devoted to the promotion of manufacturing in South Carolina.

Southern Textile Association
P.O. Box 66
Gastonia, NC 28053-0066 US
Phone: 704-215-4543
Fax: 704-215-4160

E-mail Address: *llink@southerntextile.org*
Web Address: www.southerntextile.org
The Southern Textile Association is a nonprofit organization representing people in textiles and related industries.

Supima Association of America
4141 E. Broadway Rd.
Phoenix, AZ 85040-8831 US
Phone: 602-792-6002
Fax: 602-792-6004
E-mail Address: *info@supima.com*
Web Address: www.supimacotton.org
Supima is an organization devoted to the promotion of Pima cotton.

Synthetic and Art Silk Mills' Association
Sasmira Marg, Worli
Mumbai, 400 030 India
Phone: 91-22-2493-5351-52
Fax: 91-22-2493-0225
E-mail Address: *sasmira@vsnl.com*
Web Address: www.sasmira.org
Synthetic and Art Silk Mills' Association is a cooperative venture set up by the man-made textile industry of India as a multi-functional institute to serve scientific and technological needs of the textile industry.

Synthetic Yarn and Fiber Association (SYAA)
737 Park Trail Ln.
Clover, SC 29710 US
Phone: 704-589-5895
Fax: 803-746-5566
E-mail Address: *kpettit@thesyfa.org*
Web Address: www.thesyfa.org
The Synthetic Yarn and Fiber Association (SYAA) promotes better communication of improved business practices in the textured yarn industry.

Textile Association of Los Angeles (TALA)
P.O. Box 71022
Los Angeles, CA 90071 US
Phone: 213-627-6173
Fax: 213-688-6290
E-mail Address: *info@talausa.org*
Web Address: www.talausa.org
Textile Association of Los Angeles (TALA) is professional organization for people engaged in sales of textiles and related products and services to the apparel industry.

Textile Clothing Technology Corporation (TC2)
5651 Dillard Dr.
Cary, NC 27518 US
Phone: 919-380-2156
Fax: 919-380-2181
Toll Free: 800-786-9889
Web Address: www.tc2.com
The Textile Clothing Technology Corporation (TC2) is a nonprofit consortium of fiber producers, retailers, sewn products manufacturers, labor unions, academia and the federal government. The website provides access to other TC2 sites: Size USA; 3D Body Scanning; techexchange.com and inkdropprinting.com.

Textile Council of Hong Kong
25 Kimberley Rd., Cheung Lee Comm. Bldg.
Fl. 4, Rm 401-3, TST
Kowloon, New Territories, China
Phone: 852-(2)-305-2893
Fax: 852-(2)-305-2493
E-mail Address: *sec@textilecouncil.com*
Web Address: www.textilecouncil.com.
The Textile Council of Hong Kong represents ten major textile associations and acts as a voice for the entire Hong Kong textile industry. Its affiliated associations' activities range from spinning, weaving, knitting, dyeing and finishing, garment making and the manufacture of textile goods.

Textile Institute
Oxford St.
1st Fl., St. James's Bldg.
Manchester, M1 6FQ UK
Phone: 44-161-237-1188
Fax: 44-161-236-1991
E-mail Address: *tiihq@textileinst.org.uk*
Web Address: www.textileinstitute.org
The Textile Institute is a professional organization serving the textile, clothing and footwear industry in the U.K.

Textile Society of America, Inc. (TSA)
P. O. Box 193
Middletown, DE 19709 US
Phone: 302-378-9636
Fax: 302-378-9637
E-mail Address: *tsa@textilesociety.org*
Web Address: textilesocietyofamerica.org
The Textile Society of America, Inc. (TSA) provides an international forum for the exchange and dissemination of information about textiles

worldwide, from artistic, cultural, economic, historic, political, social, and technical perspectives.

XXIII. Textile & Fabric Resources

Institute of Textile Technology (ITT)
2401 Research Dr., Box 8301
North Carolina State University, College of Textiles
Raleigh, NC 27695-8301 US
Phone: 919-513-7583
Fax: 888-348-3512
E-mail Address: *patriceh@itt.edu*
Web Address: www.itt.edu
The Institute of Textile Technology (ITT) provides private education in the textile industry aimed at continuing education and graduate students.

Ministry of Textiles, Gov. of India
Udyog Bhavan
New Delhi, 110 011 India
Phone: 91-11-23061338
Fax: 91-11-23063711
Web Address: www.texmin.nic.in
The Government of India's Ministry of Textiles is responsible for policy formulation, planning, development, export promotion and trade regulation in respect of the textile sector. This includes all natural and manmade cellulosic fibers that go into the making of textiles, clothing and handicrafts.

National Textile Center (NTC)
2 Pastern Ln.
Blue Bell, PA 19422-2429 US
Phone: 215-540-9840
Fax: 215-689-4835
E-mail Address: *marty.ntcresearch@verizon.net*
Web Address: www.ntcresearch.org
The National Textile Center (NTC) is a research consortium of eight universities that aims to enhance the knowledge base for the continuing viability of the U.S. fiber, textile, fabricated products and retail complex.

Southern Textile News
9629 Old Nations Ford Rd.
Charlotte, NC 28273 US
Phone: 704-527-5111
Toll Free: 800-738-5111
E-mail Address: *chip@mullenpublications.com*
Web Address: www.textilenews.com
Southern Textile News provides industry news and information.

TexIndex.com
No.187th Moganshan Rd.
12/F Yisheng Mansion
Hangzhou, Zhejiang 310004 China
Phone: 86-571-8767-1505
Fax: 86-571-8821-9935
E-mail Address: *succeed@texindex.com*
Web Address: www.texindex.com
TexIndex.com helps businesses find textile buyers and sellers in China, as well as promote their business online.

Textile Dictionary
E-mail Address: *fashion@ntgi.net*
Web Address: www.ntgi.net/ICCF&D/textile.htm
The Textile Dictionary, part of the Internet Centre for Canadian Fashion and Design, is a comprehensive dictionary of textile fibers.

Textile Museum
2320 South St. NW
Washington, DC 20008-4088 US
Phone: 202-667-0441
E-mail Address: *kclune@textilemuseum.org*
Web Address: www.textilemuseum.org
The Textile Museum displays a wide range of textile samples and production methods from around the world.

Textile Research Institute Princeton (TRI/Princeton)
601 Prospect Ave.
P.O. Box 625
Princeton, NJ 08542 US
Phone: 609-430-4820
E-mail Address: *info@triprinceton.org*
Web Address: www.triprinceton.org
The Textile Research Institute Princeton (TRI/Princeton) is a nonprofit organization dedicated to the competitiveness of the textile industry. It also does advanced research and education in polymers, fibers, films, human hair and porous materials.

Textile Source
6864 Indiana Ave., Ste. 202B
Riverside, CA 92506
Phone: 951-781-3000
E-mail Address: *textilepedia@aol.com*
Web Address: www.textilesource.com
The Textile Source web site provides news and information for the textile industry.

Textile World
2100 RiverEdge Pkwy., Ste. 1200
Atlanta, GA 30328 US
Phone: 770-955-5656
Fax: 770-952-0669
E-mail Address: *editor@textileworld.com*
Web Address: www.textileworld.com
Textile World, a leading magazine in the textiles industry, provides extensive coverage of business news. Users may access useful archives of textiles industry coverage online.

Textile, Clothing & Footwear (TCF) Unit-UK Trade & Investment Dept.
1 Victoria St.
London, SW1H 0ET UK
Phone: 44-20-7215-8000
E-mail Address: *andrew.strachan@ukti.gsi.gov.uk*
Web Address: www.ukti.gov.uk
The Textile, Clothing & Footwear (TCF) Unit's projects are aimed at developing the textile industry throughout the UK include; education and training; best practice; environmental issues; improvements to the supply chain and product innovation.

XXIV. Trade Associations-General

Associated Chambers of Commerce and Industry of India (ASSOCHAM)
1 Community Ctr. Zamrudpur
Kailash Colony
New Delhi, 110 048 India
Phone: 91-11-4655-0555
Fax: 91-11-4653-6481
E-mail Address: *assocham@nic.in*
Web Address: www.assocham.org
The Associated Chambers of Commerce and Industry of India (ASSOCHAM) has a membership of more than 300 chambers and trade associations and serves members from all over India. It works with domestic and international government agencies to advocate for India's industry and trade activities.

Brazilian Trade & Investment Promotion Agency (Apex-Brasil)
SBN, Quadra 02, Lote 11
Edificio Apex-Brasil
Brasilia, DF 70040-020 Brazil
Phone: 55-61-3426-0202
Web Address: www.apexbrasil.com.br
Apex-Brasil works to promote exports of Brazilian products and services, supporting some 70 industry sectors such as agribusiness, technology, civil engineering, entertainment, apparel and industrial equipment.

BUSINESSEUROPE
168 Ave. de Cortenbergh
Brussels, 1000 Belgium
Phone: 32-2-237-65-11
Fax: 32-2-231-14-45
E-mail Address: *main@businesseurope.eu*
Web Address: www.businesseurope.eu
BUSINESSEUROPE is a major European trade federation that operates in a manner similar to a chamber of commerce. Its members are the central national business federations of the 34 countries throughout Europe from which they come. Companies cannot become direct members of BUSINESSEUROPE, though there is a support group which offers the opportunity for firms to encourage BUSINESSEUROPE objectives in various ways.

Fair Trade Federation (FTF)
100 W. 10th St., Ste. 1105
Wilmington, DE 19801 US
Phone: 302-655-5203
E-mail Address: *info@fairtradefederation.org*
Web Address: www.fairtradefederation.com
The Fair Trade Federation (FTF) is an association of retailers, producers and wholesalers that seek to provide fair wages and good working conditions for artisans and farmers.

Federation of Hong Kong Industries
8 Cheung Yue St.
Fl. 31, Billion Plz.
Kowloon, Hong Kong, China
Phone: 852-2732-3188
Fax: 852-2721-3494
E-mail Address: *fhki@fhki.org.hk*
Web Address: www.industryhk.org
The Federation of Hong Kong Industries promotes the trade, investment advancement and development of opportunities for the industrial and business communities of Hong Kong. The web site hosts a trade enquiry on products and services and publishes research reports and trade publications.

United States Council for International Business (USCIB)
1212 Ave. of the Americas
New York, NY 10036 US
Phone: 212-354-4480
Fax: 212-575-0327
E-mail Address: *jhuneke@uscib.org*

Web Address: www.uscib.org
The United States Council for International Business (USCIB) promotes an open system of world trade and investment through its global network. Standard USCIB members include corporations, law firms, consulting firms and industry associations. Limited membership options are available for chambers of commerce and sole legal practitioners.

XXV. Trade Associations-Global

World Trade Organization (WTO)
Centre William Rappard
Rue de Lausanne 154
Geneva 21, CH-1211 Switzerland
Phone: 41-22-739-51-11
Fax: 41-22-731-42-06
E-mail Address: *enquiries@wto.og*
Web Address: www.wto.org
The World Trade Organization (WTO) is a global organization dealing with the rules of trade between nations. To become a member, nations must agree to abide by certain guidelines. Membership increases a nation's ability to import and export efficiently.

XXVI. Trade Resources

BrazilBiz
E-mail Address: *BrazilBiz@maxtie.com*
Web Address: www.brazilbiz.com.br
The BrazilBiz web site serves as a clearinghouse of general contact information for Brazilian companies; registration on the site is free, and allows users to search within the BrazilBiz database for firms across a range of industries.

Directory of Brazilian Exporters
Brazil SBN, Quadra 01, Bloco C
Brasilia, 70040-903 Brazil
Phone: 55-61-3317-9989
Fax: 55-61-3317-9994
Web Address: www.brazil4export.com
The Directory of Brazilian Exporters is a project of Brazil's National Confederation of Industry. The directory includes companies with annual exports of at least $100,000 (USD), and is intended to help identify potential international partnerships and to disseminate general information on Brazil's export activities. The site contains historical export information based on records from the Foreign Trade Secretariat of the Brazilian Ministry of Development, Industry and Foreign Trade.

European Union (EU) Market Access Database
Web Address:
madb.europa.eu/mkaccdb2/indexPubli.htm
The European Union (EU) Market Access Database provides information on trade tariffs, international laws and trade barriers worldwide. It recently began providing a Complaint Register, where companies may voice a concern to the European Commission regarding certain trade and investment barriers in non-EU countries.

Made-in-China.com - China Manufacturers Directory
8-12/F, Block A, Software Bldg., Xinghuo Rd.
Nanjing New & High Technology Industry Development Zone
Nanjing, Jiangsu 210061 China
Fax: 86-25-6667-0000
Web Address: www.made-in-china.com
Made-in-China.com - China Manufacturers Directory, one of the largest business to business portals in China, helps to connect Chinese manufacturers, suppliers and traders with international buyers. Made-in-China.com contains additional information on trade shows and important laws and regulations about business with China.

XXVII. U.S. Government Agencies

Bureau of Economic Analysis (BEA)
1441 L St. NW
Washington, DC 20230 US
Phone: 202-606-9900
E-mail Address: *customerservice@bea.gov*
Web Address: www.bea.gov
The Bureau of Economic Analysis (BEA), an agency of the U.S. Department of Commerce, is the nation's economic accountant, preparing estimates that illuminate key national, international and regional aspects of the U.S. economy.

Bureau of Labor Statistics (BLS)
2 Massachusetts Ave. NE
Washington, DC 20212-0001 US
Phone: 202-691-5200
Toll Free: 800-877-8339
E-mail Address: *Steinberg.Gary@bls.gov*
Web Address: stats.bls.gov
The Bureau of Labor Statistics (BLS) is the principal fact-finding agency for the Federal Government in the field of labor economics and statistics. It is an independent national statistical agency that collects, processes, analyzes and disseminates statistical data

to the American public, U.S. Congress, other federal agencies, state and local governments, business and labor. The BLS also serves as a statistical resource to the Department of Labor.

International Trade Administration (ITA)
1401 Constitution Ave. NW
U.S. Department of Commerce
Washington, DC 20230 US
Toll Free: 800-872-8723
E-mail Address: *mary.trupo@trade.gov*
Web Address: www.trade.gov
The International Trade Administration (ITA) is a U.S. Government agency that provides a wealth of information regarding imports and exports. In addition, it publishes in-depth analyses and statistics on various industry sectors and provides assistance to U.S. firms that wish to increase their exports.

National Labor Relations Board (NLRB)
1099 14th St. NW
Washington, DC 20570-0001 US
Phone: 2002-276-1000
Fax: 202-208-3013
Toll Free: 866-667-6572
E-mail Address: *publicinfo@nlrb.gov*
Web Address: www.nlrb.gov
The National Labor Relations Board (NLRB) provides case reports on labor disputes, searchable by company or union.

Office of Textiles and Apparel (OTEXA)
U.S. Department of Commerce
14th and Constitution Ave, Rm. H3100
Washington, DC 20230 US
Phone: 202-482-5078
Fax: 202-482-2331
E-mail Address: *OTEXA@trade.gov*
Web Address: otexa.ita.doc.gov
The Office of Textiles and Apparel (OTEXA) is a U.S. government agency that provides information and helps ensure fair market practices for American textile products.

U.S. Census Bureau
4600 Silver Hill Rd.
Washington, DC 20233-8800 US
Phone: 301-763-4636
Toll Free: 800-923-8282
E-mail Address: *pio.2010@census.gov*
Web Address: www.census.gov
The U.S. Census Bureau is the official collector of data about the people and economy of the U.S. Founded in 1790, it provides official social, demographic and economic information. In addition to the Population & Housing Census, which it conducts every 10 years, the U.S. Census Bureau numerous other surveys annually.

U.S. Department of Commerce (DOC)
1401 Constitution Ave. NW
Washington, DC 20230 US
Phone: 202-482-2000
E-mail Address: *TheSec@doc.gov*
Web Address: www.commerce.gov
The U.S. Department of Commerce (DOC) regulates trade and provides valuable economic analysis of the economy.

U.S. Department of Labor (DOL)
200 Constitution Ave. NW
Frances Perkins Building
Washington, DC 20210 US
Toll Free: 866-487-2365
Web Address: www.dol.gov
The U.S. Department of Labor (DOL) is the government agency responsible for labor regulations.

U.S. Department of State
2201 C St. NW
Washington, DC 20520 US
Phone: 202-647-4000
Fax: 202-647-1579
Toll Free: 800-877-8339
Web Address: www.state.gov
The Department of State is the head United States foreign affairs agency.

U.S. Securities and Exchange Commission (SEC)
100 F St. NE
Washington, DC 20549 US
Phone: 202-942-8088
E-mail Address: *publicinfo@sec.gov*
Web Address: www.sec.gov
The U.S. Securities and Exchange Commission (SEC) is a nonpartisan, quasi-judicial regulatory agency responsible for administering federal securities laws. These laws are designed to protect investors in securities markets and ensure that they have access to disclosure of all material information concerning publicly traded securities. Visitors to the web site can access the EDGAR database of corporate financial and business information.

U.S. Trade Representative (USTR)
600 17th St. NW

Washington, DC 20508 US
Phone: 202-395-3230
Fax: 202-395-4549
E-mail Address: *contactustr@ustr.eop.gov*
Web Address: www.ustr.gov
The U.S. Trade Representative (USTR) is the nation's chief trade negotiator and the principal trade policy advisor to the President.

Chapter 4

THE APPAREL & TEXTILES 350: WHO THEY ARE AND HOW THEY WERE CHOSEN

Includes Indexes by Company Name, Industry & Location

The companies chosen to be listed in PLUNKETT'S APPAREL & TEXTILES INDUSTRY ALMANAC are not the same as the "Fortune 500" or any other list of corporations. THE APPAREL & TEXTILES 350 (the actual count is 359 companies) were chosen specifically for their dominance in the many facets of the apparel and textiles industry in which they operate. Complete information about each firm can be found in the "Individual Profiles," beginning later in this chapter. These profiles are in alphabetical order by company name.

THE APPAREL & TEXTILES 350 includes companies from all parts of the United States, Canada, Europe and Asia, and from all apparel and textile industry segments. Essentially, THE APPAREL & TEXTILES 350 includes companies that are deeply involved in apparel manufacturing, textiles, retailing, design and logistics services as well as companies that provide consulting or technologies that are enabling companies in this industry to prosper.

Simply stated, the list contains 359 of the most important firms in apparel, textiles and related industries. To be included in our list, the firms had to meet the following criteria:

1) Generally, the organizations profiled are U.S.-based, for-profit corporations, subsidiaries or joint ventures. (We have also added 121 non-U.S. companies in order to better list leading firms in certain industry sectors.)
2) Prominence, or a significant presence, in apparel, textiles and supporting fields. (See the following Industry Codes section for a complete list of types of businesses that are covered).
3) The companies in THE APPAREL & TEXTILES 350 do not have to be exclusively in the apparel or textiles field.
4) Vital statistics must have been available to the editors of this book, either directly from the company being written about or from outside sources deemed reliable and accurate by the editors. A small number of companies that we would like to have included are not listed because of a lack of sufficient, objective data.

INDEXES TO THE APPAREL & TEXTILES 350, AS FOUND IN THIS CHAPTER AND IN THE BACK OF THE BOOK:

INDUSTRY LIST, WITH CODES

This book refers to the following list of unique industry codes, based on the 2012 NAIC code system (NAIC is used by many analysts as a replacement for older SIC codes because NAIC is more specific to today's industry sectors, see www.census.gov/NAICS). Companies profiled in this book are given a primary NAIC code, reflecting the main line of business of each firm.

Apparel

Apparel & Shoe Manufacturing

315190 Outerwear Kintting Mills
315210 Cut and Sew Apparel Contractors
315220 Apparel Manufacturing-Men's & Boy's Clothing
315240 Apparel Manufacturing-Women's & Children's
315280 Apparel Manufacturing-Leather & Fur
315990 Apparel Accessories and Other Apparel Manufacturing
316110 Tannery Leather Manufacturing
316210 Shoe Manufacturing

Energy

Petroleum-Refining & Manufacturing

324110 Petroleum Refineries
325110 Petrochemicals Manufacturing

Entertainment

Toys, Sporting Goods & Miscellaneous Manufacturing

339920 Sporting & Athletic Goods, Manufacturing

Food & Restaurants

Food Service

722310 Food Service Contractors

Health Care

Health Products, Manufacturing

33911 Medical/Dental/Surgical Equipment & Supplies, Manufacturing
339113A Bullet Proof Vests/Body Armor, Manufacturing

InfoTech

Computers & Electronics Manufacturing

334310 Audio & Video Equipment, Consumer Electronics

Software

511210N Computer Software, Product Lifecycle, Engineering, Design & CAD

Manufacturing

Beverage & Tobacco Manufacturing

312140 Beverages--Distilleries

Textiles Manufacturing

313110 Fiber, Yarn and Thread Mills
313210 Broadwoven Fabric Mills
313230 Textiles Manufacturing
313320 Nonwoven Fabric Mills
313240 Knit Fabric Mills
313320 Fabric Coating Mills

Home Furnishings/Textile Mills

314999 All Other Miscellaneous Textile Product Mills

Luggage Manufacturing

316998 Luggage Manufacturing

Plastics & Rubber

325220 Artificial and Sythetic Fibers and Filaments Manufacturing
3261 Plastic Products, Manufacturing

Machinery & Manufacturing Equipment

334519 Other Measuring and Controlling Device Manufacturing

Jewelry & Other Manufacturing

339993 Zippers & Fasteners

Nanotechnology

Nanotechnology

541712N Nanotechnology-General

Retailing

Furniture & Home Furnishings Stores

442110 Furniture Stores
442299 Linens/Housewares/Art/Framing Stores

Textiles

442210 Carpet & Flooring, Retail

Food & Beverage Stores

445110 Grocery Stores/Supermarkets
445120 Convenience Stores

Apparel & Accessories Stores

448110 Apparel Stores, Men's
448120 Apparel Stores, Women's
448130 Apparel Stores, Children's
448140 Apparel Stores, General/Family Clothing Stores
448210 Shoes & Accessories Stores

Sporting Goods, Hobbies, Books & Music Stores

451110 Sporting Goods Stores
451120 Toys/Hobbies/Games/Crafts Stores
451130 Sewing & Fabric Stores

Department & Discount Stores

452111 Department Stores
452112 Discount Stores

Nonstore Retailers

454111 Internet Retail Sales Sites
454113 Mail Order and Non-Store Selling, Incl. TV Shopping

Services

Personnel, Administrative & Support Services

812331 Linen/Uniform Supply

Transportation

Truck

488510 Freight Transportation Arrangement

Wholesale Distribution-Other

Distribution-Nondurable Goods

424320 Mens and Boys' Clothing and Furnishings Merchant Wholesalers

INDEX OF COMPANIES WITHIN INDUSTRY GROUPS

Company	Industry Code	2012 Sales	2012 Profits
All Other Miscellaneous Textile Product Mills			
Pacific Brands Limited	314999	1,387,216,640	-470,657,664
Apparel Accessories and Other Apparel Manufacturing			
Coach Inc	315990	4,763,180,032	1,038,910,016
Hermes International SA	315990	4,561,670,000	968,896,000
Kate Spade LLC	315990	462,000,000	
Swank Inc	315990		
Apparel Manufacturing-Leather & Fur			
Alexandra plc	315280		
Danier Leather Inc	315280	149,083,696	4,026,353
Etienne Aigner Group Inc	315280		
Superior Uniform Group Inc	315280		
Tandy Brands Accessories Inc	315280	117,601,000	-3,652,000
Apparel Manufacturing-Men's & Boy's Clothing			
China Xiniya Fashion Ltd.	315220		
Esquel Group of Companies	315220	1,400,000,000	
Gildan Activewear Inc	315220	1,948,253,056	148,464,000
Hampshire Group Ltd	315220		
Hanesbrands Inc	315220		
Oxford Industries Inc	315220	758,913,024	29,380,000
Passport Brands Inc	315220		
Perry Ellis International Inc	315220	980,592,000	25,517,000
Phat Fashions LLC	315220		
TAL Apparel Group	315220	550,000,000	
Ted Baker plc	315220		
Apparel Manufacturing-Women's & Children's			
American Apparel Inc	315240		
Bernard Chaus Inc	315240		
Bill Blass Group LLC	315240		
Burberry Group plc	315240	2,975,804,928	421,887,488
BYC Co	315240		
Carters Inc	315240		
Celebrity International Inc	315240		
Chanel SA	315240	3,500,000,000	500,000,000
Christian Dior SA	315240	11,635,399,680	1,539,655,936
CSP International Industria Calze SpA	315240		
Delta Galil Industries Ltd	315240		
Diesel SpA	315240	2,000,000,000	
Eileen Fisher Inc	315240		
Fifth & Pacific Companies Inc	315240		
Filippa K	315240		
French Connection Group plc	315240	345,136,960	8,492,228
Gerry Weber International AG	315240		
Gianni Versace SpA	315240	398,881,000	
G-III Apparel Group Ltd	315240	1,231,201,024	49,620,000

Company	Industry Code	2012 Sales	2012 Profits
Giorgio Armani SpA	315240	2,650,000,000	
Haggar Corp	315240		
Hugo Boss AG	315240		
Jaclyn Inc	315240	195,927,008	2,954,000
Jil Sander SpA	315240		
Joe's Jeans Inc	315240		
Jordache Enterprises Inc	315240		
Kellwood Co	315240		
Lacoste SA	315240	2,285,000,000	
Levi Strauss & Co	315240	4,610,000,000	144,000,000
Maidenform Inc	315240		
Marc Ecko Enterprises	315240		
MAST Global Fashions	315240		
Max Mara Fashion Group SRL	315240	2,170,000,000	
Michael Kors Holdings Ltd	315240	1,302,253,952	147,364,000
Nautica Enterprises Inc	315240		
OshKosh B'Gosh Inc	315240	283,300,000	
Paris Blues Inc	315240		
Pendleton Woolen Mills Inc	315240		
PPR S.A. (KERING)	315240		
Prada SpA Group	315240	4,320,000,000	809,300,000
Puig Beauty & Fashion Group SL	315240		
PVH Corp	315240	5,890,624,000	317,880,992
Ralph Lauren Corporation	315240	6,859,500,032	681,000,000
Spanx Inc	315240		
St John Knits International Inc	315240	420,000,000	
Sunrise Brands LLC	315240		
Tefron Ltd	315240		
Tommy Hilfiger Corp	315240	3,216,990,000	420,933,000
Tory Burch LLC	315240	800,000,000	
Valentino Fashion Group SpA	315240		
VF Corp	315240		
Wacoal Holdings Corp	315240	2,094,261,632	84,222,712
Warnaco Group Inc (The)	315240		
Wolford AG	315240		
Zegna (Ermenegildo Zegna Holditalia SpA)	315240	1,670,000,000	
Apparel Stores, Children's			
Children's Place Retail Stores Inc (The)	448130	1,715,862,016	77,225,000
Gymboree Corp (The)	448130		
Tween Brands Inc	448130	1,306,700,000	172,500,000
Apparel Stores, General/Family Clothing Stores			
Abercrombie & Fitch Co	448140	4,158,057,984	127,658,000
American Eagle Outfitters Inc	448140	3,159,817,984	151,704,992
Buckle Inc (The)	448140	1,062,945,984	151,456,000
Burlington Coat Factory	448140		
Cortefiel SA	448140		
Eddie Bauer LLC	448140	1,000,000,000	
Gap Inc (The)	448140	14,549,000,192	833,000,000
Gordmans Stores Inc	448140	558,145,984	25,173,000

Company	Industry Code	2012 Sales	2012 Profits
Guess? Inc	448140	2,688,048,128	265,500,000
H&M (Hennes & Mauritz) AB	448140	18,483,890,000	3,409,890,000
Hot Topic Inc	448140	697,934,016	-1,818,000
J Crew Group Inc	448140	2,222,770,000	96,100,000
KiK Textilien & Non-Food GmbH	448140		
Lord & Taylor LLC	448140		
lululemon Athletica Inc	448140	1,000,838,976	184,063,008
NEXT plc	448140	5,513,699,328	760,935,680
Pacific Sunwear of California Inc	448140	833,750,976	-106,423,000
Ross Stores Inc	448140	8,608,290,816	657,169,984
Tilly's Inc	448140	400,624,000	34,340,000
TJX Companies Inc (The)	448140	23,191,455,744	1,496,089,984
True Religion Apparel Inc	448140		
Uniqlo Co Ltd	448140	11,314,193,408	872,977,600
Urban Outfitters Inc	448140	2,473,800,960	185,251,008
Walking Company Holdings Inc (The)	448140		
Zara International Inc	448140		
Zumiez Inc	448140	555,873,984	37,351,000
Apparel Stores, Men's			
Casual Male Retail Group Inc	448110	397,655,008	42,663,000
Jos A Bank Clothiers Inc	448110	979,852,032	97,491,000
Men's Wearhouse Inc (The)	448110	2,382,683,904	120,601,000
Apparel Stores, Women's			
Aeropostale Inc	448120	2,342,259,968	69,515,000
Ann Inc	448120	2,212,493,056	86,566,000
Arcadia Group	448120		
Ascena Retail Group Inc	448120	3,353,299,968	162,200,000
Bebe Stores Inc	448120	530,831,008	11,721,000
Brooks Brothers Group Inc	448120	800,000,000	
Cache Inc	448120		
Cato Corp (The)	448120	931,457,984	64,834,000
Charlotte Russe Holding Inc	448120		
Charming Shoppes Inc	448120	2,000,000,000	
Chico's FAS Inc	448120	2,196,359,936	140,874,000
Christopher & Banks Corp	448120	412,796,000	-71,062,000
Claire's Stores Inc	448120	1,495,900,000	11,632,000
Deb Shops Inc	448120		
Destination Maternity Corp	448120		
Diane von Furstenberg Studio LP	448120		
Esprit Holdings Ltd	448120	3,892,308,224	112,646,616
Etam Developpement	448120		
Express Inc	448120	2,073,355,008	140,696,992
Forever 21 Inc	448120		
Francesca's Holdings Corp	448120	204,158,000	22,501,000
Frederick's of Hollywood Group Inc	448120	111,406,000	-6,432,000
J Jill Group Inc (The)	448120		
Jones Group Inc (The)	448120		
Limited Brands Inc	448120	10,364,000,256	850,000,000
Loehmann's Holdings Inc	448120		

Company	Industry Code	2012 Sales	2012 Profits
Mango/MNG Holding SL	448120	2,213,428,070	
New Look Group plc	448120		
New York & Company Inc	448120	956,456,000	-38,937,000
Simply Fashion Stores Ltd	448120		
Talbots Inc (The)	448120	1,140,000,000	-111,900,000
TOPSHOP	448120	2,000,000,000	
United Retail Group Inc	448120		
Vera Bradley Inc	448120	460,843,008	57,921,000
Victorias Secret	448120	6,574,000,000	1,188,000,000
Wet Seal Inc (The)	448120	620,097,024	15,082,000
Artificial and Sythetic Fibers and Filaments Manufacturing			
Asahi Kasei Corp	325220	19,167,031,296	679,410,304
Audio & Video Equipment, Consumer Electronics			
Samsung Group	334310		
Beverages--Distilleries			
LVMH Moet Hennessy Louis Vuitton SA	312140	36,742,300,000	7,742,080,000
Broadwoven Fabric Mills			
American Dawn Inc	313210		
Brookwood Companies Inc	313210		
Chargeurs	313210		
Cone Denim LLC	313210		
Culp Inc	313210	254,443,008	13,296,000
DOGI International Fabrics SA	313210		
Dollfus Mieg & Cie SA	313210		
Fujibo Holdings Inc	313210		
Greenwood Mills	313210		
Grupo Tavex SA	313210		
Hallwood Group Inc.	313210		
Hyosung Corporation	313210		
ICG/Holliston	313210		
International Textile Group Inc	313210		
ITOCHU Corp	313210	52,035,231,744	3,661,123,328
Johnston Textiles Inc	313210		
Klopman International	313210		
Mount Vernon Mills	313210		
Polartec LLC	313210		
Propex Operating Company LLC	313210		
Ratti SpA	313210		
Raymond Ltd	313210	656,905,664	25,791,208
Swift Galey	313210		
Toyobo Co Ltd	313210	4,258,101,760	55,884,500
Vincenzo Zucchi SpA	313210		
Worldtex Inc	313210		
Bullet Proof Vests/Body Armor, Manufacturing			
Point Blank Enterprises Inc	339113A		
Carpet & Flooring, Retail			
Mohawk Industries Inc	442210		
Computer Software, Product Lifecycle, Engineering, Design & CAD			
Lectra SA	511210N		

Company	Industry Code	2012 Sales	2012 Profits
Convenience Stores			
Seven & I Holdings Co Ltd	445120	58,313,158,656	1,581,834,752
Cut and Sew Apparel Contractors			
Benetton Group SPA	315210		
Billabong International Ltd	315210	1,535,857,280	-287,887,072
Columbia Sportswear Co	315210		
Delta Apparel Inc	315210	489,923,008	-2,447,000
Dolce & Gabbana SpA (D&G)	315210	1,500,000,000	
Everlast Worldwide Inc	315210		
Fila USA Inc	315210		
Inditex (Industria de Diseno Textil SA)	315210		
Lafuma SA	315210		
Patagonia Inc	315210	360,000,000	
Quiksilver Inc	315210		
Russell Brands LLC	315210		
Under Armour Inc	315210		
Varsity Brands Inc	315210		
Volcom Inc	315210	354,000,000	
Warnaco Swimwear Inc	315210		
Department Stores			
Barney's New York Inc	452111		
Belk Inc	452111	3,699,591,936	183,148,000
Bon-Ton Stores Inc (The)	452111	2,953,530,112	-12,128,000
Boscov's Inc	452111	1,000,000,000	
Dillard's Inc	452111	6,399,764,992	463,908,992
Elder-Beerman Stores Corp (The)	452111		
J C Penney Company Inc	452111	17,259,999,232	-152,000,000
Macy's Inc	452111	26,404,999,168	1,256,000,000
Marks & Spencer Group plc	452111	15,917,800,448	822,143,808
Neiman Marcus Group Inc (The)	452111	4,345,374,000	140,086,000
Nordstrom Inc	452111	10,876,999,680	683,000,000
Saks Inc	452111	3,013,593,088	74,790,000
Sears Canada Inc	452111	4,646,248,448	-60,450,616
Sears Holdings Corporation	452111	41,566,998,528	-3,140,000,000
Stage Stores Inc	452111	1,511,918,976	30,960,000
Stein Mart Inc	452111	1,160,366,976	19,828,000
Discount Stores			
Kmart Holding Corporation	452112	15,285,000,000	-34,000,000
Kohl's Corp	452112	18,804,000,768	1,167,000,064
Target Corporation	452112	69,864,996,864	2,928,999,936
Wal-Mart Stores Inc	452112	446,950,014,976	15,699,000,320
Fabric Coating Mills			
Nano-Tex LLC	313320		
OMNOVA Solutions Inc	313320		
W L Gore & Associates Inc	313320		
Fiber, Yarn and Thread Mills			
American & Efird Inc	313110		
Carolina Mills Inc	313110		
Chapelthorpe plc	313110		

Company	Industry Code	2012 Sales	2012 Profits
Coats plc	313110		
INVISTA	313110		
Marzotto SpA	313110		
Milliken & Co	313110		
National Spinning Company Inc	313110		
Nitto Boseki Co Ltd	313110		
Parkdale America LLC	313110		
Parkdale Mills Inc	313110		
Paul Reinhart AG	313110		
Plains Cotton Cooperative Association	313110		
Polymer Group Inc	313110		
Royal Ten Cate NV	313110		
Stowe-Pharr Mills Inc	313110		
Teijin Ltd	313110	10,408,991,744	145,942,976
Texwinca Holdings Limited	313110	1,776,258,560	123,719,400
Toho Tenax Co Ltd	313110		
Toray Industries Inc	313110	19,354,337,280	782,383,040
Unifi Inc	313110	705,086,016	11,491,000
Unitika Ltd	313110		
Weiqiao Textile Company Limited	313110		
Food Service Contractors			
ARAMARK Corporation	722310	13,250,000,000	
Freight Transportation Arrangement			
Argix Direct Inc	488510		
Fung Group	488510		
Furniture Stores			
Ryohin Keikaku Co Ltd (Muji)	442110	2,170,869,760	107,821,640
Internet Retail Sales Sites			
ASOS plc	454111	793,073,152	35,697,804
BackCountry.com	454111		
Bluefly Inc	454111		
Gilt Groupe Inc	454111	600,000,000	
Net-A-Porter LLC	454111		
YOOX SpA	454111		
Zappos.com Inc	454111	1,300,000,000	
Knit Fabric Mills			
FAB Industries Corporation	313240		
QST Industries Inc	313240		
Springs Global USA Inc	313240		
Linen/Uniform Supply			
Cintas Corp	812331	4,102,000,128	297,636,992
G & K Services Inc	812331	869,937,024	24,147,000
Johnson Service Group plc	812331		
Linens/Housewares/Art/Framing Stores			
Bed Bath & Beyond Inc	442299	9,499,889,664	989,537,024
Luggage Manufacturing			
Samsonite International SA	316998		
Mail Order and Non-Store Selling, Incl. TV Shopping			
Blair Corporation	454113		

Company	Industry Code	2012 Sales	2012 Profits
Coldwater Creek Inc	454113		
dELiA*s Inc	454113	217,152,000	-22,670,000
L L Bean Inc	454113	1,520,000,000	
Lands End Inc	454113	1,700,000,000	
Redcats USA	454113		
Sculptz Inc	454113		
Sport Supply Group Inc	454113		
Medical/Dental/Surgical Equipment & Supplies, Manufacturing			
3M Company	33911		
Aearo Technologies LLC	33911		
Alpha Pro Tech Ltd	33911		
Lakeland Industries Inc	33911	96,326,992	-376,825
Mine Safety Appliances Co	33911		
Mens and Boys' Clothing and Furnishings Merchant Wholesalers			
China Dongxiang Group	424320		
Nanotechnology-General			
Institute for Soldier Nanotechnologies	541712N		
Nonwoven Fabric Mills			
Nano-Tex LLC	313320		
OMNOVA Solutions Inc	313320		
W L Gore & Associates Inc	313320		
Other Measuring and Controlling Device Manufacturing			
Fossil Inc	334519		
Outerwear Kintting Mills			
Anvil Knitwear Inc	315190		
People's Garment PCL	315190		
Stefanel SpA	315190		
Petrochemicals Manufacturing			
Reliance Industries Ltd (Reliance Group)	325110	64,655,347,712	3,555,944,192
Petroleum Refineries			
Koch Industries Inc	324110	115,000,000,000	10,000,000,000
Plastic Products, Manufacturing			
Carpenter Co	3261		
Sewing & Fabric Stores			
Jo Ann Stores Inc	451130		
Shoe Manufacturing			
adidas AG	316210		
Allen-Edmonds Shoe Corp	316210	120,000,000	
AND 1	316210		
ASICS Corp	316210	3,018,908,416	153,715,888
Belle International Holdings Limited	316210	5,343,260,000	706,966,000
C&J Clark International Ltd	316210		
Converse Inc	316210		
Crocs Inc	316210		
Deckers Outdoor Corporation	316210		
ECCO Sko A/S	316210		
Genesco Inc	316210	2,291,986,944	81,959,000
Gucci Group NV	316210	2,117,550,000	
Iconix Brand Group Inc	316210		

Company	Industry Code	2012 Sales	2012 Profits
Kenneth Cole Productions Inc	316210	500,000,000	
Kingmaker Footwear Holdings Limited	316210	213,547,024	6,979,316
K-Swiss Inc	316210		
Lacrosse Footwear	316210		
Li Ning Company Limited	316210	1,094,600,000	-223,661,000
New Balance Athletic Shoe Inc	316210	2,175,000,000	
Nike Inc	316210	24,128,000,000	2,223,000,064
Performance + Lifestyle Group (PLG)	316210		
Phoenix Footwear Group, Inc.	316210		
PUMA SE	316210	4,308,400,000	
Red Wing Shoe Company Inc	316210		
Reebok International Ltd	316210		
RG Barry Corp	316210	155,938,000	14,549,000
Rocky Brands Inc	316210		
Salvatore Ferragamo SpA	316210		
Saucony Inc	316210		
Skechers USA Inc	316210		
Stella International Holdings Ltd	316210		
Steven Madden Ltd	316210		
Timberland Co	316210	1,440,000,000	
Tod's SpA	316210		
TOMS Shoes Inc	316210		
Weyco Group Inc	316210		
Wolverine World Wide Inc	316210		
Yue Yuen Industrial (Holdings) Limited	316210	7,300,000,000	506,200,000
Shoes & Accessories Stores			
Aldo Group Inc	448210		
Brown Shoe Company Inc	448210	2,582,823,936	24,589,000
DSW Inc	448210	2,024,328,960	174,788,000
Finish Line Inc (The)	448210	1,369,259,008	84,804,000
Foot Locker Inc	448210	5,623,000,064	278,000,000
Payless ShoeSource Inc	448210		
Shoe Carnival Inc	448210	762,534,016	26,382,000
Vans Inc	448210		
Sporting & Athletic Goods, Manufacturing			
Reebok CCM Hockey	339920		
Sporting Goods Stores			
Academy Sports & Outdoors Ltd	451110	3,000,000,000	
Bass Pro Shops Inc	451110	4,300,000,000	
Big 5 Sporting Goods Corporation	451110		
Cabela's Inc	451110		
Dick's Sporting Goods Inc	451110	5,211,802,112	263,906,000
Eastern Mountain Sports Inc	451110		
Hibbett Sports Inc	451110	732,644,992	59,060,000
Orvis Company Inc (The)	451110	245,000,000	
Recreational Equipment Inc (REI)	451110	1,930,000,000	
Sport Chalet Inc	451110	349,883,008	-5,072,000
Sports Authority Inc (The)	451110		
Sportsman's Guide Inc (The)	451110		

Company	Industry Code	2012 Sales	2012 Profits
Tannery Leather Manufacturing			
Pittards plc	316110		
Textiles Manufacturing			
ADS Inc	313230		
Albany International Corp	313230		
Cheil Industries Inc	313230		
Chori Company Ltd	313230	2,653,813,504	42,604,776
Guilford Mills Inc	313230		
Kuraray Co Ltd	313230	4,495,309,312	383,394,272
Metso Fabrics Corp	313230		
Nisshinbo Holdings Inc	313230	4,621,588,480	114,705,168
Xerium Technologies Inc	313230		
Toys/Hobbies/Games/Crafts Stores			
Motorsports Authentics LLC	451120		
Zippers & Fasteners			
Talon International Inc	339993		
Velcro Industries NV	339993		
YKK Corp	339993		

ALPHABETICAL INDEX

Filippa K
Finish Line Inc (The)
Foot Locker Inc
Forever 21 Inc
Fossil Inc
Francesca's Holdings Corp
Frederick's of Hollywood Group Inc
French Connection Group plc
Fujibo Holdings Inc
Fung Group
G & K Services Inc
Gap Inc (The)
Genesco Inc
Gerry Weber International AG
Gianni Versace SpA
G-III Apparel Group Ltd
Gildan Activewear Inc
Gilt Groupe Inc
Giorgio Armani SpA
Gordmans Stores Inc
Greenwood Mills
Grupo Tavex SA
Gucci Group NV
Guess? Inc
Guilford Mills Inc
Gymboree Corp (The)
H&M (Hennes & Mauritz) AB
Haggar Corp
Hallwood Group Inc.
Hampshire Group Ltd
Hanesbrands Inc
Hermes International SA
Hibbett Sports Inc
Hot Topic Inc
Hugo Boss AG
Hyosung Corporation
ICG/Holliston
Iconix Brand Group Inc
Inditex (Industria de Diseno Textil SA)
Institute for Soldier Nanotechnologies
Interface Inc
International Textile Group Inc
INVISTA
ITOCHU Corp
J C Penney Company Inc
J Crew Group Inc
J Jill Group Inc (The)
Jaclyn Inc
Jil Sander SpA
Jo Ann Stores Inc
Joe's Jeans Inc
Johnson Service Group plc
Johnston Textiles Inc
Jones Group Inc (The)
Jordache Enterprises Inc
Jos A Bank Clothiers Inc
Kate Spade LLC
Kellwood Co
Kenneth Cole Productions Inc
KiK Textilien & Non-Food GmbH
Kingmaker Footwear Holdings Limited
Klopman International
Kmart Holding Corporation
Koch Industries Inc
Kohl's Corp
K-Swiss Inc
Kuraray Co Ltd
L L Bean Inc
Lacoste SA
Lacrosse Footwear
Lafuma SA
Lakeland Industries Inc
Lands End Inc
Lectra SA
Levi Strauss & Co
Li Ning Company Limited
Limited Brands Inc
Loehmann's Holdings Inc
Lord & Taylor LLC
lululemon Athletica Inc
LVMH Moet Hennessy Louis Vuitton SA
Macy's Inc
Maidenform Inc
Mango/MNG Holding SL
Marc Ecko Enterprises
Marks & Spencer Group plc
Marzotto SpA
MAST Global Fashions
Max Mara Fashion Group SRL
Men's Wearhouse Inc (The)
Metso Fabrics Corp
Michael Kors Holdings Ltd
Milliken & Co
Mine Safety Appliances Co
Mohawk Industries Inc
Motorsports Authentics LLC
Mount Vernon Mills
Nano-Tex LLC
National Spinning Company Inc
Nautica Enterprises Inc
Neiman Marcus Group Inc (The)
Net-A-Porter LLC
New Balance Athletic Shoe Inc
New Look Group plc
New York & Company Inc
NEXT plc
Nike Inc
Nisshinbo Holdings Inc
Nitto Boseki Co Ltd
Nordstrom Inc
OMNOVA Solutions Inc
Orvis Company Inc (The)
OshKosh B'Gosh Inc
Oxford Industries Inc
Pacific Brands Limited
Pacific Sunwear of California Inc

Paris Blues Inc
Parkdale America LLC
Parkdale Mills Inc
Passport Brands Inc
Patagonia Inc
Paul Reinhart AG
Payless ShoeSource Inc
Pendleton Woolen Mills Inc
People's Garment PCL
Performance + Lifestyle Group (PLG)
Perry Ellis International Inc
Phat Fashions LLC
Phoenix Footwear Group, Inc.
Pittards plc
Plains Cotton Cooperative Association
Point Blank Enterprises Inc
Polartec LLC
Polymer Group Inc
PPR S.A. (KERING)
Prada SpA Group
Propex Operating Company LLC
Puig Beauty & Fashion Group SL
PUMA SE
PVH Corp
QST Industries Inc
Quiksilver Inc
Ralph Lauren Corporation
Ratti SpA
Raymond Ltd
Recreational Equipment Inc (REI)
Red Wing Shoe Company Inc
Redcats USA
Reebok CCM Hockey
Reebok International Ltd
Reliance Industries Ltd (Reliance Group)
RG Barry Corp
Rocky Brands Inc
Ross Stores Inc
Royal Ten Cate NV
Russell Brands LLC
Ryohin Keikaku Co Ltd (Muji)
Saks Inc
Salvatore Ferragamo SpA
Samsonite International SA
Samsung Group
Saucony Inc
Sculptz Inc
Sears Canada Inc
Sears Holdings Corporation
Seven & I Holdings Co Ltd
Shaw Industries
Shoe Carnival Inc
Simply Fashion Stores Ltd
Skechers USA Inc
Spanx Inc
Sport Chalet Inc
Sport Supply Group Inc
Sports Authority Inc (The)
Sportsman's Guide Inc (The)
Springs Global USA Inc
St John Knits International Inc
Stage Stores Inc
Stefanel SpA
Stein Mart Inc
Stella International Holdings Ltd
Steven Madden Ltd
Stowe-Pharr Mills Inc
Sunrise Brands LLC
Superior Uniform Group Inc
Swank Inc
Swift Galey
TAL Apparel Group
Talbots Inc (The)
Talon International Inc
Tandy Brands Accessories Inc
Target Corporation
Ted Baker plc
Tefron Ltd
Teijin Ltd
Texwinca Holdings Limited
Tilly's Inc
Timberland Co
TJX Companies Inc (The)
Tod's SpA
Toho Tenax Co Ltd
Tommy Hilfiger Corp
TOMS Shoes Inc
TOPSHOP
Toray Industries Inc
Tory Burch LLC
Toyobo Co Ltd
True Religion Apparel Inc
Tween Brands Inc
Under Armour Inc
Unifi Inc
Uniqlo Co Ltd
United Retail Group Inc
Unitika Ltd
Urban Outfitters Inc
Valentino Fashion Group SpA
Vans Inc
Varsity Brands Inc
Velcro Industries NV
Vera Bradley Inc
VF Corp
Victoria plc
Victorias Secret
Vincenzo Zucchi SpA
Volcom Inc
W L Gore & Associates Inc
Wacoal Holdings Corp
Walking Company Holdings Inc (The)
Wal-Mart Stores Inc
Warnaco Group Inc (The)
Warnaco Swimwear Inc
Weiqiao Textile Company Limited

INDEX OF U.S. HEADQUARTERS LOCATION BY STATE

To help you locate members of THE APPAREL & TEXTILES 350 geographically, the city and state of the headquarters of each company are indexed.

ALABAMA

ARKANSAS

CALIFORNIA

COLORADO

CONNECTICUT
Brooks Brothers Group Inc; Enfield

DELAWARE
W L Gore & Associates Inc; Newark

FLORIDA
Chico's FAS Inc; Fort Myers
Claire's Stores Inc; Pembroke Pines
Perry Ellis International Inc; Miami
Point Blank Enterprises Inc; Pompano Beach
Stein Mart Inc; Jacksonville
Superior Uniform Group Inc; Seminole

GEORGIA
Beaulieu of America LLC; Dalton
Carters Inc; Atlanta
Interface Inc; Atlanta
Mohawk Industries Inc; Calhoun
Oxford Industries Inc; Atlanta
Shaw Industries; Dalton
Spanx Inc; Atlanta

IDAHO
Coldwater Creek Inc; Sandpoint

ILLINOIS
Kmart Holding Corporation; Hoffman Estates
QST Industries Inc; Chicago
Sears Holdings Corporation; Hoffman Estates

INDIANA
Aearo Technologies LLC; Indianapolis
Finish Line Inc (The); Indianapolis
Shoe Carnival Inc; Evansville
Vera Bradley Inc; Fort Wayne

KANSAS
INVISTA; Wichita
Koch Industries Inc; Wichita
Payless ShoeSource Inc; Topeka

KENTUCKY
Russell Brands LLC; Bowling Green

MAINE
L L Bean Inc; Freeport

MARYLAND
Jos A Bank Clothiers Inc; Hampstead
Passport Brands Inc; Baltimore
Under Armour Inc; Baltimore

MASSACHUSETTS
Casual Male Retail Group Inc; Canton
Converse Inc; North Andover
Institute for Soldier Nanotechnologies; Cambridge
J Jill Group Inc (The); Quincy
New Balance Athletic Shoe Inc; Boston
Performance + Lifestyle Group (PLG); Lexington
Polartec LLC; Lawrence
Reebok International Ltd; Canton
Samsonite International SA; Mansfield
Saucony Inc; Lexington
Talbots Inc (The); Hingham
TJX Companies Inc (The); Framingham

MICHIGAN
Wolverine World Wide Inc; Rockford

MINNESOTA
3M Company; St. Paul
Christopher & Banks Corp; Plymouth
G & K Services Inc; Minnetonka
Red Wing Shoe Company Inc; Red Wing
Sportsman's Guide Inc (The); St. Paul
Target Corporation; Minneapolis

MISSOURI
Bass Pro Shops Inc; Springfield
Brown Shoe Company Inc; St. Louis
Everlast Worldwide Inc; Moberly
Kellwood Co; Chesterfield

NEBRASKA
Buckle Inc (The); Kearney
Cabela's Inc; Sidney
Gordmans Stores Inc; Omaha

NEVADA
Zappos.com Inc; Henderson

NEW HAMPSHIRE
Eastern Mountain Sports Inc; Peterborough
Timberland Co; Stratham

NEW JERSEY
Argix Direct Inc; Jamesburg
Bed Bath & Beyond Inc; Union
Burlington Coat Factory; Burlington
Children's Place Retail Stores Inc (The); Secaucus
Jaclyn Inc; West New York
Maidenform Inc; Iselin
United Retail Group Inc; Rochelle Park

NEW YORK
Aeropostale Inc; New York
Albany International Corp; Albany
Ann Inc; New York
Anvil Knitwear Inc; New York
Ascena Retail Group Inc; Suffern
Barney's New York Inc; New York

Bernard Chaus Inc; New York
Bill Blass Group LLC; New York
Bluefly Inc; New York
Brookwood Companies Inc; New York
Cache Inc; New York
Celebrity International Inc; New York
Coach Inc; New York
dELiA*s Inc; New York
Diane von Furstenberg Studio LP; New York
Eileen Fisher Inc; Irvington
Etienne Aigner Group Inc; New York
FAB Industries Corporation; Great Neck
Fifth & Pacific Companies Inc; New York
Fila USA Inc; New York
Foot Locker Inc; New York
G-III Apparel Group Ltd; New York
Gilt Groupe Inc; New York
Hampshire Group Ltd; New York
Iconix Brand Group Inc; New York
J Crew Group Inc; New York
Jones Group Inc (The); New York
Jordache Enterprises Inc; New York
Kate Spade LLC; New York
Kenneth Cole Productions Inc; New York
Lakeland Industries Inc; Ronkonkoma
Loehmann's Holdings Inc; Bronx
Lord & Taylor LLC; New York
Marc Ecko Enterprises; New York
Nautica Enterprises Inc; New York
New York & Company Inc; New York
Phat Fashions LLC; New York
PVH Corp; New York
Ralph Lauren Corporation; New York
Redcats USA; New York
Saks Inc; New York
Steven Madden Ltd; Long Island City
Swank Inc; New York
Tory Burch LLC; New York
Warnaco Group Inc (The); New York

NORTH CAROLINA

American & Efird Inc; Mt. Holly
Belk Inc; Charlotte
Carolina Mills Inc; Maiden
Cato Corp (The); Charlotte
Cone Denim LLC; Greensboro
Culp Inc; High Point
Guilford Mills Inc; Wilmington
Hanesbrands Inc; Winston-Salem
International Textile Group Inc; Greensboro
Motorsports Authentics LLC; Concord
National Spinning Company Inc; Washington
Parkdale America LLC; Gastonia
Parkdale Mills Inc; Gastonia
Polymer Group Inc; Charlotte
Stowe-Pharr Mills Inc; McAdenville
Swift Galey; Society Hill
Unifi Inc; Greensboro
VF Corp; Greensboro
Xerium Technologies Inc; Raleigh

OHIO

Abercrombie & Fitch Co; New Albany
Cintas Corp; Cincinnati
DSW Inc; Columbus
Express Inc; Columbus
Jo Ann Stores Inc; Hudson
Limited Brands Inc; Columbus
Macy's Inc; Cincinnati
MAST Global Fashions; Columbus
OMNOVA Solutions Inc; Fairlawn
RG Barry Corp; Pickerington
Rocky Brands Inc; Nelsonville
Tween Brands Inc; New Albany
Victorias Secret; Reynoldsburg

OREGON

Columbia Sportswear Co; Portland
Lacrosse Footwear; Portland
Nike Inc; Beaverton
Pendleton Woolen Mills Inc; Portland

PENNSYLVANIA

American Eagle Outfitters Inc; Pittsburgh
ARAMARK Corporation; Philadelphia
Blair Corporation; Warren
Bon-Ton Stores Inc (The); York
Boscov's Inc; Reading
Charming Shoppes Inc; Bensalem
Deb Shops Inc; Philadelphia
Destination Maternity Corp; Philadelphia
Dick's Sporting Goods Inc; Coraopolis
Elder-Beerman Stores Corp (The); York
Mine Safety Appliances Co; Pittsburgh
Sculptz Inc; Trevose
Urban Outfitters Inc; Philadelphia

SOUTH CAROLINA

Delta Apparel Inc; Greenville
Greenwood Mills; Greenwood
Milliken & Co; Spartanburg
Mount Vernon Mills; Mauldin
Springs Global USA Inc; Fort Mill

TENNESSEE

Dixie Group Inc; Chattanooga
Genesco Inc; Nashville
ICG/Holliston; Church Hill
Propex Operating Company LLC; Chattanooga
Varsity Brands Inc; Memphis

TEXAS

Academy Sports & Outdoors Ltd; Katy
Fossil Inc; Richardson
Francesca's Holdings Corp; Houston

INDEX OF NON-U.S. HEADQUARTERS LOCATION BY COUNTRY

To help you locate members of THE APPAREL & TEXTILES 350 that have non-U.S. headquarters, the city and country of the headquarters of such companies are included in the following index.

Gerry Weber International AG; Halle
Hugo Boss AG; Metzingen
KiK Textilien & Non-Food GmbH; Bonen
PUMA SE; Herzogenaurach

HONG KONG

Belle International Holdings Limited; Tsuen Wan
Esprit Holdings Ltd; Kowloon
Esquel Group of Companies; Hong Kong
Fung Group; Hong Kong
Kingmaker Footwear Holdings Limited; Kowloon
Michael Kors Holdings Ltd; Hong Kong
Stella International Holdings Ltd; Kowloon
TAL Apparel Group; Hong Kong
Texwinca Holdings Limited; Hong Kong
Tommy Hilfiger Corp; Hong Kong
Yue Yuen Industrial (Holdings) Limited; Hong Kong

INDIA

Raymond Ltd; Ratanagiri
Reliance Industries Ltd (Reliance Group); Mumbai

ISRAEL

Delta Galil Industries Ltd; Tel Aviv
Tefron Ltd; Misgav

ITALY

Benetton Group SPA; Ponzano
CSP International Industria Calze SpA; Ceresara
Diesel SpA; Vicenza
Dolce & Gabbana SpA (D&G); Milan
Gianni Versace SpA; Milan
Giorgio Armani SpA; Milan
Jil Sander SpA; Milan
Klopman International; Frosinone
Marzotto SpA; Vicenza
Max Mara Fashion Group SRL; Reggio Emilia
Prada SpA Group; Milano
Ratti SpA; Guanzate
Salvatore Ferragamo SpA; Florence
Stefanel SpA; Ponte di Piave
Tod's SpA; Sant'Elpidio a Mare
Valentino Fashion Group SpA; Milan
Vincenzo Zucchi SpA; Milan
YOOX SpA; Zola Predosa
Zegna (Ermenegildo Zegna Holditalia SpA); Milan

JAPAN

Asahi Kasei Corp; Tokyo
ASICS Corp; Kobe
Chori Company Ltd; Osaka
Fujibo Holdings Inc; Tokyo
ITOCHU Corp; Tokyo
Kuraray Co Ltd; Tokyo
Nisshinbo Holdings Inc; Tokyo
Nitto Boseki Co Ltd; Tokyo
Ryohin Keikaku Co Ltd (Muji); Tokyo
Seven & I Holdings Co Ltd; Tokyo
Teijin Ltd; Osaka
Toho Tenax Co Ltd; Tokyo
Toray Industries Inc; Tokyo
Toyobo Co Ltd; Osaka
Uniqlo Co Ltd; Yamaguchi City
Unitika Ltd; Osaka
Wacoal Holdings Corp; Kyoto
YKK Corp; Tokyo

KOREA

BYC Co; Seoul
Cheil Industries Inc; Seoul
Hyosung Corporation; Seoul
Samsung Group; Seoul

SPAIN

Cortefiel SA; Madrid
DOGI International Fabrics SA; Barcelona
Grupo Tavex SA; Madrid
Inditex (Industria de Diseno Textil SA); Arteixo
Mango/MNG Holding SL; Barcelona
Puig Beauty & Fashion Group SL; Barcelona
Zara International Inc; Arteixo

SWEDEN

Filippa K; Stockholm
H&M (Hennes & Mauritz) AB; Stockholm

SWITZERLAND

Paul Reinhart AG; Winterthur

THAILAND

People's Garment PCL; Bangkok

THE NETHERLANDS

Gucci Group NV; Amsterdam
Royal Ten Cate NV; Almelo

UNITED ARAB EMIRATES

Worldtex Inc; Dubai

UNITED KINGDOM

Alexandra plc; Bristol
Arcadia Group; London
ASOS plc; London
Burberry Group plc; London
C&J Clark International Ltd; Somerset
Chapelthorpe plc; Bradford
Coats plc; Uxbridge
French Connection Group plc; London
Johnson Service Group plc; Cheshire
Marks & Spencer Group plc; London
Net-A-Porter LLC; London
New Look Group plc; Weymouth
NEXT plc; Leicester

Pittards plc; Yeovil
Ted Baker plc; London
TOPSHOP; London
Victoria plc; Kidderminster

Individual Data Profiles On Each Of THE APPAREL & TEXTILES 350

3M Company

www.3m.com

NAIC Code: 33911

TYPES OF BUSINESS:

Health Care Products
Specialty Materials & Textiles
Industrial Products
Safety, Security & Protection Products
Display & Graphics Products
Consumer & Office Products
Electronics & Communications Products
Fuel Cell Technology

BRANDS/DIVISIONS/AFFILIATES:

Ceradyne Inc
Federal Signal Technologies Group
3M Canada
3M ESPE
Aearo Technologies LLC

CONTACTS: *Note: Officers with more than one job title may be intentionally listed here more than once.*

Inge G. Thulin, CEO
Inge G. Thulin, Pres.
David Meline, CFO/Sr. VP
Ian F. Hardgrove, Sr. VP-Mktg. & Sales
Marlene McGrath, Sr. VP-Human Resources
Frederick J. Palensky, Exec. VP-R&D
Frederick J. Palensky, CTO
Ivan K. Fong, General Counsel/Sr. VP-Legal Affairs
Roger H.D. Lacey, Sr. VP-Strategy & Corp. Dev.
Brad T. Sauer, Exec. VP-Health Care Bus.
Joaquin Delgado, Exec. VP-Electro & Comm. Bus.
Michael Vale, Exec. VP-Consumer & Office Bus.
George W. Buckley, Chmn.
Christopher D. Holmes, Sr. VP-Supply Chain

GROWTH PLANS/SPECIAL FEATURES:

3M Company is involved in the research, manufacturing and marketing of a variety of products. Its operations are organized into six segments: Health Care; Consumer and Office; Display and Graphics; Electro and Communications; Industrial and Transportation; and Safety, Security and Protection. The Health Care segment's products include medical and surgical supplies, skin infection prevention products, drug delivery systems, orthodontic products, health information systems and veterinary products. The Consumer and Office segment includes office supply, stationery, construction, home improvement, protective material and visual systems products. The Display and Graphics segment's products include optical film and lenses for electronic displays, touch screens and monitors, screen filters, reflective sheeting and commercial graphics systems. The Electro and Communications segment's products include packaging and interconnection devices (used in circuits), fluids used in computer chips, high-temperature and display tapes, pressure-sensitive tapes and resins and products for telecommunications systems. The Industrial and Transportation segment's products include vinyl, polyester, tapes, a variety of non-woven abrasives, adhesives, specialty materials, supply chain execution software, filtration systems, paint finishing products, engineering fluids and components for catalytic converters. The Safety, Security and Protection services segment provides products for personal protection, safety and security, energy control, commercial cleaning and protection, passports and secure cards. In 2012, the firm acquired technical ceramics manufacturer Ceradyne, Inc. and the business of Federal Signal Technologies Group.

The company offers employees medical and dental insurance, domestic partner benefits, tuition reimbursement, flexible spending accounts, disability coverage, a 401(k) and adoption assistance.

FINANCIAL DATA: *Note: Data for latest year may not have been available at press time.*

In U.S. $	2012	2011	2010	2009	2008	2007
Revenue	29,904,000,000	29,611,000,000	26,662,000,000	23,123,000,000	25,269,000,000	24,462,000,000
R&D Expense	1,634,000,000	1,570,000,000	1,434,000,000	1,293,000,000	1,404,000,000	1,368,000,000
Operating Income	6,483,000,000	6,178,000,000	5,918,000,000	4,814,000,000	5,218,000,000	6,193,000,000
Operating Margin %	21.67%	20.86%	22.19%	20.81%	20.64%	25.31%
SGA Expense	6,102,000,000	6,170,000,000	5,479,000,000	4,907,000,000	5,245,000,000	5,015,000,000
Net Income	4,444,000,000	4,283,000,000	4,085,000,000	3,193,000,000	3,460,000,000	4,096,000,000
Operating Cash Flow	5,300,000,000	5,284,000,000	5,174,000,000	4,941,000,000	4,533,000,000	4,275,000,000
Capital Expenditure	1,484,000,000	1,379,000,000	1,091,000,000	903,000,000	1,471,000,000	1,422,000,000
EBITDA	7,810,000,000	7,453,000,000	7,076,000,000	6,008,000,000	6,476,000,000	7,397,000,000
Return on Assets %	13.57%	13.86%	14.23%	12.09%	13.77%	17.81%
Return on Equity %	26.93%	27.55%	28.74%	28.20%	31.99%	37.74%
Debt to Equity	0.28	0.29	0.27	0.4	0.52	0.34

CONTACT INFORMATION:

Phone: 651 733-1110 Fax: 651 733-9973
Toll-Free: 800-364-3577
Address: 3M Center, St. Paul, MN 55144 United States

STOCK TICKER/OTHER:

Stock Ticker: MMM Exchange: NYS
Employees: 84,198 Fiscal Year Ends: 31-Dec
Parent Company:

SALARIES/BONUSES:

Top Exec. Salary: $1,228,373 Bonus: $
Second Exec. Salary: $758,013 Bonus: $

OTHER THOUGHTS:

Estimated Female Officers or Directors: 7
Hot Spot for Advancement for Women/Minorities: Y

Abercrombie & Fitch Co

NAIC Code: 448140

www.abercrombie.com

TYPES OF BUSINESS:

Casual Apparel-Young Adults, Retail
Catalog & Online Sales
Children's Apparel
Casual Adult Apparel, Retail
Lingerie Stores

BRANDS/DIVISIONS/AFFILIATES:

Abercrombie & Fitch
Gilly Hicks
Hollister Co.
abercrombie kids
Abercrombie.com
AbercrombieKids.com
HollisterCo.com

CONTACTS: *Note: Officers with more than one job title may be intentionally listed here more than once.*

Michael S. Jeffries, CEO
Jonathan E. Ramsden, CFO/Exec. VP
Ronald A. Robins, Jr., General Counsel/Sr. VP/Sec.
Leslee K. Herro, Exec. VP-Planning & Allocation
Michael S. Jeffries, Chmn.
Diane Chang, Exec. VP-Sourcing

GROWTH PLANS/SPECIAL FEATURES:

Abercrombie & Fitch Co. retails upscale casual American clothing and accessories. The Abercrombie & Fitch brand was established in 1892, and the firm became well known as a supplier of rugged, high-quality outdoor gear, outfitting expeditions, including those of Theodore Roosevelt, Ernest Hemingway, Charles Lindbergh and Richard Byrd. However, to boost sales and focus on a high-growth market segment, the company shifted its focus from outdoor gear and accessories to fashion-oriented casual wear. Abercrombie & Fitch currently operates 1,067 stores in North America, Europe and Asia through four store concepts. Abercrombie & Fitch (A&F, 278 locations), the company's main store, targets men and women, offering casual clothing, accessories and fragrances. The firm's 154 abercrombie kids stores offer A&F-style clothing for boys and girls. The Hollister Co., with 486 stores, markets Southern California and surfing-inspired clothing and accessories to teenagers. Its newest store concept, Gilly Hicks, sells women's lingerie, loungewear and personal care products at 18 retail locations. Abercrombie & Fitch also sells merchandise through a catalog and four web sites, one each for A&F, abercrombie kids, Hollister and Gilly Hicks. During 2011, direct-to-consumer sales across its brands represented 13.3% of Abercrombie & Fitch's total sales. The firm's stores generate about $500 in sales per square foot per year.

FINANCIAL DATA: *Note: Data for latest year may not have been available at press time.*

In U.S. $	2012	2011	2010	2009	2008	2007
Revenue	4,158,058,000	3,468,777,000	2,928,626,000	3,540,276,000	3,749,847,000	
R&D Expense						
Operating Income	190,030,000	231,932,000	117,912,000	439,386,000	740,497,000	
Operating Margin %	4.57%	6.68%	4.02%	12.41%	19.74%	
SGA Expense	2,325,368,000	1,990,305,000	1,779,219,000	1,931,170,000	1,782,604,000	
Net Income	127,658,000	150,283,000	254,000	272,255,000	475,697,000	
Operating Cash Flow	365,219,000	391,789,000	402,200,000	490,836,000	817,825,000	
Capital Expenditure	318,598,000	160,935,000	175,472,000	367,602,000	403,345,000	
EBITDA	427,286,000	465,523,000	364,862,000	664,720,000	886,795,000	
Return on Assets %	4.25%	5.20%		10.05%	19.75%	
Return on Equity %	6.80%	8.08%	.01%	15.71%	31.46%	
Debt to Equity	0.03	0.03	0.03	0.05		

CONTACT INFORMATION:

Phone: 614 283-6500 Fax: 614 283-6710
Toll-Free:
Address: 6301 Fitch Path, New Albany, OH 43054 United States

STOCK TICKER/OTHER:

Stock Ticker: ANF Exchange: NYS
Employees: 90,000 Fiscal Year Ends: 31-Jan
Parent Company:

SALARIES/BONUSES:

Top Exec. Salary: $1,500,000 Bonus: $
Second Exec. Salary: $977,116 Bonus: $

OTHER THOUGHTS:

Estimated Female Officers or Directors: 4
Hot Spot for Advancement for Women/Minorities: Y

Academy Sports & Outdoors Ltd

www.academy.com

NAIC Code: 451110

TYPES OF BUSINESS:

Sporting Goods Stores
Apparel
Footwear
Outdoor Sports Gear
Hunting Licenses

BRANDS/DIVISIONS/AFFILIATES:

KKR & Co LP (Kohlberg Kravis Roberts & Co)

CONTACTS: *Note: Officers with more than one job title may be intentionally listed here more than once.*

Rodney Faldyn, CEO
Rodney Faldyn, Pres.

GROWTH PLANS/SPECIAL FEATURES:

Academy Sports & Outdoors, Ltd., owned by KKR & Co LP, is one of the largest sporting goods retailers in the U.S., with annual sales in excess of $2.7 billion. The company operates over 150 stores throughout 13 states including Alabama, Florida, Georgia, Arkansas, Louisiana, Mississippi, Kansas, Missouri, Oklahoma, North Carolina, South Carolina, Tennessee and Texas. Its retail operations also include a full e-commerce retail store. Academy Sports offers a broad selection of sporting equipment, apparel and footwear. The stores, which range in size from 50,000 to 100,000 square feet, are laid out in a racetrack format with soft goods on the inside, including branded and private label athletic and casual apparel, and hard goods, such as camping, hunting, fishing, marine, footwear and fitness and sporting goods, on the outside. The company distributes merchandise to its stores from its distribution centers located in Katy, Texas and Twiggs County, Georgia. The center utilizes radio frequency identification devices (RFID), automated inventory and replenishment systems and a state-of-the-art warehouse management system to smoothly operate its large processing and inventory space.

The company offers its employees a 401(k) plan; medical, dental and vision insurance; life insurance; short- and long-term disability benefits; tuition reimbursement; merchandise discounts; bereavement leave; continuing education benefits; and business travel accident insurance.

FINANCIAL DATA: *Note: Data for latest year may not have been available at press time.*

In U.S. $	2012	2011	2010	2009	2008	2007
Revenue	3,000,000,000	2,700,000,000	2,400,000,000	2,000,000,000	2,000,000,000	2,060,000,000
R&D Expense						
Operating Income						
Operating Margin %						
SGA Expense						
Net Income						
Operating Cash Flow						
Capital Expenditure						
EBITDA						
Return on Assets %						
Return on Equity %						
Debt to Equity						

CONTACT INFORMATION:

Phone: 281-646-5200 Fax: 281-646-5000
Toll-Free: 888-922-2336
Address: 1800 N. Mason Rd., Katy, TX 77449 United States

STOCK TICKER/OTHER:

Stock Ticker: Private Exchange:
Employees: 18,000 Fiscal Year Ends: 31-Jan
Parent Company: KKR & CO LP (KOHLBERG KRAVIS ROBERTS & CO)

SALARIES/BONUSES:

Top Exec. Salary: $ Bonus: $
Second Exec. Salary: $ Bonus: $

OTHER THOUGHTS:

Estimated Female Officers or Directors:
Hot Spot for Advancement for Women/Minorities:

adidas AG

NAIC Code: 316210

www.adidas.com

TYPES OF BUSINESS:

Athletic Shoes
Golf Equipment & Accessories
Street & Sports Apparel
Personal Care Products
Eyewear
Watches

BRANDS/DIVISIONS/AFFILIATES:

adidas
Ashworth Inc
CCM
TaylorMade
Reebok International Ltd
Rockport
Adams Golf Inc
adidas Body

CONTACTS: *Note: Officers with more than one job title may be intentionally listed here more than once.*

Herbert Hainer, CEO
Glenn Bennett, Head-Global Oper.
Robin Stalker, CFO
Erich Stamminger, Pres., Global Brands
Igor Landau, Chmn.

GROWTH PLANS/SPECIAL FEATURES:

adidas AG is an athletic apparel manufacturer and one of the world's largest producers of sporting goods. The company operates more than 170 international subsidiaries and markets products under six primary brands: adidas, Reebok, Rockport, CCM, Ashworth and TaylorMade. The adidas brand develops and markets products in two divisions: the Sport Performance Division focuses on footwear and apparel for sports such as running, basketball and soccer; and the Sport Style division designs and markets products with a combination of style and performance. The company's Reebok brand develops sports apparel, footwear and athletic equipment for a variety of sports. Its divisions include Reebok, a global brand that creates and markets sports and lifestyle products in sports, fitness and women's categories; CCM, the world's largest designer, manufacturer and marketer of hockey equipment and related apparel; and Rockport, which designs and markets dress, dress casual, casual and outdoor footwear in addition to apparel and accessories. The firm's TaylorMade golf brand develops and markets a variety of products for the golf industry, including TaylorMade drivers, balls and putters; adidas Golf footwear; and Ashworth brand equiptment. In addition to these primary offerings, the company has expanded its product line to include adidas Body, a line of fragrances and personal care products; adidas eyewear, which includes sunglasses designed for a variety of sports; and adidas watches. The firm is the official soccer ball provider for the FIFA World Cup as well as the official uniform and apparel provider for the National Basketball Association (NBA), Women's National Basketball Association (WNBA) and the NBA Development League. In June 2012, TaylorMade acquired Adams Golf, Inc.

FINANCIAL DATA: *Note: Data for latest year may not have been available at press time.*

In U.S. $	2012	2011	2010	2009	2008	2007
Revenue	19,902,380,000	17,844,340,000	16,033,700,000	13,882,050,000	14,441,030,000	13,772,400,000
R&D Expense		115,000,000	102,000,000	86,000,000	81,000,000	84,000,000
Operating Income	1,230,275,000	1,351,966,000	1,195,507,000	679,326,000	1,430,864,000	1,269,056,000
Operating Margin %	6.18%	7.57%	7.45%	4.89%	9.90%	9.21%
SGA Expense		7,233,218,000	6,611,394,000	5,755,550,000	5,746,189,000	5,283,498,000
Net Income	703,396,600	895,961,500	759,561,400	327,627,700	861,192,800	742,177,000
Operating Cash Flow	1,259,695,000	1,059,107,000	1,195,507,000	1,602,033,000	664,616,200	1,043,060,000
Capital Expenditure	580,369,100	502,808,200	359,721,900	322,278,700	502,808,200	381,118,000
EBITDA	1,947,045,000	1,731,747,000	1,589,997,000	1,104,573,000	1,793,260,000	1,604,707,000
Return on Assets %	4.56%	6.10%	5.81%	2.66%	7.19%	6.59%
Return on Equity %	9.90%	13.48%	13.50%	6.82%	19.95%	18.77%
Debt to Equity		0.18	0.28	0.41	0.52	0.7

CONTACT INFORMATION:

Phone: 49 9132840 Fax: 49 9132842241
Toll-Free:
Address: Adi-Dassler-Strasse 1-2, Herzogenaurach, 91074 Germany

STOCK TICKER/OTHER:

Stock Ticker: ADDDF Exchange: PINX
Employees: 40,637 Fiscal Year Ends: 31-Dec
Parent Company:

SALARIES/BONUSES:

Top Exec. Salary: $ Bonus: $
Second Exec. Salary: $ Bonus: $

OTHER THOUGHTS:

Estimated Female Officers or Directors: 2
Hot Spot for Advancement for Women/Minorities:

ADS Inc

NAIC Code: 313230

www.adsinc.ca

TYPES OF BUSINESS:

Technical Textiles
Footwear Components
Agrotextiles & Geotextiles
Non-woven Fabrics

BRANDS/DIVISIONS/AFFILIATES:

Texel

CONTACTS:

Note: Officers with more than one job title may be intentionally listed here more than once.

Guy Drouin, CEO
Guy Drouin, Pres.
Guy Berube, VP-Oper.

GROWTH PLANS/SPECIAL FEATURES:

ADS, Inc. is a leading designer, developer and manufacturer of specialized nonwoven materials, needlepunch nonwoven textiles and polymer coated materials. Its products, which are developed and marketed primarily through subsidiary Texel, fall into six categories: industrial, filtration, medical wipes, horticulture, coated materials and geosynthetics. Industrial nonwoven products are commonly utilized for diffusion, noise and fluid absorption; footwear manufacturing; and as thermal insulators for heavy machinery such as printing equipment. The firm's specialized liquid filtration products help the filtering processes in the food processing, paint, petrochemicals and pharmaceutical industries. Its air filtration products increase the efficacy of facemasks, air purifiers and industrial filters. Its absorbent/wipes structures can be used in the medical care industry as bandages, soakers and patient wipes; in the area of home care as cleansing wipes; and in the personal care industry as cosmetic and facial wipes. Its horticulture products assist in drainage and protection of plants in nurseries and farms. The firm's coated materials products include polymer coatings on woven or nonwoven materials for use in hospital quilts, protective sheeting for civil fixtures, technical clothing such as raingear and floor coverings. The firm's geosynthetics products are used in the roads and transportation, natural resources and energy, municipal and landscape architecture and industrial and waste management industries.

FINANCIAL DATA:

Note: Data for latest year may not have been available at press time.

In U.S. $	2012	2011	2010	2009	2008	2007
Revenue						
R&D Expense						
Operating Income						
Operating Margin %						
SGA Expense						
Net Income						
Operating Cash Flow						
Capital Expenditure						
EBITDA						
Return on Assets %						
Return on Equity %						
Debt to Equity						

CONTACT INFORMATION:

Phone: 418-387-3383 Fax: 418-387-8853
Toll-Free: 800-463-8929
Address: 485 rue des Erables, Saint-Elzear, QC G0S 2J0 Canada

STOCK TICKER/OTHER:

Stock Ticker: Private
Exchange:
Employees: 225
Fiscal Year Ends: 31-Jan
Parent Company:

SALARIES/BONUSES:

Top Exec. Salary: $ Bonus: $
Second Exec. Salary: $ Bonus: $

OTHER THOUGHTS:

Estimated Female Officers or Directors:
Hot Spot for Advancement for Women/Minorities:

Aearo Technologies LLC

NAIC Code: 33911

www.aearo.com

TYPES OF BUSINESS:

Personal Protection Equipment
Safety Eyewear
Hearing Protection Devices
Shock Absorbing Composites
Respirators
Communication Headsets
Hard Hats
First Aid Kits

BRANDS/DIVISIONS/AFFILIATES:

3M Company
Aearo Technologies Inc
E-A-R
ISODAMP
TUFCOTE
CONFOR
SuperSheet
ISOLOSS

CONTACTS: *Note: Officers with more than one job title may be intentionally listed here more than once.*

Thomas R. D'Amico, Sr. VP-R&D
Thomas R. D'Amico, Sr. VP-Oper.
Inge Thulin, CEO-3M

GROWTH PLANS/SPECIAL FEATURES:

Aearo Technologies LLC, formerly Aearo Technologies, Inc., produces personal protection apparel and equipment. Its brand names include E-A-R, ISODAMP, ISOLOSS, TUFCOTE, CONFOR and SuperSheet. The E-A-R brand includes simple earplugs, custom molded earplugs, earplug dispensers, earplugs on flexible or stiff plastic connectors and earmuffs. The company first introduced the disposable polyvinyl chloride (PVC) earplug, now known as the E-A-R Classic, in 1972, and has since remained at the top of hearing conservation R&D. The Classic's yellow color, recognized worldwide, is a registered trademark in five countries. The E-A-R Specialty Composites brand manufactures various energy-absorbing materials controlling noise, vibration and shock, which are incorporated into other manufacturers' products. The firm currently specializes in a range of vinyl and urethane formulations used for shock absorption, cushioning, sound absorption and dampening materials all to reduce unwanted noise. Its products are produced in two in house facilities totaling 375,000 square feet of manufacturing space. Manufacturing capabilities include thin-sheet casting; extrusion; injection, compression and transfer molding; and fabrication including laminating, die-cutting and compositing. Its products are marketed to the industrial, automotive, electronics, mattress, athletic equipment and footwear industries. Aearo is a subsidiary of the 3M Company.

FINANCIAL DATA: *Note: Data for latest year may not have been available at press time.*

In U.S. $	2012	2011	2010	2009	2008	2007
Revenue						
R&D Expense						
Operating Income						
Operating Margin %						
SGA Expense						
Net Income						
Operating Cash Flow						
Capital Expenditure						
EBITDA						
Return on Assets %						
Return on Equity %						
Debt to Equity						

CONTACT INFORMATION:

Phone: 317-692-6666 Fax: 317-692-6772
Toll-Free:
Address: 5457 W. 79th St., Indianapolis, IN 46268 United States

STOCK TICKER/OTHER:

Stock Ticker: Subsidiary
Employees: 1,659
Parent Company: 3M COMPANY
Exchange:
Fiscal Year Ends: 30-Sep

SALARIES/BONUSES:

Top Exec. Salary: $ Bonus: $
Second Exec. Salary: $ Bonus: $

OTHER THOUGHTS:

Estimated Female Officers or Directors:
Hot Spot for Advancement for Women/Minorities:

Aeropostale Inc

www.aeropostale.com

NAIC Code: 448120

TYPES OF BUSINESS:

Teen Apparel-Retail
Online Sales
Children's Apparel-Retail

BRANDS/DIVISIONS/AFFILIATES:

Aeropostale West Inc
Aeropostale.com
P.S. from Aeropostale
ps4u.com
Jimmy Z Surf Co Inc
GoJane.com

CONTACTS:

Note: Officers with more than one job title may be intentionally listed here more than once.

Thomas P. Johnson, CEO
Michael J. Cunningham, Pres.
Marc D. Miller, CFO
Scott K. Birnbaum, Sr. VP-Mktg.
Kathy E. Gentilozzi, Sr. VP-Human Resources
Ann E. Joyce, CIO/Sr. VP
Todd Blumenthal, Sr. VP-Merch., Aeropostale
Barbara Fevelo-Hoad, Sr. VP-Prod. & Tech. Design
Edward M. Slezak, General Counsel/Sr. VP/Corp. Sec.
Barbara Pindar, Sr. VP-Planning & Allocation
Scott K. Birnbaum, Sr. VP-e-commerce
Kenneth Ohashi, VP-Media Rel.
Kenneth Ohashi, VP-Investor Rel.
Beverly House, Sr. VP-Design, Aeropostale
Mary Jo Pile, Exec. VP-Customer Engagement
Catherine E. McNeal, Sr. VP-Merch., PS from Aeropostale
Susan A. Martin, Sr. VP-Design, PS from Aeropostale
Karin Hirtler-Garvey, Chmn.
Mark A. Dorwart, Sr. VP-Logistics, Construction & Real Estate

GROWTH PLANS/SPECIAL FEATURES:

Aeropostale, Inc., together with its wholly-owned subsidiary, Aeropostale West, Inc., is a mall-based specialty designer, marketer and retailer of casual apparel and accessories, principally targeting young men and women ages 14-17. Aeropostale maintains control over its proprietary brands by designing, sourcing, marketing and selling all of its own merchandise, both through retail locations and its online store. The company's name is derived from the first transatlantic airmail carrier, Compagnie Generale Aeropostale. The firm operates over 980 stores in the U.S., Puerto Rico and Canada. It also operates 70 P.S. from Aeropostale stores in 20 states. Additionally, through a licensing agreement, there are 27 Aeropostale and PS from Aeropostale locations in the Middle East, Asia and Europe. The P.S. from Aeropostale concept offers casual clothing and accessories targeting elementary school children between the ages of 4 and 12. The firm leases a 315,000 square foot distribution and warehouse center in South River, New Jersey and a second distribution facility with 360,000 square feet of space in Ontario, California. Through its e-commerce business, the company's merchandise is available in more than 100 countries. In November 2011, it signed an agreement with FiBA Group to open 30 stores across Turkey within the next five years. In November 2012, the firm acquired GoJane.com, an apparel and footwear e-commerce site.

Aeropostale offers its employees medical, dental and vision insurance; fitness programs; tuition reimbursement; disability coverage; an employee assistance program; a 401(k); life insurance; and merchandise discounts.

FINANCIAL DATA:

Note: Data for latest year may not have been available at press time.

In U.S. $	2012	2011	2010	2009	2008	2007
Revenue	2,342,260,000	2,400,434,000	2,230,105,000	1,885,531,000	1,590,883,000	
R&D Expense						
Operating Income	113,515,000	386,794,000	382,685,000	248,299,000	202,453,000	
Operating Margin %	4.84%	16.11%	17.15%	13.16%	12.72%	
SGA Expense	494,829,000	499,368,000	464,462,000	405,883,000	345,805,000	
Net Income	69,515,000	231,339,000	229,457,000	149,422,000	129,197,000	
Operating Cash Flow	129,301,000	263,731,000	334,440,000	202,135,000	171,081,000	
Capital Expenditure	73,323,000	100,807,000	53,883,000	83,035,000	82,306,000	
EBITDA	178,628,000	445,513,000	435,536,000	296,429,000	231,321,000	
Return on Assets %	9.21%	29.55%	31.64%	25.49%	23.59%	
Return on Equity %	16.51%	53.35%	58.12%	54.10%	50.72%	
Debt to Equity	0.25					

CONTACT INFORMATION:

Phone: 646 485-5410 Fax: 646 485-5430
Toll-Free:
Address: 112 W. 34th St., 22nd Fl., New York, NY 10120 United States

STOCK TICKER/OTHER:

Stock Ticker: ARO
Exchange: NYS
Employees: 25,856
Fiscal Year Ends: 31-Jan
Parent Company:

SALARIES/BONUSES:

Top Exec. Salary: $958,269 Bonus: $
Second Exec. Salary: $637,404 Bonus: $

OTHER THOUGHTS:

Estimated Female Officers or Directors: 12
Hot Spot for Advancement for Women/Minorities: Y

Albany International Corp

www.albint.com

NAIC Code: 313230

TYPES OF BUSINESS:

Paper Machine Clothing Manufacturing
Engineered Fabrics
Specialty Materials
Filtration Products

BRANDS/DIVISIONS/AFFILIATES:

Albany Engineered Composites

CONTACTS:

Note: Officers with more than one job title may be intentionally listed here more than once.

Joseph G. Morone, CEO
Joseph G. Morone, Pres.
John Cozzolino, CFO/Treas.
Dawne H. Wimbrow, CIO/VP-Global Info. Svcs.
Robert A. Hansen, CTO/Sr. VP
Charles J. Silva, Jr., General Counsel/Sec./VP
David M. Pawlick, Controller/VP
Ralph M. Polumbo, Sr. VP/COO-Albany Engineered Composites
Daniel A. Halftermeyer, Pres., Machine Clothing
Erland E. Kailbourne, Chmn.

GROWTH PLANS/SPECIAL FEATURES:

Albany International Corp., through its subsidiaries, is a producer of man-made fibers and advanced engineered composite materials. The company divides its business operations into two segments: Machine Clothing and Albany Engineered Composites (AEC). The Machine Clothing segment specializes in fabrics used to cloth the rollers of paper and paperboard (sometimes called cardboard) machines during all stages of manufacturing, including forming, pressing and drying the paper pulp. The segment also supplies consumable fabrics used to process paper pulp as well as engineered fabrics used in a range of industries other than papermaking. These other products include belts used to make nonwovens, fiber cement building products, roofing shingles and corrugated sheets used in boxboard as well as belts used in tannery and textile applications. The AEC segment manufactures specialty materials and composites, primarily for aircraft applications in the aerospace and defense industries. Most of this segment's sales are generated from government contracts. The company maintains manufacturing facilities in the U.S., Brazil, Canada, China, France, Germany, the U.K., Italy, Mexico, South Korea and Sweden. Its conducts its primary R&D activities at facilities in Halmstad, Sweden; Manchester, England; Menasha, Wisconsin; and Rochester, New Hampshire. During 2012, the company completed the sale of its PrimaLoft Products business and its global Albany Door Systems assets.

FINANCIAL DATA:

Note: Data for latest year may not have been available at press time.

In U.S. $	2012	2011	2010	2009	2008	2007
Revenue	760,941,000	814,696,000	914,356,000	871,045,000	1,086,517,000	1,092,977,000
R&D Expense	52,962,000	56,767,000	57,163,000	54,964,000	66,486,000	
Operating Income	-44,136,000	83,073,000	81,184,000	-36,322,000	-66,917,000	37,449,000
Operating Margin %	-5.80%	10.19%	8.87%	-4.16%	-6.15%	3.42%
SGA Expense	289,509,000	179,804,000	203,159,000	205,089,000	251,506,000	316,455,000
Net Income	30,977,000	34,938,000	37,636,000	-32,380,000	-75,748,000	17,782,000
Operating Cash Flow	34,507,000	105,269,000	99,572,000	13,975,000	84,038,000	81,299,000
Capital Expenditure	37,207,000	28,680,000	31,591,000	42,496,000	140,996,000	165,238,000
EBITDA	12,987,000	148,765,000	145,859,000	83,116,000	660,000	99,475,000
Return on Assets %	2.59%	2.78%	2.87%	-2.35%	-5.16%	1.25%
Return on Equity %	6.81%	8.29%	8.84%	-7.66%	-14.85%	3.23%
Debt to Equity	0.47	0.89	0.99	1.13	1.26	0.74

CONTACT INFORMATION:

Phone: 518 445-2200 Fax: 518 445-2250
Toll-Free: 800-833-3836
Address: 1373 Broadway, Albany, NY 12204 United States

STOCK TICKER/OTHER:

Stock Ticker: AIN Exchange: NYS
Employees: 4,300 Fiscal Year Ends: 31-Dec
Parent Company:

SALARIES/BONUSES:

Top Exec. Salary: $787,250 Bonus: $
Second Exec. Salary: $602,282 Bonus: $

OTHER THOUGHTS:

Estimated Female Officers or Directors: 2
Hot Spot for Advancement for Women/Minorities:

Aldo Group Inc

www.aldoshoes.com

NAIC Code: 448210

TYPES OF BUSINESS:

Shoes, Retail
Jewelry & Accessories
Leather Goods

BRANDS/DIVISIONS/AFFILIATES:

Aldo
Spring
Globo
Call It Spring
Little Burgundy
Locale

CONTACTS:

Note: Officers with more than one job title may be intentionally listed here more than once.

Aldo Bensadoun, CEO
Rejean Dionne, Pres.
Robert Raven, VP-Finance
Norman Joskolka, VP
David Bensadoun, VP-Global Retail

GROWTH PLANS/SPECIAL FEATURES:

Aldo Group, Inc. is a retail shoe company that markets men's and women's name-brand and private-label dress and casual shoes, boots, handbags and accessories. The company has over 1,600 retail stores in Canada, the U.S., the U.K. and Ireland as well as franchises in 60 countries. Approximately 1,000 of these stores fall under the Aldo name. In addition, it markets its products through other stores within the Aldo Group, including Call It Spring, Spring, Little Burgundy and Globo stores. The firm's shoes range in price from $30-$200. Aldo's handbag products range from $12-$108 and consist of women's handbags, tote bags and clutches and men's and women's bags and wallets. The company also sells accessories, including jewelry such as necklaces, rings and earrings; sunglasses; hair accessories; outerwear, such as belts, hats, scarves, gloves and legwear; watches; and shoe care products, including shoe shine, cleaner, insoles and waterproofing spray. The company is named after owner and founder Aldo Bensadoun. Recent developments include the launch of the upscale Locale chain and the creation of a wholesale division to supply shoes to U.S. retail chains. In addition, Aldo opened stores in Cyprus, Spain, Vietnam and Hong Kong. In addition, the firm recently partnered with JC Penny to open 500 Call It Spring shop-in-shops across the U.S.

The company offers its employees an incentive package including merchandise discounts and ongoing training.

FINANCIAL DATA:

Note: Data for latest year may not have been available at press time.

In U.S. $	2012	2011	2010	2009	2008	2007
Revenue						
R&D Expense						
Operating Income						
Operating Margin %						
SGA Expense						
Net Income						
Operating Cash Flow						
Capital Expenditure						
EBITDA						
Return on Assets %						
Return on Equity %						
Debt to Equity						

CONTACT INFORMATION:

Phone: 514-747-2536 Fax: 514-747-7993
Toll-Free: 888-818-2536
Address: 2300 Emile-Belanger, Montreal, QC H4R 3J4 Canada

STOCK TICKER/OTHER:

Stock Ticker: Private Exchange:
Employees: Fiscal Year Ends: 31-Dec
Parent Company:

SALARIES/BONUSES:

Top Exec. Salary: $ Bonus: $
Second Exec. Salary: $ Bonus: $

OTHER THOUGHTS:

Estimated Female Officers or Directors:
Hot Spot for Advancement for Women/Minorities:

Alexandra plc

www.alexandra.co.uk

NAIC Code: 315280

TYPES OF BUSINESS:

Sales & Sourcing, Uniforms & Corporate Workwear
Custom Embroidery
Garment Manufacturing

BRANDS/DIVISIONS/AFFILIATES:

Alexandra BV
Alexandra Sarl
Men's Wearhouse Inc (The)
de Baer plc
Prima Corporate Wear Ltd

CONTACTS:

Note: Officers with more than one job title may be intentionally listed here more than once.

Tim Gifford, CEO
Nick Acaster, Dir.-Mktg.
Daniel Murray, Dir.-Bus. Dev.
Simon Hughes, CEO-The Men's Wearhouse, Inc. UK Group

GROWTH PLANS/SPECIAL FEATURES:

Alexandra plc, a subsidiary of The Men's Warehouse, Inc., sources and sells workwear and corporate uniforms for the catering, hospitality, transportation, medical and security industries. The firm holds a Royal Warrant of Appointment to Her Majesty the Queen as supplier of corporate uniforms and workwear; though mostly ceremonial in nature, the warrant signifies Alexandra's prominence in the British market and its reputation for quality and excellent service. The company maintains comprehensive collections of workplace items such as business wear and suiting, health care garments, chefswear and footwear, with stock inventories ready for immediate distribution. It also offers clients custom embroidery and personalization services for workplace clothing with corporate logos and trademarks. Products can be purchased through direct or retail sales channels. Originally a mail-order company, Alexandra now operates call centers, an e-commerce site and regional shops that offer clients in-person service from trained sales and support staff as well as showroom access to merchandise samples. Shops are located in Aberdeen, Blackpool, Glasgow, Manchester and Plymouth. Alexandra operates a number of sales and marketing subsidiaries: Alexandra BV in the Netherlands and Alexandra sarl in France. Sales and marketing subsidiaries in the U.K. include de Baer plc and Prima Corporate Wear Ltd. The firm owns two manufacturing sites in Glasgow and Morocco.

FINANCIAL DATA:

Note: Data for latest year may not have been available at press time.

In U.S. $	2012	2011	2010	2009	2008	2007
Revenue						
R&D Expense						
Operating Income						
Operating Margin %						
SGA Expense						
Net Income						
Operating Cash Flow						
Capital Expenditure						
EBITDA						
Return on Assets %						
Return on Equity %						
Debt to Equity						

CONTACT INFORMATION:

Phone: 44-845-155-2288 Fax: 44-845-155-2211
Toll-Free:
Address: Alexandra House, Thornbury, Bristol, BS35 2NT United Kingdom

STOCK TICKER/OTHER:

Stock Ticker: Subsidiary Exchange:
Employees: 962 Fiscal Year Ends: 31-Jan
Parent Company: MEN'S WEARHOUSE INC (THE)

SALARIES/BONUSES:

Top Exec. Salary: $ Bonus: $
Second Exec. Salary: $ Bonus: $

OTHER THOUGHTS:

Estimated Female Officers or Directors:
Hot Spot for Advancement for Women/Minorities:

Allen-Edmonds Shoe Corp

www.allenedmonds.com

NAIC Code: 316210

TYPES OF BUSINESS:

Men's Shoe Manufacturing
Footwear, Retail
Accessories

BRANDS/DIVISIONS/AFFILIATES:

Woodlore
Casual Comfort Collection
Goldner Hawn Johnson & Morrison Inc
AllenEdmonds.com

CONTACTS: *Note: Officers with more than one job title may be intentionally listed here more than once.*

Paul Grangaard, CEO
Paul Grangaard, Pres.
Colin Hall, Chief Mktg. Officer
Mark McNeill, Dir.-Merch.
Castillo Rosendo, Dir.-Quality
Al Dittrich, Sr. VP-Retail
Jim Kass, VP
Colin Hall, Gen. Mgr.-Int'l Bus.

GROWTH PLANS/SPECIAL FEATURES:

Allen-Edmonds Shoe Corp., owned by private investment firm Goldner Hawn Johnson & Morrison, Inc., manufactures and markets premium men's footwear, accessories and cedar products. The company makes its shoes from high-grade leather and full-leather linings. It crafts each pair of shoes by hand in a production process that includes over 212 production steps. Allen-Edmonds' 360-degree welt stitching binds the upper, insole and sole of a shoe all the way around. Because the method connects parts of the shoe subject to the most stress, it makes the firm's shoes last longer than other footwear and eliminates the need for a metal shank. The unique construction of Allen-Edmonds shoes allows it to offer a proprietary re-crafting service. Allen-Edmonds has three manufacturing facilities exclusively in the U.S. In order to ensure quality, the firm refuses to outsource overseas, despite the pressures of increased costs. In addition to having agents and distributors in countries around the world, Allen-Edmonds operates approximately 37 retail stores in the U.S. as well as leased departments and independent distributors in 25 international markets and an e-commerce site at AllenEdmonds.com. Wholly-owned subsidiary Woodlore manufactures and distributes cedar shoetrees, storage chests and coat hangers as well as shoe care products. The Casual Comfort Collection, consisting of shoes made from calfskin and rubber soles, is marketed toward younger customers who may not be ready to own a more formal pair of Allen-Edmonds shoes. The firm planned to open a store in Shanghai in 2012 and three in the U.S.. By the end of 2013, it plans to have a total of six Chinese stores.

FINANCIAL DATA: *Note: Data for latest year may not have been available at press time.*

In U.S. $	2012	2011	2010	2009	2008	2007
Revenue	120,000,000	100,000,000	80,000,000			
R&D Expense						
Operating Income						
Operating Margin %						
SGA Expense						
Net Income						
Operating Cash Flow						
Capital Expenditure						
EBITDA						
Return on Assets %						
Return on Equity %						
Debt to Equity						

CONTACT INFORMATION:

Phone: 262-235-6000 Fax: 262-235-7427
Toll-Free:
Address: 201 E. Seven Hills Rd., Port Washington, WI 53074 United States

STOCK TICKER/OTHER:

Stock Ticker: Private Exchange:
Employees: 600 Fiscal Year Ends: 31-Dec
Parent Company: GOLDNER HAWN JOHNSON & MORRISON INC

SALARIES/BONUSES:

Top Exec. Salary: $ Bonus: $
Second Exec. Salary: $ Bonus: $

OTHER THOUGHTS:

Estimated Female Officers or Directors:
Hot Spot for Advancement for Women/Minorities:

Alpha Pro Tech Ltd

www.alphaprotech.com

NAIC Code: 33911

TYPES OF BUSINESS:

Protective Clothing
Infection Control Products
Extended-Care Products
Pet Beds
Weatherization Products
Clean Room Clothing
Construction Materials
Plastics

BRANDS/DIVISIONS/AFFILIATES:

Alpha ProTech Engineered Products Inc
Magic Arch
REX Synfelt
Harmony Plastics Private Ltd

CONTACTS: *Note: Officers with more than one job title may be intentionally listed here more than once.*

Sheldon Hoffman, CEO
Alexander W. Millar, Pres.
Lloyd Hoffman, CFO
Danny Montgomery, Sr. VP-Eng. Prod.
Danny Montgomery, Sr. VP-Mfg.

GROWTH PLANS/SPECIAL FEATURES:

Alpha Pro Tech Ltd. develops, manufactures and markets disposable and limited-use apparel and products for the cleanroom (such as in semiconductor manufacturing plants), industrial, pharmaceutical, medical, dental, construction and pet markets. The firm offers products in three groups: disposable protective apparel, infection control and building supply. Its disposable protective apparel includes shoe covers, bouffant caps, gowns, coveralls, lab coats and frocks. The majority of these products are manufactured by a third party subcontractor in China and, to a lesser extent, a third party subcontractor in Mexico. Alpha's infection control products consist of disposable masks and eye shields. Its patented Positive Facial Lock face mask design provides a custom fit, while its Magic Arch feature holds the mask away from the nose and mouth, creating a comfortable breathing chamber. The firm's building supply segment, operated by subsidiary Alpha ProTech Engineered Products, Inc., consists of the construction of weatherization products, such as house wrap, which provides a weather resistant barrier designed to lower energy consumption costs; and synthetic roof underlayment sold under the name REX Synfelt, which is designed to resist degradation and mold growth and offers a cost reduction over standard felt-based tar paper. Alpha's house wrap and roof underlayment is manufactured in India by Harmony Plastics Private Ltd., the firm's 41.66% owned joint venture with Maple Industries; and in a facility in Valdosta, Georgia.

FINANCIAL DATA: *Note: Data for latest year may not have been available at press time.*

In U.S. $	2012	2011	2010	2009	2008	2007
Revenue	41,058,000	38,502,000	41,890,000	59,697,000	35,786,000	35,453,000
R&D Expense						
Operating Income	817,000	905,000	1,739,000	13,508,000	2,255,000	3,407,000
Operating Margin %	1.98%	2.35%	4.15%	22.62%	6.30%	9.60%
SGA Expense	12,879,000	12,323,000	13,839,000	14,701,000	12,639,000	12,541,000
Net Income	977,000	933,000	1,301,000	9,041,000	1,564,000	2,417,000
Operating Cash Flow	-1,014,000	3,766,000	-3,441,000	7,543,000	3,846,000	2,085,000
Capital Expenditure	525,000	277,000	1,143,000	548,000	1,207,000	607,000
EBITDA	1,581,000	1,723,000	2,583,000	14,531,000	2,841,000	3,901,000
Return on Assets %	2.63%	2.52%	3.36%	25.96%	5.33%	8.58%
Return on Equity %	2.79%	2.63%	3.75%	29.96%	5.87%	9.61%
Debt to Equity						

CONTACT INFORMATION:

Phone: 905 479-0654 Fax: 905 479-9732
Toll-Free:
Address: 60 Centurian Dr., Ste. 112, Markham, ON L3R 9R2 Canada

STOCK TICKER/OTHER:

Stock Ticker: APT Exchange: ASE
Employees: 110 Fiscal Year Ends: 31-Dec
Parent Company:

SALARIES/BONUSES:

Top Exec. Salary: $550,000 Bonus: $
Second Exec. Salary: $550,000 Bonus: $

OTHER THOUGHTS:

Estimated Female Officers or Directors:
Hot Spot for Advancement for Women/Minorities:

American & Efird Inc

www.amefird.com

NAIC Code: 313110

TYPES OF BUSINESS:

Sewing Threads & Supplies
Notions & Trim Products

BRANDS/DIVISIONS/AFFILIATES:

KPS Capital Partners LP
Flame-Out
Robison-Anton
Moonglow
Signature
Kevlar
Cocoon Performance Bobbin
Maxi-Lock

CONTACTS: *Note: Officers with more than one job title may be intentionally listed here more than once.*

Fred A. Jackson, CEO
Fred A. Jackson, Pres.
Craig G. Stover, Sr. VP-Finance
Michael Young, Exec.-Sales
Raymond Misiunas, VP-Info. Systems
Craig G. Stover, Sec.
Les Miller, Exec. VP-Oper.
John Eapen, VP-Environmental, Health & Safety

GROWTH PLANS/SPECIAL FEATURES:

American & Efird, Inc. (A&E), owned by KPS Capital Partners, LP, is one of the world's premier manufacturers of sewing threads and industrial yarns. The company's product segments include industrial sewing thread, embroidery thread and technical textiles, which are distributed through the industrial division; and consumer products, which are distributed through the consumer division. The industrial segment manufactures products for use by apparel, automotive, upholstery, home furnishing, footwear, bedding and mattress, industrial applications and other sewn product manufacturers. Industrial and embroidery thread products include short-term heat resistant thread, including Flame-Out and some Kevlar varieties; Performance Bobbin and Cocoon Performance Bobbin products; Robison-Anton brand embroidery thread; and denim sewing threads. The technical textile segment of this division develops mechanical rubber goods, in addition to electrical wire and cable products. The consumer division makes products for use in home sewing, including apparel sewing, crafts and quilting. Some of its brands include the Signature line, offering cotton, nylon and polyester quilting and specialty threads; the Maxi-Lock line of grisettle, buttonhole twist, topstitching and serging thread; and the Moonglow glow in the dark thread line. This division also distributes hook-and-loop tape, polypropylene webbing and YKK-brand zippers. The firm's technical support team provides multi-lingual supply-chain solutions, such as ISO stitch types, thread size cross reference charts, technical bulletins and glossaries. With 31 manufacturing centers and 76 distribution centers, the firm maintains distribution hubs in the U.S. and over 41 additional countries.

A&E offers its employees benefits including medical, dental and life insurance; a 401(k); and paid time off.

FINANCIAL DATA: *Note: Data for latest year may not have been available at press time.*

In U.S. $	2012	2011	2010	2009	2008	2007
Revenue						
R&D Expense						
Operating Income						
Operating Margin %						
SGA Expense						
Net Income						
Operating Cash Flow						
Capital Expenditure						
EBITDA						
Return on Assets %						
Return on Equity %						
Debt to Equity						

CONTACT INFORMATION:

Phone: 704-827-4311 Fax: 704-827-0974
Toll-Free: 800-453-5128
Address: 22 American St., Mt. Holly, NC 28120 United States

STOCK TICKER/OTHER:

Stock Ticker: Private Exchange:
Employees: 2,900 Fiscal Year Ends: 30-Sep
Parent Company: KPS CAPITAL PARTNERS LP

SALARIES/BONUSES:

Top Exec. Salary: $ Bonus: $
Second Exec. Salary: $ Bonus: $

OTHER THOUGHTS:

Estimated Female Officers or Directors:
Hot Spot for Advancement for Women/Minorities:

American Apparel Inc

www.americanapparel.net

NAIC Code: 315240

TYPES OF BUSINESS:

Casual Apparel Manufacturing
Apparel Stores
Apparel Wholesale
E-commerce

BRANDS/DIVISIONS/AFFILIATES:

American Apparel (USA) LLC
American Apparel Dyeing and Finishing Inc
American Apparel Retail Inc
KCL Knitting, LLC
Fresh Air Freight Inc
American Apparel Canada Wholesale Inc
American Apparel Canada Retail Inc
American Apparel (Beijing) Trading Company Ltd

CONTACTS: *Note: Officers with more than one job title may be intentionally listed here more than once.*

Dov Charney, CEO
John J. Luttrell, CFO/Exec. VP
Marty Bailey, Chief Mfg. Officer
Glenn A. Weinman, General Counsel/Exec. VP/Corp. Sec.
Adrian Taylor, Corp. Controller/VP
Dov Charney, Chmn.

GROWTH PLANS/SPECIAL FEATURES:

American Apparel, Inc. is a vertically-integrated manufacturer, distributor and retailer of branded fashion basic apparel. The company does not outsource its production, conducting all of its manufacturing in the Los Angeles metropolitan area. It primary apparel manufacturing operations are conducted from an 800,000 square foot facility in downtown Los Angeles. Its retail operations principally target fashion-conscious young adults aged 20-32, with products including basic women's and men's apparel and accessories, along with product lines for children and pets. American Apparel operates approximately 251 stores in 20 countries, including the U.S., Canada, Mexico, Brazil, the U.K., Germany, France, Italy, Spain, Israel, Australia, Japan, South Korea and China. Stores are located primarily in large metropolitan areas, emerging neighborhoods and university communities. In addition to its retail stores, American Apparel conducts online sales internationally through its e-commerce site. The firm also operates a wholesale business, supplying t-shirts and other casual wear to distributors and screen printers. American Apparel operates in four business segments: U.S. Retail, U.S. Wholesale, Canada and International. Its domestic subsidiaries include American Apparel (USA), LLC; American Apparel Retail, Inc.; American Apparel Dyeing and Finishing, Inc.; KCL Knitting, LLC; and Fresh Air Freight, Inc. It conducts business internationally through various direct and indirect subsidiaries, including American Apparel Canada Wholesale, Inc.; American Apparel Canada Retail, Inc.; and American Apparel (Beijing) Trading Company, Ltd. In 2012, the company launched e-commerce sites serving Hong Kong and Japan.

American Apparel's employees and their families are eligible for company-subsidized health insurance. Workers in the Los Angeles production facilities are offered English language classes and an onsite medical clinic, among other benefits.

FINANCIAL DATA: *Note: Data for latest year may not have been available at press time.*

In U.S. $	2012	2011	2010	2009	2008	2007
Revenue	617,310,000	547,336,000	532,989,000	558,775,000	545,050,000	387,044,000
R&D Expense						
Operating Income	962,000	-23,293,000	-50,053,000	24,415,000	36,064,000	31,122,000
Operating Margin %	.15%	-4.25%	-9.39%	4.36%	6.61%	8.04%
SGA Expense	324,774,000	313,926,000	321,365,000	294,350,000	263,051,000	184,351,000
Net Income	-37,272,000	-39,314,000	-86,315,000	1,112,000	14,112,000	15,478,000
Operating Cash Flow	23,589,000	2,305,000	-32,370,000	45,203,000	21,171,000	-5,438,000
Capital Expenditure	21,607,000	11,070,000	15,701,000	20,889,000	68,650,000	22,195,000
EBITDA	31,089,000	20,554,000	-22,269,000	83,857,000	56,132,000	46,130,000
Return on Assets %	-11.41%	-12.04%	-26.33%	.33%	4.98%	8.62%
Return on Equity %	-106.16%	-63.84%	-74.29%	.75%	14.23%	19.09%
Debt to Equity	5.11	2.05	0.07	0.45	0.77	0.01

CONTACT INFORMATION:

Phone: 213 488-0226 Fax:
Toll-Free:
Address: 747 Warehouse St., Los Angeles, CA 90021 United States

STOCK TICKER/OTHER:

Stock Ticker: APP
Exchange: ASE
Employees: 10,000
Fiscal Year Ends: 31-Dec
Parent Company:

SALARIES/BONUSES:

Top Exec. Salary: $750,000 Bonus: $
Second Exec. Salary: $378,666 Bonus: $

OTHER THOUGHTS:

Estimated Female Officers or Directors:
Hot Spot for Advancement for Women/Minorities:

American Dawn Inc

www.americandawn.com

NAIC Code: 313210

TYPES OF BUSINESS:

Towels & Linens
Laundry & Institutional Textiles
Health Care Textile Products
Salon Products
Decorating Products

BRANDS/DIVISIONS/AFFILIATES:

Classic Linen
Los Angeles Rose
Eileen West
Alexander Julian
Eileen West
Mary Jane
Vicki Payne

CONTACTS:
Note: Officers with more than one job title may be intentionally listed here more than once.

Habib Rawjee, Co-Pres.
Adnan Rawjee, Co-Pres.
Mahmud Rawjee, Chmn.
Kathleen Healy, Dir.-Sourcing

GROWTH PLANS/SPECIAL FEATURES:

American Dawn, Inc. (ADI) is a leading manufacturer, importer and distributor of textile products, serving the wholesale, commercial and retail markets. It has manufacturing and distribution plants in 12 countries. Its U.S. offices are located in California, Illinois, Georgia, Texas and New Jersey. The company's products are divided into five categories: linens & uniforms, hospitality, health care, salon and home fashion. The linens & uniforms division specializes in products intended for the restaurant industry, with a product line that consists of table linens, napkins, placemats and chefwear. Hospitality products, sold under the Classic Linen name, include bath robes, mats, towels, wash cloths, table linens and front/back house work apparel for the hotel and hospitality sector. Health care products include adult bibs, robes, baby shirts, patient gowns, scrubs, warm-up jackets, bed sheets, blankets, pillowcases, thermal spreads, aprons, diapers, surgical towels and underpads. Additionally, through subsidiary Los Angeles Rose, the firm specializes in producing fashion designed scrubs for the health care market. Salon products include barber jackets, shaving towels, haircutting cloths, shampoo towels, custom entrance mats, bath robes, blankets and tanning towels. Finally, home fashion products, sold under the brands Eileen West, Mary Jane, Vicki Payne and Alexander Julian, focus primarily on textiles for the bathroom, including bath towels, fashion print tip towels, bath rugs, seasonal accessories and others.

FINANCIAL DATA:
Note: Data for latest year may not have been available at press time.

In U.S. $	2012	2011	2010	2009	2008	2007
Revenue						
R&D Expense						
Operating Income						
Operating Margin %						
SGA Expense						
Net Income						
Operating Cash Flow						
Capital Expenditure						
EBITDA						
Return on Assets %						
Return on Equity %						
Debt to Equity						

CONTACT INFORMATION:

Phone: 310-223-2000 Fax: 310-223-2017
Toll-Free: 800-821-2221
Address: 401 W. Artesia Blvd., Compton, CA 90220 United States

STOCK TICKER/OTHER:

Stock Ticker: Private
Exchange:
Employees:
Fiscal Year Ends:
Parent Company:

SALARIES/BONUSES:

Top Exec. Salary: $ Bonus: $
Second Exec. Salary: $ Bonus: $

OTHER THOUGHTS:

Estimated Female Officers or Directors: 1
Hot Spot for Advancement for Women/Minorities:

American Eagle Outfitters Inc

www.ae.com

NAIC Code: 448140

TYPES OF BUSINESS:

Casual Apparel, Retail
Online Sales
Intimates Apparel
Children's Apparel

BRANDS/DIVISIONS/AFFILIATES:

American Eagle Outfitters
aerie by American Eagle
ae.com
aerie.com
Linmar Realty Company II LLC
Blue Heart Enterprises LLC
Blue Star Imports Ltd

CONTACTS: *Note: Officers with more than one job title may be intentionally listed here more than once.*

Robert Lee Hanson, CEO
Michael Rempell, COO
Mary Boland, CFO
Mary Boland, Chief Admin. Officer
Neil Bulman, General Counsel/Exec. VP
Dennis Parodi, Exec. VP-Store Oper.
Roger S. Markfield, Vice Chmn./Exec. Dir.-Creative
Jay L. Schottenstein, Chmn.

GROWTH PLANS/SPECIAL FEATURES:

American Eagle Outfitters, Inc. (AE) designs, markets and retails casual clothing primarily targeted at teenagers and young adults. Denim is the cornerstone, and is complemented by a variety of other fashion conscious clothing, such as graphic tees, sweaters, and accessories. The firm sells merchandise under the American Eagle Outfitters and aerie by American Eagle (aerie) brand. AE's aerie collection includes bras, underwear, camis, hoodies, robes, boxers, sweats, leggings, fitness apparel and personal care products for teen girls and young adults. AE operates approximately 911 American Eagle Outfitters stores and 158 aerie stand-alone stores in the U.S., Puerto Rico and Canada. The firm also operates ae.com and aerie.com, e-commerce sites that feature an expanded offering of sizes, colors and styles of its merchandise. It ships its products to 77 countries through its e-commerce business. The firm offers a Visa co-branded credit card, as well as a private label card issued by a third party bank. Through several franchise agreements, AE has 42 international franchise stores in 13 countries. The firm's stores generate about $470 in sales per square foot per year. In August 2012, American Eagle sold its 77kids store and brand to Ezrani 2 Corp, whose owner is the former chairman and CEO of The Children's Place.

Employees are offered medical, dental and vision insurance; life insurance; pet insurance; adoption assistance; new dependant leave; tuition reimbursement; disability coverage; merchandise discounts; a 401(k) plan; athletic club reimbursement; and an employee stock purchase plan.

FINANCIAL DATA: *Note: Data for latest year may not have been available at press time.*

In U.S. $	2012	2011	2010	2009	2008	2007
Revenue	3,159,818,000	2,967,559,000	2,990,520,000	2,988,866,000	3,055,419,000	
R&D Expense						
Operating Income	231,136,000	317,261,000	238,393,000	302,140,000	598,755,000	
Operating Margin %	7.31%	10.69%	7.97%	10.10%	19.59%	
SGA Expense	735,828,000	713,197,000	756,256,000	740,742,000	715,180,000	
Net Income	151,705,000	140,647,000	169,022,000	179,061,000	400,019,000	
Operating Cash Flow	239,256,000	381,160,000	386,462,000	302,193,000	464,270,000	
Capital Expenditure	134,322,000	84,259,000	127,419,000	265,335,000	250,407,000	
EBITDA	374,292,000	462,809,000	385,876,000	435,281,000	707,958,000	
Return on Assets %	7.92%	7.00%	8.24%	9.34%	20.75%	
Return on Equity %	10.96%	9.60%	11.31%	13.02%	29.01%	
Debt to Equity						

CONTACT INFORMATION:

Phone: 412 432-3300 Fax:
Toll-Free: 888-232-4535
Address: 77 Hot Metal St., Pittsburgh, PA 15203 United States

STOCK TICKER/OTHER:

Stock Ticker: AEO
Employees: 39,600
Parent Company:
Exchange: NYS
Fiscal Year Ends: 31-Jan

SALARIES/BONUSES:

Top Exec. Salary: $1,049,808 Bonus: $
Second Exec. Salary: $968,269 Bonus: $

OTHER THOUGHTS:

Estimated Female Officers or Directors: 2
Hot Spot for Advancement for Women/Minorities: Y

AND 1

www.and1.com

NAIC Code: 316210

TYPES OF BUSINESS:

Basketball Apparel
Street Basketball Tournaments

BRANDS/DIVISIONS/AFFILIATES:

AND 1 Mix Tape Tour
AND 1 Streetball
Brown Shoe Company Inc
American Sporting Goods Corporation
Tai Chi
Starbury
KG
Sprees

CONTACTS: *Note: Officers with more than one job title may be intentionally listed here more than once.*

Robert Purvy, Exec. Dir.
Steve Wright, Dir.-Sports Mktg.
Sue Ehnow, Dir.-Retail Oper.

GROWTH PLANS/SPECIAL FEATURES:

AND 1, often referred to as The Basketball Company, manufactures apparel branded with strategic focus on hip-hop and non-professional basketball culture. The name AND 1 refers to the free throw that is taken after someone is fouled while scoring a basket. It utilizes as its logo a faceless, raceless basketball player known as The Player. The brand was launched in 1993, initially producing printed t-shirts, and has since expanded into a complete line of basketball apparel such as performance shoes, t-shirts, shorts, pants, team duffle bags and other apparel and accessories. The company's performance Shooting Shorts line is distinctly designed by players for players. Other notable brands include Starbury, KG, Sprees and Tai Chi, which Brent Barry famously wore when he won the annual NBC dunk contest in 2000. Merchandise is sold internationally through leading sports apparel stores such as Champs, FootAction and Foot Locker as well as an e-commerce site, 1usa.com. The company has gained considerable fame in recent years with its AND 1 Mix Tape Tour, a street basketball tournament that emphasizes style, trick moves and showmanship. The Mix Tape Tours basketball team includes a group of 10 young men who promote the AND 1 brand. Additionally, the firm has licensed its brand to video game developers to produce AND 1 Streetball, a streetball video game designed to capture the hip-hop culture surrounding non-professional basketball. AND 1 is owned by American Sporting Goods Corporation, a manufacturer of athletic shoes and golf equipment, which is in turn owned by Brown Shoe Company, Inc.

FINANCIAL DATA: *Note: Data for latest year may not have been available at press time.*

In U.S. $	2012	2011	2010	2009	2008	2007
Revenue						
R&D Expense						
Operating Income						
Operating Margin %						
SGA Expense						
Net Income						
Operating Cash Flow						
Capital Expenditure						
EBITDA						
Return on Assets %						
Return on Equity %						
Debt to Equity						

CONTACT INFORMATION:

Phone: Fax: 949-756-153
Toll-Free: 800-848-8698
Address: 101 Enterprise, Ste. 100, Aliso Viejo, CA 92656 United States

STOCK TICKER/OTHER:

Stock Ticker: Subsidiary Exchange:
Employees: Fiscal Year Ends: 31-Dec
Parent Company: BROWN SHOE COMPANY INC

SALARIES/BONUSES:

Top Exec. Salary: $ Bonus: $
Second Exec. Salary: $ Bonus: $

OTHER THOUGHTS:

Estimated Female Officers or Directors: 1
Hot Spot for Advancement for Women/Minorities:

Ann Inc

NAIC Code: 448120

www.anninc.com

TYPES OF BUSINESS:

Women's Apparel, Retail
Clothing & Accessories
Shoes
Online & Catalog Sales

BRANDS/DIVISIONS/AFFILIATES:

Ann Taylor
Ann Taylor LOFT
Ann Taylor Factory Store
LOFT Outlet Store
Annco Inc
AnnTaylor Travel Inc
AnnTaylor Sourcing Far East Limited
ANN Canada

CONTACTS: *Note: Officers with more than one job title may be intentionally listed here more than once.*

Kay Krill, CEO
Kay Krill, Pres.
Michael J. Nicholson, CFO/Exec. VP
Michael J. Nicholson, Treas.
Brian Lynch, Pres., Ann Taylor Div.
Gary Muto, Pres., LOFT Div.

GROWTH PLANS/SPECIAL FEATURES:

Ann, Inc. is one of the leading national specialty retailers of high-quality women's apparel, shoes and accessories. Its stores offer a full range of career and casual separates, dresses, tops, weekend wear, shoes and accessories, coordinated as part of a total wardrobe strategy. Its target consumer is split between the Ann Taylor and Ann Taylor LOFT stores, where Ann Taylor caters to affluent career women, and LOFT targets a value conscious woman, while remaining upscale, seeking a more casual wardrobe. Through Ann Taylor Factory stores and LOFT Outlet stores, the firm offers best-selling and clearance merchandise from the other two store formats. Merchandise is marketed in 946 stores in 46 states, Washington, D.C. and Puerto Rico. Of these stores, 272 are Ann Taylor branded stores, 501 are Ann Taylor LOFT branded stores, 99 are Ann Taylor Factory stores and 74 are LOFT Outlet stores. The stores are located primarily in regional malls and upscale specialty retail centers. Ann Taylor stores average 5,200 square feet in size, while Ann Taylor LOFT stores average 5,800 square feet. The outlet stores average 7,100 (Ann Taylor) and 7,000 (LOFT) square feet. The overall trend for Ann Taylor stores is to seek a smaller footprint, be it through remodeling/downsizing or new openings. The company's merchandise is predominately developed and designed by in-house design teams exclusively for its own stores, with a small portion bought through branded vendors. During 2011, the firm sourced merchandise from approximately 138 manufacturers and vendors in 19 countries. In November 2012, Ann, Inc. announced that it has banned the use of exotic animal skins (such as ostrich, alligator or crocodile) in all of its products.

Employees are offered medical, dental and vision insurance; flexible spending accounts; adoption assistance; life insurance; a transportation reimbursement program; discounted merchandise; a stock purchase plan; and a 401(k).

FINANCIAL DATA: *Note: Data for latest year may not have been available at press time.*

In U.S. $	2012	2011	2010	2009	2008	2007
Revenue	2,212,493,000	1,980,195,000	1,828,523,000	2,194,559,000	2,396,510,000	
R&D Expense						
Operating Income	145,499,000	119,790,000	-23,954,000	-371,637,000	155,386,000	
Operating Margin %	6.57%	6.04%	-1.31%	-16.93%	6.48%	
SGA Expense	1,062,644,000	978,580,000	966,603,000	1,050,560,000	1,063,623,000	
Net Income	86,566,000	73,397,000	-18,208,000	-333,906,000	97,235,000	
Operating Cash Flow	207,828,000	164,311,000	133,703,000	172,818,000	257,197,000	
Capital Expenditure	119,168,000	63,963,000	38,573,000	110,342,000	139,998,000	
EBITDA	240,328,000	216,277,000	81,332,000	-247,738,000	280,016,000	
Return on Assets %	9.54%	8.02%	-1.95%	-28.36%	6.56%	
Return on Equity %	21.98%	17.46%	-4.36%	-53.16%	10.29%	
Debt to Equity						

CONTACT INFORMATION:

Phone: 212 541-3300 Fax: 212 541-3379
Toll-Free: 800-677-6788
Address: 7 Times Square, 15th Fl., New York, NY 10036 United States

STOCK TICKER/OTHER:

Stock Ticker: ANN Exchange: NYS
Employees: 19,900 Fiscal Year Ends: 31-Jan
Parent Company:

SALARIES/BONUSES:

Top Exec. Salary: $1,200,000 Bonus: $
Second Exec. Salary: $900,000 Bonus: $

OTHER THOUGHTS:

Estimated Female Officers or Directors: 1
Hot Spot for Advancement for Women/Minorities: Y

Anvil Knitwear Inc

NAIC Code: 315190

www.anvilknitwear.com

TYPES OF BUSINESS:

Knitwear Manufacturing
Men's, Women's & Children's Activewear

BRANDS/DIVISIONS/AFFILIATES:

AnvilSustainable
Anvil Holdings Inc
AnvilRecycled
Anvil dri-release
Towels Plus
TearAway
chromaZONE
Specrtratex Inc

CONTACTS: *Note: Officers with more than one job title may be intentionally listed here more than once.*

Anthony Corsano, CEO
Frank Ferramosca, CFO
Steve Pegg, Dir.-Human Resources
Frank Ferramosca, Exec. VP-Finance
Anthony Corsano, Chmn.

GROWTH PLANS/SPECIAL FEATURES:

Anvil Knitwear, Inc. is a leading designer, manufacturer and marketer of high-quality organic activewear and accessories. The company's clients consist primarily of customers in the imprinted or decorated segment of the U.S. apparel industry. Anvil's 70 styles of activewear products and accessories are available in over 80 colors and 17 eco friendly styles designed for men, women and children, and include long and short-sleeve tee shirts, polo shirts, caps, towels, garment-dyed tee shirts and bags. Products are manufactured from conventional cotton, certified organic cotton, recycled cotton, transitional cotton and recycled polyethylene terephthalate (PET) bottles and blends. Anvil products are available under its own brand names, including Anvil Logo, Anvil, chromaZONE, AnvilOrganic, AnvilRecycled, AnvilSustainable, Anvil dri-release, Towels Plus and TearAway, as well as private labels. ChromaZONE is a line of garment-dyed tee shirts intended to have variations in color and shading to create a unique look. AnvilOrganic is a line of 100% organic cotton tee shirts. AnvilRecycled is the company's line of tee shirts made of 70% pre-consumer recycled cotton. AnvilSustainable products are made of 55% certified organic cotton and 45% post-consumer PET (Polyethylene terephthalate) recycled polyester. Anvil's dri-release product is a performance fabric that dries four times faster than cotton and virtually eliminates odor. TearAway shirts feature a tag that can be torn off without ripping the stitches. Recent product enhancements include increased size specifications, softer fabric, designer high-gloss wood-tone buttons and ladies' satin fashion clothing labels. The firm is a wholly-owned subsidiary of Anvil Holdings, Inc. It also operates subsidiaries located around the globe, including Spectratex, Inc.; A.K.H. SA; ANNIC LLC; Star SA; Anvil GmbH; and Anvil Honduras SA.

FINANCIAL DATA: *Note: Data for latest year may not have been available at press time.*

In U.S. $	2012	2011	2010	2009	2008	2007
Revenue						
R&D Expense						
Operating Income						
Operating Margin %						
SGA Expense						
Net Income						
Operating Cash Flow						
Capital Expenditure						
EBITDA						
Return on Assets %						
Return on Equity %						
Debt to Equity						

CONTACT INFORMATION:

Phone: 212-476-0300 Fax: 212-808-4790
Toll-Free: 800-223-0332
Address: 521 Fifth Ave., 9th Fl., New York, NY 10175 United States

STOCK TICKER/OTHER:

Stock Ticker: Subsidiary Exchange:
Employees: 4,200 Fiscal Year Ends: 31-Jan
Parent Company: ANVIL HOLDINGS INC

SALARIES/BONUSES:

Top Exec. Salary: $ Bonus: $
Second Exec. Salary: $ Bonus: $

OTHER THOUGHTS:

Estimated Female Officers or Directors: 5
Hot Spot for Advancement for Women/Minorities: Y

ARAMARK Corporation

www.aramark.com

NAIC Code: 722310

TYPES OF BUSINESS:

Food Service Contractor
Facilities Management
Uniforms & Career Apparel Rental
Parks & Resorts Concessions & Facilities
Health Care Support Services
Apparel Manufacturing
Clinical Equipment Maintenance

BRANDS/DIVISIONS/AFFILIATES:

Warburg Pincus LLC
GS Capital Partners
CCMP Capital Advisors JP
J.P. Morgan Partners
Thomas H. Lee Partners LP

CONTACTS: *Note: Officers with more than one job title may be intentionally listed here more than once.*

Eric Foss, CEO
L. Frederick Sutherland, CFO/Exec. VP
Lynn B. McKee, Exec. VP-Human Resources
Stephen R. Reynolds, General Counsel/Exec. VP/Sec.
Lynn B. McKee, Exec. VP-Comm.
Joseph Neubauer, Chmn.

GROWTH PLANS/SPECIAL FEATURES:

ARAMARK Corporation is a leading food, hospitality and facilities management services company. Serving clients in 22 countries, ARAMARK offers services across five broad categories: facilities management, food service and refreshments, lodging and guest accommodations, clinical technology services and cleanroom services. ARAMARK's facilities management services include laundry, housekeeping, facilities maintenance, plant operations, landscaping, transportation, clinical equipment maintenance, grounds keeping, custodial services and construction management. Food service and refreshment offerings include on-site restaurants, convenience stores, concessions stands, banquet halls, catering and executive dining rooms. Lodging and guest accommodations are provided to parks and resorts as well as conference and convention centers. Clinical technology services include acquisitions, clinical equipment maintenance services and asset tracking. Cleanroom services are provided to the pharmaceutical, biotechnology and microelectronics manufacturing industries. ARAMARK's clients include numerous domestic and international business and industrial firms; 192 stadiums, arenas, conference rooms and convention centers; 16 state and national parks; over 1,100 health care and senior living facilities; correctional facilities; and 1,000 colleges, universities, school districts and private schools. ARAMARK is owned by an investor group led by Chairman and CEO Joseph Neubauer and investment funds managed by GS Capital Partners, CCMP Capital Advisors, J.P. Morgan Partners, Thomas H. Lee Partners and Warburg Pincus LLC.

The company offers its employees medical, dental and vision insurance; a pension plan; life and disability insurance; and an employee assistance plan.

FINANCIAL DATA: *Note: Data for latest year may not have been available at press time.*

In U.S. $	2012	2011	2010	2009	2008	2007
Revenue	13,250,000,000	13,000,000,000	12,571,676,000	12,297,869,000	12,470,200,000	12,384,300,000
R&D Expense						
Operating Income						
Operating Margin %						
SGA Expense						
Net Income			30,688,000	-6,911,000	39,500,000	30,900,000
Operating Cash Flow						
Capital Expenditure						
EBITDA						
Return on Assets %						
Return on Equity %						
Debt to Equity						

CONTACT INFORMATION:

Phone: 215-238-3000 Fax: 215-238-3333
Toll-Free: 800-272-6275
Address: 1101 Market St., ARAMARK Tower, Philadelphia, PA 19107 United States

STOCK TICKER/OTHER:

Stock Ticker: Private Exchange:
Employees: 254,000 Fiscal Year Ends: 30-Sep
Parent Company: Warburg Pincus LLC

SALARIES/BONUSES:

Top Exec. Salary: $ Bonus: $
Second Exec. Salary: $ Bonus: $

OTHER THOUGHTS:

Estimated Female Officers or Directors: 1
Hot Spot for Advancement for Women/Minorities:

Arcadia Group

NAIC Code: 448120

www.arcadiagroup.co.uk

TYPES OF BUSINESS:

Apparel Stores-Men's & Women's

BRANDS/DIVISIONS/AFFILIATES:

Dorothy Perkins
Miss Selfridge
Wallis
TOPSHOP
Evans
Burton
Outfit
BHS Limited

CONTACTS:

Note: Officers with more than one job title may be intentionally listed here more than once.

Ian Grabiner, CEO
Alana Symonds, Dir.-Public Rel. & Comm.
Philip Green, Owner-Taveta Investments
Anthony Grabiner, Chmn.

GROWTH PLANS/SPECIAL FEATURES:

Arcadia Group is one of the alrgest clothing retailers in the U.K. The firm operates several apparel chains with approximately 2,500 stores domestically. Arcadia also has 600 franchised stores in 36 countries primarily in Europe, the Middle East and Asia. The firm is owned by one of the largest privately owned retail groups in Europe, Taveta Investments. The company has eight brands: Dorothy Perkins, Miss Selfridge, Wallis, TOPSHOP, Evans, Burton, TOPMAN and BHS. Dorothy Perkins, with 600 stores in the U.K. and over 90 outlets worldwide, targets female customers age 25-35, with tall, petite and maternity ranges available. Miss Selfridge, with more than 250 stores internationally, markets to young and trendy women ages 18-24. Wallis, with over 400 standalone stores and concessions in the U.K. and Ireland, is an upscale brand that incorporates new trends and colors into its traditional glamour. TOPSHOP is a trendy, stylish and affordable brand for young women, with 300 stores across the U.K and over 140 worldwide. TOPSHOP's flagship stores are located in London, Liverpool and Leeds, U.K.; Chicago, Illinois; Las Vegas, Nevada; and on Broadway in New York. Evans retails classic clothing, footwear, lingerie and accessories for women ages 20-45 through 330 stores in Ireland and the U.K. Burton is a mid-market menswear brand for ages 25-34 with over 400 stores in Ireland and the U.K. TOPMAN is a trendy, fashion-forward and affordable brand for young men. BHS is a department store chain with over 175 stores that sell clothing and house hold items. The company also operates over 50 Outfit stores, which provide selections from the company's house brands primarily in out-of-town locations. In March 2012, the firm opened a third U.S. TOPSHOP flagship store in Las Vegas, Nevada.

Employees are offered a pension plan, a bonus plan and a 25% discount on all Arcadia Group merchandise.

FINANCIAL DATA:

Note: Data for latest year may not have been available at press time.

In U.S. $	2012	2011	2010	2009	2008	2007
Revenue		4,334,299,140	4,309,000,000	2,852,558,000		
R&D Expense						
Operating Income						
Operating Margin %						
SGA Expense						
Net Income		307,658,740				
Operating Cash Flow						
Capital Expenditure						
EBITDA						
Return on Assets %						
Return on Equity %						
Debt to Equity						

CONTACT INFORMATION:

Phone: 44-20-7636-8040 Fax: 44-20-7927-0577
Toll-Free:
Address: 70 Berners St., Colegrave House, London, W1T 3NL United Kingdom

STOCK TICKER/OTHER:

Stock Ticker: Private Exchange:
Employees: 45,000 Fiscal Year Ends: 31-Aug
Parent Company: TAVETA INVESTMENTS LTD

SALARIES/BONUSES:

Top Exec. Salary: $ Bonus: $
Second Exec. Salary: $ Bonus: $

OTHER THOUGHTS:

Estimated Female Officers or Directors: 1
Hot Spot for Advancement for Women/Minorities:

Argix Direct Inc

www.argixdirect.com

NAIC Code: 488510

TYPES OF BUSINESS:

Freight Transportation Arrangement
Direct-to-Store Delivery
Trucking
Sorting Centers

BRANDS/DIVISIONS/AFFILIATES:

CONTACTS:

Note: Officers with more than one job title may be intentionally listed here more than once.

Pete Leap, COO
Nino Silvano, CIO

GROWTH PLANS/SPECIAL FEATURES:

Argix Direct, Inc. is a specialized delivery company that focuses entirely on direct-to-store deliveries for specialty retailers. The company operates a comprehensive national delivery network through its 325,000 square foot sorting center in New Jersey which sends items to its network of 43 routing terminals across the U.S. It can deliver products throughout the continental U.S. within five days of receipt. The firm routes shipments directly from its sorting facilities to its customers, eliminating the need for distribution centers, and its out-bound trucks depart on a pre-determined schedule regardless of their load size. Its capabilities include consolidating loads from diverse sources into a single shipment and working with import vessels, such as ships or aircraft, to distribute merchandise directly from vessel to store. Argix can also provide direct to consumer shipping for catalog and e-commerce operations in collaboration with the U.S. Postal Service (USPS). Additionally, the firm maintains storage facilities for its customers. Argix uses a thorough tracking system and condensed billing for its freight, and it provides tracking information for its fleet directly through its web site. Among its key clients are retailers such as Barnes & Noble and Ann Taylor.

FINANCIAL DATA:

Note: Data for latest year may not have been available at press time.

In U.S. $	2012	2011	2010	2009	2008	2007
Revenue						
R&D Expense						
Operating Income						
Operating Margin %						
SGA Expense						
Net Income						
Operating Cash Flow						
Capital Expenditure						
EBITDA						
Return on Assets %						
Return on Equity %						
Debt to Equity						

CONTACT INFORMATION:

Phone: 732-656-2550 Fax:
Toll-Free:
Address: 100 Middlesex Center Blvd., Jamesburg, NJ 08831 United States

STOCK TICKER/OTHER:

Stock Ticker: Private
Employees:
Parent Company:
Exchange:
Fiscal Year Ends: 30-Sep

SALARIES/BONUSES:

Top Exec. Salary: $ Bonus: $
Second Exec. Salary: $ Bonus: $

OTHER THOUGHTS:

Estimated Female Officers or Directors:
Hot Spot for Advancement for Women/Minorities:

Asahi Kasei Corp

www.asahi-kasei.co.jp

NAIC Code: 325220

TYPES OF BUSINESS:

Chemical Manufacturing
Fibers & Textiles
Plastics
Synthetic Rubber
Construction Materials
Home Construction
Electronics Components
Medical Devices

BRANDS/DIVISIONS/AFFILIATES:

Asahi Kasei Fibers
Asahi Kasei Chemicals
Asahi Kasei Pharma
Asahi Kasei Construction Materials
Asahi Kasei Homes
Asahi Kasei Microdevices
Asahi Kasei Medical
ZOLL Medical

CONTACTS:

Note: Officers with more than one job title may be intentionally listed here more than once.

Taketsugu Fujiwara, Exec. Officer
Taketsugu Fujiwara, Pres.
Ichiro Itoh, Chmn.

GROWTH PLANS/SPECIAL FEATURES:

Asahi Kasei Corp. is a holdings company that provides a diverse range of chemical and materials-based solutions to the fibers, chemicals, consumer products, housing, construction, electronics and health care industries. The company operates through nine subsidiaries: Asahi Kasei Chemicals, Asahi Kasei Homes, Asahi Kasei Pharma, ZOLL Medical, Asahi Kasei Medical, Asahi Kasei Fibers, Asahi Kasei Microdevices, Asahi Kasei E-materials and Asahi Kasei Construction Materials. Asahi Kasei Chemicals, the firm's core operating company manages all chemical-related operations and produces industrial chemicals, synthetic resins and rubber; latexes; pharmaceutical and food additives; explosives; storage bags; and plastic film. Asahi Kasei Homes designs and constructs earthquake-resistant Hebel Haus houses and Hebel Maison apartments and condominiums, and also conducts real estate, remodeling and urban development operations. Asahi's health care division operates through Asahi Kasei Pharma, which produces pharmaceuticals, diagnostic reagents and nutritional products; ZOLL Medical, which produces professional defibrillators and AutoPulse, a non-invasive cardiac support pump; and Asahi Kasei Medical, which produces Planova virus removal filters, APS artificial kidneys and Sepacell leukocyte reduction filters. Asahi Kasei Fibers produces materials for apparel and industrial applications, including Roica-brand elastane, Eltas-brand spunbond fabrics, Lamous-brand artificial suede and Leona-brand nylon. Asahi Kasei Microdevices manufactures LSI products and Hall elements for electronic devices. Asahi Kasei E-materials produces electronic, photosensitive and energy materials and epoxy resins. Asahi Kasei Construction Materials manufactures Hebel AAC panels and Neoma foam insulation panels for the residential, commercial building and civil engineering markets.

FINANCIAL DATA:

Note: Data for latest year may not have been available at press time.

In U.S. $	2012	2011	2010	2009	2008	2007
Revenue	17,515,170,000	17,795,250,000	15,960,580,000	17,291,150,000	18,890,780,000	18,078,080,000
R&D Expense						
Operating Income	1,160,731,000	1,368,577,000	641,520,300	389,207,400	1,421,227,000	1,422,841,000
Operating Margin %	6.62%	7.69%	4.01%	2.25%	7.52%	7.87%
SGA Expense	3,228,677,000			3,121,018,000	3,119,182,000	3,027,677,000
Net Income	620,857,000	671,201,600	281,515,500	52,827,290	778,715,400	763,462,800
Operating Cash Flow	1,572,683,000	1,649,147,000	1,884,949,000	766,101,400	812,137,500	1,429,877,000
Capital Expenditure	750,770,900	768,016,300	1,017,112,000	1,327,418,000	848,420,700	915,476,400
EBITDA	1,972,479,000	2,079,914,000	1,525,534,000	1,149,842,000	2,049,532,000	2,129,513,000
Return on Assets %	3.93%	4.31%	1.84%	.33%	4.84%	4.83%
Return on Equity %	8.13%	9.29%	4.10%	.74%	10.66%	11.06%
Debt to Equity	0.13	0.17	0.23	0.23	0.13	0.19

CONTACT INFORMATION:

Phone: 81 332963008 Fax: 81 332963162
Toll-Free:
Address: 1-105 Kanda Jinbocho, Chiyoda-ku, Tokyo, 101-8101 Japan

STOCK TICKER/OTHER:

Stock Ticker: AHKSF Exchange: PINX
Employees: 25,016 Fiscal Year Ends: 31-Mar
Parent Company:

SALARIES/BONUSES:

Top Exec. Salary: $ Bonus: $
Second Exec. Salary: $ Bonus: $

OTHER THOUGHTS:

Estimated Female Officers or Directors:
Hot Spot for Advancement for Women/Minorities:

Ascena Retail Group Inc

NAIC Code: 448120

www.ascenaretail.com

TYPES OF BUSINESS:

Women's Apparel, Retail
Teen Fashion Stores
Fashion Accessories
Private-Label Credit Cards

BRANDS/DIVISIONS/AFFILIATES:

dressbarn
dressbarn woman
Studio Y
Industrial Exchange
Tween Brands Inc
Justice
The Dress Barn Inc
maurices

CONTACTS: *Note: Officers with more than one job title may be intentionally listed here more than once.*

David R. Jaffe, CEO
John Sullivan, COO/Exec. VP
David R. Jaffe, Pres.
Armand Correia, CFO/Exec. VP
Gene L. Wexler, General Counsel/Sr. VP
Michael W. Rayden, CEO-Justice
Elliot S. Jaffe, Chmn.

GROWTH PLANS/SPECIAL FEATURES:

Ascena Retail Group, Inc., formerly The Dress Barn, Inc., operates a national chain of value-priced specialty stores offering in-season, moderate to better quality apparel and accessories. The majority of Ascena's stores are located in strip malls. The company has two store formats: dressbarn stores, which carry junior and misses sizes, and dressbarn woman stores, which feature larger sizes; and combination stores, which carry both. The combination stores are more prevalent, though the company does have some locations that are only either dressbarn or dressbarn women. Ascena's subsidiary, maurices, offers moderately priced, up-to-date fashions designed to appeal to a younger female consumer than the dressbarn and dressbarn woman brands. maurices operates primarily within smaller cities of 25,000 to 100,000 in population. maurice's merchandise is sold under three brand names: maurices, Studio Y and Industrial Exchange. Through subsidiary Tween Brands, Inc., the company markets clothing and accessories to girls aged 7-14 through its Justice store locations. Ascena operates approximately 830 dressbarn stores, 784 maurices stores and 902 Justice stores. The company offers a dressbarn credit card operated by World Financial Network National Bank, as well as personal and corporate gift cards in a variety of values. Holders of the dressbarn credit card receive targeted promotions including directed mailings and special coupons.

The company offers its employees short-term disability; life and accidental death and dismemberment insurance; a 30% merchandise discount; career training; tuition reimbursement; and medical, dental and vision coverage.

FINANCIAL DATA: *Note: Data for latest year may not have been available at press time.*

In U.S. $	2012	2011	2010	2009	2008	2007
Revenue	3,353,300,000	2,914,000,000	2,374,571,000	1,494,236,000	1,444,165,000	1,426,607,000
R&D Expense						
Operating Income	292,600,000	289,800,000	217,457,000	104,979,000	112,614,000	154,972,000
Operating Margin %	8.72%	9.94%	9.15%	7.02%	7.79%	10.86%
SGA Expense	1,478,600,000	852,100,000	690,229,000	422,372,000	397,424,000	383,652,000
Net Income	162,200,000	170,500,000	133,378,000	69,688,000	74,088,000	101,182,000
Operating Cash Flow	361,500,000	280,800,000	231,437,000	172,691,000	145,455,000	155,232,000
Capital Expenditure	150,400,000	102,100,000	65,179,000	58,428,000	66,097,000	62,986,000
EBITDA	390,700,000	376,700,000	289,066,000	160,947,000	169,924,000	209,676,000
Return on Assets %	6.98%	9.76%	9.57%	6.46%	7.38%	11.06%
Return on Equity %	12.98%	15.68%	16.24%	11.78%	13.90%	22.03%
Debt to Equity	0.41	0.14	0.02	0.04	0.04	0.05

CONTACT INFORMATION:

Phone: 845 369-4500 Fax: 845 369-8001
Toll-Free: 800-373-7722
Address: 30 Dunnigan Dr., Suffern, NY 10901 United States

STOCK TICKER/OTHER:

Stock Ticker: ASNA Exchange: NAS
Employees: 46,000 Fiscal Year Ends: 31-Jul
Parent Company:

SALARIES/BONUSES:

Top Exec. Salary: $1,050,000 Bonus: $
Second Exec. Salary: $996,923 Bonus: $

OTHER THOUGHTS:

Estimated Female Officers or Directors: 2
Hot Spot for Advancement for Women/Minorities: Y

ASICS Corp

www.asics.com

NAIC Code: 316210

TYPES OF BUSINESS:

Athletic Footwear
Sporting Goods
Apparel
Uniforms

BRANDS/DIVISIONS/AFFILIATES:

ASICS America Corp
ASICS Europe BV
Onitsuka Tiger
Impact Guide System
Asics Gel

CONTACTS: *Note: Officers with more than one job title may be intentionally listed here more than once.*

Motoi Oyama, CEO
Motoi Oyama, Pres.
Toshiro Ikezaki, Managing Exec. Officer
Toshiyuki Sano, Managing Exec. Officer
Kazuhito Matsuo, Exec. Officer
Kousuke Hashimoto, Exec. Officer
Kiyomi Wada, Chmn.

GROWTH PLANS/SPECIAL FEATURES:

ASICS Corp. is a Japanese manufacturer of sporting and leisure goods primarily known for its footwear. The company owns and operates 10 consolidated subsidiaries and 28 overseas affiliates. The ASICS name is an acronym for a Latin phrase, Anima Sana In Corpore Sano, meaning A Sound Mind in a Sound Body. The company operates in four primary geographic regions: Japan, the Americas, Europe and the Asia Pacific region. Besides footwear, the firm produces sportswear and uniforms for a variety of sports including basketball, volleyball, cheerleading, wrestling, running and field sports as well as accessories such as hats, sports bags and protective gear. Fashion-oriented items are sold through the brand name Onitsuka Tiger. Products are sold through specialty running stores, sporting good stores, department stores and online. The company markets its shoes based on several technologies such as the Impact Guide System (I.G.S.), designed to produce a more natural heel-to-toe movement, and Gel technology for more in-shoe cushioning. ASICS operates through international offices located in the U.S., Brazil, Australia, China, Korea, Taiwan, the Netherlands, Germany, Austria, France, Poland, the U.K., Italy, Spain, Norway, Denmark, Sweden and Russia and works through distributors in numerous other countries. The firm also operates several subsidiaries, including ASICS America Corp. (North America) and Asics Europe BV. A research and development center located in Kobe, Japan develops and tests new material and construction technologies.

FINANCIAL DATA: *Note: Data for latest year may not have been available at press time.*

In U.S. $	2012	2011	2010	2009	2008	2007
Revenue	2,758,731,000	2,620,200,000	2,498,246,000	2,693,624,000		
R&D Expense						
Operating Income	218,523,500	240,177,700	195,744,900	251,923,300		
Operating Margin %	7.92%	9.16%	7.83%	9.35%		
SGA Expense	978,835,600	907,894,600	853,286,000	895,269,400		
Net Income	140,468,300	122,977,900	92,695,470	145,678,600		
Operating Cash Flow	113,993,400	106,356,000	189,064,900	209,171,600		
Capital Expenditure	35,114,280	52,370,820	33,511,090	58,471,840		
EBITDA	300,586,700	263,323,700	248,817,100	260,395,700		
Return on Assets %	6.10%	5.72%	4.62%	7.48%		
Return on Equity %	12.16%	11.09%	8.83%	14.74%		
Debt to Equity	0.22	0.23	0.15	0.16		

CONTACT INFORMATION:

Phone: 81 783032231 Fax: 81 783032241
Toll-Free: 800-678-9435
Address: 7-1-1 Minatojima-Nakamachi, Kobe, 650-8555 Japan

STOCK TICKER/OTHER:

Stock Ticker: ASCCF Exchange: GREY
Employees: 5,604 Fiscal Year Ends: 31-Mar
Parent Company:

SALARIES/BONUSES:

Top Exec. Salary: $ Bonus: $
Second Exec. Salary: $ Bonus: $

OTHER THOUGHTS:

Estimated Female Officers or Directors:
Hot Spot for Advancement for Women/Minorities:

ASOS plc

NAIC Code: 454111

www.asosplc.com

TYPES OF BUSINESS:

Online Apparel Sales

BRANDS/DIVISIONS/AFFILIATES:

ASOS White
ASOS Black
ASOS Maternity
ASOS Swim
ASOS Marketplace
Green Room (The)
ASOS Premier
ASOS Magazine

CONTACTS: *Note: Officers with more than one job title may be intentionally listed here more than once.*

Nick Robertson, CEO
Nick Beighton, Dir.-Finance
Nick Beighton, Corp. Sec.
Waheed Alli, Chmn.
Jon Kamaluddin, Dir.-Int'l

GROWTH PLANS/SPECIAL FEATURES:

ASOS plc is a U.K.-based online retailer of fashion and beauty products. The company's inventory includes about 60,000 product lines, including womenswear, menswear, footwear, accessories, jewelry and beauty items. With over 11.15 million registered users and customers throughout 190 countries, the firm's website targets the twenty-something year old demographic and attracts over 23.7 million visitors a month. The company's store websites, based in the U.K., U.S., France and Germany, provide customized shopping through localized pricing, currency rates, sizing and language options. ASOS' own private labels include ASOS White, ASOS Black, ASOS Maternity, ASOS Swim, Curve and The Green Room. Its non-proprietary brand offerings include over 850 brands, such as Ralph Lauren, American Apparel, New Look, Polo, Jack & Jones, Cheap Monday, Zadig & Voltaire, Lazy Oaf, Adidas, Espirit and Benetton. The company also generates revenues from ASOS marketplace, an online site that allows individuals to list, buy and sell new and pre-owned fashion and apparel. This site offers a basic membership, which has no monthly charge and requires a 10% commission fee on all sales; and a boutique, which requires a flat monthly rate and a 20% commission fee. The company also publishes one of the U.K.'s largest fashion magazines. Although the company is based in the U.K., it offers delivery and customer care services to Europe, the U.S. and nearly 190 other locations around world, with in-country return services available in the U.S. and Ireland. International sales account for 75% of the firm's business. Customers can purchase an ASOS Premier membership, which offers unlimited next day delivery service, free return collection and a complimentary ASOS magazine.

FINANCIAL DATA: *Note: Data for latest year may not have been available at press time.*

In U.S. $	2012	2011	2010	2009	2008	2007
Revenue	790,792,500	542,724,100	356,285,400	264,251,500	129,484,000	
R&D Expense						
Operating Income	49,846,620	25,414,600	32,450,870	22,263,940	11,123,180	
Operating Margin %	6.30%	4.68%	9.10%	8.42%	8.59%	
SGA Expense						
Net Income	35,595,140	17,333,440	23,294,460	15,991,370	8,073,175	
Operating Cash Flow	59,469,570	24,072,540	17,108,170	20,835,600	15,721,360	
Capital Expenditure	34,489,540	41,129,580	13,482,990	13,101,140	7,573,095	
EBITDA	62,746,450	33,294,460	37,758,430	25,219,690	12,661,770	
Return on Assets %	13.75%	9.78%	20.22%	20.46%	19.16%	
Return on Equity %	26.62%	18.45%	40.96%	48.05%	41.53%	
Debt to Equity						

CONTACT INFORMATION:

Phone: 44 2077561000 Fax: 44 2077561001
Toll-Free:
Address: Hampstead Rd., Greater London House, London, NW1 7FB United Kingdom

STOCK TICKER/OTHER:

Stock Ticker: ASOMF
Employees: 988
Parent Company:
Exchange: PINX
Fiscal Year Ends: 31-Aug

SALARIES/BONUSES:

Top Exec. Salary: $ Bonus: $
Second Exec. Salary: $ Bonus: $

OTHER THOUGHTS:

Estimated Female Officers or Directors: 3
Hot Spot for Advancement for Women/Minorities: Y

BackCountry.com

www.backcountry.com

NAIC Code: 454111

TYPES OF BUSINESS:

Sporting Goods-Online Retail
Outdoor Sports Equipment
Athletic Apparel
Ski Equipment

BRANDS/DIVISIONS/AFFILIATES:

HuckNroll.com
DogFunk.com
SteepAndCheap.com
Whiskey Militia
RealCyclist.com
Department of Goods

CONTACTS: *Note: Officers with more than one job title may be intentionally listed here more than once.*

Jill Layfield, CEO
John Bresee, Pres.
Scott Klossner, CFO
Dustin Robertson, Chief Mktg. Officer
CJ Singh, CTO
Mark Koppes, VP-Merch.
Brendan Quirk, VP-Customer Fulfillment
Jonathan Nielson, VP-Prod. Mgmt.
Jim Holland, Chmn.
Jeff Carter, VP-Fulfillment

GROWTH PLANS/SPECIAL FEATURES:

BackCountry.com is an online retailer of outdoor gear. The firm was founded by Jim Holland, a two-time Olympian and six-time national champion Nordic ski jumper, and John Bresee, former editor of Powder Magazine. The company offers a wide selection of goods for outdoor activities, including men's, women's and children's clothing, shoes, skis and skiing accessories, snowboards and snowboard accessories, hiking and camping gear, trail running products, snowshoeing products, technical climbing gear, kayaks and kayak accessories and outdoor travel products (backpacks, car racks, etc.). BackCountry's selection of goods encompasses nearly 50,000 products and over 250 brands, including The North Face, Oakley, Patagonia and Marmot. The company processes orders and ships from a 200,000 square foot distribution center in Salt Lake City, Utah and a 320,000 square foot distribution center in Christiansburg, Virginia. BackCountry has been expanding since its inception and has fostered four spin-off e-commerce storefronts, each focused on a specific niche of the outdoor sporting market. These Internet stores include Department of Goods, an overstock retailer; DogFunk.com, a snowboard gear retailer; SteepAndCheap.com, a discount store selling one item at a time; and WhiskeyMilitia.com, a youth-focused action sports discount store. The firm also maintains RealCyclist.com, a road cycling online retail specialty shop, and HuckNroll.com, an online mountain bike shop.

Employees are offered medical, dental and vision insurance; life and AD&D insurance; flexible spending accounts; a 401(k) plan; an employee assistance program; employee discounts; and discounted ski passes at various world-class resorts in the Park City/Salt Lake area.

FINANCIAL DATA: *Note: Data for latest year may not have been available at press time.*

In U.S. $	2012	2011	2010	2009	2008	2007
Revenue						
R&D Expense						
Operating Income						
Operating Margin %						
SGA Expense						
Net Income						
Operating Cash Flow						
Capital Expenditure						
EBITDA						
Return on Assets %						
Return on Equity %						
Debt to Equity						

CONTACT INFORMATION:

Phone: Fax: 801-973-4552
Toll-Free: 800-409-4502
Address: 2607 South 3200 W., Salt Lake City, UT 84119 United States

STOCK TICKER/OTHER:

Stock Ticker: Subsidiary Exchange:
Employees: Fiscal Year Ends: 31-Dec
Parent Company: LIBERTY INTERACTIVE CORP

SALARIES/BONUSES:

Top Exec. Salary: $ Bonus: $
Second Exec. Salary: $ Bonus: $

OTHER THOUGHTS:

Estimated Female Officers or Directors: 1
Hot Spot for Advancement for Women/Minorities:

Barney's New York Inc

NAIC Code: 452111

www.barneys.com

TYPES OF BUSINESS:

Department Stores
Apparel & Accessories
Home Furnishings
Cosmetics
Online Sales

BRANDS/DIVISIONS/AFFILIATES:

Barneys New York
Barneys CO-OP
Mrs John L Strong Stationary
Chantecaille Energy Spa
Barneys Greengrass
Completely Bare Hi-Tech Spa
Image Studio
Fred's at Barneys New York

CONTACTS:

Note: Officers with more than one job title may be intentionally listed here more than once.

Mark Lee, CEO
Kevin Dyson, Exec. VP-Store Oper.
Dawn Brown, VP-Public Rel.
Andy Watson, Chmn.

GROWTH PLANS/SPECIAL FEATURES:

Barneys New York, Inc., a wholly-owned subsidiary of Istithmar, is an upscale clothing retailer. It operates through three divisions, Barneys New York, with 9 stores; Barneys CO-OP, 13 stores; and Barneys Outlets, approximately 13 stores. The Barneys New York stores are located in upscale markets and offer a variety of luxury services including a bridal/gift registry; a concierge service; Fred's at Barneys New York, a full service restaurant located in the Madison Avenue store; the Image Studio, which is a head to toe image consultation service; personal shopping; Studio Services, which is similar to the firm's personal shopping service but is catered to stylists, celebrities and costume designers; resident makeup artists; Completely Bare Hi-Tech Spa; Mrs. John L. Strong Stationary, which offers customized high end stationary; Barney Greengrass, a full service restaurant in the Beverly Hills store; a Barneys credit card, in which customers can earn points toward purchases; and the Chantecaille Energy Spa. Barneys New York sells clothing, jewelry, handbags, shoes, cosmetics, hair products, home products and accessories for dogs. Many items are available through the company's web site, Barneys.com. Barneys CO-OP stores are located in fashion conscious areas on both coasts. The stores are centered on industrial themed edgy, urban styles at slightly lower price points than Barneys New York stores. Barneys CO-OP also sells through its web site, BarneysCoOp.com. Barneys Outlet is the company's discount clearinghouse. Outlets are located primarily on the East and West coasts, with one location in Hawaii. In addition to the permanent outlet locations, four Barneys New York warehouse sales are held annually, one each spring and fall season in New York and Santa Monica.

FINANCIAL DATA:

Note: Data for latest year may not have been available at press time.

In U.S. $	2012	2011	2010	2009	2008	2007
Revenue						
R&D Expense						
Operating Income						
Operating Margin %						
SGA Expense						
Net Income						
Operating Cash Flow						
Capital Expenditure						
EBITDA						
Return on Assets %						
Return on Equity %						
Debt to Equity						

CONTACT INFORMATION:

Phone: 212-339-7300 Fax: 212-450-8489
Toll-Free: 800-926-5393
Address: 575 5th Ave., Fl. 11, New York, NY 10017 United States

STOCK TICKER/OTHER:

Stock Ticker: Private
Exchange:
Employees:
Fiscal Year Ends: 31-Dec
Parent Company: ISTITHMAR PJSC

SALARIES/BONUSES:

Top Exec. Salary: $ Bonus: $
Second Exec. Salary: $ Bonus: $

OTHER THOUGHTS:

Estimated Female Officers or Directors: 1
Hot Spot for Advancement for Women/Minorities:

Bass Pro Shops Inc

www.basspro.com

NAIC Code: 451110

TYPES OF BUSINESS:

Sporting Goods, Retail
Sport Boats
Hunting & Fishing Equipment
Catalog & Online Sales
Outdoor Apparel
Resort Operations
Television Production

BRANDS/DIVISIONS/AFFILIATES:

Outdoor World
RedHead
American Rod & Gun
Big Cedar Lodge
Bass Pro Shop
Dogwood Canyon
Offshore Angler
White River Fly Shops

CONTACTS: *Note: Officers with more than one job title may be intentionally listed here more than once.*

James Hagale, CEO
James Hagale, Pres.
Stan Lippelman, VP-Mktg.
Leslie Weber, CIO
Larry L. Whiteley, Mgr.-Comm.
John L. Morris, Chmn.

GROWTH PLANS/SPECIAL FEATURES:

Bass Pro Shops, Inc. is a leader in sporting goods retail. The company markets its products through 57 sports superstores in 26 states across the U.S., 2 stores in Canada, a mail-order catalog and through its Internet sites. The firm is dedicated to providing outdoor recreational products, including specialty apparel, and also aims to inspire environmental conservation among its customers. The sporting goods superstores operate under the Bass Pro Shop and Outdoor World brand names. Products include boats and campers as well as fishing, hunting, camping, automobile and marine supplies. Many of these stores have a variety of unique features and attractions to draw more customers, including restaurants, snack bars, archery ranges, indoor fish tanks, waterfalls and video arcades. Aside from its stores, the company sells goods over the Internet and through catalogs/sales flyers under the Bass Pro Shops, RedHead, Offshore Angler and White River Fly Shops brand names. Its wholesale operations consist of Tracker Marine, a leader in sport boat manufacturing, and American Rod & Gun, one of the largest wholesale hunting and fishing distributors in the country. In addition to offering a variety of hunting and fishing trips and contests, Bass Pro runs the Wonders of Wildlife facility, a museum and conservation education center near its corporate headquarters, and Big Cedar Lodge, an outdoors-themed vacation spot in Missouri located near the company's own nature park, Dogwood Canyon. The company also produces a weekly television program on The Outdoor Channel and an international radio show.

FINANCIAL DATA: *Note: Data for latest year may not have been available at press time.*

In U.S. $	2012	2011	2010	2009	2008	2007
Revenue	4,300,000,000	4,000,000,000	3,830,000,000	2,800,000,000	2,650,000,000	3,000,000,000
R&D Expense						
Operating Income						
Operating Margin %						
SGA Expense						
Net Income						
Operating Cash Flow						
Capital Expenditure						
EBITDA						
Return on Assets %						
Return on Equity %						
Debt to Equity						

CONTACT INFORMATION:

Phone: 417-873-5000 Fax: 417-873-5060
Toll-Free: 800-227-7776
Address: 2500 E. Kearney St., Springfield, MO 65898 United States

STOCK TICKER/OTHER:

Stock Ticker: Private
Employees: 19,000
Parent Company:
Exchange:
Fiscal Year Ends: 31-Dec

SALARIES/BONUSES:

Top Exec. Salary: $ Bonus: $
Second Exec. Salary: $ Bonus: $

OTHER THOUGHTS:

Estimated Female Officers or Directors: 1
Hot Spot for Advancement for Women/Minorities:

Beaulieu of America LLC

NAIC Code: 314110

www.blissflooring.com

TYPES OF BUSINESS:

Carpet and Rug Mills
Yarn Extrusion
Nylon, Polypropylene & Polyester Products

BRANDS/DIVISIONS/AFFILIATES:

Access Beaulieu
Beaulieu Property Management Solutions
Hollytex Main Street Commercial
FashionSmart
BOLYU Contract
Cambridge Commercial
Aqua Hospitality
Bliss by Beaulieu

CONTACTS: *Note: Officers with more than one job title may be intentionally listed here more than once.*

Ralph Boe, CEO
Karel Vercruyssen, COO
Karel Vercruyssen, Pres.
Del Land, CFO
Gary R. Fisher, Exec. VP-Mktg.
Kathy Fisher, VP-National Accounts
Patricia Flavin, Sr. VP-Mktg.
Carl M. Bouckaert, Chmn.

GROWTH PLANS/SPECIAL FEATURES:

Beaulieu of America, LLC is one of the largest carpet manufacturers in North America. The company operates in four divisions: commercial, residential, fibers & fabrics and international. The commercial division operates in three sub-segments: Beaulieu Property Management Solutions, for the multi-family housing market; Hollytex Main Street Commercial, serving the small business market; and FashionSmart, an exclusive partnership with Armstrong Flooring Products for new home builders. Commercial brands include BOLYU Contract, Cambridge Commercial and Aqua Hospitality. The segment has directed a significant amount of its resources to research and development, which has resulted in innovations with odor-fighting features, stain-proofing and the increased incorporation of recycled materials into carpet production. Bliss by Beaulieu is the company's residential carpet brand designed by women for women in terms of features, styling and overall shopping experience. Many of the firm's carpets feature stain and wear resistant coatings, including the Hollytex Ultimate Performance carpet, which features a self-renewing odor barrier, Magic Fresh. The fibers & fabrics division of Beaulieu focuses on developing new carpet and flooring technologies, especially eco-friendly products such as Nexterra modular carpet backing and Green Smart carpet fiber. Other developments include the Silver Release antimicrobial carpet coating. The international division oversees operations outside of the U.S. In addition, Beaulieu offers Access Beaulieu, which offers customers the ability to view and purchase materials online and track orders.

Employees are offered paid vacations; an employee assistant program; a pharmacy benefit plan; hearing and vision benefits; short- and long-term disability benefits; and life, medical and dental insurance plans.

FINANCIAL DATA: *Note: Data for latest year may not have been available at press time.*

In U.S. $	2012	2011	2010	2009	2008	2007
Revenue						
R&D Expense						
Operating Income						
Operating Margin %						
SGA Expense						
Net Income						
Operating Cash Flow						
Capital Expenditure						
EBITDA						
Return on Assets %						
Return on Equity %						
Debt to Equity						

CONTACT INFORMATION:

Phone: 706-695-4624 Fax: 706-695-6237
Toll-Free: 800-227-7211
Address: 1502 Coronet Dr., Dalton, GA 30722 United States

STOCK TICKER/OTHER:

Stock Ticker: Private
Exchange:
Employees:
Fiscal Year Ends: 31-Dec
Parent Company:

SALARIES/BONUSES:

Top Exec. Salary: $ Bonus: $
Second Exec. Salary: $ Bonus: $

OTHER THOUGHTS:

Estimated Female Officers or Directors: 3
Hot Spot for Advancement for Women/Minorities: Y

Bebe Stores Inc

www.bebe.com

NAIC Code: 448120

TYPES OF BUSINESS:

Young Women's Apparel, Retail
Accessories
Shoes
Online Sales

BRANDS/DIVISIONS/AFFILIATES:

bebe
BEBE SPORT
Bebe.com
2b bebe
bbsp
bebe studio inc
bebe stores Canada
bebe Japan KK

CONTACTS: *Note: Officers with more than one job title may be intentionally listed here more than once.*

Manny Mashouf, CEO
Walter Parks, COO
Walter Parks, CFO
Hamid Mashouf, CIO
Lawrence Smith, General Counsel/Sr. VP
Ben Baum, Chief Digital Officer
Liyuan Woo, Principal Acct. Officer/Controller/VP
Susan Powers, Sr. VP-Stores
Manny Mashouf, Chmn.

GROWTH PLANS/SPECIAL FEATURES:

Bebe Stores, Inc. designs, develops and produces a line of contemporary women's apparel and accessories, marketed under the bebe, BEBE SPORT, bbsp and 2b bebe brand names. The company operates approximately 252 retail stores located in 36 states, Puerto Rico, the U.S. Virgin Islands, Canada and Japan. These stores include 205 bebe stores, 47 2b bebe stores and 18 outlet stores operated under the 2b bebe banner. The bebe line of stores averages 4,000 square feet while the 2b bebe stores average 4,500 square feet. Its stores are primarily located in regional shopping malls and freestanding street locations. In addition, the firm operates an online store at Bebe.com and has 114 international licensees in 19 countries, such as the UAE, Israel, Russia, South Korea, Turkey, Mexico and numerous countries throughout South America and South East Asia. Bebe seeks to produce a distinctive fashion that fulfills the needs of the fashion conscious woman, be it every day wear to work, weekend or party attire. With offerings including suits, tops, skirts, dresses, pants, active wear, outerwear and handbags, shoes, jewelry and other accessories. Most of its merchandise is designed and developed in-house and manufactured in conjunction with third parties. Bebe maintains an eyewear license agreement with Altair Eyewear, Inc. to manufacture and distribute bebe-branded products to be sold at bebe stores and other retailers. The company has pursued growth through new store openings and the introduction of new product categories, including denim, leather, lingerie, swimwear and footwear.

Bebe offers its employees medical, dental and vision coverage; flexible spending accounts; life and disability insurance; a 401(k) plan; a stock purchase plan; an employee assistance program; and discounts on bebe merchandise.

FINANCIAL DATA: *Note: Data for latest year may not have been available at press time.*

In U.S. $	2012	2011	2010	2009	2008	2007
Revenue	530,831,000	493,274,000	508,968,000	602,998,000	687,622,000	670,912,000
R&D Expense						
Operating Income	19,407,000	5,889,000	-11,105,000	9,884,000	79,369,000	105,257,000
Operating Margin %	3.65%	1.19%	-2.18%	1.63%	11.54%	15.68%
SGA Expense	191,666,000	185,921,000	209,012,000	233,309,000	236,044,000	216,560,000
Net Income	11,721,000	-1,779,000	-5,165,000	12,635,000	63,080,000	77,278,000
Operating Cash Flow	32,740,000	21,860,000	37,556,000	40,551,000	95,233,000	100,471,000
Capital Expenditure	42,093,000	14,723,000	14,190,000	29,418,000	41,784,000	43,243,000
EBITDA	39,686,000	27,961,000	15,213,000	36,745,000	102,044,000	125,009,000
Return on Assets %	2.62%	-.35%	-.91%	2.16%	10.47%	13.94%
Return on Equity %	3.26%	-.49%	-1.24%	2.67%	12.73%	16.93%
Debt to Equity						

CONTACT INFORMATION:

Phone: 415 715-3900 Fax:
Toll-Free: 877-232-3777
Address: 400 Valley Dr., Brisbane, CA 94005 United States

STOCK TICKER/OTHER:

Stock Ticker: BEBE
Employees: 3,294
Parent Company:
Exchange: NAS
Fiscal Year Ends: 31-Jul

SALARIES/BONUSES:

Top Exec. Salary: $600,000 Bonus: $
Second Exec. Salary: $540,384 Bonus: $

OTHER THOUGHTS:

Estimated Female Officers or Directors: 4
Hot Spot for Advancement for Women/Minorities: Y

Bed Bath & Beyond Inc

www.bedbathandbeyond.com

NAIC Code: 442299

TYPES OF BUSINESS:

Linens & Housewares, Retail
Small Appliances
Home Accessories
Health & Beauty Care
Baby & Toddler Merchandise

BRANDS/DIVISIONS/AFFILIATES:

Bed Bath & Beyond Superstores
Harmon
Christmas Tree Shops
buybuy BABY
Home & More
Harmon Face Values

CONTACTS: *Note: Officers with more than one job title may be intentionally listed here more than once.*

Steven H. Temares, CEO
Arthur Stark, Pres.
Eugene A. Castagna, CFO
Scott Hames, VP-Mktg. & Analytics
Arthur Stark, Chief Merch. Officer
Allan N. Rauch, General Counsel/VP-Legal
Richard McMahon, VP-Corp. Oper.
Richard McMahon, Chief Strategy Officer
Eugene A. Castagna, Treas.
Matthew Fiorilli, Sr. VP-Stores
G. William Waltzinger, Jr., VP-Corp. Dev./Pres., Harmon Stores, Inc.
Leonard Feinstein, Co-Chmn.
Warren Eisenberg, Co-Chmn.

GROWTH PLANS/SPECIAL FEATURES:

Bed Bath & Beyond, Inc. (BBB) is one of the nation's largest operators of domestic superstores, with approximately 1,173 stores in 50 states, as well as Washington D.C., Canada and Puerto Rico. BBB operates four retail entities: Bed Bath & Beyond; Harmon and Harmon Face Values; buybuy BABY; and Christmas Tree Shops. Bed Bath & Beyond stores offer a full line of domestic merchandise and home furnishings. The line includes items such as bed linens, bath accessories and kitchen tiles, while its home furnishings line includes a variety of cookware, dinnerware and glassware. Harmon and Harmon Face Values stores sell beauty care products in 45 stores in four states. Buybuy BABY is a retailer of infant and toddler merchandise, with 63 stores in 28 states including Florida, Georgia, Illinois, Indiana, Maryland, Michigan, New Jersey, New York, Ohio, Texas and Virginia. Christmas Tree Shops offer gifts, household items and furnishings in 75 stores in 22 states. Additionally, the company (through a joint venture) operates two stores in Mexico under the name Home & More. BBB relies on paid advertising and uses circulars and mailing pieces as its primary vehicles. In addition, the company only has three central distribution centers, since the majority of merchandise is shipped to each store from the firm's vendors.

FINANCIAL DATA: *Note: Data for latest year may not have been available at press time.*

In U.S. $	2012	2011	2010	2009	2008	2007
Revenue	9,499,890,000	8,758,503,000	7,828,793,000	7,208,340,000	7,048,942,000	6,617,429,000
R&D Expense						
Operating Income	1,568,369,000	1,288,458,000	980,687,000	673,896,000	838,022,000	889,401,000
Operating Margin %	16.50%	14.71%	12.52%	9.34%	11.88%	13.44%
SGA Expense	2,362,564,000	2,334,471,000	2,227,432,000	2,199,340,000	2,087,209,000	1,946,001,000
Net Income	989,537,000	791,333,000	600,033,000	425,123,000	562,808,000	594,244,000
Operating Cash Flow	1,225,284,000	987,407,000	905,407,000	583,970,000	614,536,000	613,632,000
Capital Expenditure	243,374,000	183,474,000	153,680,000	215,859,000	358,210,000	317,501,000
EBITDA	1,752,242,000	1,472,278,000	1,164,919,000	849,497,000	997,330,000	1,025,888,000
Return on Assets %	17.40%	14.65%	12.73%	10.48%	14.42%	16.18%
Return on Equity %	25.19%	20.86%	18.03%	15.28%	21.60%	24.19%
Debt to Equity						

CONTACT INFORMATION:

Phone: 908 688-0888 Fax: 908 810-8813
Toll-Free: 800-462-3966
Address: 650 Liberty Ave., Union, NJ 07083 United States

STOCK TICKER/OTHER:

Stock Ticker: BBBY Exchange: NAS
Employees: 48,000 Fiscal Year Ends: 28-Feb
Parent Company:

SALARIES/BONUSES:

Top Exec. Salary: $2,894,231 Bonus: $
Second Exec. Salary: $1,353,558 Bonus: $

OTHER THOUGHTS:

Estimated Female Officers or Directors:
Hot Spot for Advancement for Women/Minorities:

Belk Inc

www.belk.com

NAIC Code: 452111

TYPES OF BUSINESS:

Department Stores
Women's Accessories & Cosmetics
Jewelry

BRANDS/DIVISIONS/AFFILIATES:

Belk Merchandising Company LLC
Belk Store Services Inc
Belk Administration Company
Belk International Inc
Carmen! Carmen! Prestige Salon and Spa at Belk
Richard Joseph Studio Salon/Spa at Belk
Belk Salon and Spa

CONTACTS: *Note: Officers with more than one job title may be intentionally listed here more than once.*

Thomas M. Belk, Jr., CEO
John R. Belk, COO/Co-Pres.
Kathryn Bufano, Co-Pres.
Brian T. Marley, CFO/Exec. VP
Kathryn Bufano, Chief Merch. Officer
Ralph A. Pitts, General Counsel/Exec. VP/Sec.
H. W. McKay Belk, Vice Chmn.
Thomas M. Belk, Jr., Chmn.

GROWTH PLANS/SPECIAL FEATURES:

Belk, Inc. is the largest privately owned department store businesses in the U.S., with 301 stores in 16 states, primarily in the southern U.S. The company has operated department stores since 1888. Belk stores offer national brands of apparel, shoes and accessories for men, women and children as well as cosmetics, home furnishings, house-wares, gifts and other types of merchandise. Its aim is to provide customers with the convenience of one stop shopping, all while maintaining a uniquely southern style and sensibility. Larger Belk stores include amenities such as hair salons, spas, restaurants and optical centers. The company operates 23 hair styling salons in various locations, 17 of which offer spa services. The spas offer massage therapy, full skincare, nail and pedicure services and other specialized body treatments. Eight of the salons and spas operate under the name Carmen! Carmen! Prestige Salon and Spa at Belk, two under the name Richard Joseph Studio Salon/Spa at Belk and the remaining operating under the name Belk Salon and Spa. Although the firm operates 93 stores with more than 100,000 square feet, most stores range in size from 60,000 to 100,000 square feet. Altogether, the company has approximately 22.8 million square feet of selling space. Belk stores are typically anchor tenants in regional malls and shopping centers catering to medium and smaller markets. The firm's direct subsidiaries include BelkStore Services, Inc. and Belk International, Inc. These firms operate subsidiaries Belk Administration Company and Belk Merchandising Company, LLC respectively. In November 2012, in an effort to promote breast cancer awareness, Belk launched a mobile mammography center. This center will be of no cost to women whose Belk store is serviced.

The firm offers employees life, disability, auto, home, legal, medical, dental and vision insurance; an employee assistance program; merchandise discounts; flexible spending accounts; and a 401(k).

FINANCIAL DATA: *Note: Data for latest year may not have been available at press time.*

In U.S. $	2012	2011	2010	2009	2008	2007
Revenue	3,699,592,000	3,513,275,000	3,346,252,000	3,499,423,000	3,824,803,000	
R&D Expense						
Operating Income	300,910,000	245,981,000	147,441,000	-232,643,000	198,117,000	
Operating Margin %	8.13%	7.00%	4.40%	-6.64%	5.17%	
SGA Expense	938,008,000	914,078,000	886,263,000	947,602,000	982,425,000	
Net Income	183,148,000	127,628,000	67,136,000	-212,965,000	95,740,000	
Operating Cash Flow	251,932,000	189,241,000	387,395,000	265,301,000	215,010,000	
Capital Expenditure	143,844,000	82,409,000	42,326,000	129,282,000	202,668,000	
EBITDA	423,077,000	386,789,000	304,270,000	-62,502,000	362,473,000	
Return on Assets %	7.46%	5.13%	2.62%	-7.90%	3.35%	
Return on Equity %	15.31%	11.34%	6.31%	-17.59%	7.05%	
Debt to Equity	0.33	0.46	0.62	0.66	0.51	

CONTACT INFORMATION:

Phone: 704 357-1000 Fax: 704 357-1876
Toll-Free:
Address: 2801 W. Tyvola Rd., Charlotte, NC 28217 United States

STOCK TICKER/OTHER:

Stock Ticker: BLKIB Exchange: OTC
Employees: 23,000 Fiscal Year Ends: 31-Jan
Parent Company:

SALARIES/BONUSES:

Top Exec. Salary: $1,050,192 Bonus: $
Second Exec. Salary: $867,248 Bonus: $

OTHER THOUGHTS:

Estimated Female Officers or Directors: 4
Hot Spot for Advancement for Women/Minorities: Y

Belle International Holdings Limited

www.belleintl.com

NAIC Code: 316210

TYPES OF BUSINESS:

Shoe Manufacturing-Women's
Shoe Retail Sales
Men's Shoes
Accessories
Handbags

BRANDS/DIVISIONS/AFFILIATES:

Belle
Staccato
Teenmix
Jipi Japa
Basto
Tata
Mirabel

CONTACTS: *Note: Officers with more than one job title may be intentionally listed here more than once.*

Sheng Baijiao, CEO
Kam Kwan Leung, CFO
Kam Kwan Leung, Corp. Sec.
Sheng Fang, Sr. VP-Oper., Footwear Retail Bus.
Tang King Loy, Sr. VP-Bus. Dev.
Tang Yiu, Chmn.

GROWTH PLANS/SPECIAL FEATURES:

Belle International Holdings Limited (Belle), headquartered in Hong Kong, is one of the largest retailers of women's footwear in China. It currently operates a retail network of approximately 13,150 stores in mainland China as well as nearly 200 stores in Hong Kong and Macau. Belle directly manages its retail network and has a vertically integrated business model that encompasses design and development of products, marketing, production, distribution and retailing. The production side of the company's business includes five primary manufacturing facilities with combined annual production capacity of more than 40 million pairs of shoes. Belle's manufacturing facilities are located in Shenzhen, Dongguang, Jianhu, Suzhou and Zigui. The company typically introduces its collections on a seasonal basis, offering new product lines several times each year. The firm manages several of its own brands (including Belle, Staccato, Teenmix, Jipi Japa, Mirabell, Tata and Basto) as well as licensed brands (such as Bata, Geox, Clarks, Mephisto, BCBG and Merrell). Additionally, through its sportswear business, Belle distributes merchandise from international brands such as Nike, Adidas, PUMA and Converse. Besides women's footwear, the firm sells bags, accessories and men's footwear. The company's products are available all over China and are also sold in many other locations worldwide, including Japan, Southeast Asia, the Middle East, Africa, the U.S. and Europe.

FINANCIAL DATA: *Note: Data for latest year may not have been available at press time.*

In U.S. $	2012	2011	2010	2009	2008	2007
Revenue	5,284,242,000	4,654,761,000	3,812,278,000	3,177,980,000	2,871,490,000	1,877,017,000
R&D Expense						
Operating Income	868,870,900	846,663,700	637,232,500	454,170,000	366,547,600	282,217,800
Operating Margin %	16.44%	18.18%	16.71%	14.29%	12.76%	15.03%
SGA Expense	2,165,897,000	1,834,139,000	1,496,489,000	1,241,936,000	1,131,511,000	673,225,900
Net Income	695,543,800	681,609,800	550,718,200	406,813,900	322,119,700	318,193,100
Operating Cash Flow	706,656,200	487,473,400	425,983,600	451,103,700	379,612,800	47,070,260
Capital Expenditure		272,969,400	199,180,200	130,161,800	233,008,500	143,529,100
EBITDA	1,065,146,000	1,005,486,000	775,098,000	573,719,000	366,547,600	282,217,800
Return on Assets %	16.03%	18.29%	17.35%	14.21%	13.15%	22.00%
Return on Equity %	20.64%	23.18%	21.11%	17.62%	15.82%	27.09%
Debt to Equity						

CONTACT INFORMATION:

Phone: Fax:
Toll-Free:
Address: 918 Cheung Sha Wan Rd., 9/F Belle Tower, Tsuen Wan, Hong Kong

STOCK TICKER/OTHER:

Stock Ticker: BELLF
Exchange: PINX
Employees: 87,619
Fiscal Year Ends: 31-Dec
Parent Company:

SALARIES/BONUSES:

Top Exec. Salary: $ Bonus: $
Second Exec. Salary: $ Bonus: $

OTHER THOUGHTS:

Estimated Female Officers or Directors: 1
Hot Spot for Advancement for Women/Minorities:

Benetton Group SPA

www.benettongroup.com

NAIC Code: 315210

TYPES OF BUSINESS:

Apparel Manufacturing-Teenage & Young Adult
Retail Stores
Casual Clothing-Men's, Women's & Children's
Apparel Manufacturing
Arts Research Center
Magazine Publishing

BRANDS/DIVISIONS/AFFILIATES:

United Colors of Benetton
Sisley
Benetton International SA
COLORS
Fabrica
Undercolors of Benetton
Bencom Srl
Benetton USA Corp

CONTACTS: *Note: Officers with more than one job title may be intentionally listed here more than once.*

Biagio Chiarolanza, Co-CEO
Alberto Nathansohn, CFO
Lorenzo Zago, Head-Admin.
Andrea Pezzangora, Head-Legal & Corp. Affairs
Pietro Arnaboldi, Mgr.-Investor Rel.
Carlo Benetton, Deputy Chmn.
Franco Furno, Co-CEO
Alessandro Benetton, Chmn.

GROWTH PLANS/SPECIAL FEATURES:

Benetton Group SPA is a worldwide apparel retailer and manufacturer based in Italy. It manufactures over 150 million garments yearly at 21 factories and sells these in over 6,500 retail stores in 120 countries. Traditional western economies, U.S., U.K., Europe, Australia and Japan, account for approximately 74% of the company's sales, with developing and fast growth countries located in Asia, South America and Africa, accounting for 26% of the company's sales. Its business activities can be divided into two groups: wholesale, representing 76% of apparel sales, and retail, 24%. The company markets apparel for men, women and children under various brands: United Colors of Benetton, Undercolors of Benetton, Sisley and Playlife. Benetton's stores are characterized by their large size and prominent positions in historic and commercial centers. The company recently introduced the megastore concept to its stores. Benetton also runs a series of communications projects, including Fabrica, an arts research center; COLORS magazine, which features stories reflecting a global consciousness; and support for sports, cinema, music and other media. Of the company's 21 factories, 12 are located in Italy, two in Croatia, two in Tunisia, one in India, three in Romania and one in Hungary. The company operates numerous subsidiaries, including Bencom Srl; Benind SpA; Benetton International SA; Benetton Holding International NV SA; and Benetton USA, Corp.

FINANCIAL DATA: *Note: Data for latest year may not have been available at press time.*

In U.S. $	2012	2011	2010	2009	2008	2007
Revenue		2,695,800,000	2,899,290,000	2,748,450,000	2,889,080,000	2,780,470,000
R&D Expense						
Operating Income						
Operating Margin %						
SGA Expense						
Net Income		96,800,000	144,050,000	158,280,000	213,150,000	203,650,000
Operating Cash Flow						
Capital Expenditure						
EBITDA						
Return on Assets %						
Return on Equity %						
Debt to Equity						

CONTACT INFORMATION:

Phone: 39-0422-519036 Fax: 39-0422-519930
Toll-Free:
Address: Villa Minelli, Via Villa Minelli, 1, Ponzano, 31050 Italy

STOCK TICKER/OTHER:

Stock Ticker: Private Exchange:
Employees: 9,469 Fiscal Year Ends: 31-Dec
Parent Company:

SALARIES/BONUSES:

Top Exec. Salary: $ Bonus: $
Second Exec. Salary: $ Bonus: $

OTHER THOUGHTS:

Estimated Female Officers or Directors: 1
Hot Spot for Advancement for Women/Minorities:

Bernard Chaus Inc

www.bernardchaus.com

NAIC Code: 315240

TYPES OF BUSINESS:

Women's Careerwear
Casual Sportswear

BRANDS/DIVISIONS/AFFILIATES:

Josephine Chaus
Chaus
Cynthia Steffe
Cynthia Cynthia Steffe
Vince Camuto

CONTACTS: *Note: Officers with more than one job title may be intentionally listed here more than once.*

Josephine Chaus, CEO
William P. Runge, CFO
Biz Antrim, VP-Sales
Barton Heminover, Assistant Sec.
Barton Heminover, VP-Finance
Josephine Chaus, Chmn.

GROWTH PLANS/SPECIAL FEATURES:

Bernard Chaus, Inc. designs, arranges for the manufacture of and markets an extensive range of women's career and casual sportswear principally under the Josephine Chaus, Chaus, Cynthia Steffe and Cynthia Cynthia Steffe trademarks and other private label brand names. The firm operates in one segment, women's career and casual sportswear. Products are manufactured in both women and petite sizes and consist of accessories, suits and separates, including jackets, pants, skirts, knit tops, sweaters and dresses. The company's products are sold nationwide through Dillard's, Bon-Ton, Stage, Lord & Taylor, Nordstrom, Boscov's and Peebles. The Chaus and Josephine Chaus product lines are in the opening price points of the better career and casual clothing categories. The Cynthia Steffe and Cynthia Cynthia Steffe product lines are upscale contemporary women's apparel lines sold through department stores and specialty stores. The firm's private label product lines are sold to various customers according to their specifications. The company also sells, manufactures and distributes women's sportswear and ready-to-wear apparel under the Vince Camuto brand to specialty retailers and off-price channels in the U.S., Canada and Mexico.

FINANCIAL DATA: *Note: Data for latest year may not have been available at press time.*

In U.S. $	2012	2011	2010	2009	2008	2007
Revenue		84,818,000	100,153,000	112,096,000	118,028,000	146,772,000
R&D Expense						
Operating Income						
Operating Margin %						
SGA Expense						
Net Income		-7,857,000	-6,033,000	-9,577,000	-7,678,000	522,000
Operating Cash Flow						
Capital Expenditure						
EBITDA						
Return on Assets %						
Return on Equity %						
Debt to Equity						

CONTACT INFORMATION:

Phone: 212-354-1280 Fax: 212-863-4269
Toll-Free:
Address: 530 7th Ave., New York, NY 10018 United States

STOCK TICKER/OTHER:

Stock Ticker: Private Exchange:
Employees: 76 Fiscal Year Ends: 30-Jun
Parent Company:

SALARIES/BONUSES:

Top Exec. Salary: $ Bonus: $
Second Exec. Salary: $ Bonus: $

OTHER THOUGHTS:

Estimated Female Officers or Directors: 1
Hot Spot for Advancement for Women/Minorities: Y

Big 5 Sporting Goods Corporation

www.big5sportinggoods.com

NAIC Code: 451110

TYPES OF BUSINESS:

Sporting Goods Stores

BRANDS/DIVISIONS/AFFILIATES:

Court Casuals
Golden Bear
Big 5 Services Corp
Pacifica
Rugged Exposure
Triple Nickel

CONTACTS:

Note: Officers with more than one job title may be intentionally listed here more than once.

Steven G. Miller, CEO
Steven G. Miller, Pres.
Barry Emerson, CFO
Jeffrey Fraley, Sr. VP-Human Resources
Gary Meade, General Counsel/Sr. VP/Corp. Sec.
Shane Starr, Sr. VP-Oper.
Barry Emerson, Sr. VP/Treas.
Steven G. Miller, Chmn.
Boyd O. Clark, Sr. VP-Buying

GROWTH PLANS/SPECIAL FEATURES:

Big 5 Sporting Goods Corporation is one of the leading sporting goods retailers in the U.S., with 414 locations in 12 states. The firm's in-store products include athletic shoes; apparel and accessories for team sports; fitness equipment; camping, hunting and fishing gear; tennis and golf equipment; winter sports gear; and in-line skates. Annually, its soft goods, such as apparel and footwear, generate 45.2% of total sales, while hard goods, including sporting equipment, generate 54.8%. Big 5 sells merchandise under company owned trademark brands such as Court Casuals, Golden Bear, Pacifica, Rugged Exposure and Triple Nickel as well as national third party brands, such as adidas, Coleman, Easton, New Balance, Nike, Reebok, Spalding, Under Armour and Wilson. Stores average 11,000 square feet, which is significantly smaller than the average superstore format, allowing stores to cater to smaller markets. Store locations are normally freestanding or part of a multi-store shopping center and are chosen in collaboration with local real estate companies, mostly in markets close to those in which the company already has a presence. Big 5 owns a 953,000-square-foot distribution center in Riverside, California from which it operates all of its distribution activities. Through wholly-owned subsidiary Big 5 Services Corp., the firm manages its gift card operations.

FINANCIAL DATA:

Note: Data for latest year may not have been available at press time.

In U.S. $	2012	2011	2010	2009	2008	2007
Revenue	940,490,000	902,134,000	896,813,000	895,542,000	864,650,000	898,292,000
R&D Expense						
Operating Income	25,972,000	19,167,000	34,224,000	37,682,000	27,602,000	52,962,000
Operating Margin %	2.76%	2.12%	3.81%	4.20%	3.19%	5.89%
SGA Expense	276,797,000	272,436,000	263,488,000	260,068,000	257,883,000	256,180,000
Net Income	14,915,000	11,673,000	20,562,000	21,811,000	13,904,000	28,091,000
Operating Cash Flow	39,604,000	2,218,000	29,867,000	54,087,000	39,503,000	24,664,000
Capital Expenditure	12,901,000	12,990,000	15,628,000	5,764,000	20,447,000	20,769,000
EBITDA	44,867,000	37,711,000	52,851,000	57,082,000	46,789,000	70,698,000
Return on Assets %	3.72%	2.96%	5.42%	5.78%	3.50%	7.31%
Return on Equity %	9.29%	7.59%	14.55%	17.90%	12.58%	26.80%
Debt to Equity	0.3	0.42	0.33	0.43	0.88	0.96

CONTACT INFORMATION:

Phone: 310 536-0611 Fax: 310 297-7585
Toll-Free:
Address: 2525 E. El Segundo Blvd., El Segundo, CA 90245 United States

STOCK TICKER/OTHER:

Stock Ticker: BGFV
Exchange: NAS
Employees: 8,800
Fiscal Year Ends: 31-Jan
Parent Company:

SALARIES/BONUSES:

Top Exec. Salary: $485,000 Bonus: $200,000
Second Exec. Salary: $333,000 Bonus: $110,000

OTHER THOUGHTS:

Estimated Female Officers or Directors: 2
Hot Spot for Advancement for Women/Minorities:

Bill Blass Group LLC

www.billblass.com

NAIC Code: 315240

TYPES OF BUSINESS:

Men's & Women's Apparel
Licensing
Luxury Clothing
Design & Licensing

GROWTH PLANS/SPECIAL FEATURES:

Bill Blass Group, LLC, formerly Bill Blass, Ltd., designs and manufactures high-end men's and women's clothing. The firm is known for timeless, tailored American designs made out of fine fabrics, such as worsteds, gabardines and wool crepes from Europe. Most of the company's sales are generated from licensed merchandise featuring the Bill Blass brand, including items such as furniture, swimwear, candies, shoes, jeans, fragrances, eyewear and accessories. Its signature collections of designer fashion apparel are retailed through high-end specialty stores and high-end department stores such as Saks, Neiman Marcus and Bergdorf, and its advertising campaigns appear in several fashion magazines including Vogue, InStyle, Page Six, Glamour, W and Essence. In recent years, NexCen Brands, Inc. sold Bill Blass to Peacock International Holdings, LLC for $10 million. Upon completion of the acquisition, Peacock International changed its name to Bill Blass Group, LLC and has since sought to reinvigorate the designer's brand.

BRANDS/DIVISIONS/AFFILIATES:

Peacock International Holdings LLC
Bill Blass Ltd

CONTACTS: *Note: Officers with more than one job title may be intentionally listed here more than once.*

Cin Kim, CEO
Peter Kim, COO
Scott Patti, Pres.

FINANCIAL DATA: *Note: Data for latest year may not have been available at press time.*

In U.S. $	2012	2011	2010	2009	2008	2007
Revenue						
R&D Expense						
Operating Income						
Operating Margin %						
SGA Expense						
Net Income						
Operating Cash Flow						
Capital Expenditure						
EBITDA						
Return on Assets %						
Return on Equity %						
Debt to Equity						

CONTACT INFORMATION:

Phone: 212-689-8957 Fax: 212-213-5485
Toll-Free:
Address: 236 5th Ave., 2nd Fl., New York, NY 10018 United States

STOCK TICKER/OTHER:

Stock Ticker: Private Exchange:
Employees: Fiscal Year Ends: 31-Dec
Parent Company:

SALARIES/BONUSES:

Top Exec. Salary: $ Bonus: $
Second Exec. Salary: $ Bonus: $

OTHER THOUGHTS:

Estimated Female Officers or Directors: 1
Hot Spot for Advancement for Women/Minorities:

Billabong International Ltd

www.billabongbiz.com

NAIC Code: 315210

TYPES OF BUSINESS:

Surfing, Skateboarding & Snowboarding Apparel
Surfboards, Skateboards & Paraphernalia
Retail Sales
Surfwear Apparel
Surfing Accessories

BRANDS/DIVISIONS/AFFILIATES:

Honolua Surf Company
Element
Von Zipper
RVCA
Nixon
Sector 9
Swell.com
Surfection

CONTACTS: *Note: Officers with more than one job title may be intentionally listed here more than once.*

Launa Inman, CEO/Managing Dir.
Peter Myers, CFO
Maria Manning, Corp. Sec.
Paul Naude, Gen. Mgr.-North America
Shannan North, Gen. Mgr.-Asia Pacific
Colin Haggerty, Group Exec.-Global Retail
Ian Pollard, Chmn.
Franco Fogliato, Gen. Mgr.-Europe

GROWTH PLANS/SPECIAL FEATURES:

Billabong International, Ltd. is a leading marketer, distributor, wholesaler and retailer of surfing, skateboarding and snowboarding apparel, accessories, eyewear and wetsuits. It markets products under the Billabong, Element, Von Zipper, Honolua Surf Company, Kustom, Palmers Surf, Nixon, Xcel, Tigerlily, Sector 9, DaKine and RVCA brand names. Billabong began in 1973 when two surfing devotees, Gordon and Rena Merchant, began producing surfing supplies that were formerly completely unavailable to surfers, including triple-stitched board shorts. The firm began exporting products internationally in the 1980s, and now retails its products in more than 100 countries and approximately 11,000 stores across the globe. These venues include the company's wholly-owned stores in Australia, North America, Europe, Japan, New Zealand, South Africa and Brazil, which account for the majority of revenues, as well as licensed surf and sports dealers worldwide. Products include a wide variety of board shorts, basic tees, slim fit tees, polos, wovens, fleece, walkshorts, pants, surf tees, wetsuits, rashvests, eyewear, bags and accessories. Billabong is committed to sponsoring junior athlete development, and also sponsors professional contests held on the Gold Coast of Australia and the Jeffreys Bay in South Africa. In recent years, the company has acquired Swell.com, a U.S.-based online board sports retailer, and Surfection, a Sydney-based former joint venture held with retailer Chris Athas. Additionally, the firm and Trilantic Capital Partners established a joint venture for the purpose of accelerating the global growth of the brand Nixon.

FINANCIAL DATA: *Note: Data for latest year may not have been available at press time.*

In U.S. $	2012	2011	2010	2009	2008	2007
Revenue	1,550,872,000	1,783,504,000	1,573,580,000	1,765,293,000	1,423,425,000	1,294,796,000
R&D Expense						
Operating Income	35,711,120	171,554,100	238,032,300	265,808,300	280,366,300	250,857,400
Operating Margin %	2.30%	9.61%	15.12%	15.05%	19.69%	19.37%
SGA Expense						
Net Income	-290,701,500	125,644,900	153,960,000	161,185,200	186,011,700	176,381,000
Operating Cash Flow	83,196,930	25,664,930	197,472,100	185,278,700	161,573,300	96,189,700
Capital Expenditure	58,566,570	55,229,800	59,539,980	75,488,810	61,071,270	54,038,090
EBITDA	84,168,220	217,585,600	278,170,700	308,175,000	314,023,100	277,444,000
Return on Assets %	-12.25%	5.14%	6.58%	7.94%	11.69%	12.63%
Return on Equity %	-24.76%	9.85%	12.18%	15.50%	22.65%	22.66%
Debt to Equity	0.24	0.49	0.33	0.46	0.59	0.47

CONTACT INFORMATION:

Phone: 61 755899899 Fax: 61 755899800
Toll-Free:
Address: 1 Billabong Pl., Burleigh Heads, QLD 4220 Australia

STOCK TICKER/OTHER:

Stock Ticker: BLLAF Exchange: PINX
Employees: 6,000 Fiscal Year Ends: 30-Jun
Parent Company:

SALARIES/BONUSES:

Top Exec. Salary: $ Bonus: $
Second Exec. Salary: $ Bonus: $

OTHER THOUGHTS:

Estimated Female Officers or Directors: 5
Hot Spot for Advancement for Women/Minorities: Y

Blair Corporation

www.blair.com

NAIC Code: 454113

TYPES OF BUSINESS:

Apparel, Catalog Sales
Online Sales
Retail & Outlet Stores
Home Products

BRANDS/DIVISIONS/AFFILIATES:

Orchard Brands
Blair.com
Blair Warehouse Outlet

CONTACTS:

Note: Officers with more than one job title may be intentionally listed here more than once.

David Walde, CEO
David Walde, Pres.

GROWTH PLANS/SPECIAL FEATURES:

Blair Corporation markets women's wear, men's wear and home products through its catalogs and e-commerce site. Blair merchandise is manufactured to its specifications by independent suppliers, both domestic and foreign. The company is owned by Orchard Brands, along with other catalog and e-commerce brands, such as Willow Ridge, WinterSilks, Appleseed's and Gold Violin. Low- to middle-income customers are the primary demographic for Blair's merchandise. Women's wear merchandise includes dresses, tops, pants, skirts, lingerie, sportswear, suits, jackets, outerwear and shoes. The men's wear line includes shirts, suits, slacks, shoes and accessories. The home products line offers bedspread ensembles, draperies, furniture covers, area rugs, bath accessories, kitchenware, gifts, collectibles and personal care items. Additionally, Blair sells jewelry. In order to promote and liquidate discontinued, overstocked and returned merchandise, the company sells its products through a retail store and the Blair Warehouse Outlet store in Pennsylvania.

FINANCIAL DATA:

Note: Data for latest year may not have been available at press time.

In U.S. $	2012	2011	2010	2009	2008	2007
Revenue						
R&D Expense						
Operating Income						
Operating Margin %						
SGA Expense						
Net Income						
Operating Cash Flow						
Capital Expenditure						
EBITDA						
Return on Assets %						
Return on Equity %						
Debt to Equity						

CONTACT INFORMATION:

Phone: 814-723-3600 Fax: 814-726-6376
Toll-Free: 800-458-6057
Address: 220 Hickory St., Warren, PA 16366 United States

STOCK TICKER/OTHER:

Stock Ticker: Subsidiary Exchange:
Employees: Fiscal Year Ends: 31-Dec
Parent Company: ORCHARD BRANDS

SALARIES/BONUSES:

Top Exec. Salary: $ Bonus: $
Second Exec. Salary: $ Bonus: $

OTHER THOUGHTS:

Estimated Female Officers or Directors:
Hot Spot for Advancement for Women/Minorities:

Bluefly Inc

NAIC Code: 454111

www.bluefly.com

TYPES OF BUSINESS:

Online Apparel Retailer
Apparel & Accessories
Discount Fashions

BRANDS/DIVISIONS/AFFILIATES:

BlueFly.com
Maverick Capital Ltd
Prentice Capital Management LP
Soros Fund Management LLC
Rho Ventures LP
EyeFly
Belle & Clive
FlyandMighty.com

CONTACTS:

Note: Officers with more than one job title may be intentionally listed here more than once.

Joseph C. Park, CEO
James Gallagher, CFO
Scott A. Erdman, Sr. VP-Merch.
David Wassong, Interim Chmn.

GROWTH PLANS/SPECIAL FEATURES:

Bluefly, Inc. is a U.S. Internet retailer, which sells designer apparel, fashion accessories and home products through its web site, BlueFly.com. The company sells over 350 brands and more than 50,000 different types of items at discounts of up to 75% off retail prices. The firm acquires end-of-season and excess inventory of high-end designer fashion products. Bluefly uses a third-party warehouse and fulfillment center in Ohio to ship its orders; the majority of these orders are shipped within 24 hours of placement during the holiday season. The firm markets to existing customers through e-mails that highlight new promotions and products. Bluefly attracts new customers through print and television advertisements; through its affiliation with BravoTV.com, LifetimeTV.com and popular CW shows like Gossip Girl and America's Next Top Model; and through marketing campaigns with Facebook and YouTube, such as Closet Confessions. The company is approximately 23%-owned by affiliates of Soros Fund Management L.L.C.; Maverick Capital, Ltd. owns 13%; Prentice Capital Management, LP owns approximately 12%; and Rho Ventures, L.P. owns roughly 41%. Belle & Clive is a members-only portal that offers limited time sales events on brand name clothing products. EyeFly is an online eyewear company. The company also maintains the blog FlyandMighty.com

FINANCIAL DATA:

Note: Data for latest year may not have been available at press time.

In U.S. $	2012	2011	2010	2009	2008	2007
Revenue	93,444,000	96,282,000	88,563,000	81,222,000	95,774,000	91,493,000
R&D Expense						
Operating Income	-23,729,000	-11,085,000	-3,846,000	-2,661,000	-10,848,000	-16,070,000
Operating Margin %	-25.39%	-11.51%	-4.34%	-3.27%	-11.32%	-17.56%
SGA Expense	39,848,000	39,370,000	37,049,000	34,218,000	46,334,000	48,809,000
Net Income	-24,776,000	-10,984,000	-4,033,000	-4,369,000	-11,340,000	-15,829,000
Operating Cash Flow	-3,014,000	-8,067,000	-7,486,000	2,149,000	-3,217,000	-7,929,000
Capital Expenditure	3,233,000	4,832,000	2,139,000	375,000	2,327,000	4,110,000
EBITDA	-20,819,000	-8,789,000	-1,339,000	282,000	-8,372,000	-14,344,000
Return on Assets %	-58.18%	-24.20%	-10.36%	-11.90%	-29.21%	-32.57%
Return on Equity %	-158.78%	-39.29%	-15.31%	-21.09%	-53.21%	-48.94%
Debt to Equity					0.16	

CONTACT INFORMATION:

Phone: 212 944-8000 Fax: 212 354-3400
Toll-Free: 877-258-3359
Address: 42 W. 39th St., New York, NY 10018 United States

STOCK TICKER/OTHER:

Stock Ticker: BFLY Exchange: NAS
Employees: 94 Fiscal Year Ends: 31-Dec
Parent Company:

SALARIES/BONUSES:

Top Exec. Salary: $507,846 Bonus: $
Second Exec. Salary: $300,000 Bonus: $

OTHER THOUGHTS:

Estimated Female Officers or Directors: 1
Hot Spot for Advancement for Women/Minorities: Y

Bon-Ton Stores Inc (The)

www.bonton.com

NAIC Code: 452111

TYPES OF BUSINESS:

Department Stores
Apparel, Accessories & Footwear
Home Furnishings
Private Brand Merchandise

BRANDS/DIVISIONS/AFFILIATES:

Herberger's
Boston Store
Carson Pirie Scott
Elder-Beerman
Bergner's
Younkers
Parisian
Bon-Ton

CONTACTS: *Note: Officers with more than one job title may be intentionally listed here more than once.*

Brendan L. Hoffman, CEO
Brendan L. Hoffman, Pres.
Keith E. Plowman, CFO/Exec. VP
Luis Fernandez, Chief Mktg. Officer/Exec. VP
Dennis R. Clouser, Exec. VP-Human Resources
Dennis R. Clouser, Exec. VP-Info. Systems
Anthony J. Buccina, Pres., Merch./Vice Chmn.
Mary Kerr, VP-Public Rel.
Mary Kerr, VP-Investor Rel.
Keith E. Plowman, Principal Acct. Officer
Stephen R. Byers, Exec. VP-Stores, Visual & Loss Prevention
Byron L. Bergren, Chmn.
Dennis R. Clouser, Exec. VP-Logistics & Procurement

GROWTH PLANS/SPECIAL FEATURES:

The Bon-Ton Stores, Inc., founded in 1898, operates regional fashion department stores offering apparel, home furnishings, cosmetics, accessories and shoes. Currently, the company operates 273 stores in 24 states under the names Bon-Ton (64 locations), Younkers (49), Carson Pirie Scott (46), Elder-Beerman (42), Herberger's (42), Boston Store (14), Bergner's (13) and Parisian (3). Its primary target customers are women aged 25-60 with annual household income between $55,000 and $125,000. In many of its markets, which desire but have limited access to branded merchandise, Bon-Ton is the primary destination for fashion lines such as Calvin Klein, Chanel, Coach, Bandolino, Clarks, Estee Lauder, Jones New York, Liz Claiborne, Nine West and Ralph Lauren. The firm also offers merchandise under private-branded labels, including Laura Ashley, Victor by Victor Alfaro, Ruff Hewn, Relativity, Studio Works, Breckenridge, Living Quarters Loft, Paradise Collections, Kenneth Roberts and Cuddle Bear. In addition to its brick-and-mortar stores, Bon-Ton operates e-commerce sites at Bonton.com as well as under its other operating banners. The company offers a proprietary credit card and has over 3.8 million active card holders. Bon-Ton owns a distribution center in Rockford, Illinois and leases distribution centers in Allentown, Pennsylvania and Fairborn, Ohio.

Employees are offered medical, dental and vision insurance; life insurance; long-term disability coverage; flexible spending accounts; a 401(k); and group legal insurance.

FINANCIAL DATA: *Note: Data for latest year may not have been available at press time.*

In U.S. $	2012	2011	2010	2009	2008	2007
Revenue	2,953,530,000	3,046,485,000	3,034,937,000	3,225,415,000	3,467,659,000	
R&D Expense						
Operating Income	66,631,000	135,148,000	86,722,000	-8,990,000	125,672,000	
Operating Margin %	2.25%	4.43%	2.85%	-.27%	3.62%	
SGA Expense	936,060,000	942,660,000	963,639,000	1,033,525,000	1,065,753,000	
Net Income	-12,128,000	21,494,000	-4,055,000	-169,930,000	11,562,000	
Operating Cash Flow	99,797,000	141,135,000	194,034,000	94,206,000	135,564,000	
Capital Expenditure	67,235,000	46,268,000	32,346,000	84,810,000	109,659,000	
EBITDA	170,172,000	237,694,000	198,357,000	113,814,000	254,110,000	
Return on Assets %	-.74%	1.27%	-.22%	-8.73%	.55%	
Return on Equity %	-7.70%	13.22%	-2.93%	-68.34%	3.25%	
Debt to Equity	6.61	5	7.17	8.56	3.15	

CONTACT INFORMATION:

Phone: 717 757-7660 Fax: 717 751-3108
Toll-Free:
Address: 2801 E. Market St., York, PA 17402 United States

STOCK TICKER/OTHER:

Stock Ticker: BONT Exchange: NAS
Employees: 27,100 Fiscal Year Ends: 31-Jan
Parent Company:

SALARIES/BONUSES:

Top Exec. Salary: $1,000,000 Bonus: $
Second Exec. Salary: $842,500 Bonus: $

OTHER THOUGHTS:

Estimated Female Officers or Directors: 3
Hot Spot for Advancement for Women/Minorities: Y

Boscov's Inc

www.boscovs.com

NAIC Code: 452111

TYPES OF BUSINESS:

Department Stores
Bridal Consulting
Beauty Salons
Travel Agencies
Hearing Evaluation Centers

BRANDS/DIVISIONS/AFFILIATES:

Boscov's Department Store LLC
Boscov's Travel
Boscov's Business Travel
BoscovsTravel.com
Boscovs.com

CONTACTS: *Note: Officers with more than one job title may be intentionally listed here more than once.*

Albert Boscov, CEO
Ed McKeaney, COO
Sam Flamholz, Co-Pres.
Toni Miller, CFO/Sr. Exec. VP
Jon Holmquist, Sr. VP-Direct Mktg.
Ed Elko, Sr. VP-Human Resources
Ed McKeaney, Chief Merch. Officer-Boscov's Dept. Store LLC
Toni Miller, Chief Admin. Officer/Sr. Exec. VP
Kenneth S. Lakin, Dir-Oper.
Dean Sheaffer, Sr. VP-Financial Service
Gary Boyer, Sr. VP/Dir.-Stores
John Hlis, Sr. VP-Real Estate & Dev.
John Young, Sr. VP-Gen. Merch. Mgr.- Home
Brian Nugent, Sr. VP-Gen. Merch. Mgr.-Men's
Albert Boscov, Chmn.
Larry Bergman, Sr. VP-Supply Chain

GROWTH PLANS/SPECIAL FEATURES:

Boscov's, Inc., along with its operating partner, Boscov's Department Store, LLC, operates a string of 41 department stores in five states, many of which anchor shopping malls. A majority of the company's stores are in Pennsylvania, where it was founded, with other locations in New York, New Jersey, Maryland and Delaware. Stores feature a broad selection of products, from men's and women's apparel to electronics and appliances. Clothing brands offered include Liz Claiborne, Tommy Hilfiger, Calvin Klein, Champion and Jockey Classic. The company offers home appliances, kitchenware and decorations from brands such as Waterford Crystal, Lenox, Sauder, Whirlpool and Pfaltzgraff. Some of its newer brands, available in select stores, include Clinique, Estee Lauder and Lancome cosmetics. Shoppers also find home electronics, collectibles, sporting equipment, travel gear and gift cards. In 39 of its locations, the company has hearing evaluation and consultation centers, which sell digital hearing aids and offer free hearing tests. An affiliated travel agency, doing business as Boscov's Travel and Boscov's Business Travel, can be found at a number of Boscov's stores and through BoscovsTravel.com. Besides its retail outlets, the firm also operates an e-commerce site, Boscovs.com, which offers the company's full line of products, as well as a private label credit card that can be applied for and managed from the web site.

Boscov's offers employees an employee discount, an executive training program and summer internships.

FINANCIAL DATA: *Note: Data for latest year may not have been available at press time.*

In U.S. $	2012	2011	2010	2009	2008	2007
Revenue	1,000,000,000	960,000,000	925,000,000	1,070,000,000		
R&D Expense						
Operating Income						
Operating Margin %						
SGA Expense						
Net Income						
Operating Cash Flow						
Capital Expenditure						
EBITDA						
Return on Assets %						
Return on Equity %						
Debt to Equity						

CONTACT INFORMATION:

Phone: 610-370-0507 Fax: 610-370-3495
Toll-Free: 800-284-8155
Address: 4500 Perkiomen Ave., Reading, PA 19606 United States

STOCK TICKER/OTHER:

Stock Ticker: Private Exchange:
Employees: Fiscal Year Ends: 31-Jan
Parent Company:

SALARIES/BONUSES:

Top Exec. Salary: $ Bonus: $
Second Exec. Salary: $ Bonus: $

OTHER THOUGHTS:

Estimated Female Officers or Directors: 1
Hot Spot for Advancement for Women/Minorities:

Brooks Brothers Group Inc

www.brooksbrothers.com

NAIC Code: 448120

TYPES OF BUSINESS:

Women's & Men's Apparel, Retail
Footwear
Accessories
Fragrances

BRANDS/DIVISIONS/AFFILIATES:

Adrienne Vittadini
Carolee Designs
Brooks Brothers
Retail Brand Alliance Inc

CONTACTS: *Note: Officers with more than one job title may be intentionally listed here more than once.*

Claudio Del Vecchio, CEO
Diane Hamilton, COO
Claudio Del Vecchio, Pres.
Brian Baumann, CFO

GROWTH PLANS/SPECIAL FEATURES:

Brooks Brothers Group, Inc., formerly Retail Brand Alliance, Inc., is a private company that owns several mostly mall-based retail stores. Brooks Brothers, the oldest surviving men's clothing store in the U.S., offers men's and women's professional and casual clothes, and operates about 325 stores worldwide, with nearly 180 stores in the U.S. The company also features the brands Adrienne Vittadini and Carolee Designs. Adrienne Vittadini is an upscale retailer of women's apparel, footwear, accessories and fragrances nationwide. Carolee Designs offers women's jewelry, watches, handbags and accessories in retail locations and department stores. Carolee Designs also produces private-label products for Polo Ralph Lauren. The firm is owned by CEO Claudio del Vecchio, who acquired the company from his father's company Luxottica Group. In September 2011, the company changed its name from Retail Brand Alliance, Inc., to Brooks Brothers Group, Inc., to reflect its chief asset.

FINANCIAL DATA: *Note: Data for latest year may not have been available at press time.*

In U.S. $	2012	2011	2010	2009	2008	2007
Revenue	800,000,000	750,000,000				
R&D Expense						
Operating Income						
Operating Margin %						
SGA Expense						
Net Income						
Operating Cash Flow						
Capital Expenditure						
EBITDA						
Return on Assets %						
Return on Equity %						
Debt to Equity						

CONTACT INFORMATION:

Phone: 860-741-0771 Fax: 860-745-9714
Toll-Free:
Address: 100 Phoenix Ave., Enfield, CT 06082 United States

STOCK TICKER/OTHER:

Stock Ticker: Private Exchange:
Employees: Fiscal Year Ends: 31-Dec
Parent Company:

SALARIES/BONUSES:

Top Exec. Salary: $ Bonus: $
Second Exec. Salary: $ Bonus: $

OTHER THOUGHTS:

Estimated Female Officers or Directors: 1
Hot Spot for Advancement for Women/Minorities:

Brookwood Companies Inc

www.brookwoodcos.com

NAIC Code: 313210

TYPES OF BUSINESS:

Fabric Manufacturing
Converting & Finishing
Technical Fabrics

BRANDS/DIVISIONS/AFFILIATES:

Hallwood Group Inc (The)
Brookwood Laminating Inc
Kenyon Industries Inc
ASSAULT
STORMLITE
BARRIER-FLEX
MARATHON
ALL-TERRAIN

CONTACTS: *Note: Officers with more than one job title may be intentionally listed here more than once.*

Amber M. Brookman, CEO
Amber M. Brookman, Pres.

GROWTH PLANS/SPECIAL FEATURES:

Brookwood Companies, Inc. is a textile converter of nylon and polyester fabrics. It is also a finisher that processes technically oriented fabrics for a wide variety of end uses. The company has six operating divisions: consumer, industrial, military, Roll Goods Group, Solutions 4 and First Performance Fabric Group. The consumer group develops and sells specialty fabrics for activewear, outerwear, luggage and specialty bags. The industrial division designs high technology fabrics and textiles for industrial applications such as safety suits and hospital mattresses. The military group develops military specified fabrics, including modular sleep systems and extended cold weather clothing systems. Through the Roll Goods Group, the firm provides short lot service on rolled nylon, cotton and polyester. The Solutions 4 division provides full-service package programs, covering the production stages from initial fabric development and design to manufacturing and quality control. The First Performance Fabrics division sells nylon and polyester fabrics in coated and uncoated varieties at promotional prices. In addition to these segments, Brookwood operates two principal subsidiaries: Kenyon Industries, Inc., a leading commission finisher of woven synthetic fabrics; and Brookwood Laminating, Inc., a fabric composite laminator that laminates waterproof/breathables, non-breathables, vinyls and sail-cloth fabrics. The company's fabric brands include ASSAULT, BARRIER-FLEX, MARATHON, ALL-TERRAIN, STORMLITE and DERMA-PLUSH. The firm has a strong belief in environmentally friendly practices, recycling everything from paper to coatings and working to eliminate the use of toxic substances. The firm is a subsidiary of The Hallwood Group, a holding company engaged primarily in textiles.

FINANCIAL DATA: *Note: Data for latest year may not have been available at press time.*

In U.S. $	2012	2011	2010	2009	2008	2007
Revenue						
R&D Expense						
Operating Income						
Operating Margin %						
SGA Expense						
Net Income						
Operating Cash Flow						
Capital Expenditure						
EBITDA						
Return on Assets %						
Return on Equity %						
Debt to Equity						

CONTACT INFORMATION:

Phone: 212-551-0100 Fax: 646-472-0294
Toll-Free:
Address: 25 W. 45th St., 11th Fl., New York, NY 10036 United States

STOCK TICKER/OTHER:

Stock Ticker: Subsidiary Exchange:
Employees: 471 Fiscal Year Ends: 31-Dec
Parent Company: HALLWOOD GROUP INC (THE)

SALARIES/BONUSES:

Top Exec. Salary: $ Bonus: $
Second Exec. Salary: $ Bonus: $

OTHER THOUGHTS:

Estimated Female Officers or Directors: 1
Hot Spot for Advancement for Women/Minorities:

Brown Shoe Company Inc

NAIC Code: 448210

www.brownshoe.com

TYPES OF BUSINESS:

Shoes, Retail
Shoes, Wholesale
Online Sales

BRANDS/DIVISIONS/AFFILIATES:

Famous Footwear
Naturalizer
Dr. Scholl's Shoes
Sam Edelman
B&H Footwear Company Limited
Shoes.com Inc
American Sporting Goods Corporation
Sam & Libby

CONTACTS:

Note: Officers with more than one job title may be intentionally listed here more than once.

Diane M. Sullivan, CEO
Diane M. Sullivan, Pres.
Russ Hammer, CFO/Sr. VP
Douglas W. Koch, Chief Talent Officer/Sr. VP
Mark Schmitt, CIO
Michael I. Oberlander, General Counsel/Sr. VP/Corp. Sec.
Peggy Reilly Tharp, VP-Investor Rel.
Richard M. Ausick, Pres., Famous Footwear
John W. Schmidt, Pres., Contemporary Fashion
John Mazurk, Pres., Healthy Living Brands
Samuel Edelman, Pres., Sam Edelman Shoes
Ronald A. Fromm, Chmn.
Daniel R Friedman, Pres., Supply Chain & Global Sourcing

GROWTH PLANS/SPECIAL FEATURES:

Brown Shoe Company, Inc., founded in 1878 and incorporated in 1913, is a global retailer and wholesaler of footwear for women and men. The firm operates more than 1,300 shoe stores in the U.S., Canada, Guam and China. These stores consist of 1,054 Famous Footwear stores and 224 specialty retail stores operating under the Naturalizer, Sam Edelman, Via Spiga and Dr. Scholl's Shoes names. Additionally, through its majority-owned subsidiary, B&H Footwear Company Limited, the company operates 23 Naturalizer stores in China. Famous Footwear targets primarily female customers seeking to purchase brand-name shoes at value prices for themselves and their families. The majority of Famous Footwear stores are located in strip centers, with the remainder in outlet and regional malls. In addition to its brick-and-mortar locations, the firm maintains e-commerce sites for its brands, including FamousFootwear.com, DrSchollsShoes.com, SamEdelman.com and Naturalizer.com. Brown Shoe Company also owns online shoe retailer Shoes.com, Inc. The firm's wholesale division designs, sources and markets branded and non-branded dress, casual and athletic footwear. The company's footwear is distributed to over 2,000 retailers, including department stores, mass merchandisers, national chains, independent retailers, catalogs and online retailers throughout the U.S. and Canada as well as in 60 other countries. During 2011, retail sales accounted for roughly 66% of its revenue, while wholesale sales accounted for the remaining 34%. In 2011, Brown Shoe Company acquired American Sporting Goods Corporation, which includes shoe brands Avia, RYKA and Nevados. In September 2012, it acquired the Sam & Libby brand from The Jones Group.

Brown Shoe Company offers its employees medical, dental and vision plans; life and disability insurance; a pension plan; a 401(k); and tuition assistance.

FINANCIAL DATA:

Note: Data for latest year may not have been available at press time.

In U.S. $	2012	2011	2010	2009	2008	2007
Revenue	2,582,824,000	2,504,091,000	2,241,968,000	2,276,362,000	2,359,909,000	
R&D Expense						
Operating Income	35,550,000	72,664,000	31,523,000	-173,301,000	95,682,000	
Operating Margin %	1.37%	2.90%	1.40%	-7.61%	4.05%	
SGA Expense	937,419,000	922,976,000	859,693,000	851,893,000	847,278,000	
Net Income	24,589,000	37,233,000	9,500,000	-133,238,000	60,427,000	
Operating Cash Flow	48,086,000	-2,312,000	118,078,000	34,336,000	86,367,000	
Capital Expenditure	38,564,000	54,827,000	49,978,000	76,744,000	47,232,000	
EBITDA	94,302,000	123,189,000	82,997,000	-116,628,000	151,384,000	
Return on Assets %	2.07%	3.40%	.91%	-12.53%	5.49%	
Return on Equity %	5.94%	9.11%	2.38%	-27.97%	11.16%	
Debt to Equity	0.48	0.36	0.37	0.38	0.26	

CONTACT INFORMATION:

Phone: 314 854-4000 Fax: 314 854-4274
Toll-Free:
Address: 8300 Maryland Ave., St. Louis, MO 63105 United States

STOCK TICKER/OTHER:

Stock Ticker: BWS
Exchange: NYS
Employees: 15,300
Fiscal Year Ends: 31-Jan
Parent Company:

SALARIES/BONUSES:

Top Exec. Salary: $825,000 Bonus: $
Second Exec. Salary: $283,173 Bonus: $500,000

OTHER THOUGHTS:

Estimated Female Officers or Directors: 4
Hot Spot for Advancement for Women/Minorities: Y

Buckle Inc (The)

www.buckle.com

NAIC Code: 448140

TYPES OF BUSINESS:

Teen Apparel, Retail
Promotional Merchandise
Online Sales

BRANDS/DIVISIONS/AFFILIATES:

Buckle
Buckle [The]
Buckle Screenprinting
Buckle.com
BKE
Lucky Brand Dungarees
Big Star
Silver

CONTACTS: *Note: Officers with more than one job title may be intentionally listed here more than once.*

Dennis Nelson, CEO
Dennis Nelson, Pres.
Karen Rhoads, CFO
Kari G. Smith, VP-Sales
Patricia K. Whisler, VP-Women's Merch.
Kyle L. Hanson, General Counsel/Corp. Sec.
Karen Rhoads, VP-Finance
Brett P. Milkie, VP-Leasing
Robert M. Carlberg, VP-Men's Merch.
Thomas B. Heacock, Treas./Controller
Daniel J. Hirschfeld, Chmn.

GROWTH PLANS/SPECIAL FEATURES:

The Buckle, Inc. is a retailer of medium- to high-priced casual apparel, footwear and accessories primarily for young men and women ages 15-30. The company operates roughly 431 stores in 43 states throughout the central, northwest, southwest and southeast U.S. These stores operate under the names Buckle and The Buckle. The majority of the stores are located in regional shopping malls, although some are located in strip centers, downtown areas and lifestyle centers. Buckle markets mostly brand-name casual apparel, including denim, tops, sportswear, outerwear, accessories and footwear. Brand names such as Lucky Brand Dungarees, Big Star, Silver, Hurley, Affliction, Fossil, MEK, Billabong, G-Star RAW, Quiksilver, 7 Diamonds, OBEY and Manchester constitute roughly 68% of total sales. The remaining merchandise consists of items manufactured to the company's specifications by private labels and merchandise marketed under its exclusive BKE label. The firm emphasizes personalized attention to its customers by providing free alterations, free gift-wrapping and a frequent shopper program. Buckle Screenprinting offers promotional merchandising to outside athletic teams, organizations, clubs and individuals. The company tailors individual store inventories to reflect differences in customer buying patterns by shipping new merchandise daily to most stores through its transfer program. This assures that popular merchandise is in stock and reduces the need to lower the price of low-selling merchandise at a particular location.

Employees are offered medical and dental insurance, disability coverage, life insurance, performance bonuses, an employee assistance program, a 40% merchandise discount, a 401(k) plan and flexible spending accounts.

FINANCIAL DATA: *Note: Data for latest year may not have been available at press time.*

In U.S. $	2012	2011	2010	2009	2008	2007
Revenue	1,062,946,000	949,838,000	898,287,000	792,046,000	619,888,000	
R&D Expense						
Operating Income	236,320,000	210,767,000	199,462,000	162,196,000	109,627,000	
Operating Margin %	22.23%	22.18%	22.20%	20.47%	17.68%	
SGA Expense	232,335,000	208,362,000	201,157,000	181,292,000	144,911,000	
Net Income	151,456,000	134,682,000	127,303,000	104,409,000	75,247,000	
Operating Cash Flow	209,273,000	179,935,000	157,959,000	143,729,000	121,080,000	
Capital Expenditure	36,627,000	54,945,000	50,561,000	47,448,000	27,484,000	
EBITDA	269,089,000	240,548,000	224,597,000	183,975,000	133,897,000	
Return on Assets %	29.51%	27.38%	26.68%	22.79%	18.37%	
Return on Equity %	42.73%	38.48%	36.82%	30.91%	24.08%	
Debt to Equity						

CONTACT INFORMATION:

Phone: 308 236-8491 Fax: 308 236-4493
Toll-Free: 800-626-1255
Address: 2407 W. 24th St., Kearney, NE 68845 United States

STOCK TICKER/OTHER:

Stock Ticker: BKE Exchange: NYS
Employees: 8,600 Fiscal Year Ends: 31-Jan
Parent Company:

SALARIES/BONUSES:

Top Exec. Salary: $960,000 Bonus: $3,401,728
Second Exec. Salary: $362,000 Bonus: $680,346

OTHER THOUGHTS:

Estimated Female Officers or Directors: 4
Hot Spot for Advancement for Women/Minorities: Y

Burberry Group plc

NAIC Code: 315240

www.burberryplc.com

TYPES OF BUSINESS:

Women's, Men's & Children's Apparel
Accessories
Licensing
Outerwear
Retail Stores
Manufacturing

BRANDS/DIVISIONS/AFFILIATES:

CONTACTS: *Note: Officers with more than one job title may be intentionally listed here more than once.*

Angela Ahrendts, CEO
Stacey Cartwright, CFO/Exec. VP
Sarah Manley, Chief Mktg. Officer
Mark Taylor, Chief People Officer
John Douglas, Chief Tech. Officer
Christopher Bailey, Chief Creative Officer
Michael Mahony, General Counsel/Sr. VP-Commercial Affairs
Matt McEvoy, Sr. VP-Strategy & New Bus. Dev.
Carol Fairweather, Sr. VP-Group Finance
Emilio Foa, Sr. VP-Emerging Markets
Pascal Perrier, Pres., Asia Pacific
Andrew Maag, Pres., Europe
Stephen Gilbert, Sr. VP-Architecture
John W. Peace, Chmn.
Eugenia Ulasewicz, Pres., Americas

GROWTH PLANS/SPECIAL FEATURES:

Burberry Group plc is an English luxury brand that designs, sources, manufactures and distributes apparel and accessories through a network of retail, wholesale and licensing channels worldwide. Founded in 1856, the company's products today include apparel for women, men and children as well as accessories such as handbags, scarves, small leather goods, belts, shoes and jewelry. The firm manufactures a portion of its outerwear, with the majority of its projects sourced from third parties. It also has global licensing relationships in specialized product categories such as eyewear, fragrance and timepieces. In these arrangements, the company grants licenses for the design and manufacture of products bearing the Burberry name, while working with its partners to ensure consistency within its overall branding strategy. Burberry operates approximately 192 retail stores, 208 department store concessions and 44 outlets. Currently, retail sales account for approximately 68% of company sales, wholesale sales account for approximately 26% and licensing agreements account for approximately 6%. Burberry is responsible for the creation of gabardine fabric, the Burberry check pattern and its iconic trench coat, first worn by British officers during World War I. The company has been working to reorganize its traditional wholesale-oriented structure and to expand its retail segment, testing new store formats and accelerating store openings in new and existing markets.

FINANCIAL DATA: *Note: Data for latest year may not have been available at press time.*

In U.S. $	2012	2011	2010	2009	2008	2007
Revenue	2,967,247,000	2,398,626,000	1,893,434,000	1,919,636,000	1,590,350,000	1,358,524,000
R&D Expense						
Operating Income	602,172,900	482,665,000	345,901,900	-15,817,220	322,256,000	250,838,800
Operating Margin %	20.29%	20.12%	18.26%	-.82%	20.26%	18.46%
SGA Expense						
Net Income	420,674,200	332,960,500	130,052,700	-9,586,196	216,009,000	176,066,500
Operating Cash Flow	592,746,500	420,514,500	588,272,900	322,895,000	64,866,600	174,468,800
Capital Expenditure	244,607,800	173,190,600	111,679,200	143,633,200	77,488,420	54,801,090
EBITDA	742,131,400	582,681,000	429,461,600	63,428,670	374,500,700	295,095,100
Return on Assets %	17.70%	16.64%	7.18%	-.57%	15.87%	15.50%
Return on Equity %	32.40%	31.16%	14.18%	-1.15%	30.30%	28.13%
Debt to Equity						

CONTACT INFORMATION:

Phone: 44 2033673000 Fax: 44 2033674910
Toll-Free:
Address: Hourseferry Rd., Horseferry House, London, SW1P 2AW United Kingdom

STOCK TICKER/OTHER:

Stock Ticker: BBRYF Exchange: PINX
Employees: 9,500 Fiscal Year Ends: 31-Mar
Parent Company:

SALARIES/BONUSES:

Top Exec. Salary: $ Bonus: $
Second Exec. Salary: $ Bonus: $

OTHER THOUGHTS:

Estimated Female Officers or Directors: 6
Hot Spot for Advancement for Women/Minorities: Y

Burlington Coat Factory

www.burlingtoncoatfactory.com

NAIC Code: 448140

TYPES OF BUSINESS:

Discount Apparel Stores
Linens/Housewares
Children's Furniture
Gifts
Online Sales
Children's Apparel
Maternity Apparel
Footwear

BRANDS/DIVISIONS/AFFILIATES:

Burlington Coat Factory Warehouse Corporation
MJM Designer Shoes
Cohoes Fashions
Super Baby Depot
Bain Capital LLC

CONTACTS:

Note: Officers with more than one job title may be intentionally listed here more than once.

Thomas A. Kinsbury, CEO
Thomas A. Kingsbury, Pres.
Todd Weyhrich, CFO/Exec. VP
Paul C. Tang, General Counsel/Exec. VP/Sec.
Fred Hand, Exec. VP-Oper.
Robert L. LaPenta, Treas./VP
Fred Hand, Exec. VP-Stores

GROWTH PLANS/SPECIAL FEATURES:

Burlington Coat Factory (BCF), owned by private equity firm Bain Capital LLC, is a value priced retailer of branded apparel, baby furniture, home decor and gifts. BCF operates more than 470 stores under the Burlington Coat Factory Warehouse and MJM Designer Shoes names, two Cohoes Fashions stores and one Super Baby Depot store throughout 44 states and Puerto Rico. Burlington Coat Factory Warehouse stores offer ladies sportswear, menswear, coats, family footwear, baby furniture, accessories, home decor and gifts. MJM Designer Shoes offers men's, women's and children's moderate- to higher-priced designer and fashion shoes, sandals, boots and sneakers as well as handbags, wallets, belts, socks, hosiery and novelty gifts. Cohoes Fashions offers designer label merchandise for men and women similar to that carried in Burlington Coat Factory stores as well as decorative gifts and home furnishings. Super Baby Depot sells brand name apparel, furniture and accessories for newborns, infants and toddlers. BCF has two distribution centers in Edgewater Park, New Jersey and San Bernardino, California. The company also sells coats and baby products through its web site. Although irregular and discontinued products comprise a small portion of its merchandise, BCF primarily sells items at below-traditional prices by maintaining inventory control and using a no-frills merchandising approach to business. All of BCF's stores are located in shopping malls, strip malls or stand-alone locations, which are in close proximity to other department stores and population centers. Most of the firm's stores are relatively large, with approximately 80,000 square feet. Some stores contain departments licensed to unaffiliated parties for the sale of such items as lingerie, fragrances, shoes and jewelry.

The company offers its employees a 401(k) plan, a profit sharing plan, medical and dental insurance, life and disability insurance, an employee discount and training programs.

FINANCIAL DATA:

Note: Data for latest year may not have been available at press time.

In U.S. $	2012	2011	2010	2009	2008	2007
Revenue		3,670,000,000	3,669,602,000	3,542,000,000	3,393,417,000	3,403,000,000
R&D Expense						
Operating Income						
Operating Margin %						
SGA Expense						
Net Income			30,998,000	-15,179,000	-48,970,000	-47,199,000
Operating Cash Flow						
Capital Expenditure						
EBITDA						
Return on Assets %						
Return on Equity %						
Debt to Equity						

CONTACT INFORMATION:

Phone: 609-387-7800 Fax: 609-387-7071
Toll-Free:
Address: 1830 Route. 130 N., Burlington, NJ 08016 United States

STOCK TICKER/OTHER:

Stock Ticker: Private Exchange:
Employees: 28,700 Fiscal Year Ends: 31-Jan
Parent Company: BAIN CAPITAL LLC

SALARIES/BONUSES:

Top Exec. Salary: $ Bonus: $
Second Exec. Salary: $ Bonus: $

OTHER THOUGHTS:

Estimated Female Officers or Directors:
Hot Spot for Advancement for Women/Minorities:

BYC Co

NAIC Code: 315240

www.byc.co.kr

TYPES OF BUSINESS:

Undergarment Manufacture

BRANDS/DIVISIONS/AFFILIATES:

Scorpio
BYC

CONTACTS: *Note: Officers with more than one job title may be intentionally listed here more than once.*

Byeong Seok Kim, CEO
Seok Beom Han, Pres.

GROWTH PLANS/SPECIAL FEATURES:

BYC Co., founded in 1946, is a leader in the Korean textile industry, involved in the manufacturing of undergarments. The company brings textiles from yarn to the completed garment, with factories in Korea as well as China. It designs its garments both for functionality in the Korean climate and with an idea to incorporate the cultural legacy of traditional undergarments into the modern age. Some of its research areas include the development of garments with antibiotic, deodorizing and thermal insulating capabilities as well as being environmentally friendly. The firm's more than 2,000 fully automated knitting machines yield more than 3,300 tons of knitted fabric monthly, in over 100 different styles of fabric. BYC also processes all its own fabric, singeing the knit fabric to remove nap, increasing gloss and flexibility through a process called silket, dying it and finally steam washing it. The fabric is then machine cut and sewn by the company's 6,000 employees, producing more than 8.5 million undergarments in 3,000 different styles each month, for 100 million pieces produced annually. BYC's Chinese produced garments are exclusively exported to Japan, Africa and the Middle East, although the company is researching new products for the Chinese domestic market. Its domestic products are shipped to the same locations as its Chinese products as well as the U.S. The firm's collections include brand names such as Scorpio and BYC.

FINANCIAL DATA: *Note: Data for latest year may not have been available at press time.*

In U.S. $	2012	2011	2010	2009	2008	2007
Revenue			195,665,000	157,790,000	145,850,000	122,730,000
R&D Expense						
Operating Income						
Operating Margin %						
SGA Expense						
Net Income				17,150,000	4,360,000	7,900,000
Operating Cash Flow						
Capital Expenditure						
EBITDA						
Return on Assets %						
Return on Equity %						
Debt to Equity						

CONTACT INFORMATION:

Phone: 82-2-840-3114 Fax: 82-2-834-8142
Toll-Free:
Address: 1104-1, Daelim-Dong, Youngdeungpo-Ku, Seoul, 157-070 South Korea

STOCK TICKER/OTHER:

Stock Ticker: 1460
Exchange: Seoul
Employees: 485
Fiscal Year Ends: 31-Dec
Parent Company:

SALARIES/BONUSES:

Top Exec. Salary: $ Bonus: $
Second Exec. Salary: $ Bonus: $

OTHER THOUGHTS:

Estimated Female Officers or Directors:
Hot Spot for Advancement for Women/Minorities:

C&J Clark International Ltd

www.clarks.co.uk

NAIC Code: 316210

TYPES OF BUSINESS:

Shoe Manufacturing
Shoe Retailing

BRANDS/DIVISIONS/AFFILIATES:

Clarks
Active Air
WaveWalk
Waterproof
Soul of Africa
YoToy
Clarks Desert Boots

CONTACTS: *Note: Officers with more than one job title may be intentionally listed here more than once.*

Melissa Potter, CEO
Robin Beacham, Dir.-Finance
Roger Pedder, Chmn.

GROWTH PLANS/SPECIAL FEATURES:

C&J Clark International, Ltd. (Clark), is a leading shoe brand in the U.K. and designs, manufactures and markets shoes globally. The company produces sandals, formal, casual, and sports shoes for men, women and children. Clark is predominantly renowned for the comfort and practicality of its shoes, with style being a lesser focus for the company. The company sells its shoes under the Clarks brand, including three featured lines with certain technological advancements: Active Air, which features air circulation and cushioning; WaveWalk, which boasts a shaped sole designed to propel the body using natural momentum; and Waterproof, which combines specially treated leather with other water resistant options, including sealed seams, non-rusting trim and non-wicking laces. Additional technology ensures that all linings are fully breathable. To supplement its shoe offerings, the company markets a full line of shoe care products, including waterproofers, polishes and restorers. Clark has a line of kids shoes called YoToy, that has a secret compartment for storing toy race cars. Other lines include Clarks Desert Boots. Additionally, the firm offers its Soul of Africa line of shoes, which are hand stitched by unemployed African women and whose profits are donated to help AIDS orphans in Africa. The company's brands are sold in 150 countries worldwide and are also available to purchase online. Counting both company-owned and franchised stores, there are over 1, 200 stores in the U.K. and around the world.

Employees of the firm are offered a pension plan, a profit sharing plan, discounts on company products and health benefits.

FINANCIAL DATA: *Note: Data for latest year may not have been available at press time.*

In U.S. $	2012	2011	2010	2009	2008	2007
Revenue						
R&D Expense						
Operating Income						
Operating Margin %						
SGA Expense						
Net Income						
Operating Cash Flow						
Capital Expenditure						
EBITDA						
Return on Assets %						
Return on Equity %						
Debt to Equity						

CONTACT INFORMATION:

Phone: 44-1458-443-131 Fax: 44-1458-447-547
Toll-Free:
Address: 40 High St., Somerset, BA16 OEQ United Kingdom

STOCK TICKER/OTHER:

Stock Ticker: Private Exchange:
Employees: 11,303 Fiscal Year Ends: 31-Jan
Parent Company:

SALARIES/BONUSES:

Top Exec. Salary: $ Bonus: $
Second Exec. Salary: $ Bonus: $

OTHER THOUGHTS:

Estimated Female Officers or Directors: 2
Hot Spot for Advancement for Women/Minorities:

Cabela's Inc

NAIC Code: 451110

www.cabelas.com

TYPES OF BUSINESS:

Sporting Goods Stores
Hunting & Fishing Supplies
Camping Equipment
Outdoor Apparel
Catalog & Online Sales
Credit Card Programs

BRANDS/DIVISIONS/AFFILIATES:

World's Foremost Bank
Cabela's CLUB Visa
Cabela's Outpost Store

CONTACTS: *Note: Officers with more than one job title may be intentionally listed here more than once.*

Thomas L. Millner, CEO
Michael Copeland, COO/Exec. VP
Thomas L. Millner, Pres.
Ralph W. Castner, CFO/Exec. VP
Scott K. Williams, Chief Mktg. Officer/Exec. VP
Brian J. Linneman, Chief Merch. Officer/Exec. VP
Charles Baldwin, Chief Admin. Officer/Exec. VP
Scott K. Williams, Chief e-commerce Officer
Joe Arterburn, Mgr.-Corp. Comm.
Chris Gay, Dir.-Investor Rel.
Chris Gay, Dir.-Treasury
Sean Baker, CEO/Pres., World's Foremost Bank/Exec. VP
Richard N. Cabela, Chmn.
Douglas R. Means, Chief Supply Chain Officer/Exec. VP

GROWTH PLANS/SPECIAL FEATURES:

Cabela's, Inc. is a leading retailer of outdoor and hunting supply merchandise. Through its web site, mail-order catalogs and retail stores, the company supplies hunting, marine, automobile & ATV, fishing and camping equipment as well as brand-name casual clothing and home fashion accessories. It operates in three divisions: retail, direct and financial services. The retail division manages 37 retail stores in 24 states and three stores in Canada. The stores, which are considered tourist attractions, receive millions of visitors per year. They are designed to communicate an outdoor lifestyle environment characterized by the outdoor feel of the lighting, wood or tile flooring, cedar wood beams, open ceilings and lodge-style atmosphere. The large-format stores, roughly 150,000 square feet, contain a mountain and pond with museum-quality taxidermy and native game fish, gun libraries featuring high-quality firearms, archery training systems, virtual shooting arcades, museums or educational centers and restaurants and banquet and meeting facilities. The direct segment of Cabela's is responsible for the company's catalog and Internet sales channel operations. Annually, it sends out over 130 million catalogs to all 50 states and more than 175 countries and territories. Through its wholly-owned subsidiary World's Foremost Bank, the financial services division offers the Cabela's CLUB Visa credit card, a rewards-based program that allows customers to earn and redeem points by using their credit card. The firm currently has over 1.5 million active credit accounts. During 2012, the company opened six new stores, including its first Outpost store. The Outpost concept is a smaller store format, about 40,000 square feet, and designed for markets with less than 250,000 people. In 2013, the firm plans to open nine new stores, including two Outpost stores.

FINANCIAL DATA: *Note: Data for latest year may not have been available at press time.*

In U.S. $	2012	2011	2010	2009	2008	2007
Revenue	3,112,682,000	2,811,166,000	2,663,242,000	2,632,240,000	2,552,721,000	2,349,599,000
R&D Expense						
Operating Income	275,699,000	231,548,000	186,762,000	92,678,000	141,039,000	151,092,000
Operating Margin %	8.85%	8.23%	7.01%	3.52%	5.52%	6.43%
SGA Expense	1,046,861,000	954,125,000	895,405,000	870,147,000	871,468,000	820,121,000
Net Income	173,513,000	142,620,000	112,159,000	49,617,000	76,404,000	87,879,000
Operating Cash Flow	234,629,000	366,468,000	167,427,000	294,020,000	154,968,000	31,828,000
Capital Expenditure	214,267,000	126,740,000	75,349,000	49,113,000	91,164,000	335,644,000
EBITDA	361,106,000	310,237,000	264,034,000	170,313,000	212,616,000	219,333,000
Return on Assets %	3.18%	2.95%	3.19%	2.03%	3.31%	4.43%
Return on Equity %	13.57%	12.93%	11.16%	5.22%	8.77%	11.24%
Debt to Equity	1.56	1.11	1.2	0.35	0.41	0.45

CONTACT INFORMATION:

Phone: 308 254-5505 Fax: 308 254-4800
Toll-Free: 800-237-4444
Address: 1 Cabela Dr., Sidney, NE 69160 United States

STOCK TICKER/OTHER:

Stock Ticker: CAB
Exchange: NYS
Employees: 14,800
Fiscal Year Ends: 31-Dec
Parent Company:

SALARIES/BONUSES:

Top Exec. Salary: $989,000 Bonus: $
Second Exec. Salary: $496,971 Bonus: $

OTHER THOUGHTS:

Estimated Female Officers or Directors: 1
Hot Spot for Advancement for Women/Minorities:

Cache Inc

www.cache.com

NAIC Code: 448120

TYPES OF BUSINESS:

Women's Apparel, Retail
Fashion Accessories
Online Sales
Evening Attire
Sportswear

BRANDS/DIVISIONS/AFFILIATES:

Cache
Cache Luxe
Cache Accents

CONTACTS: *Note: Officers with more than one job title may be intentionally listed here more than once.*

Thomas Reinckens, CEO
Thomas E. Reinckens, Pres.
Margaret Feeney, CFO/Exec. VP
Margarita Croasdaile, Mgr.-Human Resources
Rabia Farhang, Gen. Merch. Officer/Exec. VP
Thomas Reinckens, Chmn.

GROWTH PLANS/SPECIAL FEATURES:

Cache, Inc. is a nationwide, mall-based specialty retailer of lifestyle sportswear and dresses targeting women aged 25-45. The company currently operates 262 Cache and Cache Luxe stores in 42 states, Puerto Rico and the U.S. Virgin Islands. Averaging 2,000 square feet, the stores are generally located in large metropolitan and suburban areas. The Cache stores carry a range of merchandise, which includes dresses for daytime and evening and accessories that complement the seasonal themes and palettes of the sportswear and dresses. Stores focus on social-occasion dressing, from informal get-togethers to formal black-tie affairs. Cache Luxe stores generally sell merchandise at higher price points than the Cache stores. Cache designs and markets three general categories of merchandise: sportswear, which consists of related tops and bottoms, with price points ranging from $35 to $280; dresses, which range from shorter lengths for cocktail and day-into-evening wear to special occasion long dresses, with price points that range from $75 to $420; and accessories, which consists primarily of jewelry, belts and handbags selected to complement the sportswear and dress selections, with price points ranging from $5 to $100. During 2011, sportswear accounted for 57% of its revenues, dresses, 35% and accessories, the remaining 8%. The company offers a loyalty discount program, Cache Accents, to its customers, providing them with a lifetime discount of 5% once the customer has spent $300. Members also receive free shipping on online and mail orders and notices of private sales at retail locations. It also offers a co-branded Visa credit card that rewards customers for shopping at Cache.

FINANCIAL DATA: *Note: Data for latest year may not have been available at press time.*

In U.S. $	2012	2011	2010	2009	2008	2007
Revenue	224,210,000	223,880,000	206,519,000	219,775,000	265,728,000	274,458,000
R&D Expense						
Operating Income	-13,436,000	2,248,000	-27,097,000	-14,389,000	-11,837,000	7,314,000
Operating Margin %	-5.99%	1.00%	-13.12%	-6.54%	-4.45%	2.66%
SGA Expense	96,229,000	93,452,000	95,888,000	97,508,000	119,392,000	119,748,000
Net Income	-12,078,000	2,081,000	-22,432,000	-8,694,000	-7,122,000	6,521,000
Operating Cash Flow	-2,253,000	5,771,000	-4,625,000	11,158,000	6,395,000	30,429,000
Capital Expenditure	8,970,000	2,817,000	3,448,000	1,976,000	10,673,000	17,804,000
EBITDA	-6,187,000	10,119,000	-17,533,000	-2,316,000	-571,000	20,315,000
Return on Assets %	-15.16%	2.43%	-22.77%	-7.54%	-5.30%	4.23%
Return on Equity %	-26.88%	4.19%	-37.82%	-11.63%	-7.93%	6.01%
Debt to Equity				0.02	0.03	0.04

CONTACT INFORMATION:

Phone: 212 575-3200 Fax:
Toll-Free: 800-788-2224
Address: 1440 Broadway, 5th Fl., New York, NY 10018 United States

STOCK TICKER/OTHER:

Stock Ticker: CACH Exchange: NAS
Employees: 3,156 Fiscal Year Ends: 31-Dec
Parent Company:

SALARIES/BONUSES:

Top Exec. Salary: $570,000 Bonus: $
Second Exec. Salary: $400,000 Bonus: $

OTHER THOUGHTS:

Estimated Female Officers or Directors: 3
Hot Spot for Advancement for Women/Minorities: Y

Carolina Mills Inc

NAIC Code: 313110

www.carolinamills.com

TYPES OF BUSINESS:

Yarn & Fiber Manufacturing
Dyeing & Finishing
Specialty Fabrics

BRANDS/DIVISIONS/AFFILIATES:

Delcron Hyrdrotec
Delcron Hydropur
4t7 Silver
TriFlon
EcoSpun
Coolmax

GROWTH PLANS/SPECIAL FEATURES:

Carolina Mills, Inc. is a supplier of raw materials to the textile industry. The company was founded in 1928 and is based in Maiden, North Carolina. The firm originally manufactured, sold and distributed yarn in domestic and international markets, but today is known for the production and distribution of specialty spun yarn. Specialty products include Delcron Hyrdrotec, Delcron Hydropur, 4t7 Silver, TriFlon, Coolmax, EcoSpun, polypropylene, heather blends, cationic disperse dye spun yarns, Soybean, Polyester, Bamboo, wool blends, packaged dyed yarns and specialty yarns. Carolina Mills has one weaving plant, three yarn plants and a commission dyeing and finishing plant; all these facilities are located near its corporate offices, in Maiden, North Carolina. The firm also has partners and production in Asia.

CONTACTS: *Note: Officers with more than one job title may be intentionally listed here more than once.*

Stephen G. Dobbins, Jr., CEO
Stephen G. Dobbins, Jr., Pres.
Bryan Beal, CFO
George Moretz, Chmn.

FINANCIAL DATA: *Note: Data for latest year may not have been available at press time.*

In U.S. $	2012	2011	2010	2009	2008	2007
Revenue						
R&D Expense						
Operating Income						
Operating Margin %						
SGA Expense						
Net Income						
Operating Cash Flow						
Capital Expenditure						
EBITDA						
Return on Assets %						
Return on Equity %						
Debt to Equity						

CONTACT INFORMATION:

Phone: 828-428-9911 Fax: 828-428-2335
Toll-Free:
Address: 618 Carolina Ave., Maiden, NC 28650 United States

STOCK TICKER/OTHER:

Stock Ticker: Private
Exchange:
Employees:
Fiscal Year Ends: 30-Sep
Parent Company:

SALARIES/BONUSES:

Top Exec. Salary: $ Bonus: $
Second Exec. Salary: $ Bonus: $

OTHER THOUGHTS:

Estimated Female Officers or Directors:
Hot Spot for Advancement for Women/Minorities:

Carpenter Co

www.carpenter.com

NAIC Code: 3261

TYPES OF BUSINESS:

Polyurethane Foam & Polyester Fiber Products
Urethane Chemistry Research & Consulting
Upholstery Cushion Materials
Memory Foam

BRANDS/DIVISIONS/AFFILIATES:

Aveena
Rely System
Renew
Sleep Better

CONTACTS: *Note: Officers with more than one job title may be intentionally listed here more than once.*

Stanley Pauley, CEO
Michael Lowery, COO
Michael Lowery, Pres.
Dan Schecter, Sr. VP-Sales & Mktg.
John Jimison, Corp. Controller
Michael Faus, Sr. VP
Stanley Pauley, Chmn.

GROWTH PLANS/SPECIAL FEATURES:

Carpenter Co. manufactures and distributes a wide variety of polyurethane foam, expanded polystyrene and polyester fiber comfort cushioning products. The company's products are primarily sold to the automotive, bedding, consumer products, floor covering and furniture industries. Clients utilize Carpenter materials in products ranging from living room sofa cushions to carpets, insulation and seat cushions. The firm's wide variety of offerings include acoustical nonwovens, air filtration media, bedding, carpet cushions, chemicals, chemical systems consulting, consumer products, expanded polystyrene products, flexible foam packaging, furniture, molded manufacturing, polyester fiber, polyurethane foam and tire products. The company's consumer products division produces pillows, mattress pads and craft fiber products designed to meet retailer needs. The firm's chemicals division manufactures a significant amount of the world's polyether polyols for internal applications in the cellular, non-cellular and rigid foam markets. Carpenter's chemical systems division also controls and produces its own urethane chemicals and utilizes its expertise to aid other companies in developing their own polyurethane foam production systems. Carpenter Tire Products are designed to prevent premature tire wear and flat tires through its Rely System, which produces tires that are nail-proof, slag-proof and bullet-proof. The company also offers Renew, a line of eco-friendly bio-based foam and fiber products. Its Sleep Better brand produces sleeping products for children and teens. Recently, the company released the new bedding material Avena, a memory foam that responds to a person's body and resets to its original shape faster than other foams.

Carpenter offers its employees medical, dental and vision coverage; life insurance; flexible spending accounts; a 401(k) plan; profit sharing; tuition assistance; product discounts; wellness programs; and personal and family counseling.

FINANCIAL DATA: *Note: Data for latest year may not have been available at press time.*

In U.S. $	2012	2011	2010	2009	2008	2007
Revenue						
R&D Expense						
Operating Income						
Operating Margin %						
SGA Expense						
Net Income						
Operating Cash Flow						
Capital Expenditure						
EBITDA						
Return on Assets %						
Return on Equity %						
Debt to Equity						

CONTACT INFORMATION:

Phone: 804-359-0800 Fax: 804-353-0694
Toll-Free:
Address: 5016 Monument Ave., Richmond, VA 23230 United States

STOCK TICKER/OTHER:

Stock Ticker: Private
Employees:
Parent Company:
Exchange:
Fiscal Year Ends: 31-Dec

SALARIES/BONUSES:

Top Exec. Salary: $ Bonus: $
Second Exec. Salary: $ Bonus: $

OTHER THOUGHTS:

Estimated Female Officers or Directors:
Hot Spot for Advancement for Women/Minorities:

Carters Inc

NAIC Code: 315240

www.carters.com

TYPES OF BUSINESS:

Babies' & Children's Apparel
Retail Stores
Online Stores

BRANDS/DIVISIONS/AFFILIATES:

Just One Year
Precious Firsts
Genuine Kids
Child of Mine
OshKosh B'gosh
Bonnie Togs
Carter's
William Carter Company

CONTACTS: *Note: Officers with more than one job title may be intentionally listed here more than once.*

Michael D. Casey, CEO
Richard F. Westenberger, CFO/Exec. VP
William G. Foglesong, Sr. VP-Mktg.
Jill Wilson, Sr. VP-Human Resources & Talent Dev.
Brendan M. Gibbons, General Counsel/Sec./Sr. VP-Legal & Corp. Affairs
Sean McHugh, VP-Investor Rel.
Sean McHugh, Treas.
James C. Petty, Pres., Retail Stores
Lisa Fitzgerald, Exec. VP/Head-OshKosh Brand
Brian J. Lynch, Exec. VP/Head-Carter's Brand
Michael D. Casey, Chmn.
Christopher W. Rork, Exec. VP-Supply Chain

GROWTH PLANS/SPECIAL FEATURES:

Carter's, Inc. is one of the largest branded marketers of apparel for babies and young children in the U.S. The company, which operates through its William Carter Company subsidiary, designs, sources and markets clothing under two primary brands: Carter's and OshKosh B'gosh. Its products are sourced through contractual arrangements with manufacturers worldwide for wholesale distribution to major domestic retailers such as Babies R Us, Macy's, JCPenny, Sears, Costco, Kohl's, Wal-Mart and Target. Additionally, the company sells its products in Carter's and OshKosh branded retail locations. There are currently 398 Carter's stores and 167 OshKosh stores throughout North America, situated predominately in strip and outlet shopping centers. Merchandise is also available through the web sites Carters.com and OshKoshBgosh.com. The retailer sells a wide array of baby and toddler goods, including bodysuits, pajamas, undershirts, bibs, blankets and layettes. The company also licenses its name to other companies to create lifestyle merchandise that appeals to older children or moms. Licensed merchandise includes hosiery, underwear, bedding, toys, shoes, diaper bags, room decor and accessories. The OshKosh division is a specialty brand designed for sizes newborn to 12. It focuses on the production of denim apparel and coordinating garments, overalls and heavier fleece or woven pieces. In addition to its two primary brands, the company produces clothing lines exclusively available at select retailers, including Just One You, Precious Firsts and Genuine Kids brands at Target and Child of Mine at Wal-Mart. In 2011, the firm acquired Canadian children's apparel retailer Bonnie Togs for $98 million.

Full time employees of Carter's receive a comprehensive benefits package that includes medical, dental and life insurance; long- and short-term disability; a college savings plan; tuition assistance; and a 401(k).

FINANCIAL DATA: *Note: Data for latest year may not have been available at press time.*

In U.S. $	2012	2011	2010	2009	2008	2007
Revenue	2,381,734,000	2,109,734,000	1,749,256,000	1,589,677,000	1,490,016,000	1,412,246,000
R&D Expense						
Operating Income	261,986,000	187,466,000	243,256,000	195,613,000	135,494,000	-6,009,000
Operating Margin %	10.99%	8.88%	13.90%	12.30%	9.09%	-.42%
SGA Expense	713,211,000	540,960,000	468,192,000	434,391,000	409,599,000	359,826,000
Net Income	161,150,000	114,016,000	146,472,000	115,640,000	75,058,000	-70,618,000
Operating Cash Flow	278,619,000	81,074,000	85,821,000	188,239,000	183,623,000	51,987,000
Capital Expenditure	83,398,000	45,495,000	39,782,000	32,980,000	37,529,000	21,876,000
EBITDA	302,234,000	220,970,000	275,558,000	228,106,000	168,288,000	26,456,000
Return on Assets %	10.62%	8.57%	11.88%	10.23%	7.41%	-6.73%
Return on Equity %	17.99%	15.34%	23.70%	23.53%	18.56%	-16.09%
Debt to Equity	0.18	0.29	0.34	0.59	0.78	0.88

CONTACT INFORMATION:

Phone: 404 745-2700 Fax: 404 892-0968
Toll-Free: 888-782-9548
Address: 1170 Peachtree St. NE, The Proscenium, Ste. 900, Atlanta, GA 30309 United States

STOCK TICKER/OTHER:

Stock Ticker: CRI Exchange: NYS
Employees: 8,684 Fiscal Year Ends: 31-Dec
Parent Company:

SALARIES/BONUSES:

Top Exec. Salary: $818,846 Bonus: $
Second Exec. Salary: $509,808 Bonus: $

OTHER THOUGHTS:

Estimated Female Officers or Directors: 4
Hot Spot for Advancement for Women/Minorities: Y

Casual Male Retail Group Inc

www.casualmale.com

NAIC Code: 448110

TYPES OF BUSINESS:

Men's Apparel-Retail
Big & Tall Apparel
Teen's Clothing & Accessories-Retail
Denim Products
Outlet Stores
Online Sales

BRANDS/DIVISIONS/AFFILIATES:

CasualMale.com
Casual Male XL
Rochester Clothing
Destination XL
True Nation
Harbor Bay
Oak Hill
DestinationXL.com

CONTACTS: *Note: Officers with more than one job title may be intentionally listed here more than once.*

David A. Levin, CEO
Dennis R. Hernreich, COO
David A. Levin, Pres.
Dennis R. Hernreich, CFO/Exec. VP
Derrick Walker, Chief Mktg. Officer/Sr. VP
Walter Sprague, Sr. VP-Human Resources
Jack McKinney, CIO/Sr. VP
Kenneth M. Ederle, Gen. Merch. Mgr., DXL/Sr. VP
Robert S. Molloy, General Counsel/Sr. VP
Brian Reaves, Sr. VP-Store Sales & Oper.
Peter Schmitz, Sr. VP-Store Dev. & Real Estate
Peter H. Stratton, Jr., Chief Acct. Officer/Sr. VP-Finance/Controller
Vickie Smith, Sr. VP-Merch. Planning & Allocation
John R. Wagner, VP/Merch. Mgr.-Tailored Clothing & Furnishings
Angela Chew, VP-Global Sourcing
Seymour Holtzman, Chmn.
Frank Chane, Sr. VP-Logistics & Dist.

GROWTH PLANS/SPECIAL FEATURES:

Casual Male Retail Group, Inc. is one of the largest specialty retailers of men's big and tall apparel, with retail operations in the U.S., Canada and Europe. The company operates 386 Casual Male XL stores and 12 Rochester Clothing stores. In addition to these stores, the firm recently launched the new concept store Destination XL, which combines all of the firm's brands under one roof. The company operates 39 of these stores and plans to have 225-250 open by 2015. The firm also has a direct business that includes several catalogs and e-commerce sites. The stores offer an extensive selection of quality sportswear, dress clothing and footwear for the big and tall customer at moderate prices. The majority of the company's merchandise is basic or fashion-neutral items, such as jeans, casual slacks, tee-shirts, polo shirts, dress shirts and suit separates from nationally-recognized brands such as Polo, Nautica, Dockers and Levi's and private label merchandise. Private label brands carried in the stores include Gold Series, Harbor Bay, Synrgy, True Nation and Oak Hill. Casual Male's clothing has features specifically designed for comfort, such as waist-relaxer pants, stretch belts, zipper ties, wide band socks, neck-relaxer shirts and clothing with comfort-stretch technology and reinforced stress points. The firm maintains the web sites ShoesXL.com, LivingXL.com, BTDirect.com, RochesterClothing.com and CasualMaleXL.com, all of which can be accessed through the portal BigandTall.com. In 2011, the company announced the launch of the e-commerce site DestinationXL.com, which, similar to the store itself, allows the customer to shop all the firm's e-commerce sites with a single, shared shopping cart. In February 2012, the firm signed a franchise agreement to open a Destination XL store in Kuwait.

Employees are offered medical, dental and vision insurance; a 401(k) plan; flexible spending accounts; tuition reimbursement; and product discounts.

FINANCIAL DATA: *Note: Data for latest year may not have been available at press time.*

In U.S. $	2012	2011	2010	2009	2008	2007
Revenue	397,655,000	393,642,000	395,168,000	444,184,000	464,128,000	
R&D Expense						
Operating Income	-6,607,000	16,249,000	8,011,000	-76,987,000	10,510,000	
Operating Margin %	-1.66%	4.12%	2.02%	-17.33%	2.26%	
SGA Expense	177,883,000	150,933,000	151,045,000	178,072,000	178,055,000	
Net Income	42,663,000	15,371,000	6,110,000	-109,257,000	414,000	
Operating Cash Flow	23,426,000	19,022,000	30,774,000	23,197,000	11,738,000	
Capital Expenditure	18,038,000	9,031,000	4,634,000	12,595,000	21,377,000	
EBITDA	5,944,000	30,025,000	24,147,000	-59,355,000	26,954,000	
Return on Assets %	20.49%	8.45%	3.19%	-41.49%	.12%	
Return on Equity %	32.11%	15.02%	7.40%	-86.10%	.20%	
Debt to Equity			0.02	0.1	0.06	

CONTACT INFORMATION:

Phone: 781 828-9300   Fax: 781 444-8999
Toll-Free: 800-767-0319
Address: 555 Turnpike St., Canton, MA 02021 United States

STOCK TICKER/OTHER:

Stock Ticker: DXLG   Exchange: NAS
Employees: 2,446   Fiscal Year Ends: 31-Jan
Parent Company:

SALARIES/BONUSES:

Top Exec. Salary: $811,200   Bonus: $
Second Exec. Salary: $621,920   Bonus: $

OTHER THOUGHTS:

Estimated Female Officers or Directors: 2
Hot Spot for Advancement for Women/Minorities:

Cato Corp (The)

NAIC Code: 448120

www.catofashions.com

TYPES OF BUSINESS:

Women's Apparel, Retail
Children's Apparel
Junior's Apparel
Footwear
Accessories

BRANDS/DIVISIONS/AFFILIATES:

Cato Fashions
Cato Plus
It's Fashion
It's Fashion Metro
Versona Accessories

CONTACTS:
Note: Officers with more than one job title may be intentionally listed here more than once.

John P. Derham Cato, CEO
John P. Derham Cato, Pres.
John R. Howe, CFO/Exec. VP
Robert C. Brummer, Sr. VP-Human Resources
Sally J. Almason, Exec. VP/Gen. Merch. Mgr.
Gordon D. Smith, Chief Real Estate & Store Dev. Officer/Exec. VP
Miachael T. Greer, Exec. VP/Dir.-Stores
John P. Derham Cato, Chmn.

GROWTH PLANS/SPECIAL FEATURES:

The Cato Corp. is a women's specialty fashion retailer. The firm operates under the names Cato, Cato Plus, Cato Fashions, It's Fashion and It's Fashion Metro. The company has 1,288 specialty stores in 31 states, primarily located in non-metropolitan markets in the southern and southeastern U.S. The Cato Corp.'s stores feature a broad assortment of apparel and accessories, including dressy, career and casual sportswear; dresses; coats; and accessories such as shoes, lingerie, costume jewelry and handbags. Stores operating under the Cato brand primarily market products in junior or missy, plus sizes and girls sizes 7 to 16. The firm's It's Fashion stores market apparel, shoes and accessories for juniors, junior plus, young men, boys' sizes 8 to 20 and girls' sizes 7 to 16. It's Fashion Metro stores carry apparel in juniors and junior plus sizes, men and big men's sizes, boys' and girls' sizes, infants, toddlers and newborn sizes as well as jewelry, shoes and accessories. A major portion of the company's merchandise is sold under its private labels and is produced by various vendors in accordance with the company's specifications. Most of its merchandise is purchased from about 100 primary vendors. The Cato Corp. offers its own credit card and layaway plan; these represent approximately 9.5% of the firm's retail sales. Most stores range in size from 4,500 to 10,000 square feet (with an average of roughly 4,300 square feet) and are located primarily in strip shopping centers anchored by national discounters or market-dominant grocery stores. In determining store placement, the company uses site-selection and demographic criteria. Recently, the company established the Versona Accessories division, which offers fashion jewelry and accessories accented by key apparel items.

FINANCIAL DATA:
Note: Data for latest year may not have been available at press time.

In U.S. $	2012	2011	2010	2009	2008	2007
Revenue	931,458,000	925,528,000	883,995,000	857,718,000	846,437,000	
R&D Expense						
Operating Income	100,271,000	90,809,000	68,914,000	45,445,000	41,024,000	
Operating Margin %	10.76%	9.81%	7.79%	5.29%	4.84%	
SGA Expense	238,982,000	251,121,000	245,483,000	227,645,000	210,892,000	
Net Income	64,834,000	57,739,000	45,765,000	33,634,000	32,319,000	
Operating Cash Flow	81,341,000	79,476,000	84,689,000	71,567,000	74,164,000	
Capital Expenditure	35,890,000	19,559,000	9,960,000	19,443,000	18,330,000	
EBITDA	122,117,000	112,668,000	90,809,000	75,235,000	71,454,000	
Return on Assets %	12.08%	11.51%	9.98%	7.85%	7.57%	
Return on Equity %	18.73%	18.71%	16.54%	13.21%	12.33%	
Debt to Equity						

CONTACT INFORMATION:

Phone: 704 554-8510 Fax: 704 551-7200
Toll-Free:
Address: 8100 Denmark Rd., Charlotte, NC 28273 United States

STOCK TICKER/OTHER:

Stock Ticker: CATO Exchange: NYS
Employees: 9,500 Fiscal Year Ends: 31-Jan
Parent Company:

SALARIES/BONUSES:

Top Exec. Salary: $1,092,100 Bonus: $
Second Exec. Salary: $382,692 Bonus: $

OTHER THOUGHTS:

Estimated Female Officers or Directors: 1
Hot Spot for Advancement for Women/Minorities:

Celebrity International Inc

www.vitaminsbaby.com

NAIC Code: 315240

TYPES OF BUSINESS:

Children's & Babies' Clothing
Apparel for Preemies
Children's Books

BRANDS/DIVISIONS/AFFILIATES:

Vitamins Baby
Baby V
Whistle & Wink
Early Start
Educational Publishing LLC
My Skidders
Skidders Footwear Inc

CONTACTS:

Note: Officers with more than one job title may be intentionally listed here more than once.

Eli Matalon, CEO
Eli Matalon, Pres.

GROWTH PLANS/SPECIAL FEATURES:

Celebrity International, Inc. designs children's apparel, marketing licensed, private label and branded children's wear in sizes 0-4T. The firm's brands include Vitamins Baby, Baby V, Whistle & Wink, Early Start and My Skidders. The Vitamins Baby brand, featuring European-influenced, 100% cotton designs, is produced for babies 0-24 months and preemie babies from 0-5 lbs. Its products include playwear sets, blankets, socks, sleepwear and shoes for both girls and boys. The Vitamins brand typically targets upper moderate consumers, such as mothers, grandmothers and gift givers, and finds space in upscale department stores, including Macy's, Dillard's, Bloomingdale's and Nordstrom. Celebrity also markets Baby V, a line of playwear and active wear, for value-conscious consumers. Through the Whistle & Wink brand, the firm designs and markets products for nurseries, including furniture, bedding, decor and wall art. The Early Start brand, through Educational Publishing LLC, produces interactive story books and other reading materials for children learning to read. The My Skidders brand, produced by subsidiary Skidders Footwear, Inc., offers hybrid socks/shoes for kids with a cotton sock upper portion and a no-slip rubber sole that provides traction. It also markets a line of gripper socks for women.

FINANCIAL DATA:

Note: Data for latest year may not have been available at press time.

In U.S. $	2012	2011	2010	2009	2008	2007
Revenue	3,350,000,000	3,340,000,000	3,375,700,000	3,307,900,000	3,442,000,000	3,035,400,000
R&D Expense						
Operating Income						
Operating Margin %						
SGA Expense						
Net Income			122,600,000	88,300,000	-60,000,000	50,400,000
Operating Cash Flow						
Capital Expenditure						
EBITDA						
Return on Assets %						
Return on Equity %						
Debt to Equity						

CONTACT INFORMATION:

Phone: 212-279-1616 Fax: 212-868-3995
Toll-Free:
Address: 100 W. 33rd St., Ste. 800, New York, NY 10001 United States

STOCK TICKER/OTHER:

Stock Ticker: Private
Exchange:
Employees: 800
Fiscal Year Ends: 31-Dec
Parent Company:

SALARIES/BONUSES:

Top Exec. Salary: $ Bonus: $
Second Exec. Salary: $ Bonus: $

OTHER THOUGHTS:

Estimated Female Officers or Directors:
Hot Spot for Advancement for Women/Minorities:

Chanel SA

NAIC Code: 315240

www.chanel.com

TYPES OF BUSINESS:

Women's Fashion Apparel
Retail-Fashion Boutiques
Fine Jewelry & Watches
Beachwear & Lingerie
Fragrances & Cosmetics

BRANDS/DIVISIONS/AFFILIATES:

Chanel Fragrance
Chanel Beauty
Chanel No. 5
Eres
CC
Paraffection
Robert Goossens

CONTACTS: *Note: Officers with more than one job title may be intentionally listed here more than once.*

Maureen Chiquet, CEO
Francoise Montenay, Pres.
Alain Wertheimer, Chmn.

GROWTH PLANS/SPECIAL FEATURES:

Chanel SA is an haute couture landmark in fashion, cosmetics and fragrances, with such memorable creations as the little black dress, Chanel No. 5, the tweed suit, the jersey dress, the chain belt and two-tone shoes. The company is also famous for its CC trademark, found on everything from cosmetic compacts to handbag closures. Founded in 1913 by stylist Gabrielle Chanel, known as Coco, the firm floundered for years until it was taken over by Karl Lagerfeld, who updated the style for a younger market. The house's style is known for its simplicity and pure lines as well as several recurrent themes, including the camellia, the star, the faceted rectangle (inspired by the gardens of the Place Vendome) and matelassee (the lozenge design quilted into racecourse jockey jackets). The company also creates watches and fine jewelry and has a single license with Luxxotica for eyewear. Chanel operates more than 310 boutiques worldwide, several of which are located in Japan. These boutiques are divided into watches and fine jewelry, fashion, eyewear, and fragrance and beauty. Some locations carry only one or a selection of these product lines. The firm also owns Eres, a French swimwear and lingerie label with several boutiques in France, the U.S. and Italy. Other Chanel brands include Paraffection and Robert Goossens.

FINANCIAL DATA: *Note: Data for latest year may not have been available at press time.*

In U.S. $	2012	2011	2010	2009	2008	2007
Revenue	3,500,000,000	3,200,000,000	2,485,228,000			
R&D Expense						
Operating Income						
Operating Margin %						
SGA Expense						
Net Income	500,000,000	400,000,000	385,079,000			
Operating Cash Flow						
Capital Expenditure						
EBITDA						
Return on Assets %						
Return on Equity %						
Debt to Equity						

CONTACT INFORMATION:

Phone: 33-1-46-43-40-00 Fax: 33-1-47-47-60-34
Toll-Free: 800-550-0005
Address: 135 Ave. Charles De Gaulle, Neuilly-sur-Seine, 92521 France

STOCK TICKER/OTHER:

Stock Ticker: Private
Exchange:
Employees: 1,220
Fiscal Year Ends: 31-Dec
Parent Company:

SALARIES/BONUSES:

Top Exec. Salary: $ Bonus: $
Second Exec. Salary: $ Bonus: $

OTHER THOUGHTS:

Estimated Female Officers or Directors: 1
Hot Spot for Advancement for Women/Minorities:

Chapelthorpe plc

NAIC Code: 313110

www.chapelthorpe.com

TYPES OF BUSINESS:

Polypropylene Fiber
Specialty Coatings

BRANDS/DIVISIONS/AFFILIATES:

Drake Extrusion Inc
Drake Extrusion Ltd
Asota GmbH
International Fibres Group Limited (The)

CONTACTS:

Note: Officers with more than one job title may be intentionally listed here more than once.

Peter Gyllenhammar, Chmn.

GROWTH PLANS/SPECIAL FEATURES:

Chapelthorpe plc is a leading manufacturer and distributor of colored polypropylene staple fiber and filament yarn. Fibers, the company's sole operating segment, supplies colored polypropylene fiber and filament yarns to the automotive, floor coverings, technical textiles, geotextiles and home furnishings industries in the U.K., Austria and the U.S. Subsidiary The International Fibres Group Limited (IFG) is responsible for operating the firm's fiber business segment. IFG maintains its operations through three trading subsidiaries: Drake Extrusion, Ltd., based in the U.K.; Austria-based Asota GmbH; and U.S.-based Drake Extrusion, Inc. IFG remains one of the largest suppliers of polypropylene staple fiber in Europe, the sole provider of polypropylene staple fiber in the U.K. and a top supplier for the U.S. automotive industry, which uses the fibers in a variety of applications such as parcel shelves, kick panels, sound deadening materials, speaker covers, door panels and boot panels. Applications in other industries include durable and colorfast fibers for carpets; smooth fibers for home furnishings; and high tenacity fibers or geotextiles for the construction, engineering and environmental industries.

FINANCIAL DATA:

Note: Data for latest year may not have been available at press time.

In U.S. $	2012	2011	2010	2009	2008	2007
Revenue	155,000,000	150,000,000	143,300,000	138,840,000	161,860,000	157,780,000
R&D Expense						
Operating Income						
Operating Margin %						
SGA Expense						
Net Income			1,700,000	-1,130,000	-1,460,000	-22,210,000
Operating Cash Flow						
Capital Expenditure						
EBITDA						
Return on Assets %						
Return on Equity %						
Debt to Equity						

CONTACT INFORMATION:

Phone: 44-113-285-9020 Fax: 44-113-285-9033
Toll-Free:
Address: Old Mills, Whitehall Grove, Drighlington, Bradford, BD11 1BY United Kingdom

STOCK TICKER/OTHER:

Stock Ticker: Subsidiary Exchange:
Employees: 331 Fiscal Year Ends: 31-Mar
Parent Company: BRONSSTADET AB

SALARIES/BONUSES:

Top Exec. Salary: $ Bonus: $
Second Exec. Salary: $ Bonus: $

OTHER THOUGHTS:

Estimated Female Officers or Directors:
Hot Spot for Advancement for Women/Minorities:

Chargeurs

NAIC Code: 313210

www.chargeurs.fr

TYPES OF BUSINESS:

Wool Textiles
Wool Processing
Interlining
Protective Films

BRANDS/DIVISIONS/AFFILIATES:

Lainier de Picardie
DHJ International
Intissel
DHJ Industries

CONTACTS:

Note: Officers with more than one job title may be intentionally listed here more than once.

Eduardo Malone, CEO
Martine Odillard, COO
Federico Paullier, Managing Dir.-Chargeurs Wool
Stephane Rigaut, Managing Dir.-Chargeurs Interlining
Laurent Derolez, Managing Dir.-Chargeurs Protective Films
Eduardo Malone, Chmn.

GROWTH PLANS/SPECIAL FEATURES:

Chargeurs, managed as an investment fund, owns several manufacturing and services businesses. Chargeurs has three operating business segments: wool, interlining and protective films. The wool segment processes wool tops for its customer base of spinners. Wool tops are strands of combed wool that have been scoured for impurities and blended to create a more consistent product from which to make yarn. The interlining segment produces woven and non-woven interlining through its subsidiaries and their brand names, including Lainier de Picardie, DHJ International, Intissel and DHJ Industries. Interlining is a material produced by a fusible fixing process that is positioned between layers within garments to enhance their shape and/or feel. The interlining segment also produces technical textiles such as fire-retardant and ion-exchanging materials. The protective films segment is a designer, producer and vendor of self-adhesive films for the temporary protection of surfaces. These films are utilized by the building, household appliance and transportation industries to ensure that the materials covered emerge from machining, assembly, transportation or storage without damage to the exterior finishing. The company has operations in 33 countries worldwide. Most of the firm's sales are generated in Europe, with 46% of revenue. The Asia Pacific and Africa region generates 32% of revenues and the Americas are responsible for 22%.

FINANCIAL DATA:

Note: Data for latest year may not have been available at press time.

In U.S. $	2012	2011	2010	2009	2008	2007
Revenue		732,400,000	670,500,000	653,130,000	866,590,000	1,231,940,000
R&D Expense						
Operating Income						
Operating Margin %						
SGA Expense						
Net Income		13,900,000	16,900,000	-66,380,000	-86,490,000	23,560,000
Operating Cash Flow						
Capital Expenditure						
EBITDA						
Return on Assets %						
Return on Equity %						
Debt to Equity						

CONTACT INFORMATION:

Phone: 33-1-71-72-33-50 Fax: 33-1-71-72-33-51
Toll-Free:
Address: 29-31 rue Washington, Paris, 75008 France

STOCK TICKER/OTHER:

Stock Ticker: CRI
Exchange: Paris
Employees: 2,215
Fiscal Year Ends: 31-Dec
Parent Company:

SALARIES/BONUSES:

Top Exec. Salary: $ Bonus: $
Second Exec. Salary: $ Bonus: $

OTHER THOUGHTS:

Estimated Female Officers or Directors: 1
Hot Spot for Advancement for Women/Minorities:

Charlotte Russe Holding Inc

www.charlotte-russe.com

NAIC Code: 448120

TYPES OF BUSINESS:

Women's Apparel, Retail

BRANDS/DIVISIONS/AFFILIATES:

Charlotte Russe
blu chic
Refuge
Advent International Corporation
Charlotte-Russe.com

CONTACTS:

Note: Officers with more than one job title may be intentionally listed here more than once.

Jenny Ming, CEO
Edward Wong, COO
Jenny Ming, Pres.
Frederick G. Silny, CFO/Exec. VP
Jennifer Mitchell, Sr. VP-Merch.
Frederick G. Silny, Treas.
Kara Stangl, Sr. VP-Merch.
Jennifer C. Salopek, Chmn.

GROWTH PLANS/SPECIAL FEATURES:

Charlotte Russe Holding, Inc., privately owned by Advent International, is a mall-based specialty retailer of fashionable, value-priced apparel and accessories targeting young women between ages 14-35. The stores offer a broad assortment of merchandise centered on styles that are affordable and feminine and reflect the latest fashion trends. There are over 560 Charlotte Russe stores located in 46 states and Puerto Rico. These stores reflect established fashion trends and rely on in-store graphics and window displays to convey a fashion-forward orientation. Charlotte Russe offers ready-to-wear apparel, such as tops, dresses, shorts, pants and skirts, as well as seasonal items, such as prom dresses and outerwear. The majority of merchandise sold at Charlotte Russe stores is under the company's proprietary labels Charlotte Russe, Refuge and blu Chic. The company markets this merchandise to both younger career women and to teenagers, building brand awareness through a national print marketing campaign. Charlotte Russe stores are located predominantly in high-visibility, center court mall locations in spaces that average 7,100 square feet. Additionally, the firm offers online shopping at Charlotte-Russe.com. The company maintains a distribution center in Ontario, California.

FINANCIAL DATA:

Note: Data for latest year may not have been available at press time.

In U.S. $	2012	2011	2010	2009	2008	2007
Revenue						
R&D Expense						
Operating Income						
Operating Margin %						
SGA Expense						
Net Income						
Operating Cash Flow						
Capital Expenditure						
EBITDA						
Return on Assets %						
Return on Equity %						
Debt to Equity						

CONTACT INFORMATION:

Phone: 858-587-1500 Fax: 858-587-0902
Toll-Free: 877-266-9327
Address: 575 Florida St., San Francisco, CA 94110 United States

STOCK TICKER/OTHER:

Stock Ticker: Private Exchange:
Employees: 8,961 Fiscal Year Ends: 30-Sep
Parent Company: ADVENT INTERNATIONAL CORPORATION

SALARIES/BONUSES:

Top Exec. Salary: $ Bonus: $
Second Exec. Salary: $ Bonus: $

OTHER THOUGHTS:

Estimated Female Officers or Directors: 5
Hot Spot for Advancement for Women/Minorities: Y

Charming Shoppes Inc

www.charmingshoppes.com

NAIC Code: 448120

TYPES OF BUSINESS:

Women's Apparel, Retail
Plus-Size Women's Apparel
Fashion Accessories
Food & Gifts

BRANDS/DIVISIONS/AFFILIATES:

Lane Bryant
Catherine's Plus Sizes
Cacique
Lane Bryant Outlet
Figi's
Fashion Bug
Fashion Bug Plus

CONTACTS: *Note: Officers with more than one job title may be intentionally listed here more than once.*

Anthony M. Romano, CEO
Anthony M. Romano, Pres.
Eric M. Specter, CFO/Exec. VP
Frederick Lamster, Exec. VP-Human Resources
Colin D. Stern, General Counsel/Exec. VP/Sec.
John Lee, Chief Accounting Officer/Sr. VP
Brian P. Woolf, Pres., Lane Bryant, Cacique And Lane Bryant Outlet
MaryEllen MacDowell, Pres., Fashion Bug
Carol Williams, Pres., Catherines
Michael Goldstein, Chmn.
Bryan Q. Eshelman, Chief Supply Chain Officer/Exec. VP

GROWTH PLANS/SPECIAL FEATURES:

Charming Shoppes, Inc. operates women's plus size specialty apparel stores, including Lane Bryant, Fashion Bug and Catherine's Plus Sizes. Charming Shoppes maintains 685 Lane Bryant stores, catering to women ages 35-55, in 46 states; these stores average 5,800 square feet. Lane Bryant stores with an attached intimate apparel section under the Cacique brand name have a floor space of 7,200 square feet and account for 136 of total Lane Bryant stores. Through private labels, such as Cacique and Lane Bryant, the company offers fashionable and sophisticated apparel in sizes 12-32, including intimate apparel, wear-to-work and casual sportswear, as well as accessories. Lane Bryant Outlet operates 117 stores in 37 states, and is the only national chain offering women's plus-size apparel in the outlet sales channel. The company's 620 Fashion Bug and Fashion Bug Plus stores specialize in selling a wide variety of plus size, misses and junior apparel, accessories, intimate apparel and footwear. It targets customers 30-50 years old (sizes 6-30) who shop in the low-to-moderate price range. Fashion Bug's stores are located in 40 states, primarily in strip shopping centers, and average approximately 8,600 square feet. The firm's 435 Catherine's Plus Sizes stores, catering primarily to women ages 40-65, specialize in plus-sized classic apparel and accessories for career and casual lifestyles. The chain is well known for its extended sizes (28-34). Located in 44 states, the stores, averaging approximately 4,100 square feet, are primarily in strip shopping centers in the Southeast, Mid-Atlantic and Eastern Central U.S. Some of Charming Shoppes' products are sold online or through catalogues. Its Figi's catalog offers food and gifts. Plus-size apparel represents approximately 84% of Charming Shoppes' sales. In June 2012, the firm was acquired by Ascena Retail Group for $900 million.

Employees are offered a 40% discount on company merchandise; a 401(k); a stock purchase plan; medical, prescription, dental, life, AD&D, disability and vision insurance; flexible spending accounts; credit union membership; adoption assistance; and tuition reimbursement.

FINANCIAL DATA: *Note: Data for latest year may not have been available at press time.*

In U.S. $	2012	2011	2010	2009	2008	2007
Revenue	2,000,000,000	2,061,819,000	2,064,602,000	2,474,898,000	2,722,462,000	2,751,845,000
R&D Expense						
Operating Income						
Operating Margin %						
SGA Expense						
Net Income		-53,982,000	-77,962,000	-255,273,000	-88,290,000	108,923,000
Operating Cash Flow						
Capital Expenditure						
EBITDA						
Return on Assets %						
Return on Equity %						
Debt to Equity						

CONTACT INFORMATION:

Phone: 215-245-9100 Fax: 215-633-4640
Toll-Free:
Address: 450 Winks Lane, Bensalem, PA 19020 United States

STOCK TICKER/OTHER:

Stock Ticker: Subsidiary Exchange:
Employees: 27,000 Fiscal Year Ends: 31-Jan
Parent Company: Ascena Retail Group Inc

SALARIES/BONUSES:

Top Exec. Salary: $ Bonus: $
Second Exec. Salary: $ Bonus: $

OTHER THOUGHTS:

Estimated Female Officers or Directors: 4
Hot Spot for Advancement for Women/Minorities: Y

Cheil Industries Inc

www.cii.samsung.com

NAIC Code: 313230

TYPES OF BUSINESS:

Fabric Manufacturing
Chemical Manufacturing
Electronic Chemical Manufacturing

BRANDS/DIVISIONS/AFFILIATES:

STARON
Bean Pole
Nine West
Pal Zileri
FUBU
Steve Madden
Samsung Group

CONTACTS: *Note: Officers with more than one job title may be intentionally listed here more than once.*

Jong Woo Park, CEO
Jang-jae Lee, Exec. VP-Chemical Business Group
Seoung-ku Lee, Exec. VP-Mgmt. Support Div.

GROWTH PLANS/SPECIAL FEATURES:

Cheil Industries, Inc., a subsidiary of the Samsung Group, is a Korean manufacturer of textiles, chemicals and electronics. The company is comprised of three divisions: chemicals, ECM (electronic chemical materials) and fashion. Its chemicals division develops and produces plastics such as flame-retardant ABS (Acrylonitrile Butadiene Styrene); flame-retardant HIP (High Impact Polystyrene); sheet ABS; transparent ABS; heat-resistant ABS; STARON, a proprietary artificial marble material; and functional polymers. These chemicals are used in products such as automobiles, refrigerators and electronics such as TVs and mobile phones. Its ECM division develops and produces display materials, such as diffusion plates, CR (color resist) and ACF (anisotropic conductive film) used in LCD screens; semiconductor materials such as EMCs (epoxy molding compounds), which protect semiconductors from moisture, shock, heat and electric charge; and CMP (Chemical Mechanical Polishing) Slurries, used to polish and planarize insulating layers between metal layers or wafers; and other functional materials such as paste, a material used both for insulation and conduction inside electronic parts and plasma display panels. Its fashion business is comprised mainly of the Bean Pole brand, a leading fashion brands in Korea, producing casual wear for men, women and kids as well as golf clothes, men's formal wear, jeans and a line of accessories. The firm also operates a textile business sector comprised of the Gumi Plant, a producer of woolen fabrics, as well as sales, planning and management teams. It has six facilities in Korea: an R&D center, the two fashion group offices and three plants; and 18 other offices in the UAE, Slovakia, Hungry, Germany, the U.S., Japan, China and Taiwan. The firm also owns several international brands including Nine West, FUBU, Steve Madden and Pal Zileri.

FINANCIAL DATA: *Note: Data for latest year may not have been available at press time.*

In U.S. $	2012	2011	2010	2009	2008	2007
Revenue	5,000,000,000	4,913,500,000	4,741,070,000	3,534,090,000	3,413,320,000	2,294,150,000
R&D Expense						
Operating Income						
Operating Margin %						
SGA Expense						
Net Income		228,100,000	248,620,000	96,150,000	142,290,000	111,500,000
Operating Cash Flow						
Capital Expenditure						
EBITDA						
Return on Assets %						
Return on Equity %						
Debt to Equity						

CONTACT INFORMATION:

Phone: 82-54-468-2114 Fax: 82-31-596-3088
Toll-Free:
Address: 290 Kongdan-Dong, Seoul, 730-710 South Korea

STOCK TICKER/OTHER:

Stock Ticker: 1300 Exchange: Seoul
Employees: 3,853 Fiscal Year Ends: 31-Dec
Parent Company: SAMSUNG GROUP

SALARIES/BONUSES:

Top Exec. Salary: $ Bonus: $
Second Exec. Salary: $ Bonus: $

OTHER THOUGHTS:

Estimated Female Officers or Directors:
Hot Spot for Advancement for Women/Minorities:

Chico's FAS Inc

NAIC Code: 448120

www.chicos.com

TYPES OF BUSINESS:

Women's Apparel, Retail
Online & Catalog Sales
Franchising

BRANDS/DIVISIONS/AFFILIATES:

Chico's
White House/Black Market
Soma Intimates
Boston Proper Inc
Passport Club

CONTACTS: *Note: Officers with more than one job title may be intentionally listed here more than once.*

David F. Dyer, CEO
Kent Kleeberger, COO/Exec. VP
David F. Dyer, Pres.
Pamela K. Knous, CFO
Celia Rao Visconti, Sr. VP-Mktg.
Sara K. Stensrud, Chief Human Resources Officer/Exec. VP
Gary A. King, CIO/Exec. VP
Lee Eisenberg, Chief Creative Officer/Exec. VP
A. Alexander (Sandy) Rhodes, General Counsel/Sec./Exec. VP
Lynn Willis, VP-Public Rel.
Todd Vogensen, VP-Investor Rel.
Pamela K. Knous, Chief Acct. Officer
Mori MacKenzie, Chief Stores Officer/Exec. VP
Donna M. Colaco, Pres., White House/Black Market Brand
Cynthia S. Murray, Pres., Chico's Brand
Laurie Van Brunt, Pres., Soma Intimates Brand
Ross E. Roeder, Chmn.

GROWTH PLANS/SPECIAL FEATURES:

Chico's FAS, Inc. retails exclusively designed, private label, casual-to-formal clothing, complementary accessories, intimate apparel and gift items for women under the Chico's, White House/Black Market, Soma Intimates and Boston Proper brands. Chico's currently operates 1,256 retail stores in 48 states, Washington, D.C., the U.S. Virgin Islands and Puerto Rico as well as e-commerce sites for all its brands. Chico's primarily targets women aged 30 and over with moderate and high income levels. There are currently 601 Chico's boutique stores and 83 Chico's outlet stores. The majority of the firm's products are designed and developed by its Product Development Team, headquartered in Fort Myers, Florida. The company offers Passport Club membership to women who have spent at least $500 over time, allowing them perks such as a permanent 5% discount on future purchases. Chico's also operates the White House/Black Market chain of 364 women's clothing stores and 27 outlet locations that focuses on women aged 25 and older who lead active work and social lives with moderate to high income levels. White House/Black Market offers clothes predominately in shades of white and black, although the stores do offer a line of denim jeans as well. The White House/Black Market brand contributed 31.7% of the company's total net sales in 2011. The Soma Intimates intimate wear line currently consists of 164 boutique-style stores and 17 outlet locations. Soma Intimates offers foundation products in intimate apparel, sleepwear, bodywear and active wear. The Chico's and Soma brands accounted for 66.5% of 2011 net sales. In 2011, the firm acquired Boston Proper, Inc., a women's apparel and accessories retailer.

Chico's employee benefits include merchandise discounts, a retirement savings plan and a stock purchase plan.

FINANCIAL DATA: *Note: Data for latest year may not have been available at press time.*

In U.S. $	2012	2011	2010	2009	2008	2007
Revenue	2,196,360,000	1,904,954,000	1,713,150,000	1,582,405,000	1,714,326,000	
R&D Expense						
Operating Income	222,377,000	177,082,000	108,153,000	-39,594,000	121,458,000	
Operating Margin %	10.12%	9.29%	6.31%	-2.50%	7.08%	
SGA Expense	998,861,000	889,625,000	836,562,000	835,422,000	847,603,000	
Net Income	140,874,000	115,394,000	69,646,000	-19,137,000	93,421,000	
Operating Cash Flow	255,181,000	239,626,000	215,370,000	99,446,000	208,647,000	
Capital Expenditure	131,757,000	73,045,000	67,920,000	104,615,000	235,119,000	
EBITDA	321,807,000	271,195,000	204,525,000	57,978,000	213,437,000	
Return on Assets %	9.91%	8.43%	5.47%	-1.54%	8.09%	
Return on Equity %	13.58%	11.27%	7.39%	-2.10%	10.88%	
Debt to Equity						

CONTACT INFORMATION:

Phone: 239 277-6200 Fax: 239 277-5237
Toll-Free: 888-550-5559
Address: 11215 Metro Pkwy., Fort Myers, FL 33966 United States

STOCK TICKER/OTHER:

Stock Ticker: CHS Exchange: NYS
Employees: 19,800 Fiscal Year Ends: 31-Jan
Parent Company:

SALARIES/BONUSES:

Top Exec. Salary: $950,000 Bonus: $
Second Exec. Salary: $332,418 Bonus: $485,000

OTHER THOUGHTS:

Estimated Female Officers or Directors: 8
Hot Spot for Advancement for Women/Minorities: Y

Children's Place Retail Stores Inc (The)

www.childrensplace.com

NAIC Code: 448130

TYPES OF BUSINESS:

Apparel-Children's, Retail
Online Sales

BRANDS/DIVISIONS/AFFILIATES:

Children's Place (The)
ChildrensPlace.com

CONTACTS:

Note: Officers with more than one job title may be intentionally listed here more than once.

Jane Elfers, CEO
Eric Bauer, COO
Jane Elfers, Pres.
Steve Baginski, CFO/Sr. VP
Larry McClure, Sr. VP-Human Resources
Bill Melvin, CIO/Sr. VP
Natalie Levy, Exec. VP-Merch. & Design
Bradley P. Cost, General Counsel/Sr. VP/Sec.
Jason McAndrew, Sr. VP-Planning & Allocation
Jane Singer, VP-Investor Rel.
Greg Poole, Sr. VP-Global Sourcing
Michael Giannelli, Sr. VP-Design
Adrienne Gernand, Sr. VP-Store Dev.
Norman Matthews, Chmn.
James Bruce Marshall, Sr. VP/Managing Dir.-Int'l Bus. Dev.
Liam Devoy, Sr. VP-Logistics & Dist.

GROWTH PLANS/SPECIAL FEATURES:

The Children's Place Retail Stores, Inc. is a leading retailer of children's apparel and accessories. The firm designs, contracts to manufacture and sells high-quality, value-priced merchandise that is sold exclusively at its retail stores and on its web site, ChildrensPlace.com. The company sells clothes, shoes, socks, jewelry, belts, pajamas and various other seasonal items to boys and girls in three major size categories: zero to 24 months; six months to 5T; and four to 14. Currently, the firm operates 1,049 Children's Place stores in the U.S., Puerto Rico and Canada, 135 of which are outlet stores. The company stores are designed around four concepts in an effort to make the merchandise appealing to both parents and their children. The Apple-Maple store concept consists of light wood floors and trim, along with floor-to-ceiling glass windows that are ideal for displaying the company's colorful merchandise. The Technicolor concept store features bright colors within a boutique setting. The Tech2 (Tech Squared) stores have the brand aesthetics of a Technicolor store with the functionality of an Apple-Maple location and feature darker ceilings and floors, with white floor-wall fixtures. Outlet stores are approximately 7,500 square feet and strategically located to allow for quick liquidation of overstocked merchandise. In June 2012, the company announced plans to open retail stores in Saudi Arabia and the UAE, Kuwait, Qatar, Bahrain and Oman through franchise agreements.

Employees are offered tuition assistance and health, dental and vision insurance.

FINANCIAL DATA:

Note: Data for latest year may not have been available at press time.

In U.S. $	2012	2011	2010	2009	2008	2007
Revenue	1,715,862,000	1,673,999,000	1,643,587,000	1,630,323,000	2,162,559,000	
R&D Expense						
Operating Income	110,007,000	136,336,000	130,072,000	122,397,000	-88,698,000	
Operating Margin %	6.41%	8.14%	7.91%	7.50%	-4.10%	
SGA Expense	477,076,000	452,459,000	455,782,000	471,302,000	710,610,000	
Net Income	77,225,000	83,124,000	88,354,000	82,370,000	-59,567,000	
Operating Cash Flow	156,103,000	174,511,000	155,177,000	218,378,000	-1,223,000	
Capital Expenditure	79,764,000	83,945,000	62,217,000	51,693,000	199,766,000	
EBITDA	185,546,000	207,976,000	201,519,000	197,687,000	-5,767,000	
Return on Assets %	9.05%	9.73%	9.85%	8.50%	-6.15%	
Return on Equity %	12.69%	13.89%	15.54%	16.14%	-11.98%	
Debt to Equity				0.1		

CONTACT INFORMATION:

Phone: 201 558-2400 Fax: 201 227-0321
Toll-Free: 877-752-2387
Address: 500 Plaza Dr., Secaucus, NJ 07094 United States

STOCK TICKER/OTHER:

Stock Ticker: PLCE Exchange: NAS
Employees: 18,900 Fiscal Year Ends: 31-Jan
Parent Company:

SALARIES/BONUSES:

Top Exec. Salary: $1,080,769 Bonus: $
Second Exec. Salary: $633,346 Bonus: $

OTHER THOUGHTS:

Estimated Female Officers or Directors: 5
Hot Spot for Advancement for Women/Minorities: Y

China Dongxiang Group

http://www.dxsport.com

NAIC Code: 424320

TYPES OF BUSINESS:

Mens and Boys' Clothing and Furnishings Merchant Wholesalers

BRANDS/DIVISIONS/AFFILIATES:

Phenix Co Ltd
Kappa
PHENIX
X-NIX
INHABITANT

CONTACTS:

Note: Officers with more than one job title may be intentionally listed here more than once.

Chen Yihong, CEO
Qin Dazhong, COO
Sun Jianjun, Head- Phenix Co., Ltd.
Chen Yihong, Chmn.

GROWTH PLANS/SPECIAL FEATURES:

China Dongxiang Group (CDG) operates as a designer, developer, marketer and wholesaler of branded sportswear in China. The firm's primary brands are Kappa and PHENIX, through Phenix Co., Ltd. The Kappa brand, which CDG owns the rights for in China and Macau, produces items designed for both sports and leisure wear and targets high income earners of the 18 to 35 demographic. The apparel under the Kappa brand are specifically designed to match together with each other and cover a diverse line of items including football, classic, national, gym, lifestyle and others. The Phenix Co. is one of the largest producers of ski-related items in Japan as well as having the rights to the Kappa brand in Japan. In addition to the Kappa brand, Phenix Co. also operates the sports brands of PHENIX, devoted to extreme outdoor sports like skiing; X-NIX, a snowboarding apparel brand; and INHABITANT, a brand whose products cover extreme sports such as surfing, wakeboarding, BMX and skateboarding. Outside of its Japanese locations, Phenix Co. has international sales outlets located in China, Korea, Australia, Europe, Canada and the U.S. CDG has adopted an asset-light business model, which focuses on brand management, product design and research, supply chain management and expansion of distribution networks in different segments.

FINANCIAL DATA:

Note: Data for latest year may not have been available at press time.

In U.S. $	2012	2011	2010	2009	2008	2007
Revenue	284,935,900	440,928,500	685,332,800	638,503,300	534,267,700	275,159,300
R&D Expense						
Operating Income	11,124,750	15,869,610	280,041,800	272,945,500	214,150,300	116,546,500
Operating Margin %	3.90%	3.59%	40.86%	42.74%	40.08%	42.35%
SGA Expense	128,063,100	171,712,000	165,699,800	131,187,800	110,158,400	47,600,790
Net Income	31,166,720	16,991,300	235,427,700	234,765,800	219,951,100	117,969,200
Operating Cash Flow	43,892,700	1,451,844	232,154,100	238,673,800	149,455,000	105,552,000
Capital Expenditure	1,851,149	3,276,297	6,361,385	4,596,916	3,637,650	13,248,480
EBITDA	50,363,120	24,694,530	311,178,000	287,253,200	225,645,300	116,546,500
Return on Assets %	2.39%	1.28%	17.95%	19.20%	20.23%	20.74%
Return on Equity %	2.57%	1.42%	19.68%	20.74%	21.67%	23.62%
Debt to Equity						

CONTACT INFORMATION:

Phone: (8610) 6783-6666 Fax: (8610) 6785-6626
Toll-Free:
Address: Building 21, No. 2 Jingyuanbei Street, Beijing, 100176 China

STOCK TICKER/OTHER:

Stock Ticker: CDGXF Exchange: PINX
Employees: 429 Fiscal Year Ends: 31-Dec
Parent Company:

SALARIES/BONUSES:

Top Exec. Salary: $ Bonus: $
Second Exec. Salary: $ Bonus: $

OTHER THOUGHTS:

Estimated Female Officers or Directors:
Hot Spot for Advancement for Women/Minorities:

China Xiniya Fashion Ltd.

http://www.xiniya.com

NAIC Code: 315220

TYPES OF BUSINESS:

Mens and Boys' Cut and Sew Suit, Coat, and Overcoat Manufacturing

BRANDS/DIVISIONS/AFFILIATES:

Xiniya

CONTACTS:

Note: Officers with more than one job title may be intentionally listed here more than once.

Qiming Xu, CEO
Kangkai Zeng, COO
Chee Jiong Ng, CFO
Mingjiang Liu, Dir.-Mktg. & Sales/Deputy Gen. Mgr.
Qifa Zhang, Chief Prod. Officer
Qiwen (Zi Yang) Yang, Chief Designer
Qiming Xu, Chmn.

GROWTH PLANS/SPECIAL FEATURES:

China Xiniya Fashion Ltd (CXF) is a provider of men's business casual and business formal apparel and accessories in China. The firm markets its products through its Xiniya brand. Through its distribution network of 29 distributers, the Xiniya brand is made available to customers through over 1,700 owned and managed third party authorized retail outlets in 21 provinces, four autonomous regions and four municipalities in China. The authorized retail outlets are designed to have a uniform look and feel that fits the brands image, with in-store displays that accentuate the quality and style of the products. Each outlet is contracted to sell only CXF products. CXF's marketing strategy is to focus on second and lower-tier cities whose recent increase in affluence has led to an improvement in living standards, but where competing international brands have yet to establish a significant presence. The brands target demographic is male professionals between the ages of 25 and 45. In 2012, the firm sold approximately 6.5 million garment units with business casual items accounting for 92.6%; business formal, 5.8%; and accessories, 1.6%. In November 2012, CXF announced that it had opened its sixth and seventh flagship stores in Urumqi City, Xinjiang Province and Shenyang City, Liaoning Province.

FINANCIAL DATA:

Note: Data for latest year may not have been available at press time.

In U.S. $	2012	2011	2010	2009	2008	2007
Revenue	222,518,500	189,768,300	144,613,800	108,080,200	77,145,040	
R&D Expense						
Operating Income	34,854,860	49,768,910	46,436,970	35,773,920	23,178,200	
Operating Margin %	15.66%	26.22%	32.11%	33.09%	30.04%	
SGA Expense	39,339,210	15,345,190	3,555,152	1,872,216	3,656,627	
Net Income	28,215,270	40,478,590	40,581,190	31,253,560	20,265,510	
Operating Cash Flow	20,287,700	33,493,560	30,808,420	16,044,420	18,970,460	
Capital Expenditure	2,697,200	4,264,831	322			
EBITDA	35,636,110	50,033,610	46,499,520	35,857,230	23,261,340	
Return on Assets %	11.17%	19.41%	36.36%	76.68%	57.17%	
Return on Equity %	13.30%	22.50%	40.99%	103.42%	85.49%	
Debt to Equity						

CONTACT INFORMATION:

Phone: 86-0595-88886166 Fax: 86-0595-83005066
Toll-Free:
Address: Xiniya Industry Mansion, Jinjiang, Fujian Province China

STOCK TICKER/OTHER:

Stock Ticker: XNY Exchange: NYS
Employees: Fiscal Year Ends: 31-Dec
Parent Company:

SALARIES/BONUSES:

Top Exec. Salary: $ Bonus: $
Second Exec. Salary: $ Bonus: $

OTHER THOUGHTS:

Estimated Female Officers or Directors:
Hot Spot for Advancement for Women/Minorities:

Chori Company Ltd

NAIC Code: 313230

www.chori.co.jp

TYPES OF BUSINESS:

Textiles Manufacturing
Specialty Chemicals
Metals
Machines
Fine Chemicals
Life Sciences
Real Estate
Apparel

BRANDS/DIVISIONS/AFFILIATES:

Toray Industries Inc
Chori America Inc
e-ryoshizai
Textilenet

CONTACTS:

Note: Officers with more than one job title may be intentionally listed here more than once.

Shuji Yamazaki, CEO
Shuji Yamazaki, Pres.

GROWTH PLANS/SPECIAL FEATURES:

Chori Company Ltd., a subsidiary of Toray Industries, Inc., manufactures chemicals, textiles and machinery for a variety of uses. Founded in 1861, the firm has expanded from a silk wholesaler to a multifaceted conglomerate operating numerous businesses, from fabrics to chemicals. The firm is divided into four segments: textile materials, textile goods, chemicals and machinery and overseas businesses. The textile materials segment procures, produces and sells synthetic fiber and automotive fabrics; knitted products, including raw materials for knit apparel, socks and tights; nonwoven sheets with soundproofing, shock proofing and water-proofing capability; and textile-garments such as SBX-10 Suede, Air Concept fabrics and Thobes. The textile goods segment provides apparel and related services, including men's and women's wear; uniforms, including uniform planning; intimate apparel; and sportswear, especially golf wear. This division also generates revenues through its two online e-commerce clothing material sites: e-ryoshizai, used for individual order placement; and Textilenet, used for commercial business-to-business (B2B) order placement. The chemicals and machinery segment has a variety of operations, including the import and export of organic chemicals; the import of inorganic fine chemicals for medical, agricultural and organic applications; the manufacture of industrial chemicals, including glass substrate materials; the marketing of petrochemical raw materials and glass substrates; the import of life science products, such as protein- and sweetener-based food chemicals; and the export of cars, motorcycles and specialty vehicles and parts manufactures in Japan. Chori's overseas division operates globally through a network of subsidiaries, with a large presence in China's trading and consumer markets.

FINANCIAL DATA:

Note: Data for latest year may not have been available at press time.

In U.S. $	2012	2011	2010	2009	2008	2007
Revenue	2,425,101,000	2,472,406,000	2,128,110,000	2,482,894,000		
R&D Expense						
Operating Income	65,864,330	57,636,850	52,214,960	65,040,460		
Operating Margin %	2.71%	2.33%	2.45%	2.61%		
SGA Expense	157,012,300	151,868,700	151,145,000	157,157,000		
Net Income	38,932,990	59,807,840	32,486,840	35,971,540		
Operating Cash Flow	-2,928,046	-12,224,310	40,213,310	-24,704,690		
Capital Expenditure	2,126,452	1,692,255	1,425,057	3,529,242		
EBITDA	70,685,030	61,956,560	58,416,180	66,231,730		
Return on Assets %	5.30%	8.84%	5.23%	5.94%		
Return on Equity %	12.89%	22.83%	14.64%	17.51%		
Debt to Equity						

CONTACT INFORMATION:

Phone: Fax:
Toll-Free:
Address: 1-7-3, Awajimachi, Chuo-ku, Osaka, 540-8603 Japan

STOCK TICKER/OTHER:

Stock Ticker: CCLDF Exchange: GREY
Employees: 1,113 Fiscal Year Ends:
Parent Company: TORAY INDUSTRIES INC

SALARIES/BONUSES:

Top Exec. Salary: $ Bonus: $
Second Exec. Salary: $ Bonus: $

OTHER THOUGHTS:

Estimated Female Officers or Directors:
Hot Spot for Advancement for Women/Minorities:

Christian Dior SA

www.dior.com

NAIC Code: 315240

TYPES OF BUSINESS:

Luxury Apparel
Leather Goods
Perfume & Cosmetics
Wines & Spirits
Watches & Jewelry
Online Sales
Cruise Lines

BRANDS/DIVISIONS/AFFILIATES:

Christian Dior Couture
Donna Karan International
LVMH Moet Hennessey Louis Vuitton SA
Dom Perignon
Veuve Clicquot
Hennessy
Sephora
Fendi

CONTACTS: *Note: Officers with more than one job title may be intentionally listed here more than once.*

Sidney Toledano, CEO
Eric Guerlain, Vice Chmn.
Pierre Gode, Group Managing Dir.
Bernard Arnault, Chmn.

GROWTH PLANS/SPECIAL FEATURES:

Christian Dior SA is a major international manufacturer and retailer of luxury goods. It owns the Dior fashion brand, and has a 40.9% stake in LVMH Moet Hennessy Louis Vuitton S.A. The company's business is divided into six segments: Christian Dior Couture; wines and spirits; fashion and leather goods; watches and jewelry; perfumes and cosmetics; and selective retailing. The company operates approximately 3,263 stores across these businesses. Christian Dior Couture creates individual haute couture clothing designs for celebrity clients as well as luxury ready-to-wear clothing and accessories that are sold both through other retailers and at the firm's international boutiques. In addition to Dior-branded fashions for men and women in its couture segment, the company owns a number of other fashion businesses, such as Givenchy, Donna Karan International, Fendi, Celine, Marc Jacobs and Kenzo. The wines and spirits segment produces brands such as Moet & Chandon, Veuve Clicquot, Dom Perignon and Hennessy. The fashion and leather goods segment includes the Louis Vuitton, Celine, Loewe, Marc Jacobs, Pucci and Berluti brands, among others. The watches and jewelry segment includes the TAG Heuer, Bulgari, Hublot and Zenith brands. Sephora, which is part of the firm's selective retailing and perfumes and cosmetics segments, operates 1,313 cosmetic boutiques in 27 countries as well as an online retail web site. The perfumes and cosmetics segment also includes BeneFit Cosmetics, an American brand that has expanded into the U.K., and carries Guerlain, Parfums Givenchy, Parfums Kenzo and other lines. The selective retailing additionally owns and operates Starboard Cruise Services in Miami, Florida as well as Cruise Line Holdings Co. in Delaware. The U.S. generates 22% of the firm's revenues; Japan, 8%; Asia outside of Japan, 30%; France, 11%; Europe outside of France, 19%; and other markets, 10%.

FINANCIAL DATA: *Note: Data for latest year may not have been available at press time.*

In U.S. $	2012	2011	2010	2009	2008	2007
Revenue	12,025,940,000	32,933,940,000	28,246,860,000	23,729,610,000	23,981,010,000	23,060,980,000
R&D Expense						
Operating Income	2,312,115,000	6,957,743,000	5,579,032,000	4,231,078,000	4,637,604,000	4,671,035,000
Operating Margin %	19.22%	21.12%	19.75%	17.83%	19.33%	20.25%
SGA Expense	5,423,910,000	14,624,230,000	12,484,620,000	10,653,920,000	10,707,410,000	10,129,710,000
Net Income	1,591,335,000	1,710,350,000	4,371,490,000	2,543,461,000	2,974,057,000	3,113,132,000
Operating Cash Flow	1,032,362,000	5,306,232,000	5,470,714,000	2,933,939,000	3,016,849,000	3,268,253,000
Capital Expenditure	661,941,700	2,453,865,000	1,474,993,000		1,432,201,000	1,370,687,000
EBITDA	2,952,661,000	8,260,230,000	7,728,002,000	5,385,130,000	5,604,440,000	5,520,193,000
Return on Assets %	.75%	2.76%	3.26%	1.94%	2.26%	2.60%
Return on Equity %	1.51%	5.74%	7.06%	4.42%	5.45%	9.31%
Debt to Equity		0.21	0.21	0.32	0.3	0.24

CONTACT INFORMATION:

Phone: 33 140735444 Fax:
Toll-Free:
Address: 30 Ave. Montaigne, Paris, 75008 France

STOCK TICKER/OTHER:

Stock Ticker: CHDRF Exchange: PINX
Employees: 86,818 Fiscal Year Ends: 31-Dec
Parent Company:

SALARIES/BONUSES:

Top Exec. Salary: $ Bonus: $
Second Exec. Salary: $ Bonus: $

OTHER THOUGHTS:

Estimated Female Officers or Directors: 2
Hot Spot for Advancement for Women/Minorities:

Christopher & Banks Corp

NAIC Code: 448120

www.christopherandbanks.com

TYPES OF BUSINESS:

Women's Business Apparel, Retail
Private-Label Merchandise
Accessories

BRANDS/DIVISIONS/AFFILIATES:

C.J. Banks
ChristopherandBanks.com
CJBanks.com
Christopher & Banks

CONTACTS: *Note: Officers with more than one job title may be intentionally listed here more than once.*

LuAnn Via, CEO
LuAnn Via, Pres.
Peter G Michielutti, CFO/Sr. VP
Monica L. Dahl, Sr. VP-Multi-Channel Mktg.
Cindy J. Stemper, VP-Human Resources
Lisa M. Klein, VP-IT, Planning & Allocation
Luke R. Komarek, General Counsel/Sr. VP/Corp. Sec.
Michelle Rice, Sr. VP-Store Oper.
Monica L. Dahl, Sr. VP-Bus. Strategy
Monica L. Dahl, Sr. VP-Investor Rel.
Michael J. Lyftogt, Chief Acct. Officer
Donna J. Fauchald, VP-Real Estate & Construction
Paul L. Snyder, Chmn.
Michael J. Tripp, VP-Supply Chain & Logistics

GROWTH PLANS/SPECIAL FEATURES:

Christopher & Banks Corp. (C&B) is a Minnesota-based specialty retailer of women's apparel. The company operates 640 stores in 44 states under the names Christopher & Banks (388 stores) and C.J. Banks (172) as well as 55 dual-concept stores and 25 outlet stores. The stores are generally mall-based and located in small to mid-sized markets. Sportswear and sweaters account for the bulk of the company's sales, though the firm has been shifting merchandise focus away from sweaters and expanding its offering of novelty jackets and fashion-knit tops. The principal store concept, Christopher & Banks, emphasizes style, quality and value in casual sportswear and sweaters exclusively designed for working women ages 45-55. The company's plus size store concept, C.J. Banks, offers similar apparel in sizes 14-24, and is often paired with an existing Christopher & Banks store. It also offers petite sizes at most of its Christopher & Banks locations. In addition to its brick-and-mortar locations, C&B maintains e-commerce sites for both of its lines at ChristopherandBanks.com and CJBanks.com. To keep its fashions fresh, the company has begun testing new product categories, including sunglasses, scarves, handbags, swimwear and outerwear. The company has also upgraded its customer service, offering merchandise on wooden hangers, receipts placed in envelopes and purchases wrapped in tissue and placed in drawstring bags. In late 2011, the company planned to close 103 underperforming stores, and by February 2012, it had closed 95 stores. During this span, it reduced its overall full-time and part-time store sales associate positions by 14% and its corporate office headcount by 15%.

FINANCIAL DATA: *Note: Data for latest year may not have been available at press time.*

In U.S. $	2012	2011	2010	2009	2008	2007
Revenue	412,796,000	448,130,000	455,402,000	530,742,000	575,780,800	
R&D Expense						
Operating Income	-71,773,000	-14,559,000	-1,367,000	-14,108,290	21,835,850	
Operating Margin %		-3.24%	- .30%	-2.65%	3.79%	
SGA Expense	131,259,000	142,461,000	138,711,000	172,295,300	166,361,700	
Net Income	-71,062,000	-22,167,000	158,000	-12,750,500	17,018,150	
Operating Cash Flow	-25,880,000	7,793,000	35,946,000	22,110,550	62,029,030	
Capital Expenditure	11,742,000	8,428,000	5,969,000	18,430,790	35,901,000	
EBITDA	-51,571,000	10,177,000	24,618,000	12,255,720	44,439,200	
Return on Assets %		-8.84%	.05%	-4.23%	5.49%	
Return on Equity %		-12.38%	.08%	-6.08%	7.65%	
Debt to Equity						

CONTACT INFORMATION:

Phone: 763 551-5000 Fax: 763 551-5198
Toll-Free:
Address: 2400 Xenium Ln. N., Plymouth, MN 55441 United States

STOCK TICKER/OTHER:

Stock Ticker: CBK Exchange: NYS
Employees: 6,400 Fiscal Year Ends: 31-Jan
Parent Company:

SALARIES/BONUSES:

Top Exec. Salary: $ Bonus: $
Second Exec. Salary: $ Bonus: $

OTHER THOUGHTS:

Estimated Female Officers or Directors: 9
Hot Spot for Advancement for Women/Minorities: Y

Cintas Corp

www.cintas.com

NAIC Code: 812331

TYPES OF BUSINESS:

Linen & Uniform Supply
Uniform Rental, Sales & Cleaning
Uniform Design & Manufacturing
Outsourcing Services
Dust Control Services
Restroom Cleaning Services
Document Shredding & Management
First Aid & Safety Products

BRANDS/DIVISIONS/AFFILIATES:

CONTACTS: *Note: Officers with more than one job title may be intentionally listed here more than once.*

Scott D. Farmer, CEO
William C. Gale, CFO/Sr. VP
Thomas Frooman, General Counsel/VP/Corp. Sec.
J. Michael Hansen, Treas./VP
Robert J. Kohlhepp, Chmn.

GROWTH PLANS/SPECIAL FEATURES:

Cintas Corp. is a leading uniform supplier in the U.S. that designs, manufactures and implements corporate identity uniform programs. The company's products include entrance mats, restroom supplies, hygiene service supplies, first aid and safety products, fire protection products, document shredding and storage, cleanroom resources and flame resistant clothing. Cintas supplies products and services to more than 900,000 businesses. The firm operates through four segments: Rental Uniforms and Ancillary Products, which accounted for 71% of the company's 2011 revenue; Uniform Direct Sales, 11%; First Aid, Safety and Fire Protection Services, 10%; and Document Management Services, 7%. The Rental Uniforms and Ancillary Products segment controls the rental and servicing of uniforms and other garments, as well as supplying mats, mops, shop towels and restroom and hygiene products and services. Rental services include the cleaning of uniforms and providing on-going uniform replacements as required to each customer. The Uniform Direct Sales unit consists of the direct sale of uniforms and related items as well as branded promotional products. The First Aid, Safety and Fire Protection Services segment consists of on-site defibrillators and accessories, first aid kits and full fire protection systems and maintenance. Lastly, the Document Management Services division offers document destruction, document imaging and document retention services. The company also has specialized services tailored to the requirements of a number of additional industries, including automotive, casino, food processing, healthcare, lawn and garden, lodging, pest control, restaurant, supermarket and veterinary. The firm operates six manufacturing plants and eight distribution centers across North America, as well as more than 400 other facilities in nearly 300 cities nationwide. Cintas supplies its products and services through a national distribution network and approximately 7,800 local delivery routes.

Benefits that Cintas offers to its employees include medical insurance, a 401(k) plan, profit sharing, paid vacations and holidays, banking discounts and direct deposit.

FINANCIAL DATA: *Note: Data for latest year may not have been available at press time.*

In U.S. $	2012	2011	2010	2009	2008	2007
Revenue	4,102,000,000	3,810,384,000	3,547,339,000	3,774,685,000	3,937,900,000	3,706,900,000
R&D Expense						
Operating Income	539,627,000	440,343,000	390,809,000	409,065,000	577,455,000	577,397,000
Operating Margin %	13.15%	11.55%	11.01%	10.83%	14.66%	15.57%
SGA Expense	1,198,981,000	1,168,944,000	1,109,888,000	1,082,709,000	1,104,145,000	1,003,958,000
Net Income	297,637,000	246,989,000	215,620,000	226,357,000	335,405,000	334,538,000
Operating Cash Flow	469,862,000	340,886,000	561,572,000	523,522,000	544,543,000	449,391,000
Capital Expenditure	160,802,000	182,592,000	111,078,000	160,092,000	190,333,000	180,824,000
EBITDA	697,400,000	593,259,000	544,563,000	611,935,000	775,430,000	759,803,000
Return on Assets %	6.99%	5.93%	5.60%	6.01%	9.09%	9.56%
Return on Equity %	13.40%	10.21%	8.79%	9.79%	15.17%	15.72%
Debt to Equity	0.49	0.55	0.3	0.33	0.41	0.4

CONTACT INFORMATION:

Phone: 513 459-1200 Fax: 513 573-4030
Toll-Free:
Address: 6800 Cintas Blvd., Cincinnati, OH 45262 United States

STOCK TICKER/OTHER:

Stock Ticker: CTAS Exchange: NAS
Employees: 30,000 Fiscal Year Ends: 31-May
Parent Company:

SALARIES/BONUSES:

Top Exec. Salary: $769,153 Bonus: $
Second Exec. Salary: $562,277 Bonus: $

OTHER THOUGHTS:

Estimated Female Officers or Directors:
Hot Spot for Advancement for Women/Minorities:

Claire's Stores Inc

NAIC Code: 448120

www.clairestores.com

TYPES OF BUSINESS:

Teenage Apparel & Accessories, Retail
Jewelry
Novelty Items
Piercing Services
Franchising
Accessories
Seasonal Merchandise
Headwear

BRANDS/DIVISIONS/AFFILIATES:

Claire's
Icing
RSI
Apollo Advisors LP

CONTACTS: *Note: Officers with more than one job title may be intentionally listed here more than once.*

James D. Fielding, CEO
J. Per Brodin, CFO/Sr. VP
Dina C. Keenan, Chief Mktg. Officer/Sr. VP
Joseph Defalco, Sr. VP-Human Resources
Robert G. Cole, Sr. VP-IT
Rebecca Orand, General Counsel/Sr. VP
Colleen Collins, Sr. VP-Store Oper., USA
Andrea Guthrie, Sr. VP-Strategic New Bus.
Michael R. Basler, Sr. VP-Finance
Beatrice Lafon, Pres., Europe
Denise Vujovich, Exec. VP/Gen. Mgr.-Merch., North America
Shena MacDonald, Exec. VP/Gen. Mgr.-Merch., Europe
William Hoeller, Exec. VP-Int'l & Global Franchising
Kirk Hanselman, Exec. VP-Supply Chain, Allocation & Planning, U.S.

GROWTH PLANS/SPECIAL FEATURES:

Claire's Stores, Inc., owned by the private equity firm Apollo Advisors, LP., is a specialty retailer of value-priced costume jewelry and other fashion accessories. The company operates two store concepts: Claire's, which caters to fashion-conscious girls and teens age 3-18, and Icing, which caters to young women age 19-27; Icing only operates in North America. The company has 3071 stores, 1,953 of which operate in all 50 U.S. states, Canada, Puerto Rico and the U.S. Virgin Islands. The other 1,118 stores are located in the U.K., Switzerland, Netherlands, Portugal, Czech Republic, Hungary, Austria, Germany, France, Spain, Holland, Belgium and Ireland. In addition, the firm operates 381 franchised stores in the Middle East, Russia, South Africa, Poland, Greece, Guatemala, Malta and Turkey through various merchandising and licensing agreements. All stores share a similar format, with different store concepts and trade-names allowing for multiple store locations within a single mall. The company selects its merchandise from over 8,000 stock-keeping units (SKUs) and also provides ear-piercing services. Claire's stores are primarily located in shopping malls; and the company estimates that it has a store in approximately 90% of all U.S. shopping malls and 32 countries outside of the U.S. The firm purchases merchandise, which is largely imported, from approximately 650 suppliers and then ships to distribution centers located in Illinois, England, Switzerland and Austria. The company's distributor is based in Hong Kong and is responsible for buying, merchandise development and overseas quality control.

Employee benefits include life and supplemental life insurance; medical, dental and AD&D coverage; income protection programs; several forms of paid time off; and a 401(k) plan.

FINANCIAL DATA: *Note: Data for latest year may not have been available at press time.*

In U.S. $	2012	2011	2010	2009	2008	2007
Revenue	1,495,900,000	1,426,397,000	1,342,389,000	1,412,960,000	1,085,932,000	1,480,987,000
R&D Expense						
Operating Income						
Operating Margin %						
SGA Expense						
Net Income	11,632,000	4,323,000	-10,402,000	-643,592,000	641,000	188,762,000
Operating Cash Flow						
Capital Expenditure						
EBITDA						
Return on Assets %						
Return on Equity %						
Debt to Equity						

CONTACT INFORMATION:

Phone: 954-433-3900 Fax: 954-433-3999
Toll-Free: 800-252-4737
Address: 3 SW 129th Ave., Pembroke Pines, FL 33027 United States

STOCK TICKER/OTHER:

Stock Ticker: Private Exchange:
Employees: 18,300 Fiscal Year Ends: 31-Jan
Parent Company: APOLLO ADVISORS LP

SALARIES/BONUSES:

Top Exec. Salary: $ Bonus: $
Second Exec. Salary: $ Bonus: $

OTHER THOUGHTS:

Estimated Female Officers or Directors: 12
Hot Spot for Advancement for Women/Minorities: Y

Coach Inc

www.coach.com

NAIC Code: 315990

TYPES OF BUSINESS:

Leather Accessories
Online & Catalog Sales
Outlet Stores
Purses

BRANDS/DIVISIONS/AFFILIATES:

Coach.com
Coach Japan, Inc.

CONTACTS: *Note: Officers with more than one job title may be intentionally listed here more than once.*

Lew Frankfort, CEO
Jerry Stritzke, COO
Jerry Stritzke, Pres.
Jane Neilson, CFO/Exec. VP
Sarah Dunn, Sr. VP-Human Resources
Todd Kahn, General Counsel/Sr. VP/Sec.
Michael Tucci, Pres., Retail Div.-North American Retail
Reed Krakoff, Exec. Creative Dir.
Lew Frankfort, Chmn.

GROWTH PLANS/SPECIAL FEATURES:

Coach, Inc., founded in 1949, is a designer, producer and marketer of fine accessories and gifts for men and women, including handbags, women's and men's accessories, footwear, outerwear, business luggage and travel accessories, cases, eyewear, watches, jewelry and fragrance. The firm also licenses its name for watches, shoes and eyewear. The company sells its products through a number of direct channels, accounting for approximately 87% of its sales. Various Coach brand products are available at 3,000 department store locations in the U.S. and 211 international department stores, retail store and duty-free shop locations in over 20 countries. Approximately 63% of the company's sales come from handbags, roughly 27% come from accessories (wristlets, cosmetic cases, money pieces, etc.) and the final 10% of sales derive from all other products (sunglasses, watches, fragrance, etc.). Over the last several years, Coach has successfully transformed itself from a manufacturer of classic leather products to a marketer of more modern, fashionable handbags and accessories, using a broader range of fabrics and materials. Through the firm's web site Coach.com, customers can shop for new products and locate a store nearest to them. At approximately 850 company-owned stores, Coach's average sales per square foot of store space per year were an estimated $3,000 in 2012. The company has been expanding aggressively with new stores in China.

Coach employees receive medical, dental, life, short- and long-term disability, accidental death and dismemberment, dependent life and care, vision and prescription drug insurance; health care flexible spending accounts; an employee assistance program; a profit-sharing plan; a 401(k); an employee stock purchase plan; paid holiday, vacation and sick leave; product discounts; tuition assistance; transit discounts; and entertainment, travel and shopping discounts.

FINANCIAL DATA: *Note: Data for latest year may not have been available at press time.*

In U.S. $	2012	2011	2010	2009	2008	2007
Revenue	4,763,180,000	4,158,507,000	3,607,636,000	3,230,468,000	3,180,757,000	2,612,456,000
R&D Expense						
Operating Income	1,511,989,000	1,304,924,000	1,150,171,000	971,913,000	1,147,129,000	993,397,000
Operating Margin %	31.74%	31.37%	31.88%	30.08%	36.06%	38.02%
SGA Expense	1,954,089,000	1,718,617,000	1,483,520,000	1,350,697,000	1,259,974,000	1,029,589,000
Net Income	1,038,910,000	880,800,000	734,940,000	623,369,000	783,055,000	663,665,000
Operating Cash Flow	1,221,689,000	1,033,271,000	990,877,000	809,152,000	923,361,000	779,102,000
Capital Expenditure	184,309,000	147,744,000	81,116,000	240,329,000	174,720,000	140,872,000
EBITDA	1,644,898,000	1,430,030,000	1,276,915,000	1,094,927,000	1,247,833,000	1,074,284,000
Return on Assets %	36.20%	34.52%	29.21%	25.76%	33.15%	32.56%
Return on Equity %	57.62%	56.50%	45.91%	38.81%	45.71%	42.82%
Debt to Equity		0.01	0.01	0.01		

CONTACT INFORMATION:

Phone: 212 594-1850 Fax: 212 594-1682
Toll-Free: 888-262-6224
Address: 516 W. 34th St., New York, NY 10001 United States

STOCK TICKER/OTHER:

Stock Ticker: COH Exchange: NYS
Employees: 8,200 Fiscal Year Ends: 30-Jun
Parent Company:

SALARIES/BONUSES:

Top Exec. Salary: $2,850,650 Bonus: $1,101,475
Second Exec. Salary: $1,500,000 Bonus: $

OTHER THOUGHTS:

Estimated Female Officers or Directors: 5
Hot Spot for Advancement for Women/Minorities: Y

Coats plc

NAIC Code: 313110

www.coats.com

TYPES OF BUSINESS:

Fiber Manufacturing
Thread & Specialty Threads
Craft Products
Zippers & Trim

BRANDS/DIVISIONS/AFFILIATES:

Guiness Peat Group plc
Coats Sewing Solutions
Opti
Sylko
Epic
Anchor
Red Heart
Coats Insectiban

CONTACTS:

Note: Officers with more than one job title may be intentionally listed here more than once.

Paul Forman, CEO
Roger Bevan, CFO
Andy Speak, Chief Human Resources Officer
Richard Cammish, CIO
Rajiv Sharma, CEO-Industrial Div.
Michael Schofer, CEO-Crafts Div.
Frazer Smith, Chief Commercial Officer
Mike Allen, Chmn.
Chris Butcher, Chief Supply Chain Officer

GROWTH PLANS/SPECIAL FEATURES:

Coats plc, formerly Coats Group, Ltd., owned by London-based investment firm Guinness Peat Group plc (GPG), focuses exclusively on the industrial and craft thread markets. Coats has two businesses: Industrial and Crafts. The Industrial business, Coats Sewing Solutions, produces thread for apparel, embroidery and specialty purposes as well as for Opti brand zippers. Apparel thread is sold under the brands Epic, which features high seam strength and high temperature stability; Dual Duty, a durable mixed polyester and cotton thread; Gramax, a softer polyester thread; and Astra, a thread that resists degradation from chemicals and sunlight. Embroidery threads come in Sylko and Alcazar brands. The Industrial division also produces engineered specialty yarns for fiber optics, the medical industry and food applications, among others. Craft products include knitting, crocheting, quilting, stitching and sewing products sold in 70 countries, mainly in Europe as well as in North and South America, Asia and the Middle East. Its brands include Anchor, Red Heart, Maia, Patons, Rowan, Gedifra, Regia and Opti. In 2013, the firm released two new products Coats Insectiban, the first line of sewing threat and zippers designed to repel bed bugs; and Coats Protect an anti-microbial thread line.

FINANCIAL DATA:

Note: Data for latest year may not have been available at press time.

In U.S. $	2012	2011	2010	2009	2008	2007
Revenue		1,701,600,000	1,583,600,000	1,408,300,000	1,645,400,000	1,681,200,000
R&D Expense						
Operating Income						
Operating Margin %						
SGA Expense						
Net Income		71,200,000	59,900,000	-4,500,000	-4,000,000	67,000,000
Operating Cash Flow						
Capital Expenditure						
EBITDA						
Return on Assets %						
Return on Equity %						
Debt to Equity						

CONTACT INFORMATION:

Phone: 44-20-8210-5000 Fax: 44-20-8210-5025
Toll-Free:
Address: 1 The Square, Stockley Park, Uxbridge, Middlesex UB11 1TD United Kingdom

STOCK TICKER/OTHER:

Stock Ticker: Subsidiary Exchange:
Employees: 22,000 Fiscal Year Ends: 31-Dec
Parent Company: GUINNESS PEAT GROUP PLC

SALARIES/BONUSES:

Top Exec. Salary: $ Bonus: $
Second Exec. Salary: $ Bonus: $

OTHER THOUGHTS:

Estimated Female Officers or Directors:
Hot Spot for Advancement for Women/Minorities:

Coldwater Creek Inc

www.coldwatercreek.com

NAIC Code: 454113

TYPES OF BUSINESS:

Upscale Apparel-Women's
Catalog Sales
Gifts, Jewelry & Accessories
Retail & Outlet Stores
Online Sales

BRANDS/DIVISIONS/AFFILIATES:

Coldwater Creek, The Spa
Gifts-To-Go
ColdWaterCreek.com

CONTACTS: *Note: Officers with more than one job title may be intentionally listed here more than once.*

Dennis C. Pence, CEO
James A. Bell, COO/Exec. VP
Jill Brown Dean, Pres.
James A. Bell, CFO
Tim Dilworth, Sr. VP-Mktg.
John E. Hayes, III, Sr. VP-Human Resources
David Edwards, CIO/Sr. VP
Jill Brown Dean, Chief Merch. Officer
John E. Hayes, III, General Counsel
Jeffrey Parisian, Chief Strategy Officer/Sr. VP-Real Estate
Lyn Walther, VP-Investor Rel.
Jerome Jessup, Pres./Chief Creative Officer
Kim Sorenson, Sr. VP-Stores
Nancy Binger, Sr. VP-Brand & Creative Svcs.
Deanna Emsley, VP-Inventory Planning & Allocation
Dennis C. Pence, Chmn.

GROWTH PLANS/SPECIAL FEATURES:

Coldwater Creek, Inc. retails women's apparel, jewelry, footwear, gift items and home merchandise through three sales channels: a traditional catalog business, an e-commerce web site and retail stores. The company targets professional women who are 35 years of age and older, with household incomes in excess of $75,000. The Coldwater Creek catalog features merchandise available in the company's stores and is designed to encourage customers to shop in store. Coldwater Creek also sends periodic specialty mailings, such as the Gifts-To-Go holiday catalog. The firm's e-commerce site, ColdWaterCreek.com, features its entire full-priced, first-line merchandise collection. The company currently has 355 full-line stores and 38 factory outlet stores across the U.S. Additionally, the firm operates nine day spas under the name Coldwater Creek The Spa, which offer massages, facials, body treatments, manicures and pedicures and carry an assortment of relevant apparel as well as lines of personal care products. Coldwater Creek also maintains a customer contact and technical center in Coeur d'Alene, Idaho, comprising 60,000 square feet, and a 960,000-square-foot customer contact and distribution center in Mineral Wells, West Virginia.

Employees are offered medical, dental and life insurance; a 401(k) plan; an employee stock purchase plan; and discounts on Coldwater Creek products.

FINANCIAL DATA: *Note: Data for latest year may not have been available at press time.*

In U.S. $	2012	2011	2010	2009	2008	2007
Revenue	773,021,000	981,101,000	1,038,581,000	1,024,221,000	1,151,472,000	
R&D Expense						
Operating Income	-96,160,000	-43,379,000	-45,178,000	-46,212,000	-10,669,000	
Operating Margin %	-12.43%	-4.42%	-4.34%	-4.51%	-.92%	
SGA Expense	320,272,000	346,733,000	378,852,000	395,320,000	460,852,000	
Net Income	-99,694,000	-44,111,000	-56,132,000	-25,963,000	-2,488,000	
Operating Cash Flow	-44,224,000	-989,000	24,625,000	95,533,000	52,359,000	
Capital Expenditure	8,895,000	31,084,000	21,681,000	81,215,000	121,263,000	
EBITDA	-39,417,000	20,942,000	19,314,000	17,854,000	43,479,000	
Return on Assets %	-21.67%	-8.09%	-9.26%	-4.14%	-.41%	
Return on Equity %	-64.43%	-20.58%	-21.67%	-8.88%	-.80%	
Debt to Equity	0.22	0.06	0.04	0.04	0.04	

CONTACT INFORMATION:

Phone: 208-263-2266 Fax: 208-263-1582
Toll-Free: 800-510-2808
Address: 1 Coldwater Creek Dr., Sandpoint, ID 83864 United States

STOCK TICKER/OTHER:

Stock Ticker: CWTR Exchange: NAS
Employees: 9,198 Fiscal Year Ends: 31-Jan
Parent Company:

SALARIES/BONUSES:

Top Exec. Salary: $576,923 Bonus: $312,000
Second Exec. Salary: $726,923 Bonus: $

OTHER THOUGHTS:

Estimated Female Officers or Directors: 6
Hot Spot for Advancement for Women/Minorities: Y

Columbia Sportswear Co

NAIC Code: 315210

www.columbia.com

TYPES OF BUSINESS:

Sportswear & Outdoor Apparel
Sports Accessories & Equipment
Footwear
Retail Sales
Bicycles

BRANDS/DIVISIONS/AFFILIATES:

Columbia
Montrail
Mountain Hardwear
Sorel

CONTACTS: *Note: Officers with more than one job title may be intentionally listed here more than once.*

Timothy P. Boyle, CEO
Bryan L. Timm, COO/Exec. VP
Timothy P. Boyle, Pres.
Thomas B. Cusick, CFO
Daniel G. Hanson, VP-Mktg.
Susan G. Popp, VP-Human Resources
Frederick R. Pond, CIO/VP
Mark Nenow, VP-Global Footwear Merch.
Patrick J. Werner, VP-Global Apparel Mfg.
Peter J. Bragdon, General Counsel/Sec./Sr. VP-Legal & Corp. Affairs
Lisa A. Kulok, VP-Customer Oper. & Marketplace Planning
Ron Parham, Sr. Dir.-Corp. Comm.
Ron Parham, Sr. Dir.-Investor Rel.
Thomas B. Cusick, Sr. VP-Finance
Christopher A. Gaylord, Pres., Mountain Hardwear
Kerry Barnes, VP-Retail
C. Mitchell Fields, VP-Global Apparel Sales
Adrienne L. Moser, VP-Apparel & Design
Gertrude Boyle, Chmn.
William Tung, VP-Latin America & Asia Pacific
Daniel A. Dougherty, VP-Global Dist.

GROWTH PLANS/SPECIAL FEATURES:

Columbia Sportswear Co. designs, sources, markets and distributes active outdoor apparel and footwear. The firm is one of the largest outerwear companies in the world and a leading seller of skiwear in the U.S. The company's products are sold in more than 100 countries across North America, Europe and Asia. Its apparel offerings include outerwear and other accessories designed to protect the user from weather in everyday use and in outdoor activities including skiing, hiking, hunting and fishing; and sportswear, which includes fishing and hunting shirts, hiking shorts, fleece and pile products, sweaters, chinos, knit shirts, sweats and jeans. Footwear consists of seasonal outdoor footwear for adults and youth, designed with the outdoor consumer in mind. Related accessories and equipment include hats, caps, scarves, gloves and headbands, tents and camping gear. Additionally, the firm licenses its brand name for products such as socks, packs and travel bags, leather goods, watches, cutlery and home furnishings. Columbia's brand names include the Columbia, Montrail, Mountain Hardwear and Sorel. The company's products are sold in specialty and department stores throughout the world, with its U.S. sales accounting for 56.7% of 2012 revenue. The firm operates approximately 150 branded and outlet retail stores in the U.S. and internationally. In August 2012, the firm established a joint venture with Swire Resources Ltd. to expand its business in China.

Columbia employees receive medical, dental and vision insurance; a 401(k) plan; education assistance; employee discounts; an employee volunteer program; and paid time off.

FINANCIAL DATA: *Note: Data for latest year may not have been available at press time.*

In U.S. $	2012	2011	2010	2009	2008	2007
Revenue	1,669,563,000	1,693,985,000	1,483,524,000	1,244,023,000	1,317,835,000	1,356,039,000
R&D Expense						
Operating Income	133,528,000	136,406,000	103,327,000	87,762,000	118,706,000	199,139,000
Operating Margin %	7.99%	8.05%	6.96%	7.05%	9.00%	14.68%
SGA Expense	596,635,000	614,658,000	534,068,000	444,715,000	430,350,000	385,769,000
Net Income	99,859,000	103,479,000	77,037,000	67,021,000	95,047,000	144,452,000
Operating Cash Flow	148,668,000	63,796,000	23,517,000	214,372,000	144,930,000	124,522,000
Capital Expenditure	50,491,000	78,404,000	28,838,000	33,074,000	47,580,000	34,289,000
EBITDA	174,420,000	179,966,000	141,757,000	124,015,000	157,464,000	238,517,000
Return on Assets %	7.02%	7.73%	6.14%	5.67%	8.21%	13.16%
Return on Equity %	8.91%	9.96%	7.70%	6.90%	9.93%	16.04%
Debt to Equity						

CONTACT INFORMATION:

Phone: 503 985-4000 Fax:
Toll-Free: 800-622-6953
Address: 14375 NW Science Park Dr., Portland, OR 97229 United States

STOCK TICKER/OTHER:

Stock Ticker: COLM
Exchange: NAS
Employees: 4,161
Fiscal Year Ends: 31-Dec
Parent Company:

SALARIES/BONUSES:

Top Exec. Salary: $529,615 Bonus: $
Second Exec. Salary: $515,673 Bonus: $

OTHER THOUGHTS:

Estimated Female Officers or Directors: 5
Hot Spot for Advancement for Women/Minorities: Y

Cone Denim LLC

www.conedenim.com

NAIC Code: 313210

TYPES OF BUSINESS:

Fabrics, Manufacturing
Denim Production
Finishing & Dyeing Services

BRANDS/DIVISIONS/AFFILIATES:

International Textile Group Inc
Cone Denim (Jiaxing) Limited
White Oak
Sustainblue
Selvage
S Gene
S Gene 2 Bi-stretch Denim

CONTACTS: *Note: Officers with more than one job title may be intentionally listed here more than once.*

Kenneth Kunberger, Pres., ITG Apparel & Specialty Fabrics Group
Joseph L. Gorga, Pres./CEO-ITG

GROWTH PLANS/SPECIAL FEATURES:

Cone Denim LLC is one of the world's leading producers of denim fabrics in North America. A subsidiary of International Textile Group (ITG), the firm operates in the U.S., Mexico and China, with additional sales operations in Japan, Italy, Korea and Australia. The company produces denim in Mexico through Cone Mexico; and in China through Cone Denim (Jiaxing) Limited, a joint venture. Its denims are usually yarn-dyed, meaning the yarn is dyed before the fabric is woven, which produces variations of color in the resulting products. These fabrics are marketed to premium and vintage jeanswear markets, where styling and innovation are important factors, as well as to the fashion and better basic jeanswear markets, cut and sew contractors and apparel wholesalers. The firm's product developers and designers work directly with customers to develop and provide differentiated denim products. Cone Denim also markets the White Oak collection, produced in the U.S., which combines vintage designs with new materials and processes. Other brands include Sustainblue, Selvage, S Gene and S Gene 2 Bi-stretch Denim. Cone Denim and Burlington Worldwide, another ITG company, comprise the ITG Apparel & Specialty Fabrics business unit.

FINANCIAL DATA: *Note: Data for latest year may not have been available at press time.*

In U.S. $	2012	2011	2010	2009	2008	2007
Revenue						
R&D Expense						
Operating Income						
Operating Margin %						
SGA Expense						
Net Income						
Operating Cash Flow						
Capital Expenditure						
EBITDA						
Return on Assets %						
Return on Equity %						
Debt to Equity						

CONTACT INFORMATION:

Phone: 336-379-6220 Fax: 336-379-6287
Toll-Free:
Address: 804 Green Valley Rd., Ste. 300, Greensboro, NC 27408 United States

STOCK TICKER/OTHER:

Stock Ticker: Subsidiary Exchange:
Employees: 3,200 Fiscal Year Ends: 31-Dec
Parent Company: INTERNATIONAL TEXTILE GROUP INC

SALARIES/BONUSES:

Top Exec. Salary: $ Bonus: $
Second Exec. Salary: $ Bonus: $

OTHER THOUGHTS:

Estimated Female Officers or Directors:
Hot Spot for Advancement for Women/Minorities:

Converse Inc

NAIC Code: 316210

www.converse.com

TYPES OF BUSINESS:

Athletic Shoes
Athletic Apparel

BRANDS/DIVISIONS/AFFILIATES:

Nike Inc
Chuck Taylor All Star
Jack Purcell
Converse by John Varvatos
One Star
All Star
Star Chevron
DC Comics

CONTACTS: *Note: Officers with more than one job title may be intentionally listed here more than once.*

Michael Spillane, CEO
Geoff Cottrill, Chief Mktg. Officer
Keith Gulla, Mgr.-Communications

GROWTH PLANS/SPECIAL FEATURES:

Converse, Inc., a wholly-owned subsidiary of Nike Inc., is a leading designer and producer of athletic shoes and sports apparel. The company, headquartered in Massachusetts, is most famous for its iconic Chuck Taylor All Star canvas basketball shoes. The company has sold over 800 million pairs of this model of shoe alone. The shoe was introduced in 1917 and remained the only high-performance basketball shoe mass-produced for over 50 years after its introduction. Converse divides its products into three categories: Sports Performance, which includes footwear and apparel designed for athletic activities such as basketball and skateboarding; the Sports Lifestyle collection, including sport-inspired casual footwear and clothing; and Sports Classics, which in addition to the Chuck Taylor brand includes the Jack Purcell line of sneakers. Through the Converse web site, customers can order products from original and new collections or choose to personally design their own unique footwear. In addition, the company designs and sells shoes, apparel and accessories under the All Star, One Star, Star Chevron, DC Comics, Dr. Seuss and Converse by John Varvatos brand names. Aside from direct product sales, the firm also generates revenues by licensing its name to third party sports apparel manufacturers.

FINANCIAL DATA: *Note: Data for latest year may not have been available at press time.*

In U.S. $	2012	2011	2010	2009	2008	2007
Revenue		1,130,000,000	983,000,000	915,000,000	729,000,000	550,000,000
R&D Expense						
Operating Income						
Operating Margin %						
SGA Expense						
Net Income						
Operating Cash Flow						
Capital Expenditure						
EBITDA						
Return on Assets %						
Return on Equity %						
Debt to Equity						

CONTACT INFORMATION:

Phone: 978-983-3300 Fax: 978-983-3502
Toll-Free: 888-792-3307
Address: 1 High St., North Andover, MA 01845-2601 United States

STOCK TICKER/OTHER:

Stock Ticker: Subsidiary Exchange:
Employees: Fiscal Year Ends: 31-May
Parent Company: NIKE INC

SALARIES/BONUSES:

Top Exec. Salary: $ Bonus: $
Second Exec. Salary: $ Bonus: $

OTHER THOUGHTS:

Estimated Female Officers or Directors:
Hot Spot for Advancement for Women/Minorities:

Cortefiel SA

www.grupocortefiel.com

NAIC Code: 448140

TYPES OF BUSINESS:

Men's & Women's Apparel, Retail
Lingerie Stores
Perfume & Cosmetics
Online Sales

BRANDS/DIVISIONS/AFFILIATES:

Cortefiel
Springfield
Women'Secret
Pedro del Hierro
Fifty Factory
Antonio Miro
WomenSecret.com
Cortefiel.com

CONTACTS: *Note: Officers with more than one job title may be intentionally listed here more than once.*

Juan Garcia-Bosque, CEO
Ignacio Garcia, CFO
Joaquin G.Q. Rodriguez, Corp. Sec.
Ignacio Garcia, Dir.-Investor Rel.
Alex Cara, Managing Dir.-Global Franchise
Gonzalo H.F. de Angulo, Chmn.

GROWTH PLANS/SPECIAL FEATURES:

Cortefiel SA operates retail chains that specialize in clothing, perfume and lingerie. The company operates 1,855 stores in 63 countries, including 546 franchises mostly under the Cortefiel, Pedro del Hierro, Springfield and Women'Secret brands. Cortefiel is a women's clothing brand with a modern fashionable look that is continually updated with in-store releases of new color proposals, combinations and styles every two weeks. Pedro del Hierro, or PdH, is a pret-a-porter (ready-to-wear) chain of independent stores and has locations within Cortefiel stores. Springfield is a men's fashion brand with a cosmopolitan, urban and youthful feel, targeted to young men ages 20-30. Women'Secret is a lingerie and casualwear chain targeted to women ages 18-35 that retails online at WomenSecret.com and worldwide through its franchised stores. In addition, the firm operates an outlet chain for all the Cortefiel brands under the name Fifty Factory. Cortifiel owns six factories, including two in Spain, three in Morocco and one in Hungary. Recently, Cortefiel announced the launch of Cortefiel.com, completing the firms desire to make all its stores and brands accessible online.

FINANCIAL DATA: *Note: Data for latest year may not have been available at press time.*

In U.S. $	2012	2011	2010	2009	2008	2007
Revenue						
R&D Expense						
Operating Income						
Operating Margin %						
SGA Expense						
Net Income						
Operating Cash Flow						
Capital Expenditure						
EBITDA						
Return on Assets %						
Return on Equity %						
Debt to Equity						

CONTACT INFORMATION:

Phone: 34-902-45-3545 Fax: 34-913-87-3874
Toll-Free:
Address: Ave. Del Llano Castellano, 51, Madrid, 28034 Spain

STOCK TICKER/OTHER:

Stock Ticker: Private
Exchange:
Employees:
Fiscal Year Ends: 28-Feb
Parent Company:

SALARIES/BONUSES:

Top Exec. Salary: $ Bonus: $
Second Exec. Salary: $ Bonus: $

OTHER THOUGHTS:

Estimated Female Officers or Directors: 1
Hot Spot for Advancement for Women/Minorities:

Crocs Inc

NAIC Code: 316210

www.crocs.com

TYPES OF BUSINESS:

Footwear
Apparel
Accessories

BRANDS/DIVISIONS/AFFILIATES:

Crocs
Croslite
YOU by Crocs
Crocs Work
Jibbitz
Crocs RX
Ocean Minded

CONTACTS:

Note: Officers with more than one job title may be intentionally listed here more than once.

John McCarvel, CEO
Scott Crutchfield, COO
John McCarvel, Pres.
Jeff Lasher, CFO
Mike DeBell, Chief Sales Officer
Dale Bathum, Chief Prod. Officer
Dan Hart, Chief Admin. Officer/Exec. VP
Dan Hart, Chief Legal Officer
Vince Gunn, VP-Europe
Dave Thielen, VP-Asia

GROWTH PLANS/SPECIAL FEATURES:

Crocs, Inc. is a designer, manufacturer and marketer of footwear for men, women and children under the Crocs brand. All of the company's footwear products incorporate its proprietary closed-cell resin material, known as Croslite, which enables the production of soft, lightweight, non-marking and odor-resistant shoes. To provide maximum comfort and functionality, Crocs shoes are designed around several structural features such as Italian-designed orthotic heels, built-in arch supports and tarsal bars. These properties make the footwear appropriate for casual wear as well as for recreational uses such as boating, hiking, fishing or gardening. Apart from its Crocs line of footwear, the company produces the Crocs Work line for the hospital, restaurant and hospitality markets; Crocs RX for the general foot care and diabetic-needs markets; Ocean Minded sandals for beachwear; and YOU by Crocs, a fashion line for women. Crocs' more than 300 models are available in thousands of stores nationwide, with retail price points ranging from $20 to $60. The firm operates 537 retail locations and mall kiosks throughout 90 countries. In addition to footwear, Crocs markets a line of apparel and accessory items, including Jibbitz accessories, which are specifically suited to snap on to Crocs shoes. The firm also produces a line of lightweight kneepads made from Croslite, which it markets to gardening retailers and hardware stores; golf shoes; sunglasses; and other apparel. Since Crocs' start in 2002, the company has sold more than 200 million pairs of shoes.

Employees are offered medical, dental and vision insurance; flexible spending accounts; a 401(k); and complimentary shoes.

FINANCIAL DATA:

Note: Data for latest year may not have been available at press time.

In U.S. $	2012	2011	2010	2009	2008	2007
Revenue	1,123,301,000	1,000,903,000	789,695,000	645,767,000	721,589,000	847,350,000
R&D Expense						
Operating Income	146,174,000	136,505,000	81,035,000	-51,184,000	-188,282,000	237,767,000
Operating Margin %	13.01%	13.63%	10.26%	-7.92%	-26.09%	28.06%
SGA Expense	460,393,000	404,803,000	342,121,000	311,592,000	368,800,000	259,882,000
Net Income	131,343,000	112,788,000	67,726,000	-42,078,000	-185,076,000	168,228,000
Operating Cash Flow	128,356,000	142,376,000	104,274,000	61,109,000	72,863,000	8,938,000
Capital Expenditure	60,836,000	41,640,000	45,105,000	27,027,000	66,218,000	73,904,000
EBITDA	183,079,000	174,806,000	118,508,000	-10,455,000	-150,267,000	261,713,000
Return on Assets %	17.22%	18.11%	14.12%	-9.72%	-34.16%	36.29%
Return on Equity %	23.68%	25.99%	20.40%	-14.64%	-50.61%	51.57%
Debt to Equity						

CONTACT INFORMATION:

Phone: 303 848-7000 Fax: 303 468-4266
Toll-Free: 866-306-3179
Address: 6328 Monarch Park Pl., Niwot, CO 80503 United States

STOCK TICKER/OTHER:

Stock Ticker: CROX Exchange: NAS
Employees: 4,157 Fiscal Year Ends: 31-Dec
Parent Company:

SALARIES/BONUSES:

Top Exec. Salary: $840,000 Bonus: $
Second Exec. Salary: $498,750 Bonus: $

OTHER THOUGHTS:

Estimated Female Officers or Directors:
Hot Spot for Advancement for Women/Minorities:

CSP International Industria Calze SpA

www.cspinternational.it

NAIC Code: 315240

TYPES OF BUSINESS:

Underwear & Hosiery

BRANDS/DIVISIONS/AFFILIATES:

Oroblu
Sanpellegrino
Le Bourget
Lepel
Belsole Lepel
Claudia Lemes
6teen
Textiles Well SA

CONTACTS: *Note: Officers with more than one job title may be intentionally listed here more than once.*

Simone Ruffoni, Head-Investor Rel.
Maria G. Bertoni, Vice Chmn.
Mario Bertoni, Exec. Dir.
Francesco Bertoni, Chmn.

GROWTH PLANS/SPECIAL FEATURES:

CSP International Industria Calze SpA, established in 1873, is one of Italy's largest manufacturers of hosiery and underwear. The company distributes products to 40,000 wholesalers and retailers in over 40 countries. CSP's proprietary brands include premium-priced Oroblu, a hosiery, corsetry and seamless underwear brand found mainly in Western Europe; Sanpellegrino, a value-oriented tights and seamless underwear brand; Le Bourget, a leading French hosiery brand; and Lepel, a leading Italian bra and corsetry brand. Through the Lepel subsidiary, the firm also owns the Belsole Lepel, 6teen, Miss Lepel and Claudia Lemes sub-brands. CSP's seamless underwear collections include Sanpellegrino Comodo and Oroblu Dolce Vita. The company offers a casual tights collection for fashion-conscious teenagers under the Oroblu On Line brand. In addition, the firm produces men's underwear and licenses its brands for men's socks, swimwear and lingerie. CSP also distributes various private-label products. For the Sixty Group, CSP distributes underwear, swimwear and hosiery collections for women under the brand name Miss Sixty and under the name Energie for men; for designer Sergio Tacchini, it distributes swimwear, t-shirts and accessories; and for Puma, it distributes sports socks in the Italian market. CSP also operates a production unit in Poland, Sanpellegrino Polska. In recent years, the firm acquired the French hosiery company Textiles Well SA.

FINANCIAL DATA: *Note: Data for latest year may not have been available at press time.*

In U.S. $	2012	2011	2010	2009	2008	2007
Revenue			141,800,000	132,300,000	149,700,000	157,000,000
R&D Expense						
Operating Income						
Operating Margin %						
SGA Expense						
Net Income			7,900,000	7,800,000	7,300,000	10,900,000
Operating Cash Flow						
Capital Expenditure						
EBITDA						
Return on Assets %						
Return on Equity %						
Debt to Equity						

CONTACT INFORMATION:

Phone: 39-0376-8101 Fax: 39-0376-87573
Toll-Free:
Address: Via Piubega, 5C, Ceresara, 46040 Italy

STOCK TICKER/OTHER:

Stock Ticker: CSP Exchange: Milan
Employees: 910 Fiscal Year Ends: 31-Dec
Parent Company:

SALARIES/BONUSES:

Top Exec. Salary: $ Bonus: $
Second Exec. Salary: $ Bonus: $

OTHER THOUGHTS:

Estimated Female Officers or Directors: 3
Hot Spot for Advancement for Women/Minorities: Y

Culp Inc

NAIC Code: 313210

www.culpinc.com

TYPES OF BUSINESS:

Upholstery Fabrics
Mattress Fabrics
Yarn

BRANDS/DIVISIONS/AFFILIATES:

Culp Home Fashions
Culp Express
Embrace

CONTACTS:

Note: Officers with more than one job title may be intentionally listed here more than once.

Franklin N. Saxon, CEO
Franklin N. Saxon, Pres.
Kenneth R. Bowling, CFO/Treas.
Kenneth R. Bowling, Corp. Sec.
Thomas B. Gallagher, Jr., Controller
Robert G. Culp, IV, Pres., Culp Home Fashions
Robert G. Culp, III, Chmn.

GROWTH PLANS/SPECIAL FEATURES:

Culp, Inc. is a leading manufacturer of mattress fabrics used for covering mattresses and box springs, and upholstery fabrics primarily for use in the production of residential and commercial upholstered furniture. Culp's fabrics go into mattress sets by Sealy, Serta and Simmons; and into sofas, recliners, chairs, love seats, sectionals, sofa-beds and office seating manufactured by companies and brands like La-Z-Boy, Lane, Broyhill, Basset and Flexsteel. The company is divided into two segments: mattress fabrics, representing 57.2% of sales; and upholstery fabrics, which accounts for 42.8%. Subsidiary Culp Home Fashions produces the company's mattress fabrics, which encompass woven jacquards (floral and other intricate designs woven on complex looms), knitted fabric (designs produced on special-width circular machines utilizing a variety of synthetic and natural yarns) and some upholstery type fabrics. The upholstery fabrics division includes woven jacards, woven dobbies, velvets and suede fabrics under the Culp Express and Embrace product lines. Culp Express offers a wide range of colors, styles and prints, from solid to patterns, while the Embrace line focuses on the production of recycled leather fabric. The company has twelve manufacturing facilities located in North and South Carolina; Poznan, Poland; Quebec, Canada; and Shanghai, China. Culp also sources fabrics from other manufacturers, located primarily in China and Turkey, with almost all of such fabrics being produced specifically for Culp and created by Culp designers. In recent years, the portion of the company's sales represented by fabrics produced outside of the U.S. and Canada has increased significantly, especially in the company's upholstery fabrics segment, where more than 85% of its sales consist of fabrics produced in China.

FINANCIAL DATA:

Note: Data for latest year may not have been available at press time.

In U.S. $	2012	2011	2010	2009	2008	2007
Revenue	254,443,000	216,806,000	206,416,000	203,938,000	254,046,000	250,533,000
R&D Expense						
Operating Income	14,706,000	15,743,000	16,342,000	-4,570,000	8,300,000	641,000
Operating Margin %	5.77%	7.26%	7.91%	-2.24%	3.26%	.25%
SGA Expense	25,026,000	21,069,000	22,805,000	19,751,000	23,973,000	27,030,000
Net Income	13,296,000	16,164,000	13,188,000	-38,842,000	5,385,000	-1,316,000
Operating Cash Flow	12,003,000	14,821,000	21,577,000	22,768,000	16,364,000	11,465,000
Capital Expenditure	5,890,000	6,352,000	7,431,000	1,970,000	4,846,000	6,262,000
EBITDA	20,086,000	20,757,000	20,188,000	2,676,000	13,739,000	8,779,000
Return on Assets %	9.67%	13.32%	12.68%	-31.92%	3.49%	-.82%
Return on Equity %	15.70%	22.54%	23.74%	-57.80%	6.51%	-1.71%
Debt to Equity	0.07	0.11	0.24	0.24	0.16	0.27

CONTACT INFORMATION:

Phone: 336 889-5161 Fax: 336 889-8339
Toll-Free:
Address: 1823 Eastchester Dr., High Point, NC 27265 United States

STOCK TICKER/OTHER:

Stock Ticker: CFI
Exchange: NYS
Employees: 1,143
Fiscal Year Ends: 30-Apr
Parent Company:

SALARIES/BONUSES:

Top Exec. Salary: $359,625 Bonus: $
Second Exec. Salary: $282,562 Bonus: $

OTHER THOUGHTS:

Estimated Female Officers or Directors:
Hot Spot for Advancement for Women/Minorities:

Danier Leather Inc

www.danier.com

NAIC Code: 315280

TYPES OF BUSINESS:

Leather & Suede Goods
Men's & Women's Leather Apparel
Leather Accessories
Retail Stores
Online Sales

BRANDS/DIVISIONS/AFFILIATES:

Canada Sportswear

CONTACTS:

Note: Officers with more than one job title may be intentionally listed here more than once.

Jeffrey Wortsman, CEO
Jeffrey Wortsman, Pres.
Bryan Tatoff, CFO/Sr. VP
Elizabeth Gill, VP-Mktg.
Phil Cutter, CIO/VP-IT
Olga Koel, Chief Merch. Officer/Exec. VP
Bryan Tatoff, Corp. Sec.
Cris Ruivo, VP-Store Oper.
Bruce Aitken, VP-Planning & Allocation
Karen Marshall, VP-Logistics & Dist.
Michael Watson, Sr.VP-Retail Oper.
Edwin F. Hawken, Chmn.
Guia Lopez, Chief Sourcing Officer/VP

GROWTH PLANS/SPECIAL FEATURES:

Danier Leather, Inc. designs, manufactures and retails leather and suede apparel and accessories. The company's products include men's and women's leather jackets, pants, tops, coats and blazers as well as business and travel accessories for both sexes. The firm's products are sold in approximately 90 locations in Canada including shopping malls, store fronts and power centers. At certain locations, customers can also receive alteration services. Additionally, the company offers corporate promotional and incentive products, including briefcases, laptop bags, portfolios, wallets and notebook and passport covers, through its corporate sales division, Canada Sportswear. For the creation of one single garment, an average of 87 sewing operations and 18 pattern-making operations are used. The firm considers its core customers to be middle to upper-income females between the ages of 35-55. Danier has increased its brand-oriented advertising through mediums such as magazines, billboards, transit shelters and fashion-oriented brochures as well as through more frequent and better quality newspaper inserts, and aims to increase the sales of its accessory products. Outerwear products account for approximately 60% of Danier's total sales and sportswear represents about 8%, with accessories sales accounting for approximately 32%. In 2012, the firm opened up a second accessories sourcing office in China and aims to increase accessory sales to 40%-50% of total revenue by next year.

FINANCIAL DATA:

Note: Data for latest year may not have been available at press time.

In U.S. $	2012	2011	2010	2009	2008	2007
Revenue	150,354,000	159,891,500	166,582,500	164,441,100	165,905,900	160,376,400
R&D Expense						
Operating Income	5,426,050	11,001,220	11,106,720	-3,165,957	-2,431,528	2,205,316
Operating Margin %	3.60%	6.88%	6.66%	-1.92%	-1.46%	1.37%
SGA Expense	72,384,860	76,587,540	76,959,830	75,802,390	79,713,940	77,459,930
Net Income	4,060,661	7,748,022	7,322,987	-2,342,260	17,790,630	1,676,811
Operating Cash Flow	10,270,850	4,588,152	14,204,710	10,928,180	5,832,826	13,611,280
Capital Expenditure	4,083,993	2,424,427	2,710,489	3,627,511	3,947,048	2,906,269
EBITDA	9,391,357	15,100,430	15,412,860	2,100,832	3,351,593	8,383,039
Return on Assets %	5.16%	10.14%	10.08%	-3.24%	22.65%	2.01%
Return on Equity %	6.24%	13.00%	12.94%	-3.95%	32.22%	3.42%
Debt to Equity						0.01

CONTACT INFORMATION:

Phone: 416 762-8175 Fax: 416 762-4570
Toll-Free: 877-932-6437
Address: 2650 St. Clair Ave. W., Toronto, ON M6N 1M2 Canada

STOCK TICKER/OTHER:

Stock Ticker: DL Exchange: TSE
Employees: 1,125 Fiscal Year Ends: 30-Jun
Parent Company:

SALARIES/BONUSES:

Top Exec. Salary: $ Bonus: $
Second Exec. Salary: $ Bonus: $

OTHER THOUGHTS:

Estimated Female Officers or Directors: 5
Hot Spot for Advancement for Women/Minorities: Y

Deb Shops Inc

www.debshops.com

NAIC Code: 448120

TYPES OF BUSINESS:

Women's Apparel, Retail
Outlet Stores
Teen Apparel
Plus-Size Apparel

GROWTH PLANS/SPECIAL FEATURES:

Deb Shops, Inc. operates over 310 women's specialty apparel retail stores in 42 states under the DEB and Tops â€˜N Bottoms names. The firm is owned by private equity company Lee Equity Partners LLC. In addition, the company offers its merchandise on its e-commerce web site DebShops.com. Deb's stores offer moderately priced, coordinated women's sportswear, dresses, coats, lingerie, accessories and shoes for junior and plus-sizes. DEB merchandise consists of clothing and accessories for female consumers from size 0-26 aimed primarily at the 13-25 year age range. Stores tend to be located in either enclosed or strip malls. Deb uses an electronic point of sale merchandise data system to keep a close eye on changing purchase trends. Slow-moving products are often given an early discount.

BRANDS/DIVISIONS/AFFILIATES:

DEB
Lee Equity Partners LLC
DebShops.com

CONTACTS: *Note: Officers with more than one job title may be intentionally listed here more than once.*

Thomas Haskell Lee, Pres., Lee Equity Partners LLC

FINANCIAL DATA: *Note: Data for latest year may not have been available at press time.*

In U.S. $	2012	2011	2010	2009	2008	2007
Revenue						
R&D Expense						
Operating Income						
Operating Margin %						
SGA Expense						
Net Income						
Operating Cash Flow						
Capital Expenditure						
EBITDA						
Return on Assets %						
Return on Equity %						
Debt to Equity						

CONTACT INFORMATION:

Phone: 215-676-6000 Fax: 215-969-2830
Toll-Free: 866-803-4803
Address: 9401 Blue Grass Rd., Philadelphia, PA 19144 United States

STOCK TICKER/OTHER:

Stock Ticker: Private Exchange:
Employees: Fiscal Year Ends: 31-Jan
Parent Company: LEE EQUITY PARTNERS LLC

SALARIES/BONUSES:

Top Exec. Salary: $ Bonus: $
Second Exec. Salary: $ Bonus: $

OTHER THOUGHTS:

Estimated Female Officers or Directors:
Hot Spot for Advancement for Women/Minorities:

Deckers Outdoor Corporation

www.deckers.com

NAIC Code: 316210

TYPES OF BUSINESS:

Footwear-Outdoor & Sports
Direct Sales, Footwear
Sandals
Hiking Boots
Trail Running Shoes

BRANDS/DIVISIONS/AFFILIATES:

Teva
UGG Australia
Sanuk
Tsubo
Mozo
Ahnu
Fraid Not
Sidewalk Surfers

CONTACTS: *Note: Officers with more than one job title may be intentionally listed here more than once.*

Angel R. Martinez, CEO
Zohar Ziv, COO
Angel R. Martinez, Pres.
Thomas A. George, CFO
Jessica Buttimer, Chief Mktg. Officer
Graciela Montgomery, Chief Human Resources Officer
Yul Vanek, VP-IT
David Lafitte, General Counsel
Dave Powers, Pres., Global Direct to Consumer
Constance X. Rishwain, Pres., UGG Australia
Joel Heath, Pres., Teva
Jake Brandman, Pres., Sanuk
Jim Van Dine, Pres., Ahnu & Tsubo
Angel R. Martinez, Chmn.
Peter K. Worley, Pres., Deckers Asia Pacific
Mark N. Fegley, Sr. VP-Supply Chain

GROWTH PLANS/SPECIAL FEATURES:

Deckers Outdoor Corporation designs and produces both casual and outdoor active footwear for men, women and children. Additionally, the company sells accessories such as outerwear and handbags through its channels of domestic and international distributors, retailers and direct to consumers using its retail stores and web sites. The firm's proprietary brands include UGG, Teva and Sanuk. UGG Australia offers luxury sheepskin footwear as well as cold weather accessories, apparel and handbags. Recently, UGG has focused on increasing its merchandising capabilities by creating products targeted towards men. UGG sells its products in its 56 concept stores and 21 retail outlets, but primarily its products are sold in higher-end department stores, specialty retailers and online. Teva is the company's outdoor performance and lifestyle brand, known for its proprietary sport sandal product offering. Teva markets its open and closed-toe footwear through sporting goods retailers like Dick's Sporting Goods, The Sports Authority and REI. Sanuk is an action sport footwear brand rooted in the surf culture. Its products include Yoga Mat sandals, Fraid Not sandals and Sidewalk Surfers shoes and are sold at sport retailers, high-end outlets and national chain stores. Additionally, the company's brand offerings include Mozo, a line of footwear primarily targeting individuals who spend long hours on their feet; Ahnu, an outdoor lifestyle and performance footwear brand; and Tsubo, a high-end casual footwear brand. During 2012, Deckers opened seven UGG stores in the U.S. and 23 internationally.

FINANCIAL DATA: *Note: Data for latest year may not have been available at press time.*

In U.S. $	2012	2011	2010	2009	2008	2007
Revenue	1,414,398,000	1,377,283,000	1,000,989,000	813,177,000	689,445,000	448,929,000
R&D Expense						
Operating Income	186,948,000	284,838,000	249,088,000	181,247,000	116,919,000	105,553,000
Operating Margin %	13.21%	20.68%	24.88%	22.28%	16.95%	23.51%
SGA Expense	445,206,000	394,157,000	253,850,000	188,843,000	152,574,000	101,918,000
Net Income	128,866,000	199,052,000	158,235,000	116,786,000	73,948,000	66,437,000
Operating Cash Flow	163,906,000	30,091,000	139,922,000	185,474,000	53,276,000	61,054,000
Capital Expenditure	66,533,000	59,563,000	22,489,000	13,699,000	22,337,000	6,385,000
EBITDA	186,948,000	284,838,000	249,088,000	181,247,000	122,927,000	114,323,000
Return on Assets %	11.63%	20.36%	22.47%	21.57%	17.32%	21.43%
Return on Equity %	16.36%	26.73%	27.65%	26.67%	21.65%	25.90%
Debt to Equity						

CONTACT INFORMATION:

Phone: 805 967-7611 Fax: 805 967-9722
Toll-Free:
Address: 495-A S. Fairview Ave., Goleta, CA 93117 United States

STOCK TICKER/OTHER:

Stock Ticker: DECK Exchange: NAS
Employees: 1,900 Fiscal Year Ends: 31-Dec
Parent Company:

SALARIES/BONUSES:

Top Exec. Salary: $1,200,000 Bonus: $
Second Exec. Salary: $655,080 Bonus: $

OTHER THOUGHTS:

Estimated Female Officers or Directors: 6
Hot Spot for Advancement for Women/Minorities: Y

dELiA*s Inc

NAIC Code: 454113

www.deliasinc.com

TYPES OF BUSINESS:

Apparel Sales-Catalogs
Apparel Stores
Young Women's Apparel
Cosmetics & Accessories
Online Retailing

BRANDS/DIVISIONS/AFFILIATES:

dELiA*s
Alloy Inc
DELiAs.com
Alloy.com

CONTACTS: *Note: Officers with more than one job title may be intentionally listed here more than once.*

Walter Killough, CEO
Dyan Jozwick, Pres., dELiA*s Brand
David J. Dick, CFO
Marc Schuback, General Counsel/Sec./VP
David J. Dick, Treas./Sr. VP
Carter S. Evans, Chmn.

GROWTH PLANS/SPECIAL FEATURES:

dELiA*s, Inc. is a direct marketer and retailer of apparel and fashion accessories primarily marketed towards teenage girls and young women. The company operates two brands: dELiA*s and Alloy. Its catalogs feature a broad assortment of merchandise, including jeans, dresses, accessories, shoes and outerwear. The firm offers items from its own line as well as products from recognized and emerging brands. In 2011, the firm circulated approximately 38.8 million direct mail catalogs, of which approximately 19.8 million were dELiA*s catalogs and 19.0 million were Alloy catalogs. The company's online stores, Delias.com and Alloy.com, offer apparel, cosmetics, accessories, footwear and other products designed to complement the catalog merchandise. All catalog and online merchandise is shipped from a single, company-owned distribution facility located in Hanover, Pennsylvania. The firm also operates 113 dELiA*s mall-based stores in 33 states, ranging in size from 2,600 to 5,100 square feet, with an average store size of 3,800 square feet.

FINANCIAL DATA: *Note: Data for latest year may not have been available at press time.*

In U.S. $	2012	2011	2010	2009	2008	2007
Revenue	217,152,000	220,697,000	223,866,000	215,620,000	274,256,000	
R&D Expense						
Operating Income	-22,942,000	-29,427,000	-15,867,000	-19,182,000	-2,117,000	
Operating Margin %	-10.56%	-13.33%	-7.08%	-8.89%	-.77%	
SGA Expense	92,740,000	95,746,000	94,939,000	95,560,000	105,007,000	
Net Income	-22,670,000	-21,643,000	-10,424,000	17,159,000	-2,335,000	
Operating Cash Flow	4,367,000	-7,765,000	-36,094,000	-1,724,000	1,526,000	
Capital Expenditure	4,015,000	5,819,000	12,571,000	12,158,000	19,558,000	
EBITDA	-11,496,000	-18,758,000	-5,774,000	-10,388,000	5,469,000	
Return on Assets %	-17.77%	-14.00%	-5.57%	9.39%	-1.48%	
Return on Equity %	-31.02%	-22.79%	-9.42%	16.14%	-2.39%	
Debt to Equity					0.02	

CONTACT INFORMATION:

Phone: 212-590-6200 Fax:
Toll-Free:
Address: 50 W. 23rd St., New York, NY 10010 United States

STOCK TICKER/OTHER:

Stock Ticker: DLIA Exchange: NAS
Employees: 2,131 Fiscal Year Ends: 31-Jan
Parent Company:

SALARIES/BONUSES:

Top Exec. Salary: $525,000 Bonus: $
Second Exec. Salary: $303,040 Bonus: $22,000

OTHER THOUGHTS:

Estimated Female Officers or Directors:
Hot Spot for Advancement for Women/Minorities:

Delta Apparel Inc

www.deltaapparelinc.com

NAIC Code: 315210

TYPES OF BUSINESS:

T-Shirts & Sports Clothing
Headwear

BRANDS/DIVISIONS/AFFILIATES:

Junkfood Clothing Company
MJ Soffe LLC
To The Game LLC
Art Gun LLC
Delta Pro-Weight
Delta Magnum Weight
Junk Food
Salt Life

CONTACTS: *Note: Officers with more than one job title may be intentionally listed here more than once.*

Robert W. Humphreys, CEO
Steven E. Cochran, COO
Steven E. Cochran, Pres.
Deborah H. Merrill, CFO/VP/Treas.
Martha M. Watson, Chief Human Resources Officer/VP
Justin M. Grow, General Counsel/Assistant Sec.
Nancy P. Bubanich, Corp. Controller/Assistant Treas.
Robert W. Humphreys, Chmn.

GROWTH PLANS/SPECIAL FEATURES:

Delta Apparel, Inc. designs, manufactures, sources and markets branded and private label casual and athletic apparel through several wholly-owned subsidiaries, including Junkfood Clothing Company; M.J. Soffe, LLC; To The Game, LLC; and Art Gun, LLC. The firm's operations are divided into two segments: basics and branded. The company's basics segment consists of the Delta Catalog and FunTees business units, primarily focused on garment styles that are characterized by low fashion risk. The division markets, distributes and manufactures unembellished knit apparel under the main brand names Delta Pro-Weight and Delta Magnum Weight. These products are primarily sold to screen printing companies or under private labels for retailers, corporate industry programs, sports licensed apparel marketers and major branded sportswear companies. The branded apparel segment consists of the business units primarily focused on more specialized apparel garments and headwear designed to meet consumer preferences and fashion trends. The division sells embellished and unembellished products through specialty and boutique stores, high-end and mid-tier retail chains, sporting goods stores, college bookstores and to the U.S. military. The products in this segment are marketed under the brands Soffe, The Cotton Exchange, Junk Food, Salt Life and others. The company's manufacturing and cut & sew operations are conducted in North Carolina, El Salvador, Honduras and Mexico.

FINANCIAL DATA: *Note: Data for latest year may not have been available at press time.*

In U.S. $	2012	2011	2010	2009	2008	2007
Revenue	489,923,000	475,236,000	424,411,000	355,197,000	322,034,000	312,438,000
R&D Expense						
Operating Income	-6,222,000	25,296,000	20,162,000	12,147,000	4,887,000	12,388,000
Operating Margin %	-1.26%	5.32%	4.75%	3.41%	1.51%	3.96%
SGA Expense	89,973,000	91,512,000	80,695,000	64,388,000	59,898,000	59,187,000
Net Income	-2,447,000	17,327,000	12,187,000	6,456,000	-508,000	6,343,000
Operating Cash Flow	-19,101,000	2,890,000	32,269,000	21,731,000	-12,107,000	20,450,000
Capital Expenditure	6,626,000	7,966,000	6,965,000	3,058,000	16,590,000	10,915,000
EBITDA	1,270,000	32,553,000	26,950,000	19,225,000	11,218,000	17,572,000
Return on Assets %	-.77%	6.15%	4.79%	2.48%	-.20%	2.91%
Return on Equity %	-1.74%	12.94%	10.24%	5.94%	-.48%	6.19%
Debt to Equity	0.79	0.59	0.49	0.76	0.91	0.67

CONTACT INFORMATION:

Phone: 864 232-5200 Fax: 678 775-6992
Toll-Free:
Address: 322 S. Main St., Greenville, SC 29601 United States

STOCK TICKER/OTHER:

Stock Ticker: DLA
Exchange: ASE
Employees: 7,200
Fiscal Year Ends: 30-Jun
Parent Company:

SALARIES/BONUSES:

Top Exec. Salary: $733,125 Bonus: $
Second Exec. Salary: $333,365 Bonus: $

OTHER THOUGHTS:

Estimated Female Officers or Directors: 5
Hot Spot for Advancement for Women/Minorities: Y

Delta Galil Industries Ltd

NAIC Code: 315240

www.deltagalil.com

TYPES OF BUSINESS:

Intimate Apparel & Underwear
Baby Apparel
Socks
Retail Stores
Loungewear
Active-wear

BRANDS/DIVISIONS/AFFILIATES:

Delta
Taga
Yodfat
Schiesser AG
LittleMissMatched

CONTACTS:

Note: Officers with more than one job title may be intentionally listed here more than once.

Issac Dabah, CEO
Noam Lautman, Chmn.

GROWTH PLANS/SPECIAL FEATURES:

Delta Galil Industries, Ltd. manufactures and markets apparel for men, women and children. Its products include lingerie, underwear for men and women, bras and socks, baby clothes and casual wear. It produces private-label garments for retailers in Europe and the U.S.; manufactures branded garments under license; supplies garments for leading brands; and sells products directly in Israel under the brand names Delta, Taga and Yodfat. The company's private label retail customers include Marks & Spencer, Wal-Mart, Dillard's, Carrefour, Target, Victoria's Secret and Kohl's. Its licensed products fall under the brand names Wilson, Converse, Maidenform, Nicole Miller, Tommy Hilfiger, Kenneth Cole and others. In Israel, Delta products are available through various retailers as well as the firm's 115 chain stores in the country. During 2012, the company acquired Schiesser AG, a manufacturer of underwear in Germany, and the trademarks of LittleMissMatched.

FINANCIAL DATA:

Note: Data for latest year may not have been available at press time.

In U.S. $	2012	2011	2010	2009	2008	2007
Revenue			620,074,000	572,534,000	622,834,000	647,409,000
R&D Expense						
Operating Income			31,066,000	15,552,000	-15,726,000	4,623,000
Operating Margin %			5.00%	2.71%	-2.52%	.71%
SGA Expense			87,007,000	88,298,000	105,656,000	97,109,000
Net Income			21,060,000	7,650,000	-26,400,000	-12,951,000
Operating Cash Flow			14,260,000	40,824,000	15,790,000	10,397,000
Capital Expenditure			5,217,000	4,072,000	7,781,000	8,181,000
EBITDA			42,096,000	15,552,000	-4,490,000	17,902,000
Return on Assets %			5.29%	2.02%	-6.36%	-2.87%
Return on Equity %			11.36%	4.73%	-16.32%	-7.10%
Debt to Equity			0.31	0.03		0.04

CONTACT INFORMATION:

Phone: 972 35193636 Fax: 972 35193790
Toll-Free:
Address: 2 Kaufman St., Textile Bldg., Tel Aviv, 68012 Israel

STOCK TICKER/OTHER:

Stock Ticker: DELTY Exchange: PINX
Employees: 7,130 Fiscal Year Ends: 31-Dec
Parent Company:

SALARIES/BONUSES:

Top Exec. Salary: $ Bonus: $
Second Exec. Salary: $ Bonus: $

OTHER THOUGHTS:

Estimated Female Officers or Directors:
Hot Spot for Advancement for Women/Minorities:

Destination Maternity Corp

www.destinationmaternity.com

NAIC Code: 448120

TYPES OF BUSINESS:

Women's Apparel, Retail
Maternity Apparel
Online Sales
Leased Maternity Departments
Spas

BRANDS/DIVISIONS/AFFILIATES:

Motherhood Maternity
Destination Maternity
A Pea in the Pod
Oh Baby by Motherhood
Edamame The Maternity Spa

CONTACTS: *Note: Officers with more than one job title may be intentionally listed here more than once.*

Edward M. Krell, CEO
Christopher F. Daniel, Pres.
Judd P. Tirnauer, CFO/Exec. VP
Ronald J. Masciantonio, Chief Admin. Officer
Ronald J. Masciantonio, General Counsel/Exec. VP
Arnaud Ajdler, Chmn.

GROWTH PLANS/SPECIAL FEATURES:

Destination Maternity Corp. is a leading designer, manufacturer and retailer of maternity apparel. It currently operates 625 stores under the Motherhood Maternity, A Pea in the Pod and Destination Maternity names as well as 1,383 leased maternity departments in department stores and baby specialty stores. The firm's stores, located throughout all 50 states, Canada and Puerto Rico, offer career, casual, exercise and special-occasion maternity apparel. Motherhood Maternity, serving the value-priced market, is the broadest of the firm's lines, featuring casual and career wear, formal attire, lingerie, sportswear and outerwear as well as accessories and plus sizes. The A Pea in the Pod brand features upscale maternity fashions, including career and casual merchandise as well as exclusive designer labels. Destination Maternity superstores typically carry the Motherhood and Pea brands and are located in regional malls and lifestyle centers. Seven of its superstores also feature its Edamame-branded spas. Additionally, its Oh Baby by Motherhood collection is offered exclusively at Kohl's stores. The company's leased departments are located in stores such as Macy's, Sears, Bloomingdale's and Gordman's. In addition to brick-and-mortar locations, the firm markets its brands through direct marketing and several e-commerce sites. Additionally, the company has been expanding internationally through various franchise arrangements in India, the Middle East and South Korea. Currently, its international presence consists of 119 franchised locations, including 103 shop-in-shop locations and 16 Destination Maternity branded stores. In 2011, the company closed all 293 of its Kmart leased department locations and entered the South Korean market through a franchise arrangement with Agabang & Company. In October 2012, the firm stopped operating leased departments in Babies â€˜R Us Stores.

Destination Maternity offers employees a healthcare plan, dental plan, an on-site fitness center, an employee assistance program, credit union membership and flexible spending accounts.

FINANCIAL DATA: *Note: Data for latest year may not have been available at press time.*

In U.S. $	2012	2011	2010	2009	2008	2007
Revenue	541,476,000	545,394,000	531,192,000	531,251,000	564,602,000	581,371,000
R&D Expense						
Operating Income	33,105,000	38,244,000	31,433,000	-29,259,000	5,072,000	18,709,000
Operating Margin %	6.11%	7.01%	5.91%	-5.50%	.89%	3.21%
SGA Expense	255,623,000	257,421,000	251,653,000	259,552,000	274,508,000	281,507,000
Net Income	19,372,000	22,988,000	16,829,000	-40,682,000	-1,389,000	-393,000
Operating Cash Flow	42,697,000	21,443,000	25,974,000	42,525,000	27,822,000	27,398,000
Capital Expenditure	9,521,000	12,583,000	10,741,000	12,981,000	16,031,000	15,453,000
EBITDA	45,569,000	51,009,000	44,329,000	-14,400,000	21,192,000	26,511,000
Return on Assets %	9.72%	11.38%	8.39%	-17.99%	-.52%	-.13%
Return on Equity %	19.60%	27.98%	27.72%	-58.42%	-1.56%	-.46%
Debt to Equity		0.3	0.54	1.01	0.85	1.03

CONTACT INFORMATION:

Phone: 215 873-2200 Fax:
Toll-Free:
Address: 456 N. 5th St., Philadelphia, PA 19123 United States

STOCK TICKER/OTHER:

Stock Ticker: DEST
Exchange: NAS
Employees: 4,500
Fiscal Year Ends: 30-Sep
Parent Company:

SALARIES/BONUSES:

Top Exec. Salary: $750,000 Bonus: $
Second Exec. Salary: $525,000 Bonus: $

OTHER THOUGHTS:

Estimated Female Officers or Directors: 1
Hot Spot for Advancement for Women/Minorities: Y

Diane von Furstenberg Studio LP

www.dvf.com

NAIC Code: 448120

TYPES OF BUSINESS:

Fashion Apparel--Design & Retail
Jewelry
Home Furnishings
Handbags
Ready-to-wear Clothing
Footwear
Eyewear
Luggage

BRANDS/DIVISIONS/AFFILIATES:

CONTACTS: *Note: Officers with more than one job title may be intentionally listed here more than once.*

Paula Sutter, Pres.

GROWTH PLANS/SPECIAL FEATURES:

Diane von Furstenberg Studio, LP (DVF) is a global luxury brand fashion house that retails complete lines of its ready-to-wear women's apparel. The company's eponymous apparel line started with the innovative wrap dress that is still sold in boutiques and retailers throughout the world. DVF is headquartered in New York City and markets its products towards women in need of versatile apparel. DVF retails collections of accessories, beachwear, dresses, coats, tops, jackets, books, jewelry, sweaters, watches, shoes, luggage and handbags. In addition, the company retails sportswear, cosmetics, fragrances and home products. Consumers may purchase items from the collection in 45 freestanding DVF boutiques located throughout the U.S., China, Japan, Indonesia, Russia, South Korea, Spain, the U.K., Belgium, Greece, Malaysia, Singapore, Taiwan and France. The DVF line is also sold in high end retail stores such as Henri Bendel, Bergdorf Goodman, Nordstrom, Barney's New York, Intermix and others in over 70 countries. Products are also available for sale online. Additionally, the company operates a creative studio that serves as a lab for young budding artists to showcase and develop their work. DVF also collaborates with other luxury brands, such as the Diane von Furstenberg for H. Stern jewelry collection and DVF signature rugs for The Rug Company.

FINANCIAL DATA: *Note: Data for latest year may not have been available at press time.*

In U.S. $	2012	2011	2010	2009	2008	2007
Revenue						
R&D Expense						
Operating Income						
Operating Margin %						
SGA Expense						
Net Income						
Operating Cash Flow						
Capital Expenditure						
EBITDA						
Return on Assets %						
Return on Equity %						
Debt to Equity						

CONTACT INFORMATION:

Phone: 646-486-4800 Fax: 212-929-3971
Toll-Free: 888-472-2383
Address: 874 Washington St., New York, NY 10014 United States

STOCK TICKER/OTHER:

Stock Ticker: Private Exchange:
Employees: Fiscal Year Ends:
Parent Company:

SALARIES/BONUSES:

Top Exec. Salary: $ Bonus: $
Second Exec. Salary: $ Bonus: $

OTHER THOUGHTS:

Estimated Female Officers or Directors: 3
Hot Spot for Advancement for Women/Minorities: Y

Dick's Sporting Goods Inc

www.dickssportinggoods.com

NAIC Code: 451110

TYPES OF BUSINESS:

Sporting Goods Stores
Outdoor Apparel
Footwear
Hunting & Fishing Supplies
Golf Supplies
Bicycles
Online Sales

BRANDS/DIVISIONS/AFFILIATES:

Dick's Sporting Goods
Golf Galaxy
Field & Stream
Top-Flite
Ativa
Power Bolt
Walter Hagen
Fitness Gear

CONTACTS: *Note: Officers with more than one job title may be intentionally listed here more than once.*

Edward W. Stack, CEO
Joseph H. Schmidt, COO
Joseph H. Schmidt, Pres.
Timothy E. Kullman, CFO/Exec. VP-Finance
Lauren Hobart, Chief Mktg. Officer/Sr. VP
Kathy Sutter, Sr. VP-Human Resources
Matthew J. Lynch, CIO/Sr. VP
John G. Duken, Exec. VP-Global Merch.
Timothy E. Kullman, Exec. VP-Admin.
David I. Mosse, General Counsel/Sr. VP/Corp. Sec.
Don Germano, Sr. VP-Oper.
David I. Mosse, Chief Strategy Officer
Michele B. Willoughby, Sr. VP-e-Commerce
Anne-Marie Megela, Dir.-Investor Rel.
Joe Oliver, Chief Acct. Officer/Sr. VP
David G. Stanchak, Sr. VP-Real Estate
Lee J. Belitsky, Chief Risk & Compliance Officer/Sr. VP
Nancy Goldfarb, Sr. VP-Prod. Dev. & Global Sourcing
Moussa Coulibaly, Sr. VP-Planning & Allocation
Edward W. Stack, Chmn.

GROWTH PLANS/SPECIAL FEATURES:

Dick's Sporting Goods, Inc. (DSG) is a retail sporting goods chain that operates 518 Dick's Sporting Goods stores in 44 states as well as 81 Golf Galaxy stores in 30 states. The company offers a broad assortment of sporting goods equipment, footwear and apparel under national and private-label brands, including its own Ativa, Power Bolt, Walter Hagen, Fitness Gear, Maxfli and Acuity brands. Each of the Dick's Sporting Goods locations typically contains six store-within-a-store specialty units: footwear, team sports, outdoor lodge, golf, fitness and athletic apparel. In addition to apparel and equipment sales, its stores offer a variety of services such as golf club fitting, repair and grip replacement; tennis and lacrosse racket stringing; ice skate sharpening; bicycle repair and servicing; scope mounting and bore sighting; and CO2 tank filling for paintball. The Golf Galaxy stores are designed for serious golf enthusiasts and include artificial bent grass putting greens, golf simulators and private lessons. In addition to its retail locations, the firm maintains separate catalog and e-commerce sites for its chains. During 2012, DSG purchased a stake in JJB Sports plc, a leading U.K. sports retailer; and acquired the intellectual property rights to the Top-Flite brand from Callaway Golf Company. In March 2013, the firm disclosed plans to open its first Field & Stream retail concept store, which will focus on outdoor equipment and accessories.

FINANCIAL DATA: *Note: Data for latest year may not have been available at press time.*

In U.S. $	2012	2011	2010	2009	2008	2007
Revenue	5,211,802,000	4,871,492,000	4,412,835,000	4,130,128,000	3,888,422,000	
R&D Expense						
Operating Income	432,020,000	309,249,000	225,571,000	30,380,000	268,817,000	
Operating Margin %	8.28%	6.34%	5.11%	.73%	6.91%	
SGA Expense	1,162,861,000	1,139,781,000	981,252,000	944,442,000	889,246,000	
Net Income	263,906,000	182,077,000	135,359,000	-35,094,000	155,036,000	
Operating Cash Flow	410,421,000	389,967,000	401,329,000	159,811,000	262,834,000	
Capital Expenditure	201,807,000	159,067,000	140,269,000	216,530,000	172,366,000	
EBITDA	562,475,000	421,921,000	328,667,000	123,609,000	345,435,000	
Return on Assets %	9.43%	7.51%	6.42%	-1.75%	8.71%	
Return on Equity %	17.61%	14.88%	13.68%	-3.93%	20.54%	
Debt to Equity	0.09	0.1	0.13	0.2	0.2	

CONTACT INFORMATION:

Phone: 724 273-3400 Fax:
Toll-Free: 877-846-9997
Address: 345 Court St., Coraopolis, PA 15108 United States

STOCK TICKER/OTHER:

Stock Ticker: DKS Exchange: NYS
Employees: 28,400 Fiscal Year Ends: 31-Jan
Parent Company:

SALARIES/BONUSES:

Top Exec. Salary: $605,096 Bonus: $750,000
Second Exec. Salary: $1,019,231 Bonus: $

OTHER THOUGHTS:

Estimated Female Officers or Directors: 5
Hot Spot for Advancement for Women/Minorities: Y

Diesel SpA

NAIC Code: 315240

www.diesel.com

TYPES OF BUSINESS:

Apparel Design
Retail Stores
Men's, Women's & Children's Casual Clothing
Accessories
Sportswear
Hotel Operation
Licensing
Wine

BRANDS/DIVISIONS/AFFILIATES:

Diesel
Diesel Black Gold
Diesel Kids
55DSL
Pelican Hotel (The)
Diesel Farm
Fiat 500 by Diesel
Only The Brave

CONTACTS: *Note: Officers with more than one job title may be intentionally listed here more than once.*

Renzo Rosso, CEO
Sebastiano Zanolli, Mgr.-Sales

GROWTH PLANS/SPECIAL FEATURES:

Diesel SpA designs and retails premium jeans and casual clothing as well as sportswear and accessories for men, women and children. Diesel's main product lines are Diesel, a collection focused mainly on denim for men and women; Diesel Black Gold, a high-end collection pioneering in design, fabrics and product innovation; Diesel Kids, clothing for young people with bright colors and modern lines; and 55DSL, a sportswear collection for emerging extreme and action sports. Its products are sold in over 80 countries through high-end department stores and specialty retailers as well as some 400 company-owned stores (with 50 located in the U.S.). These stores include four planet stores (located in Milan, New York, Tokyo and Hong Kong), which offer on-site tailoring, messenger service, storewide Wi-Fi, e-commerce portals for shopping at Diesel.com, wireless cash stations for purchases, a multilingual staff and VIP services, along with its full clothing and accessory lines. The firm also markets its goods through catalogs and e-commerce sites. Additionally, Diesel owns and operates The Pelican Hotel, a 28-room, 1950s-style hotel on South Beach with rooms designed to feel like movie sets; and Diesel Farm, a winery. Through a collaboration with Italian automaker Fiat, the company designed the 500 by Diesel, a limited edition automobile featuring the brand's styling.

FINANCIAL DATA: *Note: Data for latest year may not have been available at press time.*

In U.S. $	2012	2011	2010	2009	2008	2007
Revenue	2,000,000,000	1,800,000,000	1,500,000,000			
R&D Expense						
Operating Income						
Operating Margin %						
SGA Expense						
Net Income						
Operating Cash Flow						
Capital Expenditure						
EBITDA						
Return on Assets %						
Return on Equity %						
Debt to Equity						

CONTACT INFORMATION:

Phone: 39-0424-477555 Fax: 39-0424-708492
Toll-Free:
Address: Via dell Industria 7, Molvena, Vicenza, 36060 Italy

STOCK TICKER/OTHER:

Stock Ticker: Private Exchange:
Employees: Fiscal Year Ends: 31-Dec
Parent Company: Only The Brave

SALARIES/BONUSES:

Top Exec. Salary: $ Bonus: $
Second Exec. Salary: $ Bonus: $

OTHER THOUGHTS:

Estimated Female Officers or Directors:
Hot Spot for Advancement for Women/Minorities:

Dillard's Inc

www.dillards.com

NAIC Code: 452111

TYPES OF BUSINESS:

Department Stores
Online Sales
Catalogs
Construction & Remodeling
Real Estate Investment
Insurance

BRANDS/DIVISIONS/AFFILIATES:

Dillards.com
CDI Contractors LLC
CDI Contractors Inc

CONTACTS:

Note: Officers with more than one job title may be intentionally listed here more than once.

William T. Dillard, II, CEO
Alex Dillard, Pres.
James I. Freeman, CFO/Sr. VP
Paul J. Schroeder, Jr., General Counsel/VP
Mike Dillard, Exec. VP
Drue Matheny, Exec. VP
William T. Dillard, II, Chmn.

GROWTH PLANS/SPECIAL FEATURES:

Dillard's, Inc. is one of the U.S.'s largest retail department stores offering apparel, cosmetics and home furnishings. Its 304 department stores are located in 29 states, primarily in the southwestern, mid-western and southeastern U.S. It sells name-brand and private-label merchandise including women's, petites, juniors, men's and children's apparel in addition to accessories, lingerie, cosmetics, furniture, electronics, shoes and other gifts and products. Most Dillard's stores are located in suburban shopping malls. The firm also sells merchandise through catalogs, various mail promotions and its web site, Dillards.com, which offers an online wedding, baby and gift registry service. Dillard's catalogs include the Men's; Women's; Shoes; Home and Entertainment; Handbags and Accessories; and Gifts categories. Many of Dillard's stores also offer day spas as well as nail and hair care services. Dillard's also offers proprietary credit cards to its customers. The firm has six company owned distribution centers in Arkansas, Arizona Georgia, Kansas, North Carolina and Texas as well as an Internet fulfillment center in Nashville, Tennessee. CDI Contractors, LLC and CDI Contractors, Inc. are full service construction companies based in Little Rock, Arkansas that construct stores for the company.

FINANCIAL DATA:

Note: Data for latest year may not have been available at press time.

In U.S. $	2012	2011	2010	2009	2008	2007
Revenue	6,399,765,000	6,253,535,000	6,226,628,000	6,988,440,000	7,370,806,000	
R&D Expense						
Operating Income	396,669,000	268,716,000	84,525,000	-117,829,000	159,949,000	
Operating Margin %	6.19%	4.29%	1.35%	-1.68%	2.17%	
SGA Expense	1,679,017,000	1,676,838,000	1,702,454,000	1,994,213,000	2,125,275,000	
Net Income	463,909,000	179,620,000	68,531,000	-241,065,000	53,761,000	
Operating Cash Flow	501,140,000	512,922,000	554,007,000	350,005,000	254,449,000	
Capital Expenditure	115,651,000	98,184,000	75,089,000	189,579,000	396,337,000	
EBITDA	728,195,000	605,903,000	423,291,000	-5,000,000	457,052,000	
Return on Assets %	10.68%	4.00%	1.46%	-4.78%	1.00%	
Return on Equity %	22.41%	8.18%	3.00%	-10.11%	2.10%	
Debt to Equity	0.4	0.43	0.42	0.34	0.31	

CONTACT INFORMATION:

Phone: 501 376-5200 Fax: 501 376-5917
Toll-Free: 800-345-5273
Address: 1600 Cantrell Rd., Little Rock, AR 72201 United States

STOCK TICKER/OTHER:

Stock Ticker: DDS Exchange: NYS
Employees: 38,900 Fiscal Year Ends: 31-Jan
Parent Company:

SALARIES/BONUSES:

Top Exec. Salary: $950,000 Bonus: $
Second Exec. Salary: $920,000 Bonus: $

OTHER THOUGHTS:

Estimated Female Officers or Directors: 2
Hot Spot for Advancement for Women/Minorities:

Dixie Group Inc

www.thedixiegroup.com

NAIC Code: 314110

TYPES OF BUSINESS:

Carpet and Rug Mills
Soft Flooring
Yarn

BRANDS/DIVISIONS/AFFILIATES:

Fabrica International
Masland Carpets
Masland Contract
Masland Residential
Dixie Home
Avant Contract

CONTACTS: *Note: Officers with more than one job title may be intentionally listed here more than once.*

Daniel K. Frierson, CEO
D. Kennedy Frierson, Jr., COO/VP
Jon A. Faulkner, CFO/VP
W. Derek Davis, VP-Human Resources
Starr Klein, Sec.
D. Eugene Lasater, Controller
Paul B. Comiskey, Pres., Residential/VP
Daniel K. Frierson, Chmn.

GROWTH PLANS/SPECIAL FEATURES:

Dixie Group, Inc. markets, manufactures and sells carpet and rugs. The group markets its products to high-end residential and commercial customers through its Fabrica International, Masland Carpets, Avant Contract and Dixie Home brands. Fabrica manufactures high-end carpets and custom rugs primarily for interior decorators and designers; selected retailers and furniture stores; luxury home builders; and manufacturers of luxury motor coaches and yachts. Masland Carpets is divided into two segments: Masland Contract and Masland Residential. Masland Contract manufactures broadloom and modular carpet tiles for the commercial market. Its products are marketed primarily through either the architectural community or specialty floor covering retailers to commercial end-users. Masland Residential manufactures specialty carpets and rugs for the high-end residential market. Its products are marketed primarily through the interior design community, as well as through some specialty floor covering retailers. Avant Contract focuses on the corporate office market through multi-line sales agents. Dixie Home provides an array of stylishly designed, differentiated tufted broadloom residential carpets to selected retailers and home centers under the Dixie Home and private label brands. Residential floor coverings account for approximately 75% of the company's total revenue, while commercial flooring products represent 25%. In November 2012, the firm acquired Lineage PCR, Inc's. Colormaster carpet dyeing facility in Calhoun, Georgia.

Employees are offered medical, dental and life insurance; a 401(k) plan; health care plans; educational benefits; and scholarships for children of employees.

FINANCIAL DATA: *Note: Data for latest year may not have been available at press time.*

In U.S. $	2012	2011	2010	2009	2008	2007
Revenue	266,372,000	270,110,000	231,322,000	203,480,000	282,710,000	320,795,000
R&D Expense						
Operating Income	1,815,000	5,668,000	-2,570,000	-45,323,000	-28,396,000	16,724,000
Operating Margin %	.68%	2.09%	-1.11%	-22.27%	-10.04%	5.21%
SGA Expense	63,690,000	61,038,000	58,918,000	64,516,000	78,432,000	80,952,000
Net Income	-927,000	986,000	-4,654,000	-42,098,000	-31,403,000	6,266,000
Operating Cash Flow	-4,712,000	5,054,000	3,918,000	26,510,000	8,399,000	19,766,000
Capital Expenditure	3,386,000	6,740,000	1,771,000	2,436,000	9,469,000	16,659,000
EBITDA	11,488,000	15,075,000	8,722,000	-31,638,000	-14,318,000	29,759,000
Return on Assets %	-.48%	.54%	-2.60%	-19.50%	-11.56%	2.20%
Return on Equity %	-1.44%	1.55%	-7.16%	-48.08%	-25.15%	4.51%
Debt to Equity	1.25	1.01	0.93	0.87	0.79	0.56

CONTACT INFORMATION:

Phone: 423 510-7000 Fax:
Toll-Free: 866-274-9776
Address: 104 Nowlin Ln., Ste. 101, Chattanooga, TN 37421 United States

STOCK TICKER/OTHER:

Stock Ticker: DXYN
Exchange: NAS
Employees: 1,150
Fiscal Year Ends: 31-Dec
Parent Company:

SALARIES/BONUSES:

Top Exec. Salary: $560,000 Bonus: $109,072
Second Exec. Salary: $260,000 Bonus: $50,641

OTHER THOUGHTS:

Estimated Female Officers or Directors: 1
Hot Spot for Advancement for Women/Minorities:

DOGI International Fabrics SA

www.dogi.com

NAIC Code: 313210

TYPES OF BUSINESS:

Textile Manufacturing
Elastic Fabrics & Fibers
Specialty & Nanotechnology Fabrics

BRANDS/DIVISIONS/AFFILIATES:

Sari
Winn
Dazzling
Liquid
Cristal
Invisible
Textiles ATA

CONTACTS:

Note: Officers with more than one job title may be intentionally listed here more than once.

Carlos Jaime Miguel Schroder Quijano, Managing Dir.
Raimon Pages Vida, Dir.-Global Oper.
Josep Abellan Pecheco, CFO
Kerim Helalli, Dir.-Sales
Roser Bonancia Canellas, Global Mgr.-Human Resources
Antonio Penarroja Castell, Corp. Sec.
James Robbins, Dir.-EFA
Josep Domenech Gimenez, Chmn.
Sergio Domenech Alvaro, Exec. Dir.-Global Mktg.
Eduardo Domenech Alvaro, Exec. Dir.-Global Purchases

GROWTH PLANS/SPECIAL FEATURES:

DOGI International Fabrics SA, based in Spain, is a leading textile manufacturer specializing in elastic fabrics and related yarns for use in intimate apparel, swimwear and technical sportswear. The intimate apparel offerings focus on comfort and style, and including the Sari line of silky fabrics and the Winn line, infused with Vitamin E and Aloe Vera. The company's swimwear division focuses on the modern fashion of swimwear and develops materials that are both comfortable and fashionable. DOGI's sportswear offerings focus on durability, comfort and practicality, developing fabrics that are both elastic and breathable. The firm's brands include: Dazzling, an iridescent and delicate material; Liquid, a plain smooth product that can be used for night or day wear; Invisible; a transparent and supportive fabric suitable for big sizes; and Cristal, a luxurious garment intended for elegant usage. The company performs manufacturing operations in three factories located in Spain, China and the U.S. The firm has commercial offices in Spain, France, the U.S., Hong Kong and China as well as strategic alliances with representatives in other parts of the world. Through Mexican subsidiary Textiles ATA, DOGI manufactures and sells fabric throughout North America.

FINANCIAL DATA:

Note: Data for latest year may not have been available at press time.

In U.S. $	2012	2011	2010	2009	2008	2007
Revenue		52,800,000	86,630,000	95,450,000	170,370,000	238,200,000
R&D Expense						
Operating Income						
Operating Margin %						
SGA Expense						
Net Income		-20,200,000	18,660,000	-56,840,000	-85,620,000	-34,600,000
Operating Cash Flow						
Capital Expenditure						
EBITDA						
Return on Assets %						
Return on Equity %						
Debt to Equity						

CONTACT INFORMATION:

Phone: 34-93-462-80-00 Fax: 34-93-462-80-34
Toll-Free:
Address: Pintor Domenech Farre 13-15, El Masnou, Barcelona, 08320 Spain

STOCK TICKER/OTHER:

Stock Ticker: DGI
Exchange: Madrid
Employees: 703
Fiscal Year Ends: 31-Dec
Parent Company:

SALARIES/BONUSES:

Top Exec. Salary: $ Bonus: $
Second Exec. Salary: $ Bonus: $

OTHER THOUGHTS:

Estimated Female Officers or Directors:
Hot Spot for Advancement for Women/Minorities: Y

Dolce & Gabbana SpA (D&G)

www.dolcegabbana.com

NAIC Code: 315210

TYPES OF BUSINESS:

Men's, Women's & Children's Clothing
Designer Apparel
Retail Stores
Leather Goods
Accessories
Knitwear
Underwear
Beachwear

BRANDS/DIVISIONS/AFFILIATES:

Sicily
Alta Moda
One (The)
Gold
Light Blue

CONTACTS: *Note: Officers with more than one job title may be intentionally listed here more than once.*

Alfonso Dolce, CEO
Stefano Gabbana, Pres.
Domenico Dolce, Chmn.

GROWTH PLANS/SPECIAL FEATURES:

Dolce & Gabbana SpA creates, markets and distributes high-end designer men's and women's collections of knitwear, leather goods, scarves, ties, shoes, underwear and beachwear. The Dolce & Gabbana brand has a cosmopolitan and elegant look that can be either distinctly masculine or feminine. Through various licensing agreements, the firm creates and distributes fragrances, eyewear and footwear for both of its brands. Its products are sold through company-owned boutiques as well as high-end shops and department stores in over 80 countries around the world. The company has developed an anti-counterfeiting system, which includes a safety hologram produced by the Italian State Printing Works and Mint and found on the inside of every article of clothing. In addition to its clothing lines, the company has perfume lines called The One, Sicily and Light Blue; a line of cosmetics; and a restaurant in Milan called Gold. Recently, the designers behind the company, Domenico Dolce and Stefano Gabbana, formed a partnership with Madonna's MG ICON brand (a joint venture with Iconix Brand Group, Inc.) to design a line of sunglasses in a first ever co-branded project from the company. In July 2012, the company debuted its first couture line, Atla Moda, consisting of 73 Sicilian inspired dresses.

FINANCIAL DATA: *Note: Data for latest year may not have been available at press time.*

In U.S. $	2012	2011	2010	2009	2008	2007
Revenue	1,500,000,000	1,480,000,000				
R&D Expense						
Operating Income						
Operating Margin %						
SGA Expense						
Net Income						
Operating Cash Flow						
Capital Expenditure						
EBITDA						
Return on Assets %						
Return on Equity %						
Debt to Equity						

CONTACT INFORMATION:

Phone: 39-02-774271 Fax:
Toll-Free:
Address: Via Goldoni,10, Milan, 20129 Italy

STOCK TICKER/OTHER:

Stock Ticker: Private Exchange:
Employees: Fiscal Year Ends: 31-Mar
Parent Company:

SALARIES/BONUSES:

Top Exec. Salary: $ Bonus: $
Second Exec. Salary: $ Bonus: $

OTHER THOUGHTS:

Estimated Female Officers or Directors:
Hot Spot for Advancement for Women/Minorities:

Dollfus Mieg & Cie SA

www.dmc.com

NAIC Code: 313210

TYPES OF BUSINESS:

Textiles Manufacturing
Yarn & Fiber
Sportswear Fabrics
Craft Stores

BRANDS/DIVISIONS/AFFILIATES:

DMC Club
Loisirs et Creation
SAIC Velcorex

CONTACTS: *Note: Officers with more than one job title may be intentionally listed here more than once.*

Dominique Poile, Pres.
Francois Demoulin, Chmn.

GROWTH PLANS/SPECIAL FEATURES:

Dollfus Mieg & Cie SA (DMC), based in Paris, is a major producer of threads, fabrics and craft supplies. DMC offers dozens of patterns for cross stitching and embroidery; ready-to-stitch products such as fabrics, bookmarks, bags and stuffed animals with images of cartoon characters and holiday images; over 100 tapestry designs; more than 200 fabrics; craft publications; craft accessories from magnifying lamps and storage kits to scissors and needles; cotton and wool threads specialized for cross stitching and embroidery; knitting and crocheting; tapestries and friendship bracelets; and other crafts. DMC supplies a variety of fabrics intended as a base for cross stitching and embroidery projects including Impressions printed fabrics, pre-cut fabrics meant for pillows or other small projects, fabrics by the meter, woven fabrics for afghans, tablecloths, ribbon-like bands of cloth, canvases and dissolving stitching guides. The DMC Club is a craft information hub with links to patterns, instructions and advice on embroidery, crochet, cross-stitch and tapestry projects. U.S.-based points of sale include retailers such as Jo Ann Fabrics, Michael's and Hobby Lobby. The sportswear division, SAIC Velcorex, jointly-owned by Pakistani firm Kohinoor, produces fabric for various sportswear and casual applications. The group also owns a chain of stores, Loisirs et Creation, based in France.

FINANCIAL DATA: *Note: Data for latest year may not have been available at press time.*

In U.S. $	2012	2011	2010	2009	2008	2007
Revenue						
R&D Expense						
Operating Income						
Operating Margin %						
SGA Expense						
Net Income						
Operating Cash Flow						
Capital Expenditure						
EBITDA						
Return on Assets %						
Return on Equity %						
Debt to Equity						

CONTACT INFORMATION:

Phone: 33-3-89-32-44 Fax:
Toll-Free:
Address: 13 Rue de Pfastatt, Mulhouse, 68200 France

STOCK TICKER/OTHER:

Stock Ticker: Private Exchange:
Employees: 1,181 Fiscal Year Ends: 31-Dec
Parent Company:

SALARIES/BONUSES:

Top Exec. Salary: $ Bonus: $
Second Exec. Salary: $ Bonus: $

OTHER THOUGHTS:

Estimated Female Officers or Directors:
Hot Spot for Advancement for Women/Minorities:

DSW Inc

www.dswshoes.com

NAIC Code: 448210

TYPES OF BUSINESS:

Discount Shoe Stores
Women's Merchandise
Men's Merchandise
Handbags
Luxury Designers
Accessories
Clearance Merchandise
Online Services

BRANDS/DIVISIONS/AFFILIATES:

DSW Rewards
DSW.com
Retail Venture Inc

CONTACTS: *Note: Officers with more than one job title may be intentionally listed here more than once.*

Michael R. MacDonald, CEO
Michael R. MacDonald, Pres.
Douglas J. Probst, CFO/Exec. VP
Derek Ungless, Chief Mktg. Officer/Exec. VP
Deborah L. Ferree, Chief Merch. Officer/Vice Chmn.
William L. Jordan, General Counsel/Exec. VP
Carrie McDermott, Exec. VP-Oper. & Stores
Jay L. Schottenstein, Chmn.
Harris Mustafa, Exec. VP-Supply Chain & Merch. Planning

GROWTH PLANS/SPECIAL FEATURES:

DSW, Inc. is a leading retailer of specialty branded footwear in the U.S. DSW operates 364 stores in 41 states, Washington, D.C. and Puerto Rico. The stores offer a wide selection of brand name and designer shoes in dress, casual and athletic categories. Each store offers approximately 24,000 shoes, in more than 2,000 styles. In addition, the retailer operates 346 leased shoe departments for Gordmans, Stein Mart and Frugal Fannies. The stores are largely self-service, allowing customers to try on shoes and check size availability without sales associates. New styles are located on the main floor, clearance in the rear and accessories and impulse buys are featured in the front. The firm operates a 700,000-square-foot distribution center in Columbus, Ohio. The company maintains a loyalty program called DSW Rewards. The retailer uses this program to classify members by frequency, and uses direct mail and on-line communication to stimulate further sales and traffic. The company's most recent estimate showed 18 million members enrolled in the DSW Rewards loyalty program; furthermore, 88% of DSW store and DSW.com net sales were generated by shoppers in the program.

DSW employees are offered competitive wages; comprehensive medical, vision, prescription and dental insurance; life insurance programs; long- term disability insurance; flexible spending accounts; a 401(k) plan; an employee discount; and others.

FINANCIAL DATA: *Note: Data for latest year may not have been available at press time.*

In U.S. $	2012	2011	2010	2009	2008	2007
Revenue	2,024,329,000	1,822,376,000	1,602,605,000	1,462,944,000	1,405,615,000	
R&D Expense						
Operating Income	151,450,000	173,583,000	93,455,000	42,813,000	81,321,000	
Operating Margin %	7.48%	9.52%	5.83%	2.92%	5.78%	
SGA Expense	448,583,000	392,098,000			288,814,000	
Net Income	174,788,000	107,624,000	54,741,000	26,902,000	53,775,000	
Operating Cash Flow	214,788,000	140,949,000	164,453,000	97,109,000	70,860,000	
Capital Expenditure	74,707,000	46,735,000	23,080,000	82,191,000	98,971,000	
EBITDA	205,310,000	226,140,000	139,565,000	81,415,000	113,642,000	
Return on Assets %	15.76%	11.57%	6.96%	3.80%	8.25%	
Return on Equity %	24.49%	18.46%	11.05%	5.98%	13.30%	
Debt to Equity						

CONTACT INFORMATION:

Phone: 614 237-7100 Fax: 614 238-4200
Toll-Free: 866-379-7463
Address: 810 DSW Dr., Columbus, OH 43219 United States

STOCK TICKER/OTHER:

Stock Ticker: DSW Exchange: NYS
Employees: 10,500 Fiscal Year Ends: 31-Jan
Parent Company:

SALARIES/BONUSES:

Top Exec. Salary: $1,062,500 Bonus: $
Second Exec. Salary: $996,250 Bonus: $

OTHER THOUGHTS:

Estimated Female Officers or Directors: 5
Hot Spot for Advancement for Women/Minorities: Y

Eastern Mountain Sports Inc

www.ems.com

NAIC Code: 451110

TYPES OF BUSINESS:

Sporting Goods Stores
Outdoor Apparel
Outdoor Gear
Outdoor Sports Instruction

BRANDS/DIVISIONS/AFFILIATES:

EMS Climbing School
EMS Kayak School

CONTACTS: *Note: Officers with more than one job title may be intentionally listed here more than once.*

William O. Manzer, CEO
Robert Mayerson, COO
Robert Mayerson, Pres.
Ralph Lucarelli, Dir.-Visual Merch.
Ralph Lucarelli, Dir.-Store Planning
Joe Lentini, Dir.-Climbing & Adventure School

GROWTH PLANS/SPECIAL FEATURES:

Eastern Mountain Sports (EMS) operates 69 outdoor specialty retail stores in the Northeast. The stores provide an extensive selection of clothing and gear for activities such as rock and ice climbing, mountaineering, kayaking and canoeing, camping, cycling and fitness. EMS stocks most popular brand names in the outdoor industry, including Patagonia, Sierra Designs, Gregory, Kelty, Thule, The North Face, Marmot and Teva. In addition, the company designs and offers its own line of outdoor gear and clothing under the EMS name. The firm utilizes a large store format that includes comprehensive product offerings, allowing it to provide lower prices than those offered by local specialty outdoor retailers. EMS has a liberal return policy that allows customers to exchange, return or repair any item with a web invoice or original store receipt, even if the item has been damaged from use. Many of the firm's products are tested and demonstrated at the firm's 11 outdoor skills training schools. The EMS Climbing School in New Hampshire offers beginner, intermediate and advanced courses in rock climbing, ice climbing and mountaineering throughout the White Mountains. The EMS Kayak School in Massachusetts provides classes for every level through day and overnight trips throughout the Northeast.

FINANCIAL DATA: *Note: Data for latest year may not have been available at press time.*

In U.S. $	2012	2011	2010	2009	2008	2007
Revenue						
R&D Expense						
Operating Income						
Operating Margin %						
SGA Expense						
Net Income						
Operating Cash Flow						
Capital Expenditure						
EBITDA						
Return on Assets %						
Return on Equity %						
Debt to Equity						

CONTACT INFORMATION:

Phone: 603-924-9571 Fax: 603-924-9138
Toll-Free:
Address: 1 Vose Farm Rd., Peterborough, NH 03458 United States

STOCK TICKER/OTHER:

Stock Ticker: Private Exchange:
Employees: Fiscal Year Ends: 31-Jan
Parent Company:

SALARIES/BONUSES:

Top Exec. Salary: $ Bonus: $
Second Exec. Salary: $ Bonus: $

OTHER THOUGHTS:

Estimated Female Officers or Directors:
Hot Spot for Advancement for Women/Minorities:

ECCO Sko A/S

NAIC Code: 316210

www.ecco.com

TYPES OF BUSINESS:

Casual Footwear
Retail Sales
Running Shoes
Golf Shoes
Walking Shoes
Outdoor Shoes
Leather Production

BRANDS/DIVISIONS/AFFILIATES:

ECCO Shoes International AG
ECCO Leather
Comfort Fibre System
Hydromax
Cambrelle

CONTACTS: *Note: Officers with more than one job title may be intentionally listed here more than once.*

Dieter Kasprzak, CEO
Michael Hauge Sorensen, COO
Dieter Kasprzak, Pres.
Steen Borgholm, CFO
Andreas Wortmann, Exec. VP-Prod. & Brand
Steen Borgholm, Exec. VP-Finance
Panagiotis Mytaros, Exec. VP-Prod.
Hanni T. Kasprzak, Chmn.
Michel Krol, Exec. VP-Global Sales

GROWTH PLANS/SPECIAL FEATURES:

ECCO Sko A/S is a Danish designer and manufacturer of footwear for men, women and children. Besides dress and casual shoes, the company produces athletic shoes, including shoes for golfing, running, walking and outdoor activities. ECCO's hallmark is its unique direct injection technology, wherein a polyurethane and thermoplastic urethane-based spray is applied directly onto the molded upper part of the shoe to form the sole. ECCO feels this special direct injection process helps provide its shoes with higher quality, flexibility and lightness than other shoes. It operates shoe factories in China, Indonesia, Portugal, Thailand and Slovakia as well as product development facilities in Denmark and Portugal. The firm also produces accessories through its Swiss subsidiary, ECCO Shoes International AG. The company's leather supply operations, managed by ECCO Leather, include tanneries in China, Thailand, Indonesia and the Netherlands, which supply ECCO's shoe factories and other shoe manufactures as well as manufacturers of gloves, bags and airplane and car seats. Annually, ECCO sells over 33% of its leather production to external customers. The company has invented numerous products, including the Comfort Fibre System vegetable tanned, metal-free insole; the Hydromax leather treatment; and the Cambrelle removable, washable felt insole. ECCO manufactures over 20 million pairs of shoes each year, making it one of the largest shoe manufacturers in the world. The company's shoes are sold in 91 markets, including approximately 1,095 ECCO retail stores worldwide. In June 2012, the firm purchased a 56 square mile plot of land in Thailand for a new production facility.

FINANCIAL DATA: *Note: Data for latest year may not have been available at press time.*

In U.S. $	2012	2011	2010	2009	2008	2007
Revenue		1,234,570,000	1,130,182,000	908,210,000	955,010,000	927,540,000
R&D Expense						
Operating Income						
Operating Margin %						
SGA Expense						
Net Income		96,594,000	75,826,000	53,910,000	93,720,000	95,530,000
Operating Cash Flow						
Capital Expenditure						
EBITDA						
Return on Assets %						
Return on Equity %						
Debt to Equity						

CONTACT INFORMATION:

Phone: 45-74-91-16-25 Fax: 45-74-71-03-60
Toll-Free:
Address: Industrivej 5, Bredebro, 6261 Denmark

STOCK TICKER/OTHER:

Stock Ticker: Private Exchange:
Employees: 19,759 Fiscal Year Ends: 31-Dec
Parent Company:

SALARIES/BONUSES:

Top Exec. Salary: $ Bonus: $
Second Exec. Salary: $ Bonus: $

OTHER THOUGHTS:

Estimated Female Officers or Directors: 3
Hot Spot for Advancement for Women/Minorities: Y

Eddie Bauer LLC

www.eddiebauer.com

NAIC Code: 448140

TYPES OF BUSINESS:

Apparel Stores
Catalogs
Online Sales
Outlet Stores
Outdoor Wear
High-Performance Mountaineering Wear
Private Label Credit Cards

BRANDS/DIVISIONS/AFFILIATES:

Eddie Bauer Credit Card
EddieBauer.com
Golden Gate Capital

CONTACTS: *Note: Officers with more than one job title may be intentionally listed here more than once.*

Michael R. Egeck, CEO
Daniel E. Templin, COO
Michael R. Egeck, Pres.
Daniel E. Templin, CFO
Kathleen Boyer, Chief Merchandising Officer/Sr. VP
Steven M. Newman, Pres., Apparel Bus.
Joshua Olshanksky, Owner
Engle E. Saez, Chief Creative Officer/Sr. VP
Marvin Edward Toland, Sr. VP-Direct
David Chamberlain, Chmn.

GROWTH PLANS/SPECIAL FEATURES:

Eddie Bauer, LLC, owned by Golden Gate Capital, is a leading multi-channel specialty retailer offering casual men's, women's and children's apparel as well as accessories and home furnishings. Founded in 1920, the company offers its apparel and related accessories through its Eddie Bauer apparel stores, outlet stores, mail-order catalogs and through its web site, EddieBauer.com. The company has approximately 370 stores in the U.S. and Canada (48 U.S. states and 8 Canadian provinces), consisting of 251 retail stores and 119 outlet stores as well as joint ventures in Germany and Japan. The firm's retail stores are generally located in upscale regional malls or in high-traffic metropolitan areas. Eddie Bauer's outlet stores are located predominately in outlet malls, value strip centers and other areas not serviced by its core specialty retail stores. The company also offers Eddie Bauer Credit Cards to its customers. Eddie Bauer partners with several companies and organizations that share the company's spirit of adventure and love of the outdoors, such as Airstream, American Mountaineering Museum and the Leave No Trace Center for Outdoor Ethics.

The company offers its corporate and full-time store employees medical, dental and vision coverage; prescription plans; life insurance; a 401(k) plan; flexible spending accounts; an employee assistance program; and discounts on Eddie Bauer merchandise.

FINANCIAL DATA: *Note: Data for latest year may not have been available at press time.*

In U.S. $	2012	2011	2010	2009	2008	2007
Revenue	1,000,000,000					
R&D Expense						
Operating Income						
Operating Margin %						
SGA Expense						
Net Income						
Operating Cash Flow						
Capital Expenditure						
EBITDA						
Return on Assets %						
Return on Equity %						
Debt to Equity						

CONTACT INFORMATION:

Phone: Fax:
Toll-Free: 800-426-8020
Address: 10401 NE 8th St., Ste. 500, Bellevue, WA 98004 United States

STOCK TICKER/OTHER:

Stock Ticker: Private Exchange:
Employees: Fiscal Year Ends: 31-Dec
Parent Company: GOLDEN GATE CAPITAL

SALARIES/BONUSES:

Top Exec. Salary: $ Bonus: $
Second Exec. Salary: $ Bonus: $

OTHER THOUGHTS:

Estimated Female Officers or Directors: 2
Hot Spot for Advancement for Women/Minorities: Y

Eileen Fisher Inc

NAIC Code: 315240

www.eileenfisher.com

TYPES OF BUSINESS:

Women's Apparel
Retail Stores
Online Sales

BRANDS/DIVISIONS/AFFILIATES:

Eileen Fisher Home
Eco Collection
EileenFisher.com

CONTACTS:

Note: Officers with more than one job title may be intentionally listed here more than once.

Eileen Fisher, Pres.
Kenneth Pollak, CFO
Ann Gilligan, Dir.-Mktg. Planning
Celeste Thompson, VP-Human Resources
Jim Gundell, VP-e-Commerce & Retail
Hayley Gluck, Controller
Karen Gray, Dir.-Stores
Amy Hall, Dir.-Social Consciousness
Ellie Thoren, Dir.-Visual Advertising

GROWTH PLANS/SPECIAL FEATURES:

Eileen Fisher, Inc. manufactures and retails women's business and casual clothing with an emphasis on comfort, versatility and style. Offerings include tanks and camis; tops and tees; sweaters and cardigans; jackets and vests; pants; skirts and dresses; sleepwear; outerwear; shoes; the Eco Collection; and accessories such as scarves, handbags and belts. Its products are sold through more than 58 company owned boutiques and outlets in 21 states. Eileen Fisher also distributes its products nationwide and in Canada through department and specialty stores, including Nordstrom, Saks Fifth Avenue and Neiman Marcus, and online, through EileenFisher.com. In addition to clothing, Eileen Fisher designs organic cotton and linen sheets and other home textiles out of natural fabrics for Garnet Hill. Through the Global Organic Textile Standard (GOTS) program, the firm buys fair-traded cotton from farmers and artisans in Peru.

Employees are offered medical and dental coverage, flexible spending accounts, life insurance, an employee assistance program, educational assistance, parental and partner leave, domestic partner coverage, adoption assistance, flexible work arrangements, a clothing allowance, sample sales and summer hours. Additionally, Eileen Fisher contributes to and supports programs that influence women's health, well-being, independence and empowerment. Wellness activities are offered at corporate offices and include massages, yoga classes and foot reflexology.

FINANCIAL DATA:

Note: Data for latest year may not have been available at press time.

In U.S. $	2012	2011	2010	2009	2008	2007
Revenue	345,000,000	340,000,000	325,000,000	260,000,000	262,000,000	250,000,000
R&D Expense						
Operating Income						
Operating Margin %						
SGA Expense						
Net Income						
Operating Cash Flow						
Capital Expenditure						
EBITDA						
Return on Assets %						
Return on Equity %						
Debt to Equity						

CONTACT INFORMATION:

Phone: 914-591-5700 Fax: 914-591-8824
Toll-Free: 800-445-1603
Address: 2 Bridge St., Irvington, NY 10533-1527 United States

STOCK TICKER/OTHER:

Stock Ticker: Private
Exchange:
Employees:
Fiscal Year Ends: 31-Dec
Parent Company:

SALARIES/BONUSES:

Top Exec. Salary: $ Bonus: $
Second Exec. Salary: $ Bonus: $

OTHER THOUGHTS:

Estimated Female Officers or Directors: 8
Hot Spot for Advancement for Women/Minorities: Y

Elder-Beerman Stores Corp (The)

www.elder-beerman.com

NAIC Code: 452111

TYPES OF BUSINESS:

Department Stores
Furniture Stores

BRANDS/DIVISIONS/AFFILIATES:

Bon-Ton Stores Inc (The)
Elder-Beerman Furniture Stores

CONTACTS: *Note: Officers with more than one job title may be intentionally listed here more than once.*

Brendan L. Hoffman, CEO/Pres., Bon-Ton Stores, Inc.
Byron L. Bergen, Chmn.

GROWTH PLANS/SPECIAL FEATURES:

The Elder-Beerman Stores Corp. is a department store retailer, originally founded in Ohio in 1847. The firm is a wholly-owned subsidiary of The Bon-Ton Stores, Inc., which operates roughly 272 department stores in 23 states under the Bon-Ton, Elder-Beerman, Bergner's, Boston Store, Carson Pirie Scott, Parisian, Herberger's and Younkers brand names. Elder-Beerman's stores sell a wide range of moderate- to better-branded merchandise, including women's, men's and children's apparel and accessories; cosmetics; home furnishings; and other consumer goods. Elder-Beerman maintains approximately 55 department stores principally located in the smaller Midwestern markets of Ohio, West Virginia, Indiana, Illinois, Michigan, Wisconsin and Kentucky. These stores typically feature innovations such as centralized customer service centers, bridal registry kiosks and self-select cosmetics. The firm's historical competitive advantage is its niche in medium-sized and small cities.

The company offers employees benefits including medical, dental, vision and life insurance; long-term disability coverage; flexible spending accounts; and a 401(k) plan with company matching and retirement contribution.

FINANCIAL DATA: *Note: Data for latest year may not have been available at press time.*

In U.S. $	2012	2011	2010	2009	2008	2007
Revenue						
R&D Expense						
Operating Income						
Operating Margin %						
SGA Expense						
Net Income						
Operating Cash Flow						
Capital Expenditure						
EBITDA						
Return on Assets %						
Return on Equity %						
Debt to Equity						

CONTACT INFORMATION:

Phone: 717-757-7660 Fax: 717-751-3196
Toll-Free:
Address: 2801 E. Market St., York, PA 17402 United States

STOCK TICKER/OTHER:

Stock Ticker: Subsidiary Exchange:
Employees: Fiscal Year Ends: 31-Jan
Parent Company: BON-TON STORES INC (THE)

SALARIES/BONUSES:

Top Exec. Salary: $ Bonus: $
Second Exec. Salary: $ Bonus: $

OTHER THOUGHTS:

Estimated Female Officers or Directors:
Hot Spot for Advancement for Women/Minorities: Y

Esprit Holdings Ltd

www.esprit.com

NAIC Code: 448120

TYPES OF BUSINESS:

Apparel Stores
Footwear
Accessories
Women's Clothing
Men's Clothing
Children's Clothing
Home Furnishings

BRANDS/DIVISIONS/AFFILIATES:

Esprit
Red Earth

GROWTH PLANS/SPECIAL FEATURES:

Esprit Holdings Ltd. is a fashion retailer, selling clothing, footwear and accessories for men, women and children, and retailing a home furnishings line. Headquartered in Hong Kong with additional corporate offices in Germany and New York, the firm operates over 1,100 stores and has over 12,000 wholesale outlets internationally. Under the brand names Esprit and Red Earth, the company sells its products at freestanding stores, department store shop-in-shops and indoor boutiques. The retailer has 12 established product lines featuring women's wear, kid's wear and shoes and accessories. Esprit focuses on providing its customers with stylish and inexpensive clothing and accessories. In 2011 and 2012, the company opened 141 new retail stores and closed 94.

CONTACTS:

Note: Officers with more than one job title may be intentionally listed here more than once.

Jose Manuel Martinez Gutierrez, CEO
Tanya Todd, COO
Thomas Tang Wing Yung, CFO
Ernst-Peter Vogel, Exec. VP-IT
Jan Nord, Dir.-Brand & New Bus.
Colin Henry, Chief Product Officer
Ng Wai Yin Florence, Company Sec.
Patrick Lau, Investor Rel.
Ernst-Peter Vogel, Exec. VP-Global Finance
Christophe Bezu, Chief Wholesale Officer
Gert Van De Weerdhof, Chief Retail Officer
Raymond Or Ching Fai, Chmn.
John Gunn, Pres., Americas

FINANCIAL DATA:

Note: Data for latest year may not have been available at press time.

In U.S. $	2012	2011	2010	2009	2008	2007
Revenue	3,891,103,000	4,355,740,000	4,351,483,000	4,448,357,000	4,802,059,000	3,823,381,000
R&D Expense						
Operating Income	151,052,000	89,263,830	488,371,200	739,006,500	995,962,400	807,373,200
Operating Margin %	3.88%	2.04%	11.22%	16.61%	20.74%	21.11%
SGA Expense	881,931,800	884,511,700	770,223,000	670,639,700		
Net Income	112,611,700	10,190,520	545,128,500	612,076,400	832,011,100	668,188,800
Operating Cash Flow	94,165,600	236,704,000	698,115,300	680,056,300	770,094,000	758,613,600
Capital Expenditure	183,171,400	185,235,300	194,651,900	259,406,900	174,399,900	79,331,300
EBITDA	242,895,700	209,744,200	824,658,400	839,105,800	1,093,353,000	920,372,000
Return on Assets %	3.44%	.30%	19.10%	22.82%	33.14%	34.67%
Return on Equity %	5.48%	.48%	27.72%	31.26%	46.03%	48.89%
Debt to Equity	0.06	0.09	0.12			

CONTACT INFORMATION:

Phone: 852 27654321 Fax: 852 23625576
Toll-Free:
Address: 39 Wang Chiu Rd., Enterprise Sq. 3, 43/F, Kowloon, Kowloon, Hong Kong Hong Kong

STOCK TICKER/OTHER:

Stock Ticker: ESHDF
Exchange: PINX
Employees: 12,455
Fiscal Year Ends: 30-Jun
Parent Company:

SALARIES/BONUSES:

Top Exec. Salary: $ Bonus: $
Second Exec. Salary: $ Bonus: $

OTHER THOUGHTS:

Estimated Female Officers or Directors: 1
Hot Spot for Advancement for Women/Minorities:

Esquel Group of Companies

www.esquel.com

NAIC Code: 315220

TYPES OF BUSINESS:

Manufacturing-Apparel
Cotton Shirts
Retail Stores
Textile Research
Cotton Farming
Spinning, Dyeing & Knitting

BRANDS/DIVISIONS/AFFILIATES:

PYE

CONTACTS:

Note: Officers with more than one job title may be intentionally listed here more than once.

John Cheh, CEO
Teresa Yang, Vice Chmn.
Marjorie Yang, Chmn.

GROWTH PLANS/SPECIAL FEATURES:

Esquel Group of Companies is one of the world's largest producers of cotton shirts. The firm's vertically integrated operation starts in China where it oversees cotton farms, which supply spinning, dyeing and knitting facilities that produce high-quality cotton fabric. On behalf of brands including Banana Republic, Tommy Hilfiger, Hugo Boss, Brooks Brothers, Abercrombie and Fitch, Nike, Nordstrom and Lands' End as well as private-label retailers including Marks and Spencer, Esquel companies transform its fabric into premium men's and women's wear at plants in China, Hong Kong, Malaysia, Mauritius, Sri Lanka and Vietnam. The company also sells its products through its proprietary brand, PYE, which markets high-end cotton apparel in China with a flagship luxury retail store in Beijing. Esquel maintains group sales offices in select locations around the world, including U.S. offices in New York City and regional locations convenient to key accounts. Unlike many textile companies, the firm has an expressed strategy of in-company resource development, seeking to ensure high standards of quality and consistency across all of its product lines. The company has partnered with the Zhejiang Institute of Science and Technology to establish a joint Eco-Textile Research Center that will promote scientific research on eco-textile technologies.

FINANCIAL DATA:

Note: Data for latest year may not have been available at press time.

In U.S. $	2012	2011	2010	2009	2008	2007
Revenue	1,400,000,000	1,200,000,000	970,000,000			
R&D Expense						
Operating Income						
Operating Margin %						
SGA Expense						
Net Income			100,000,000			
Operating Cash Flow						
Capital Expenditure						
EBITDA						
Return on Assets %						
Return on Equity %						
Debt to Equity						

CONTACT INFORMATION:

Phone: 852-2811-8077 Fax: 852-2960-6988
Toll-Free:
Address: 25 Harbour Rd., 12/F Harbour Ctr., Hong Kong, Hong Kong

STOCK TICKER/OTHER:

Stock Ticker: Private
Exchange:
Employees: 50,000
Fiscal Year Ends:
Parent Company:

SALARIES/BONUSES:

Top Exec. Salary: $ Bonus: $
Second Exec. Salary: $ Bonus: $

OTHER THOUGHTS:

Estimated Female Officers or Directors: 2
Hot Spot for Advancement for Women/Minorities: Y

Etam Developpement

www.etamdeveloppement.com

NAIC Code: 448120

TYPES OF BUSINESS:

Women's Apparel Stores
Retail Lingerie
Retail Accessories

BRANDS/DIVISIONS/AFFILIATES:

1.2.3
Undiz
Etam Lingerie
Etam Ready-to-Wear

CONTACTS: *Note: Officers with more than one job title may be intentionally listed here more than once.*

Pierre Milchior, Interim-Mgmt. Chmn.
Bernard Duche, Managing Dir.-Human Resources
Claude Fradin-Lasabatie, Dir.-Info. Systems & Methods
Vanessa Michior, Gen. Partner
Gerard Billaud, Chmn.
Jean-Marie Fersing, CEO-China

GROWTH PLANS/SPECIAL FEATURES:

Etam Developpement, headquartered in France, is an international group that focuses on women's fashions including ready-to-wear, lingerie and accessories. The company has over 4,461 outlets in 40 countries, including 738 in France; 174 outlets in the rest of Europe; 3,286 in China; and 128 franchised outlets in Eastern Europe, Asia-Oceania, Africa and the Middle East. The outlets operate under the Etam Lingerie, Etam Ready-to Wear, 1.2.3 and Undiz names. Its products cover much of the market in women's clothing, both in terms of age and price range. The company engages in the design and retailing of fashion products through three complementary brands: Etam, 1.2.3 and Undiz. Each brand has its own team of stylists, designers and product managers. Once each collection is designed, the company uses a global network of subcontractors and textile manufacturers for production, mainly in Asia, Southern and Eastern Europe and the Mediterranean basin. The Etam brand covers ready-to-wear, lingerie and accessories through three main categories: city wear, casual chic and weekend wear. Designers create a number of mini-collections each season, and retail stock is regularly rotated. The 1.2.3 brand aims to make upscale fashion available to a wider range of customers with broad, coordinated collections. The Undiz brand specializes in women's lingerie, offering a broad range of mix-and-match styles targeted at young urban women. The brand also offers a small collection of underwear basics for men.

FINANCIAL DATA: *Note: Data for latest year may not have been available at press time.*

In U.S. $	2012	2011	2010	2009	2008	2007
Revenue		1,533,230,000	1,493,580,000	1,394,610,000	1,322,100,000	1,488,000,000
R&D Expense						
Operating Income						
Operating Margin %						
SGA Expense						
Net Income		53,736,000	69,968,000	25,700,000	8,200,000	40,300,000
Operating Cash Flow						
Capital Expenditure						
EBITDA						
Return on Assets %						
Return on Equity %						
Debt to Equity						

CONTACT INFORMATION:

Phone: 33-1-55-90-70-70 Fax: 33-1-55-90-70-01
Toll-Free:
Address: 67/73, rue de Rivoli, Paris, 75001 France

STOCK TICKER/OTHER:

Stock Ticker: TAM Exchange: Paris
Employees: 19,890 Fiscal Year Ends: 31-Dec
Parent Company:

SALARIES/BONUSES:

Top Exec. Salary: $ Bonus: $
Second Exec. Salary: $ Bonus: $

OTHER THOUGHTS:

Estimated Female Officers or Directors: 1
Hot Spot for Advancement for Women/Minorities:

Etienne Aigner Group Inc

www.etienneaigner.com

NAIC Code: 315280

TYPES OF BUSINESS:

Leather Goods Manufacturing
Shoes
Belts
Accessories
Jewelry
Online & Catalog Sales

BRANDS/DIVISIONS/AFFILIATES:

CONTACTS:

Note: Officers with more than one job title may be intentionally listed here more than once.

Jim Piotti, Pres.
Richard Schaefer, CFO
Terry McCormick, Exec. VP-Mktg.
Tony Cheng, Exec. VP-Human Resources
Ilene Yaroslawitz, Corp. Sec.
Robert Payne, Dir.-Electronic Commerce
Kristina Brochu, Dir.-Corp. Comm.
Daniela Anastasio, Creative Dir.
Sudesh Conker, Controller
Udo Dengs, Chmn.

GROWTH PLANS/SPECIAL FEATURES:

Etienne Aigner Group, Inc. (Etienne) is a leader in the design, manufacture and distribution of fine leather goods sold in department and specialty stores across the U.S. as well as through catalogs and the Internet. Etienne's products include handbags, jewelry, accessories and small leather goods. Additionally, through licensing partnerships with leading manufacturers, the Etienne brand is found on footwear, gloves and belts. The firm's products are sold throughout North and South America and the Caribbean. The firm owns over 60 retail boutiques in Europe and Asia where its goods can be purchased, as well as several airports and duty-free shops in China, Denmark, Egypt, Germany, Korea, Kuwait, Malaysia, Taiwan, Thailand and the UAE. Etienne's products can also be purchased at various online stores including Zappos.com, OnlineShoes.com, Shopping.com, Shopzilla.com and eBags.com. However, starting with the Holiday 2012 collection, Etienne products will only be available for purchase through the company's website and various specialty retailers for the spring collection starting in 2014.

FINANCIAL DATA:

Note: Data for latest year may not have been available at press time.

In U.S. $	2012	2011	2010	2009	2008	2007
Revenue						
R&D Expense						
Operating Income						
Operating Margin %						
SGA Expense						
Net Income						
Operating Cash Flow						
Capital Expenditure						
EBITDA						
Return on Assets %						
Return on Equity %						
Debt to Equity						

CONTACT INFORMATION:

Phone: 212-239-7788 Fax: 212-239-7981
Toll-Free:
Address: 19 W. 34th St., Ste. 302, New York, NY 10001 United States

STOCK TICKER/OTHER:

Stock Ticker: Private Exchange:
Employees: Fiscal Year Ends: 31-Mar
Parent Company:

SALARIES/BONUSES:

Top Exec. Salary: $ Bonus: $
Second Exec. Salary: $ Bonus: $

OTHER THOUGHTS:

Estimated Female Officers or Directors: 3
Hot Spot for Advancement for Women/Minorities: Y

Everlast Worldwide Inc

www.everlast.com

NAIC Code: 315210

TYPES OF BUSINESS:

Athletic Apparel
Apparel
Sporting Goods
Boxing Equipment
Work Boots & Safety Footwear

BRANDS/DIVISIONS/AFFILIATES:

Brands Holdings
Everlast

CONTACTS:

Note: Officers with more than one job title may be intentionally listed here more than once.

Neil Morton, CEO
Adam Geisler, Pres.
Gary J. Dailey, CFO
Angelo V. Giusti, Sr. VP-Sales
Gerard J. Delisser, Chief Merch. Officer
Angelo V. Giusti, Sec.
Angelo V. Giusti, Pres., Sporting Equipment Bus.

GROWTH PLANS/SPECIAL FEATURES:

Everlast Worldwide, Inc., a subsidiary of Brands Holdings, designs, manufactures, markets and sells active wear, sportswear and outerwear, with a particular focus on boxing related products, such as gloves, heavy bags, speed bags, trunks and other gym equipment. The company is a leading brand for boxing and mixed martial arts enthusiasts, selling everything from training equipment to heavy duty competition cages. The firm also licenses the Everlast trademark to companies that produce women's and children's apparel, underwear, footwear, eyewear, hats, fragrances, batteries, jewelry and nutritional products, fitness toys, heart rate monitors and other products. The Everlast brand is currently present in over 101 countries and has 88 worldwide licenses. Everlast maintains a graphic arts department to develop advertising campaigns, provide brand management, packaging solutions, retail advertising and catalogs for all of the company's product lines. The firm has received continued exposure in print, television and movie media as well as product placement on the Academy Award winning movie Million Dollar Baby, the reality TV drama The Contender, the movie Cinderella Man and the EA Sports Fight Night videogame. Everlast has partnered with several health and fitness businesses, including 24 Fitness, Crunch, Lifetime Fitness, Xtreme Couture MMA, Title Boxing, Gleason's Gym, Five Points Academy and Everlast Sports Nutrition. The company's strategy is to expand its network of retailers carrying Everlast products, focusing on department stores, specialty stores, sporting goods stores, catalog operations and better mass merchandisers. Recently, the firm signed a deal with Sears Holdings to distribute its products in Sears and Kmart stores. In 2012, Everlast signed mixed martial arts champion Ben Henderson to an exclusive contract.

FINANCIAL DATA:

Note: Data for latest year may not have been available at press time.

In U.S. $	2012	2011	2010	2009	2008	2007
Revenue						
R&D Expense						
Operating Income						
Operating Margin %						
SGA Expense						
Net Income						
Operating Cash Flow						
Capital Expenditure						
EBITDA						
Return on Assets %						
Return on Equity %						
Debt to Equity						

CONTACT INFORMATION:

Phone: 660-263-4381 Fax:
Toll-Free: 800-821-7930
Address: 1900 Highway DD, Moberly, MO 65270 United States

STOCK TICKER/OTHER:

Stock Ticker: Subsidiary Exchange:
Employees: Fiscal Year Ends: 31-Dec
Parent Company: BRANDS HOLDINGS

SALARIES/BONUSES:

Top Exec. Salary: $ Bonus: $
Second Exec. Salary: $ Bonus: $

OTHER THOUGHTS:

Estimated Female Officers or Directors:
Hot Spot for Advancement for Women/Minorities:

Express Inc

www.express.com

NAIC Code: 448120

TYPES OF BUSINESS:

Women's Apparel, Retail
Men's Apparel, Retail
Private Label Credit Cards

BRANDS/DIVISIONS/AFFILIATES:

Express
Express Men
Express Credit Card

CONTACTS: *Note: Officers with more than one job title may be intentionally listed here more than once.*

Michael A. Weiss, CEO
Matt Moellering, COO/Exec. VP
David G. Kornberg, Pres.
D. Paul Dascoli, CFO/Sr. VP
Lisa A. Gavales, Chief Mktg. Officer/Exec. VP
Michael Keane, Exec. VP-Human Resources
John J. Rafferty, Exec. VP-Planning & Allocation
D. Paul Dascoli, Treas.
Jeanne L. St. Pierre, Exec. VP-Stores
Doug H. Tilson, Exec. VP-Real Estate
Michael A. Weiss, Chmn.
Colin Campbell, Exec. VP-Sourcing & Prod.

GROWTH PLANS/SPECIAL FEATURES:

Express, Inc. is a specialty retailer of women's and men's private-label apparel, sportswear and accessories. Its internationally-inspired styles cater to fashion-conscious casual, professional and urban women and men between the ages of 20 and 30. Express' women's apparel includes tops, shirts, hoodies, jackets, suit jackets, outerwear, skirts, dresses, jeans, pants and accessories. Women's merchandise accounted for 65% of its total sales. Men's apparel, which accounted for the remaining 35%, includes shirts, graphic tees, crewnecks, track jackets, outerwear, jeans, pants and accessories. It currently operates 609 women's, men's and dual gender stores in the U.S., Puerto Rico and Canada, each averaging about 8,700 gross square feet. Its in-house design studio, located on 5th Avenue in New York City, is responsible for designing substantially all of the firm's merchandise. The company's web site allows customers to browse and purchase its most up-to-date apparel styles and trends for the coming season as well as gain information about current and forthcoming storewide sales. The web site also offers Internet-only product specials, in addition to offering a radio playlist of the music playing in Express retail stores. The entire catalog is available on Facebook as well. The company also offers an Express credit card that gives holders certain privileges including a rewards program, invitations to special events and other special offers. In July 2012, the company announced that two flagship stores will open in 2013: one store in Times Square in New York City, New York and one store in Union Square in San Francisco, California.

The company offers employees benefits including medical, dental, vision and pharmacy programs; life insurance; short- and long-term disability benefits; flexible spending accounts; 401(k) retirement plan; tuition reimbursement; adoption assistance; employee assistance programs; and merchandise discounts.

FINANCIAL DATA: *Note: Data for latest year may not have been available at press time.*

In U.S. $	2012	2011	2010	2009	2008	2007
Revenue	2,073,355,000	1,905,814,000	1,721,066,000	1,737,010,000	1,796,346,000	
R&D Expense						
Operating Income	270,946,000	199,251,000	126,837,000	3,914,000	3,691,000	
Operating Margin %	13.06%	10.45%	7.36%	.22%	.20%	
SGA Expense	483,823,000	461,073,000	409,198,000	447,071,000	445,250,000	
Net Income	140,697,000	127,388,000	75,307,000	-29,036,000	-10,457,000	
Operating Cash Flow	212,609,000	219,958,000	200,721,000	35,234,000	328,104,000	
Capital Expenditure	77,236,000	54,843,000	26,873,000	51,801,000	38,146,000	
EBITDA	339,471,000	269,792,000	202,199,000	89,151,000	3,691,000	
Return on Assets %	16.31%	14.70%	8.70%	-3.07%		
Return on Equity %	68.41%	116.06%	80.77%	-8.15%		
Debt to Equity	0.7	2.81	4.64	4.34		

CONTACT INFORMATION:

Phone: 614 474-4001 Fax: 212 884-3221
Toll-Free: 888-397-1980
Address: 1 Express Dr., Columbus, OH 43230 United States

STOCK TICKER/OTHER:

Stock Ticker: EXPR Exchange: NYS
Employees: 18,000 Fiscal Year Ends: 31-Jan
Parent Company:

SALARIES/BONUSES:

Top Exec. Salary: $1,413,461 Bonus: $
Second Exec. Salary: $604,423 Bonus: $100,000

OTHER THOUGHTS:

Estimated Female Officers or Directors: 4
Hot Spot for Advancement for Women/Minorities: Y

FAB Industries Corporation

www.fab-industries.com

NAIC Code: 313240

TYPES OF BUSINESS:

Textiles Manufacturing
Bedding & Home Furnishings Fabrics
Heat Resistant Fabrics
Health care Fabrics

BRANDS/DIVISIONS/AFFILIATES:

Lida Stretch Fabrics
ShopCeliaRachel.com
Celia Rachel
Mohican Mills Inc

CONTACTS: *Note: Officers with more than one job title may be intentionally listed here more than once.*

Steven Myers, COO
Steven Myers, Pres.
Jerry Deese, CFO/VP

GROWTH PLANS/SPECIAL FEATURES:

FAB Industries Corporation manufactures warp and circular knit fabrics, Raschel laces, laminated fabrics and bedding products. Its textiles are used in apparel, home furnishings and bedding, industrial and health care markets. The company sells its fabrics, primarily in piece-dyed form, to a variety of manufacturers of ready-to-wear and intimate apparel, sportswear and sleepwear. The company's manufacturing operations are carried out at its 750,000 square foot facility, Mohican Mills, Inc. in North Carolina. FAB Industries' Raschel knitting division, supplied by the Raval Lace division, provides laces to the intimate apparel, hosiery and home furnishing markets. The company's industrial textiles division makes polyurethane-coated, fire-resistant fabric often used in fire fighting materials. FAB Industries has developed an entire line of proprietary knitted fabrics for sale to manufacturers of men's, women's and children's apparel in both domestic and foreign markets. The firm's Lida Stretch Fabrics division specializes in circular knit products blending nylon, cotton, polyester and rayon with spandex to create stretch fabrics in a variety of constructions. These fabrics are sold as piece and yarn dyes to the ready-to-wear, aerobic wear, swimwear and intimate apparel markets. The company's broad range of products includes jerseys, ribs, jacquards and failles. Through Celia Rachel and its e-commerce site, ShopCeliaRachel.com, the firm markets finished textile products such as lingerie, sleepwear and bedding directly to consumers.

FINANCIAL DATA: *Note: Data for latest year may not have been available at press time.*

In U.S. $	2012	2011	2010	2009	2008	2007
Revenue						
R&D Expense						
Operating Income						
Operating Margin %						
SGA Expense						
Net Income						
Operating Cash Flow						
Capital Expenditure						
EBITDA						
Return on Assets %						
Return on Equity %						
Debt to Equity						

CONTACT INFORMATION:

Phone: 516-498-3200 Fax: 516-829-0783
Toll-Free:
Address: 98 Cutter Mill Rd., Ste. 412 N., Great Neck, NY 11021 United States

STOCK TICKER/OTHER:

Stock Ticker: Private
Employees: 500
Parent Company:
Exchange:
Fiscal Year Ends: 30-Nov

SALARIES/BONUSES:

Top Exec. Salary: $ Bonus: $
Second Exec. Salary: $ Bonus: $

OTHER THOUGHTS:

Estimated Female Officers or Directors:
Hot Spot for Advancement for Women/Minorities:

Fifth & Pacific Companies Inc

fifthandpacific.com

NAIC Code: 315240

TYPES OF BUSINESS:

Women's Apparel Manufacturing
Apparel Marketing
Jewelry Design & Marketing

BRANDS/DIVISIONS/AFFILIATES:

Juicy Couture
kate spade
Lucky Brand
Adelington Design Group
Liz Claiborne New York
Mexx
Kate Spade LLC
Kate Spade Japan Co Ltd

CONTACTS: *Note: Officers with more than one job title may be intentionally listed here more than once.*

William L. McComb, CEO
George M. Carrara, CFO
Evon L. Jones, CIO/Sr. VP
Nicholas Rubino, General Counsel/Corp. Sec./Chief Legal Officer
John S. Moroz, Sr. VP-Global Oper.
Jane Randel, Sr. VP-Corp. Comm. & Brand Svcs.
Michael Rinaldo, Chief Acct. Officer/Corp. Controller/VP
Tim Gunn, Fashion Dean
Robert Vill, Treas./Sr. VP-Finance

GROWTH PLANS/SPECIAL FEATURES:

Fifth & Pacific Companies, Inc., formerly Liz Claiborne, Inc., designs and markets a portfolio of retail-based, premium, global lifestyle brands including Juicy Couture, kate spade and Lucky Brand. Fifth & Pacific also owns the Adelington Design Group, a private brand jewelry design and development group, which markets brands through department stores and serves jcpenney via exclusive supplier agreements for the Liz Claiborne and Monet jewelry lines and Kohl's via an exclusive supplier agreement for Dana Buchman jewelry. In addition, the company has licenses for the Liz Claiborne New York brand, available at QVC and Lizwear, which is distributed through the club store channel. Fifth & Pacific maintains an 18.75% stake in Mexx, a European and Canadian apparel and accessories retail-based brand. The company was formed in May 2012 after selling the Liz Claiborne brand to J.C. Penney Corporation, owner of jcpenney department stores. In July 2012, the firm announced that its kate spade unit, Kate Spade LLC, exercised an option to buy out joint venture partner Sanei International Co., Ltd.'s 51% share of Kate Spade Japan Co., Ltd.

Fifth & Pacific offers employees medical, dental and vision coverage; life insurance; short- and long-term disability; supplemental maternity leave; flexible spending accounts; a 401(k) and profit sharing; product discounts; and back-up child care.

FINANCIAL DATA: *Note: Data for latest year may not have been available at press time.*

In U.S. $	2012	2011	2010	2009	2008	2007
Revenue	1,505,094,000	1,518,721,000	2,500,072,000	3,011,859,000	3,984,946,000	4,577,251,000
R&D Expense						
Operating Income	-34,451,000	-96,252,000	-179,514,000	-334,584,000	-733,780,000	-425,813,000
Operating Margin %	-2.28%	-6.33%	-7.18%	-11.10%	-18.41%	-9.30%
SGA Expense	877,426,000	904,619,000	1,415,441,000	1,715,327,000	1,943,963,000	2,104,420,000
Net Income	-74,505,000	-171,687,000	-251,467,000	-305,729,000	-951,811,000	-372,798,000
Operating Cash Flow	11,358,000	-17,028,000	150,641,000	207,290,000	158,361,000	273,876,000
Capital Expenditure	82,792,000	73,653,000	77,369,000	65,332,000	184,260,000	173,032,000
EBITDA	70,103,000	-10,283,000	-10,029,000	-175,027,000	-582,318,000	-268,094,000
Return on Assets %	-8.04%	-15.55%	-17.56%	-17.41%	-36.79%	-11.02%
Return on Equity %			-261.43%	-84.90%	-94.27%	-20.45%
Debt to Equity				2.38	1.25	0.55

CONTACT INFORMATION:

Phone: 212-354-4900 Fax:
Toll-Free:
Address: 1441 Broadway, New York, NY 10018 United States

STOCK TICKER/OTHER:

Stock Ticker: FNP Exchange: NYS
Employees: 6,100 Fiscal Year Ends: 31-Dec
Parent Company:

SALARIES/BONUSES:

Top Exec. Salary: $1,300,000 Bonus: $
Second Exec. Salary: $500,000 Bonus: $

OTHER THOUGHTS:

Estimated Female Officers or Directors: 1
Hot Spot for Advancement for Women/Minorities:

Fila USA Inc

NAIC Code: 315210

www.fila.com

TYPES OF BUSINESS:

Sports Apparel & Accessories
Footwear
Casual Wear

BRANDS/DIVISIONS/AFFILIATES:

Fila Korea Ltd
True Sportsman
Heritage
Baseline
Center Court
Collezione
Essenza

CONTACTS:

Note: Officers with more than one job title may be intentionally listed here more than once.

Jon Epstein, Pres.
Young-Chan Cho, CFO
Kelly Macmanus Funke, Sr. Mgr.-Mktg.
Sandy Hamilton, Dir.-IT
Jarita Bridges, Mgr.-Prod. Placement & Entertainment Mktg.
Kelly Macmanus Funke, Sr. Mgr.-e-commerce
Jarita Bridges, Head-Investor Rel.
Gene Yoon, CEO/Chmn.-Fila Korea Ltd.
Rob Baker, Dir.-Sportstyle Mktg.

GROWTH PLANS/SPECIAL FEATURES:

Fila USA, Inc., a subsidiary of Fila Korea Ltd., produces athletic footwear, apparel and accessories for men, women and children. Its products are segmented into categories based on sport applications, with specific lines tailored for basketball, tennis, soccer, yoga, fitness and training. These products are designed to complement athletes' movements. Its tennis line features a number of individual collections, from the classic True Sportsman and Heritage collections to the updated Baseline, Center Court, Collezione and Essenza lines. The company has also introduced athletic apparel lines in collaboration with athletes including James Blake, Janko Tipsarevic and Dmitry Tursunov. In addition to athletic wear, Fila offers a line of lifestyle clothing designed for a younger, more style conscious shopper, featuring vintage styling and updated classics from Fila's history. Fila's original focus revolved around the textiles business before going on to specialize in knitwear production. The company entered the sports industry in 1973, after consolidating its expertise in the Italian textiles industry. Fila established itself in the tennis apparel industry with its innovative tubular manufacturing process that, until then, was only used for other products. It was also Fila that challenged the white-only tradition on tennis courts by creating a line of colored tennis apparel. The firm is committed to using cutting-edge technology and innovative materials. In 2012, Fila signed endorsement agreements with the women's tennis players Jelena Jankovic, a WTA Tour star and former number one, and Vera Zvonareva, a Grand Slam finalist.

Employees are offered a health, dental and vision plan; short- and long-term disability; additional coverage through Aflac; a flexible spending account; a 401 (k) savings plan; and an employee discount.

FINANCIAL DATA:

Note: Data for latest year may not have been available at press time.

In U.S. $	2012	2011	2010	2009	2008	2007
Revenue						
R&D Expense						
Operating Income						
Operating Margin %						
SGA Expense						
Net Income						
Operating Cash Flow						
Capital Expenditure						
EBITDA						
Return on Assets %						
Return on Equity %						
Debt to Equity						

CONTACT INFORMATION:

Phone: 212-726-5900 Fax:
Toll-Free: 800-845-3452
Address: 340 Madison Ave, Fl. 3, New York, NY 10173 United States

STOCK TICKER/OTHER:

Stock Ticker: Subsidiary
Exchange:
Employees:
Fiscal Year Ends: 31-Dec
Parent Company: FILA KOREA LTD

SALARIES/BONUSES:

Top Exec. Salary: $ Bonus: $
Second Exec. Salary: $ Bonus: $

OTHER THOUGHTS:

Estimated Female Officers or Directors: 3
Hot Spot for Advancement for Women/Minorities: Y

Filippa K

www.filippa-k.com

NAIC Code: 315240

TYPES OF BUSINESS:

Men's & Women's Apparel
Apparel Stores

BRANDS/DIVISIONS/AFFILIATES:

Filippa K

CONTACTS:

Note: Officers with more than one job title may be intentionally listed here more than once.

Jon Abrahamsson, CEO
Magdalena Rodell Andersson, CFO
Eva Boding, Dir.-Mktg.
Birgitta Andersson, Mgr.-Human Resources
Susanna Liljeqvist, Mgr.-IT
Monica Nygren, Mgr.-Comm. & PR
Elin Larsson, Mgr.-Corp. Responsibility
Nina Bogstedt, Dir.-Range/Head-Design
Krister Nylund, Mgr.-Wholesale
Elin Mohlander, Mgr.-Australia & U.K.
Sten Lunden, Mgr.-Supply Chain

GROWTH PLANS/SPECIAL FEATURES:

Filippa K is a Swedish company engaged in the design, manufacture and retailing of fashion apparel and accessories for men and women. The company's products include blouses, shirts, dresses, jackets, jeans, knits, outerwear, shoes, skirts, tights, trousers and accessories such as belts, wallets, handbags, hats, neckties, scarves and gloves. Filippa K's line typically targets young urbanites. The company markets its products through 50 brand stores and 13 shop-in-shops in Scandinavia, Europe and the U.S., with its U.S. flagship store located in San Francisco. Filippa K operates offices in Norway, Denmark, Germany and Holland and offers its apparel via agreements with over 700 international retailers in 20 markets. Some of these retailers include Selfridges in the U.K., Barney's in New York, Ten Over Six in Los Angeles, Greta & Luis in Berlin and Frenchtrotters in Paris. Recently, the company announced the opening of a new concept store in Helsinki. In May 2012, Filippa K opened two new stores in Copenhagen, Denmark.

FINANCIAL DATA:

Note: Data for latest year may not have been available at press time.

In U.S. $	2012	2011	2010	2009	2008	2007
Revenue						
R&D Expense						
Operating Income						
Operating Margin %						
SGA Expense						
Net Income						
Operating Cash Flow						
Capital Expenditure						
EBITDA						
Return on Assets %						
Return on Equity %						
Debt to Equity						

CONTACT INFORMATION:

Phone: 46-8-615-7000 Fax:
Toll-Free:
Address: Soder Malarstrand 65, 9tr, Stockholm, 118 25 Sweden

STOCK TICKER/OTHER:

Stock Ticker: Private
Exchange:
Employees:
Fiscal Year Ends:
Parent Company:

SALARIES/BONUSES:

Top Exec. Salary: $ Bonus: $
Second Exec. Salary: $ Bonus: $

OTHER THOUGHTS:

Estimated Female Officers or Directors: 10
Hot Spot for Advancement for Women/Minorities: Y

Finish Line Inc (The)

NAIC Code: 448210

www.finishline.com

TYPES OF BUSINESS:

Athletic Shoes, Retail
Activewear
Athletic Accessories

BRANDS/DIVISIONS/AFFILIATES:

Finish Line
FinishLine.com
m.FinishLine.com
Running Company (The)
Run On!

CONTACTS: *Note: Officers with more than one job title may be intentionally listed here more than once.*

Glenn S. Lyon, CEO
Steven J. Schneider, COO
Steven J. Schneider, Pres.
Edward W. Wilhelm, CFO/Exec. VP
Jay Custard, VP-Digital Mktg.
Harvey Bierman, VP-Digital Tech.
Scott Hoffman, Chief Merch. Officer/Exec. VP
Michael L. Marchetti, Exec. VP-Store Oper.
Mark S. Landau, Chief Bus. Dev. Officer/Exec. VP
Christopher S. Ladd, Chief Digital Officer/Exec. VP
George S. Sanders, Exec. VP-Store Dev. & Real Estate
Samuel M. Sato, Pres., Finish Line Brand
Amber Vanes, VP-Planning & Allocation
Jeffrey Sherry, VP-Bus. Dev. & Corp. Strategy
Glenn S. Lyon, Chmn.

GROWTH PLANS/SPECIAL FEATURES:

The Finish Line, Inc. is a mall-based specialty retailer of men's, women's and children's brand-name athletic, outdoor and casual footwear, activewear and accessories in the U.S. The company owns and operates 647 Finish Line stores in 47 states, which average approximately 5,400 square feet. The firm attracts its core customers, typically males between 18 and 24, with a full range of leading athletic branded merchandise, including Nike, adidas, Puma, New Balance, Asics, Converse, Lacoste, K-Swiss, Reebok and Under Armour. Footwear products are categorized into sections by application, including basketball, running, sport style, fitness and outdoor. Most categories are available in men's, women's and children's styles. Footwear accounts for approximately 86% of the firm's net sales. Its stores also offer a range of softgoods products and licensed apparel, socks, athletic bags, backpacks, sunglasses, watches and shoe-care products, along with core basics apparel under a private label. The company purchases products from approximately 120 suppliers and manufacturers, the largest of which, Nike, represents around 60% of annual purchasing. The Finish Line maintains an e-commerce business through FinishLine.com and a mobile commerce site at M.FinishLine.com. In addition to Finish Line-branded locations, the firm operates 24 specialty running stores in seven states under The Running Company and Run On! brands. These locations specialize in men's and women's performance running shoes, apparel and accessories. In September 2012, the firm announced an arrangement with Macy's to operate more than 450 Finish Line-branded athletic footwear shops as leased departments within Macy's stores across the U.S. and at Macys.com. The company will also manage the athletic footwear assortment and inventory at the remaining 225 Macy's locations.

The Finish Line offers its employees an education reimbursement plan; employee discounts; a stock purchase plan; a 401(k) savings plan; profit sharing; basic life insurance; and medical, dental, vision and prescription drug coverage plans.

FINANCIAL DATA: *Note: Data for latest year may not have been available at press time.*

In U.S. $	2012	2011	2010	2009	2008	2007
Revenue	1,369,259,000	1,229,002,000	1,172,415,000	1,262,263,000	1,277,162,000	1,338,207,000
R&D Expense						
Operating Income	134,335,000	109,633,000	72,058,000	10,175,000	-67,813,000	51,179,000
Operating Margin %	9.81%	8.92%	6.14%	.80%	-5.30%	3.82%
SGA Expense	344,820,000	303,068,000	300,030,000	335,160,000	342,234,000	342,160,000
Net Income	84,804,000	68,834,000	35,672,000	3,758,000	-60,812,000	32,364,000
Operating Cash Flow	94,740,000	108,574,000	157,458,000	59,402,000	40,509,000	58,986,000
Capital Expenditure	29,681,000	19,088,000	8,454,000	14,734,000	29,405,000	73,993,000
EBITDA	161,362,000	136,592,000	102,035,000	47,671,000	-26,656,000	90,750,000
Return on Assets %	12.32%	10.79%	5.90%	.60%	-9.35%	5.03%
Return on Equity %	16.63%	14.76%	8.23%	.88%	-13.97%	7.37%
Debt to Equity						

CONTACT INFORMATION:

Phone: 317 899-1022 Fax: 317 895-2884
Toll-Free: 888-777-3949
Address: 3308 N. Mitthoeffer Rd., Indianapolis, IN 46235 United States

STOCK TICKER/OTHER:

Stock Ticker: FINL Exchange: NAS
Employees: 11,800 Fiscal Year Ends: 28-Feb
Parent Company:

SALARIES/BONUSES:

Top Exec. Salary: $950,000 Bonus: $
Second Exec. Salary: $609,000 Bonus: $

OTHER THOUGHTS:

Estimated Female Officers or Directors: 3
Hot Spot for Advancement for Women/Minorities: Y

Foot Locker Inc

NAIC Code: 448210

www.footlocker-inc.com

TYPES OF BUSINESS:

Athletic Shoes, Retail
Athletic Apparel
Catalogs
Online Sales
Athletic Equipment

BRANDS/DIVISIONS/AFFILIATES:

Foot Locker
Lady Foot Locker
Kids Foot Locker
Champs Sports
Eastbay Inc
Footaction
CCS
SIX:02

CONTACTS:

Note: Officers with more than one job title may be intentionally listed here more than once.

Ken C. Hicks, CEO
Richard Johnson, COO/Exec. VP
Ken C. Hicks, Pres.
Lauren B. Peters, CFO/Exec. VP
Laurie J. Petrucci, Sr. VP-Human Resources
Peter D. Brown, CIO/Sr. VP
Gary M. Bahler, General Counsel/Sr. VP/Sec.
John A. Maurer, VP-Investor Rel.
Giovanna Cipriano, Chief Acct. Officer/Sr. VP
Jeffrey L. Berk, Sr. VP-Real Estate
John A. Maurer, Treas.
Stephen Jacobs, Pres./CEO-Foot Locker U.S. & Lady Foot Locker
Ken C. Hicks, Chmn.

GROWTH PLANS/SPECIAL FEATURES:

Foot Locker, Inc. is one of the largest athletic footwear and apparel retailers in the world, operating approximately 3,372 stores in the U.S., and world wide. The company operates several store formats, including Foot Locker, Lady Foot Locker, Kids Foot Locker, Champs Sports, Footaction and CCS. Foot Locker stores offer athletic-inspired performance products, manufactured primarily by the leading athletic brands. These stores offer products for activities such as running, basketball and cross-training. Lady Foot Locker is a U.S. retailer of athletic footwear, apparel and accessories for active women. Kids Foot Locker offers brand name athletic footwear, apparel and accessories for infants, boys and girls. Champs Sports is a mall-based retailer of specialty athletic footwear and apparel as well as an assortment of sporting equipment. Footaction stores offer athletic footwear and clothing with a street-inspired fashion style. The primary customers of Footaction are young males. CCS is a specialty skateboard footwear and apparel store catering to 12-18 year old males. In addition to its retail stores, the direct-to-customers segment of business, through affiliates such as Eastbay, Inc., sells merchandise through catalogs and web sites, including Amazon.com. In 2012, the company introduced a new concept, SIX:02. This concept is designed to find the right product for women, of any activity, style or body type.

FINANCIAL DATA:

Note: Data for latest year may not have been available at press time.

In U.S. $	2012	2011	2010	2009	2008	2007
Revenue	5,623,000,000	5,049,000,000	4,854,000,000	5,237,000,000	5,437,000,000	
R&D Expense						
Operating Income	437,000,000	262,000,000	70,000,000	156,000,000	78,000,000	
Operating Margin %	7.77%	5.18%	1.44%	2.97%	1.43%	
SGA Expense	1,244,000,000	1,138,000,000	1,099,000,000	1,174,000,000	1,176,000,000	
Net Income	278,000,000	169,000,000	48,000,000	-80,000,000	51,000,000	
Operating Cash Flow	497,000,000	326,000,000	345,000,000	383,000,000	283,000,000	
Capital Expenditure	152,000,000	97,000,000	89,000,000	146,000,000	148,000,000	
EBITDA	552,000,000	368,000,000	195,000,000	46,000,000	137,000,000	
Return on Assets %	9.35%	5.91%	1.68%	-2.61%	1.56%	
Return on Equity %	13.44%	8.50%	2.47%	-3.81%	2.23%	
Debt to Equity	0.06	0.06	0.07	0.07	0.09	

CONTACT INFORMATION:

Phone: 212 720-3700 Fax: 212 553-7026
Toll-Free: 800-991-6815
Address: 112 W. 34th St., New York, NY 10120 United States

STOCK TICKER/OTHER:

Stock Ticker: FL
Exchange: NYS
Employees: 39,157
Fiscal Year Ends: 31-Jan
Parent Company:

SALARIES/BONUSES:

Top Exec. Salary: $1,100,000 Bonus: $
Second Exec. Salary: $837,500 Bonus: $

OTHER THOUGHTS:

Estimated Female Officers or Directors: 5
Hot Spot for Advancement for Women/Minorities: Y

Forever 21 Inc

NAIC Code: 448120

www.forever21.com

TYPES OF BUSINESS:

Apparel-Women's & Men's, Retail
Accessories
Footwear
Cosmetics
Lingerie
Online Sales

BRANDS/DIVISIONS/AFFILIATES:

Forever XXI
Love 21 Cotemporary
Forever + 21 Plus Size
Heritage 1981
Heritage1981.com
Forever21.com
Love & Beauty

CONTACTS: *Note: Officers with more than one job title may be intentionally listed here more than once.*

Do Won Chang, CEO
Lawrence Meyer, CFO/Sr. VP
Linda Chang, Sr. Mgr.-Mktg.

GROWTH PLANS/SPECIAL FEATURES:

Forever 21, Inc. operates approximately 480 largely mall-based clothing stores in the U.S., Canada, Europe, the Middle East and Asia. The company's business model is fast fashion, that is, to quickly imitate and manufacture the latest fashions and sell them at a low price. Traditional Forever 21 stores offer competitively priced trendy clothing, lingerie, cosmetics and accessories for women and junior girls. Forever XXI is the company's flagship store concept. Love 21 Contemporary is the firm's low priced sophisticated contemporary line. Forever 21 + Plus Sizes is the firm's plus size line. Heritage 1981 sells casual wear for girls and guys. Love & Beauty is a beauty collection featuring lip gloss, mascara, eye shadow and eyeliner. In addition to its retail locations, customers can buy clothing through the firm's online web sites, Forever21.com and Heritage1981.com. Forever 21 opens an average of 30 new stores each year. Taking advantage of low lease rates on large, empty spaces, the company announced plans to open a 90,000 square foot store in New York's Times Square; a 33,000 square foot store in a former Saks location in Charleston, South Carolina; an 85,000 square foot store in Cerritos, California; and a 120,000 square foot store in Las Vegas. As it is moving into larger spaces, the firm has been adding new product categories, such as men's and children's clothing. In January 2012, the firm opened its first store location in Paris, France.

FINANCIAL DATA: *Note: Data for latest year may not have been available at press time.*

In U.S. $	2012	2011	2010	2009	2008	2007
Revenue	2,800,000,000	2,600,000,000	2,500,000,000	2,300,000,000	1,800,000,000	1,260,000,000
R&D Expense						
Operating Income						
Operating Margin %						
SGA Expense						
Net Income						
Operating Cash Flow						
Capital Expenditure						
EBITDA						
Return on Assets %						
Return on Equity %						
Debt to Equity						

CONTACT INFORMATION:

Phone: 213-741-5100 Fax: 213-741-5161
Toll-Free:
Address: 2001 S. Alameda St., Los Angeles, CA 90058 United States

STOCK TICKER/OTHER:

Stock Ticker: Private Exchange:
Employees: 27,000 Fiscal Year Ends: 28-Feb
Parent Company:

SALARIES/BONUSES:

Top Exec. Salary: $ Bonus: $
Second Exec. Salary: $ Bonus: $

OTHER THOUGHTS:

Estimated Female Officers or Directors: 1
Hot Spot for Advancement for Women/Minorities:

Fossil Inc

www.fossil.com

NAIC Code: 334519

TYPES OF BUSINESS:

Watches and Parts (except crystals) Manufacturing
Accessories
Online Sales
Leather Goods
Belts
Handbags
Jewelry
Retail Stores

BRANDS/DIVISIONS/AFFILIATES:

Fossil
Relic
Zodiac
Skagen Designs

CONTACTS: *Note: Officers with more than one job title may be intentionally listed here more than once.*

Kosta N. Kartsotis, CEO
Mike L. Kovar, CFO/Exec. VP
Darren Hart, Exec. VP-Human Resources
Mike L. Kovar, Treas.
Livio Galanti, Exec. VP
Mark D. Quick, Vice Chmn.
Jennifer Pritchard, Pres., Retail Div.
Kosta N. Karsotis, Chmn.

GROWTH PLANS/SPECIAL FEATURES:

Fossil, Inc. designs, develops, markets and distributes fashion accessories. The company's principal offerings include a line of men's and women's watches and jewelry sold under proprietary and licensed brands, handbags, leather goods, sunglasses and apparel. In the watch and jewelry product category, Fossil has a diverse portfolio of globally recognized brands such as Fossil, Relic and Zodiac. Also, through license agreements, the company utilizes prestigious brand names such as Burberry, DKNY, Michael Kors, Marc Jacobs, Adidas and Emporio Armani. The company distributes products through various channels including wholesale and export as well as directly to the consumer. Domestically, the company sells its products through a distribution network that includes Neiman Marcus, Nordstrom, Macy's, Dillard's, JCPenney, Kohl's, Sears, Wal-Mart and Target. The firm also sells its products through a network of company-owned stores, which includes 123 retail and 74 outlet stores. Additionally, the company offers an extensive collection of Fossil brand products through its catalog and web site as well as proprietary and licensed watch and jewelry brands through other managed and affiliate web sites. Internationally, products are sold to department stores and specialty stores in over 120 countries through 22 company-owned foreign sales subsidiaries and through approximately 60 independent distributors. In 2012, Fossil acquired Skagen Designs, a company offering Danish design watches, jewelry, sunglasses and clocks internationally.

FINANCIAL DATA: *Note: Data for latest year may not have been available at press time.*

In U.S. $	2012	2011	2010	2009	2008	2007
Revenue	2,857,508,000	2,567,302,000	2,030,690,000	1,548,093,000	1,583,242,000	1,432,984,000
R&D Expense						
Operating Income	488,840,000	471,991,000	376,414,000	211,627,000	205,770,000	186,485,000
Operating Margin %	17.10%	18.38%	18.53%	13.67%	12.99%	13.01%
SGA Expense	1,117,703,000	967,195,000	778,750,000	633,223,000	645,381,000	555,546,000
Net Income	343,401,000	294,702,000	255,205,000	139,188,000	138,097,000	123,261,000
Operating Cash Flow	451,600,000	251,267,000	209,177,000	265,988,000	108,954,000	131,964,000
Capital Expenditure	122,804,000	131,496,000	46,812,000	38,072,000	87,598,000	47,080,000
EBITDA	488,840,000	471,991,000	376,414,000	211,627,000	227,626,000	221,212,000
Return on Assets %	19.70%	18.94%	18.60%	11.77%	12.49%	12.48%
Return on Equity %	29.35%	27.41%	25.43%	15.77%	17.54%	17.94%
Debt to Equity	0.06					

CONTACT INFORMATION:

Phone: 972 234-2525 Fax: 972 234-4669
Toll-Free: 800-449-3056
Address: 2280 N. Greenville Ave., Richardson, TX 75082 United States

STOCK TICKER/OTHER:

Stock Ticker: FOSL Exchange: NAS
Employees: 13,100 Fiscal Year Ends: 31-Dec
Parent Company:

SALARIES/BONUSES:

Top Exec. Salary: $601,539 Bonus: $
Second Exec. Salary: $475,385 Bonus: $125,000

OTHER THOUGHTS:

Estimated Female Officers or Directors: 4
Hot Spot for Advancement for Women/Minorities: Y

Francesca's Holdings Corp

www.francescascollections.com

NAIC Code: 448120

TYPES OF BUSINESS:

Ladies Apparel Stores

BRANDS/DIVISIONS/AFFILIATES:

Francesca's Collection Inc
FrancescasCollections.com
Francesca's Collections

CONTACTS:
Note: Officers with more than one job title may be intentionally listed here more than once.

John De Meritt, CEO
Theresa Backers, COO/Exec. VP
Neill Davis, Pres.
Cynthia Thomassee, Interim CFO
Kal Malik, General Counsel/Exec. VP
Cynthia Thomassee, Controller
Greg Brenneman, Chmn.

GROWTH PLANS/SPECIAL FEATURES:

Francesca's Holdings Corp. operates a national chain of 283 retail stores designed and merchandised to feel like upscale, independently owned boutiques. The firm conducts its business through the wholly-owned subsidiary Francesca's Collection, Inc. In addition to the retail stores, Francesca's also operates an e-commerce site at FrancescasCollections.com. Through its stores and e-commerce site, the company offers its customers a mix of fashion-forward women's apparel, jewelry, accessories and gift items. Keeping with its fashion-forward theme, the firm carries a broad selection of items in limited numbers and adds new items on a fairly regular basis to create a sense of scarcity and newness. Francesca's tailors its assortment of goods to appeal to its core target group, fashion conscious women between the ages of 18 and 35. Francesca's goal is to have 900 retail operations in the U.S and is poised to expand over 200 stores in the next four years.

FINANCIAL DATA:
Note: Data for latest year may not have been available at press time.

In U.S. $	2012	2011	2010	2009	2008	2007
Revenue	204,158,000	135,176,000	79,367,000			
R&D Expense						
Operating Income	43,531,000	29,643,000	17,482,000			
Operating Margin %	21.32%	21.92%	22.02%			
SGA Expense	63,262,000	40,525,000	24,641,000			
Net Income	22,501,000	16,895,000	10,604,000			
Operating Cash Flow	46,471,000	21,020,000	13,277,000			
Capital Expenditure	16,894,000	16,208,000	5,538,000			
EBITDA	48,467,000	32,020,000	18,737,000			
Return on Assets %	34.23%	37.40%	-165.57%			
Return on Equity %						
Debt to Equity	1.3					

CONTACT INFORMATION:

Phone: 713 864-1358 Fax:
Toll-Free:
Address: 3480 West 12th St., Houston, TX 77008 United States

STOCK TICKER/OTHER:

Stock Ticker: FRAN Exchange: NAS
Employees: 1,970 Fiscal Year Ends: 31-Jan
Parent Company:

SALARIES/BONUSES:

Top Exec. Salary: $375,000 Bonus: $175,000
Second Exec. Salary: $325,000 Bonus: $

OTHER THOUGHTS:

Estimated Female Officers or Directors: 4
Hot Spot for Advancement for Women/Minorities: Y

Frederick's of Hollywood Group Inc

www.fohgroup.com

NAIC Code: 448120

TYPES OF BUSINESS:

Lingerie & Sleepwear
Loungewear
Formal Apparel
Jewelry & Accessories
Shoes

BRANDS/DIVISIONS/AFFILIATES:

Seduction by Frederick's of Hollywood
Frederick's of Hollywood
Hollywood Exxtreme Cleavage
Get Cheeky

CONTACTS: *Note: Officers with more than one job title may be intentionally listed here more than once.*

Thomas J. Lynch, CEO
Don Jones, COO
Don Jones, Pres.
Thomas Rende, CFO/Exec. VP
Thomas J. Lynch, Chmn.

GROWTH PLANS/SPECIAL FEATURES:

Frederick's of Hollywood Group, Inc. designs, manufactures, markets and sells an extensive line of ladies' intimate apparel. Its product line includes bras, panties, lingerie, corsets, hosiery, shape wear, sleepwear, dresses, tops, pants, shorts, skirts, formal wear, shoes, fragrances, personal care products, jewelry and bridal undergarments. The company operates in two divisions: retail and direct to consumer. The retail division manages 118 mall-based Frederick's of Hollywood retail stores in 29 U.S. states (approximately one-third of these are located in California). These stores range in size from 900 square feet to 3,200, with its flagship store in Hollywood, California at 5,700 square feet. The direct to consumer business handles the company's e-commerce and catalog marketing operations. In 2012, the company sent out 10.8 million catalogs to 5 million households. Some of its brands and trademarks include Hollywood Exxtreme Cleavage, Get Cheeky and Seduction by Frederick's of Hollywood. In addition to its intimate apparel, the firm has licensed the Frederick's of Hollywood brand to manufacturers of swimwear, Halloween costumes, jewelry and accessories. Recently, the company announced the construction of Frederick's of Hollywood retail stores throughout the Middle East through a licensing agreement with Emirates Associated Business Group. As of October 2012, the firm operated three stores in the UAE.

FINANCIAL DATA: *Note: Data for latest year may not have been available at press time.*

In U.S. $	2012	2011	2010	2009	2008	2007
Revenue	111,406,000	119,615,000	133,855,000	176,310,000	182,233,000	63,493,000
R&D Expense						
Operating Income	-4,133,000	-8,713,000	-7,109,000	-32,384,000	-13,181,000	1,020,000
Operating Margin %	-3.70%	-7.28%	-5.31%	-18.36%	-7.23%	1.60%
SGA Expense	45,757,000	49,771,000	55,079,000	74,496,000	80,108,000	17,434,000
Net Income	-6,432,000	-12,055,000	-21,164,000	-34,047,000	-15,383,000	146,000
Operating Cash Flow	-6,571,000	-11,137,000	2,171,000	5,538,000	-4,400,000	-137,000
Capital Expenditure	312,000	96,000	816,000	3,882,000	7,165,000	260,000
EBITDA	-1,673,000	-5,591,000	-2,902,000	-26,493,000	-8,201,000	1,382,000
Return on Assets %	-14.81%	-22.86%	-31.61%	-37.37%	-14.54%	.64%
Return on Equity %		-210.25%	-245.76%	-149.49%	-39.09%	1.02%
Debt to Equity		47.33		2.12	0.31	

CONTACT INFORMATION:

Phone: 323 466-5151 Fax:
Toll-Free:
Address: 8 W. 38th St., Ste. 802, Hollywood, CA United States

STOCK TICKER/OTHER:

Stock Ticker: FOHL Exchange: PINX
Employees: 926 Fiscal Year Ends: 31-Jul
Parent Company:

SALARIES/BONUSES:

Top Exec. Salary: $540,000 Bonus: $
Second Exec. Salary: $334,615 Bonus: $

OTHER THOUGHTS:

Estimated Female Officers or Directors:
Hot Spot for Advancement for Women/Minorities: Y

French Connection Group plc

www.frenchconnection.com

NAIC Code: 315240

TYPES OF BUSINESS:

Apparel Design, Manufacture & Distribution
Accessories
Shoes
Cosmetics & Fragrances
Home Furnishings
Online & Mail Order Sales
Franchising
Retail Stores

BRANDS/DIVISIONS/AFFILIATES:

YMC
Great Plains
Toast
French Connection U.K.
fcuk
YMC

CONTACTS:

Note: Officers with more than one job title may be intentionally listed here more than once.

Stephen Marks, CEO
Neil Williams, Dir.-Oper.
Roy Naismith, Dir.-Finance
Stephen Marks, Chmn.

GROWTH PLANS/SPECIAL FEATURES:

French Connection Group plc is a youth-oriented apparel company that retails and wholesales fashion clothing and accessories in more than 30 countries worldwide. Generally, the firm targets men and women ages 18-35. Its products are sold through its retail stores in the U.K., Europe, U.S., Australia, the Middle East and Canada; mail order service in the U.K. and U.S.; and on the Internet. The company has licensed French Connection operators in Australia, Asia and the Middle East. The firm also operates wholesale operations in London, Hong Kong, Paris, Toronto and New York. It owns four brands: French Connection; Great Plains; Toast; and the menswear brand, YMC. French Connection designs, produces and distributes its own brands of men's and women's products, including outerwear, dresses, skirts, jeans, footwear, casual wear, suits, denim, accessories and children's apparel, and is supported by unique advertising featuring the acronym for French Connection U.K., fcuk. Accessories such as hats, sunglasses, backpacks, fragrances and the company's new cosmetics line complement the brand's fashions. A range of homewares are also available through some of its stores. Fcuk is available worldwide. It is represented in the U.K., Europe, the U.S. and through the firm's joint ventures in Asia. The Great Plains brand provides a range of fashionable mid-priced basic clothing for men and women and is available across the U.K. and in Europe. The Toast brand, represented only in the U.K., provides a range of garments and products that are available through mail order, online and at some retail locations.

FINANCIAL DATA:

Note: Data for latest year may not have been available at press time.

In U.S. $	2012	2011	2010	2009	2008	2007
Revenue	344,144,400	341,588,100	339,511,100	341,268,600	377,216,800	385,524,900
R&D Expense						
Operating Income	5,272,408	5,911,488	-4,793,098	-2,716,089	2,236,779	5,272,408
Operating Margin %	1.53%	1.73%	-1.41%	- .79%	.59%	1.36%
SGA Expense						
Net Income	8,467,807	-3,834,479	-39,782,720	-26,521,810	2,236,779	159,770
Operating Cash Flow	-2,716,089	-1,757,469	2,077,009	-639,080	5,432,178	1,597,699
Capital Expenditure	2,556,319	1,597,699	3,355,169	9,745,966	3,994,249	7,349,418
EBITDA	9,745,966	11,663,210	2,875,859	10,065,510	15,018,370	20,929,860
Return on Assets %	4.35%	-1.93%	-17.17%	-10.12%	.80%	.05%
Return on Equity %	7.21%	-3.33%	-28.75%	-15.16%	1.17%	.08%
Debt to Equity						

CONTACT INFORMATION:

Phone: 44 2070367200 Fax: 44 2070367201
Toll-Free:
Address: 39 Camden St., Centro 1, London, NW1 0DX United Kingdom

STOCK TICKER/OTHER:

Stock Ticker: FCOGF
Exchange: PINX
Employees: 2,558
Fiscal Year Ends: 31-Jan
Parent Company:

SALARIES/BONUSES:

Top Exec. Salary: $ Bonus: $
Second Exec. Salary: $ Bonus: $

OTHER THOUGHTS:

Estimated Female Officers or Directors: 1
Hot Spot for Advancement for Women/Minorities:

Fujibo Holdings Inc

www.fujibo.co.jp

NAIC Code: 313210

TYPES OF BUSINESS:

Textile Manufacturing
Specialty Fabrics
Machinery
Health Products
Specialty Chemicals
Real Estate

BRANDS/DIVISIONS/AFFILIATES:

BVD
Chitopoly
Fujibo Apparel Corp
Fujibo Electronics Corp
Fujibo Wakayama Corp
Fujibo Ehime Corp

CONTACTS: *Note: Officers with more than one job title may be intentionally listed here more than once.*

Miki Yasufumi, Managing Dir.
Nakano Mitsuo, Pres.

GROWTH PLANS/SPECIAL FEATURES:

Fujibo Holdings, Inc., based in Tokyo, is a holding company that primarily manufactures, processes and sells textiles. Its operations are divided into four segments: Fibers, Abrasive Materials, Chemical Products and Other Business. The Fibers division oversees the production of a diverse range of textiles. Products include Spandex-brand synthetic fabrics, specialized yarn, cellulose, specialized knit/woven material, raw cotton, flame-resistant fabrics, recycled polyester, silk-blended fiber and wool-blended fiber. Some unique products include Chitopoly, a fabric consisting of chitosan, a uniquely Japanese material made from crustacean shells, blended with polynosic, a plant-based fiber. Chitopoly is said to be antibacterial and anti-odor and to cause almost no irritation of the skin. The Abrasive Materials segment offers non woven fabrics, functional films and ultra-precise abrasive materials. Fujibo's Chemical Products division manufactures chemicals, largely for use in its textiles. The firm has developed a seaweed extract that, when incorporated into a fabric, can purportedly stimulate fat-dissolving enzymes in the wearer. Other Fujibo chemicals can, when used with fabrics, replenish vitamins needed for the skin, retain moisture and generate heat, among other applications. The segment also manufactures various other types of cosmetics and pharmaceutical products. The Other Business segment has interests in the automotive sector; it is focused on refining and the production of electronic products as well as some forms of industrial equipment; and engages in the sale, rental and management of real estate. A key subsidiary, Fujibo Apparel Corp., markets BVD branded undergarments for Japanese consumers. Other subsidiaries include Fujibo Electronics Corp., Fujibo Wakayama Corp. and Fujibo Ehime Corp.

FINANCIAL DATA: *Note: Data for latest year may not have been available at press time.*

In U.S. $	2012	2011	2010	2009	2008	2007
Revenue		461,700,000	414,640,000	423,650,000	523,980,000	557,290,000
R&D Expense						
Operating Income						
Operating Margin %						
SGA Expense						
Net Income		11,800,000	9,300,000	4,110,000	8,370,000	5,290,000
Operating Cash Flow						
Capital Expenditure						
EBITDA						
Return on Assets %						
Return on Equity %						
Debt to Equity						

CONTACT INFORMATION:

Phone: 81-3-3665-7777 Fax:
Toll-Free:
Address: 1-18-12, Ningyou-machi, Nihonbashi, Chuo-ku, Tokyo, 103-0013 Japan

STOCK TICKER/OTHER:

Stock Ticker: 3104
Employees: 1,479
Parent Company:
Exchange: Tokyo
Fiscal Year Ends: 31-Mar

SALARIES/BONUSES:

Top Exec. Salary: $ Bonus: $
Second Exec. Salary: $ Bonus: $

OTHER THOUGHTS:

Estimated Female Officers or Directors:
Hot Spot for Advancement for Women/Minorities:

Fung Group

NAIC Code: 488510

www.funggroup.com

TYPES OF BUSINESS:

Freight Transportation Arrangement
Supply Chain Management
Distribution & Warehousing
Ground Transport
Healthcare & Consumer Products, Manufacturing
Convenience Stores
Toys, Retail
Clothing, Distribution & Retail

BRANDS/DIVISIONS/AFFILIATES:

Li & Fung Ltd
Fung Retailing Limited
Circle K
Convenience Retail Asia Limited
Trinity Limited
Bluedog
Gieves & Hawkes
LEO

CONTACTS: *Note: Officers with more than one job title may be intentionally listed here more than once.*

William K. Fung, Group Deputy Chmn.
Victor K. Fung, Chmn.

GROWTH PLANS/SPECIAL FEATURES:

Fung Group through its subsidiaries, is engaged in logistics, trading, distribution and retailing. Li & Fung Ltd., the publicly traded export trading, logistics and distribution branch of the company, manages the logistics of producing and exporting an extensive range of private-label consumer goods. It operates through over 15,000 suppliers that deliver its goods in 40 countries around the world. Garments comprise the majority of Li & Fung's trading business, though it also deals in fashion accessories, home furnishings, toys, sporting goods and travel goods. Fung Group's distribution division is also carried out by Li & Fung Ltd, whose distribution business operates in three segments: private labels, proprietary brands and licensing of recognized brands. The firm has over 300 distribution centers and offices. The firm's retail division is carried out through subsidiary Fung Retailing Limited, which operates roughly 600 retail outlets under the Circle K and Saint Honore brands in China through Convenience Retail Asia Limited. Through Trinity Limited, Fung Retailing offers high-end men's apparel at 520 locations under six international brands: Cerruti 1881, Altea, Gieves & Hawkes, D'urban, Intermezzo and Kent & Curwen. Branded Lifestyle Holdings Limited operates over 1,000 outlets focused on casual fashion brands like Hang Ten, Roots, Arnold Palmer, H:Connect and LEO. Fung Kids (Holdings) Limited consists of the Stride Rite and Toonsland children's shoes and apparel brands throughout the Asia pacific region. The Suhyang Networks operates 218 shop-in-shop stores focused on infant and child apparel under brands including Bluedog, Minkmui and Lulabee. Through Toysâ€˜R'US (Asia), Fung Retailing operates 150 Toys R Us Asia locations under license. In 2012, Trinity Limited acquired the Gieves & Hawkes fashion brand from Wing Tai Properties Limited.

FINANCIAL DATA: *Note: Data for latest year may not have been available at press time.*

In U.S. $	2012	2011	2010	2009	2008	2007
Revenue	20,221,810,000	20,030,270,000	15,947,550,000	13,472,260,000	14,286,580,000	11,856,970,000
R&D Expense						
Operating Income	790,703,000	879,937,000	726,749,100	514,512,500	392,823,800	461,667,200
Operating Margin %	3.91%	4.39%	4.55%	3.81%	2.74%	3.89%
SGA Expense	2,452,622,000	2,192,148,000	1,516,512,000	1,125,367,000	1,153,518,000	300,586,000
Net Income	622,564,000	681,404,000	549,985,200	434,000,400	312,730,800	392,536,500
Operating Cash Flow	586,121,000	924,223,000	747,008,400	870,831,400	461,621,400	515,494,300
Capital Expenditure	187,832,000	118,055,000	82,747,270	84,328,030	66,019,190	40,973,130
EBITDA	1,096,924,000	1,127,200,000	855,547,600	582,627,200	392,823,800	461,667,200
Return on Assets %	5.38%	6.66%	7.37%	8.48%	7.00%	11.35%
Return on Equity %	13.59%	17.97%	18.55%	21.65%	20.86%	33.59%
Debt to Equity	0.24	0.31	0.68	0.36	0.43	

CONTACT INFORMATION:

Phone: 852 23002300 Fax: 852 23002000
Toll-Free:
Address: 888 Cheung Sha Wan Rd., LiFung Tower, 11th Fl., Hong Kong, Hong Kong

STOCK TICKER/OTHER:

Stock Ticker: LFUGF Exchange: PINX
Employees: 27,000 Fiscal Year Ends: 31-Dec
Parent Company:

SALARIES/BONUSES:

Top Exec. Salary: $ Bonus: $
Second Exec. Salary: $ Bonus: $

OTHER THOUGHTS:

Estimated Female Officers or Directors:
Hot Spot for Advancement for Women/Minorities: Y

G & K Services Inc

www.gkservices.com

NAIC Code: 812331

TYPES OF BUSINESS:

Linen & Uniform Supply
Uniform Rental
Cleanroom Apparel
Facility Services
Customized Apparel

BRANDS/DIVISIONS/AFFILIATES:

Best Uniform Rental Inc
Lion Uniform Group
Grand Rapids Coat & Apron Service Inc
Alltex Uniform Rental Service Inc
Nanoclean Ltd

CONTACTS: *Note: Officers with more than one job title may be intentionally listed here more than once.*

Douglas A. Milroy, CEO
Jeffrey L. Wright, CFO/Exec. VP
David A. Euson, VP-Mktg. & Sales
Randall R. Ross, VP-Human Resources
Ian G. Davis, CIO/VP
Jeffrey L. Cotter, General Counsel/VP/Corp. Sec.
Richard J. Stutz, Sr. VP-Oper.
Timothy N. Curran, Sr. VP-U.S. Field
M. Lenny Pippin, Chmn.
Robert G. Wood, Pres., G&K Svcs. Canada, Inc.
Richard J. Stutz, Sr. VP-Sourcing

GROWTH PLANS/SPECIAL FEATURES:

G&K Services, Inc. provides corporate uniforms and facility services to more than 165,000 clients in the North American industrial, service and high-technology industries. The company rents and leases branded identity apparel, including uniforms, corporate casual wear and executive apparel. It offers customized apparel for individual companies based on brand images, safety and security needs. G&K Services accommodates a variety of industries, as it includes flame-resistant and food safety specializations in its uniform offerings. The company operates out of approximately 160 locations in North America and Europe and serves 85 of the top 100 metropolitan markets in the U.S. and Canada. Through its Lion Uniform Group division, it provides direct purchase services to customers over the Internet or through catalogs. In addition to customized apparel programs for its clients, G&K Services offers value-added services such as corporate identity and security, brand awareness, image management, employee retention, worker protection and product protection. It also provides services for cleanroom operations, such as microelectronic, pharmaceutical, biotechnology, defense and automotive businesses as well as for situations requiring food-safety apparel. The company's facility services include dust control, maintenance, hand care and hygiene products and services. G&K Services owns Grand Rapids Coat & Apron Service, Inc.; Alltex Uniform Rental Service, Inc.; Best Uniform Rental, Inc.; and Nanoclean, Ltd., an Irish cleanroom business. The firm manufactures approximately 55% of the work apparel it markets, primarily out of its facility located in the Dominican Republic.

Employees receive health, dental and life insurance; a flexible spending account; an employee assistance program; tuition reimbursement; and disability coverage.

FINANCIAL DATA: *Note: Data for latest year may not have been available at press time.*

In U.S. $	2012	2011	2010	2009	2008	2007
Revenue	869,937,000	828,861,000	833,592,000	935,965,000	1,002,395,000	929,542,000
R&D Expense						
Operating Income	42,271,000	65,375,000	59,621,000	-76,043,000	90,523,000	79,362,000
Operating Margin %	4.85%	7.88%	7.15%	-8.12%	9.03%	8.53%
SGA Expense	220,108,000	196,076,000	191,705,000	226,115,000	229,987,000	203,614,000
Net Income	24,147,000	33,160,000	28,612,000	-72,464,000	46,079,000	43,190,000
Operating Cash Flow	72,973,000	67,004,000	72,710,000	103,180,000	103,058,000	80,389,000
Capital Expenditure	34,026,000	20,670,000	16,710,000	23,330,000	27,057,000	31,515,000
EBITDA	76,254,000	102,975,000	99,809,000	-31,791,000	138,927,000	124,957,000
Return on Assets %	2.77%	3.94%	3.43%	-7.61%	4.50%	4.44%
Return on Equity %	5.26%	6.75%	6.32%	-14.56%	8.01%	7.58%
Debt to Equity	0.54	0.18	0.34	0.51	0.5	0.25

CONTACT INFORMATION:

Phone: 952 912-5500 Fax: 952 912-5999
Toll-Free: 800-452-2737
Address: 5995 Opus Pkwy., Minnetonka, MN 55343 United States

STOCK TICKER/OTHER:

Stock Ticker: GK Exchange: NAS
Employees: 7,500 Fiscal Year Ends: 30-Jun
Parent Company:

SALARIES/BONUSES:

Top Exec. Salary: $613,478 Bonus: $
Second Exec. Salary: $425,770 Bonus: $

OTHER THOUGHTS:

Estimated Female Officers or Directors:
Hot Spot for Advancement for Women/Minorities: Y

Gap Inc (The)

NAIC Code: 448140

www.gapinc.com

TYPES OF BUSINESS:

Casual Apparel Stores
Online Retailing
Shoes & Accessories
Personal Care Products
Children's Clothing
Babies' Clothing
Maternity Clothing
Exercise Clothing

BRANDS/DIVISIONS/AFFILIATES:

Piperlime
GapKids
babyGap
Banana Republic
Old Navy Clothing Co
Gap Outlet
Gap Inc Direct
Athleta

CONTACTS: *Note: Officers with more than one job title may be intentionally listed here more than once.*

Glenn K. Murphy, CEO
Sabrina Simmons, CFO/Exec. VP
Eva Sage-Gavin, Exec. VP-Global Human Resources
Tom Keiser, CIO/Exec. VP
Michelle Banks, General Counsel/Exec. VP/Chief Compliance Officer
Eva Sage-Gavin, Exec. VP-Corp. Affairs
Jack Calhoun, Pres., Banana Republic
Toby Lenk, Pres., Gap, Inc. Direct
Art Peck, Pres., Gap North America
Stefan Larsson, Pres., Old Navy
Glenn K. Murphy, Chmn.
Stephen Sunnucks, Pres., International
Colin Funnell, Exec. VP-Global Supply Chain

GROWTH PLANS/SPECIAL FEATURES:

The Gap, Inc. (Gap) is a retailer of clothing, accessories and personal care products for adults, kids and babies. It has over 3,263 stores in the U.S., Canada, Mexico, the U.K., Japan, Germany, Ireland and France. It has more than 100 Gap and Banana Republic stores through franchise agreements in Asia, Europe, Latin America and the Middle East. The Gap brand includes GapKids, babyGap, GapBody and Gap Outlet as well as Banana Republic and Old Navy. Gap stores typically embrace all Gap-branded units, such as Gap, babyGap and GapKids under one roof but segregate them into different areas, each with its own street/mall entrance. The firm also sells its products via Gap.com, BananaRepublic.com and OldNavy.com. Gap has two online-only brands: Athleta, which sells women's exercise and casual apparel, footwear and accessories; and Piperlime, which sells apparel, footwear and handbags for women, men and kids. Gap stores sell classically styled, high-quality casual apparel. Old Navy operates stores in the U.S and Canada and offers a broad selection of apparel, shoes and accessories for adults, children and infants at more affordable prices. Banana Republic has locations in the U.S., Canada, the U.K. and Japan and sells men's and women's clothing, shoes, accessories, personal care products, intimate apparel and jewelry to a higher price-point market. In late 2011, Gap announced a major cutback. By the end of 2013, it plans to shrink its North American store base by nearly 200 units. Meanwhile, it hopes to expand in emerging nations such as China, where it already has a minor presence. This firm's stores generate about $400 in sales per square foot per year.

Gap offers its employees benefits including medical, dental and vision insurance; life, disability, AD&D and business travel accident insurance; an employee assistance program; travel, moving and relocation assistance; tuition reimbursement; a 401(k) plan with matching; an employee stock purchase plan; and merchandise discounts at all Gap-affiliated stores.

FINANCIAL DATA: *Note: Data for latest year may not have been available at press time.*

In U.S. $	2012	2011	2010	2009	2008	2007
Revenue	14,549,000,000	14,664,000,000	14,197,000,000	14,526,000,000	15,763,000,000	
R&D Expense						
Operating Income	1,438,000,000	1,968,000,000	1,815,000,000	1,548,000,000	1,315,000,000	
Operating Margin %	9.88%	13.42%	12.78%	10.65%	8.34%	
SGA Expense						
Net Income	833,000,000	1,204,000,000	1,102,000,000	967,000,000	833,000,000	
Operating Cash Flow	1,363,000,000	1,744,000,000	1,928,000,000	1,412,000,000	2,081,000,000	
Capital Expenditure	548,000,000	557,000,000	334,000,000	431,000,000	682,000,000	
EBITDA	2,035,000,000	2,630,000,000	2,477,000,000	2,153,000,000	1,979,000,000	
Return on Assets %	11.50%	15.99%	14.17%	12.55%	10.16%	
Return on Equity %	24.37%	26.84%	23.75%	22.32%	17.63%	
Debt to Equity	0.92	0.21			0.01	

CONTACT INFORMATION:

Phone: 650 952-4400 Fax: 650 952-4407
Toll-Free: 800-333-7899
Address: 2 Folsom St., San Francisco, CA 94105 United States

STOCK TICKER/OTHER:

Stock Ticker: GPS Exchange: NYS
Employees: 132,000 Fiscal Year Ends: 31-Jan
Parent Company:

SALARIES/BONUSES:

Top Exec. Salary: $1,500,000 Bonus: $
Second Exec. Salary: $921,383 Bonus: $

OTHER THOUGHTS:

Estimated Female Officers or Directors: 5
Hot Spot for Advancement for Women/Minorities: Y

Genesco Inc

www.genesco.com

NAIC Code: 316210

TYPES OF BUSINESS:

Shoes, Retail
Retail Stores
Men's Accessories
Wholesale Operations
Hats, Retail
Catalog & Online Operations
Athletic Team Products

BRANDS/DIVISIONS/AFFILIATES:

Johnston & Murphy
Journeys
Underground Station
Journeys Kidz
Schuh Group Ltd
Dockers
Lids
Jarman

CONTACTS: *Note: Officers with more than one job title may be intentionally listed here more than once.*

Robert J. Dennis, CEO
Robert J. Dennis, Pres.
James S. Gulmi, CFO/Sr. VP-Finance
Roger G. Sisson, General Counsel/Corp. Sec./Sr. VP
Mimi E. Vaughn, Sr. VP-Strategy & Shared Svcs.
Paul D. Williams, Chief Acct. Officer/VP
Kenneth J. Kocher, Sr. VP/Pres., Hat World
James C. Estepa, Sr. VP/CEO/Pres., Genesco Retail
Jonathan D. Caplan, Sr. VP/CEO-Genesco Branded Group
Jonathan D. Caplan, Pres., Johnston & Murphy
Robert J. Dennis, Chmn.

GROWTH PLANS/SPECIAL FEATURES:

Genesco, Inc. is a leading specialty retailer of footwear, headwear, licensed sportswear and accessories through 2,309 retail stores in the U.S., Canada and Puerto Rico. The company operates six business segments: Journeys Group, Underground Station Group, Schuh Group, Lids Sports Group, Johnston & Murphy Group and Licensed Brands. The Journey's Group segment operates 1,017 stores, including Journeys, Journeys Kidz and Shi by Journeys, and markets products through a catalog and Journeys.com. Journeys stores target customers in the 13-22 year age group. Shi carries shoes targeted at women in their mid 20s to early 30s. The Underground Station Group segment operates 137 stores, including Underground Station and Jarman. The Schuh Group segment operates 64 stores averaging 4,575 square feet and 14 footwear concessions in Republic apparel stores. Schuh stores targets men and women between 15 and 30 years of age. The Lids Sports Group segment operates 1,002 stores, including Hat World, Lids, Hat Shack, Hat Zone, Head Quarters, Cap Connection, Lids Kids and Lids Locker Room, and also markets products online. The Johnston & Murphy Group segment operates 153 retail and factory stores, as well as JohnstonMurphy.com, and sells footwear, luggage and accessories primarily targeting male business and professional customers. Retail prices for Johnston & Murphy footwear generally range from $100 to $275. The Licensed Brands segment is comprised primarily of Dockers footwear, marketed under a license from Levi Strauss & Company. In June 2011, the firm acquired Schuh Group Ltd., a United Kingdom-based casual and athletic footwear retailer.

Employee benefits include educational assistance, flexible spending accounts, adoption assistance, scholarships, healthcare and dental plans, company discounts, an employee credit association, child care and fitness classes.

FINANCIAL DATA: *Note: Data for latest year may not have been available at press time.*

In U.S. $	2012	2011	2010	2009	2008	2007
Revenue	2,291,987,000	1,789,839,000	1,574,352,000	1,551,562,000	1,502,119,000	
R&D Expense						
Operating Income	143,870,000	86,083,000	60,422,000	263,192,000	45,161,000	
Operating Margin %	6.27%	4.80%	3.83%	16.96%	3.00%	
SGA Expense	1,007,502,000	807,197,000	722,087,000	713,365,000	696,352,000	
Net Income	81,959,000	53,211,000	28,813,000	152,636,000	6,885,000	
Operating Cash Flow	144,960,000	102,608,000	142,096,000	179,103,000	23,943,000	
Capital Expenditure	49,456,000	29,299,000	33,825,000	49,420,000	80,662,000	
EBITDA	197,672,000	133,829,000	101,951,000	310,271,000	90,419,000	
Return on Assets %	7.45%	5.83%	3.42%	18.81%	.86%	
Return on Equity %	12.32%	8.89%	5.65%	35.58%	1.63%	
Debt to Equity	0.04			0.26	0.37	

CONTACT INFORMATION:

Phone: 615 367-7000 Fax: 615 367-8278
Toll-Free:
Address: 1415 Murfreesboro Rd., Nashville, TN 37217 United States

STOCK TICKER/OTHER:

Stock Ticker: GCO Exchange: NYS
Employees: 15,200 Fiscal Year Ends: 31-Jan
Parent Company:

SALARIES/BONUSES:

Top Exec. Salary: $794,500 Bonus: $
Second Exec. Salary: $566,500 Bonus: $

OTHER THOUGHTS:

Estimated Female Officers or Directors: 2
Hot Spot for Advancement for Women/Minorities:

Gerry Weber International AG

www.gerryweber-ag.de

NAIC Code: 315240

TYPES OF BUSINESS:

Women's Apparel, Retail
Plus-Size Apparel
Accessories
Brand Licensing

BRANDS/DIVISIONS/AFFILIATES:

GERRY WEBER
TAIFUN
SAMOON
GERRY WEBER EDITION
G.W.
House of Gerry Weber
CASTRO Deutschland GmbH & Co

CONTACTS: *Note: Officers with more than one job title may be intentionally listed here more than once.*

Gerhard Weber, CEO
Thomas Kronefeld, Dir.-Retail
Corinna Korte, Dir.-Public Rel. & Corp. Comm.
Jorg Stuber, Dir.-Investor Rel.
David Frink, Managing Dir.
Doris Stratker, Managing Dir.
Ernst F. Schroder, Chmn.

GROWTH PLANS/SPECIAL FEATURES:

Gerry Weber International AG retails women's ready-to-wear clothing under three proprietary brands throughout Germany as well as internationally. The firm's brands are GERRY WEBER, an upper-mid-priced, complementary knit, trouser and outdoor women's wear line; TAIFUN, a mid-priced, internationally influenced business and casual fashion line that is more colorful and designed for a younger consumer; and SAMOON, a plus-size line. The company licenses its GERRY WEBER brand for ladies' accessories, including handbags, shoes, scarves, watches, jewelry and eyewear. In addition, the GERRY WEBER line includes two sub-labels: GERRY WEBER EDITION, a single item collection consisting of knits, shirts, blouses, pants, skirts and outdoor jackets; and G.W., offering trendy fashion items at lower prices. Instead of offering the traditional collection schedule, the firm offers bi-monthly collections for all of its brands, allowing it to more efficiently and effectively respond to customer demand and changing trends. Products are sold through more than 560 House of Gerry Weber branded stores as well as 2,400 shop-in-shops worldwide, e-commerce sites and several TAIFUN and SAMOON single branded shops. GERRY WEBER products accounted for approximately 78% of company revenues, TAIFUN products represented 17% and SAMOON products accounted for 5% of total revenue, with less than 1% of revenue stemming from miscellaneous sources. The company's export business has grown in recent years, and sales outside of Germany now account for approximately 40% of annual revenues. Major export markets include the U.K., Ireland, Belgium, the Netherlands, Luxembourg, Scandinavia, Austria and Switzerland, while additional exports go to locations such as Eastern Europe, Asia, the Middle East and Canada. Recently, the firm acquired German retailer CASTRO Deutschland GmbH & Co.

FINANCIAL DATA: *Note: Data for latest year may not have been available at press time.*

In U.S. $	2012	2011	2010	2009	2008	2007
Revenue	1,072,867,000	939,728,900	831,697,300	794,397,800	762,239,000	678,073,500
R&D Expense						
Operating Income	155,032,100	133,201,300	111,383,800	95,264,080	83,901,990	69,195,180
Operating Margin %	14.45%	14.17%	13.39%	11.99%	11.00%	10.20%
SGA Expense						
Net Income	105,417,100	89,646,700	72,158,640	57,535,460	52,706,800	36,059,310
Operating Cash Flow	105,670,500	93,065,940	81,299,820	123,264,200	51,258,360	79,355,440
Capital Expenditure	75,438,620	57,751,000	31,817,330	20,244,720	28,697,510	
EBITDA	176,008,600	148,837,500	126,665,500	111,333,100	98,693,490	69,195,180
Return on Assets %	17.54%	18.08%	17.41%	14.56%	13.83%	10.53%
Return on Equity %	23.29%	25.56%	29.21%	25.31%	24.16%	19.67%
Debt to Equity	0.02	0.04	0.1			0.27

CONTACT INFORMATION:

Phone: 49 52011850 Fax: 49 520110931
Toll-Free:
Address: Neulehenstrasse 8, Halle, 33790 Germany

STOCK TICKER/OTHER:

Stock Ticker: GRYIF Exchange: PINX
Employees: 3,260 Fiscal Year Ends: 31-Oct
Parent Company:

SALARIES/BONUSES:

Top Exec. Salary: $ Bonus: $
Second Exec. Salary: $ Bonus: $

OTHER THOUGHTS:

Estimated Female Officers or Directors: 2
Hot Spot for Advancement for Women/Minorities:

Gianni Versace SpA

www.versace.com

NAIC Code: 315240

TYPES OF BUSINESS:

Designer Apparel
Retail Boutiques
Accessories & Eyewear
Home Furnishings
Beauty Products & Fragrances
Hotels & Resorts

BRANDS/DIVISIONS/AFFILIATES:

Versace Boutique
Versace Home Collection
Versus
Versace Jeans Couture
Sunland Group Ltd
Palazzo Versace Luxury Hotel and Resort
Palazzo Versace Dubai
Versace Collection

CONTACTS: *Note: Officers with more than one job title may be intentionally listed here more than once.*

Gian Giacomo Ferraris, CEO
Francesco Buccola, CFO
Donatella Versace, Dir.-Creative
Santo Versace, Chmn.

GROWTH PLANS/SPECIAL FEATURES:

Gianni Versace SpA, founded in 1978, is one of the world's leading international fashion houses. The company designs, manufactures, distributes and retails luxury fashion and lifestyle products, clothing, jewelry, accessories, fragrances, cosmetics and home furnishings under a variety of brand names. The company's portfolio of brands covers two primary segments of the market. The higher end is addressed by the Gianni Versace collection and includes apparel, jewelry, watches, accessories, fragrances, cosmetics and home furnishings, all of which are distributed through a world-wide retail network including 82 free-standing boutiques in principal cities. The second segment is addressed towards a younger target consumer and includes the Versace Collection, Versace Sport, Versace Jeans Couture and Versus brands. These items are distributed through free-standing stores. The company's products are also sold through its 56 branded stores and approximately 1,800 corners in multi-brand/department stores. In partnership with Sunland Group Ltd., an Australian development company, the company owns and operates the Palazzo Versace Luxury Hotel and Resort, a 205-room hotel located on the Gold Coast of Australia; and Palazzo Versace Dubai, located in the UAE. The firm, owned entirely by the Versace family, is currently undergoing a complete restructuring of its brands and retail strategy.

FINANCIAL DATA: *Note: Data for latest year may not have been available at press time.*

In U.S. $	2012	2011	2010	2009	2008	2007
Revenue	398,881,000	390,110,000	382,796,000			
R&D Expense						
Operating Income						
Operating Margin %						
SGA Expense						
Net Income						
Operating Cash Flow						
Capital Expenditure						
EBITDA						
Return on Assets %						
Return on Equity %						
Debt to Equity						

CONTACT INFORMATION:

Phone: 39-02-76-09-31 Fax: 39-02-76-00-41-22
Toll-Free:
Address: Via Manzoni, 38, Milan, 20121 Italy

STOCK TICKER/OTHER:

Stock Ticker: Private
Exchange:
Employees:
Fiscal Year Ends: 31-Dec
Parent Company:

SALARIES/BONUSES:

Top Exec. Salary: $ Bonus: $
Second Exec. Salary: $ Bonus: $

OTHER THOUGHTS:

Estimated Female Officers or Directors: 1
Hot Spot for Advancement for Women/Minorities: Y

G-III Apparel Group Ltd

NAIC Code: 315240

www.g-iii.com

TYPES OF BUSINESS:

Apparel Importing & Manufacturing
Outerwear & Sportswear
Handbags & Luggage
Leather Accessories

BRANDS/DIVISIONS/AFFILIATES:

Andrew Marc
G-III
Black Rivet
Jessica Howard
Vince Camuto
Wilsons Leather
Vilebrequin International SA
Eliza J

CONTACTS: *Note: Officers with more than one job title may be intentionally listed here more than once.*

Morris Goldfarb, CEO
Wayne S. Miller, COO
Jeanette Nostra-Katz, Pres.
Neal S. Nackman, CFO
Wayne S. Miller, Corp. Sec.
Jeffrey Goldfarb, Dir.-Strategic Planning
Neal S. Nackman, Treas.
Sammy Aaron, Vice Chmn.
Morris Goldfarb, Chmn.

GROWTH PLANS/SPECIAL FEATURES:

G-III Apparel Group, Ltd. designs, manufactures, imports and markets an extensive range of licensed and private-label branded outerwear and sportswear, including coats, jackets, pants, dresses and other sportswear items as well as handbags and luggage. G-III operates in three segments: licensed apparel, which accounted for roughly 65% of its net sales in 2011; non-licensed apparel, 22%; and retail, 13%. Among its many licensed brands are Calvin Klein, Guess?, Kenneth Cole, Cole Haan, Tommy Hilfiger, Levi's, Dockers, Jessica Simpson, Sean John, Jones New York, Nine West and Ellen Tracy as well as products containing trademarks for the NFL, NBA, MLB and NHL sports leagues and numerous U.S. colleges and universities. The company-owned labels sold by the non-licensed apparel segment include G-III, Black Rivet, Marc New York, Andrew Marc, Jessica Howard and Eliza J. The retail segment operates 134 Wilsons Leather retail outlet stores and 4 outlet stores under the Andrew Marc banner. This segment also includes a joint venture with the Camuto Group that operates 11 footwear and accessory retail outlet stores under the Vince Camuto banner. G-III's products are generally sold by department, specialty and mass merchant retail stores. The firm imports its products from independent manufacturers located primarily in China and, to a lesser extent, in Vietnam, India, Indonesia, Thailand, Sri Lanka, Taiwan and Central and South America. In November 2011, G-III entered into a license agreement to operate Calvin Klein Performance retail stores in the U.S. and China, opening the first location in Scottsdale, Arizona in February 2012. In March 2012, it entered into a joint venture agreement with Finity Apparel Retail Limited to open and operate Calvin Klein Performance retail stores in China and Hong Kong. In August 2012, the firm acquired Vilebrequin International SA, a leading global provider of luxury swimwear, accessories and resort-wear.

FINANCIAL DATA: *Note: Data for latest year may not have been available at press time.*

In U.S. $	2012	2011	2010	2009	2008	2007
Revenue	1,231,201,000	1,063,404,000	800,864,000	711,146,000	518,868,000	
R&D Expense						
Operating Income	86,224,000	96,932,000	56,207,000	-3,877,000	32,355,000	
Operating Margin %	7.00%	9.11%	7.01%	-.54%	6.23%	
SGA Expense	277,019,000	248,380,000	205,281,000	164,098,000	101,669,000	
Net Income	49,620,000	56,682,000	31,718,000	-14,029,000	17,490,000	
Operating Cash Flow	6,776,000	-28,649,000	43,999,000	22,518,000	10,623,000	
Capital Expenditure	17,410,000	19,407,000	1,477,000	2,411,000	1,445,000	
EBITDA	92,426,000	102,665,000	61,587,000	3,070,000	37,782,000	
Return on Assets %	9.89%	14.37%	10.34%	-5.40%	8.50%	
Return on Equity %	15.00%	21.16%	16.08%	-8.34%	12.08%	
Debt to Equity						

CONTACT INFORMATION:

Phone: 212 403-0500 Fax: 212 403-0551
Toll-Free:
Address: 512 7th Ave., New York, NY 10018 United States

STOCK TICKER/OTHER:

Stock Ticker: GIII Exchange: NAS
Employees: 2,592 Fiscal Year Ends: 31-Jan
Parent Company:

SALARIES/BONUSES:

Top Exec. Salary: $1,000,000 Bonus: $4,930,213
Second Exec. Salary: $750,000 Bonus: $3,218,333

OTHER THOUGHTS:

Estimated Female Officers or Directors: 1
Hot Spot for Advancement for Women/Minorities:

Gildan Activewear Inc

www.gildan.com

NAIC Code: 315220

TYPES OF BUSINESS:

Wholesale Apparel
T-Shirts & Activewear
Socks & Undergarments

BRANDS/DIVISIONS/AFFILIATES:

Gildan
Anvil
Gold Toe
CanAm Yarns LLC
Anvil Holdings Inc

CONTACTS: *Note: Officers with more than one job title may be intentionally listed here more than once.*

Glenn J. Chamandy, CEO
Glenn J. Chamandy, Pres.
Laurence G. Sellyn, CFO/Exec. VP
Benito Masi, Exec. VP-Mfg.
Laurence G. Sellyn, Chief Admin. Officer
Genevieve Gosselin, Dir.-Corp. Comm.
Sophie Argiriou, Dir.-Investor Comm.
Michael R. Hoffman, Pres., Printwear
Eric R. Lehman, Pres., Branded Apparel
Peter Iliopoulos, Sr. VP-Corp & Public Affairs
William D. Anderson, Chmn.

GROWTH PLANS/SPECIAL FEATURES:

Gildan Activewear, Inc., based in Quebec, manufactures and markets basic, non-fashion apparel and activewear for the casual and wholesale imprinted apparel market. Its core product line consists of blank shirts and sweatshirts in various fabric weights, sizes, colors and styles, made of 100% cotton or of blends of cotton and synthetic fibers. The firm divides its operations into two segments: Printwear and Branded Apparel. The Printwear segment, headquartered in Christ Church, Barbados, designs, manufactures, sources and distributes undecorated activewear products in large quantities primarily to wholesale distributors in printwear markets in over 30 countries across North America, Europe and the Asia Pacific region. The products sold through this segment consist mainly of undecorated, or blank, T-shirts, fleece and sport shirts, marketed primarily under the company-owned Gildan and Anvil brands, as well as licensed brands such as New Balance. The Branded Apparel segment, headquartered in Charleston, South Carolina, designs, manufactures, sources and distributes branded family apparel, which includes socks, underwear and activewear products, mainly to U.S. retailers. Brands in this segment include company-owned Gold Toe and licensed brands Under Armour and New Balance, for which Gildan is the exclusive sock licensee in the U.S. The company has significant manufacturing infrastructure in two main hubs in Central America and the Caribbean Basin, and a small vertically-integrated factory in Bangladesh. During 2012, Gildan acquired Anvil Holdings, Inc., a supplier of basic T-shirts and sport shirts. The company also recently acquired the 50% outstanding stake in CanAm Yarns LLC, its former joint venture with Frontier Spinning Mills, Inc.

FINANCIAL DATA: *Note: Data for latest year may not have been available at press time.*

In U.S. $	2012	2011	2010	2009	2008	2007
Revenue	1,948,253,000	1,726,041,000	1,311,463,000	1,038,319,000	1,249,711,000	964,429,000
R&D Expense						
Operating Income	155,128,000	230,151,000	200,878,000	89,349,000	186,445,000	170,158,000
Operating Margin %	7.96%	13.33%	15.31%	8.60%	14.91%	17.64%
SGA Expense	226,035,000	199,132,000	154,674,000	134,785,000	151,453,000	110,979,000
Net Income	148,464,000	239,904,000	198,245,000	95,329,000	144,592,000	130,020,000
Operating Cash Flow	219,582,000	181,550,000	301,551,000	169,179,000	238,904,000	91,163,000
Capital Expenditure	76,755,000	164,722,000	127,881,000	44,938,000		134,282,000
EBITDA	249,691,000	309,042,000	267,578,000	154,337,000	245,640,000	172,373,000
Return on Assets %	7.84%	14.94%	16.49%	8.72%	14.63%	16.26%
Return on Equity %	10.78%	19.65%	19.57%	11.06%	19.59%	21.76%
Debt to Equity	0.12	0.15			0.06	0.08

CONTACT INFORMATION:

Phone: 514 735-2023 Fax: 514 735-6810
Toll-Free: 866-755-2023
Address: 600 de Maisonneuve Blvd. W., 33rd Fl., MontrÃ©al, QC H3A 3J2 Canada

STOCK TICKER/OTHER:

Stock Ticker: GIL
Employees: 28,000
Parent Company:
Exchange: NYS
Fiscal Year Ends: 30-Sep

SALARIES/BONUSES:

Top Exec. Salary: $795,875 Bonus: $
Second Exec. Salary: $500,201 Bonus: $

OTHER THOUGHTS:

Estimated Female Officers or Directors: 3
Hot Spot for Advancement for Women/Minorities: Y

Gilt Groupe Inc

NAIC Code: 454111

www.gilt.com

TYPES OF BUSINESS:

Online Retailing-Fashion Apparel

BRANDS/DIVISIONS/AFFILIATES:

Gilt
Gilt City
Jetsetter
Little Gilt
Gilt Taste

CONTACTS:

Note: Officers with more than one job title may be intentionally listed here more than once.

Michelle Peluso, CEO
Andy Page, Pres.
Tom Sansone, CFO
Elizabeth Francis, Chief Marketing Officer
Kathy Leo, Chief People Officer
Steve Jacobs, CIO
Michael Bryzek, CTO
Kathy Leo, Chief Legal Officer
Tom Sansone, Head-Corp. Dev.
Jyothi Rao, Exec. VP-Fashion
Sarah Chubb, Pres./Gen. Mgr.-Gilt City
Alexis Maybank, Chief Strategy Officer
Alexandra Wilkis Wilson, Head-Nat'l Sales-Gilt City
Susan Lyne, Chmn.
Peter Glusker, CEO-Gilt Japan & Gilt City Japan
Chris Halkyard, Chief Supply Chain Officer

GROWTH PLANS/SPECIAL FEATURES:

Gilt Groupe Inc. operates Gilt.com, an online members-only retailer of discounted designer home and fashion merchandise. The site has over 6 million members. Members of the site receive discounts of up to 60% on men's, women's and children's apparel in addition to home goods including kitchenware, furniture and decor. Gilt Taste, a separate section on Gilt.com, offers discounts on gourmet meats, wines, cheeses, sweets, coffee, tea and other food items as well as high-end kitchen items. Gilt members can shop for fashion items by type of clothing, sales of the day or ages for children. Women and Men's apparel also includes the Handbag Shop, for women's bags by style; the Tailored Shop, with men's dress apparel and shoes; and the Casual Shop, with men's casual apparel and shoes. The firm offers travel discounts on hotels and cruises through Jetsetter; members can explore these sales by destination, by perusing the Flash Sales page or through the themed Travel Ideas page. Jetsetter also offers a featured Sleep of the Week Sale on a particular location. Gilt City offers discounts of up to 75% on events in 12 U.S. cities, Washinton D.C. and National Events. Events can include city specific store discounts, salon specials, restaurant coupons, fitness classes, theater and musical events, parties and concerts. In September 2012, the firm launched its own collection of Baby & Kids clothing, Little Gilt, featuring 17 styles of baby apparel at initial launch.

FINANCIAL DATA:

Note: Data for latest year may not have been available at press time.

In U.S. $	2012	2011	2010	2009	2008	2007
Revenue	600,000,000	450,000,000	300,000,000			
R&D Expense						
Operating Income						
Operating Margin %						
SGA Expense						
Net Income						
Operating Cash Flow						
Capital Expenditure						
EBITDA						
Return on Assets %						
Return on Equity %						
Debt to Equity						

CONTACT INFORMATION:

Phone: Fax:
Toll-Free: 877-445-8692
Address: 2 Park Ave. 4th Fl., New York, NY 10016 United States

STOCK TICKER/OTHER:

Stock Ticker: Private
Exchange:
Employees:
Fiscal Year Ends:
Parent Company:

SALARIES/BONUSES:

Top Exec. Salary: $ Bonus: $
Second Exec. Salary: $ Bonus: $

OTHER THOUGHTS:

Estimated Female Officers or Directors: 9
Hot Spot for Advancement for Women/Minorities: Y

Giorgio Armani SpA

www.giorgioarmani.com

NAIC Code: 315240

TYPES OF BUSINESS:

Designer Apparel
Men's & Women's Suits & Apparel
Children's Apparel
Home Furnishings & Accessories
Retail Apparel Stores
Jewelry, Cosmetics & Fragrances
Watches
Online Sales

BRANDS/DIVISIONS/AFFILIATES:

Giorgio Armani
Armani Collezioni
Emporio Armani
Armani Junior
Armani Exchange
Armani Casa
Armani Hotel Milano
Armani Hotels & Resorts

CONTACTS: *Note: Officers with more than one job title may be intentionally listed here more than once.*

Giorgio Armani, CEO
Giorgio Armani, Pres.
Bruno Laguardia, Head-Giorgio Armani Corp.
Livio Proli, Gen. Mgr.
Giorgio Armani, Chmn.

GROWTH PLANS/SPECIAL FEATURES:

Giorgio Armani SpA is one of the world's leading fashion and luxury goods groups, best known for its men's and women's suits. It designs, manufactures, distributes and retails fashion and lifestyle products, including apparel, accessories, glasses, watches, jewelry, home interiors, fragrances and cosmetics under a variety of brand names. The company also licenses its name for perfume, watches and accessories. The firm's brands include the men's and women's apparel collections Giorgio Armani, Armani Collezioni and Emporio Armani; sportswear brand EA7; Armani Jeans; children's brand Armani Junior, which includes boy's, girl's and baby/toddler lines; and a home collection called Armani Casa, which includes furniture, home accessories, dishware and linens lines. Armani retails its products through over 609 free-standing stores in 46 countries as well as through ArmaniExchange.com. The company has approximately 183 Emporio Armani stores, 180 Armani Exchange stores, 85 Giorgio Armani stores, 28 Armani Casa stores, 22 Armani Jeans stores and 10 Giorgio Armani Accessori stores. Armani also owns a chain of chocolate stores Armani Dolci, with 19 locations throughout Europe, the Middle East and Asia. The company also owns manufacturing companies Intai Spa, Antinea Srl and Simint Spa. The company established Armani Hotels & Resorts through an agreement with Emaar Properties to develop, own and operate a collection of hotels, resorts and residences in major world cities and holiday destinations, including a luxury hotel in Dubai, in the Burj Khalifa, the world's tallest building; and a luxury hotel in Milan.

FINANCIAL DATA: *Note: Data for latest year may not have been available at press time.*

In U.S. $	2012	2011	2010	2009	2008	2007
Revenue	2,650,000,000	2,350,000,000	2,050,000,000			
R&D Expense						
Operating Income						
Operating Margin %						
SGA Expense						
Net Income						
Operating Cash Flow						
Capital Expenditure						
EBITDA						
Return on Assets %						
Return on Equity %						
Debt to Equity						

CONTACT INFORMATION:

Phone: 39-02-723-18-1 Fax: 39-02-723-18-549
Toll-Free:
Address: Via Borgonuovo 11, Milan, 20121 Italy

STOCK TICKER/OTHER:

Stock Ticker: Private
Exchange:
Employees:
Fiscal Year Ends: 31-Dec
Parent Company:

SALARIES/BONUSES:

Top Exec. Salary: $ Bonus: $
Second Exec. Salary: $ Bonus: $

OTHER THOUGHTS:

Estimated Female Officers or Directors:
Hot Spot for Advancement for Women/Minorities:

Gordmans Stores Inc

www.gordmans.com

NAIC Code: 448140

TYPES OF BUSINESS:

Apparel Stores
Off-Price Retail

BRANDS/DIVISIONS/AFFILIATES:

CONTACTS: *Note: Officers with more than one job title may be intentionally listed here more than once.*

Jeffrey J. Gordman, CEO
Michael E. Wirkkala, COO
Jeffrey J. Gordman, Pres.
Michael D. James, CFO/Sr. VP
Roger Glenn, Sr. VP-Human Resources
Richard H. Heyman, CIO/Sr. VP
Johanna K. Lewis, Chief Merch. Officer/Exec. VP
Jeffrey J. Gordman, Sec.
Michael E. Wirkkala, Exec. VP-Oper.
Michael S. Morand, Exec. VP-Planning, Allocation & Analysis
Michael D. James, Treas.
Tracie L. Wickenhauser, Sr. VP-Stores
Thomas V. Taylor, Chmn.

GROWTH PLANS/SPECIAL FEATURES:

Gordmans Stores, Inc. is a discount apparel and accessories retailer operating throughout the Midwest and Southeast U.S. The firm currently operates 83 stores in 18 states, all under the Gordmans name. Its stores are located in a variety of shopping center developments, including regional enclosed shopping malls, lifestyle centers and power centers. The firm's merchandise assortment includes apparel for all ages, which accounted for 54% of its 2011 revenue; home fashions, 28%; and accessories, 18%. Additionally, the company maintains a licensing agreement with Destination Maternity Corporation to provide maternity products in each of its stores. It also has a similar arrangement with DSW, Inc. for its footwear departments. Gordmans is aligned in the off-price retail segment, offering nationally branded merchandise at prices below larger department store competitors. Unlike similar retailers in the off-price segment, Gordmans utilizes merchandising techniques, visual displays, a departmental floor layout, fixture systems, signing and graphics akin to those of department and specialty stores. The firm, through an agreement with Alliance Data Systems, offers a private label Gordmans credit card.

FINANCIAL DATA: *Note: Data for latest year may not have been available at press time.*

In U.S. $	2012	2011	2010	2009	2008	2007
Revenue	558,146,000	523,322,000	463,212,000	432,670,000		
R&D Expense						
Operating Income	40,895,000	25,960,000	26,193,000	5,524,000		
Operating Margin %	7.32%	4.96%	5.65%	1.27%		
SGA Expense	201,084,000	198,302,000	167,842,000	170,533,000		
Net Income	25,173,000	15,598,000	15,868,000	2,387,000		
Operating Cash Flow	27,116,000	12,383,000	40,070,000	15,774,000		
Capital Expenditure	31,164,000	8,959,000	3,865,000	10,411,000		
EBITDA	45,202,000	28,595,000	27,950,000	10,940,000		
Return on Assets %	17.49%	14.35%	18.67%	3.06%		
Return on Equity %	38.79%	39.99%	62.61%	9.64%		
Debt to Equity		0.01	0.02			

CONTACT INFORMATION:

Phone: 402 691-4000 Fax: 402 691-4269
Toll-Free:
Address: 12100 W. Center Rd., Omaha, NE 68144 United States

STOCK TICKER/OTHER:

Stock Ticker: GMAN Exchange: NAS
Employees: 4,900 Fiscal Year Ends: 31-Jan
Parent Company:

SALARIES/BONUSES:

Top Exec. Salary: $635,000 Bonus: $
Second Exec. Salary: $303,000 Bonus: $

OTHER THOUGHTS:

Estimated Female Officers or Directors: 2
Hot Spot for Advancement for Women/Minorities:

Greenwood Mills

www.greenwoodmills.com

NAIC Code: 313210

TYPES OF BUSINESS:

Textiles Manufacturing
Cotton Fabrics & Yarns
Real Estate Development
Golf Courses

BRANDS/DIVISIONS/AFFILIATES:

Greenwood Development Corporation
SingleSource Apparel
Palmetto Dunes Oceanfront Resort
Shelter Cove Harbour Resort

CONTACTS: *Note: Officers with more than one job title may be intentionally listed here more than once.*

James C. Self, III, COO
James C. Self, III, Pres.
Thomas J. Davis, CFO/Exec. VP
Gary Niederauer, VP-Sales
Warren L. Moore, VP-Human Resources
Glenn Langrehr, Dir.-IT Svcs.
W. Eddie Gaither, VP-Greige Mfg.
William M. Self, Chmn.
H. Doyle Kidd, VP-Purchasing

GROWTH PLANS/SPECIAL FEATURES:

Greenwood Mills, founded 1889, is a producer of fine woven fabrics. The firm's plants produce unfinished, unbleached and un-dyed textiles to sell to various finishers and manufacturers. Greenwood Mills operates two plants in its home state, South Carolina; a joint venture plant in Aguascalientes, Mexico; and a sales office in Huntington, New York. The company manufactures specialty fabrics for the industrial, apparel, home furnishings and military markets. It also produces garments, offering full-package manufacturing to prominent American clothing brands such as Levi Strauss, Abercrombie & Fitch and Hollister through its SingleSource Apparel subsidiary. Single Source produces woven jeans, pants, skirts and jackets. Greenwood Mills also oversees wholly-owned subsidiary Greenwood Development Corporation, which develops real estate primarily in Hilton Head Island. Developments include Palmetto Dunes Oceanfront Resort and Shelter Cove Harbour Resort both located in Hilton Head as well as several residential communities located throughout the Southwest U.S. Founded by James Self, the firm is still owned and operated by the Self family.

FINANCIAL DATA: *Note: Data for latest year may not have been available at press time.*

In U.S. $	2012	2011	2010	2009	2008	2007
Revenue						
R&D Expense						
Operating Income						
Operating Margin %						
SGA Expense						
Net Income						
Operating Cash Flow						
Capital Expenditure						
EBITDA						
Return on Assets %						
Return on Equity %						
Debt to Equity						

CONTACT INFORMATION:

Phone: 864-227-2121 Fax:
Toll-Free:
Address: 300 Morgan Ave., Greenwood, SC 29646 United States

STOCK TICKER/OTHER:

Stock Ticker: Private
Exchange:
Employees: 700
Fiscal Year Ends: 31-Dec
Parent Company:

SALARIES/BONUSES:

Top Exec. Salary: $ Bonus: $
Second Exec. Salary: $ Bonus: $

OTHER THOUGHTS:

Estimated Female Officers or Directors:
Hot Spot for Advancement for Women/Minorities:

Grupo Tavex SA

NAIC Code: 313210

www.grupotavex.com

TYPES OF BUSINESS:

Textiles Manufacturing
Denim Fabric
Protective Garments & Construction Apparel
Sports Apparel
Geotextiles
Medical Supplies

BRANDS/DIVISIONS/AFFILIATES:

Tavex Corporation
Ombu-Aire Libre
Santista Textil SA
Tavex International SA
Tavemex SA de CV
Industrial Textil Puebla
Tavex USA Inc
Camargo Correa SA

CONTACTS: *Note: Officers with more than one job title may be intentionally listed here more than once.*

Ricardo Antonio Weiss, CEO
Marcio Garcia de Souza, Chmn.

GROWTH PLANS/SPECIAL FEATURES:

Grupo Tavex SA, also known as Tavex Corporation, is a multinational textile group that specializes in the manufacture of denim. Established in 2006 through the merger of Tavex Algodonera (a Spainsh textile company founded in 1846) and Santista Textil (a Brazilian firm established in 1929), Tavex is currently one of the largest producers of denim in Europe and Latin America. The company exports over 80% of its products to clients in more than 50 countries. The firm manufactures textiles for the home furnishings and apparel industries, operating through eight factories on three continents and over 30 subsidiaries around the world. The company's products are used by many prominent designers and apparel brands, including Dolce & Gabbana, Guess, Levi's, Hugo Boss, Gap, Giorgio Armani and Tommy Hilfiger, among others. Besides denim cloth, it produces raincoats, canvas, polyester blends, linen, lycra, Teflon HT, laminates, tapestries, geotextiles, protective garments, construction apparel and medical supplies. Under the Santista brand, the firm is a leading supplier of workwear to the Latin American market. Additionally, it markets a proprietary line of casual apparel under the trademark Ombu-Aire Libre. Subsidiaries of the company include Santista Textil SA; Tavex International SA; Tavex Polska; Tavex Peru; Alginet Textil SA; Textile and Garment Sourcing SL; Settavex SA; Tavemex SA de CV; Industrial Textil Puebla; Tejidos Hogar Decoracion; Sanpere Logistica SA; Sanpere Textile Hogar; and Tavex USA, Inc. The Brazilian firm Camargo Correa SA owns roughly 49% of Tavex.

FINANCIAL DATA: *Note: Data for latest year may not have been available at press time.*

In U.S. $	2012	2011	2010	2009	2008	2007
Revenue		634,200,000	584,400,000	412,700,000	461,000,000	465,900,000
R&D Expense						
Operating Income						
Operating Margin %						
SGA Expense						
Net Income		30,600,000	-25,900,000	-49,100,000	-35,700,000	1,100,000
Operating Cash Flow						
Capital Expenditure						
EBITDA						
Return on Assets %						
Return on Equity %						
Debt to Equity						

CONTACT INFORMATION:

Phone: 34-91-391-13-50 Fax: 34-91-319-99-18
Toll-Free:
Address: Calle Rosario Pino 14-16, Madrid, 28020 Spain

STOCK TICKER/OTHER:

Stock Ticker: ASA
Exchange: Madrid
Employees: 6,109
Fiscal Year Ends: 31-Dec
Parent Company:

SALARIES/BONUSES:

Top Exec. Salary: $ Bonus: $
Second Exec. Salary: $ Bonus: $

OTHER THOUGHTS:

Estimated Female Officers or Directors: 1
Hot Spot for Advancement for Women/Minorities:

Gucci Group NV

www.gucci.com

NAIC Code: 316210

TYPES OF BUSINESS:

High-End Apparel
Luxury Goods
Leather Goods & Shoes
Retail Stores
Beauty Products
Cosmetics
Ties, Scarves & Accessories
Perfume

BRANDS/DIVISIONS/AFFILIATES:

Yves Saint Laurent
Sergio Rossi
Boucheron
Rive Gauche
Bottega Veneta
Alexander McQueen
Stella McCartney
Balenciaga

CONTACTS: *Note: Officers with more than one job title may be intentionally listed here more than once.*

Francois-Henri Pinault, CEO
Frida Giannini, Creative Dir.
Alexis Babeau, Deputy-CEO
Paul Deneve, CEO-Yves Saint Deneve
Jonathan Akeroyd, CEO/Pres., Alexander McQueen

GROWTH PLANS/SPECIAL FEATURES:

Gucci Group NV is one of the world's largest multi-brand designers, producers and distributors of luxury goods and apparel. The firm operates in four segments: clothing, accessories, watches and luggage. The Gucci group owns and manages brands such as Alexander McQueen, Boucheron, Bottega Veneta, Balenciaga, Gucci, Sergio Rossi, Stella McCartney and Yves Saint Laurent. Recently, Alexander McQueen added a men's collection to its lineup. The Boucheron brand is a maker of jewelry, perfumes and watches. In addition Boucheron offers a selection of clocks designed with various materials, including precious and semiprecious stones, gold, silver and stainless steel. The Bottega Veneta brand makes signature leather goods, including handbags, luggage and leather wallets. The Balenciaga brand makes designer apparel, handbags, shoes and fragrances. The Gucci brand produces luggage, small leather goods, shoes, timepieces, ties, jewelry, scarves, eyewear and perfume. The Sergio Rossi brand makes designer Italian shoes and handbags. The Stella McCartney brand makes designer clothing and fragrances as well as organic skincare under the CARE product line. The Yves Saint Laurent brand creates women's and men's ready-to-wear and leather goods, including shoes sold under the Rive Gauche label. It also provides men's apparel, accessories and eyewear. The YSL Beaute segment offers Yves Saint Laurent brand perfumes, skincare and cosmetics. The firm operates over 680 stores in major markets around the world. The company sells its products through directly operated stores, franchise stores, duty-free boutiques, department stores and specialty stores.

FINANCIAL DATA: *Note: Data for latest year may not have been available at press time.*

In U.S. $	2012	2011	2010	2009	2008	2007
Revenue	2,117,550,000	1,654,655,000				
R&D Expense						
Operating Income						
Operating Margin %						
SGA Expense						
Net Income						
Operating Cash Flow						
Capital Expenditure						
EBITDA						
Return on Assets %						
Return on Equity %						
Debt to Equity						

CONTACT INFORMATION:

Phone: 31 204621700 Fax: 31 204653569
Toll-Free:
Address: 1 Amstelplein, Rembrandt Tower, Amsterdam, 1096 HA Netherlands

STOCK TICKER/OTHER:

Stock Ticker: GUCG Exchange: GREY
Employees: Fiscal Year Ends:
Parent Company: Kering

SALARIES/BONUSES:

Top Exec. Salary: $ Bonus: $
Second Exec. Salary: $ Bonus: $

OTHER THOUGHTS:

Estimated Female Officers or Directors: 1
Hot Spot for Advancement for Women/Minorities: Y

Guess? Inc

www.guessinc.com

NAIC Code: 448140

TYPES OF BUSINESS:

Casual Clothing Stores
Accessories
Fragrances
Footwear
Online Sales
Jeans

BRANDS/DIVISIONS/AFFILIATES:

Baby GUESS
Guess Jeans
Triangle Design
Question Mark
GUESS by Marciano
GUESS Kids
GUESS U.S.A.
YES

CONTACTS: *Note: Officers with more than one job title may be intentionally listed here more than once.*

Paul Marciano, CEO
J. Michael Prince, COO
Dennis Secor, CFO/Sr. VP
Amber Tarshis, Sr. VP-Mktg.
Chet Kuchinad, Chief People Officer
Michael Relich, CIO/Exec. VP
Maurice Marciano, Chmn.
Frank Rescigna, Pres., Licensing-Worldwide

GROWTH PLANS/SPECIAL FEATURES:

Guess?, Inc. designs, markets, distributes and licenses collections of contemporary apparel and accessories for men, women and children that reflect the American lifestyle and European fashion sensibilities. The company's apparel is marketed under trademarks including GUESS?, GUESS U.S.A., GUESS Jeans, Triangle Design, Question Mark, a stylized G, GUESS Kids, Baby GUESS, YES and GUESS by Marciano. The lines include full collections of denim and cotton clothing, including jeans, pants, overalls, skirts, dresses, shorts, blouses, shirts, jackets and knitwear. The firm also grants licenses to manufactures and distributes a broad range of products that complement its apparel lines, including eyewear, watches, handbags, footwear, infants' and children's apparel, leather apparel, fragrance, jewelry and other fashion accessories. Guess products are sold through three primary distribution channels: its own stores in the U.S. and Canada, a network of wholesale accounts in the U.S., and via the Internet. The company operates 481 stores in the U.S. and Canada, consisting of 192 full-price retail stores, 118 factory outlet stores, 54 G by GUESS stores, 57 GUESS by MARCIANO stores and 60 Guess Accessories stores. Guess also directly operates 125 stores in Europe and the Middle East, 43 stores in Asia and 10 stores in Central and South America. The firm's European headquarters is located in Lugano, Switzerland.

Guess offers its employees a 401(k), paid vacation and healthcare.

FINANCIAL DATA: *Note: Data for latest year may not have been available at press time.*

In U.S. $	2012	2011	2010	2009	2008	2007
Revenue	2,688,048,000	2,487,294,000	2,128,466,000	2,093,390,000	1,749,916,000	
R&D Expense						
Operating Income	397,235,000	404,633,000	358,816,000	328,787,000	309,139,000	
Operating Margin %	14.77%	16.26%	16.85%	15.70%	17.66%	
SGA Expense	747,587,000	680,311,000	577,963,000	569,398,000	483,630,000	
Net Income	265,500,000	289,508,000	242,761,000	213,562,000	186,472,000	
Operating Cash Flow	364,494,000	346,374,000	358,210,000	228,625,000	175,460,000	
Capital Expenditure	124,870,000	123,065,000	82,286,000	89,971,000	100,178,000	
EBITDA	480,629,000	490,082,000	428,479,000	384,635,000	365,484,000	
Return on Assets %	15.04%	18.00%	17.48%	17.55%	15.71%	
Return on Equity %	23.80%	27.90%	27.02%	29.80%	28.38%	
Debt to Equity		0.01	0.01	0.01	0.02	

CONTACT INFORMATION:

Phone: 213 765-3100 Fax: 213 744-7838
Toll-Free: 800-224-8377
Address: 1444 S. Alameda St., Los Angeles, CA 90021 United States

STOCK TICKER/OTHER:

Stock Ticker: GES Exchange: NYS
Employees: 15,000 Fiscal Year Ends: 31-Jan
Parent Company:

SALARIES/BONUSES:

Top Exec. Salary: $1,000,000 Bonus: $1,400,000
Second Exec. Salary: $1,000,000 Bonus: $

OTHER THOUGHTS:

Estimated Female Officers or Directors: 2
Hot Spot for Advancement for Women/Minorities: Y

Guilford Mills Inc

www.guilfordtextiles.com

NAIC Code: 313230

TYPES OF BUSINESS:

Engineered Fabrics
Synthetic Fibers
Warp Knit Fabrics
Specialty Fabrics
Health Care Supplies
Automotive Fabrics

BRANDS/DIVISIONS/AFFILIATES:

Lear Corporation

CONTACTS:
Note: Officers with more than one job title may be intentionally listed here more than once.

Matthew J. Simoncini, Chmn./CEO/Pres., Lear Corporation
Ray Scott, Pres., Seating-Lear Corporation

GROWTH PLANS/SPECIAL FEATURES:

Guilford Mills, Inc., a subsidiary of Lear Corporation, designs and produces value-added engineered fabrics. The company has domestic locations in Michigan and North Carolina as well as international locations in Mexico, the U.K., France, Germany, Spain, China and Japan. Guilford Mills is one of the world's largest and most diversified warp knit textile companies and one of the largest manufacturers of bodycloth for the automotive industry. Manufacturing technologies include warp, weft and circular knitting, bodycloth printing, flat and velour weaving and double needle bars. Guilford Mills also offers a wide variety of specialty fabrics for consumer and industrial applications including casket linings, loop/closure systems (commonly known as Velcro), reverse osmosis, medical products (such as incontinence pads, braces and surgical towels), blankets, mattress pad coverings, lamination, shoe linings, gaming table fabrics and window coverings. Moreover, the firm is an integrated designer and producer of warp knit fabrics for activewear, team sports, performance, cap and gown, loungewear/lingerie, ready-to-wear and choral robes. With regard to automotive products, Guilford Mills develops and fabricates body cloth for many aspects of car upholstery, including seat fabrics, door linings and headliners. In April 2012, Guilford Mills announced that its previous owner, Cerberus Capital Management, had sold the firm to Lear Corporation, a supplier of automotive seating and electrical power management systems.

FINANCIAL DATA:
Note: Data for latest year may not have been available at press time.

In U.S. $	2012	2011	2010	2009	2008	2007
Revenue						
R&D Expense						
Operating Income						
Operating Margin %						
SGA Expense						
Net Income						
Operating Cash Flow						
Capital Expenditure						
EBITDA						
Return on Assets %						
Return on Equity %						
Debt to Equity						

CONTACT INFORMATION:

Phone: 910-794-5800 Fax:
Toll-Free:
Address: 1001 Military Cutoff Rd., Ste. 300, Wilmington, NC 28405 United States

STOCK TICKER/OTHER:

Stock Ticker: Subsidiary
Employees:
Parent Company: Lear Corporation
Exchange:
Fiscal Year Ends: 30-Sep

SALARIES/BONUSES:

Top Exec. Salary: $ Bonus: $
Second Exec. Salary: $ Bonus: $

OTHER THOUGHTS:

Estimated Female Officers or Directors:
Hot Spot for Advancement for Women/Minorities: Y

Gymboree Corp (The)

www.gymboree.com

NAIC Code: 448130

TYPES OF BUSINESS:

Apparel-Children's, Retail
Parent-Child Developmental Play Programs
Online Sales

BRANDS/DIVISIONS/AFFILIATES:

Gymboree
Gymboree Outlet
Janie & Jack
Crazy 8
ShadeClothing.com
Gymboree Play & Music
JanieAndJack.com
BAIN CAPITAL LLC

CONTACTS: *Note: Officers with more than one job title may be intentionally listed here more than once.*

Kip M. Garcia, Interim CEO
Yvonne Hao, Interim COO
Kip M. Garcia, Pres.
Lynda Gustafson, Acting CFO
Lynda Gustafson, Corp. Controller/VP

GROWTH PLANS/SPECIAL FEATURES:

The Gymboree Corp., a subsidiary of Bain Capital, LLC, is a specialty retailer of children's apparel and accessories. The company operates 1,288 retail stores in the U.S., Canada, Puerto Rico and Australia under the Gymboree, Gymboree Outlet, Janie and Jack and Crazy 8 brands. Online shopping is provided to customers through Gymboree.com, JanieandJack.com, Crazy8.com and ShadeClothing.com (which offers women's apparel). The firm currently has 634 Gymboree stores (586 in the U.S., 41 in Canada, 6 in Australia and 1 in Puerto Rico). Stores are primarily located in regional shopping malls and in select suburban and urban locations. Gymboree apparel and accessories are characterized by bright colors, patterns, whimsical graphics and complex embellishments, and are designed for both functionality and durability for newborns to children age 12. The company operates 156 Gymboree Outlet stores selling similar merchandise at lower prices. The majority of this merchandise is developed and manufactured exclusively for the Gymboree Outlet stores. The Janie and Jack division, with 130 stores, offers higher-end clothing for newborn children through age eight. These stores are designed in an old-fashioned mercantile boutique style. The company's Crazy 8 division operates 308 stores, marketing towards children under fourteen. Gymboree Play & Music offers a variety of developmental classes, toys, books and music for children. The company operates directed parent-child developmental play programs at 712 franchised and company-operated Gymboree Play & Music centers throughout the U.S. and 40 countries. Additionally, third party overseas partners operate 24 Gymboree stores in the Middle East and South Korea.

FINANCIAL DATA: *Note: Data for latest year may not have been available at press time.*

In U.S. $	2012	2011	2010	2009	2008	2007
Revenue	1,200,000,000	1,100,000,000	1,001,527,000	987,859,000	909,410,000	781,172,000
R&D Expense						
Operating Income						
Operating Margin %						
SGA Expense						
Net Income				93,480,000	80,331,000	60,250,000
Operating Cash Flow						
Capital Expenditure						
EBITDA						
Return on Assets %						
Return on Equity %						
Debt to Equity						

CONTACT INFORMATION:

Phone: 415-278-7000 Fax: 707-678-1315
Toll-Free: 877-449-6932
Address: 500 Howard St., San Francisco, CA 94105 United States

STOCK TICKER/OTHER:

Stock Ticker: Private Exchange:
Employees: 12,400 Fiscal Year Ends: 31-Jan
Parent Company: BAIN CAPITAL LLC

SALARIES/BONUSES:

Top Exec. Salary: $ Bonus: $
Second Exec. Salary: $ Bonus: $

OTHER THOUGHTS:

Estimated Female Officers or Directors: 2
Hot Spot for Advancement for Women/Minorities:

H&M (Hennes & Mauritz) AB

www.hm.com

NAIC Code: 448140

TYPES OF BUSINESS:

Apparel Stores, Moderately Priced
Cosmetics
Men's & Women's Clothing
Accessories
Catalog & Online Sales

BRANDS/DIVISIONS/AFFILIATES:

H&M
Cheap Monday
H&M Rowells
COS (Collection of Style)
H&M Home
Cheap Monday
Monki
Weekday

CONTACTS: *Note: Officers with more than one job title may be intentionally listed here more than once.*

Karl-Johan Persson, CEO
Jyrki Tervonen, Dir.-Finance
Anna Tillberg Pantzar, Dir.-Mktg.
Sanna Lindberg, Dir.-Human Resources
Kjell-Olof Nilsson, Dir.-IT
Karl Gunnar Fagerlin, Dir.-Prod.
Bjorn Magnusson, Dir.-Bus. Dev.
Kristina Stenvinkel, Dir-Comm.
Nils Vinge, Dir-Investor Rel.
Anders Jonasson, Dir.-Accounts
Ann-Sofie Johansson, Dir.-Design
Stefan Larsson, VP-Sales
Jonas Guldstrand, Dir.-Logistics
Pernilla Wohlfahrt, Dir.-New Bus.
Stefan Persson, Chmn.
Madeleine Persson, Dir.-Buying

GROWTH PLANS/SPECIAL FEATURES:

H&M (Hennes & Mauritz) AB is a Sweden-based fashion retailer that operates approximately 2,700 stores in 48 countries, including the U.S. H&M's largest market exists in Sweden, Germany and the U.K. The company offers clothing for women, men, teenagers and children, concentrating predominately on customers aged 18-45. H&M also sells accessories and an in-house cosmetics line. Production takes place in Europe and Asia. In some regions, the firm also sells its products via catalogs and the Internet (which are operated by H&M Rowells). While H&M sees the Internet as an important factor in its growth, it foresees that the bulk of its revenue will be earned at its stores, with a growth target of increasing its stores by 10-15% per year. To that end, a key component of the company's success is the selection of superior store locations, primarily areas with high foot traffic. The firm's growth strategy is focused mainly on expanding into new markets and choosing attractive store locations. In particular, the company sees growth potential in the U.S., China, Canada, Slovenia, Germany, France, Portugal, Spain, the U.K., Poland and the Czech Republic. H&M recently opened stores in Mexico, Bulgaria and Kuala Lumpur. In addition to H&M's primary operations, the company operates several other brands, including the COS (Collection of Style) store chain, which is similar to H&M, but at higher price points; H&M Home, which offers textile fashions for the home through Internet and catalog sales and a showroom in Stockholm; Monki; Weekday; and Cheap Monday. In 2012, H&M opened 42 stores in North America, including a 25,000 square foot flagship store in New York City's Time Warner Center. Also in 2012, the firm opened its first stores in Thailand, Latvia, Mexico and Malaysia.

FINANCIAL DATA: *Note: Data for latest year may not have been available at press time.*

In U.S. $	2012	2011	2010	2009	2008	2007
Revenue	18,605,900,000	16,942,450,000	16,708,950,000	15,616,920,000	13,636,020,000	12,067,140,000
R&D Expense						
Operating Income	3,350,630,000	3,138,848,000	3,798,069,000	3,333,688,000	3,101,728,000	2,831,263,000
Operating Margin %	18.00%	18.52%	22.73%	21.34%	22.74%	23.46%
SGA Expense	7,719,203,000	7,049,354,000	6,716,972,000	6,288,786,000	5,287,632,000	4,538,307,000
Net Income	2,597,917,000	2,436,808,000	2,877,316,000	2,523,523,000	2,355,638,000	2,092,873,000
Operating Cash Flow	2,911,047,000	2,683,092,000	3,363,568,000	2,768,267,000	2,767,189,000	2,369,038,000
Capital Expenditure	1,051,519,000	785,982,700	763,803,300	875,778,500	731,150,200	555,717,300
EBITDA	4,003,845,000	3,728,759,000	4,324,368,000	3,841,504,000	3,604,153,000	3,232,802,000
Return on Assets %	28.02%	26.50%	32.90%	31.02%	32.89%	35.16%
Return on Equity %	38.36%	35.84%	44.06%	42.24%	44.30%	45.39%
Debt to Equity						

CONTACT INFORMATION:

Phone: 46 87965500 Fax: 46 8208094
Toll-Free:
Address: Regeringsgatan 48, Stockholm, SE-106-38 Sweden

STOCK TICKER/OTHER:

Stock Ticker: HMRZF
Exchange: PINX
Employees: 94,000
Fiscal Year Ends: 30-Nov
Parent Company:

SALARIES/BONUSES:

Top Exec. Salary: $ Bonus: $
Second Exec. Salary: $ Bonus: $

OTHER THOUGHTS:

Estimated Female Officers or Directors: 14
Hot Spot for Advancement for Women/Minorities: Y

Haggar Corp

NAIC Code: 315240

www.haggar.com

TYPES OF BUSINESS:

Men's & Women's Apparel
Retail Stores

BRANDS/DIVISIONS/AFFILIATES:

Haggar Clothing Company
Infinity Associates LLC
LK Life Khaki
Haggar Apparel Ltd
Haggar Canada Co

CONTACTS:

Note: Officers with more than one job title may be intentionally listed here more than once.

Michael Stitt, CEO
Tim Lyons, Pres., Haggar Clothing Co.
Frank Brackenfd, Pres./COO-Haggar Clothing Co.

GROWTH PLANS/SPECIAL FEATURES:

Haggar Corp., along with its subsidiaries (including Haggar Clothing Co., its primary subsidiary), designs, manufactures, imports and markets casual and dress men's and women's apparel. The company is based in Dallas, Texas and markets clothing and accessories throughout the U.S., Canada, Mexico and the U.K. The firm is famous for first coining the term slacks in 1938 and for the introduction of wrinkle free cotton casual pants, pre-cuffed pants and expandable waist pants. Today, its products include pants, shorts, suits, sport coats, sweaters, shirts, dresses, skirts and vests. Recently, the company launched a new line of eco-friendly slacks called the LK Life Khaki brand. Hagar Clothing Company is also an official clothing partner of both the Pro Football Hall of Fame and the Hockey Hall of Fame. The company conducts foreign operations through Haggar Apparel, Ltd. and Haggar Canada Co. Haggar Corp. is owned by private equity group Infinity Associates LLC.

FINANCIAL DATA:

Note: Data for latest year may not have been available at press time.

In U.S. $	2012	2011	2010	2009	2008	2007
Revenue						
R&D Expense						
Operating Income						
Operating Margin %						
SGA Expense						
Net Income						
Operating Cash Flow						
Capital Expenditure						
EBITDA						
Return on Assets %						
Return on Equity %						
Debt to Equity						

CONTACT INFORMATION:

Phone: 214-352-8481 Fax: 214-956-4446
Toll-Free:
Address: 11511 Luna Rd., Dallas, TX 75234 United States

STOCK TICKER/OTHER:

Stock Ticker: Private Exchange:
Employees: Fiscal Year Ends: 30-Sep
Parent Company: INFINITY ASSOCIATES LLC

SALARIES/BONUSES:

Top Exec. Salary: $ Bonus: $
Second Exec. Salary: $ Bonus: $

OTHER THOUGHTS:

Estimated Female Officers or Directors:
Hot Spot for Advancement for Women/Minorities:

Hallwood Group Inc.

http://www.hallwood.com

NAIC Code: 313210

TYPES OF BUSINESS:

Broadwoven Fabric Mills

BRANDS/DIVISIONS/AFFILIATES:

Brookwood Companies Incorporated
Kenyon Industries Inc
Brookwood Laminating Inc

CONTACTS: *Note: Officers with more than one job title may be intentionally listed here more than once.*

Anthony J. Gumbiner, CEO
William L. Guzzetti, COO
William L. Guzzetti, Pres.
Richard Kelley, CFO
Richard Kelley, Sec.
Joseph T. Koenig, Treas.
Amber M. Brookman, CEO/Pres., Brookwood Companies Inc.
Anthony J. Gumbiner, Chmn.

GROWTH PLANS/SPECIAL FEATURES:

Hallwood Group Inc. operates as a holding company with interests in the textile products industry. Textile products operations are conducted through the firm's wholly owned subsidiary, Brookwood Companies Incorporated. Brookwood is an integrated textile firm that develops and produces innovative fabrics and related products through finishing, treating and coating processes. It operates as a converter, finisher and laminator in the textile industry, which processes fabrics at its plants, located in Rhode Island and Connecticut, or by contracting with independent finishers. Brookwood is one of the largest coaters of woven nylons in the United States of America and is known for its extensive, in-house expertise in high-tech fabric development and is a major supplier of specialty fabric to U.S. military contractors, which represented over 52% of 2012 revenues. Brookwood has two principle subsidiaries: Kenyon Industries, Inc. and Brookwood Laminating Inc. Kenyon provides quality finishing services for fabrics used in a variety of markets, such as military, luggage and knapsacks, flag and banner, apparel, industrial and sailcloth. Brookwood Laminating provides laminating services for fabrics used in military clothing and equipment, sailcloth, medical equipment, industrial applications and consumer apparel.

FINANCIAL DATA: *Note: Data for latest year may not have been available at press time.*

In U.S. $	2012	2011	2010	2009	2008	2007
Revenue	130,524,000	139,499,000	168,354,000	179,554,000	162,237,000	132,497,000
R&D Expense						
Operating Income	-18,410,000	-9,009,000	16,156,000	25,632,000	15,767,000	7,250,000
Operating Margin %	-14.10%	-6.45%	9.59%	14.27%	9.71%	5.47%
SGA Expense	36,742,000	32,643,000	26,317,000	25,110,000	22,675,000	20,329,000
Net Income	-17,943,000	-6,331,000	9,880,000	17,055,000	1,398,000	-32,825,000
Operating Cash Flow	-21,108,000	-5,437,000	23,350,000	8,637,000	34,760,000	4,457,000
Capital Expenditure	1,750,000	2,343,000	7,089,000	3,102,000	3,207,000	2,358,000
EBITDA	-16,133,000	-6,501,000	18,429,000	27,993,000	6,082,000	-46,179,000
Return on Assets %	-22.44%	-7.26%	11.37%	21.61%	1.74%	-33.09%
Return on Equity %	-35.76%	-10.16%	16.32%	36.34%	3.21%	-50.19%
Debt to Equity	0.34	0.03	0.03	0.11	0.27	0.35

CONTACT INFORMATION:

Phone: +1 214 528-5588 Fax: +1 214 522-9254
Toll-Free:
Address: 3710 Rawlins, Dallas, TX 75219 United States

STOCK TICKER/OTHER:

Stock Ticker: HWG Exchange: ASE
Employees: 458 Fiscal Year Ends: 31-Dec
Parent Company:

SALARIES/BONUSES:

Top Exec. Salary: $996,000 Bonus: $
Second Exec. Salary: $317,539 Bonus: $300,000

OTHER THOUGHTS:

Estimated Female Officers or Directors: 2
Hot Spot for Advancement for Women/Minorities:

Hampshire Group Ltd

www.hamp.com

NAIC Code: 315220

TYPES OF BUSINESS:

Men's Apparel

BRANDS/DIVISIONS/AFFILIATES:

Hampshire Brands Inc
scott james LLC
Rio Garment SA
Keynote Services Ltd

CONTACTS: *Note: Officers with more than one job title may be intentionally listed here more than once.*

Paul M. Buxbaum, CEO
Maura M. Langley, CFO/VP/Treas.
Robert Stec, Pres., Hampshire Brands
David L. Gren, Pres., Rio Garment
Peter H. Woodward, Chmn.

GROWTH PLANS/SPECIAL FEATURES:

Hampshire Group, Ltd. is a holding company with subsidiaries involved in the design, manufacture and marketing of apparel in the men's sweaters, woven and knit categories. Its business consists of three primary subsidiaries: Hampshire Brands, Inc.; scott james, LLC; and Rio Garment SA. Hampshire Brands offers sweaters under the licensed names of Geoffrey Beene and Dockers as well as certain private labels of its customers. Hampshire has an exclusive arrangement with JC Penney for a full sportswear line under the licensed JOE Joseph Abboud label, which includes sweaters, knits, woven tops, blazers and a range of men's bottoms. scott james is a men's specialty retailer and wholesale provider of apparel that operates one store, located in Boston. Rio Garment SA is an apparel manufacturer based in Honduras that designs, manufactures and sources knit tops for men, women and children. Customers include Kohl's Department Stores, Target Stores, Macy's, JC Penney and Federated Department Stores. Another subsidiary, Keynote Services, Ltd., assists in Chinese sourcing and quality control operations. During 2012, Hampshire entered into licensing agreements to produce men's tops under the Dockers brand and a full line of lifestyle apparel for men under the Panama Jack brand.

FINANCIAL DATA: *Note: Data for latest year may not have been available at press time.*

In U.S. $	2012	2011	2010	2009	2008	2007
Revenue	117,560,000	86,148,000	134,482,000	165,178,000	240,901,000	310,793,000
R&D Expense						
Operating Income	-11,155,000	-21,106,000	-11,528,000	-11,734,000	-17,554,000	122,000
Operating Margin %	-9.48%	-24.49%	-8.57%	-7.10%	-7.28%	.03%
SGA Expense	33,965,000	28,156,000	33,484,000	41,768,000	58,598,000	68,613,000
Net Income	-11,739,000	-9,960,000	-9,693,000	-6,046,000	-29,897,000	-4,000
Operating Cash Flow	-12,416,000	-12,888,000	5,244,000	-594,000	5,646,000	-23,476,000
Capital Expenditure	797,000	409,000	167,000	359,000	10,291,000	4,107,000
EBITDA	-7,199,000	-18,267,000	-8,830,000	-9,385,000	-12,889,000	3,159,000
Return on Assets %	-14.38%	-11.74%	-11.71%	-6.51%	-23.43%	
Return on Equity %	-30.59%	-21.69%	-17.90%	-9.77%	-34.89%	
Debt to Equity	0.13					

CONTACT INFORMATION:

Phone: 1-212-840-5666 Fax: 864 225-4421
Toll-Free:
Address: 114 W. 41st St., New York, NY 10036 United States

STOCK TICKER/OTHER:

Stock Ticker: HAMP
Exchange: PINX
Employees: 1,100
Fiscal Year Ends: 31-Dec
Parent Company:

SALARIES/BONUSES:

Top Exec. Salary: $ Bonus: $
Second Exec. Salary: $ Bonus: $

OTHER THOUGHTS:

Estimated Female Officers or Directors: 1
Hot Spot for Advancement for Women/Minorities:

Hanesbrands Inc

www.hanesbrands.com

NAIC Code: 315220

TYPES OF BUSINESS:

Underwear
Outerwear
Hosiery
Internet Sales
Catalogs

BRANDS/DIVISIONS/AFFILIATES:

Champion
Hanes
Playtex
barely there
Just My Size
Wonderbra
Outer Banks
Hanes Beefy-T

CONTACTS: *Note: Officers with more than one job title may be intentionally listed here more than once.*

Richard A. Noll, CEO
Gerald W. Evans, Jr., Co-COO
Richard D. Moss, CFO
Kevin W. Oliver, Chief Human Resources Officer
Joia M. Johnson, General Counsel/Chief Legal Officer/Corp. Sec.
William J. Nictakis, Co-COO
Richard A. Noll, Chmn.

GROWTH PLANS/SPECIAL FEATURES:

Hanesbrands, Inc. designs, manufactures, sources and sells apparel including t-shirts, bras, panties, men's underwear, kids' underwear, socks, hosiery, casual wear and active wear. The company is organized into four operating segments: Innerwear, which accounted for 52% of its 2012 revenue; Outerwear, 29%; Direct-to-Consumer, 8%; and International, 11%. Hanesbrands' innerwear segment produces bras, panties, men's underwear, kids' underwear and socks marketed under the Hanes, Playtex, Bali, barely there, Just My Size and Wonderbra brands. Hanesbrands also maintains a licensing agreement with Polo Ralph Lauren to produce underwear. The outerwear segment produces t-shirts, fleece, athletic uniforms, thermals sleepwear and casual wear marketed under the Hanes, Champion, Outer Banks, Hanes Beefy-T and Just My Size brands. The direct-to-consumer segment operates 201 outlet stores as well as catalogs and e-commerce channels. The international segment includes sales in Asia, Latin America, Canada and Australia with some of the company's largest international markets located in Japan, Canada, Mexico, Brazil and Australia. During 2012, sales in the U.S. accounted for 89% of its revenue, while international sales accounted for the remainder. Its largest customers are Wal-Mart Stores, Inc., Target Corporation and Kohl's Corporation, which accounted for 28%, 18% and 6%, respectively, of total revenue. Hanesbrands engages in manufacturing through both company-owned and operated facilities and third-party contractors. The firm's design, research and product development activities are primarily located in a North Carolina facility, with some activities in New York City and Lenexa, Kansas.

Employees are offered medical, dental and vision insurance; life insurance; disability coverage; a retirement savings plan; adoption assistance; and an employee assistance program.

FINANCIAL DATA: *Note: Data for latest year may not have been available at press time.*

In U.S. $	2012	2011	2010	2009	2008	2007
Revenue	4,525,721,000	4,637,143,000	4,326,713,000	3,891,275,000	4,248,770,000	4,474,537,000
R&D Expense						
Operating Income	440,115,000	478,281,000	404,188,000	270,856,000	317,480,000	388,569,000
Operating Margin %	9.72%	10.31%	9.34%	6.96%	7.47%	8.68%
SGA Expense	979,932,000	1,062,090,000	1,010,581,000	940,530,000	1,009,607,000	1,040,754,000
Net Income	164,681,000	266,688,000	211,293,000	51,283,000	127,169,000	126,127,000
Operating Cash Flow	548,902,000	167,957,000	133,054,000	414,504,000	177,397,000	359,040,000
Capital Expenditure	40,994,000	90,099,000	106,240,000	126,825,000	186,957,000	96,626,000
EBITDA	492,836,000	569,006,000	490,800,000	367,611,000	439,291,000	521,485,000
Return on Assets %	4.29%	6.81%	5.93%	1.49%	3.64%	3.66%
Return on Equity %	21.00%	42.88%	47.09%	19.72%	53.65%	43.65%
Debt to Equity	1.48	2.65	3.53	5.16	11.5	8.01

CONTACT INFORMATION:

Phone: 336 519-8080   Fax: 312 726-3712
Toll-Free:
Address: 1000 E. Hanes Mill Rd., Winston-Salem, NC 27105 United States

STOCK TICKER/OTHER:

Stock Ticker: HBI   Exchange: NYS
Employees: 53,300   Fiscal Year Ends: 31-Dec
Parent Company:

SALARIES/BONUSES:

Top Exec. Salary: $1,100,000   Bonus: $
Second Exec. Salary: $491,667   Bonus: $275,000

OTHER THOUGHTS:

Estimated Female Officers or Directors: 3
Hot Spot for Advancement for Women/Minorities: Y

Hermes International SA

www.hermes.com

NAIC Code: 315990

TYPES OF BUSINESS:

Leather Goods
Luxury Scarves & Ties
Accessories
Luxury Goods
Housewares
Retail Stores
Equestrian Gear
Apparel

BRANDS/DIVISIONS/AFFILIATES:

CONTACTS:
Note: Officers with more than one job title may be intentionally listed here more than once.

Patrick Thomas, Exec. Chmn.
Mireille Maury, Managing Dir.-Finance
Beatriz Gonzales-Cristobal Poyo, Managing Dir.-Mktg.
Guillaume De Seynes, Managing Dir.-Mfg. Div.
Mireille Maury, Managing Dir.-Admin.
Pierre-Alexis Dumas, Managing Dir.-Mktg.
Patrick Albaladejo, Deputy Managing Dir.-Strategic Dev. & Corp. Image
Bertrand Puech, Chmn.-Emile Hermes SARL
Guillaume De Seynes, Managing Dir.-Equity Investments
Pierre-Alexis Dumas, Managing Dir.-Artistic
Eric de Seynes, Chmn.

GROWTH PLANS/SPECIAL FEATURES:

Hermes International SA, founded in 1837 as a leather goods company that catered to equestrians, has since grown to include other articles of clothing. Current product lines include leather goods, including handbags, luggage, diaries, small leather goods and equestrian gear; perfumes; watches; tableware; clothing and accessories, including men's wear, women's wear, belts, hats, gloves, shoes and leather and enamel jewelry; and silk goods, such as scarves, ties and other accoutrements. Leather goods and saddle equipment account for approximately 47% of the firm's revenue; clothing and accessories, 20%; silks and textiles, 12%; perfumes, 6%; watches, 5%; tableware, 2%; other Hermes sectors 4% and other items, 7%. Hermes distributes its products through 341 retail stores in more than 47 countries. The company also sells its products through a network of specialty stores located in airports and on board aircraft. Hermes owns a 21% stake in Swiss watchmaker Vaucher Manufacture Fleurier; a 39.5% stake in Les Tissages Perrin; and a 32% stake in Joseph Erard Holding. Customers for the most expensive Hermes handbags, costing $7,000 or more, may find themselves on waiting lists that last years until their turn arrives to receive one of these limited edition items.

FINANCIAL DATA:
Note: Data for latest year may not have been available at press time.

In U.S. $	2012	2011	2010	2009	2008	2007
Revenue		3,799,412,000	3,210,484,000	2,559,909,000	2,359,722,000	2,173,175,000
R&D Expense						
Operating Income		1,183,739,000	893,554,400	619,015,800	600,695,400	566,595,300
Operating Margin %		31.15%	27.83%	24.18%	25.45%	26.07%
SGA Expense		1,264,643,000	1,072,747,000	883,391,300	818,668,100	762,770,800
Net Income		806,900,200	577,293,400	395,025,400	394,891,700	392,885,800
Operating Cash Flow		984,086,700	887,670,500	611,794,600	404,653,700	434,340,700
Capital Expenditure		247,659,800	184,942,500	266,648,800	214,763,300	160,337,000
EBITDA		1,332,308,000	1,023,402,000	728,269,600	700,321,000	652,982,100
Return on Assets %		19.27%	15.73%	12.11%	13.21%	14.36%
Return on Equity %		26.47%	21.25%	17.01%	19.00%	19.95%
Debt to Equity				0.01	0.01	0.01

CONTACT INFORMATION:

Phone: 33 140174920 Fax: 33 140174921
Toll-Free:
Address: 24 rue du Faubourg Saint-Honore, Paris, 75008 France

STOCK TICKER/OTHER:

Stock Ticker: HESAF Exchange: PINX
Employees: 10,118 Fiscal Year Ends: 31-Dec
Parent Company:

SALARIES/BONUSES:

Top Exec. Salary: $ Bonus: $
Second Exec. Salary: $ Bonus: $

OTHER THOUGHTS:

Estimated Female Officers or Directors: 4
Hot Spot for Advancement for Women/Minorities: Y

Hibbett Sports Inc

www.hibbett.com

NAIC Code: 451110

TYPES OF BUSINESS:

Sporting Goods Stores
Sports Apparel
Athletic Shoes
Training Equipment

BRANDS/DIVISIONS/AFFILIATES:

Hibbett Sports
Sports & Co
Sports Additions
Hibbett Team Sales Inc

CONTACTS: *Note: Officers with more than one job title may be intentionally listed here more than once.*

Jeffry O. Rosenthal, CEO
Jeffry O. Rosenthal, Pres.
Gary A. Smith, CFO/Sr. VP
Terry Mayfield, CIO/VP
Rebecca A. Jones, Sr. VP-Merch.
Cathy E. Pryor, Sr. VP-Oper.

GROWTH PLANS/SPECIAL FEATURES:

Hibbett Sports, Inc. is an operator of sporting goods stores in small to mid-sized markets predominantly in the Southeast, Southwest, Midwest and Mid-Atlantic U.S. Its stores offer a broad assortment of athletic equipment, footwear and apparel. The company's merchandise assortment features a broad selection of brand name merchandise emphasizing team sports complemented by localized apparel and accessories designed to appeal to a wide range of customers. Hibbett operates a total of 670 stores, as well as 19 Sports Additions stores and 1 Sports & Co. superstores, throughout 26 states. The firm's primary retail format is Hibbett Sports, a 5,000-square-foot store model located in strip centers frequently within in close proximity to Wal-Mart stores. Hibbett's Sports Additions stores are small, mall-based stores averaging 2,500 square feet with roughly 90% of merchandise consisting of athletic footwear and the remainder consisting of caps and a limited assortment of apparel. Sports Additions stores offer a broader assortment of athletic footwear, with a greater emphasis on fashion than the athletic footwear choices offered by traditional Hibbett Sports stores. The firm's Sports & Co. superstores average 25,000 square feet and offer a broader assortment of athletic footwear, apparel and equipment than the Hibbett Sports stores. Subsidiary Hibbett Team Sales, Inc. supplies customized athletic apparel, equipment and footwear to school, athletic and youth programs in Alabama, Georgia, Mississippi and Florida. It sells its merchandise directly to educational institutions and youth associations.

Employees are offered life, disability, medical and dental insurance; a vision care plan; a stock purchase plan; a 401(k) plan; employee discounts; and a 529 college savings plan.

FINANCIAL DATA: *Note: Data for latest year may not have been available at press time.*

In U.S. $	2012	2011	2010	2009	2008	2007
Revenue	732,645,000	664,954,000	593,492,000	564,188,000	520,720,000	
R&D Expense						
Operating Income	93,531,000	73,547,000	52,407,000	47,972,000	48,227,000	
Operating Margin %	12.76%	11.06%	8.83%	8.50%	9.26%	
SGA Expense	155,672,000	143,232,000	129,888,000	123,075,000	108,463,000	
Net Income	59,060,000	46,400,000	32,549,000	29,448,000	30,329,000	
Operating Cash Flow	54,921,000	61,918,000	36,914,000	38,997,000	48,022,000	
Capital Expenditure	12,997,000	10,476,000	9,605,000	13,697,000	16,376,000	
EBITDA	106,761,000	87,212,000	66,372,000	62,337,000	60,963,000	
Return on Assets %	18.81%	15.70%	12.71%	13.03%	14.12%	
Return on Equity %	29.24%	24.73%	20.88%	23.03%	23.72%	
Debt to Equity	0.01	0.01				

CONTACT INFORMATION:

Phone: 205 942-4292 Fax: 205 912-7290
Toll-Free:
Address: 451 Industrial Ln., Birmingham, AL 35211 United States

STOCK TICKER/OTHER:

Stock Ticker: HIBB Exchange: NAS
Employees: 6,700 Fiscal Year Ends: 31-Jan
Parent Company:

SALARIES/BONUSES:

Top Exec. Salary: $440,000 Bonus: $
Second Exec. Salary: $400,000 Bonus: $

OTHER THOUGHTS:

Estimated Female Officers or Directors: 3
Hot Spot for Advancement for Women/Minorities: Y

Hot Topic Inc

NAIC Code: 448140

www.hottopic.com

TYPES OF BUSINESS:

Music-Related Apparel Stores
Online Sales
Accessories
Cosmetics
Jewelry
Lingerie

BRANDS/DIVISIONS/AFFILIATES:

Hot Topic
Torrid
HotTopic.com
Torrid.com
Blackheart
BlackHeartLingerie.com

CONTACTS: *Note: Officers with more than one job title may be intentionally listed here more than once.*

Lisa Harper, CEO
James J. McGinty, CFO
Don Hendricks, CIO
Kate Horton, VP/Gen. Merch. Mgr., Torrid
Jonathan Block, Corp. Sec.
Mark Mizicko, Chief Planning Officer
George Wehlitz, Jr., VP-Finance
Lisa Harper, Chmn.

GROWTH PLANS/SPECIAL FEATURES:

Hot Topic, Inc. is a mall and Internet-based specialty retailer that operates under the Hot Topic and Torrid brands. The firm owns approximately 620 Hot Topic stores throughout the U.S., Canada and Puerto Rico and 186 Torrid stores in 33 states and Puerto Rico. Hot Topic stores are approximately 1,770 square feet on average, while Torrid locations average roughly 2,493 square feet. The firm also sells merchandise similar to that in its two stores through two distinct web sites: HotTopic.com and Torrid.com. Hot Topic stores target males and females primarily between the ages of 12-22, selling a selection of music-inspired apparel and accessories. Torrid targets plus-size females between the ages of 15-29, selling apparel, lingerie, shoes and accessories. The firm has developed a strategy focused on offering music-related merchandise in the mall environment. It tracks alternative and rock music trends by visiting nightclubs, attending concerts and monitoring new music, music videos and radio airplay. In November 2012, the company introduced a new lingerie concept, Blackheart, launching an e-commerce site (BlackHeartLingerie.com) and opening five initial Blackheart retail locations in California and Texas.

The firm offers employees medical, dental and vision insurance; life insurance; disability coverage; a 401(k) plan; an employee stock purchase plan; tuition assistance; and an employee assistance program.

FINANCIAL DATA: *Note: Data for latest year may not have been available at press time.*

In U.S. $	2012	2011	2010	2009	2008	2007
Revenue	697,934,000	708,244,000	736,710,000	761,074,000	728,121,000	
R&D Expense						
Operating Income	-3,455,000	-13,762,000	19,247,000	30,822,000	24,297,000	
Operating Margin %	-.49%	-1.94%	2.61%	4.04%	3.33%	
SGA Expense	236,308,000	247,089,000	237,010,000	242,483,000	227,147,000	
Net Income	-1,818,000	-8,235,000	11,880,000	19,742,000	16,012,000	
Operating Cash Flow	45,896,000	35,018,000	45,994,000	77,004,000	51,591,000	
Capital Expenditure	23,891,000	31,031,000	24,332,000	23,257,000	48,766,000	
EBITDA	32,274,000	27,164,000	57,564,000	69,150,000	65,880,000	
Return on Assets %	-.61%	-2.39%	3.18%	5.61%	4.92%	
Return on Equity %	-.90%	-3.33%	4.43%	7.99%	7.01%	
Debt to Equity						

CONTACT INFORMATION:

Phone: 626 839-4681 Fax: 626 839-4686
Toll-Free:
Address: 18305 E. San Jose Ave., City of Industry, CA 91748 United States

STOCK TICKER/OTHER:

Stock Ticker: HOTT Exchange: NAS
Employees: 8,800 Fiscal Year Ends: 31-Jan
Parent Company:

SALARIES/BONUSES:

Top Exec. Salary: $589,904 Bonus: $600,000
Second Exec. Salary: $413,462 Bonus: $620,000

OTHER THOUGHTS:

Estimated Female Officers or Directors: 3
Hot Spot for Advancement for Women/Minorities: Y

Hugo Boss AG

group.hugoboss.com

NAIC Code: 315240

TYPES OF BUSINESS:

Men's Apparel, Retail
Women's Apparel
Fragrances
Home Accessories
Leather Goods
Eyewear

BRANDS/DIVISIONS/AFFILIATES:

Valentino Fashion Group SpA
BOSS Black
BOSS Orange
BOSS Selection
BOSS Green
HUGO
Hugo Boss

CONTACTS: *Note: Officers with more than one job title may be intentionally listed here more than once.*

Claus-Dietrich Lahrs, CEO
Mark Langer, CFO
Claus-Dietrich Lahrs, Dir.-Sales
Mark Langer, Dir.-Human Resources & Labor Rel.
Mark Langer, Dir.-IT
Christoph Auhagen, Dir.-Mfg.
Mark Langer, Dir.-Legal Affairs
Claus-Dietrich Lahrs, Dir.-Comm.
Mark Langer, Dir.-Investor Rel.
Mark Langer, Dir.-Finance & Controlling
Christoph Auhagen, Chief Brand Officer
Hellmut Albrecht, Chmn.
Mark Langer, Dir.-Logistics

GROWTH PLANS/SPECIAL FEATURES:

Hugo Boss AG is an international designer and licenser of clothes, accessories and fragrances. The firm's products are sold through approximately 622 company-owned stores, over 1,000 Hugo Boss franchised locations and roughly 6,300 third party retailers in 124 countries. The firm, which sells under the brands BOSS and HUGO, was formerly known for its men's wear, but has moved into women's wear, fragrances and other offerings, such as home accessories. BOSS comes in four labels: BOSS Black, BOSS Selection, BOSS Orange and BOSS Green. BOSS Black focuses on office, formal attire and leisurewear; the line includes shoes, bags, accessories, eyewear, watches and fragrances for men and women as well as bodywear and skincare for men. BOSS Selection is the premium tier of the company's menswear brands, featuring high-quality materials and expert tailoring. BOSS Orange is a more casual and unconventional leisurewear line for men and women with an urban sportswear theme. It includes men's and women's shoes, bags and accessories as well as men's bodywear and fragrances. BOSS Green is a men's collection designed for athletics and active outdoor performance, with an emphasis on golf. The HUGO brand includes business and leisurewear for young men and women with an emphasis on avant-garde yet appropriate style. HUGO brand products include shoes, bags, accessories, bodywear, eyewear and fragrances. The designing process for the company's new lines takes place at the Metzingen headquarters in Germany. The firm also has manufacturing facilities in Poland, the U.S., Turkey and Italy. Valentino Fashion Group SpA, an Italian textile maker, owns a majority of the company.

Hugo Boss offers apprenticeships, internships, traineeships and even dissertation writing as gateways to employment.

FINANCIAL DATA: *Note: Data for latest year may not have been available at press time.*

In U.S. $	2012	2011	2010	2009	2008	2007
Revenue	3,137,007,000	2,753,197,000	2,312,712,000	2,088,611,000	2,315,950,000	2,182,426,000
R&D Expense			47,286,000			
Operating Income	579,300,600	526,999,200	352,959,400	211,881,500	254,236,400	294,588,100
Operating Margin %	18.46%	19.14%	15.26%	10.14%	10.97%	13.49%
SGA Expense	1,364,060,000	1,164,412,000	929,275,200	924,966,500		
Net Income	411,064,400	389,097,400	253,046,300	139,078,600	149,832,800	206,150,000
Operating Cash Flow	524,741,900	403,450,100	412,321,500	467,385,700	220,413,200	226,012,300
Capital Expenditure	221,719,700	145,064,200	74,330,030	64,638,940	158,840,600	
EBITDA	702,242,600	637,511,400	455,526,900	307,349,600	317,523,400	294,588,100
Return on Assets %	20.26%	20.28%	15.35%	9.34%	10.18%	15.55%
Return on Equity %	52.94%	64.33%	66.03%	51.82%	30.04%	29.60%
Debt to Equity						0.22

CONTACT INFORMATION:

Phone: 49 07123940 Fax: 49 07123942014
Toll-Free:
Address: Dieselstrasse 12, Metzingen, 72555 Germany

STOCK TICKER/OTHER:

Stock Ticker: BOSSY Exchange: GREY
Employees: 2,600 Fiscal Year Ends: 31-Dec
Parent Company: VALENTINO FASHION GROUP SPA

SALARIES/BONUSES:

Top Exec. Salary: $ Bonus: $
Second Exec. Salary: $ Bonus: $

OTHER THOUGHTS:

Estimated Female Officers or Directors:
Hot Spot for Advancement for Women/Minorities:

Hyosung Corporation

www.hyosung.com

NAIC Code: 313210

TYPES OF BUSINESS:

Manufacturing-Textiles
Information & Communications Systems
Electrical Equipment Manufacturing
Chemicals Manufacturing
Construction
Trading-Metal & Chemicals
Transportation & Logistics
Car Dealerships

BRANDS/DIVISIONS/AFFILIATES:

FIREX
Creora
Hyosung GoodSprings
Hyosung Ebara Engineering Co Ltd
Hyosung Transworld Co Ltd
Class Hyosung Co Ltd (The)
Nautilus Hyosung Inc
Hyosung Information Systems Co Ltd

CONTACTS: *Note: Officers with more than one job title may be intentionally listed here more than once.*

Seok Rae Cho, CEO
Sang-Woon Lee, COO/Vice Chmn.
Yoon-Taik Chung, CFO/Exec. VP
Hyun-Joon Cho, Pres., Corp. Strategy Center
Hyun-Joon Cho, Pres., Textile PG & Info. & Comm. PG
Seok Rae Cho, Chmn.

GROWTH PLANS/SPECIAL FEATURES:

Hyosung Corporation is a diversified manufacturer operating in seven performance groups (PGs), which are further subdivided into additional performance units (PUs). The Textile PG has three PUs engaged in weaving, dyeing and manufacturing fibers such as nylon, spandex and polyester. Its products include flame-retardant FIREX fabrics and Creora brand spandex. Within the Industrial Materials PG, the Interior PU manufactures carpet yarn and commercial and automotive carpets; the Tire & Industrial Reinforcements PU manufactures nylon, polyester and steel cords used to reinforce tires; and the Technical Yarn PU produces high-performance fibers, including nylon 66, used in airbags. The Chemicals PG manufactures plastic bottles, nylon and polyester films; optical films used in LCD screens; polypropylene; and terephthalic acid (TPA, used to manufacture polyester). The Power & Industrial Systems PG's operations include the Power Systems PU, which manufactures equipment, including distribution transformers and welding machines; and the Industrial Machinery PU, which produces motors, wind turbine generating systems and test facilities. Subsidiary Hyosung GoodSprings, which manufactures heavy industrial pumps, also operates as part of this division. The Construction PG consists of the Construction PU, mainly involved in residential building, and subsidiary Hyosung Ebara Engineering Co., Ltd., which builds municipal infrastructure. The Trading PG has three PUs, which trade in steel, metal and chemicals. Hyosung Transworld Co., Ltd., a transportation and logistics company, is also included in this group. The Information & Communication PG comprises subsidiaries Nautilus Hyosung, Inc., offering computerized banking hardware and software; Hyosung Information Systems Co., Ltd., manufacturing large-capacity data storage devices; and Galaxia Photonics, Galaxia Electronics, Galaxia Communications and Galaxia Device. Additional companies not included in one of the seven PGs are The Class Hyosung Co., Ltd., a Mercedes dealer; Hyosung Toyota, a leading dealer of Toyota vehicles in Korea; and Hyosung Capital Co., Ltd., offering general corporate loans, financing and leasing.

FINANCIAL DATA: *Note: Data for latest year may not have been available at press time.*

In U.S. $	2012	2011	2010	2009	2008	2007
Revenue		12,340,535,000	10,730,900,000	9,266,050,000	8,887,340,000	3,931,030,000
R&D Expense						
Operating Income						
Operating Margin %						
SGA Expense						
Net Income		165,268,000	246,670,000	299,420,000	109,690,000	120,800,000
Operating Cash Flow						
Capital Expenditure						
EBITDA						
Return on Assets %						
Return on Equity %						
Debt to Equity						

CONTACT INFORMATION:

Phone: 82-2-707-7000 Fax: 82-2-707-7799
Toll-Free:
Address: Hyosung Bldg. 450 Gongdeok-dong, Mapo-gu, Seoul, 121-720 South Korea

STOCK TICKER/OTHER:

Stock Ticker: 4800 Exchange: Seoul
Employees: 7,786 Fiscal Year Ends: 31-Dec
Parent Company:

SALARIES/BONUSES:

Top Exec. Salary: $ Bonus: $
Second Exec. Salary: $ Bonus: $

OTHER THOUGHTS:

Estimated Female Officers or Directors:
Hot Spot for Advancement for Women/Minorities:

ICG/Holliston

www.icgholliston.com

NAIC Code: 313210

TYPES OF BUSINESS:

Coverings, Manufacturing
Binding & Packaging Supplies
Office & Industrial Supplies

BRANDS/DIVISIONS/AFFILIATES:

Luminaire
Versa Capital Management Inc
Impressions
Pinnacle

CONTACTS: *Note: Officers with more than one job title may be intentionally listed here more than once.*

Keith A. Polak, COO
Steve W. Lister, Sr. VP-Sales & Mktg.
Charles H. Crowley, Dir.-Quality & Process Improvement
Gary F. Sweeney, Sr. VP-Oper.
Gary F. Sweeney, Sr. VP-Bus. Dev.
Steve W. Lister, VP-Finance
William D. Chapman, VP-Technical Sales
William C. Waldron, III, VP-Security Prod.
Gregory W. Stanley, Dir.-Oper.
Sheldon Sillyman, Oper. Controller
Russell L. McAlister, Dir.-Supply Chain & Customer Service

GROWTH PLANS/SPECIAL FEATURES:

ICG/Holliston designs and manufactures coverings for bookbinding, box covering, passports and security products, packaging and the mechanical binding industry. Additional product lines are designed for use with office supplies, tags and labels, building products and as separator material. With over 115 years of experience, the firm has been an industry leader in this niche market for a considerable amount of time. Some covering products include coated and starch-filled natural book cloth; value-added rayon materials; leather-like coated cloth; buckram library cloth; and Luminaire, a pearlescent cloth used in bookbinding and in all forms of packaging. This product group also includes exceptionally tough materials for passports and other security documents as well as bindery supplies. ICG's office and industrial supply products include tape stock, ID tags, clothing labels, drill cloth, box tape and a variety of cloth products used in the office. Additionally, the firm produces the Impressions and Pinnacle brands of printable cloth fabrics, which can be used for short-run book covers or other forms of art. The company is owned by private equity firm Versa Capital Management, Inc.

FINANCIAL DATA: *Note: Data for latest year may not have been available at press time.*

In U.S. $	2012	2011	2010	2009	2008	2007
Revenue						
R&D Expense						
Operating Income						
Operating Margin %						
SGA Expense						
Net Income						
Operating Cash Flow						
Capital Expenditure						
EBITDA						
Return on Assets %						
Return on Equity %						
Debt to Equity						

CONTACT INFORMATION:

Phone: 423-357-6141 Fax: 423-357-8840
Toll-Free: 800-251-0451
Address: 905 Holliston Mills Rd., Church Hill, TN 37642 United States

STOCK TICKER/OTHER:

Stock Ticker: Private Exchange:
Employees: 300 Fiscal Year Ends: 31-Dec
Parent Company: VERSA CAPITAL MANAGEMENT INC

SALARIES/BONUSES:

Top Exec. Salary: $ Bonus: $
Second Exec. Salary: $ Bonus: $

OTHER THOUGHTS:

Estimated Female Officers or Directors:
Hot Spot for Advancement for Women/Minorities:

Iconix Brand Group Inc

www.iconixbrand.com

NAIC Code: 316210

TYPES OF BUSINESS:

Brand Management

BRANDS/DIVISIONS/AFFILIATES:

Candie's
Joe Boxer
Mudd
London Fog
Mossimo
Ocean Pacific
Danskin
Sharper Image (The)

CONTACTS: *Note: Officers with more than one job title may be intentionally listed here more than once.*

Neil Cole, CEO
Neil Cole, Pres.
Warren Clamen, CFO/Exec. VP
Dari Marder, Chief Mktg. Officer
Lanie List, Chief Merch. Officer
Andrew Tarshis, General Counsel/Exec. VP
David Blumberg, Head-Strategic Dev./Exec. VP
Seth Horowitz, Pres., Men's Div.
Neil Cole, Chmn.

GROWTH PLANS/SPECIAL FEATURES:

Iconix Brand Group, Inc. is a brand management company engaged in the licensing, marketing and trend direction of its owned consumer brands. These brands include Mossimo, Badgley Mischka, Candie's, Bongo, Joe Boxer, Rampage, Mudd, London Fog, Ocean Pacific/OP, Danskin, Rocawear, Fieldcrest, Cannon, Royal Velvet, Charisma, Zoo York, The Sharper Image, Starter and Waverly. Mossimo is a contemporary lifestyle brand; it is licensed in the U.S. exclusively by Target for casual sportswear, denim, swimwear, body wear, watches, handbags and other fashion accessories. Badgley Mischka is a premier couture eveningwear brand. Candie's is a junior lifestyle brand in the footwear, apparel and accessories categories. Bongo offers a range of women's and children's casual apparel and accessories. Joe Boxer is an underwear, sleepwear and loungewear brand. Rampage is a contemporary women's and junior's sportswear brand. Mudd is a junior apparel brand, with a primary presence in the denim and footwear categories. London Fog is a classic brand for outerwear, cold weather accessories, umbrellas, luggage and travel products. Ocean Pacific and OP are global action-sports lifestyle brands that have roots as a 1960's surfboard label. Danskin is a brand of women's active wear and fitness equipment. Rocawear is an urban lifestyle apparel brand. Fieldcrest, Cannon, Royal Velvet and Charisma manufacture home textiles. Zoo York is an East Coast graffiti inspired action lifestyle brand. The Sharper Image offers innovative home products and consumer electronics. Starter is a brand of athletic apparel and footwear. The Waverly brand focuses on home furnishings and lifestyle products. Additionally, the firm owns interests in Hardy Way LLC, the owner of the Ed Hardy brands; and the Ecko portfolio of brands, including Mark Ecko. In 2012, the firm acquired global lifestyle brand Lee Cooper and sports apparel and footwear brand Umbro. In 2013, the firm acquired a 51% interest in the Buffalo David Bitton brand.

FINANCIAL DATA: *Note: Data for latest year may not have been available at press time.*

In U.S. $	2012	2011	2010	2009	2008	2007
Revenue	353,818,000	369,845,000	332,559,000	232,058,000	216,761,000	160,004,000
R&D Expense						
Operating Income	215,450,000	228,766,000	209,715,000	152,565,000	142,052,000	121,789,000
Operating Margin %	60.89%	61.85%	63.06%	65.74%	65.53%	76.11%
SGA Expense	138,368,000	140,985,000	138,532,000	79,356,000	73,816,000	44,254,000
Net Income	109,408,000	126,105,000	98,847,000	75,111,000	70,153,000	63,755,000
Operating Cash Flow	202,169,000	177,609,000	165,723,000	118,718,000	89,243,000	83,687,000
Capital Expenditure	2,228,000	3,810,000	3,278,000	3,873,000	7,701,000	349,000
EBITDA	237,360,000	272,719,000	216,151,000	166,485,000	166,379,000	138,770,000
Return on Assets %	4.71%	6.13%	5.26%	4.66%	5.09%	6.25%
Return on Equity %	9.21%	11.33%	10.16%	9.86%	12.29%	12.83%
Debt to Equity	0.72	0.26	0.52	0.62	0.96	1.23

CONTACT INFORMATION:

Phone: 212 730-0030 Fax: 212 391-2057
Toll-Free:
Address: 1450 Broadway, 3rd Fl., New York, NY 10018 United States

STOCK TICKER/OTHER:

Stock Ticker: ICON Exchange: NAS
Employees: 133 Fiscal Year Ends: 31-Dec
Parent Company:

SALARIES/BONUSES:

Top Exec. Salary: $1,000,000 Bonus: $3,000,000
Second Exec. Salary: $378,125 Bonus: $200,000

OTHER THOUGHTS:

Estimated Female Officers or Directors: 2
Hot Spot for Advancement for Women/Minorities: Y

Inditex (Industria de Diseno Textil SA)

www.inditex.com

NAIC Code: 315210

TYPES OF BUSINESS:

Fashion Apparel
Retail Stores
Textiles
Home Furnishings
Retail Stores

BRANDS/DIVISIONS/AFFILIATES:

Zara
Pull & Bear
Massimo Dutti
Bershka
Stradivarius
Oysho
Zara Home
Uterque

CONTACTS: *Note: Officers with more than one job title may be intentionally listed here more than once.*

Pablo Isla Alvarez de Tejera, CEO
Antonio A. Abadin, General Counsel/Corp. Sec.
Pablo Isla Alvarez de Tejera, Chmn.

GROWTH PLANS/SPECIAL FEATURES:

Inditex (Industria de Diseno Textil SA) is a Spanish conglomerate of over 100 companies engaged in textile design, production, distribution and related businesses. As one of the world's largest fashion distribution companies, it owns or leases approximately 6,009 retail stores in 86 countries under the names Zara, Pull & Bear, Massimo Dutti, Bershka, Stradivarius, Oysho, Zara Home and Uterque. Zara is the firm's largest fashion brand with 1,751 stores. Pull & Bear, which focuses on clothing for urban youth ages 14-28, has approximately 816 locations in 59 countries. Massimo Dutti, with 630 stores located in 60 countries, is designed to appeal to younger professionals. Bershka is aimed at a younger target market, with 885 stores open in 62 countries. Stradivarius boasts 780 stores in 52 countries, stocking more creative designs, fabrics and accessories. Oysho is a chain of stores focused primarily on women's lingerie and undergarments as well as casual outerwear and accessories; it currently has 524 locations in 35 countries. Zara Home is the Inditex Group's chain specializing in home furnishings, with a focus on textiles such as bed, table and bathroom linen, complemented with cutlery, glassware and decorative items. Zara Home currently features 357 locations in 35 countries. Uterque, the firm's newest store type, features fashion accessories and a select range of textile and leather garments all designed by an in-house creative team, with 92 store locations in 18 countries. The firm is famous for its lean inventory and fast-fashion strategy. It is able to get a new item of clothing from its design team, through its manufacturing plants and into its stores in as little as two weeks. Much of its manufacturing is done in Europe to reduce shipping time. In October 2012, Massimo Dutti expanded its operations to the U.S. with a flagship store in New York City. In January 2013, the firm acquired the four Zara Stores in Finland.

FINANCIAL DATA: *Note: Data for latest year may not have been available at press time.*

In U.S. $	2012	2011	2010	2009	2008	2007
Revenue	18,444,250,000	16,751,260,000	14,821,490,000	13,916,770,000		
R&D Expense						
Operating Income	3,372,568,000	3,062,943,000	2,311,297,000	2,151,024,000		
Operating Margin %	18.28%	18.28%	15.59%	15.45%		
SGA Expense	6,578,400,000					
Net Income	2,583,964,000	2,328,537,000	1,768,036,000	1,687,102,000		
Operating Cash Flow	3,220,996,000	3,418,311,000	3,103,567,000	2,373,865,000		
Capital Expenditure	1,609,620,000	923,697,500	665,335,600	1,243,703,000		
EBITDA	4,298,779,000	3,971,538,000	3,146,696,000	2,912,414,000		
Return on Assets %	18.59%	19.07%	16.31%	16.11%		
Return on Equity %	27.84%	29.36%	25.97%	26.39%		
Debt to Equity						

CONTACT INFORMATION:

Phone: 34 981185400 Fax: 34 981185544
Toll-Free:
Address: Avenida de la Diputacion, Edificio Inditex, Arteixo, 15142 Spain

STOCK TICKER/OTHER:

Stock Ticker: IDEXF
Exchange: PINX
Employees: 100,140
Fiscal Year Ends:
Parent Company:

SALARIES/BONUSES:

Top Exec. Salary: $ Bonus: $
Second Exec. Salary: $ Bonus: $

OTHER THOUGHTS:

Estimated Female Officers or Directors: 2
Hot Spot for Advancement for Women/Minorities: Y

Institute for Soldier Nanotechnologies

web.mit.edu/isn

NAIC Code: 541712N

TYPES OF BUSINESS:

Nanotechnology-Military Applications
Nanotech-Based Battle Suits
Nanotech Fabrics
Medical Nanotechnologies
Ballistic & Blast Protection Systems
Chemical & Biological Defensive Systems
Advanced Nanoscale Communications Devices
Fiber & Fabric Nanotechnologies

BRANDS/DIVISIONS/AFFILIATES:

Massachusetts Institute of Technology (MIT)
US Army Research Office

CONTACTS:

Note: Officers with more than one job title may be intentionally listed here more than once.

William A. Peters, Exec. Dir.
Amy Tatem-Bannister, Mgr.-Lab
Steven Kooi, Research Eng.
Josh Freedman, Assistant Dir.-Admin.
Franklin Hadley, Dir.-Outreach & Comm.
Josh Freedman, Assistant Dir.-Finance
Paula T. Hammond, Dir.-HBCU-MI Program
Ivan Celanovic, Research Eng.
Kurt Keville, SDC Coordinator/Research Specialist
Donna Johnson, Research Support Associate

GROWTH PLANS/SPECIAL FEATURES:

Institute for Soldier Nanotechnologies (ISN) was established in 2002 as an interdepartmental research center at MIT through a five-year, $50-million contract from the U.S. Army Research Office. ISN consists of faculty members from more than a dozen academic departments; students and post-doctoral candidates; staff members; and visitors from industry, government and academia. The goal of the initiative is to employ nanotechnology to improve the survivability of soldiers. ISN's projected goal is to create a new type of lightweight battlesuit that offers superior technology, durability and enhanced performance, replacing bulky nylon/cotton outfits and heavy packs with automatically adapting clothing the thickness of spandex and offering integrated real-time medical monitoring, high-tech communications and protection from ballistic, chemical and biological assaults. Current nanotechnologies will enable the researchers to create miniaturized communication, surveillance and sensory systems with powerful coatings for waterproofing and durable materials for enhanced protection. Research is conducted in five Strategic Research Areas (SRAs): light weight, multifunctional nanostructured fibers and materials (SRA 1); battle suit medicine (SRA 2); blast and ballistic protection (SRA 3); nuclear, analyte and IED detection and prevention (SRA 4); and nanosystems integration (SRA 5). SRA 1 focuses on embedding diverse, nano-enabled functionalities into materials for clothing and other gear, especially sensors. SRA 2 focuses on far-forward medical treatment, such as in field hospitals, utilizing nano-enabled materials and devices, including drug delivery systems, to be ultimately incorporated into a battle suit. SRA 3 researches blast interactions with human tissue and various anthropogenic energy absorbing materials, including cloth-like armors of metallic nanostructures reinforced with nano-trusses. SRA 4 focuses on enabling the detection of hazardous substances in the environment and developing means to protect soldiers from hazardous substances. SRA 5 focuses on nanoelectromechanical devices, including nanoelectronics, developing advanced nanoscale communications devices and sensing equipment.

FINANCIAL DATA:

Note: Data for latest year may not have been available at press time.

In U.S. $	2012	2011	2010	2009	2008	2007
Revenue						
R&D Expense						
Operating Income						
Operating Margin %						
SGA Expense						
Net Income						
Operating Cash Flow						
Capital Expenditure						
EBITDA						
Return on Assets %						
Return on Equity %						
Debt to Equity						

CONTACT INFORMATION:

Phone: 617-324-4700 Fax:
Toll-Free:
Address: 77 Massachusetts Ave., Bldg. NE47-4F, Cambridge, MA 02139 United States

STOCK TICKER/OTHER:

Stock Ticker: Private
Employees:
Parent Company:
Exchange:
Fiscal Year Ends: 30-Jun

SALARIES/BONUSES:

Top Exec. Salary: $ Bonus: $
Second Exec. Salary: $ Bonus: $

OTHER THOUGHTS:

Estimated Female Officers or Directors: 7
Hot Spot for Advancement for Women/Minorities: Y

Interface Inc

NAIC Code: 314110

www.interfaceglobal..com

TYPES OF BUSINESS:

Carpet and Rug Mills
Modular Carpets
Broadloom Carpet
Antimicrobial Chemicals
Carpet Installation Services

BRANDS/DIVISIONS/AFFILIATES:

Dominus Capital
TacTiles
Convert
Heuga
FLOR
InterfaceSERVICES
Intersept

CONTACTS: *Note: Officers with more than one job title may be intentionally listed here more than once.*

Daniel T. Hendrix, CEO
Daniel T. Hendrix, Pres.
Patrick C. Lynch, CFO/Sr. VP
Sanjay Lall, CIO/Sr. VP
Raymond S. Willoch, Sr. VP-Admin.
Raymond S. Willoch, General Counsel/Sec.
John R. Wells, Sr. VP-Americas Floorcovering
Robert A. Coombs, Sr. VP-Asia Pacific Floorcovering
Lindsey K. Parnell, Sr. VP-Europe Floorcoverings
Erin Meezah, VP-Sustainability
Nigel Stansfield, Chief Innovations Officer/VP

GROWTH PLANS/SPECIAL FEATURES:

Interface, Inc. is a worldwide designer, producer and seller of modular carpet for the commercial, institutional and residential markets. Its modular carpet system, marketed under the established global brands Interface and Heuga, utilizes carpet tiles cut in precise, dimensionally stable 50 cm by 50 cm squares to produce a floorcovering that combines the appearance and texture of traditional soft floorcovering with the advantages of a modular carpet system. The FLOR line specifically targets modular carpet sales to the residential market segment. Its GlasBacRE technology employs a fiberglass-reinforced polymeric composite backing that allows the tiles to be installed and remain flat on the floor without adhesives or fasteners. The Convert collection of carpet tile is designed and manufactured with yarn containing varying degrees of post-consumer nylon, depending on the style and color. Interface also manufactures and sells two-meter roll goods that are structure-backed and are often used in conjunction with carpet tiles to create special design effects. Additional products include the TacTiles carpet tile installation system; a variety of traditional adhesives and products for carpet installation and maintenance; Intercell, a branded raised/access flooring product sold in Europe; and Intersept, a proprietary antimicrobial chemical compound which the firm licenses out to other companies. Through its InterfaceSERVICES division, the firm also offers carpet installation and reclamation services for national accounts and other large customers. Interface produces its carpet products through three U.S. facilities and additional plants in the Netherlands, the U.K., Thailand and China. Products are marketed in over 110 countries. In August 2012, the company sold its Bentley Prince Street segment to Dominus Capital, ending its presence in the broadloom carpet industry.

The firm offers employees partial medical and dental plans, vision and prescription drug plans, short- and long-term disability coverage, life insurance, a 401(k), tuition reimbursement and flexible spending accounts.

FINANCIAL DATA: *Note: Data for latest year may not have been available at press time.*

In U.S. $	2012	2011	2010	2009	2008	2007
Revenue		1,057,051,000	961,827,000	859,888,000	1,082,344,000	1,081,273,000
R&D Expense						
Operating Income						
Operating Margin %						
SGA Expense						
Net Income		38,721,000	8,283,000	10,918,000	-40,873,000	-10,812,000
Operating Cash Flow						
Capital Expenditure						
EBITDA						
Return on Assets %						
Return on Equity %						
Debt to Equity						

CONTACT INFORMATION:

Phone: 770-437-6800 Fax: 770-319-0070
Toll-Free:
Address: 2859 Paces Ferry Rd., Ste. 2000, Atlanta, GA 30339 United States

STOCK TICKER/OTHER:

Stock Ticker: IFSIA
Employees: 3,566
Parent Company:
Exchange: NAS
Fiscal Year Ends: 28-Feb

SALARIES/BONUSES:

Top Exec. Salary: $ Bonus: $
Second Exec. Salary: $ Bonus: $

OTHER THOUGHTS:

Estimated Female Officers or Directors: 3
Hot Spot for Advancement for Women/Minorities: Y

International Textile Group Inc

NAIC Code: 313210

www.itg-global.com

TYPES OF BUSINESS:

Medical Textiles
Apparel Fabrics
Upholstery Fabrics
Fabric Finishing
Chemical Treatments
Denim
Automotive Safety Fabrics

BRANDS/DIVISIONS/AFFILIATES:

Narricot Industries LLC
Burlington Worldwide
Carlisle Finishing
Cone Denim
Burlington Contract Fabrics
Safety Components

CONTACTS:

Note: Officers with more than one job title may be intentionally listed here more than once.

Joseph L. Gorga, CEO
Joseph L. Gorga, Pres.
Robert E. Garren, VP-Human Resources
Robert E. Garren, VP-Corp. Comm.

GROWTH PLANS/SPECIAL FEATURES:

International Textile Group, Inc. (ITG) manufactures and supplies textiles and fabrics for automotive safety, apparel, barrier products (such as surgical masks) and interior furnishings. Operations are divided among six companies: Burlington WorldWide Apparel, Carlisle Finishing, Cone Denim, Burlington Contract Fabrics, Narricot Industries LLC and Safety Components. Burlington WorldWide provides a variety of synthetic, wool, cotton and cotton blend fabrics to the apparel industry. Carlisle Finishing provides finishing services to the apparel, military and furniture industries on a commission basis. Cone Denim is a leading producer and supplier of denim to apparel companies, with operations in the U.S., Mexico, China, Nicaragua, Turkey and India. Burlington Contract Fabrics, a manufacturer of piece-dyed fabrics from yarn to completed fabric, produces branded and unbranded antimicrobial, fire retardant and other specialty fabrics. Narricot Industries LLC manufactures woven narrow fabrics, including seatbelt webbing and other polyester and nylon webbing products for industrial, military, fall protection and commercial applications. Safety Components develops and manufactures automotive air bag fabrics and technical fabrics for fire-fighting, military and other specialty applications. ITG has manufacturing facilities in the U.S., Mexico, South Africa, Romania, Poland, Germany, Vietnam, China, the Czech Republic and Nicaragua.

FINANCIAL DATA:

Note: Data for latest year may not have been available at press time.

In U.S. $	2012	2011	2010	2009	2008	2007
Revenue	619,075,000	694,372,000	616,130,000	673,121,000	995,132,000	1,010,853,000
R&D Expense						
Operating Income	17,325,000	-8,914,000	2,921,000	-48,154,000	-186,835,000	-47,016,000
Operating Margin %	2.79%	-1.28%	.47%	-7.15%	-18.77%	-4.65%
SGA Expense	46,079,000	48,486,000	50,792,000	75,044,000	105,807,000	118,342,000
Net Income	-64,673,000	-59,140,000	-37,582,000	-202,680,000	-241,026,000	-80,160,000
Operating Cash Flow	45,923,000	-19,952,000	-14,827,000	14,107,000	-36,453,000	-78,338,000
Capital Expenditure	2,037,000	3,366,000	3,770,000	5,578,000	52,882,000	170,521,000
EBITDA	28,360,000	4,424,000	21,097,000	-106,929,000	-139,329,000	4,554,000
Return on Assets %	-23.38%	-18.39%	-13.40%	-37.49%	-29.89%	-13.44%
Return on Equity %						-78.27%
Debt to Equity						5.68

CONTACT INFORMATION:

Phone: 336 379-6220 Fax: 336 379-6287
Toll-Free:
Address: 804 Green Valley Rd., Ste. 300, Greensboro, NC 27408 United States

STOCK TICKER/OTHER:

Stock Ticker: ITXN Exchange: PINX
Employees: 8,800 Fiscal Year Ends: 31-Dec
Parent Company:

SALARIES/BONUSES:

Top Exec. Salary: $ Bonus: $
Second Exec. Salary: $ Bonus: $

OTHER THOUGHTS:

Estimated Female Officers or Directors:
Hot Spot for Advancement for Women/Minorities: Y

INVISTA

www.invista.com

NAIC Code: 313110

TYPES OF BUSINESS:

Fibers & Textiles
Polymers & Resins

BRANDS/DIVISIONS/AFFILIATES:

Koch Industries Inc
COOLMAX
CORDURA
LYCRA
DACRON
STAINMASTER
ANTRON
Ashburn Hill Corp

CONTACTS: *Note: Officers with more than one job title may be intentionally listed here more than once.*

Jeff Gentry, CEO
David Trerotola, Pres., INVISTA Apparels
David Duncan, Pres., Performance Surfaces and Materials
Warren Primeaux, Pres., INVISTA Intermediates
Robert Francois, Pres., INVISTA Specialty Materials
Jeff Gentry, Chmn.

GROWTH PLANS/SPECIAL FEATURES:

INVISTA, a subsidiary of Koch Industries, Inc., is a global producer and marketer of premium integrated fibers, resins, polymers and intermediates. The company, which operates in over 20 countries, has four segments: INVISTA Apparel; INVISTA Intermediates; INVISTA Performance Surfaces and Materials; and INVISTA Performance Technologies. INVISTA Apparel markets fibers and fabrics for ready-to-wear, intimate apparel, swimwear, activewear, legwear and career apparel. The segment's brands include LYCRA, which has recently introduced four-way stretch technology denim and improved chlorine durability for its swimwear; COOLMAX; and CORDURA. INVISTA Intermediates creates specialty intermediates that serve as the backbone in applications that utilize polyamides, polyesters, polyurethanes, plasticizers and coatings under the Terate polyols, Terathane PTMEG, Dytek Amines and DBE Dibasic Esters brand names. INVISTA Performance Surfaces and Materials provides solutions for area rugs, bedding, sheets, pillows, comforters, airbags, automotive flooring and industrial nylon, including such brands as DACRON, STAINMASTER, ANTRON and CORDURA. INVISTA Performance Technologies handles the company's technology licensing activities. In December 2012, the firm acquired a manufacturing facility from VSL Born BV. In March 2013, the firm acquired flame resistant apparel manufacturer, Ashburn Hill Corp.

FINANCIAL DATA: *Note: Data for latest year may not have been available at press time.*

In U.S. $	2012	2011	2010	2009	2008	2007
Revenue						
R&D Expense						
Operating Income						
Operating Margin %						
SGA Expense						
Net Income						
Operating Cash Flow						
Capital Expenditure						
EBITDA						
Return on Assets %						
Return on Equity %						
Debt to Equity						

CONTACT INFORMATION:

Phone: 316-828-1000 Fax: 316-828-1801
Toll-Free: 877-446-8478
Address: 4123 E. 37th St. N., Wichita, KS 67220 United States

STOCK TICKER/OTHER:

Stock Ticker: Subsidiary Exchange:
Employees: Fiscal Year Ends: 31-Dec
Parent Company: KOCH INDUSTRIES INC

SALARIES/BONUSES:

Top Exec. Salary: $ Bonus: $
Second Exec. Salary: $ Bonus: $

OTHER THOUGHTS:

Estimated Female Officers or Directors:
Hot Spot for Advancement for Women/Minorities: Y

ITOCHU Corp

NAIC Code: 313210

www.itochu.co.jp

TYPES OF BUSINESS:

Textiles & Fabrics Manufacturing
Brand Development
Machinery
Mining & Processing Minerals
Food Production
Electronics Components
Construction Equipment
Financial Services

BRANDS/DIVISIONS/AFFILIATES:

METSA FIBRE Oy
SKINS
Toyo Advanced Technologies Co Ltd
Agromate Holdings Sdn Bhd
Bramhope Group Holdings Ltd
CSC ESI Sdn Bhd
CSC Automated Pte Ltd
HyLife Group Holdings Ltd

CONTACTS: *Note: Officers with more than one job title may be intentionally listed here more than once.*

Masahiro Okafuji, CEO
Masahiro Okafuji, Pres.
Tadayuki Seki, CFO
Fumihiko Kobayashi, Gen. Mgr.-Human Resources Div.
Toru Matsushima, Chief Admin. Officer
Koji Takayanagi, Chief Strategy Officer
Kimio Kitamura, Gen. Mgr.-Gen. Acct. Control Div.
Ichiro Nakamura, Pres., Metals & Minerals Company
Hitoshi Okamoto, Pres., Textile Company
Takao Shiomi, Pres., Machinery Company
Masahiro Imai, CEO-ITOCHU Europe PLC
Eizo Kobayashi, Chmn.
Eiichi Yonekura, CEO/Pres., ITOCHU Int'l, Inc.

GROWTH PLANS/SPECIAL FEATURES:

ITOCHU Corp. is a globally integrated Japanese trading company. It is one of the world's largest corporations, with 9 offices in Japan and 114 overseas offices in approximately 66 countries. The firm has seven division companies which operate approximately 170 domestic and 243 overseas subsidiaries and associates. The Textile division markets raw materials for textile fabrics and garments; manufactures men's and women's casual wear, underwear, uniforms and sportswear; provides commodity materials for industrial applications; and owns the Japanese trademark rights for Converse-branded apparel, accessories and shoes. The Machinery division provides equipment for power plants, ships, automobiles and construction equipment. The Energy & Chemicals division facilitates the development and trade of crude oil and natural gas in addition to handling battery-related materials, synthetic resin, basic chemicals and other chemical products. The Metals & Minerals division trades, mines and/or produces basic materials made from iron, aluminum, coal and nuclear fuels. The Food division handles all aspects of food production, including foodstuffs procurement through processing, distribution and retail, and deals in grains, ingredients for animal feed, oilseed, vegetable oil, sugar, fruit juice, coffee beans and livestock products. The ICT (Information & Communication Technology) General Products & Realty Company division is involved in IT solutions; internet-related services; green technology; building products; soft wood pulp production; rubber; and the development, sale and operation/management of residential houses, industrial facilities, office buildings and commercial facilities. In 2012, the firm acquired 24.9% of METSA FIBRE Oy; the trade mark for the SKINS brand in five regions in Asia; a majority stake in Toyo Advanced Technologies Co., Ltd.; 25% of Agromate Holdings Sdn Bhd; and Bramhope Group Holdings Ltd. In 2013, ITOCHU Corp. acquired CSC ESI Sdn Bhd of Malaysia; CSC Automated Pte. Ltd. of Singapore; Dole Food Company's packaged fresh and Asia fresh produce businesses; and a 33.4% stake in HyLife Group Holdings Ltd.

FINANCIAL DATA: *Note: Data for latest year may not have been available at press time.*

In U.S. $	2012	2011	2010	2009	2008	2007
Revenue	47,550,710,000	40,632,710,000	38,038,290,000	38,065,270,000	31,854,580,000	29,471,970,000
R&D Expense						
Operating Income	3,035,148,000	2,851,026,000	1,641,198,000	2,741,085,000	2,967,279,000	3,001,024,000
Operating Margin %	6.38%	7.01%	4.31%	7.20%	9.31%	10.18%
SGA Expense	8,382,249,000	8,637,301,000	8,571,570,000	8,551,619,000	8,053,829,000	7,115,496,000
Net Income	3,345,598,000	1,792,175,000	1,426,760,000	1,841,329,000	2,433,562,000	1,971,243,000
Operating Cash Flow	2,369,490,000	3,750,437,000	3,288,496,000	3,082,286,000	729,807,000	2,626,524,000
Capital Expenditure	1,478,730,000	906,592,000	1,059,028,000	1,460,560,000	1,322,631,000	881,297,200
EBITDA	4,913,439,000	3,167,911,000	2,968,259,000	3,551,018,000	4,500,740,000	4,474,989,000
Return on Assets %	4.93%	2.88%	2.40%	3.16%	4.15%	3.51%
Return on Equity %	23.86%	14.28%	13.15%	18.09%	23.31%	21.81%
Debt to Equity	1.65	1.87	1.91	2.27	1.93	2

CONTACT INFORMATION:

Phone: 81 662412121 Fax:
Toll-Free:
Address: 5-1 Kita-Aoyama 2-chome, Minato-ku, Tokyo, 107-8077 Japan

STOCK TICKER/OTHER:

Stock Ticker: ITOCF Exchange: GREY
Employees: 62,635 Fiscal Year Ends: 31-Mar
Parent Company:

SALARIES/BONUSES:

Top Exec. Salary: $ Bonus: $
Second Exec. Salary: $ Bonus: $

OTHER THOUGHTS:

Estimated Female Officers or Directors:
Hot Spot for Advancement for Women/Minorities:

J C Penney Company Inc

www.jcpenney.com

NAIC Code: 452111

TYPES OF BUSINESS:

Department Stores
Online & Catalog Sales
Optometry
Photography Services
Salons
Custom Decorating

BRANDS/DIVISIONS/AFFILIATES:

Original Arizona Jean Company (The)
JCPenney Optical Services
JCPenney Portraits
JCPenney Salon
JCPenney Custom Decorating
Monet
Worthington
St. John's Bay

CONTACTS: *Note: Officers with more than one job title may be intentionally listed here more than once.*

Ron Johnson, CEO
Michael Kramer Michael Kramer, COO
Ken Hannah, CFO
Dan Walker, Chief Talent Officer
Kristen Blum, CTO/Exec. VP
Ken Mangone, Exec. VP-Prod. Dev., Design & Sourcing
Janet L. Dhillon, General Counsel/Exec. VP/Sec.
Benjamin Fay, Exec. VP-Real Estate, Store Design & Dev.
Laurie Beja Miller, Exec. VP-The Square
Liz Sweney, Chief Merchant
Tim Nichols, Exec. VP-JC Penney Stores
Thomas Engibous, Chmn.
John Singleton, Sr. VP-Supply Chain

GROWTH PLANS/SPECIAL FEATURES:

J.C. Penney Company, Inc. is a holding company for J.C. Penney Corp., Inc., a department store retailer. J.C. Penney provides merchandise and services through department stores, catalogs and the Internet. The company operates 1,102 JCPenney department stores in 49 U.S. states and Puerto Rico. The firm's major products include family apparel, jewelry, shoes, accessories and home furnishings. J.C. Penney's entire product offering is available online at JCPenny.com. Local stores receive revenue credit for online and catalog sales made within their regions. Company brands include Every Day Matters, Worthington, St. John's Bay, Okie Dokie, she said, The Original Arizona Jean Company, Ambrielle, J. Ferrar, JCPenney Home Collection, Studio by JCPenney Home Collection, Decree, Linden Street, Article 365 and Stafford. Other services offered by the company are JCPenney Optical Services, JCPenney Portraits, JCPenney Salon and JCPenney Custom Decorating all of which can be found within JCPenney department stores. Additionally, the firm sells beauty products through Sephora stores at 231 J.C. Penney locations. J.C. Penney purchases its merchandise from more than 2,500 U.S.-based and international suppliers. The company has 13 merchandise distribution facilities in 13 states. It also maintains five regional warehouses, five furniture distribution centers and four Internet/catalog order fulfillment centers.

FINANCIAL DATA: *Note: Data for latest year may not have been available at press time.*

In U.S. $	2012	2011	2010	2009	2008	2007
Revenue	17,260,000,000	17,759,000,000	17,556,000,000	18,486,000,000	19,860,000,000	
R&D Expense						
Operating Income	-2,000,000	832,000,000	663,000,000	1,135,000,000	1,888,000,000	
Operating Margin %	-.01%	4.68%	3.77%	6.13%	9.50%	
SGA Expense	5,251,000,000	5,617,000,000	5,747,000,000	5,336,000,000	5,403,000,000	
Net Income	-152,000,000	389,000,000	251,000,000	572,000,000	1,111,000,000	
Operating Cash Flow	820,000,000	592,000,000	1,574,000,000	1,155,000,000	1,252,000,000	
Capital Expenditure	634,000,000	499,000,000	600,000,000	969,000,000	1,243,000,000	
EBITDA	516,000,000	1,323,000,000	1,158,000,000	1,604,000,000	2,302,000,000	
Return on Assets %	-1.24%	3.03%	2.04%	4.34%	8.23%	
Return on Equity %	-3.21%	7.59%	5.61%	12.08%	23.14%	
Debt to Equity	0.71	0.56	0.62	0.84	0.65	

CONTACT INFORMATION:

Phone: 972 431-1000 Fax: 972 591-9322
Toll-Free: 800-322-1189
Address: 6501 Legacy Dr., Plano, TX 75024 United States

STOCK TICKER/OTHER:

Stock Ticker: JCP Exchange: NYS
Employees: 156,000 Fiscal Year Ends: 31-Jan
Parent Company:

SALARIES/BONUSES:

Top Exec. Salary: $624,621 Bonus: $2,000,000
Second Exec. Salary: $1,500,000 Bonus: $

OTHER THOUGHTS:

Estimated Female Officers or Directors: 7
Hot Spot for Advancement for Women/Minorities: Y

J Crew Group Inc

www.jcrew.com

NAIC Code: 448140

TYPES OF BUSINESS:

Apparel, Retail
Catalog Sales
Online Sales
Outlet Stores
Private-Label Credit Card

BRANDS/DIVISIONS/AFFILIATES:

J.Crew
crewcuts
Madewell
Leonard Green & Partners LP
TPG Capital
jcrew.com
madewell.com
jcrew.com/factory

CONTACTS: *Note: Officers with more than one job title may be intentionally listed here more than once.*

Millard Drexler, CEO
Jenna Lyons, Pres./Exec. Creative Dir.
Stuart Haselden, CFO
Lynda Markoe, Exec. VP-Human Resources
James S. Scully, Chief Admin. Officer
Jennifer L. O'Connor, General Counsel/Sr. VP/Sec.
Libby Wadle, Exec. VP-J. Crew
Millard Drexler, Chmn.

GROWTH PLANS/SPECIAL FEATURES:

J.Crew Group, Inc. is a multi-channel retailer of apparel and accessories for men, women and children. Its target customer base consists primarily of affluent, college-educated, professional and fashion-conscious men and women. Men's offerings include sweaters, shirts, suits, ties, underwear, coats, swimwear, belts, shoes and accessories; women's offerings are similar with the additions of dresses, skirts and sleepwear. The firm's crewcuts brand features similar J.Crew styles designed for the children's market, from toddler size 2 to children size 14. J.Crew also sells women's merchandise under the Madewell brand, which focuses on vintage-inspired denim, apparel and accessories. Its operations are split into two primary sales channels: retail and direct. J.Crew's retail operations include 279 retail stores (including eight crewcuts and 39 Madewell stores), 97 factory outlet stores (including three crewcuts factory stores) and three clearance stores throughout the U.S. and Canada. Stores are generally located in regional malls, lifestyle centers, shopping centers and street locations. During 2011, the retail division accounted for approximately 70% of its revenue. The company's direct marketing division consists of its catalog and e-commerce operations. J.Crew publishes about 20 editions of its catalog per year, with an annual circulation of approximately 36 million. Its e-commerce business includes jcrew.com, madewell.com and jcrew.com/factory. Additionally, the firm offers a private-label credit card through an agreement with Comenity Bank. In 2011, sales on J.Crew credit cards accounted for 17% of its revenue. The company utilizes a 282,000-square-foot distribution center in Asheville, North Carolina, servicing retail operations, and a 262,000-square-foot customer contact center in Lynchburg, Virginia, for J.Crew direct operations. Most company merchandise is produced abroad, primarily in Asia. In 2011, J.Crew was acquired by affiliates of private equity firms TPG Capital and Leonard Green & Partners LP for $3 billion.

FINANCIAL DATA: *Note: Data for latest year may not have been available at press time.*

In U.S. $	2012	2011	2010	2009	2008	2007
Revenue	2,222,770,000	1,854,988,000	1,722,227,000	1,427,970,000	1,334,723,000	1,152,100,000
R&D Expense						
Operating Income						
Operating Margin %						
SGA Expense						
Net Income	96,100,000	51,514,000	121,505,000	54,117,000	97,075,000	77,782,000
Operating Cash Flow						
Capital Expenditure						
EBITDA						
Return on Assets %						
Return on Equity %						
Debt to Equity						

CONTACT INFORMATION:

Phone: 212-209-2500 Fax: 212-209-2666
Toll-Free: 800-562-0258
Address: 770 Broadway, New York, NY 10003 United States

STOCK TICKER/OTHER:

Stock Ticker: Private Exchange:
Employees: 8,000 Fiscal Year Ends: 31-Jan
Parent Company: TPG CAPITAL

SALARIES/BONUSES:

Top Exec. Salary: $ Bonus: $
Second Exec. Salary: $ Bonus: $

OTHER THOUGHTS:

Estimated Female Officers or Directors: 4
Hot Spot for Advancement for Women/Minorities: Y

J Jill Group Inc (The)

www.jjill.com

NAIC Code: 448120

TYPES OF BUSINESS:

Women's Apparel, Retail
Online & Catalog Sales
Retail Stores
Accessories
Footwear
Credit Cards

BRANDS/DIVISIONS/AFFILIATES:

Golden Gate Capital
J Jill
J Jill Compassion Fund
JJill.com

CONTACTS: *Note: Officers with more than one job title may be intentionally listed here more than once.*

Paula Bennett, CEO
Paula Bennett, Pres.
Hilary Chasin, Chief Mktg. Officer/Sr. VP
Mark Campton, VP-Visual Merch.
Mara D. Calame, General Counsel/VP

GROWTH PLANS/SPECIAL FEATURES:

The J. Jill Group, Inc. (J. Jill), owned by Golden Gate Capital, is a multi-channel specialty retailer of women's apparel, accessories and footwear, ranging from career to weekend wear, targeting a customer base of active, affluent women aged 35 and older. The firm operates approximately 200 retail locations including both department and outlet center locations throughout the U.S. Consumers have the option of making purchases through the store locations, catalogue or online at JJill.com. J. Jill's products include sweaters, shirts, knit tees and tops, camisoles, pants, shorts, skirts, jackets, vests, dresses, shoes, jewelry, handbags and belts. In order to assure that all sizes and shapes of women are served, it carries sizes such as petites, misses, woman and tall. The brand also offers a private label credit card for direct and retail purchases. The merchandise, sold predominantly under the company's private label, is designed to be simple and comfortable with an emphasis on natural fibers, neutral colors and creative details. The firm gives back to the community through donations to the J. Jill Compassion Fund, which primarily funds organizations geared toward aiding women and their children. The company is headquartered in Quincy, Massachusetts and has a distribution center in Tilton, New Hampshire.

Employees are offered medical, dental, life and disability coverage; a 401(k); flexible spending accounts; service awards; paid time off; and employee assistance programs.

FINANCIAL DATA: *Note: Data for latest year may not have been available at press time.*

In U.S. $	2012	2011	2010	2009	2008	2007
Revenue						
R&D Expense						
Operating Income						
Operating Margin %						
SGA Expense						
Net Income						
Operating Cash Flow						
Capital Expenditure						
EBITDA						
Return on Assets %						
Return on Equity %						
Debt to Equity						

CONTACT INFORMATION:

Phone: 617-376-4300 Fax: 603-266-2802
Toll-Free: 800-343-5700
Address: 4 Batterymarch Park, Quincy, MA 02169 United States

STOCK TICKER/OTHER:

Stock Ticker: Private Exchange:
Employees: Fiscal Year Ends: 31-Dec
Parent Company: GOLDEN GATE CAPITAL

SALARIES/BONUSES:

Top Exec. Salary: $ Bonus: $
Second Exec. Salary: $ Bonus: $

OTHER THOUGHTS:

Estimated Female Officers or Directors: 3
Hot Spot for Advancement for Women/Minorities: Y

Jaclyn Inc

NAIC Code: 315240

www.jaclyninc.com

TYPES OF BUSINESS:

Women's & Children's Apparel
Handbags & Accessories

BRANDS/DIVISIONS/AFFILIATES:

Topsville

CONTACTS:

Note: Officers with more than one job title may be intentionally listed here more than once.

Robert Chestnov, CEO
Robert Chestnov, Pres.
Anthony Christon, CFO
Amanda Cahill, VP-Merch. & Prod. Dev., Jaclyn Apparel Div.
Jaclyn Hartstein, Sec.
Anthony Christon, Chief Acct. Officer
Peggy Hoy, Pres., Premium Incentives Div.
Lori Nitzberg, Exec. VP-Topsville Div.
Gina Labartino, VP-Prod. Dev., Bonnie Int'l Div.
Allan Ginsburg, Chmn.

GROWTH PLANS/SPECIAL FEATURES:

Jaclyn, Inc. is engaged in the design, distribution and sale of women's and children's apparel as well as vinyl, leather and fabric handbags, sports bags, backpacks, cosmetic bags and related products. Jaclyn operates through four divisions. The Jaclyn Apparel division develops fabrics as well as designing men's and women's sleepwear, including bathrobes and loungewear, for department stores, mid-level retailers and national chains. The Topsville division designs and merchandises private label children's clothing for newborns to size 16 or 20, depending on gender. Topsville also produces licensed clothing from Disney, Angelina Ballerina, Thomas and Friends, Wishes and Kisses and Messages from the Heart. The Bonnie International division designs and markets private label handbags, backpacks, accessories and related products. Its clients have included leading retailers such as Macy's, Wal-Mart, JC Penny, Target and Kohl's. The firm's Premium Incentive division partners with businesses to create rewards programs, including firms in the cosmetics, medical, retail, accessory, apparel and fragrance industries. The division designs packaging for promotional materials and offers other creative consultations. Its partners have included leading beauty brands such as Estee Lauder, L'Oreal and Roche Diagnostics. Apparel and accessory lines are marketed to department and other retail stores as well as to major mail-order catalogs and other specialty retailers. All of the company's products are manufactured by outside contractors.

FINANCIAL DATA:

Note: Data for latest year may not have been available at press time.

In U.S. $	2012	2011	2010	2009	2008	2007
Revenue	195,927,000	196,020,000	106,167,000	166,913,000	151,304,000	154,507,000
R&D Expense						
Operating Income	1,328,000	3,306,000	136,000	3,449,000	873,000	5,687,000
Operating Margin %	.67%	1.68%		2.06%	.57%	3.68%
SGA Expense	35,431,000	38,324,000	26,664,000	35,570,000	32,052,000	30,573,000
Net Income	2,954,000	1,940,000	71,000	2,010,000	362,000	709,000
Operating Cash Flow	16,856,000	-15,975,000	7,445,000	3,315,000	-8,539,000	9,891,000
Capital Expenditure	517,000	229,000	277,000	646,000	263,000	3,158,000
EBITDA	5,995,000	4,336,000	818,000	4,645,000	1,904,000	2,971,000
Return on Assets %	5.09%	2.95%		3.95%	.84%	1.77%
Return on Equity %	12.30%	8.63%		11.00%	1.98%	3.70%
Debt to Equity		0.03		0.1	0.12	0.12

CONTACT INFORMATION:

Phone: 201 868-9400 Fax: 201 854-7202
Toll-Free:
Address: 197 W. Spring Valley Ave., West New York, NJ 07607 United States

STOCK TICKER/OTHER:

Stock Ticker: JCLY
Exchange: PINX
Employees: 161
Fiscal Year Ends: 30-Jun
Parent Company:

SALARIES/BONUSES:

Top Exec. Salary: $ Bonus: $
Second Exec. Salary: $ Bonus: $

OTHER THOUGHTS:

Estimated Female Officers or Directors: 5
Hot Spot for Advancement for Women/Minorities: Y

Jil Sander SpA

www.jilsander.com

NAIC Code: 315240

TYPES OF BUSINESS:

Apparel, Retail
Women's & Men's Apparel
Eyewear
Accessories
Fragrances
Footwear
Jewelry
Small Leather Goods

BRANDS/DIVISIONS/AFFILIATES:

Onward Holdings Co Ltd

CONTACTS:

Note: Officers with more than one job title may be intentionally listed here more than once.

Alessandro Cremonesi, CEO
Jil Sander, Creative Dir.

GROWTH PLANS/SPECIAL FEATURES:

Jil Sander SpA is a fashion house that designs, manufactures and markets high-fashion men's and women's clothing lines. It is currently owned by Japanese apparel conglomerate Onward Holdings Co., Ltd. With foundations in Germany in the late â€˜60s, the company is known for its finely tailored, minimalist approach to fashion design. Its products include men's and women's fashions, handbags, belts, jewelry, sunglasses, fragrances and footwear. The firm has recently launched a new line of women's wear, Jil Sander Navy, with a focus on more causal pieces and a more moderate price point than its flagship collection. In addition to its ready-to-wear collections, the firm provides sartorial, made-to-measure services for men's suiting and dress shirts. The firm sells its products through flagship stores, freestanding stores and store-within-a-store concepts inside department stores. Additionally, the firm operates an e-commerce web site, JilSander.com. Jil Sander has a presence in Germany, France, the U.K., Switzerland, Italy, Taiwan, Singapore, China, Hong Kong, South Korea, Japan and the U.S. Through strategic licenses with various manufacturers, the company sells jewelry, eyewear, underwear and beachwear. In February 2012, Jil Sander, who founded the company in 1968, returned to the company as creative director after eight years away from the firm.

FINANCIAL DATA:

Note: Data for latest year may not have been available at press time.

In U.S. $	2012	2011	2010	2009	2008	2007
Revenue						
R&D Expense						
Operating Income						
Operating Margin %						
SGA Expense						
Net Income						
Operating Cash Flow						
Capital Expenditure						
EBITDA						
Return on Assets %						
Return on Equity %						
Debt to Equity						

CONTACT INFORMATION:

Phone: 39-02-80-69-131 Fax: 39-02-80-69-13-364
Toll-Free:
Address: Via Beltrami, 5, Milan, 20121 Italy

STOCK TICKER/OTHER:

Stock Ticker: Subsidiary Exchange:
Employees: Fiscal Year Ends: 31-Dec
Parent Company: ONWARD HOLDINGS CO LTD

SALARIES/BONUSES:

Top Exec. Salary: $ Bonus: $
Second Exec. Salary: $ Bonus: $

OTHER THOUGHTS:

Estimated Female Officers or Directors: 1
Hot Spot for Advancement for Women/Minorities:

Jo Ann Stores Inc

NAIC Code: 451130

www.joann.com

TYPES OF BUSINESS:

Fabrics & Sewing Accessories, Retail
Crafts Equipment
Home Decorating Items
Seasonal Merchandise
Instructional Classes
Online Sales

BRANDS/DIVISIONS/AFFILIATES:

Jo-Ann Fabric and Craft
Leonard Green & Partners LP

CONTACTS: *Note: Officers with more than one job title may be intentionally listed here more than once.*

Travis Smith, CEO
Travis Smith, Pres.
James Kerr, CFO/Exec. VP
Riddiane Kline, Sr. VP-Mktg. & Advertising
Thomas Williams, Sr. VP-Human Resources
Matt Susz, CIO/Sr. VP
Keith Kotel, VP-Prod. Dev. & Sourcing
David Goldston, General Counsel/Sr. VP/Sec.
Kenneth Haverkost, Exec. VP-Store Oper.
Richard Chapman, VP-Store Dev. & Facilities
Margaret Skinner, Dir.-Corp. Communications
Renee Jefferson, Treas./VP
Carolyn Tackett, Regional VP-California Region
Diana Young, Regional VP- Northeast Region
Roger Hawkins, Regional VP-Southeast Region
Edward Dann, Regional VP-Central Region
Darrell Webb, Chmn.
Scott Clark, VP-Supply Chain

GROWTH PLANS/SPECIAL FEATURES:

Jo-Ann Stores, Inc. owned by Leonard Green & Partners L.P., is one of the largest fabric and craft retailers in the U.S. The company's 790 retail locations operate under the Jo-Ann Fabric and Craft name. The stores, located throughout 49 states, offer a wide variety of competitively priced items, including fashion, decorator, quilting and craft fabrics; notions; patterns; craft components; and silk and dried flowers. Larger Jo-Ann's stores, offer an expanded and more comprehensive product assortment, including such services as custom framing and educational programs. The company divides its products in two categories: sewing, which consists of the firm's principal fabric business; and non-sewing components, which include yarn and knitting accessories, paper crafting components, brand name fine art materials such as painting supplies, floral products, an assortment of craft related books, frames and framing equipment and home decor accessories. As a way to draw in new customers and maintain customer loyalty, the company also offers classes in sewing and other crafts. In addition to its retail stores, Jo-Ann operates a web site, which features a community forum, craft projects and ideas and exclusive products only available online.

The company offers employees medical, dental and vision insurance; healthcare and dependent care spending accounts; an employee assistance program; disability coverage; life insurance; discounted home and auto insurance; family and medical leave; a 401(k) plan; and a college savings plan.

FINANCIAL DATA: *Note: Data for latest year may not have been available at press time.*

In U.S. $	2012	2011	2010	2009	2008	2007
Revenue	2,150,000,000	2,079,000,000	1,990,700,000	1,901,100,000	1,878,800,000	1,850,600,000
R&D Expense						
Operating Income						
Operating Margin %						
SGA Expense						
Net Income		93,100,000	66,600,000	21,900,000	15,400,000	-1,900,000
Operating Cash Flow						
Capital Expenditure						
EBITDA						
Return on Assets %						
Return on Equity %						
Debt to Equity						

CONTACT INFORMATION:

Phone: 330-656-2600 Fax: 330-463-6760
Toll-Free: 888-739-4120
Address: 5555 Darrow Rd., Hudson, OH 44236 United States

STOCK TICKER/OTHER:

Stock Ticker: Private Exchange:
Employees: 21,453 Fiscal Year Ends: 31-Jan
Parent Company: LEONARD GREEN & PARTNERS LP

SALARIES/BONUSES:

Top Exec. Salary: $ Bonus: $
Second Exec. Salary: $ Bonus: $

OTHER THOUGHTS:

Estimated Female Officers or Directors: 6
Hot Spot for Advancement for Women/Minorities: Y

Joe's Jeans Inc

www.joesjeans.com

NAIC Code: 315240

TYPES OF BUSINESS:

Apparel Design-Jeans & Denim

BRANDS/DIVISIONS/AFFILIATES:

Rocker
Shirt (The)
Joe's
Chelsea
Cigarette
Honey
Ex-Lover
Twiggy

CONTACTS: *Note: Officers with more than one job title may be intentionally listed here more than once.*

Marc B. Crossman, CEO
Marc B. Crossman, Pres.
Hamish S. Sandhu, CFO
Joseph Dahan, Dir.-Creative
Samuel J. (Sam) Furrow, Chmn.
Sarah Choufani, VP-Intl. Sales

GROWTH PLANS/SPECIAL FEATURES:

Joe's Jeans, Inc. designs, develops and markets Joe's brand denim jeans, casual wear and accessories. The company was formed through the merger of Innovo, Inc. and JD Holdings. Joe's Jeans relies on third parties in Mexico, the U.S., Morocco, Hong Kong, India, Portugal, Peru, Korea and China to manufacture its apparel products. The firm's products include women's, men's and children's denim jeans, pants, sweaters and jackets. Its women's product line accounts for the largest source of revenue and consists primarily of denim jeans in a variety of different fits, fabrics, washes and detailing. Joe's Jeans has branded its different women's fit styles, which include Chelsea (ultra slim), Cigarette (straight and narrow), Visionaire (high waist bootcut), Stardust (super flare), Honey (curvy), Ex-Lover (relaxed), Icon Muse (high waist), Provocateur (petite), Rocker (lean flare), Socialite (classic bootcut), Starlet (slim bootcut) and Twiggy (tall). The company has also branded its men's fit styles, which include Rebel, Brixton, Outsider, Classic and the Rocker. Children's product offerings include basic denim bottoms, tops, t-shirts and jackets for infants', toddlers', girls' and boys' sizes. Additionally, the company offers a line of unisex woven shirts marketed under The Shirt brand name. Product development for the Joe's brand is managed internally by a team of designers led by Joe Dahan. Joe's Jeans sells its products to retailers such as Bloomingdale's, Macy's, Neiman Marcus, Nordstrom, Saks Fifth Avenue, Dillard's, American Rag, Anthropologie, Bergdorf Goodman, Scoop NYC, Galleries Lafayette, Le Bon Marche, Barney's Japan and Gio Moretti. The company also operates 11 Joe's stores and 19 outlet stores. In February 2012, the company launched a new, affordable denim line marketed for young women called else, exclusively sold in Macy's store across the U.S.

FINANCIAL DATA: *Note: Data for latest year may not have been available at press time.*

In U.S. $	2012	2011	2010	2009	2008	2007
Revenue	118,642,000	95,420,000	98,176,000	80,116,000	69,174,000	62,767,000
R&D Expense						
Operating Income	10,717,000	-565,000	6,021,000	8,523,000	6,120,000	3,186,000
Operating Margin %	9.03%	- .59%	6.13%	10.63%	8.84%	5.07%
SGA Expense	43,997,000	41,617,000	39,349,000	30,726,000	26,161,000	23,085,000
Net Income	5,565,000	-1,365,000	2,601,000	24,520,000	4,902,000	2,254,000
Operating Cash Flow	5,782,000	10,606,000	-4,149,000	10,599,000	4,587,000	-7,471,000
Capital Expenditure	2,779,000	2,055,000	3,402,000	873,000	2,508,000	790,000
EBITDA	12,173,000	603,000	6,864,000	9,059,000	6,470,000	3,532,000
Return on Assets %	6.69%	-1.68%	3.22%	35.71%	9.31%	7.58%
Return on Equity %	8.15%	-2.10%	4.11%	50.25%	14.74%	13.37%
Debt to Equity						

CONTACT INFORMATION:

Phone: 323 837-3700 Fax:
Toll-Free: 877-528-5637
Address: 2340 S. Eastern Ave., Commerce, CA 90040 United States

STOCK TICKER/OTHER:

Stock Ticker: JOEZ
Employees: 320
Parent Company:
Exchange: NAS
Fiscal Year Ends: 30-Nov

SALARIES/BONUSES:

Top Exec. Salary: $462,000 Bonus: $384,000
Second Exec. Salary: $317,000 Bonus: $

OTHER THOUGHTS:

Estimated Female Officers or Directors: 3
Hot Spot for Advancement for Women/Minorities: Y

Johnson Service Group plc

www.jsg.com

NAIC Code: 812331

TYPES OF BUSINESS:

Uniforms & Linen Rentals
Apparel Manufacturing
Hospitality Support Services
Dry Cleaning
Facilities Management

BRANDS/DIVISIONS/AFFILIATES:

Johnsons Apparelmaster
Johnson Cleaners
Stalbridge Linen Services
SGP Property & Facilities Management Limited
Jeeves of Belgravia
Cannon Textile Care
Nickleby & Co

CONTACTS: *Note: Officers with more than one job title may be intentionally listed here more than once.*

John Talbot, CEO
Yvonne Monaghan, Dir.-Finance
Yvonne Monaghan, Corp. Sec.
Christopher Sander, Managing Dir.-Johnson Apparelmaster
Kevin P. Elliott, Managing Dir.-SGP Property & Facilities Mgmt.
John Talbot, Chmn.

GROWTH PLANS/SPECIAL FEATURES:

Johnson Service Group plc, based in the U.K., deals in a range of managed products and services relating to the textile industry. From a traditional base in textile-related services, including work-wear and linen rental, garment sourcing and dry cleaning, the company has developed an expanded offering in the growing hospitality services sector. The firm offers three primary services: dry cleaning, textile rental and facilities management. Its dry cleaning services include Johnson Cleaners, one of the nation's largest dry cleaning, specialist textile and garment aftercare service chains, with stores located across the U.K.; and Jeeves of Belgravia, a London-based premium provider of men's and women's personal valet services, including apparel laundering, curtain restoration, repairs and alterations, rug and carpet cleaning, shoe and luggage care, stain removal and haute-couture clothing care. Johnson's textile rental services are conducted through Johnsons Apparelmaster, Britain's leading work wear rental business, which provides sourcing supply and aftercare services in addition to a wide range of workplace clothing; and Stalbridge Linen Services, a provider of laundry services to customers in the hospitality, catering and leisure markets. The company's facilities management services are offered through SGP Property & Facilities Management Limited, which provides integrated property, building, maintenance and facilities management services to clients in the retail, hospitality, health care, education and corporate sectors, managing over 40,000 locations throughout the U.K. and Ireland. In mid-2012, after completing a strategic review of its businesses, Johnson disclosed plans to close approximately 100 of its dry cleaning retail locations. In December 2012, the company sold Alex Reid Ltd., a provider of wholesale cleaning agents, machinery and equipment for the clothes care sector. In the same year, it acquired Cannon Textile Care, a textile rental company, and Nickleby & Co., a facilities management firm.

FINANCIAL DATA: *Note: Data for latest year may not have been available at press time.*

In U.S. $	2012	2011	2010	2009	2008	2007
Revenue		386,292,000	372,250,000	375,360,000	397,130,000	453,284,000
R&D Expense						
Operating Income						
Operating Margin %						
SGA Expense						
Net Income		16,416,000	5,240,000	18,660,000	-9,990,000	-47,360,000
Operating Cash Flow						
Capital Expenditure						
EBITDA						
Return on Assets %						
Return on Equity %						
Debt to Equity						

CONTACT INFORMATION:

Phone: 44-1928-704-600 Fax: 44-1928-704-620
Toll-Free:
Address: Abbots Park, Monks Way, Preston Brook, Cheshire, WA7 3GH United Kingdom

STOCK TICKER/OTHER:

Stock Ticker: JSG
Exchange: London
Employees: 5,577
Fiscal Year Ends: 31-Dec
Parent Company:

SALARIES/BONUSES:

Top Exec. Salary: $ Bonus: $
Second Exec. Salary: $ Bonus: $

OTHER THOUGHTS:

Estimated Female Officers or Directors: 1
Hot Spot for Advancement for Women/Minorities:

Johnston Textiles Inc

www.johnstontextiles.com

NAIC Code: 313210

TYPES OF BUSINESS:

Textiles Manufacturing
Specialty & Industrial Fabrics

BRANDS/DIVISIONS/AFFILIATES:

Southern Phenix Textiles
Shawmut Finishing Complex
Omega
Meriturn Partners
TexTest
Wellington Decorative Fabrics
Chef Check
Meriturn Partners

CONTACTS: *Note: Officers with more than one job title may be intentionally listed here more than once.*

Tommy Cooper, CEO
Tommy Cooper, Pres.
Grover Elliott, CFO/Exec. VP

GROWTH PLANS/SPECIAL FEATURES:

Johnston Textiles, Inc. manufactures value-added textiles for a vast array of markets. The company's manufacturing operations are located exclusively in the state of Alabama. Its facilities include Southern Phenix Textiles in Phenix City, Alabama; and the Shawmut Finishing Complex and distribution operation in Valley, Alabama. Johnston's brands include Wellington Decorative Fabrics, Omega and Chef Check. Services include TexTest, a comprehensive apparel testing laboratory. As a niche manufacturer, Johnston is somewhat removed from competition with direct importers. Its different fabrics include decoratives, technical wovens and fabrics made for industrial and career use. These fabrics can be used in bedding and upholstery, as linens and other home products. Johnston also produces technical fabrics for use in a variety of applications including automotive fabrics for seat suspensions and hood linings; rubber coated fabrics for conveyor belts; coated and laminated fabrics for book binding, banners and awnings; filtration fabrics; and specialty finishing including latex and fire retardant treatments. Johnston is owned by Meriturn Partners, an equity management firm.

FINANCIAL DATA: *Note: Data for latest year may not have been available at press time.*

In U.S. $	2012	2011	2010	2009	2008	2007
Revenue						
R&D Expense						
Operating Income						
Operating Margin %						
SGA Expense						
Net Income						
Operating Cash Flow						
Capital Expenditure						
EBITDA						
Return on Assets %						
Return on Equity %						
Debt to Equity						

CONTACT INFORMATION:

Phone: 334-298-9351 Fax: 334-291-3002
Toll-Free:
Address: 300 General Colin Powell Pkwy., Phenix City, AL 36868 United States

STOCK TICKER/OTHER:

Stock Ticker: Private
Exchange:
Employees: 500
Fiscal Year Ends: 31-Dec
Parent Company: Meriturn Partners

SALARIES/BONUSES:

Top Exec. Salary: $ Bonus: $
Second Exec. Salary: $ Bonus: $

OTHER THOUGHTS:

Estimated Female Officers or Directors:
Hot Spot for Advancement for Women/Minorities:

Jones Group Inc (The)

NAIC Code: 448120

www.jonesgroupinc.com

TYPES OF BUSINESS:

Women's Apparel, Retail
Wholesale Operations
Jewelry
Footwear
Accessories

BRANDS/DIVISIONS/AFFILIATES:

Jones New York
Evan-Picone
Nine West
Kasper
l.e.i.
ShoeWoo
Kurt Geiger
Brian Atwood

CONTACTS: *Note: Officers with more than one job title may be intentionally listed here more than once.*

Wesley R. Card, CEO
Cynthia DiPietrantonio, COO
Richard Dickson, Pres.
John T. McClain, CFO
Susie Rieland, Chief Merch. Officer
Ira M. Dansky, General Counsel/Exec. VP/Sec.
Christopher R. Cade, Chief Acct. Officer/Controller/Exec. VP
Richard Dickson, CEO-Branded Bus.
Stefani Greenfield, Chief Creative Officer
Kathy Nedorostek, Pres., Global Footwear & Accessories
George Sharp, Exec. VP-Design
Sidney Kimmel, Chmn.
Scott Bowman, Pres., Global Retail & Int'l Dev.

GROWTH PLANS/SPECIAL FEATURES:

The Jones Group, Inc. designs and markets branded apparel, footwear and accessories. The company's more than 35 nationally recognized brands include Jones New York, Evan-Picone, l.e.i., Energie, Nine West, Easy Spirit, Bandolino, Anne Klein, Le Suit, Kasper and Judith Jack. The firm also markets Givenchy costume jewelry, Dockers Woman footwear and Rachel Roy apparel through various license arrangements. The Jones Group operates in six segments: domestic wholesale sportswear, domestic wholesale jeanswear, domestic wholesale footwear and accessories, domestic retail, international wholesale and international retail. The domestic wholesale sportswear group markets career and casual sportswear, jeans, dresses, suits and related accessories that cover a broad array of categories for the women's, juniors and girls markets. The domestic wholesale jeanswear group consists of jeanswear, casual sportswear and dresses covering the women's, juniors and girl's markets. The segment's brands include Jones New York, Gloria Vanderbilt, Energie and l.e.i. The domestic wholesale footwear and accessories group sells brand name and private-label footwear, handbags, small leather goods, luggage and costume, semi-precious, sterling silver and marcasite jewelry. The domestic retail segment operates 672 stores under four specialty banners (Nine West, Easy Spirit, Stuart Weitzman and ShoeWoo) and five outlet formats (Nine West, Jones New York, Easy Spirit, Kasper and Stuart Weitzman). It also includes six e-commerce stores to support its various brands. International wholesale operations include the sale of brand-name footwear, handbags, small leather goods, jewelry and apparel to distributors and retailers in over 75 countries outside the U.S. The international retail segment operates 315 retail locations in the U.K., Ireland and Spain, including 66 specialty retail stores, 39 outlet stores and 210 concession locations. In 2011, The Jones Group announced the acquisition of U.K. luxury shoe retailer Kurt Geiger. In June 2012, it acquired the Brian Atwood brand.

FINANCIAL DATA: *Note: Data for latest year may not have been available at press time.*

In U.S. $	2012	2011	2010	2009	2008	2007
Revenue	3,798,100,000	3,785,300,000	3,642,700,000	3,327,400,000	3,616,400,000	3,848,500,000
R&D Expense						
Operating Income	74,700,000	140,500,000	144,900,000	-12,100,000	-731,400,000	-27,000,000
Operating Margin %	1.96%	3.71%	3.97%	- .36%	-20.22%	- .70%
SGA Expense	1,226,900,000	1,173,200,000	1,073,000,000	1,008,700,000	1,069,200,000	1,100,400,000
Net Income	-56,100,000	50,700,000	53,800,000	-86,600,000	-765,400,000	311,100,000
Operating Cash Flow	112,700,000	271,700,000	141,300,000	349,000,000	175,500,000	159,500,000
Capital Expenditure	76,500,000	98,000,000	41,000,000	30,000,000	71,200,000	111,200,000
EBITDA	167,900,000	233,400,000	237,400,000	64,200,000	-643,000,000	83,500,000
Return on Assets %	-2.11%	2.00%	2.46%	-3.88%	-27.02%	8.85%
Return on Equity %	-5.35%	4.55%	4.82%	-7.61%	-48.15%	14.78%
Debt to Equity	0.95	0.78	0.47	0.48	0.44	0.38

CONTACT INFORMATION:

Phone: 212 642-3860 Fax: 215 785-1228
Toll-Free:
Address: 1411 Broadway, New York, NY 10018 United States

STOCK TICKER/OTHER:

Stock Ticker: JNY Exchange: NYS
Employees: 6,690 Fiscal Year Ends: 31-Dec
Parent Company:

SALARIES/BONUSES:

Top Exec. Salary: $1,600,000 Bonus: $
Second Exec. Salary: $1,099,452 Bonus: $

OTHER THOUGHTS:

Estimated Female Officers or Directors: 6
Hot Spot for Advancement for Women/Minorities: Y

Jordache Enterprises Inc

www.jordachecorporate.com

NAIC Code: 315240

TYPES OF BUSINESS:

Jeans & Denim Apparel
Real Estate Investments
Airlines
Hotels & Resorts

BRANDS/DIVISIONS/AFFILIATES:

Dittos
US Polo Association
Gasoline
Halutza
Jordache
Gas
KIKIT Jeans
Love Lola

CONTACTS: *Note: Officers with more than one job title may be intentionally listed here more than once.*

Joseph Nakash, CEO
Liz Berlinger, Pres.
Joe Taylor, CFO
Charles Flores, VP-Mktg.
Charles Flores, VP-e-commerce
Joseph Nakash, Chmn.

GROWTH PLANS/SPECIAL FEATURES:

Jordache Enterprises, Inc. manufactures and markets denim, casual wear and accessories. Jordache Enterprises' brands include Jordache, Gas, KIKIT Jeans, Dittos, Love Lola, XOXO and Gasoline. The licensed sportswear brand, U.S. Polo Association, is one of Jordache's fastest-growing brands, offering a mid-priced line of jeans, t-shirts, knit shirts and accessories through department stores and discount big-box retailers. Additionally, Jordache operates several U.S. Polo Association-branded retail shops in the U.S. The company operates manufacturing facilities in Jordan, Mexico, Kenya, Madagascar, Morocco and Bangladesh, with U.S. distribution centers situated close to major clients. The company manufactures private label denim for Gap, Tommy Hilfiger, Levi's and Abercrombie & Fitch, among others. The company also invests in worldwide residential and commercial real estate; owns a number of hotel and resort properties, including five hotels in Miami and several in Israel; holds interests in a private jet charter service; and owns Halutza, an award-winning olive oil company.

FINANCIAL DATA: *Note: Data for latest year may not have been available at press time.*

In U.S. $	2012	2011	2010	2009	2008	2007
Revenue						
R&D Expense						
Operating Income						
Operating Margin %						
SGA Expense						
Net Income						
Operating Cash Flow						
Capital Expenditure						
EBITDA						
Return on Assets %						
Return on Equity %						
Debt to Equity						

CONTACT INFORMATION:

Phone: 212-944-1330 Fax: 212-239-0063
Toll-Free:
Address: 1400 Broadway, 15th Fl., New York, NY 10018 United States

STOCK TICKER/OTHER:

Stock Ticker: Private
Exchange:
Employees:
Fiscal Year Ends: 31-Dec
Parent Company:

SALARIES/BONUSES:

Top Exec. Salary: $ Bonus: $
Second Exec. Salary: $ Bonus: $

OTHER THOUGHTS:

Estimated Female Officers or Directors: 1
Hot Spot for Advancement for Women/Minorities:

Jos A Bank Clothiers Inc

NAIC Code: 448110

www.josbank.com

TYPES OF BUSINESS:

Business Apparel-Men's, Retail
Catalog Sales
Online Retail
Tailoring Services
Footwear
Casual Apparel
Accessories

BRANDS/DIVISIONS/AFFILIATES:

Jos. A. Bank Executive Collection
Signature Collection
Signature Gold Collection
JosBankFormal.com

CONTACTS: *Note: Officers with more than one job title may be intentionally listed here more than once.*

R. Neal Black, CEO
R. Neal Black, Pres.
David E. Ullman, CFO/Exec. VP
Robert B. Hensley, Exec. VP-Human Resources
James W. Thorne, Chief Merch. Officer/Exec. VP
Charles D. Frazer, General Counsel/Sr. VP/Corp. Sec.
Gary M. Merry, Exec. VP-Store & Catalog Oper.
Robert B. Hensley, Exec. VP-Real Estate & Loss Prevention
Frank J. Barbarino, Sr. VP/Gen. Merch. Mgr.
Robert N. Wildrick, Chmn.

GROWTH PLANS/SPECIAL FEATURES:

Jos. A. Bank Clothiers, Inc., established in 1905, is a leading retailer and direct marketer of men's tailored and casual clothing, footwear, golf apparel and accessories. The company sells all of its products under the Jos. A. Bank label, targeting the male career professional. Products are offered on three levels, which include the basic Jos. A. Bank Executive collection and the more refined Signature and Signature Gold collections. The company designs its own tailored products and selects, and sometimes purchases, raw materials, such as finished wool. Merchandise is manufactured by third party domestic and international contract manufacturers to the firm's specifications. Jos. A. Bank operates 568 stores, including 26 company-owned factory and outlet stores and 15 stores owned by franchisees, in 44 states and Washington, D.C. Stores are located primarily in lifestyle centers, malls and strip centers. The company's newer stores operate under a format that integrates the firm's corporate casual and sportswear divisions with its other products. Most of the company's stores feature a tailor shop that provides a range of tailoring services. The firm also offers corporate customers the Corporate Card, which provides users with merchandise discounts. In addition to its retail stores, the firm operates a nationwide catalog, including four seasonal editions, and an e-commerce site. In September 2012, the company launched JosBankFormal.com, a new site for tuxedo rentals.

The company offers its employees medical, dental and vision insurance; flexible spending accounts; prescription drug coverage; merchandise discounts; a 401(k); tuition reimbursement; an employee assistance program; and legal services.

FINANCIAL DATA: *Note: Data for latest year may not have been available at press time.*

In U.S. $	2012	2011	2010	2009	2008	2007
Revenue	979,852,000	858,128,000	770,316,000	695,908,000	604,010,000	
R&D Expense						
Operating Income	159,407,000	141,607,000	117,403,000	95,489,000	82,751,000	
Operating Margin %	16.26%	16.50%	15.24%	13.72%	13.70%	
SGA Expense	448,868,000	395,936,000	354,720,000	335,465,000	295,895,000	
Net Income	97,491,000	85,799,000	71,155,000	58,408,000	50,168,000	
Operating Cash Flow	91,816,000	106,239,000	84,748,000	73,540,000	63,712,000	
Capital Expenditure	37,531,000	29,352,000	16,333,000	35,105,000	27,696,000	
EBITDA	185,855,000	166,675,000	140,160,000	116,954,000	103,165,000	
Return on Assets %	13.21%	14.08%	13.58%	12.54%	12.41%	
Return on Equity %	18.26%	19.58%	19.90%	20.03%	21.37%	
Debt to Equity					0.19	

CONTACT INFORMATION:

Phone: 410 239-2700 Fax: 410 239-5700
Toll-Free:
Address: 500 Hanover Pike, Hampstead, MD 21074 United States

STOCK TICKER/OTHER:

Stock Ticker: JOSB Exchange: NAS
Employees: 4,998 Fiscal Year Ends: 31-Jan
Parent Company:

SALARIES/BONUSES:

Top Exec. Salary: $783,138 Bonus: $
Second Exec. Salary: $492,450 Bonus: $

OTHER THOUGHTS:

Estimated Female Officers or Directors:
Hot Spot for Advancement for Women/Minorities:

Kate Spade LLC

www.katespade.com

NAIC Code: 315990

TYPES OF BUSINESS:

Clothing & Accessories-Retail
Fashion Accessories
Shoes
Baby Items
Online Sales
Retail Stores
Stationary

BRANDS/DIVISIONS/AFFILIATES:

Kate Spade
Liz Claiborne Inc
Kate's Pretty Papers
Jack Spade
Kate Spade Saturday
Fifth & Pacific Comp;anies, Inc.

CONTACTS: *Note: Officers with more than one job title may be intentionally listed here more than once.*

Craig Leavitt, CEO
Deborah J. Lloyd, Pres.
Kyle Andrew, Sr. VP-Brand Mktg.
Sydney Price, Sr. VP-Global Merch.

GROWTH PLANS/SPECIAL FEATURES:

Kate Spade LLC, formerly a subsidiary of Liz Claiborne, Inc.., and now part of Fifth & Pacific Companies, Inc., is a designer clothing and accessories chain founded through a partnership between Kate Spade and Andy Spade in 1993. Since then, the company has grown dramatically, expanding from its New York City base. Originally a women's handbag designer, the firm now offers a variety of accessories, such as wallets, hats, hair accessories and umbrellas as well as shoes, baby items, sunglasses, stationary and beauty products. The firm's stationary line, Kate's Pretty Papers, is manufactured by Crane & Co. and includes nine designs for wedding invitations, correspondence cards and thank-you notes. The line also offers accessories such as organizers, refills, pencil cases, frames, albums and desk calendars. The firm also offers a line of high fashion women's clothing. The firm has a total of 800 stores in the U.S., and products are distributed to over 400 stores worldwide. The company's Jack Spade line, geared towards men, includes messenger bags, clothing, laptop cases, brief cases, travel bags and small goods. The firm also runs Kate Spade Saturday, a casual clothing and accessory line focused on bright colors and graphic prints at a lower price range than the Kate Spade merchandise. Internationally, Kate Spade has extensive operations in Japan, Hong Kong, Canada, China and Australia as well as shops in Turkey, Belgium, Brazil, the Bahamas, Thailand, Singapore, Indonesia, the Philippines, Korea, Macau, Malaysia and Guam. Customers are able to shop and order products or gift certificates from the company web site.

FINANCIAL DATA: *Note: Data for latest year may not have been available at press time.*

In U.S. $	2012	2011	2010	2009	2008	2007
Revenue	462,000,000					
R&D Expense						
Operating Income						
Operating Margin %						
SGA Expense						
Net Income						
Operating Cash Flow						
Capital Expenditure						
EBITDA						
Return on Assets %						
Return on Equity %						
Debt to Equity						

CONTACT INFORMATION:

Phone: 212-739-6550 Fax: 212-739-6544
Toll-Free: 866-999-5283
Address: 48 W. 25th St., New York, NY 10010 United States

STOCK TICKER/OTHER:

Stock Ticker: Subsidiary Exchange:
Employees: Fiscal Year Ends: 31-Jul
Parent Company: Fifth & Pacific Companies Inc

SALARIES/BONUSES:

Top Exec. Salary: $ Bonus: $
Second Exec. Salary: $ Bonus: $

OTHER THOUGHTS:

Estimated Female Officers or Directors: 2
Hot Spot for Advancement for Women/Minorities: Y

Kellwood Co

NAIC Code: 315240

www.kellwood.com

TYPES OF BUSINESS:

Apparel Design
Women's Sportswear
Intimate Apparel
Infant Apparel
Children's Apparel

BRANDS/DIVISIONS/AFFILIATES:

Sun Capital Partners Inc
Jolt
Sag Harbor
Rebecca Taylor
Briggs New York
Vince
Rewind
My Michelle

GROWTH PLANS/SPECIAL FEATURES:

Kellwood Co., owned by Sun Capital Partners, Inc., is a marketer of apparel and soft goods primarily for women, offering branded as well as private-label products to all channels of distribution. The apparel maker designs, merchandises and sells designer and lifestyle sportswear for women. Its products (including blazers, dresses, sweaters, blouses, vests, skirts, jeans and pants) consists of branded goods sold at popular to moderate price points as well as some upper-price-point lines sold to specialty stores, department stores and catalog houses. Brands in this division include Rebecca Taylor, XOXO, David Meister, My Michelle, Briggs NY, Sag Harbor and Zobha. The firm's Vince label designs clothing and accessories for men and women. Its children's segment consists of brands such as Jolt and Rewind.

CONTACTS: *Note: Officers with more than one job title may be intentionally listed here more than once.*

Jill Granoff, CEO
Michael M. Saunders, COO/Sr. VP
Christina Cassar, Interim Chief Mktg. Officer/ VP-Mktg.
Patricia (Patty) Burman, VP-Digital

FINANCIAL DATA: *Note: Data for latest year may not have been available at press time.*

In U.S. $	2012	2011	2010	2009	2008	2007
Revenue						
R&D Expense						
Operating Income						
Operating Margin %						
SGA Expense						
Net Income						
Operating Cash Flow						
Capital Expenditure						
EBITDA						
Return on Assets %						
Return on Equity %						
Debt to Equity						

CONTACT INFORMATION:

Phone: 314-576-3100 Fax: 314-576-3460
Toll-Free:
Address: 600 Kellwood Pkwy., Chesterfield, MO 63017 United States

STOCK TICKER/OTHER:

Stock Ticker: Private Exchange:
Employees: Fiscal Year Ends: 31-Jan
Parent Company: SUN CAPITAL PARTNERS INC

SALARIES/BONUSES:

Top Exec. Salary: $ Bonus: $
Second Exec. Salary: $ Bonus: $

OTHER THOUGHTS:

Estimated Female Officers or Directors: 3
Hot Spot for Advancement for Women/Minorities: Y

Kenneth Cole Productions Inc

www.kennethcole.com

NAIC Code: 316210

TYPES OF BUSINESS:

Shoes, Manufacturing
Footwear & Handbags
Men's & Women's Apparel
Retail Stores
Catalog & Online Sales

BRANDS/DIVISIONS/AFFILIATES:

Kenneth Cole New York
Kenneth Cole Reaction
Unlisted
Le Tigre
Gentle Souls
Kenneth Cole
Kenneth Cole Outlet

CONTACTS: *Note: Officers with more than one job title may be intentionally listed here more than once.*

Kenneth D. Cole, Chief Creative Officer
Kenneth D. Cole, Chmn.

GROWTH PLANS/SPECIAL FEATURES:

Kenneth Cole Productions, Inc. (KCP) designs, sources, markets and licenses a broad range of fashion footwear, handbags, apparel and accessories under the Kenneth Cole New York, Kenneth Cole Reaction, Unlisted, Le Tigre and Gentle Souls brand names. The firm's products are targeted to appeal to modern, fashion-conscious consumers who seek accessible designer fashion that reflects a metropolitan lifestyle. Its operations are divided into three segments: wholesale, consumer direct and licensing. The wholesale segment includes sales to department stores, independent specialty stores, national chains and mass market retailers. Its products are currently distributed to over 4,700 U.S. retail shops as well as select locations in Canada, Europe, Australia, Malaysia and the Middle East. This segment also designs, develops and sources private label footwear and handbags for selected retailers. The consumer direct segment is responsible for the firm's full-priced retail stores and outlets as well as its e-commerce operations. It currently operates 97 retail locations, which include 24 Kenneth Cole full-priced retail stores and 73 outlet stores under the Kenneth Cole name. The licensing segment handles the firm's relationships with its licensing partners. Though KCP maintains in-house design teams for the development of footwear, handbags and men's sportswear collections, all other product categories sold under its various brands are produced by licensee partners. Licensed product categories include men's tailored clothing, neckwear, belts, socks and jewelry; women's apparel, outerwear, jewelry and swimwear; children's apparel; and men's and women's watches, optical frames, sunglasses and fragrances. In September 2012, Kenneth D. Cole acquired the company via KCP Holdco, Inc, an entity under his control that was created specifically for the acquisition. Upon the completion of the acquisition, the company was delisted from the New York Stock Exchange.

FINANCIAL DATA: *Note: Data for latest year may not have been available at press time.*

In U.S. $	2012	2011	2010	2009	2008	2007
Revenue	500,000,000	478,900,000	411,652,000	370,955,000	448,578,000	466,405,000
R&D Expense						
Operating Income						
Operating Margin %						
SGA Expense						
Net Income		-2,900,000	2,084,000	-63,238,000	-14,800,000	7,083,000
Operating Cash Flow						
Capital Expenditure						
EBITDA						
Return on Assets %						
Return on Equity %						
Debt to Equity						

CONTACT INFORMATION:

Phone: 212-265-1500 Fax: 212-315-8279
Toll-Free:
Address: 603 W. 50th St., New York, NY 10019 United States

STOCK TICKER/OTHER:

Stock Ticker: Private
Exchange:
Employees: 1,800
Fiscal Year Ends: 31-Dec
Parent Company:

SALARIES/BONUSES:

Top Exec. Salary: $ Bonus: $
Second Exec. Salary: $ Bonus: $

OTHER THOUGHTS:

Estimated Female Officers or Directors:
Hot Spot for Advancement for Women/Minorities: Y

KiK Textilien & Non-Food GmbH

www.kik-textilien.com

NAIC Code: 448140

TYPES OF BUSINESS:

Clothing & Apparel Retail Stores
General Merchandise Stores

BRANDS/DIVISIONS/AFFILIATES:

Tengelmann Group

GROWTH PLANS/SPECIAL FEATURES:

KiK Textilien & Non-Food GmbH, a member of the Tengelmann Group, is a German discount retailer of general merchandise, operating 3,164 stores in Germany (2,591), Austria (253), the Czech Republic (145), Hungary (57), Slovakia (54), Slovenia (44) and Croatia (20). KiK focuses primarily on inexpensive apparel for men, women and children, offering items including hats, blouses, sweaters, jackets, mittens, pants, skirts, socks and shoes. The firm's non-apparel merchandise comprises toys, kitchen items, jewelry and stationary as well as pet supplies, tools, perfumes and decorative pieces. The company averages 200 store openings per year and is pushing to reach 5,000 stores across Europe. In 2012, KiK opened its first store in Poland.

CONTACTS: *Note: Officers with more than one job title may be intentionally listed here more than once.*

Heinz Speet, CEO

FINANCIAL DATA: *Note: Data for latest year may not have been available at press time.*

In U.S. $	2012	2011	2010	2009	2008	2007
Revenue						
R&D Expense						
Operating Income						
Operating Margin %						
SGA Expense						
Net Income						
Operating Cash Flow						
Capital Expenditure						
EBITDA						
Return on Assets %						
Return on Equity %						
Debt to Equity						

CONTACT INFORMATION:

Phone: 49-2383-95-4116 Fax:
Toll-Free:
Address: Siemensstrasse 21, Bonen, 59199 Germany

STOCK TICKER/OTHER:

Stock Ticker: Private Exchange:
Employees: Fiscal Year Ends:
Parent Company:

SALARIES/BONUSES:

Top Exec. Salary: $ Bonus: $
Second Exec. Salary: $ Bonus: $

OTHER THOUGHTS:

Estimated Female Officers or Directors:
Hot Spot for Advancement for Women/Minorities:

Kingmaker Footwear Holdings Limited www.kingmaker-footwear.com

NAIC Code: 316210

TYPES OF BUSINESS:

Mens Footwear Manufacturing
Womens Footwear Manufacturing
Athletic Shoes Manufacturing

GROWTH PLANS/SPECIAL FEATURES:

Kingmaker Footwear Holdings Limited is engaged in the manufacture of rugged-casual, dress-casual, work and sports footwear. Its product range covers rugged all-terrain footwear, hiking shoes, boots, steel-toe shoes and waterproof footwear. Partnered brands include Skechers, Clarks, Stride Rite, Rockport, New Balance and Fiona's Prince. Kingmaker Footwear has operations in China, Hong Kong, Taiwan and Cambodia, including 10 manufacturing plants.

BRANDS/DIVISIONS/AFFILIATES:

CONTACTS: *Note: Officers with more than one job title may be intentionally listed here more than once.*

Mickey Ming-Hsiung Chen, CEO
Mickey Ming-Hsiung Chen, Pres.
Helen Hsiu-Duan Huang, Exec. Dir.-Finance
Phillip Brian Kimmel, Gen. Mgr.-Mktg./Exec. Dir.
Helen Hsiu-Duan Huang, Exec. Dir.-Personnel
Mickey Ming-Hsiung Chen, Chmn.

FINANCIAL DATA: *Note: Data for latest year may not have been available at press time.*

In U.S. $	2012	2011	2010	2009	2008	2007
Revenue	213,480,900	193,989,900	166,361,500	188,824,500	169,995,600	149,202,900
R&D Expense						
Operating Income	5,705,920	15,206,070	15,825,110	5,143,377	6,305,613	3,064,381
Operating Margin %	2.67%	7.83%	9.51%	2.72%	3.70%	2.05%
SGA Expense	24,836,240	23,948,510	20,854,970	19,866,360	19,514,340	15,848,850
Net Income	6,977,155	15,776,480	13,547,080	6,862,093	6,501,039	4,358,836
Operating Cash Flow	5,922,114	455,091	18,728,500	21,996,700	14,829,150	1,541,091
Capital Expenditure	14,382,700	3,040,130	4,028,998	4,857,655	4,721,438	2,125,434
EBITDA	13,556,490	23,663,040	23,191,050	13,596,220	14,028,480	8,059,931
Return on Assets %	3.92%	9.56%	9.09%	4.91%	4.96%	3.56%
Return on Equity %	5.60%	13.57%	13.21%	7.20%	7.15%	4.97%
Debt to Equity						

CONTACT INFORMATION:

Phone: 852-2723-1858 Fax:
Toll-Free:
Address: Empress Plaza, 17-19 Chatham Rd. S., Fl. 17, Kowloon, Hong Kong

STOCK TICKER/OTHER:

Stock Ticker: KMSWF
Employees: 12,000
Parent Company:
Exchange: PINX
Fiscal Year Ends: 31-Mar

SALARIES/BONUSES:

Top Exec. Salary: $ Bonus: $
Second Exec. Salary: $ Bonus: $

OTHER THOUGHTS:

Estimated Female Officers or Directors: 2
Hot Spot for Advancement for Women/Minorities:

Klopman International

www.klopman.com

NAIC Code: 313210

TYPES OF BUSINESS:

Specialty Fabrics
Uniforms, Manufacturing
Protective Wear

BRANDS/DIVISIONS/AFFILIATES:

MW Corp
MW Unitexx SA
Firestrike 245
Klopman Bravo
Sungrazer
Luminex C50
Klopman Marketing Services Department

CONTACTS: *Note: Officers with more than one job title may be intentionally listed here more than once.*

Giuseppe Abbruzzesi, Gen. Mgr.
Pasquale Salemi, Dir.-Finance
Holger Beule, Dir.-Sales
Giovanni Ricci, Dir.-Human Resources
Pasquale Salemi, Dir.-Admin.
Amaury Sartorius, Dir.-Sales
Alfonso Marra, Managing Dir.

GROWTH PLANS/SPECIAL FEATURES:

Klopman International is a manufacturer of specialty fabrics and uniforms. Its products are used by professionals in a variety of industries, including automotive, health care, law enforcement, utilities, agriculture, the military, hospitality, tourism and transportation. The company maintains offices in Italy, India, France, Russia, Poland, Spain and Sweden. Klopman has developed several fabrics, made from either cotton or cotton/polyester blends, which provide protection and comfort for its clients' employees. Additional fabrics include Klopman Bravo, made with organic Fairtrade Certified Cotton; Sungrazer, which combines flame retardancy and anti-staticity for use in multi risk situations; Luminex C50, a 50/50 cotton and polyester blend; and Firestrike 245, a fabric combining permanent, inherent flame retardancy with anti-static and protection from liquid chemical splash. The firm also develops specialized fabrics for use in furnishings. The company manages Klopman Marketing Services Department, which is responsible for providing customers with tailor-made marketing support solutions that include advertising, public relations, exhibitions, literature, videos, point-of-sale material and presentations. Klopman is owned by MW Unitexx SA, a subsidiary of MW Corp., based in Mumbai. In November 2012, the firm expanded its operations to Asia through a joint venture with ARGO Manunggal Textile, Klopman-AMT.

FINANCIAL DATA: *Note: Data for latest year may not have been available at press time.*

In U.S. $	2012	2011	2010	2009	2008	2007
Revenue						
R&D Expense						
Operating Income						
Operating Margin %						
SGA Expense						
Net Income						
Operating Cash Flow						
Capital Expenditure						
EBITDA						
Return on Assets %						
Return on Equity %						
Debt to Equity						

CONTACT INFORMATION:

Phone: 39-0775-298-382 Fax: 39-0775-294-679
Toll-Free:
Address: Loc. ta Mola dei Frati, Frosinone, 03100 Italy

STOCK TICKER/OTHER:

Stock Ticker: Subsidiary Exchange:
Employees: Fiscal Year Ends: 30-Sep
Parent Company: MW CORP

SALARIES/BONUSES:

Top Exec. Salary: $ Bonus: $
Second Exec. Salary: $ Bonus: $

OTHER THOUGHTS:

Estimated Female Officers or Directors:
Hot Spot for Advancement for Women/Minorities:

Kmart Holding Corporation

www.kmart.com

NAIC Code: 452112

TYPES OF BUSINESS:

Discount Department Stores
Online Sales
Pharmacy Services
Grocery Sales
Video Rentals
Optical Shops

BRANDS/DIVISIONS/AFFILIATES:

Sears Holdings Corporation
Sears Auto Centers
Kmart
Kmart Super Center
Kenmore
DieHard
Craftsman

CONTACTS: *Note: Officers with more than one job title may be intentionally listed here more than once.*

Mark Snyder, Chief Mktg. Officer
Louis J. D'Ambrosio, CEO/Pres., Sears Holding Corporation
Robert A. Schriesheim, CFO-Sears Holding Corporation
Imran Jooma, Pres., Online, Mktg., Pricing & Financial Svcs.
Dane A. Drobny, Sr. VP/General Counsel-Sears Holding Corporation

GROWTH PLANS/SPECIAL FEATURES:

Kmart Holding Corporation, a subsidiary of Sears Holdings Corporation, is a mass merchandise retailer. The firm has 1,305 stores in 49 states, Guam, Puerto Rico and the U.S. Virgin Islands. This store count includes the traditional discount stores, ranging from 80,000 to 110,000 square feet; and the Kmart Super Centers, which are between 140,000 to 190,000 square feet. The traditional Kmart discount stores are one floor, free standing units that carry a wide assortment of general merchandise and often include a pharmacy. These stores carry between 60,000 and 80,000 different products. The Kmart Super Centers store format focuses on three areas of retail: home fashions, children's apparel and consumable products. These stores stock between 100,000 and 150,000 different items. The company sells certain proprietary Sears brand products, such as Kenmore, Craftsman and DieHard products, and offers services for these products within certain Kmart stores. About 268 Kmart stores sell an assortment of major home appliances, including Kenmore-branded products. The Super Centers feature a full line of general merchandise and groceries as well as a variety of ancillary services including video rentals, dry cleaning, hair care and optical and floral shops, and full grocery, deli and bakery operations, open 24 hours a day. Kmart also sells its products online, offering over 220,000 items of Kmart merchandise, including appliances and outdoor equipment. In addition, there are 22 Sears Auto Centers operating in certain Kmart stores. Sears Auto Centers offer a variety of professional automotive repair and maintenance services as well as a full assortment of automotive accessories.

FINANCIAL DATA: *Note: Data for latest year may not have been available at press time.*

In U.S. $	2012	2011	2010	2009	2008	2007
Revenue	15,285,000,000	15,593,000,000	15,743,000,000	16,219,000,000	17,256,000,000	18,647,000,000
R&D Expense						
Operating Income						
Operating Margin %						
SGA Expense						
Net Income	-34,000,000	353,000,000	190,000,000	172,000,000	402,000,000	
Operating Cash Flow						
Capital Expenditure						
EBITDA						
Return on Assets %						
Return on Equity %						
Debt to Equity						

CONTACT INFORMATION:

Phone: 847-286-2500 Fax: 847-286-5500
Toll-Free:
Address: 3333 Beverly Rd., Hoffman Estates, IL 60179 United States

STOCK TICKER/OTHER:

Stock Ticker: Subsidiary Exchange:
Employees: 312,000 Fiscal Year Ends: 31-Jan
Parent Company: SEARS HOLDINGS CORPORATION

SALARIES/BONUSES:

Top Exec. Salary: $ Bonus: $
Second Exec. Salary: $ Bonus: $

OTHER THOUGHTS:

Estimated Female Officers or Directors:
Hot Spot for Advancement for Women/Minorities:

Koch Industries Inc

NAIC Code: 324110

www.kochind.com

TYPES OF BUSINESS:

Petroleum Refining
Chemicals
Textiles
Pipelines
Fertilizer Production
Chemical Equipment
Asphalt & Paving Supplies
Beef Production

BRANDS/DIVISIONS/AFFILIATES:

Flint Hills Resources
Koch Carbon LLC
Koch Fertilizer, LLC
Matador Cattle Company (The)
Koch Chemical Technology Group LLC
J&H Bunn Limited
INVISTA BV
Georgia-Pacific LLC

CONTACTS: *Note: Officers with more than one job title may be intentionally listed here more than once.*

Charles G. Koch, CEO
David L. Robertson, COO
David L. Robertson, Pres.
David H. Koch, Exec. VP
Charles G. Koch, Chmn.

GROWTH PLANS/SPECIAL FEATURES:

Koch Industries, Inc. is a diversified group of companies with operations in markets such as refining and chemicals; process and pollution control equipment and technologies; minerals and fertilizers; polymers and fibers; commodity and financial trading and services; and forest and consumer products. It also conducts operations in venture capital investments, municipal finance, capital market investments and business development. Subsidiary Flint Hills Resources operates petroleum refineries in Alaska, Minnesota and Texas, with a combined crude oil processing capacity of more than 800,000 barrels per day. These plants produce aromatics, olefins, polymers and intermediate chemicals. The subsidiary obtains its supply from purchasing offices in Houston, Denver and Calgary, Canada. Subsidiary Koch Carbon, LLC supplies coal and petroleum coke as well as cement, pulp and paper, sulfur and related products internationally. Koch Fertilizer, LLC and its affiliates own or have interests in nitrogen fertilizer plants capable of marketing and distributing more than 13 million metric tons of nitrogen products annually. Koch's ranching subsidiary, The Matador Cattle Company, a division of Koch Agriculture Company, operates three ranches in Kansas, Montana and Texas totaling 425,000 acres. Subsidiary Koch Pipeline Company, LP and its affiliates operate a 4,000-mile network of pipelines. Subsidiary Koch Chemical Technology Group, LLC and its affiliates design, manufacture, sell, install and service process and pollution control equipment. Subsidiary INVISTA BV is a global producer and marketer of polymers and fibers, primarily for nylon, spandex and polyester applications. Subsidiary Georgia-Pacific LLC manufactures and markets tissue, packaging, paper, building products, related chemicals and fluff, filter and market pulp under such brand names as Quilted Northern, Angel Soft, Brawny, Sparkle and Dixie. In March 2011, the firm's Koch Fertilizer, LLC subsidiary acquired J&H Bunn Limited, a UK-based fertilizer blender and distributor.

FINANCIAL DATA: *Note: Data for latest year may not have been available at press time.*

In U.S. $	2012	2011	2010	2009	2008	2007
Revenue	115,000,000,000	110,000,000,000	105,000,000,000	90,000,000,000	94,000,000,000	98,000,000,000
R&D Expense						
Operating Income						
Operating Margin %						
SGA Expense						
Net Income	10,000,000,000	8,500,000,000	8,000,000,000	7,000,000,000	7,000,000,000	8,000,000,000
Operating Cash Flow						
Capital Expenditure						
EBITDA						
Return on Assets %						
Return on Equity %						
Debt to Equity						

CONTACT INFORMATION:

Phone: 316-828-5500 Fax: 316-828-5739
Toll-Free:
Address: 4111 E. 37th St. N., Wichita, KS 67220 United States

STOCK TICKER/OTHER:

Stock Ticker: Private
Employees: 60,000
Parent Company:
Exchange:
Fiscal Year Ends: 31-Dec

SALARIES/BONUSES:

Top Exec. Salary: $ Bonus: $
Second Exec. Salary: $ Bonus: $

OTHER THOUGHTS:

Estimated Female Officers or Directors:
Hot Spot for Advancement for Women/Minorities:

Kohl's Corp

www.kohls.com

NAIC Code: 452112

TYPES OF BUSINESS:

Discount Department Stores
Online Sales

BRANDS/DIVISIONS/AFFILIATES:

Kohls.com

CONTACTS:

Note: Officers with more than one job title may be intentionally listed here more than once.

Kevin Mansell, CEO
Kevin Mansell, Pres.
Wes McDonald, CFO/Sr. Exec. VP
Donald Brennan, Chief Merch. Officer
John Worthington, Chief Admin. Officer
Peggy Eskenasi, Sr. Exec. VP-Prod. Dev.
Kevin Mansell, Chmn.

GROWTH PLANS/SPECIAL FEATURES:

Kohl's Corp. operates family-oriented specialty department stores. The company currently operates 1,134 stores in 49 states, with three store formats: prototype (approximately 88,000 square feet of retail space), small (approximately 68,000 square feet of retail space) and urban (approximately 125,000 square feet of retail space). Kohl's stores offer apparel, shoes and accessories for women, children and men; soft home products, such as sheets and pillows; and other home products, such as small electronics and luggage. The stores' brands include Dockers, Lee, Levi's, Jockey, Candie's, Nike, Dana Buchman, LC Lauren Conrad, Simply Vera Vera Wang, Dana Buchman Food Network and FILA Sport. Approximately 31% of Kohl's' merchandise is marketed towards women; 19% is marketed towards men; 19% is comprised of home furnishing products; 13% is designed for children; 10% is comprised of accessories; and 8% is comprised of footwear. In addition to its physical retail locations, the company also markets its products through its e-commerce site, Kohls.com. The company maintains 11 distribution centers located in Ohio, Texas, Virginia, Missouri, New York, California, Georgia and Illinois, and e-commerce fulfillment centers in Monroe, Ohio; Edgewood, Maryland; and San Bernardino, California. In 2012, Kohl's launched a solar initiative by installing solar panels in 30 stores, with plans to have 150 solar locations by the end of the year to offset its traditional energy usage. This firm's stores generate about $225 in sales per square foot per year.

The company offers its employees medical, dental and vision insurance; long-term disability and life insurance; a 401(k) plan; tuition reimbursement; on-site fitness classes, farmers market, child care, dry cleaning and food service; adoption assistance; parental leave; vacation; and merchandise discounts.

FINANCIAL DATA:

Note: Data for latest year may not have been available at press time.

In U.S. $	2012	2011	2010	2009	2008	2007
Revenue	18,804,000,000	18,391,000,000	17,178,000,000	16,389,000,000	16,473,730,000	
R&D Expense						
Operating Income	2,158,000,000	1,914,000,000	1,712,000,000	1,536,000,000	1,804,477,000	
Operating Margin %	11.47%	10.40%	9.96%	9.37%	10.95%	
SGA Expense	4,243,000,000	4,462,000,000	4,196,000,000	3,978,000,000	3,757,563,000	
Net Income	1,167,000,000	1,114,000,000	991,000,000	885,000,000	1,083,851,000	
Operating Cash Flow	2,143,000,000	1,676,000,000	2,234,000,000	1,701,000,000	1,234,363,000	
Capital Expenditure	927,000,000	761,000,000	666,000,000	1,014,000,000	1,541,683,000	
EBITDA	2,940,000,000	2,579,000,000	2,313,000,000	2,100,000,000	2,277,644,000	
Return on Assets %	8.43%	8.33%	8.09%	8.08%	11.05%	
Return on Equity %	15.97%	13.96%	13.58%	13.78%	18.51%	
Debt to Equity	0.63	0.2	0.26	0.3	0.33	

CONTACT INFORMATION:

Phone: 262 703-7000 Fax: 262 703-6373
Toll-Free:
Address: N56 W17000 Ridgewood Dr., Menomonee Falls, WI 53051 United States

STOCK TICKER/OTHER:

Stock Ticker: KSS Exchange: NYS
Employees: 136,000 Fiscal Year Ends: 31-Jan
Parent Company:

SALARIES/BONUSES:

Top Exec. Salary: $1,329,300 Bonus: $
Second Exec. Salary: $920,250 Bonus: $

OTHER THOUGHTS:

Estimated Female Officers or Directors: 3
Hot Spot for Advancement for Women/Minorities: Y

K-Swiss Inc

www.kswiss.com

NAIC Code: 316210

TYPES OF BUSINESS:

Footwear Manufacturing
Athletic Footwear & Apparel
Online Sales

BRANDS/DIVISIONS/AFFILIATES:

K-Swiss Classic
Palladium
Form Athletics
E-Land World Limited

CONTACTS: *Note: Officers with more than one job title may be intentionally listed here more than once.*

Steven Nichols, CEO
Steven Nichols, Pres.
George Powlick, CFO/VP-Finance/Sec.
George Powlick, Chief Admin. Officer
Kimberly Scully, Corp. Controller
Brian Sullivan, VP-National Accounts
David Nichols, Exec. VP
Steven Nichols, Chmn.

GROWTH PLANS/SPECIAL FEATURES:

K-Swiss, Inc. designs, develops and markets an array of athletic footwear for high-performance sports use, fitness activities and casual wear under the K-Swiss brand. The firm is best known for its original K-Swiss Classic shoe, an all-white, all-weather, all-leather tennis shoe, which has undergone only slight modifications since its introduction in 1966. In addition to footwear, the K-Swiss brand is used to market a line of branded apparel and accessories, including high-tech tennis apparel that consists of skirts, shorts, tops, polos, dresses and warm-ups for both men and women. The firm's apparel is targeted to a variety of markets, from consumers wanting performance apparel and accessories to casual sportstyle consumers. Other brands owned by the company include Palladium, which markets all-terrain hiking footwear, and Form Athletics, a marketer of mixed martial arts apparel. The company's products are manufactured by independent suppliers in China, Thailand, Vietnam and Indonesia and are sold in department and specialty retail stores as well as directly through the firm's web site. As a marketing strategy, the company endeavors to use classic styling instead of fashion-oriented footwear in order to reduce the impact of changes in consumer taste and prolong the life cycle of its products. In January 2013, K-Swiss agreed to be acquired by E-Land World Limited, a South Korean apparel distributor.

FINANCIAL DATA: *Note: Data for latest year may not have been available at press time.*

In U.S. $	2012	2011	2010	2009	2008	2007
Revenue	222,851,000	268,357,000	216,987,000	240,729,000	340,160,000	410,432,000
R&D Expense						
Operating Income	-30,770,000	-63,990,000	-57,395,000	-36,962,000	-12,510,000	32,361,000
Operating Margin %	-13.80%	-23.84%	-26.45%	-15.35%	-3.67%	7.88%
SGA Expense	107,895,000	153,626,000	142,474,000	118,303,000	147,869,000	157,498,000
Net Income	-34,779,000	-70,471,000	-68,212,000	-27,962,000	20,885,000	39,073,000
Operating Cash Flow	-572,000	-89,665,000	-46,735,000	-17,531,000	16,288,000	42,515,000
Capital Expenditure	1,563,000	2,704,000	2,377,000	1,444,000	11,065,000	10,489,000
EBITDA	-26,961,000	-53,857,000	-56,125,000	-32,046,000	28,272,000	34,664,000
Return on Assets %	-17.55%	-27.75%	-21.57%	-7.56%	4.96%	9.18%
Return on Equity %	-23.37%	-35.15%	-25.41%	-8.92%	5.89%	10.70%
Debt to Equity						

CONTACT INFORMATION:

Phone: 818 706-5100 Fax: 818 773-2390
Toll-Free: 800-938-8000
Address: 31248 Oak Crest Dr., Westlake Village, CA 91361 United States

STOCK TICKER/OTHER:

Stock Ticker: KSWS Exchange: NAS
Employees: 609 Fiscal Year Ends: 31-Dec
Parent Company:

SALARIES/BONUSES:

Top Exec. Salary: $937,267 Bonus: $
Second Exec. Salary: $305,623 Bonus: $

OTHER THOUGHTS:

Estimated Female Officers or Directors: 1
Hot Spot for Advancement for Women/Minorities:

Kuraray Co Ltd

www.kuraray.co.jp

NAIC Code: 313230

TYPES OF BUSINESS:

Textiles Manufacturing
Chemicals
Advanced Materials
Synthetic Textiles
Dental Reconstruction Materials
Fine Resins
High Performance Materials
Molded Plastic Fasteners

BRANDS/DIVISIONS/AFFILIATES:

KurarayKuraFlex Co Ltd
EVAL
Kuraray Fastening Co Ltd
Genestar
Kuraray Noritake Dental Inc
Clarino
Magic Tape
Magilock

CONTACTS: *Note: Officers with more than one job title may be intentionally listed here more than once.*

Fumio Ito, Pres.
Nobuo Fujii, Sr. Exec.-Human Resources
Setsuo Yamashita, Chief Officer-R&D
Setsuo Yamashita, CTO
Setsuo Yamashita, Chief Officer.-New Bus. Dev. Div.
Kohei Maeda, Gen. Mgr.-Acct. & Finance
Kazuhiro Tenkumo, Pres., Fibers & Textiles Company
Tomoyuki Aya, Pres., Isoprene Company
Keiji Murakami, Pres., Vinyl Acetate Company
Sadaaki Matsuyama, Pres., Kuraray Noritake Dental, Inc.
Yasuaki Wakui, Chmn.
George Avdey, Pres., KAI
Nobuo Fujii, Sr. Exec.-Purchasing & Logistics

GROWTH PLANS/SPECIAL FEATURES:

Kuraray Co. Ltd. manufactures chemicals, resins, fibers, textiles and other high-performance materials. Its operations are divided among four primary areas: Vinyl Acetate; Functional Materials; Fibers and Textiles; and Isoprene. The Vinyl Acetate segment produces poval resin and poval film, used in the information, chemical and IT sectors; EVAL, its proprietary ethylene vinyl alcohol (EVOH) copolymer resin, used in food packaging; and polyvinyl butyral (PVB) film, used in laminated safety glass. The Functional Materials segment consists of three sub-divisions. The methacrylate division produces methacrylate monomers and resins used as synthetic marble, optical parts, vehicle parts and raw material for other plastics. The medical division, or Kuraray Noritake Dental, Inc., develops bond cement, composite resin and other bone filling materials for the field of regenerative medicine. Lastly, the synthetic leather division produces alternative man-made leather, under the name Clarino, that possesses the same qualities as natural leather. The unit's plant in Okayama produces approximately 15.5 million square meters per year through four production lines. The Fibers and Textiles segment operates in two divisions. The fibers and industrial materials division sells industrial polyvinyl fibers used as cement reinforcers and in fishing nets, automotive brake hoses, ropes and knitted fabrics. The KuraFlex division, operating through KurarayKuraFlex Co., Ltd., produces nonwoven fabric used in medical, agricultural, industrial and electronic component applications. The fastening division, which operates through Kuraray Fastening Co., Ltd., produces molded plastic hook-and-loop fasteners under the Magic Tape and Magilock brand names. The Isoprene segment consists of three divisions. The isoprene chemicals division produces commodity chemicals, functional specialty chemicals, aroma chemicals, cosmetics raw materials and pharmaceutical and agricultural chemical intermediates. The elastomer division produces and markets thermoplastic rubbers such as Septon polymers for use in adhesives, sealants, asphalt and more. The Genestar division produces Genestar, a high-resistant polyamide resin.

FINANCIAL DATA: *Note: Data for latest year may not have been available at press time.*

In U.S. $	2012	2011	2010	2009	2008	2007
Revenue	4,107,892,000	4,043,498,000	3,706,037,000	4,194,754,000	4,649,258,000	4,289,476,000
R&D Expense						
Operating Income	609,356,400	591,120,100	339,018,700	325,981,700	535,843,500	447,779,500
Operating Margin %	14.83%	14.61%	9.14%	7.77%	11.52%	10.43%
SGA Expense	753,754,700				826,644,100	770,187,300
Net Income	350,352,400	319,992,000	181,639,000	144,554,200	284,499,200	249,518,500
Operating Cash Flow	474,120,800	774,996,900	896,650,000	522,361,100	628,538,900	433,773,800
Capital Expenditure	409,893,000	208,158,400	240,912,500	431,747,600	518,631,500	362,665,700
EBITDA	918,916,400	896,961,700	678,404,800	644,938,200	800,024,500	684,661,700
Return on Assets %	6.10%	5.69%	3.34%	2.69%	5.11%	4.52%
Return on Equity %	8.92%	8.47%	4.96%	4.01%	7.48%	6.42%
Debt to Equity	0.14	0.12	0.16	0.15	0.06	0.04

CONTACT INFORMATION:

Phone: 81 367011000 Fax: 81 367011005
Toll-Free:
Address: Omeda-Hankyu Bldg. Tower, 8-1, Kakudacho, Kita-ku, Tokyo, 530-8611 Japan

STOCK TICKER/OTHER:

Stock Ticker: KURRY
Exchange: PINX
Employees: 6,544
Fiscal Year Ends: 31-Mar
Parent Company:

SALARIES/BONUSES:

Top Exec. Salary: $ Bonus: $
Second Exec. Salary: $ Bonus: $

OTHER THOUGHTS:

Estimated Female Officers or Directors:
Hot Spot for Advancement for Women/Minorities:

L L Bean Inc

NAIC Code: 454113

www.llbean.com

TYPES OF BUSINESS:

Outdoor Apparel, Catalogs
Camping & Fishing Equipment
Footwear
Housewares
Online Sales
Outdoor Activity Classes

BRANDS/DIVISIONS/AFFILIATES:

Bean Boot
Outdoor Discovery Schools
Walk-On Adventures
L.L. Bean Home Store
L.L. Bean
L.L. Bean Outlet Store

CONTACTS: *Note: Officers with more than one job title may be intentionally listed here more than once.*

Christopher J. McCormick, CEO
Christopher J. McCormick, Pres.
Mark Fasold, CFO
Edward R. Howell, Chief Retail Officer
Leon A. Gorman, Chmn.

GROWTH PLANS/SPECIAL FEATURES:

L.L. Bean, Inc. is a retailer of more than 20,000 outdoor-related products, including apparel for men, women and children; footwear; luggage; housewares; and equipment for outdoor activities such as camping, hiking and fishing. With initial mail-order operations dating to 1912, the company now offers products through its retail stores, e-commerce sites and approximately 50 annual catalogs distributed to more than 160 countries. Approximately 90% of all merchandise sold carries the L.L. Bean label. The company currently operates retail stores in the U.S., Japan and China, including its Freeport, Maine flagship store, a 200,000-square foot facility open 24-hours-a-day, year-round. U.S. locations include approximately 18 L.L. Bean stores and 10 L.L. Bean Outlet stores in 10 states, primarily in New England and the Mid-Atlantic Region. The company has opened a total of 30 stores in China in 2011, with expectations for more to open. Additionally, the company operates the L.L. Bean Home Store, a 33,000 square-foot store that features vintage, handcrafted and one of a kind house wares. L.L. Bean's merchandise includes the Bean Boot, one of the company's signature products; boating gear; tote bags; and customized corporate and business items. The company manufactures its products at two facilities in Brunswick and Lewiston, Maine. In addition to retailing, the firm offers Outdoor Discovery Schools and Walk-On Adventures, programs that assist customers in discovering a variety of outdoor activities.

L.L. Bean offers employees benefits including health coverage; flexible spending accounts; an employee assistance program; life and disability insurance; a 401(k) plan and a company-paid pension plan; tuition reimbursement; employee discounts; onsite fitness programs and subsidies for gym memberships; the use of company vacation properties; and access to the Employee Use Room, which allows employees to borrow gear for outdoor activities.

FINANCIAL DATA: *Note: Data for latest year may not have been available at press time.*

In U.S. $	2012	2011	2010	2009	2008	2007
Revenue	1,520,000,000	1,450,000,000	1,400,000,000	1,360,000,000	1,500,000,000	1,620,000,000
R&D Expense						
Operating Income						
Operating Margin %						
SGA Expense						
Net Income						
Operating Cash Flow						
Capital Expenditure						
EBITDA						
Return on Assets %						
Return on Equity %						
Debt to Equity						

CONTACT INFORMATION:

Phone: 207-552-3028 Fax: 207-552-3080
Toll-Free: 800-441-5713
Address: 3 Campus Dr., Freeport, ME 04033 United States

STOCK TICKER/OTHER:

Stock Ticker: Private
Employees: 5,000
Parent Company:
Exchange:
Fiscal Year Ends: 28-Feb

SALARIES/BONUSES:

Top Exec. Salary: $ Bonus: $
Second Exec. Salary: $ Bonus: $

OTHER THOUGHTS:

Estimated Female Officers or Directors:
Hot Spot for Advancement for Women/Minorities:

Lacoste SA

www.lacoste.com

NAIC Code: 315240

TYPES OF BUSINESS:

Sportswear Design & Manufacture
Men's, Women's & Children's Apparel Stores
Apparel-Manufacturing
Apparel-Accessories
Home Textiles
Fragrances

BRANDS/DIVISIONS/AFFILIATES:

Lacoste Champions
Devanlay SA
Charment Group
Maus Freres SA
Movado Group

CONTACTS: *Note: Officers with more than one job title may be intentionally listed here more than once.*

Michel Lacoste, CEO
Felipe Oliveira Baptista, Artistic Dir.
Marc Lumet, Sr. Exec. VP
Jean-Claude Fauvet, Sr. Exec. VP
Sofie Lacoste Dournel, Chmn.

GROWTH PLANS/SPECIAL FEATURES:

Lacoste SA is a manufacturer and retailer of fashionable sport and street clothes for young men and women. Lacoste operates as a subsidiary of Devanlay SA, which is itself owned by the Swedish company Maus Freres SA. The company revolutionized sportswear by placing Lacoste's trademark crocodile logo on its line of simple, lightweight cotton polo shirts. Lacoste continues to market its clothing for amateurs and professionals of golf, tennis and yachting (its traditional clientele), but it has also branched out into more general club and casual lines. In addition to a variety of apparel and shoes for casual, sport and club settings, Lacoste offers a kaleidoscope of accessories, including leather goods, belts, watches, footwear, key chains, linens, umbrellas, eyewear and its own line of fragrances. Though some clothing is manufactured in France, the firm owns several international production facilities. The company markets its clothing in 114 countries, with major markets in the U.S., France, Italy, the U.K. and Spain. It distributes its products through more than 1,165 Lacoste boutiques and more than 2,000 locations within department stores and outlet malls. Devanlay owns the exclusive rights to manufacture Lacoste brand apparel; however, several companies were granted licensing agreements, such as the Charment Group and Movado Group, which produce Lacoste brand eyewear and watches. The company also granted the manufacturing rights of its fragrance line to Proctor & Gamble. In addition, Lacoste offers sponsorships to the Lacoste Champions, a group of professional golfers, yachters and tennis players who endorse the company's clothing. The Lacoste family still maintains creative control over the brand. In November 2012, Devanlay agreed to sell its 28% stake in Lacoste to Maus Freres.

FINANCIAL DATA: *Note: Data for latest year may not have been available at press time.*

In U.S. $	2012	2011	2010	2009	2008	2007
Revenue	2,285,000,000	2,095,410,000	1,850,000,000			
R&D Expense						
Operating Income						
Operating Margin %						
SGA Expense						
Net Income						
Operating Cash Flow						
Capital Expenditure						
EBITDA						
Return on Assets %						
Return on Equity %						
Debt to Equity						

CONTACT INFORMATION:

Phone: 33-1-44-58-12-12 Fax: 33-1-44-58-12-19
Toll-Free: 800-452-2678
Address: 8 rue de Castiglione, Paris, 75001 France

STOCK TICKER/OTHER:

Stock Ticker: Subsidiary Exchange:
Employees: Fiscal Year Ends:
Parent Company: DEVANLAY SA

SALARIES/BONUSES:

Top Exec. Salary: $ Bonus: $
Second Exec. Salary: $ Bonus: $

OTHER THOUGHTS:

Estimated Female Officers or Directors: 1
Hot Spot for Advancement for Women/Minorities:

Lacrosse Footwear

www.lacrossefootwear.com

NAIC Code: 316210

TYPES OF BUSINESS:

Footwear Manufacturing

BRANDS/DIVISIONS/AFFILIATES:

Danner
LaCrosse
LaCrosse Europe ApS
ABC-MART

CONTACTS:

Note: Officers with more than one job title may be intentionally listed here more than once.

Joseph P. Schneider, CEO
Richard Rosenthal, Chmn.

GROWTH PLANS/SPECIAL FEATURES:

LaCrosse Footwear, Inc. is a developer and marketer of branded, premium and innovative footwear for work and outdoor users. Its Danner and LaCrosse brands are distributed in the U.S. through a nationwide network of retailers and distributors, and internationally through its Danish subsidiary, LaCrosse Europe ApS, and through distributors and retailers in Asia, Europe and Canada. Work customers include people in law enforcement, transportation, mining, oil and gas, military services and other occupations that need high-performance and protective footwear as a critical tool for the job. Outdoor customers include people active in hunting, hiking and other outdoor recreational activities. The company sources approximately two-thirds of the products it sells through a network of international contract manufacturers, mainly in China and Thailand, with the remaining one-third manufactured domestically in its 36,000 square foot facility in Portland, Oregon. LaCrosse Footwear markets its two brands through four channels of distribution: wholesale (formerly known as retail), government, consumer direct and international. In 2012, the firm was acquired by Japanese footwear retailer ABC-MART for approximately $138 million.

FINANCIAL DATA:

Note: Data for latest year may not have been available at press time.

In U.S. $	2012	2011	2010	2009	2008	2007
Revenue		131,321,000	150,542,000	139,152,000	127,956,000	118,179,000
R&D Expense						
Operating Income						
Operating Margin %						
SGA Expense						
Net Income		3,000,000	6,881,000	5,510,000	6,167,000	7,300,000
Operating Cash Flow						
Capital Expenditure						
EBITDA						
Return on Assets %						
Return on Equity %						
Debt to Equity						

CONTACT INFORMATION:

Phone: 503 262-0110 Fax: 503 782-8190
Toll-Free:
Address: 17634 NE Airport Way, Portland, OR 97230 United States

STOCK TICKER/OTHER:

Stock Ticker: Subsidiary
Exchange:
Employees: 499
Fiscal Year Ends: 31-Dec
Parent Company: ABC-MART Inc

SALARIES/BONUSES:

Top Exec. Salary: $ Bonus: $
Second Exec. Salary: $ Bonus: $

OTHER THOUGHTS:

Estimated Female Officers or Directors:
Hot Spot for Advancement for Women/Minorities:

Lafuma SA

www.groupe-lafuma.fr

NAIC Code: 315210

TYPES OF BUSINESS:

Outdoor Apparel
Outdoor Equipment & Apparel
Backpacks
Footwear
Camping Furniture
Technical Mountaineering Equipment
Hunting & Fishing Gear

BRANDS/DIVISIONS/AFFILIATES:

Lafuma
Eider
Le Chameau
Millet
Oxbow
Ober

CONTACTS: *Note: Officers with more than one job title may be intentionally listed here more than once.*

Felix Sulzberger, CEO
Gilles Venet, CFO
Emmannuel Durand, Head-Human Resources
Olivier Salivet, Head-IT

GROWTH PLANS/SPECIAL FEATURES:

Lafuma SA specializes in sports and outdoor apparel and equipment. The company markets its products under six primary brands: Lafuma, Ober, Millet, Eider, Oxbow and Le Chameau. It organizes its products into four segments: great outdoor, mountain, boardsports and country. The great outdoor segment is composed mostly of the Lafuma brand, which specializes in hiking, trail running, skiing and camping gear. The mountain segment focuses on high altitude and cold weather attire and includes the Millet and Eider brands. Millet is a mountaineering, bouldering and skiing brand, and Eider offers technical gear, including extreme cold weather gear designed for Himalayan and arctic expeditions. The boardsports segment is made up of Oxbow, which offers apparel and equipment designed for surfing, windsurfing, kite surfing, snowboarding and skiing. The country sector, comprised of Le Chameau, specializes in horseback riding, hunting and country apparel and accessories. The firm operates a number of flagship Lafuma and Oxbow retail stores in France, as well as international Lafuma shops in England, Spain, Belgium and Germany.

FINANCIAL DATA: *Note: Data for latest year may not have been available at press time.*

In U.S. $	2012	2011	2010	2009	2008	2007
Revenue		332,600,000	321,300,000	343,220,000	348,150,000	332,740,000
R&D Expense						
Operating Income						
Operating Margin %						
SGA Expense						
Net Income		5,200,000	-4,700,000	2,290,000	2,130,000	3,980,000
Operating Cash Flow						
Capital Expenditure						
EBITDA						
Return on Assets %						
Return on Equity %						
Debt to Equity						

CONTACT INFORMATION:

Phone: 33-4-75-31-31-31 Fax: 33-4-75-31-57-26
Toll-Free:
Address: 6, Rue Victor Lafuma, Anneyron, 26140 France

STOCK TICKER/OTHER:

Stock Ticker: LAF Exchange: Paris
Employees: 1,925 Fiscal Year Ends: 30-Sep
Parent Company:

SALARIES/BONUSES:

Top Exec. Salary: $ Bonus: $
Second Exec. Salary: $ Bonus: $

OTHER THOUGHTS:

Estimated Female Officers or Directors:
Hot Spot for Advancement for Women/Minorities:

Lakeland Industries Inc

www.lakeland.com

NAIC Code: 33911

TYPES OF BUSINESS:

Safety Clothing
Reusable Industrial & Medical Apparel
Chemical Protection Clothing
Specialty Safety Gloves
Heat Resistant Clothing
Disposable Protective Garments

BRANDS/DIVISIONS/AFFILIATES:

TyChem
Fyrepel
Kevlar
Tyvek

CONTACTS: *Note: Officers with more than one job title may be intentionally listed here more than once.*

Christopher J. Ryan, CEO
Christopher J. Ryan, Pres.
Gary Pokrassa, CFO
Paul C. Smith, VP-Sales
Gregory D. Pontes, VP-Mfg.
Christopher J. Ryan, Corp. Sec.
Stephen M. Bachelder, Chmn./National Sales Mgr.
Charles D. Robseron, VP-Int'l Sales

GROWTH PLANS/SPECIAL FEATURES:

Lakeland Industries, Inc. manufactures and sells a comprehensive line of safety garments and accessories for industrial safety and protective clothing industries. In North America, the firm's products are sold through a network of over 1,200 safety and mill supply distributors. Lakeland's products are sold in over 30 countries internationally. The company's major product areas include disposable and limited-use protective clothing, chemical protective suits, fire fighting and heat protective apparel, gloves and arm guards for the food service market, reusable woven garments and high visibility clothing for traffic and public safety officials. The firm's disposable and reusable garments protect the wearer from contaminants or irritants, such as chemicals, pesticides, fertilizers, paint, grease and dust; from viruses and bacteria; and from limited exposure to hazardous waste and toxic chemicals including acids, asbestos, lead and hydro-carbons (PCBs). Lakeland's products are also used to prevent human contamination of manufacturing processes in clean-room environments. Disposable clothing products include coveralls, lab coats, hoods, aprons, sleeves and smocks. Lakeland's heat protective gear is used by firefighters (under the Fyrepel brand) as well as for maintenance of extreme high-temperature industrial equipment (up to 1,500 degrees Fahrenheit ambient) and for crash and rescue operations. The company's high-end chemical protective suits protect wearers from highly concentrated and powerful chemical and biological toxins such as toxic wastes at Super Fund sites, accidental toxic chemical spills or biological discharges, the handling of chemical or biological warfare weapons and the cleaning and maintenance of chemical, petrochemical and nuclear facilities. Lakeland buys most of its raw materials for manufacturing from DuPont, including those required for the TyChem, Kevlar and Tyvek brands.

FINANCIAL DATA: *Note: Data for latest year may not have been available at press time.*

In U.S. $	2012	2011	2010	2009	2008	2007
Revenue	96,326,990	101,236,300	94,140,820	102,268,100	95,740,060	100,170,900
R&D Expense						
Operating Income	1,437,567	3,453,965	2,457,507	6,272,310	4,983,474	6,721,905
Operating Margin %	1.49%	3.41%	2.61%	6.13%	5.20%	6.71%
SGA Expense	27,348,230	26,314,390	22,948,240	21,696,880	17,373,880	17,553,970
Net Income	-376,825	971,999	1,029,792	4,549,468	3,290,862	5,104,199
Operating Cash Flow	-4,459	-836,813	18,677,710	1,636,950	-2,119,202	8,225,873
Capital Expenditure	4,871,932	1,698,500	1,192,251	2,371,914	3,427,458	912,651
EBITDA	3,457,410	3,940,479	4,291,222	8,524,874	6,381,906	8,334,757
Return on Assets %	-.36%	1.01%	1.07%	4.88%	4.14%	6.96%
Return on Equity %	-.51%	1.32%	1.46%	6.63%	4.87%	8.04%
Debt to Equity	0.22	0.17	0.02	0.37	0.12	0.05

CONTACT INFORMATION:

Phone: 631 981-9700 Fax: 631 981-9751
Toll-Free: 800-645-9291
Address: 701 Koehler Ave., Ste. 7, Ronkonkoma, NY 11779 United States

STOCK TICKER/OTHER:

Stock Ticker: LAKE Exchange: NAS
Employees: 1,840 Fiscal Year Ends: 31-Jan
Parent Company:

SALARIES/BONUSES:

Top Exec. Salary: $368,000 Bonus: $17,250
Second Exec. Salary: $207,036 Bonus: $

OTHER THOUGHTS:

Estimated Female Officers or Directors:
Hot Spot for Advancement for Women/Minorities:

Lands End Inc

www.landsend.com

NAIC Code: 454113

TYPES OF BUSINESS:

Casual Apparel, Catalogs
Retail & Outlet Stores
Online Sales
Luggage
Home Products

BRANDS/DIVISIONS/AFFILIATES:

Sears Holdings Corporation
Lands' End Canvas
Durable Goods

CONTACTS: *Note: Officers with more than one job title may be intentionally listed here more than once.*

Edgar Huber, CEO
Tim Martin, COO
Edgar Huber, Pres.
Tim Martin, CFO/Sr. VP
Susan Sachatello, Chief Mktg. Officer/Sr. VP
Kelly Ritchie, Sr. VP-Employee & Customer Svcs.
Chris Kolbe, Pres., Brand
Harry S. Schutte, VP-Int'l Oper.
Harry S. Schutte, VP-Dist.

GROWTH PLANS/SPECIAL FEATURES:

Lands' End, Inc., a subsidiary of Sears Holdings Corporation, is a leading direct retailer of traditionally-styled casual clothing for men, women and children as well as outerwear, swimwear, home products and wheeled and soft luggage. The company's products are offered through regular mailings of its primary and specialty catalogs, retail and outlet stores and via the Internet, at LandsEnd.com. The firm also operates several Inlet stores in Wisconsin, Illinois and Minnesota. The firm additionally offers Not Quite Perfect clothing and accessories at up to 70% off original catalog prices. A selection of Lands' End merchandise can also be found in Lands' End Shops in select Sears retail locations. In addition, Lands' End has developed web sites for customers in the U.K., Japan, Austria, Germany and France as well as a business outfitters web site. The firm recently launched the Lands' End Canvas collection, which features updated designs and fits of classic casual apparel at higher price points than its regular collection. In March 2012, the firm introduced Durable Goods, a new line for men made in the U.S.

Lands' End offers its employees health insurance, a retirement savings plan, employee discounts and emergency child care.

FINANCIAL DATA: *Note: Data for latest year may not have been available at press time.*

In U.S. $	2012	2011	2010	2009	2008	2007
Revenue	1,700,000,000	1,600,000,000				
R&D Expense						
Operating Income						
Operating Margin %						
SGA Expense						
Net Income						
Operating Cash Flow						
Capital Expenditure						
EBITDA						
Return on Assets %						
Return on Equity %						
Debt to Equity						

CONTACT INFORMATION:

Phone: 608-935-6170 Fax: 800-332-0103
Toll-Free: 800-963-4816
Address: 1 Lands' End Ln., Dodgeville, WI 53595 United States

STOCK TICKER/OTHER:

Stock Ticker: Subsidiary Exchange:
Employees: Fiscal Year Ends: 31-Dec
Parent Company: SEARS HOLDINGS CORPORATION

SALARIES/BONUSES:

Top Exec. Salary: $ Bonus: $
Second Exec. Salary: $ Bonus: $

OTHER THOUGHTS:

Estimated Female Officers or Directors: 2
Hot Spot for Advancement for Women/Minorities: Y

Lectra SA

NAIC Code: 511210N

www.lectra.com

TYPES OF BUSINESS:

Computer Software
Consulting

BRANDS/DIVISIONS/AFFILIATES:

LectraOnline Exchange

CONTACTS:

Note: Officers with more than one job title may be intentionally listed here more than once.

Daniel Harari, CEO
Jerome Viala, CFO
Edouard Macquin, Dir.-Sales
Laurent Alt, Dir.-Software R&D
Veronique Zoccoletto, CIO/Human Capital Officer
Laurence Jacquot, Dir.-Mfg. & Hardware R&D
Daniel Dufag, General Counsel
Myriam Akoun-Brunet, Dir.-Comm.
Didier Teiller, Dir.-Svcs.
Jean-Maurice Ferauge, Dir.-Professional Svcs.
Roy Shurling, Managing Dir.-North America
Adriana Vono Papavero, Managing Dir.-South America
Andre Harari, Chmn.
Andreas Kim, Managing Dir.-China

GROWTH PLANS/SPECIAL FEATURES:

Lectra SA provides a range of high-technology products and services to customers using soft materials in the fashion, apparel, retail, textile, furniture, footwear, luggage, leather goods, automotive, aerospace, marine and industrial fabrics markets. It organizes its products and services into four segments: software, accounting for approximately 28% of revenues; computer-aided design and computer-aided manufacturing (CAD/CAM) equipment, 26% of revenues; services (including training, consulting, hardware maintenance and on-line services), 23%; and spare parts and consumables, 23%. These technologies are developed differently to meet the needs of each individual industry and for each individual material. The software segment is designed to aid customers in the design, development and manufacturing processes. The firm also offers product lifecycle management (PLM) programs. The CAD/CAM software and equipment programs are designed to handle 3D body measurement, pattern plotting, digital fabric printing, spreading and automated cutting. Lectra's services are provided by a team of professionals offering training, consulting, installation, hardware maintenance, engineering services, on-line assistance and support through the company's six international call centers. This support includes the shipping of spare parts, another core business. The LectraOnline Exchange Internet-based platform permits secure data communication and collaboration. Lectra is a global company with approximately 23,000 customers, 31 subsidiaries and operations in more than 100 countries. During 2012, approximately 47% of its new orders were derived from customers in Europe; 26% from the Americas; 6% from the Asia/Pacific region; and 21% from the rest of the world. The fashion industry is its largest market, which accounted for 49% of its 2012 revenues, while automotive made up 37%; furniture, 7%; and other industries, 7%.

FINANCIAL DATA:

Note: Data for latest year may not have been available at press time.

In U.S. $	2012	2011	2010	2009	2008	2007
Revenue	265,359,700	275,371,800	254,466,400	204,850,200	264,954,500	289,602,800
R&D Expense	11,536,000	11,463,000	9,547,000	8,673,000	10,607,000	14,225,000
Operating Income	26,496,390	38,676,120	33,547,740	-6,248,997	9,308,639	13,654,720
Operating Margin %	9.98%	14.04%	13.18%	-3.05%	3.51%	4.71%
SGA Expense	151,927,000	138,974,300	138,674,800	137,007,200	153,527,700	159,887,700
Net Income	18,245,520	25,671,300	20,924,040	-4,844,879	4,330,035	7,770,795
Operating Cash Flow	21,824,020	23,685,480	49,692,430	14,704,470	-1,636,801	-4,006,419
Capital Expenditure	6,771,864	5,005,349	3,139,877	3,119,818	4,918,428	6,910,939
EBITDA	35,132,390	46,243,650	41,147,360	7,781,493	21,144,690	21,487,030
Return on Assets %	8.07%	11.44%	9.52%	-2.16%	1.84%	3.39%
Return on Equity %	22.00%	38.12%	46.90%	-13.72%	11.90%	11.79%
Debt to Equity	0.01	0.28	0.65	1.7	1.75	1.85

CONTACT INFORMATION:

Phone: 33 0153644200 Fax: 33 0153644300
Toll-Free:
Address: 16-18, Rue Chalgrin, Paris, 75016 France

STOCK TICKER/OTHER:

Stock Ticker: LCTSF Exchange: GREY
Employees: 1,326 Fiscal Year Ends: 31-Dec
Parent Company:

SALARIES/BONUSES:

Top Exec. Salary: $ Bonus: $
Second Exec. Salary: $ Bonus: $

OTHER THOUGHTS:

Estimated Female Officers or Directors: 3
Hot Spot for Advancement for Women/Minorities: Y

Levi Strauss & Co

www.levistrauss.com

NAIC Code: 315240

TYPES OF BUSINESS:

Jeans & Casual Wear
Footwear & Accessories
Retail Stores
COM Franchised Stores

BRANDS/DIVISIONS/AFFILIATES:

501 Original
Levi's
Dockers
Levi Strauss Signature
Levi's Workwear by Filson
Denizen

CONTACTS: *Note: Officers with more than one job title may be intentionally listed here more than once.*

Chip Bergh, CEO
Chip Bergh, Pres.
Harmit Singh, CFO/Exec. VP
Varun Bhatia, Chief Human Resources Officer/Sr. VP
Tom Peck, CIO/Sr. VP
Seth Jaffe, Chief Counsel/Sr. VP
Jill Nash, Chief Comm. Officer/VP-Corp. Affairs
Seth Ellison, Exec. VP/Pres., Global Dockers Brand
Aaron Boey, Exec. VP/Pres., Asia Pacific
James Curleigh, Exec. VP/Pres., The Levi's Brand
Anne Rohosy, Exec. VP/Pres., Commercial Oper. & Europe
Stephen Neal, Chmn.
Robert Hanson, Pres., Global Levi's Brand
David Love, Chief Supply Chain Officer/Sr. VP

GROWTH PLANS/SPECIAL FEATURES:

Levi Strauss & Co. is one of the world's largest and best-recognized name brands in the apparel industry, with retail outlets in over 110 countries. Best known for its trademark 501 Original jeans, the company has recently begun transforming its clothing line, expanding its core Levi's and Dockers brands and introducing its new Levi Strauss Signature brand for value-conscious consumers in North America, Europe and Asia. In addition to jeans, the company's product line includes an array of tops, footwear, outerwear and accessories. The firm is organized into three divisions: Levi Strauss North America, based in the company's San Francisco main office; Levi Strauss Europe, based in Brussels; and the Asia Pacific Division, based in Singapore. The company's products are sold in approximately 50,000 locations worldwide, including 2,300 locations that are company-owned or franchised. Levi Strauss was one of the first international corporations to establish a comprehensive ethical code of conduct for its manufacturing and finishing contractors, ensuring proper working conditions for its overseas workforce. Levi Strauss recently launched Denizen, a line of jeans marketed towards middle class consumers in the Asia Pacific. Recently, the firm has partnered with Filson, a Seattle-based outdoor clothing retailer, to introduce Levi's Workwear by Filson, a line of menswear.

Levi Strauss offers employees emergency financial assistance; apparel, technology and business discounts; and medical, dental, life and vision insurance.

FINANCIAL DATA: *Note: Data for latest year may not have been available at press time.*

In U.S. $	2012	2011	2010	2009	2008	2007
Revenue	4,610,000,000	4,762,000,000	4,410,000,000	4,105,766,000	4,400,914,000	4,360,929,000
R&D Expense						
Operating Income						
Operating Margin %						
SGA Expense						
Net Income	144,000,000	138,000,000	157,000,000	151,875,000	229,285,000	460,385,000
Operating Cash Flow						
Capital Expenditure						
EBITDA						
Return on Assets %						
Return on Equity %						
Debt to Equity						

CONTACT INFORMATION:

Phone: 415-501-6000 Fax: 415-501-7112
Toll-Free:
Address: 1155 Battery St., San Francisco, CA 94111 United States

STOCK TICKER/OTHER:

Stock Ticker: Private
Employees: 17,000
Parent Company:
Exchange:
Fiscal Year Ends: 30-Nov

SALARIES/BONUSES:

Top Exec. Salary: $ Bonus: $
Second Exec. Salary: $ Bonus: $

OTHER THOUGHTS:

Estimated Female Officers or Directors: 6
Hot Spot for Advancement for Women/Minorities: Y

Li Ning Company Limited

NAIC Code: 316210

www.lining.com

TYPES OF BUSINESS:

Apparel & Shoe Manufacturing

BRANDS/DIVISIONS/AFFILIATES:

LI-NING
Double Happiness
Kason
AIGLE
Lotto
TPG

CONTACTS: *Note: Officers with more than one job title may be intentionally listed here more than once.*

Li Ning, CEO
Catherine Teng, Chief Mktg. Officer/VP
Edwin Alexander Jonkers, Chief Prod. & Merch. Officer/VP
Jin-Goon Kim, Dir.-Daily Oper.
Jin-Goon Kim, Dir.-Dev.
Zhang Zhi Yong, Exec. Dir.
Li Ning, Chmn.
Deng Hongbing, Chief Supply Chain Officer/VP

GROWTH PLANS/SPECIAL FEATURES:

Li Ning Company Limited is a leading China-based sports apparel company focused on branding, research and development, design, manufacturing, distribution and retail. The company's products are designed with a focus on five athletic areas: basketball, running, soccer, tennis and general fitness. Its production and supply chain operations are outsourced to third-party manufacturing and franchising agents. Li Ning sells sports footwear, apparel, equipment and accessories for sports and leisure under the LI-NING brand with an emphasis on soccer and tennis. Li Ning has additionally established a joint venture with AIGLE (Aigle International SA) for the manufacture, marketing and distribution of AIGLE branded outdoor and casual wear. The company maintains a 57.5% controlling interest in subsidiary Double Happiness, which brands, develops and manufactures equipment for table tennis. Furthermore, Li Ning manufactures and sells licensed sports fashion items under the brand name Lotto, which is an Italian sports fashion company. Badminton equipment is developed and produced under the Kason brand. The firm both franchises and directly manages its LI-NING retail outlets throughout China and has research and development centers in mainland China, Hong Kong and the U.S. The company has been going through a three-year restructuring plan. This included the closing of about 1,800 stores during 2012, reducing the total to 6,434. The company has also revitalized its inventory and written off obsolete merchandise. Private equity firm TPG has been helping with this effort.

FINANCIAL DATA: *Note: Data for latest year may not have been available at press time.*

In U.S. $	2012	2011	2010	2009	2008	2007
Revenue	1,083,722,000	1,435,847,000	1,524,296,000	1,348,746,000	1,075,869,000	699,346,600
R&D Expense						
Operating Income	-256,072,200	101,467,600	248,745,600	215,797,900	154,417,300	98,074,230
Operating Margin %	-23.62%	7.06%	16.31%	15.99%	14.35%	14.02%
SGA Expense	693,286,000	583,276,800	503,265,400	443,059,800	373,904,100	45,407,420
Net Income	-314,458,800	66,090,250	182,065,200	155,883,600	116,961,200	76,226,140
Operating Cash Flow	-149,741,900	2,503,900	159,351,400	210,132,700	112,404,800	63,188,330
Capital Expenditure	34,952,160	62,361,580	43,797,500	43,765,660	41,879,290	39,087,530
EBITDA	-220,222,600	145,798,200	285,096,400	215,797,900	154,417,300	98,074,230
Return on Assets %	-29.62%	5.55%	18.56%	19.44%	20.26%	19.15%
Return on Equity %	-72.27%	10.68%	34.52%	38.15%	37.62%	30.12%
Debt to Equity	0.35					0.03

CONTACT INFORMATION:

Phone: 8610-8080-0808 Fax: 8610-8080-0000
Toll-Free:
Address: #8 Xing Guang 5th St., Opto-Mechanics Indistrial Park, Tongzhou Dist., Beijing, 101111 China

STOCK TICKER/OTHER:

Stock Ticker: LNNGF Exchange: PINX
Employees: 4,001 Fiscal Year Ends: 31-Dec
Parent Company:

SALARIES/BONUSES:

Top Exec. Salary: $ Bonus: $
Second Exec. Salary: $ Bonus: $

OTHER THOUGHTS:

Estimated Female Officers or Directors: 1
Hot Spot for Advancement for Women/Minorities: Y

Limited Brands Inc

www.limitedbrands.com

NAIC Code: 448120

TYPES OF BUSINESS:

Apparel, Retail
Contract Manufacturing
Apparel Importing
Catalog & Online Sales
Lingerie
Cosmetics
Fragrances
Candles

BRANDS/DIVISIONS/AFFILIATES:

Limited, Inc. (The)
Victoria's Secret
Henri Bendel
Bath & Body Works
White Barn Candle Company
C.O. Bigelow
La Senza Corporation
Mast Global

CONTACTS: *Note: Officers with more than one job title may be intentionally listed here more than once.*

Leslie H. Wexner, CEO
Charles C. McGuigan, COO
Stuart Burgdoerfer, CFO/Exec. VP
Jane L. Ramsey, Exec. VP-Human Resources
Sharen J. Turney, CEO/Pres., Victoria's Secret Megabrand
Nicholas Coe, CEO-Bath & Body Works
Leslie H. Wexner, Chmn.

GROWTH PLANS/SPECIAL FEATURES:

Limited Brands, Inc., formerly The Limited, Inc., is an apparel, lingerie, personal care products, accessories and fragrances retailer under brand names including Victoria's Secret and Bath & Body Works. The company operates approximately 2,941 retail stores. Victoria's Secret and Victoria's Secret Beauty, with approximately 1,036 stores in the U.S. and Canada, are specialty retailers of women's intimate apparel, beauty products and related accessories. Victoria's Secret's direct marketing segment is in charge of the catalog, which reaches over 390 million customers a year. Bath & Body Works, with 1,656 stores in the U.S. and Canada, features personal care products and also operates White Barn Candle Company to supply its candle needs and C.O. Bigelow in Chicago for body products. Limited Brands also retains full ownership of the approximately 19 Henri Bendel women's personal care, apparel and accessory stores. In addition, the firm owns a 49% interest in Mast Global, a contract manufacturer and apparel importer, which supplies merchandise to Victoria's Secret, Express and Limited Stores. Finally, the company operates La Senza, a Canadian lingerie company. La Senza operates approximately 300 stores. Limited Brands runs four e-commerce sites: VictoriasSecret.com, HenriBendel.com, LaSenza.com and BathandBodyWorks.com. In November 2011, private equity firm Sycamore Partners acquired a 51% stake in Mast Global.

Limited Brands offers its employees medical, dental, prescription, life and vision insurance; short- and long-term disability coverage; an employee assistance program; flexible spending; a retirement plan; a 401(k); a stock purchase plan; product discounts; tuition assistance; paid time off; and adoption assistance.

FINANCIAL DATA: *Note: Data for latest year may not have been available at press time.*

In U.S. $	2012	2011	2010	2009	2008	2007
Revenue	10,364,000,000	9,613,000,000	8,632,001,000	9,043,000,000	10,134,000,000	
R&D Expense						
Operating Income	1,238,000,000	1,284,000,000	868,000,000	589,000,000	1,110,000,000	
Operating Margin %	11.94%	13.35%	10.05%	6.51%	10.95%	
SGA Expense	2,698,000,000	2,341,000,000	2,166,000,000	2,311,000,000	2,662,000,000	
Net Income	850,000,000	805,000,000	448,000,000	220,000,000	718,000,000	
Operating Cash Flow	1,266,000,000	1,284,000,000	1,174,000,000	954,000,000	765,000,000	
Capital Expenditure	426,000,000	274,000,000	202,000,000	479,000,000	749,000,000	
EBITDA	1,864,000,000	1,853,000,000	1,280,000,000	977,000,000	1,630,000,000	
Return on Assets %	13.53%	11.81%	6.33%	3.05%	9.88%	
Return on Equity %	105.39%	44.00%	22.08%	10.75%	27.75%	
Debt to Equity	25.4	1.69	1.24	1.54	1.3	

CONTACT INFORMATION:

Phone: 614 415-7000 Fax: 614 479-7440
Toll-Free:
Address: 3 Limited Pkwy., Columbus, OH 43230 United States

STOCK TICKER/OTHER:

Stock Ticker: LTD
Exchange: NYS
Employees: 96,500
Fiscal Year Ends: 31-Jan
Parent Company:

SALARIES/BONUSES:

Top Exec. Salary: $1,924,000 Bonus: $
Second Exec. Salary: $1,380,769 Bonus: $

OTHER THOUGHTS:

Estimated Female Officers or Directors: 4
Hot Spot for Advancement for Women/Minorities: Y

Loehmann's Holdings Inc

www.loehmanns.com

NAIC Code: 448120

TYPES OF BUSINESS:

Discount Apparel Stores
Designer Apparel

BRANDS/DIVISIONS/AFFILIATES:

Back Room (The)
Insider Club (The)
Istithmar PJSC

CONTACTS: *Note: Officers with more than one job title may be intentionally listed here more than once.*

Jerry Politzer, CEO
Joseph M. Melvin, COO
Robert Glass, Pres.
Nancy Straface, Sr. VP-Human Resources
William Thayer, CIO/Sr. VP-IT
Lesley Heller, VP-Merch.
Arthur Reiner, Chmn.
William Thayer, Sr. VP-Logistics

GROWTH PLANS/SPECIAL FEATURES:

Loehmann's Holdings, Inc. is a leading national specialty retailer of well-known designer and brand-name women's fashion apparel, accessories and shoes. The firm operates 40 stores in 11 states and Washington, D.C. Certain stores also carry a wide range of men's apparel, including dress shirts, ties, belts, suspenders, sweaters and jackets. Loehmann's products are offered at prices that are typically 30% to 65% below department store prices. The company's target customers are relatively affluent women aged 30-55 who are attracted to designer merchandise offered at bargain prices. The firm's brand-name merchandise includes offerings from Donna Karan, Calvin Klein, Ralph Lauren, Diane Von Furstenberg, Kenneth Cole, Dolce & Gabbana, Valentino, Nicole Miller, Anne Klein and Michael Kors. The company's simple and functional store formats are designed to project the image of deep discounts and exceptional value. Loehmann's stores have a large, open selling area with wall-to-wall merchandise and a smaller, more intimate area called The Back Room. The Back Room displays designer, bridge and import merchandise, including evening dresses and suits. The majority of the firm's advertising efforts involve direct-mail announcements to members of The Insider Club, a $25 per year membership program that gives members birthday discounts and additional savings with a points program. The company is owned by private equity firm Istithmar PJSC, based in Dubai. Recently, the firm emerged from bankruptcy protection by securing $45 million in exit financing from Wells Fargo Bank and Whippoorwill Associates.

The firm offers employees life, short- and long-term disability, medical, dental and vision insurance; employee assistance programs; a 401(k) plan; a flexible spending plan; commuter benefits; and merchandise discounts.

FINANCIAL DATA: *Note: Data for latest year may not have been available at press time.*

In U.S. $	2012	2011	2010	2009	2008	2007
Revenue						
R&D Expense						
Operating Income						
Operating Margin %						
SGA Expense						
Net Income						
Operating Cash Flow						
Capital Expenditure						
EBITDA						
Return on Assets %						
Return on Equity %						
Debt to Equity						

CONTACT INFORMATION:

Phone: 718-409-2000 Fax: 718-518-2766
Toll-Free: 800-366-5634
Address: 2500 Halsey St., Bronx, NY 10461 United States

STOCK TICKER/OTHER:

Stock Ticker: Private Exchange:
Employees: Fiscal Year Ends: 31-Jan
Parent Company: ISTITHMAR PJSC

SALARIES/BONUSES:

Top Exec. Salary: $ Bonus: $
Second Exec. Salary: $ Bonus: $

OTHER THOUGHTS:

Estimated Female Officers or Directors: 2
Hot Spot for Advancement for Women/Minorities:

Lord & Taylor LLC

www.lordandtaylor.com

NAIC Code: 448140

TYPES OF BUSINESS:

Department Stores

BRANDS/DIVISIONS/AFFILIATES:

Hudson's Bay Company
LordAndTaylor.com

CONTACTS:

Note: Officers with more than one job title may be intentionally listed here more than once.

Bonnie Brooks, CEO
Bonnie Brooks, Pres.
Don Watros, COO-Hudson's Bay Company
Richard A. Baker, Chmn.

GROWTH PLANS/SPECIAL FEATURES:

Lord & Taylor LLC operates approximately 47 department stores in nine states in the northeastern U.S., ranging in size from 70,000 to 160,000 square feet. The firm also operates an e-commerce web site, LordAndTaylor.com. Brands sold by the company include Seven For All Mankind, Diesel, Iisli, Tracy Reese, Bryan by Bryan Bradley, Cole Haan, Lacoste and Kate Spade. Sportswear, denim, dresses and jewelry are some of the firm's best-performing categories. Additional products sold include shoes, handbags, fragrance, beauty products and accessories. The company's 650,000 square foot flagship store on Fifth Avenue in New York is famous for its annual Christmas window display. Lord & Taylor's biggest customers are suburban women in their mid-40s with roughly $125,000 in annual household income; 88% of the firm's customers own a home, 68% are female, more than 80% are married and the majority have more than two children. Annual sales per square foot of store space were approximately $220 in 2013.

Employees are offered medical, dental and vision benefits; a 401(k) plan; and store-wide discounts.

FINANCIAL DATA:

Note: Data for latest year may not have been available at press time.

In U.S. $	2012	2011	2010	2009	2008	2007
Revenue						
R&D Expense						
Operating Income						
Operating Margin %						
SGA Expense						
Net Income						
Operating Cash Flow						
Capital Expenditure						
EBITDA						
Return on Assets %						
Return on Equity %						
Debt to Equity						

CONTACT INFORMATION:

Phone: 212-391-3344 Fax: 212-768-0743
Toll-Free: 800-223-7440
Address: 424 5th Ave., New York, NY 10018 United States

STOCK TICKER/OTHER:

Stock Ticker: Subsidiary Exchange:
Employees: Fiscal Year Ends:
Parent Company: HUDSON'S BAY COMPANY

SALARIES/BONUSES:

Top Exec. Salary: $ Bonus: $
Second Exec. Salary: $ Bonus: $

OTHER THOUGHTS:

Estimated Female Officers or Directors: 1
Hot Spot for Advancement for Women/Minorities:

lululemon Athletica Inc

NAIC Code: 448140

www.lululemon.com

TYPES OF BUSINESS:

Athletic Apparel
Organic Products
Accessories

BRANDS/DIVISIONS/AFFILIATES:

lululemon athletica

CONTACTS:
Note: Officers with more than one job title may be intentionally listed here more than once.

Christine Day, CEO
John E. Currie, CFO
Delaney Schweitzer, Exec. VP-Retail Oper., North America
Dennis J. Wilson, Chmn.

GROWTH PLANS/SPECIAL FEATURES:

lululemon athletica, Inc. is a designer and retailer of technical athletic apparel in North America, specializing in yoga-inspired apparel and accessories. The company offers a comprehensive line of performance apparel and accessories for both women and men. The apparel assortment, including items such as fitness pants, shorts, tops and jackets, is designed for healthy lifestyle activities such as yoga, dance, running and general fitness. Stores also host self-defense classes, yoga classes and goal-setting workshops led by community ambassadors. The firm's fitness-related accessories include an array of items such as bags, socks, hats, underwear, balance balls, yoga mats, yoga bricks, towels, water bottles and headbands. Lululemon's clothing line consists of several branded fabrics, including luon, made of nylon and lycra, which wicks away moisture, stretches in four directions and is designed to eliminate irritation; silverescent, which incorporates silver directly into the fabric as a way to reduce odors due to certain antibacterial properties of silver; and luxtreme, a silky and lightweight wicking fabric that is incorporated in the firm's running lines. The company also offers a line of casual, organic products made with organically grown, spun and dyed cotton as well as natural, sustainable and/or recycled materials, such as vitasea, derived from a seaweed compound. In addition, customers can bring in lululemon pants to any location for a complimentary hemming. Lululemon sells its products through 186 corporate-owned stores in North America, including 135 in the U.S. and 51 in Canada as well as its e-commerce site.

The company hires sales people who are fit, well educated and enthusiastic about exercise and healthy life styles. The stores attempt to develop deep relationships with customers through yoga classes, an engaging sales style and other special activities. Merchandise is typically sold at full price.

FINANCIAL DATA:
Note: Data for latest year may not have been available at press time.

In U.S. $	2012	2011	2010	2009	2008	2007
Revenue	1,000,839,000	711,704,000	452,898,000	353,488,200	274,713,300	
R&D Expense						
Operating Income	286,958,000	180,391,000	86,546,000	56,564,170	50,124,800	
Operating Margin %	28.67%	25.34%	19.10%	16.00%	18.24%	
SGA Expense	282,312,000	212,784,000	136,161,000	118,098,300	96,177,340	
Net Income	184,063,000	121,847,000	58,281,000	39,363,230	30,842,440	
Operating Cash Flow	203,615,000	179,995,000	117,960,000	46,437,780	38,090,560	
Capital Expenditure	122,311,000	42,839,000	16,307,000	40,530,460	29,675,940	
EBITDA	317,217,000	205,005,000	107,378,000	72,387,120	58,465,540	
Return on Assets %	29.83%	30.21%	22.46%	21.46%	27.18%	
Return on Equity %	37.11%	39.08%	30.04%	29.49%	41.28%	
Debt to Equity						

CONTACT INFORMATION:

Phone: 604 732-6124 Fax: 604 874-6124
Toll-Free: 877-263-9300
Address: 400-1818 Cornwall Ave., Vancouver, BC V6J 1C7 Canada

STOCK TICKER/OTHER:

Stock Ticker: LULU
Exchange: NAS
Employees: 3,219
Fiscal Year Ends: 31-Jan
Parent Company:

SALARIES/BONUSES:

Top Exec. Salary: $646,771 Bonus: $
Second Exec. Salary: $428,461 Bonus: $

OTHER THOUGHTS:

Estimated Female Officers or Directors: 5
Hot Spot for Advancement for Women/Minorities: Y

LVMH Moet Hennessy Louis Vuitton SA

www.lvmh.com

NAIC Code: 312140

TYPES OF BUSINESS:

Luxury Goods, Manufacture & Retail
Wines & Spirits, Manufacturing
Fashion & Leather Goods, Manufacturing & Retail
Perfumes & Cosmetics, Manufacturing & Retail
Duty-Free Retail
Online Cosmetics & Apparel Retail
Watches & Jewelry, Manufacturing
Magazine Publishing & Advertising Services

BRANDS/DIVISIONS/AFFILIATES:

Dom Perignon
Aqua Di Parma
BeneFit Cosmetics
Christian Dior SA
Le Bon Marche
Fresh
Make Up For Ever
Louis Vuitton

CONTACTS: *Note: Officers with more than one job title may be intentionally listed here more than once.*

Bernard Arnault, CEO
Chantal Gaemperie, Dir.-Human Resources
Nicolas Bazire, Dir.-Dev. & Acquisitions
Jean-Jacques Guiony, Dir.-Finance
Antonio Belloni, Group Managing Dir.
Pierre-Yves Roussel, Dir.-Fashion Div.
Francesco Trapani, Dir.-Watches & Jewelry
Christophe Navarre, Dir.-Wines & Spirits
Bernard Arnault, Chmn.

GROWTH PLANS/SPECIAL FEATURES:

LVMH Moet Hennessy Louis Vuitton SA is a luxury goods company controlling a portfolio of approximately 60 luxury brands, operating in five sectors: Wines & Spirits, Fashion & Leather Goods, Watches & Jewelry, Perfumes & Cosmetics and Selective Retailing. The company's Wines & Spirits brands include champagnes, such as Dom Perignon, Moet & Chandon, Krug, Mercier, Ruinart and Veuve Clicquot; several vineyards, including Chateau d'Yquem, Cloudy Bay and Cape Mentelle; Hennessy cognac; and several brands of whiskey, vodka, rum and sparkling wines. The firm's Fashion & Leather Goods division includes the Donna Karan, Marc Jacobs, Givenchy, Kenzo, Berluti, Fendi and Louis Vuitton brands, while its Watches & Jewelry segment includes brands such as TAG Heuer, Zenith, Dior Watches, Chaumet, De Beers and FRED. The firm's perfumes include Aqua di Parma, Guerlain, Kenzo and Givenchy, and its cosmetics brands include Fresh, Make Up For Ever and BeneFit Cosmetics. The company's Selective Retailing division operates several retail store brands, including Sephora cosmetics stores and Sephora.com; Le Bon Marche and La Samaritaine department stores; Miami Cruiseline Services, which offers duty-free luxury shops aboard cruise ships; and DFS Galleria liquor, tobacco, cosmetics and fragrances duty-free stores, primarily located in the Asia-Pacific region. Outside of its luxury brands, LVMH owns Groupe Les Echos, a French media conglomerate that produces 21 publications with corresponding web sites, including Les Echos, Serie Limitee and Investir. Christian Dior SA owns a 42.4% stake in the firm.

FINANCIAL DATA: *Note: Data for latest year may not have been available at press time.*

In U.S. $	2012	2011	2010	2009	2008	2007
Revenue	37,580,900,000	31,638,140,000	27,173,040,000	22,804,230,000	22,991,440,000	22,039,320,000
R&D Expense						
Operating Income	7,674,512,000	6,892,217,000	5,575,020,000	4,227,066,000	4,660,337,000	4,585,450,000
Operating Margin %	20.42%	21.78%	20.51%	18.53%	20.26%	20.80%
SGA Expense	16,401,440,000	13,779,090,000	11,787,910,000	10,078,900,000	10,100,290,000	9,548,007,000
Net Income	4,578,764,000	4,633,592,000	4,438,352,000	2,638,406,000	3,099,759,000	3,117,144,000
Operating Cash Flow	3,308,371,000	5,224,659,000	5,414,550,000	3,923,509,000	3,046,269,000	3,286,975,000
Capital Expenditure		2,338,861,000	1,339,930,000	1,000,267,000	1,389,409,000	1,323,883,000
EBITDA	9,411,608,000	8,106,446,000	7,649,104,000	5,124,365,000	5,557,636,000	5,378,444,000
Return on Assets %	7.05%	7.27%	8.75%	5.51%	6.50%	6.80%
Return on Equity %	13.92%	14.69%	18.38%	12.67%	16.54%	18.24%
Debt to Equity	0.14	0.17	0.18	0.27	0.28	0.21

CONTACT INFORMATION:

Phone: 33 144132222 Fax: 33 144132119
Toll-Free:
Address: 22 Ave. Montaigne, Paris, 75008 France

STOCK TICKER/OTHER:

Stock Ticker: LVMHF Exchange: PINX
Employees: 83,542 Fiscal Year Ends: 31-Dec
Parent Company:

SALARIES/BONUSES:

Top Exec. Salary: $ Bonus: $
Second Exec. Salary: $ Bonus: $

OTHER THOUGHTS:

Estimated Female Officers or Directors: 4
Hot Spot for Advancement for Women/Minorities: Y

Macy's Inc

NAIC Code: 452111

www.macysinc.com

TYPES OF BUSINESS:

Department Stores
Bridal & Formalwear Stores
Direct Marketing
Online Sales
Catalogs
Wedding Planning & Bridal Registries
Credit Services
Furniture Stores

BRANDS/DIVISIONS/AFFILIATES:

Holiday Lane
Bloomingdale's By Mail
FDS Bank
Macy's Credit & Customer Service Inc
Macy's Systems and Technology Inc
Macy's Merchandising Group Inc
Macy's Logistics & Operations
Alfani

CONTACTS: *Note: Officers with more than one job title may be intentionally listed here more than once.*

Terry J. Lundgren, CEO
Terry J. Lundgren, Pres.
Karen M. Hoguet, CFO
Martine Reardon, Chief Mktg. Officer
Jeffrey Gennette, Chief Merch. Officer
Thomas L. Cole, Chief Admin. Officer
Jeffrey A. Kantor, Chmn.-Macys.com
Michael Gould, CEO/Chmn.-Bloomingdales
Timothy M. Adams, Chief Private Brand Officer
Julie Greiner, Chief Merch. Planning Officer
Peter Sachse, Chief Stores Officer
Terry J. Lundgren, Chmn.

GROWTH PLANS/SPECIAL FEATURES:

Macy's, Inc. is a U.S. operator of full-line department stores. The firm currently has approximately 840 stores in 45 states, Washington D.C., Guam and Puerto Rico. Macy's department stores, which operate under the brand names Macy's and Bloomingdale's, offer men's, women's and children's apparel and accessories; cosmetics; home furnishings; and other consumer goods. Through its stores, the firm conducts direct-to-customer catalog and e-commerce businesses under the names Bloomingdale's By Mail, Bloomingdales.com and Macys.com. Additionally, the company offers online bridal registry and gift purchase facilities. The firm's owned brands include Charter Club; Alfani; Epic Threads; first impressions; and Holiday Lane. It also licenses brands American Rag and Martha Stewart Collection from third parties. Feminine accessories, intimate apparel, shoes and cosmetics generate approximately 37% of Macy's revenues; feminine apparel accounts for 25%; 23% is derived from men's and children's apparel; and home and miscellaneous products account for 15%. The company conducts many of its support functions through direct/indirect subsidiary firms and divisions. These include FDS Bank and Macy's Credit and Customer Service, Inc., which both offer credit processing, certain collections, customer service and credit marketing services; Macy's Systems and Technology, Inc., which offers operational electronic data processing and management information services; Macy's Merchandising Group, Inc., which designs, develops and markets the firm's private label brands; and Macy's Logistics and Operations, which operates warehouses and distribution services.

Macy's offers its employees medical and dental coverage; flexible spending accounts; wellness programs, including health screenings and Weight Watchers; disability protection; a 401(k) plan; employee assistance; tuition assistance; adoption assistance; paid vacation; leave of absence; life, AD&D and travel insurance; and a merchandise discount.

FINANCIAL DATA: *Note: Data for latest year may not have been available at press time.*

In U.S. $	2012	2011	2010	2009	2008	2007
Revenue	26,405,000,000	25,003,000,000	23,489,000,000	24,892,000,000	26,313,000,000	
R&D Expense						
Operating Income	2,411,000,000	1,894,000,000	1,063,000,000	-4,378,000,000	1,863,000,000	
Operating Margin %	9.13%	7.57%	4.52%	-17.58%	7.08%	
SGA Expense	8,281,000,000	8,260,000,000	8,062,000,000	8,481,000,000	8,554,000,000	
Net Income	1,256,000,000	847,000,000	350,000,000	-4,803,000,000	893,000,000	
Operating Cash Flow	2,093,000,000	1,506,000,000	1,750,000,000	1,879,000,000	2,238,000,000	
Capital Expenditure	764,000,000	505,000,000	460,000,000	897,000,000	1,105,000,000	
EBITDA	3,500,000,000	3,049,000,000	2,279,000,000	-3,099,000,000	3,172,000,000	
Return on Assets %	5.87%	4.03%	1.61%	-19.23%	3.11%	
Return on Equity %	21.91%	16.55%	7.48%	-66.00%	8.05%	
Debt to Equity	1.12	1.26	1.79	1.87	0.91	

CONTACT INFORMATION:

Phone: 513-579-7000 Fax:
Toll-Free: 800-261-5385
Address: 7 W. 7th St., Cincinnati, OH 45202 United States

STOCK TICKER/OTHER:

Stock Ticker: M
Exchange: NYS
Employees: 166,000
Fiscal Year Ends:
Parent Company:

SALARIES/BONUSES:

Top Exec. Salary: $1,591,667 Bonus: $
Second Exec. Salary: $1,000,000 Bonus: $

OTHER THOUGHTS:

Estimated Female Officers or Directors: 7
Hot Spot for Advancement for Women/Minorities: Y

Maidenform Inc

www.maidenform.com

NAIC Code: 315240

TYPES OF BUSINESS:

Intimate Apparel
Online Sales
Outlet Stores

BRANDS/DIVISIONS/AFFILIATES:

Flexees
Lilyette
Maidenform
Sweet Nothings
Bodymates
Control It!
Self Expressions
Inspirations

CONTACTS:

Note: Officers with more than one job title may be intentionally listed here more than once.

Maurice Reznik, CEO
Chris Vieth, COO/Exec. VP
Chris Vieth, CFO
Mitch Kauffman, Sr. VP-Sales
Bob Russo, CIO
Steven Castellano, Sr. VP-Merch. & Design Oper.
Nanci Prado, General Counsel/Exec. VP/Corp. Sec.
Karen Rose, Chmn.
Patricia J. Royak, Sr. VP/Managing Dir.-Int'l
Anthony D'Onofrio, Chief Supply Chain Officer

GROWTH PLANS/SPECIAL FEATURES:

Maidenform, Inc. is an international designer and marketer of a wide range of branded intimate apparel, including bras, panties, full figure apparel, shapewear, sleepwear and lingerie. Maidenform operates 77 outlet stores in 63 countries and territories. The firm's primary brand names include Maidenform, Flexees and Lilyette, with additional brand offerings including Control It!, Bodymates, Self Expressions, Sweet Nothings and Inspirations The Maidenform brand markets its collection of shapewear, bras and panties as providing everyday comfort for career women with active lifestyles. Flexees brand sells shapewear designed with shape defining properties to provide anywhere from light to firm body shape contouring. Lilyette brand targets full-figured women with bra cup sizes larger than C. The company's product offerings are sold through a variety of distribution channels such as department stores, national chains, mass merchants and warehouse clubs, specialty stores, outlet stores, kiosks and over the Internet. The particular distribution channel for a given product depends largely on its brand. For example, the Maidenform, Control It!, Flexees and Lilyette brands are primarily sold in department stores such as Macy's and Bloomingdale's as well as in national chains like Kohl's and JC Penney, while the Self Expressions, Sweet Nothings and Bodymates names appear more in mass merchant stores like Target, Wal-Mart and Costco. Maidenform sources the majority of its products from the Asia Pacific region. The firm has two distribution centers: one in Fayetteville, North Carolina, which serves its U.S. and Canadian operations; and one in Shannon, Ireland, which serves its European markets.

FINANCIAL DATA:

Note: Data for latest year may not have been available at press time.

In U.S. $	2012	2011	2010	2009	2008	2007
Revenue	600,277,000	606,327,000	556,709,000	466,255,000	413,542,000	422,164,000
R&D Expense						
Operating Income	52,288,000	52,475,000	76,363,000	53,173,000	45,913,000	67,919,000
Operating Margin %	8.71%	8.65%	13.71%	11.40%	11.10%	16.08%
SGA Expense	133,193,000	132,202,000	123,982,000	107,810,000	109,781,000	97,880,000
Net Income	33,459,000	33,192,000	45,280,000	36,993,000	24,720,000	34,204,000
Operating Cash Flow	31,718,000	4,362,000	31,316,000	45,193,000	32,973,000	40,420,000
Capital Expenditure	7,307,000	8,585,000	6,884,000	5,894,000	3,197,000	6,371,000
EBITDA	58,740,000	58,271,000	81,128,000	57,611,000	50,685,000	75,275,000
Return on Assets %	8.05%	8.86%	12.71%	11.26%	8.55%	13.04%
Return on Equity %	14.01%	16.04%	25.33%	25.82%	22.54%	39.19%
Debt to Equity	0.26	0.3	0.36	0.51	0.72	0.88

CONTACT INFORMATION:

Phone: 732 621-2500 Fax:
Toll-Free: 888-888-9328
Address: 485F US Hwy 1 S., Iselin, NJ 08830 United States

STOCK TICKER/OTHER:

Stock Ticker: MFB
Exchange: NYS
Employees: 1,230
Fiscal Year Ends: 31-Dec
Parent Company:

SALARIES/BONUSES:

Top Exec. Salary: $700,000 Bonus: $
Second Exec. Salary: $600,000 Bonus: $

OTHER THOUGHTS:

Estimated Female Officers or Directors: 5
Hot Spot for Advancement for Women/Minorities: Y

Mango/MNG Holding SL

www.mango.es

NAIC Code: 448120

TYPES OF BUSINESS:

Women's Apparel, Retail
Apparel Manufacturing
Fast Fashion

BRANDS/DIVISIONS/AFFILIATES:

M by MNG
MngJEANS
MANGO
HE
MANGO Touch

CONTACTS: *Note: Officers with more than one job title may be intentionally listed here more than once.*

Enric Casi, Gen. Dir.
Isak Andic, Pres.
Nahman Andic, Exec. VP
Isak Andic, Chmn.
Salvador Valles, Head-Purchasing & Quality

GROWTH PLANS/SPECIAL FEATURES:

Mango/MNG Holding SL is a multinational company engaged in the design, manufacture and marketing of clothing and accessories for young men and women. It has approximately 2,415 women's clothing boutiques in 140 countries. The company specializes in fast fashion, a retail trend where stores continually fill their racks with a steady stream of new, trendy merchandise. Mango's clothing promotes a party-girl image and is famous for carrying clothing in an eclectic mix of body-hugging styles. The firm is also famous for its store atmosphere and layout. Mango mandates that store racks hang only one of each size of each garment, creating a sense of urgency in the customer. The company has several private-label brands including M by MNG and MngJEANS, over which it maintains tight control. Having manufacturing facilities in Europe allows Mango to make last-minute changes on pieces. Mango currently has stores under three names MANGO, its main brand stores; HE, its men's clothing stores known for their black and gold decor; and MANGO TOUCH, stores with a feminine edge known for their mixture of classic and modern furnishings. Every Mango store also has its own set of traits dependent upon climate, shop location and whether large- or small-size clothing sells best at the store. The company's proprietary computer system matches the personality traits of its clothing to its stores, allowing the firm to dictate what pieces and styles should go to which stores. Mango expects to have 3,300 stores by 2015.

FINANCIAL DATA: *Note: Data for latest year may not have been available at press time.*

In U.S. $	2012	2011	2010	2009	2008	2007
Revenue	2,213,428,070	1,924,870,000	1,673,197,010			
R&D Expense						
Operating Income						
Operating Margin %						
SGA Expense						
Net Income						
Operating Cash Flow						
Capital Expenditure						
EBITDA						
Return on Assets %						
Return on Equity %						
Debt to Equity						

CONTACT INFORMATION:

Phone: 34-93-860-24-24 Fax: 34-93-860-22-07
Toll-Free:
Address: Mercaders 9-11, P.I. Riera de Caldes, Apt. 280, Barcelona, 08184 Spain

STOCK TICKER/OTHER:

Stock Ticker: Private
Employees:
Parent Company:
Exchange:
Fiscal Year Ends: 31-Dec

SALARIES/BONUSES:

Top Exec. Salary: $ Bonus: $
Second Exec. Salary: $ Bonus: $

OTHER THOUGHTS:

Estimated Female Officers or Directors:
Hot Spot for Advancement for Women/Minorities:

Marc Ecko Enterprises

www.marceckoenterprises.com

NAIC Code: 315240

TYPES OF BUSINESS:

Apparel Manufacturing-Men's & Boy's
Apparel Manufacturing-Women's
Apparel Retailing
Magazine Publishing
Video Games
Skateboards

BRANDS/DIVISIONS/AFFILIATES:

*ecko unltd
Marc Ecko Cut & Sew
ecko red
Zoo York
Complex Magazine
Marc Echo Entertainment
Avirex
Complex Media Network

CONTACTS: *Note: Officers with more than one job title may be intentionally listed here more than once.*

Seth Gerszberg, CEO
Effy Zinkin, Pres.
Marc Ecko, Chief Creative Officer
Marc Ecko, Chmn.

GROWTH PLANS/SPECIAL FEATURES:

Marc Ecko Enterprises (MEE) is a private clothing designer, manufacturer and retailer focusing on urban, fashion forward customers. The company retails its products through department and specialty stores in the U.S. and through full price and outlet stores in over 80 international countries. MEE has eight primary brands: *ecko unltd., a clothing line geared toward young men; Marc Ecko Cut & Sew, a menswear line; Zoo York, an action sports inspired clothing and accessories line as well as a line of skateboard and skate accessories; Complex Media, which includes Complex Magazine, with a circulation of approximately 300,000, and the Complex Media Network, which consists of 14 web sites that generate over 40 million page views; ecko red, a juniors apparel line; Avirex, a sportswear line; Marc Ecko Entertainment, a videogame development and production firm; and ecko TV, an in-store and online video channel that showcases film projects from customers and independent film makers. MEE licensing agreements include footwear via the Skechers brand, watches through Callanen and kidswear through the Kids Headquarters line. Through the various MEE brands, the firm is able to offer a wide variety of merchandise including men's and women's apparel, formalwear, children's wear, outerwear, footwear, watches, eyewear, underwear, accessories and small leather goods. The company also developed a graffiti-themed video game with Atari, Inc. called Getting Up: Contents Under Pressure. The company maintains an e-commerce web site, ShopEcko.com, which has all MEE brands available.

FINANCIAL DATA: *Note: Data for latest year may not have been available at press time.*

In U.S. $	2012	2011	2010	2009	2008	2007
Revenue						
R&D Expense						
Operating Income						
Operating Margin %						
SGA Expense						
Net Income						
Operating Cash Flow						
Capital Expenditure						
EBITDA						
Return on Assets %						
Return on Equity %						
Debt to Equity						

CONTACT INFORMATION:

Phone: 917-262-1002 Fax:
Toll-Free:
Address: 40 W. 23rd St, New York, NY 10010 United States

STOCK TICKER/OTHER:

Stock Ticker: Private
Exchange:
Employees:
Fiscal Year Ends:
Parent Company:

SALARIES/BONUSES:

Top Exec. Salary: $ Bonus: $
Second Exec. Salary: $ Bonus: $

OTHER THOUGHTS:

Estimated Female Officers or Directors: 1
Hot Spot for Advancement for Women/Minorities: Y

Marks & Spencer Group plc

www.marksandspencer.com

NAIC Code: 452111

TYPES OF BUSINESS:

Department Stores
Grocery Stores
Food & Housewares
Financial Services
Food Courts
Franchising

BRANDS/DIVISIONS/AFFILIATES:

Marks & Spencer
Simply Food
Portfolio
Autograph
Indigo
Per Una
M&S Direct
Classic

CONTACTS: *Note: Officers with more than one job title may be intentionally listed here more than once.*

Marc Bolland, CEO
Alan Stewart, CFO
Steven Sharp, Exec. Dir.-Mktg.
Tanith Dodge, Dir.-Human Resources
Darrell Stein, Dir.-IT
John Dixon, Exec. Dir.-Gen. Merch.
Amanda Mellor, Group Sec./Dir.-Corp. Governance
Patrick Bousquet-Chavanne, Corp. Dir.-Strategy Implementation & Bus. Dev.
Laura Wade-Gery, Exec. Dir.-Multi-Channel e-commerce
Dominic Fry, Dir-Comm.
Dominic Fry, Dir.-Investor Rel.
Steve Rowe, Exec. Dir.-Food
Clem Constantine, Dir.-Property
Nayna McIntosh, Dir.-Store Environment & Product Presentation
Robert Swannel, Chmn.
Jan Heere, Dir.-Int'l
Darrell Stein, Dir.-Logistics

GROWTH PLANS/SPECIAL FEATURES:

Marks & Spencer Group plc serves as a holding company for the Marks & Spencer group of companies. Marks & Spencer is one of the leading mass merchandise retailers in the U.K., with an average of more than 21 million customers per week. The company has products sourced from more than 2,000 global suppliers. Marks & Spencer sells women's, men's and children's apparel as well as housewares and women's lingerie. The company's most notable apparel brands include Classic, Portfolio, Per Una, Limited Collection, Indigo and Autograph. The company also retails food products that include fresh produce, staple foods and ready made meals. Overall, 49% of sales are derived from apparel and homeware, and 51% from food sales. The company has 703 stores throughout the U.K. that include Marks & Spencer flagship stores measuring up to 170,000 square feet and the company's Simply Food stores that measure approximately 7,000 square feet. In addition, the company has 400 international and franchised stores in roughly 44 countries throughout Europe, Asia and the Middle East. M&S Direct is the company's e-commerce site, which includes the company's home catalogue and flower and wine delivery. In addition, the web site provides Christmas hamper delivery, which provides in-store food ordering services and lunchtime platter delivery services.

The company offers employees benefits including a pension plan, a sharesave plan, subsidized healthcare and 20% employee discount.

FINANCIAL DATA: *Note: Data for latest year may not have been available at press time.*

In U.S. $	2012	2011	2010	2009	2008	2007
Revenue	15,872,030,000	15,562,070,000	15,236,620,000	14,478,510,000	14,414,440,000	13,721,200,000
R&D Expense						
Operating Income	1,192,682,000	1,337,115,000	1,361,240,000	1,391,117,000	1,935,293,000	1,671,034,000
Operating Margin %	7.51%	8.59%	8.93%	9.60%	13.42%	12.17%
SGA Expense						
Net Income	819,779,600	977,792,100	840,869,200	811,631,300	1,312,830,000	1,054,322,000
Operating Cash Flow	1,704,905,000	1,683,176,000	1,702,509,000	1,747,084,000	1,567,183,000	1,833,360,000
Capital Expenditure	1,151,462,000	725,036,000	686,211,900	1,058,316,000	1,628,056,000	1,139,799,000
EBITDA	1,959,099,000	2,084,039,000	2,044,895,000	2,044,576,000	2,442,723,000	2,122,703,000
Return on Assets %	7.02%	8.44%	7.30%	7.04%	13.10%	12.40%
Return on Equity %	18.80%	25.16%	24.55%	24.99%	45.49%	46.27%
Debt to Equity						

CONTACT INFORMATION:

Phone: 44 2079354422 Fax: 44 2074872670
Toll-Free:
Address: 35 N. Wharf Rd., Waterside House, London, W2 1NW United Kingdom

STOCK TICKER/OTHER:

Stock Ticker: MAKSF Exchange: PINX
Employees: 78,000 Fiscal Year Ends: 31-Mar
Parent Company:

SALARIES/BONUSES:

Top Exec. Salary: $ Bonus: $
Second Exec. Salary: $ Bonus: $

OTHER THOUGHTS:

Estimated Female Officers or Directors: 6
Hot Spot for Advancement for Women/Minorities: Y

Marzotto SpA

www.marzotto.it

NAIC Code: 313110

TYPES OF BUSINESS:

Yarns & Fabrics
Home Decor
Blankets

BRANDS/DIVISIONS/AFFILIATES:

Manifattura Lane Gaetano Marzotto & Figli
Lanerossi Filati
Linificio e Canapificio Nazionale
Guabello
Marlane
Tessuti di Sondrio
Lanerossi
Marzotto

CONTACTS:

Note: Officers with more than one job title may be intentionally listed here more than once.

Sergio Tamborini, CEO/Gen. Mgr.
Andrea D. Dalle Rose, Vice Chmn.
Antonio Favrin, Chmn.

GROWTH PLANS/SPECIAL FEATURES:

Marzotto SpA, formerly Manifattura Lane Gaetano Marzotto & Figli, is a leading international manufacturer of pure wool and linen fabrics, yarns and blankets. The company divides its operations into three primary divisions: yarns, fabrics and home decor. The yarns division consists of two companies: Lanerossi Filati and Linificio e Canapificio Nazionale (National Linen and Hemp Mill) SpA. Lanerossi Filati produces outer knitwear yarns that combine natural fibers such as wool, cashmere, angora and alpaca wool, mohair, linen, silk and cotton. Linificio e Canapificio Nazionale is one of the largest makers of long line flax yarns in the world. The fabrics division includes four brands. Marzotto, Guabello and Marlane focus on wool fabric production, and Tessuti di Sondrio makes high-quality cotton fabrics. Lastly, the home decor division consists of two distinct brands, Lanerossi and Marzotto, as well as a special supplies division that provides blankets to hotel chains and major airline companies, including American Airlines and Alitalia. The firm supplies companies with finished products as well as raw materials. The company also owns and operates the Valentino haute couture fashion house, which is now Valentino Fashion Group, and owns a controlling share in Hugo Boss.

FINANCIAL DATA:

Note: Data for latest year may not have been available at press time.

In U.S. $	2012	2011	2010	2009	2008	2007
Revenue						
R&D Expense						
Operating Income						
Operating Margin %						
SGA Expense						
Net Income						
Operating Cash Flow						
Capital Expenditure						
EBITDA						
Return on Assets %						
Return on Equity %						
Debt to Equity						

CONTACT INFORMATION:

Phone: 39-0445-429411 Fax: 39-0445-402000
Toll-Free:
Address: Largo S. Margherita 1, Valdagno, Vicenza, 36078 Italy

STOCK TICKER/OTHER:

Stock Ticker: Private
Exchange:
Employees: 3,438
Fiscal Year Ends: 31-Dec
Parent Company:

SALARIES/BONUSES:

Top Exec. Salary: $ Bonus: $
Second Exec. Salary: $ Bonus: $

OTHER THOUGHTS:

Estimated Female Officers or Directors:
Hot Spot for Advancement for Women/Minorities: Y

MAST Global Fashions

NAIC Code: 315240

www.mast.com

TYPES OF BUSINESS:

Contract Apparel Manufacturing

BRANDS/DIVISIONS/AFFILIATES:

Limited Brands Inc
Sycamore Partners LLC
Mast Industries

GROWTH PLANS/SPECIAL FEATURES:

MAST Global Fashions, previously MAST Industries, Inc., is a merchandise supply chain management and sourcing company, which is currently 51% owned by Sycamore Partners LLC and 49% owned by Limited Brands. It is one of the largest contract manufacturers, importers and distributors of men's, women's and children's apparel in the world. Its logistics services include product development, production, and distribution. MAST maintains a global network of manufacturing partners, bringing decades of experience to every new partnership, which in turn helps the company to consistently meet demands for quality, quantity and price.

CONTACTS: *Note: Officers with more than one job title may be intentionally listed here more than once.*

James Schwartz, CEO
James Schwartz, Pres.

FINANCIAL DATA: *Note: Data for latest year may not have been available at press time.*

In U.S. $	2012	2011	2010	2009	2008	2007
Revenue						
R&D Expense						
Operating Income						
Operating Margin %						
SGA Expense						
Net Income						
Operating Cash Flow						
Capital Expenditure						
EBITDA						
Return on Assets %						
Return on Equity %						
Debt to Equity						

CONTACT INFORMATION:

Phone: 614-337-5600 Fax: 614-337-5080
Toll-Free:
Address: 8655 E. Broad St., Columbus, OH 43230 United States

STOCK TICKER/OTHER:

Stock Ticker: Joint Venture Exchange:
Employees: 600 Fiscal Year Ends: 31-Jan
Parent Company: SYCAMORE PARTNERS LLC

SALARIES/BONUSES:

Top Exec. Salary: $ Bonus: $
Second Exec. Salary: $ Bonus: $

OTHER THOUGHTS:

Estimated Female Officers or Directors:
Hot Spot for Advancement for Women/Minorities:

Max Mara Fashion Group SRL

www.maxmarafashiongroup.com

NAIC Code: 315240

TYPES OF BUSINESS:

Luxury Men's & Women's Clothing
Apparel Manufacturing
Retail Stores
Perfume

BRANDS/DIVISIONS/AFFILIATES:

Max Mara
Sportmax
Pennyblack
Marella
Max & Co
Marina Renaldi
Max Mara Hosiery
Persona

CONTACTS: *Note: Officers with more than one job title may be intentionally listed here more than once.*

Ignazio Maramotti, Managing Dir.
Laura Lusuardi, Dir.-Design
Luigi Maramotti, Chmn.

GROWTH PLANS/SPECIAL FEATURES:

Max Mara Fashion Group SRL is one of the largest Italian clothing manufacturers and retailers of designer apparel for men and women. It has nine brands, including Max Mara, a collection of suits, coats and classic garments made from top quality fabrics; Sportmax, a collection designed to appeal to younger people; Pennyblack, which offers casual, elegant and romantic lines; Marella, a lower priced line; Persona, a collection of modern garments and accessories, which targets young women of all sizes; and Marina Rinaldi, a collection of full-sized fashions. The family-owned and -operated company sells its clothing through high-end boutiques and in over 2,300 mono-brand and over 10,000 multi-brand stores in 105 countries under the names Max Mara, for luxury clothing and Max & Co., for trendy clothing. The company's clothing is also sold through luxury retailers including Neiman Marcus, Saks Fifth Avenue and Barneys. All Max Mara clothing is produced at the company's production facility in Italy under heavy quality supervision. Max Mara also offers a perfume line and the Max Mara Hosiery brand.

FINANCIAL DATA: *Note: Data for latest year may not have been available at press time.*

In U.S. $	2012	2011	2010	2009	2008	2007
Revenue	2,170,000,000	2,000,000,000	1,700,000,000	1,500,000,000	1,500,000,000	
R&D Expense						
Operating Income						
Operating Margin %						
SGA Expense						
Net Income						
Operating Cash Flow						
Capital Expenditure						
EBITDA						
Return on Assets %						
Return on Equity %						
Debt to Equity						

CONTACT INFORMATION:

Phone: 39-05-22-3991 Fax: 39-05-22-399-3254
Toll-Free:
Address: Via Giulia Maramotti 4, Reggio Emilia, 42124 Italy

STOCK TICKER/OTHER:

Stock Ticker: Private
Employees: 5,000
Parent Company:
Exchange:
Fiscal Year Ends: 31-Dec

SALARIES/BONUSES:

Top Exec. Salary: $ Bonus: $
Second Exec. Salary: $ Bonus: $

OTHER THOUGHTS:

Estimated Female Officers or Directors: 1
Hot Spot for Advancement for Women/Minorities:

Men's Wearhouse Inc (The)

www.menswearhouse.com

NAIC Code: 448110

TYPES OF BUSINESS:

Men's Apparel, Retail
Men's Suits
Shoes & Accessories
Business Casual Wear
Sportswear
Shoes & Accessories
Ladies' Career Apparel
Dry Cleaning & Laundry

BRANDS/DIVISIONS/AFFILIATES:

K&G
Moores Clothing for Men
MW Tux
Perfect Fit
Men's Wearhouse and Tux
MW Cleaners
Dimensions
Alexandra

CONTACTS: *Note: Officers with more than one job title may be intentionally listed here more than once.*

Douglas S. Ewert, CEO
Douglas S. Ewert, Pres.
Diana M. Wilson, Interim CFO/Exec. VP
Charles Bresler, Exec. VP-Mktg.
Charles Bresler, Exec. VP-Human Resources
Diana M. Wilson, Treas./Principal Financial Officer
David H. Edwab, Vice Chmn.
George A. Zimmer, Chmn.

GROWTH PLANS/SPECIAL FEATURES:

The Men's Wearhouse, Inc. is a leading specialty retailer of men's suits and provider of tuxedo rental products in the U.S. and Canada. The company conducts business through two segments: retail and corporate apparel. The retail segment comprises more than 1,049 stores in 49 U.S. states and Washington, D.C., primarily operating under the brand names of Men's Wearhouse and K&G. Men's Wearhouse apparel stores target middle- and upper-middle income men and offer designer, brand name and private label merchandise at discounted prices through 607 locations. Merchandise offered includes suits, sport coats, slacks, formalwear, business casual, sportswear, outerwear, dress shirts, shoes and accessories. Most of these stores also offer tuxedo rental products in a section branded as MW Tux. In addition, tuxedo rentals are available in 343 stores under the brand name of Men's Wearhouse and Tux in 35 states. The Men's Wearhouse's 99 K&G branded stores target more price-sensitive customers and include ladies' career apparel at 44 locations. In Canada, retail operations include 117 retail apparel stores in 10 provinces operating under the brand name of Moores Clothing for Men. The Moores brand targets middle- and upper-middle income men. The corporate apparel segment provides corporate clothing uniforms and workwear, with operations provided by Twin Hill in the U.S., and by a U.K. holding company operating under the Dimensions and Alexandra brands in the U.K. The firm also conducts retail dry cleaning and laundry operations through MW Cleaners in the Houston, Texas area. The company offers a private label credit card as well as a Perfect Fit loyalty program to its customers.

The Men's Wearhouse offers employees benefits including medical, dental, life and long-term disability insurance; a wellness program; 401(k) savings plan with company matching; employee stock discount program; tuition reimbursement; and merchandise discounts.

FINANCIAL DATA: *Note: Data for latest year may not have been available at press time.*

In U.S. $	2012	2011	2010	2009	2008	2007
Revenue	2,382,684,000	2,102,664,000	1,909,575,000	1,972,418,000	2,112,558,000	
R&D Expense						
Operating Income	185,432,000	101,671,000	68,204,000	90,471,000	228,652,000	
Operating Margin %	7.78%	4.83%	3.57%	4.58%	10.82%	
SGA Expense	861,453,000	790,908,000	710,049,000	760,041,000	741,405,000	
Net Income	120,601,000	67,697,000	45,508,000	58,844,000	147,041,000	
Operating Cash Flow	162,797,000	169,947,000	163,155,000	129,490,000	204,877,000	
Capital Expenditure	91,820,000	58,868,000	56,912,000	88,225,000	126,116,000	
EBITDA	261,824,000	177,984,000	155,206,000	183,728,000	357,002,000	
Return on Assets %	8.84%	5.30%	3.76%	4.81%	12.49%	
Return on Equity %	12.12%	7.22%	5.21%	7.09%	18.73%	
Debt to Equity			0.04	0.07	0.11	

CONTACT INFORMATION:

Phone: 281 776-7000 Fax:
Toll-Free: 800-776-7848
Address: 6380 Rogerdale Rd., Houston, TX 77072 United States

STOCK TICKER/OTHER:

Stock Ticker: MW Exchange: NYS
Employees: 16,600 Fiscal Year Ends: 31-Jan
Parent Company:

SALARIES/BONUSES:

Top Exec. Salary: $932,000 Bonus: $
Second Exec. Salary: $395,000 Bonus: $200,000

OTHER THOUGHTS:

Estimated Female Officers or Directors: 3
Hot Spot for Advancement for Women/Minorities: Y

Metso Fabrics Corp

www.metso.com

NAIC Code: 313230

TYPES OF BUSINESS:

Technical Textiles
Filter Products
Paper Machine Clothing

BRANDS/DIVISIONS/AFFILIATES:

Valstone Control Inc
MW Power
Metso Oyj
Jacobs Engineering Group
Shaoriu Heavy Industries
ExperTune

CONTACTS: *Note: Officers with more than one job title may be intentionally listed here more than once.*

Matti Kahkonen, CEO
Matti Kahkonen, Pres.
Harri Nikunen, Exec.-Finance
Kalle Reponen, Sr. VP-Strategy & Mergers & Acquisitions
Andrew Benko, Pres., Mining & Construction
Perttu Louhiluoto, Pres., Automation
Pasi Laine, Pres., Pulp, Paper & Power/Exec. VP
Matti Kahkonen, Chmn.

GROWTH PLANS/SPECIAL FEATURES:

Metso Fabrics Corp., owned by Metso Oyj, is one of the world's leading developers and marketers of technical textiles used in filtration manufacturing processes. Through more than 50 owned and operated subsidiaries operating in 50 countries, Metso offers its services to customers in process industries, such as mining and construction, energy, automation, recycling, pulp and paper, food, oil and gas and pharmaceuticals as well as commercial laundries. The company's operations are divided into three divisions: mining and construction, offering mineral processing solutions, equipment for crushing and screening as well as recycling; automation, consisting of process automation systems and flow control business lines; and service lines for paper, fiber, power. In 2012, the mining and construction segment accounted for 44% of total revenue; pulp, paper and power accounted for 40%; automation accounted for 12%; and other businesses, 4%. In 2012, the firm acquired Valstone Control, Inc.; MW Power; Jacobs Engineering Group; and Shaoriu Heavy Industries. Additionally, it created a joint-venture with LiuGong Group Corp. to establish a screening and crushing business in China. In January 2013, the company acquired ExperTune, a U.S. company with experience in industrial maintenance software.

FINANCIAL DATA: *Note: Data for latest year may not have been available at press time.*

In U.S. $	2012	2011	2010	2009	2008	2007
Revenue		8,598,784,000	7,183,335,936	6,489,843,712	8,280,502,272	8,086,427,648
R&D Expense						
Operating Income						
Operating Margin %						
SGA Expense						
Net Income		460,602,944	332,513,920	194,074,272	503,299,296	492,948,640
Operating Cash Flow						
Capital Expenditure						
EBITDA						
Return on Assets %						
Return on Equity %						
Debt to Equity						

CONTACT INFORMATION:

Phone: 358 20484100 Fax: 358 20484101
Toll-Free:
Address: Yrittajankatu 21, Helsinki, 33710 Finland

STOCK TICKER/OTHER:

Stock Ticker: Subsidiary Exchange:
Employees: 29,000 Fiscal Year Ends: 31-Dec
Parent Company: METSO OYJ

SALARIES/BONUSES:

Top Exec. Salary: $ Bonus: $
Second Exec. Salary: $ Bonus: $

OTHER THOUGHTS:

Estimated Female Officers or Directors: 3
Hot Spot for Advancement for Women/Minorities: Y

Michael Kors Holdings Ltd

www.michaelkors.com

NAIC Code: 315240

TYPES OF BUSINESS:

Men's & Women's Fashion Apparel
Designer Accessories
Retail Stores

BRANDS/DIVISIONS/AFFILIATES:

MichaelKors.com
MICHAEL
KORS
MICHAEL Michael Kors
Michael Kors Lifestyle

CONTACTS: *Note: Officers with more than one job title may be intentionally listed here more than once.*

John D. Idol, CEO
Joseph B. Parsons, COO
Joseph B. Parsons, CFO/Exec. VP
Lee S. Sporn, General Counsel/Sec.
Lee S. Sporn, Sr. VP-Bus. Affairs
Joseph B. Parsons, Treas.
Michael Kors, Chief Creative Officer/Honorary Chmn.
John D. Idol, Chmn.

GROWTH PLANS/SPECIAL FEATURES:

Michael Kors Holdings, Ltd. is a designer, manufacturer and retailer of luxury American sportswear and accessories. The label was founded in 1981 by American fashion designer Michael Kors, and has since expanded from a few boutiques offering women's ready-to-wear fashions to include full lines of handbags, shoes, accessories, beauty products and men's fashions. In total, Michael Kors Holdings operates approximately 181 stores in North America, and 46 international stores in Europe and Japan. It has a presence in 85 countries. The company sells its products wholesale in 2,677 locations: 2,027 North American department and specialty stores; and 650 international department and specialty stores. The company operates a small number of Michael Kors collection boutiques that feature the designer's most luxurious collection of sportswear and accessories. More numerous are the Michael Kors Lifestyle stores, which offer mainly shoe and handbag collections from the Michael Kors and MICHAEL Michael Kors collections and a limited selection of ready-to-wear pieces. These stores are located throughout the U.S., primarily in high-end malls and shopping centers. The firm also operates an e-commerce store at MichaelKors.com. Apart from the firm's owned boutiques and shops, the Michael Kors collections are sold in luxury department stores such as Neiman Marcus, Saks and Bloomingdale's. In addition to women's and men's fashion apparel and accessories, the company produces several lines of fragrances for men and women. The company is comprised of three segments: retail, wholesale and licensing. In December 2011, Michael Kors Holdings filed its initial public offering.

FINANCIAL DATA: *Note: Data for latest year may not have been available at press time.*

In U.S. $	2012	2011	2010	2009	2008	2007
Revenue	1,302,254,000	803,339,000	508,099,000			
R&D Expense						
Operating Income	247,682,000	136,866,000	56,174,000			
Operating Margin %	19.01%	17.03%	11.05%			
SGA Expense	464,568,000	279,822,000	191,717,000			
Net Income	147,364,000	72,506,000	39,248,000			
Operating Cash Flow	115,290,000	110,308,000	28,592,000			
Capital Expenditure	88,187,000	57,830,000	32,175,000			
EBITDA	285,236,000	162,409,000	75,017,000			
Return on Assets %	23.49%	16.69%				
Return on Equity %	43.37%	65.25%				
Debt to Equity		0.81				

CONTACT INFORMATION:

Phone: 852 39285563 Fax: 852 39285697
Toll-Free:
Address: Miramar Tower, Unit 1001, 10/F, 132 Nathan Rd., Hong Kong, Hong Kong

STOCK TICKER/OTHER:

Stock Ticker: KORS Exchange: NYS
Employees: 2,945 Fiscal Year Ends: 31-Mar
Parent Company:

SALARIES/BONUSES:

Top Exec. Salary: $ Bonus: $
Second Exec. Salary: $ Bonus: $

OTHER THOUGHTS:

Estimated Female Officers or Directors: 1
Hot Spot for Advancement for Women/Minorities:

Milliken & Co

www.milliken.com

NAIC Code: 313110

TYPES OF BUSINESS:

Textile Manufacturing
Yarn
Specialty Chemicals
Textile Research

BRANDS/DIVISIONS/AFFILIATES:

Milliken Visa
Millitron
Millad
Agilon
Pegasus
Contex
SiVance LLC

CONTACTS: *Note: Officers with more than one job title may be intentionally listed here more than once.*

Joe Salley, CEO
Joe Salley, Pres.
Jim McNulty, CFO
Roger Milliken, Chmn.

GROWTH PLANS/SPECIAL FEATURES:

Milliken & Co., owned and operated by the Milliken family, is one of the largest privately held textile and chemical manufacturers in the world. The company makes a variety of fabric, yarn and chemical products that are present in homes and businesses around the globe. The firm has developed several branded materials and processes, including Milliken Visa, a water-absorbing polyester; Millitron, a computer-controlled carpet inkjet printer; Millad, which makes Tupperware clear; Agilon, which gives nylon mechanical stretch; Pegasus, a process of digitally creating a two- or three-dimensional design in fabrics; and Contex, which allows fabric to be uniformly coated without affecting the fabric's flexibility or softness. Milliken has extensive facilities for research, testing and analysis, including an analytical laboratory, a microscopy facility, a textile testing laboratory and the Milliken Research Library, coupled with resources like university partnerships and pilot plant facilities. The company currently has more than 39 facilities worldwide, including facilities in Brazil, Europe and Asia. Recently, the company acquired SiVance LLC, a siloxane intermediates, silicone and silane provider.

The firm offers its employees flexible spending accounts, medical and dental coverage, employee and dependant life insurance, a 401(k) and service awards.

FINANCIAL DATA: *Note: Data for latest year may not have been available at press time.*

In U.S. $	2012	2011	2010	2009	2008	2007
Revenue						
R&D Expense						
Operating Income						
Operating Margin %						
SGA Expense						
Net Income						
Operating Cash Flow						
Capital Expenditure						
EBITDA						
Return on Assets %						
Return on Equity %						
Debt to Equity						

CONTACT INFORMATION:

Phone: 864-503-2020 Fax: 864-503-2100
Toll-Free:
Address: 920 Milliken Rd., Spartanburg, SC 29303 United States

STOCK TICKER/OTHER:

Stock Ticker: Private Exchange:
Employees: Fiscal Year Ends: 30-Nov
Parent Company:

SALARIES/BONUSES:

Top Exec. Salary: $ Bonus: $
Second Exec. Salary: $ Bonus: $

OTHER THOUGHTS:

Estimated Female Officers or Directors:
Hot Spot for Advancement for Women/Minorities:

Mine Safety Appliances Co

NAIC Code: 33911

www.msanet.com

TYPES OF BUSINESS:

Equipment/Supplies-Manufacturer
Safety & Health Equipment
Personal Protective Products
Respiratory Protective Equipment
Combat Helmets
Thermal Imaging Cameras
Security Sensors & Systems

BRANDS/DIVISIONS/AFFILIATES:

General Monitors

CONTACTS: *Note: Officers with more than one job title may be intentionally listed here more than once.*

William M. Lambert, CEO
William M. Lambert, Pres.
Dennis L. Zeitler, CFO/Sr. VP
Paul R. Uhler, VP-Global Human Resources
Markus Weber, CIO/VP
Thomas Muschter, VP-Global Prod. Leadership
Douglas K. McClaine, General Counsel/Sec./VP
Steven C. Blanco, VP-Global Oper. Excellence
Dennis L. Zeitler, Treas.
Joseph A. Bigler, VP/Pres., MSA North America
Kerry M. Bove, Pres., MSA Asia, Australia, Africa & Latin America
John T. Ryan, III, Chmn.
Ronald N. Herring, Jr., Pres., MSA Europe, Russia, Middle East & India

GROWTH PLANS/SPECIAL FEATURES:

Mine Safety Appliances Co. (MSA) develops, manufactures and supplies safety and health equipment. The firm divides its product offerings by the geographic areas of its customers: North America, Europe and International. The company's safety products are used by workers around the world in fire service, homeland security, the military, construction services and other industries. MSA's principle product offerings fall under five categories: respiratory protection; thermal imaging cameras; portable and permanent gas detection instruments; eye, face, hearing and head protection; and fall protection. The company produces numerous respiratory systems, such as Self Contained Breathing Apparatuses (SCBA); escape hoods, which allow workers to escape from dangerous gases; and air-purifying respirators, such as military full face gas masks, half-mask respirators, powered-air purifying respirators and pollen masks. Hand-held thermal imaging cameras are used by firefighters to see victims through dense smoke and to detect a fire's source. MSA's portable and permanent gas detection products include single- and multiple-gas hand-held detectors; multi-point permanently installed gas detectors; and flame and open-path infrared gas detectors. Eye, face, hearing, and head protection is used in work environments with hazards such as dust, metal fragments, chemicals, extreme glare, optical radiation and items dropped from above. The company produces industrial hard hats, fire helmets, military helmets, hearing protection products, protective eyewear and face shields. MSA's fall protection equipment includes confined space equipment, harnesses, fall arrest equipment, lanyards and lifelines. The firm devotes significant resources to R&D and produces safety products that exceed worldwide government and industry standards. In October 2011, the firm sold its Paraclete Armor and Equipment business to Protective Products Enterprises, Inc. In July 2012, MSA sold its North American ballistic helmet business to Revision Military.

MSA offers its employees medical, dental and vision coverage; flexible spending accounts; an assistance program; tuition reimbursement; adoption benefits; and credit union access.

FINANCIAL DATA: *Note: Data for latest year may not have been available at press time.*

In U.S. $	2012	2011	2010	2009	2008	2007
Revenue	1,179,895,000	1,178,608,000	982,668,000	915,851,000	1,139,543,000	1,007,648,000
R&D Expense	40,900,000	39,245,000	32,784,000	28,781,000	35,020,000	30,196,000
Operating Income	134,290,000	104,818,000	57,349,000	65,340,000	128,324,000	115,969,000
Operating Margin %	11.38%	8.89%	5.83%	7.13%	11.26%	11.50%
SGA Expense	321,234,000	306,367,000	262,940,000	230,006,000	270,584,000	241,138,000
Net Income	90,637,000	69,852,000	38,104,000	43,295,000	70,422,000	67,588,000
Operating Cash Flow	150,476,000	85,258,000	31,606,000	120,846,000	59,785,000	41,345,000
Capital Expenditure	32,209,000	30,390,000	25,024,000	25,737,000	44,450,000	32,884,000
EBITDA	177,353,000	151,801,000	95,248,000	99,782,000	149,028,000	140,464,000
Return on Assets %	8.14%	6.04%	3.67%	4.94%	7.44%	7.05%
Return on Equity %	20.37%	15.91%	8.65%	10.51%	16.60%	15.14%
Debt to Equity	0.59	0.77	0.81	0.18	0.24	0.22

CONTACT INFORMATION:

Phone: 724 776-8600 Fax:
Toll-Free: 800-672-2222
Address: 1000 Cranberry Woods Dr., Pittsburgh, PA 16066 United States

STOCK TICKER/OTHER:

Stock Ticker: MSA
Employees: 5,300
Parent Company:
Exchange: NYS
Fiscal Year Ends: 31-Dec

SALARIES/BONUSES:

Top Exec. Salary: $703,500 Bonus: $
Second Exec. Salary: $370,325 Bonus: $

OTHER THOUGHTS:

Estimated Female Officers or Directors: 1
Hot Spot for Advancement for Women/Minorities:

Mohawk Industries Inc

www.mohawkind.com

NAIC Code: 442210

TYPES OF BUSINESS:

Floor Covering Stores
Tile & Stone Products
Extrusion
Laminate Flooring Technology
Roofing Systems

BRANDS/DIVISIONS/AFFILIATES:

Mohawk Carpet LLC
Dal-Tile Corporation
Unilin BVBA
American Orlean
Aladdin
Pergo
Spano Invest NV
Marazzi Group

CONTACTS: *Note: Officers with more than one job title may be intentionally listed here more than once.*

Jeffrey S. Lorberbaum, CEO
W. Christopher Wellborn, COO
W. Christopher Wellborn, Pres.
Frank H. Boykin, CFO
James T. Lucke, General Counsel/VP/Sec.
James F. Brunk, Chief Acct. Officer/Controller
Brian M. Carson, Pres., Mohawk Flooring
Bernard P. Thiers, Pres., Unilin BVBA
John C. Turner, Jr., Pres., Dal-Tile Corporation
Jeffrey S. Lorberbaum, Chmn.

GROWTH PLANS/SPECIAL FEATURES:

Mohawk Industries, Inc. is a leading producer of floor-covering products for residential and commercial applications in the U.S. and residential applications in Europe. Its primary operating subsidiaries are Mohawk Carpet LLC, Dal-Tile Corporation and Unilin BVBA. The Mohawk segment manufactures, markets and distributes carpet, rugs, pad, ceramic tile, hardwood and resilient product lines through its network of regional distribution centers and satellite warehouses. Products are sold through over 24,000 customers, including floor covering retailers, home centers, mass merchandisers, department stores, independent distributors, commercial dealers and commercial end users. Some of the company's products are also marketed through private labeling programs. Its brand names include Mohawk, Aladdin, Horizon, SmartStrand, Lees, Bigelow and Durkan. Dal-Tile manufactures, markets and distributes ceramic tile, porcelain tile and stone products through its network of regional distribution centers and approximately 250 company-operated sales service centers. Dal-Tile's products, marketed under the Dal-Tile and American Orlean brands, are sold primarily to floor covering retailers, home centers, independent distributors, tile specialty dealers and tile contractors. The Unilin segment manufactures and distributes laminate flooring, wood flooring, medium-density fiberboard, roofing systems and other wood products for use in residential remodeling and new construction projects in the U.S. and Europe. The segment has developed several new technologies, including the patented UNICLIC glueless installation system. Unilin sells its flooring products under the Quick-Step, Columbia Flooring, Century Flooring and Mohawk brands. In January 2013, the firm acquired Pergo, a manufacturer of laminate flooring in the U.S. and Europe. Also in January, Mohawk agreed to acquire Belgian panel board manufacturer Spano Invest NV. In April 2013, it completed the acquisition of the Marazzi Group for $1.5 billion.

Employees are offered medical, dental and vision insurance; prescription drug coverage; a 401(k); tuition reimbursement; flexible spending accounts; and discounts on company products.

FINANCIAL DATA: *Note: Data for latest year may not have been available at press time.*

In U.S. $	2012	2011	2010	2009	2008	2007
Revenue	5,787,980,000	5,642,258,000	5,319,072,000	5,344,024,000	6,826,348,000	7,586,018,000
R&D Expense						
Operating Income	379,508,000	315,542,000	314,169,000	43,730,000	-1,124,134,000	750,106,000
Operating Margin %	6.55%	5.59%	5.90%	.81%	-16.46%	9.88%
SGA Expense	1,110,550,000	1,101,337,000	1,088,431,000	1,188,500,000	1,318,501,000	1,364,678,000
Net Income	250,258,000	173,922,000	185,471,000	-5,499,000	-1,458,228,000	706,814,000
Operating Cash Flow	587,590,000	300,993,000	319,712,000	672,205,000	570,034,000	875,077,000
Capital Expenditure	208,294,000	275,573,000	156,180,000	108,925,000	217,824,000	163,076,000
EBITDA	659,498,000	599,225,000	622,572,000	352,322,000	-856,062,000	1,065,023,000
Return on Assets %	4.00%	2.82%	2.96%	- .08%	-19.28%	8.38%
Return on Equity %	7.01%	5.20%	5.73%	- .17%	-37.09%	16.78%
Debt to Equity	0.35	0.35	0.39	0.56	0.58	0.42

CONTACT INFORMATION:

Phone: 706 629-7721 Fax: 706 625-3851
Toll-Free: 800-241-4494
Address: 160 S. Industrial Blvd., Calhoun, GA 30701 United States

STOCK TICKER/OTHER:

Stock Ticker: MHK Exchange: NYS
Employees: 26,200 Fiscal Year Ends: 31-Dec
Parent Company:

SALARIES/BONUSES:

Top Exec. Salary: $1,030,000 Bonus: $
Second Exec. Salary: $890,000 Bonus: $

OTHER THOUGHTS:

Estimated Female Officers or Directors: 2
Hot Spot for Advancement for Women/Minorities:

Motorsports Authentics LLC

www.motorsports-authentics.com

NAIC Code: 451120

TYPES OF BUSINESS:

Toys & Collectibles
Auto Racing Collectibles & Memorabilia
Motorsports Apparel & Merchandise

BRANDS/DIVISIONS/AFFILIATES:

International Speedway Corp
Speedway Motorsports Inc
Chase Authentics
Minichamp

CONTACTS: *Note: Officers with more than one job title may be intentionally listed here more than once.*

Mark Dyer, CEO
Mark Dyer, Pres.
Pat Harvey, CFO

GROWTH PLANS/SPECIAL FEATURES:

Motorsports Authentics LLC, a joint venture between International Speedway Corp. and Speedway Motorsports, Inc., is a leading designer of licensed NASCAR, NHRA, Formula One, IRL and World of Outlaws die-cast scaled replicas of vehicles, apparel and memorabilia. The company was formed in 2005 through the combining of Action Performance Companies, Inc., the original exclusive license owner of Dale Earnheardt, and Team Caliber, the original official licensed manufacturer of die-cast. The firm's business plan is focused on harboring long-term licensing agreements with motorsports drivers and team owners, providing the company with the exclusive rights to design and develop products bearing the team and driver's name, likeness and image. The company has agreements with NASCAR drivers such as Dale Earnhardt, Jr., Jeff Gordon, Jimmie Johnson, Tony Stewart, Kevin Harvick, Carl Edwards and Kasey Kahne. In addition to its die cast memorabilia, the firm produces and markets a line of officially licensed NASCAR apparel under the Chase Authentics brand. This apparel includes a line of jackets, hats, t-shirts and jerseys featuring the names, likenesses and color schemes of popular NASCAR drivers. The company distributes its products through a variety of channels, including NASCAR events, specialty retail stores, team and driver shops, track retail stores and the Internet. Additionally, the company has agreements with wholesale distributors, such as QVC, as well as throughout Germany, where Motorsports Authentics markets its Minichamps collectibles die-cast brand. Motorsports designs every die-cast collectible that it manufactures. The company also manages an online Die-cast Identification Number (DIN) Program, which allows collectors to manage their die-cast collection virtually.

FINANCIAL DATA: *Note: Data for latest year may not have been available at press time.*

In U.S. $	2012	2011	2010	2009	2008	2007
Revenue						
R&D Expense						
Operating Income						
Operating Margin %						
SGA Expense						
Net Income						
Operating Cash Flow						
Capital Expenditure						
EBITDA						
Return on Assets %						
Return on Equity %						
Debt to Equity						

CONTACT INFORMATION:

Phone: 704-454-4000 Fax: 704-454-4006
Toll-Free: 888-332-4000
Address: 6301 Performance Dr., Concord, NC 28027 United States

STOCK TICKER/OTHER:

Stock Ticker: Joint Venture Exchange:
Employees: Fiscal Year Ends: 30-Sep
Parent Company: INTERNATIONAL SPEEDWAY CORP

SALARIES/BONUSES:

Top Exec. Salary: $ Bonus: $
Second Exec. Salary: $ Bonus: $

OTHER THOUGHTS:

Estimated Female Officers or Directors:
Hot Spot for Advancement for Women/Minorities:

Mount Vernon Mills

www.mvmills.com

NAIC Code: 313210

TYPES OF BUSINESS:

Textiles Manufacturing
Denim
Bedding Products
Medical Supplies
Upholstery Fabrics
Trucking
Specialty Fabrics
Textile Chemicals

BRANDS/DIVISIONS/AFFILIATES:

RB Pamplin Corp
Apollo Chemicals
Brentex
LaFrance Industries
PhilChem
Smith & Waters Inc
Mount Vernon Chemicals LLC
FCI Technology

CONTACTS: *Note: Officers with more than one job title may be intentionally listed here more than once.*

Roger W. Chastain, CEO
David Hastings, COO
David Hastings, Pres.
William E. Duncan, CFO
James A. Knight, Treas.
Robert Boisseau (Bob) Pamplin, Jr., Chmn.

GROWTH PLANS/SPECIAL FEATURES:

Mount Vernon Mills (MVM), owned by R. B. Pamplin Corp., manufactures textile, chemical and related products for the apparel, industrial, institutional and commercial markets. Its business encompasses six areas: Brentex, apparel fabrics, LaFrance Industries, consumer products, chemicals and transportation. The Brentex division manufactures pocketings and other apparel trim items as well as fabric for numerous end uses, including apparel, home furnishings, book bindings, military apparel components, electronic composites, flags and boat covers. The group operates textile facilities in Cuero, Texas and Columbus, Mississippi. MVM's largest group, apparel fabrics, operates one of the world's largest denim manufacturing facilities, in Trion, Georgia, which produces washed, over-dyed and stretch denim as well as twills, drills, duck and plain weave fabrics. The latter products are piece-dyed for career apparel, work clothing and sportswear. This group also produces flame-retardant fabrics and military uniform fabrics. LaFrance Industries produces fabrics primarily for paint rollers and seating surfaces, focusing on the automotive, transportation, residential and commercial markets. MVM's consumer products group produces infant bedding products; surgical towels; and tablecloths, napkins, sheets and towels for hotels and restaurants. The group sells its products through RiegelLinen.com. The chemicals group, operating through subsidiary Mount Vernon Chemicals LLC, offers products through three divisions: Apollo Chemicals, which produces lubricants and softeners for textile dyeing and finishing; PhilChem, a supplier of textile sizing and other dry chemical products; and FCI Technology, which provides chemicals to the personal care, textiles, paper, food and metal working industries. Through its subsidiary Smith & Waters, Inc., the transportation group provides nationwide truck hauling services and supporting logistics for clients as well as the company's products. MVM also maintains a corporate division that provides support services for its business groups.

FINANCIAL DATA: *Note: Data for latest year may not have been available at press time.*

In U.S. $	2012	2011	2010	2009	2008	2007
Revenue						
R&D Expense						
Operating Income						
Operating Margin %						
SGA Expense						
Net Income						
Operating Cash Flow						
Capital Expenditure						
EBITDA						
Return on Assets %						
Return on Equity %						
Debt to Equity						

CONTACT INFORMATION:

Phone: 864-688-7100 Fax: 864-370-2315
Toll-Free:
Address: 503 S. Main St., Mauldin, SC 29662 United States

STOCK TICKER/OTHER:

Stock Ticker: Subsidiary Exchange:
Employees: 5,500 Fiscal Year Ends: 31-Dec
Parent Company: RB PAMPLIN CORP

SALARIES/BONUSES:

Top Exec. Salary: $ Bonus: $
Second Exec. Salary: $ Bonus: $

OTHER THOUGHTS:

Estimated Female Officers or Directors:
Hot Spot for Advancement for Women/Minorities:

Nano-Tex LLC

NAIC Code: 313320

www.nano-tex.com

TYPES OF BUSINESS:

Textile Treatments-Nanotechnology
Nanotextiles

BRANDS/DIVISIONS/AFFILIATES:

Resists Spills
Coolest Comfort
Resists Spills and Releases Stains
Resists Static
Dry Inside
Neutralizer
Vitalize DP
Aquapel

CONTACTS: *Note: Officers with more than one job title may be intentionally listed here more than once.*

Jim Curley, CEO
Jim Curley, Pres.
Mark Brutten, Sr. VP-Sales & Mktg.
Kelvin Chen, Sr. VP-R&D
Kelvin Chen, Sr. VP-Tech. Oper.

GROWTH PLANS/SPECIAL FEATURES:

Nano-Tex LLC specializes in applying nanotechnology to textiles in order to produce a range of breathable fabrics resistant to water, spills, stains and static. These nanotextile fabrics are treated with a process designed to attach these specialized enhancements directly to the molecules in the fabric. The company does not engage in the manufacturing of the fabrics, but instead provides the chemical enhancements and technologies to a network of licensed mills, which then make the fabrics and manufacture garments made from them. Nano-Tex produces several types of fabric technologies, including Resists Spills, Coolest Comfort, Resists Static, Resists Spills and Releases Stains, Dry Inside, Neutralizer, Speed Dry, Releases Stains and All Conditions. The Resists Spills fabric protection product attaches tiny whiskers, less than 1/1000th the size of a cotton fiber and designed to repel water directly, to the fibers of the material. This process can be applied to many different fabrics, such as cotton, synthetics, wool, silk, rayon and polypropylene. The firm's Coolest Comfort treatment gives materials a super-absorbency, which pulls perspiration away from the skin and keeps the body cool. The Resist Static enhancement reduces static cling and repels statically attractive substances. The Resists Spills and Releases Stains technology allows fabrics to resist stains soaking in as well as loosening the stains up from the fibers during a normal wash cycle. Dry Inside or performance cotton keeps moisture away from skin and dries faster than normal cotton. Neutralizer isolates and neutralizes odor. Speed Dry reduces water absorption to quicken drying times. Releases Stains fights tough stains by allowing stains to wash out easily and also provides protection against future stains. All Conditions is designed to resist moisture and liquid in outdoor environments. Recently, Nano-Tex introduced new technologies Vitalize DP, a wrinkle resistant solution; and Aquapel, a fluorocarbon free water-resistant solution.

FINANCIAL DATA: *Note: Data for latest year may not have been available at press time.*

In U.S. $	2012	2011	2010	2009	2008	2007
Revenue						
R&D Expense						
Operating Income						
Operating Margin %						
SGA Expense						
Net Income						
Operating Cash Flow						
Capital Expenditure						
EBITDA						
Return on Assets %						
Return on Equity %						
Debt to Equity						

CONTACT INFORMATION:

Phone: 510-434-2700 Fax: 510-420-3764
Toll-Free:
Address: 2220 Livington St., Ste. 201, Oakland, CA 94606 United States

STOCK TICKER/OTHER:

Stock Ticker: Private
Exchange:
Employees:
Fiscal Year Ends:
Parent Company:

SALARIES/BONUSES:

Top Exec. Salary: $ Bonus: $
Second Exec. Salary: $ Bonus: $

OTHER THOUGHTS:

Estimated Female Officers or Directors:
Hot Spot for Advancement for Women/Minorities:

National Spinning Company Inc

www.natspin.com

NAIC Code: 313110

TYPES OF BUSINESS:

Yarn, Fabrics & Fiber
Dyeing

GROWTH PLANS/SPECIAL FEATURES:

National Spinning Company, Inc. provides a variety of yarns for apparel, home fashions and industrial end-uses. The firm's yarn for manufacturers includes types such as performance polyester, wool and wool blends, acrylic yarns and dyed filament yarns. Performance polyester is spun for moisture management, insulation and outdoor performance. The firm's performance offerings include UltraSpun performance polyester; Coolmax polyester; Thermolite polyester; Hydrotec polyester; Naturespun recycled polyester; Naturespun wool and polyester blend; and Ingeo recycled polyester, which uses vegetable- and corn-based fiber. Its wool and wool blend offerings include both long and short staple varieties with varying blend levels. Acrylic yarns include longstaple/high bulk acrylic, ring spun short staple acrylic, open end short staple acrylic, MicroDenier acrylic, low pill/high bulk acrylic and grey heather yarns. The dyed filament polyester yarns are offered in semi-dull round, bright delta, post-consumer recycled, antimony free and fire retardant varieties. Recently, the company sold its subsidiary Caron International to Spinrite LP.

BRANDS/DIVISIONS/AFFILIATES:

CONTACTS: *Note: Officers with more than one job title may be intentionally listed here more than once.*

Jim Chestnut, CEO
Jim Booterbaugh, COO
Linda Fanton, CFO
Chase Stallings, Head-Human Resources
Harry Wetmore, VP-Dev.
Morgan Miller, Vice-Chmn.
Bob Miller, Exec. VP
Jim Chestnut, Chmn.
Bob Miller, Head-Purchasing

FINANCIAL DATA: *Note: Data for latest year may not have been available at press time.*

In U.S. $	2012	2011	2010	2009	2008	2007
Revenue						
R&D Expense						
Operating Income						
Operating Margin %						
SGA Expense						
Net Income						
Operating Cash Flow						
Capital Expenditure						
EBITDA						
Return on Assets %						
Return on Equity %						
Debt to Equity						

CONTACT INFORMATION:

Phone: 252-975-7111 Fax:
Toll-Free: 800-868-7104
Address: 1481 W. 2nd St., Washington, NC 27889 United States

STOCK TICKER/OTHER:

Stock Ticker: Private
Exchange:
Employees: 750
Fiscal Year Ends: 31-Dec
Parent Company:

SALARIES/BONUSES:

Top Exec. Salary: $ Bonus: $
Second Exec. Salary: $ Bonus: $

OTHER THOUGHTS:

Estimated Female Officers or Directors: 2
Hot Spot for Advancement for Women/Minorities:

Nautica Enterprises Inc

NAIC Code: 315240

www.nautica.com

TYPES OF BUSINESS:

Casual Clothing for Men & Women
Retail Stores
Retailing
Fragrances
Accessories
Home Furnishings

BRANDS/DIVISIONS/AFFILIATES:

VF Corp
Nautica
Nautica Jeans Company
Nautica Spa
Nautica Home Furniture

CONTACTS:

Note: Officers with more than one job title may be intentionally listed here more than once.

Eric C. Wiseman, CEO/Pres./Chmn.-VF Corp.

GROWTH PLANS/SPECIAL FEATURES:

Nautica Enterprises, Inc., a subsidiary of VF Corp., is a designer, producer, distributor and retailer of men's, women's and children's fashion sportswear, denim bottoms, footwear and sleepwear. The company's target is men and women between the ages of 25-44. Types of apparel for men, women and children include Polo shirts, T-shirts, fleece and activewear, sweaters, pants and shorts, dresses, skirts, jeans, outerwear and blazers, swimwear, underwear and sleepwear. All fabric is preshrunk and seams are taped to prevent abrasion. Recently, the company introduced a new line of women's sportswear. Additionally, the company licenses Nautica brand home furnishings and accessories through the Nautica Home Furniture business segment. Each collection's products include unique comforter sets, sheet collections and pillowcases, quilts, decorative pillows, towels and bath rugs. Accessories include watches, fragrances, belts, scarves, cufflinks, sunglasses, hats, socks, ties, wallets, umbrellas, backpacks, towel bags and duffel bags. Company-owned brands include Nautica, Nautica Spa and Nautica Jeans Company. The Nautica Spa collection is only available for individual in-store purchase. Nautica is the principal lifestyle brand in VF Corp.'s Sportswear Coalition business segment, which operates Nautica retail outlet stores in outlet malls across the U.S. In addition to outlet stores, Nautica products are sold wholesale, primarily to leading department and specialty stores, and can be purchased by individuals at chain retail locations as well as online. Customers receive free shipping on all online purchases exceeding $100. Independent licensees also operate 170 Nautica-branded stores worldwide. Nautica apparel is sold in over 64 countries.

Parent company VF Corp. offers its employees medical, life, disability, AD&D, dental and vision coverage; tuition assistance; an employee assistance program; flexible spending accounts; product discounts; wellness programs; and a 401(k) plan.

FINANCIAL DATA:

Note: Data for latest year may not have been available at press time.

In U.S. $	2012	2011	2010	2009	2008	2007
Revenue						
R&D Expense						
Operating Income						
Operating Margin %						
SGA Expense						
Net Income						
Operating Cash Flow						
Capital Expenditure						
EBITDA						
Return on Assets %						
Return on Equity %						
Debt to Equity						

CONTACT INFORMATION:

Phone: 212-541-5757 Fax: 212-887-8136
Toll-Free: 866-376-4184
Address: 40 W. 57th St., New York, NY 10019 United States

STOCK TICKER/OTHER:

Stock Ticker: Subsidiary
Exchange:
Employees:
Fiscal Year Ends: 28-Feb
Parent Company: VF CORP

SALARIES/BONUSES:

Top Exec. Salary: $ Bonus: $
Second Exec. Salary: $ Bonus: $

OTHER THOUGHTS:

Estimated Female Officers or Directors:
Hot Spot for Advancement for Women/Minorities:

Neiman Marcus Group Inc (The)

www.neimanmarcus.com

NAIC Code: 452111

TYPES OF BUSINESS:

Upscale Department Stores
Online & Catalog Sales
Corporate Gifts
Fine Jewelry
Fine Apparel
Cosmetics
Housewares & Linens

BRANDS/DIVISIONS/AFFILIATES:

InCircle
Horchow
Bergdorf Goodman
Cusp
Last Call
Warburg Pincus LLC
TPG Capital

CONTACTS:

Note: Officers with more than one job title may be intentionally listed here more than once.

Karen W. Katz, CEO
James E. Skinner, COO/Exec. VP
Karen W. Katz, Pres.
James E. Skinner, CFO
Wanda Gierhart, Chief Mktg. Officer/Sr. VP
Joseph Weber, Chief Human Resources Officer/Sr. VP
Michael R. Kingston, CIO/Sr. VP
Wayne Hussey, Sr. VP-Store Dev. & Properties
John Koryl, Pres., Neiman Marcus Direct
Ginger Reeder, VP-Corp. Comm.
Dale Stapleton, Chief Acct. Officer/Sr. VP
Stacie Shirley, Sr. VP-Finance/Treas.
James Gold, Pres., Specialty Retail
Thomas Lind, Sr. VP-Program Mgmt.
Burton M. Tansky, Chmn.

GROWTH PLANS/SPECIAL FEATURES:

The Neiman Marcus Group, Inc. specializes in high-end retail sales, focusing on a largely affluent customer base. The company's 42 Neiman Marcus stores, 33 Last Call stores and two Bergdorf Goodman stores showcase global designer brands such as Prada, Armani, Chanel and Gucci. In addition to apparel, the firm offers distinctive and unusual upscale merchandise such as fine jewelry, home furnishings and decor, chocolates, perfumes and accessories. The company also operates six new concept stores under the name Cusp. These smaller stores are located in Virginia, Massachusetts, Illinois, California and Washington D.C. and are designed to appeal to fashion-forward women aged 21-45. Neiman Marcus is well-known for its emphasis on maintaining its customer base through personal contact, including invitations to events, thank-you notes, birthday greetings and personal correspondence alerting customers that new merchandise has arrived. The firm's highly successful rewards program, InCircle, helps drive sales, representing 35% of the group's yearly revenue. Additionally, to complement its brick-and-mortar operations, the firm's Neiman Marcus Direct division conducts online and catalog sales of fashion apparel, accessories and home furnishings through the Neiman Marcus brand; online and catalog sales of home furnishings and accessories through the Horchow brand; and online sales of fashion apparel and accessories through the Bergdorf Goodman brand. The company is owned by an investor group led by TPG Capital and Warburg Pincus LLC.

Neiman Marcus offers its employees a range of benefits including federal credit union membership; an education assistance plan; medical, disability, life, AD&D and dental coverage; same-sex domestic partner benefits; employee assistance programs; adoption benefits; flexible spending accounts; a scholarship award program for dependent children; prenatal/well baby care programs; a medical leave program; a retirement savings plan; a computer purchase program; and discounts on store and catalog merchandise.

FINANCIAL DATA:

Note: Data for latest year may not have been available at press time.

In U.S. $	2012	2011	2010	2009	2008	2007
Revenue	4,345,374,000	4,000,000,000	3,690,000,000	3,640,000,000	4,600,536,000	4,390,076,000
R&D Expense						
Operating Income						
Operating Margin %						
SGA Expense						
Net Income	140,086,000	31,600,000	-1,800,000,000	-668,000,000	142,813,000	111,932,000
Operating Cash Flow						
Capital Expenditure						
EBITDA						
Return on Assets %						
Return on Equity %						
Debt to Equity						

CONTACT INFORMATION:

Phone: 214-741-6911 Fax: 214-573-5320
Toll-Free: 888-888-4757
Address: 1618 Main St., 1 Marcus Square, Dallas, TX 75201 United States

STOCK TICKER/OTHER:

Stock Ticker: Private
Employees: 14,900
Parent Company: TPG CAPITAL
Exchange:
Fiscal Year Ends: 31-Jul

SALARIES/BONUSES:

Top Exec. Salary: $ Bonus: $
Second Exec. Salary: $ Bonus: $

OTHER THOUGHTS:

Estimated Female Officers or Directors: 5
Hot Spot for Advancement for Women/Minorities: Y

Net-A-Porter LLC

NAIC Code: 454111

www.net-a-porter.com

TYPES OF BUSINESS:

Online High Fashion Sales

BRANDS/DIVISIONS/AFFILIATES:

Net-A-Porter.com
theOutnet.com
Financiere Richemont SA
MrPorter.com

CONTACTS:

Note: Officers with more than one job title may be intentionally listed here more than once.

Mark Sebba, CEO
Alison Loehnis, VP-Mktg. & Sales
Naomi Hewitt, Dir.-Human Resources
Richard Lloyd-Williams, Dir.-IT
Paul Layton, Dir.-Oper.
Nick Cust, Dir.-e-commerce
Tenley Zinke, Head-Communications
Mario Muttenthaler, Head-Sales & Mktg., Mr. Porter
James Hudson, Head-Global Recruitment
Steve Crease, Head-Oper., Asia Pacific
Natalie Massenet, Chmn.
Rotimi Akinyemiju, Head-U.S. Oper.

GROWTH PLANS/SPECIAL FEATURES:

Net-A-Porter LLC is an online retailer of high-fashion, designer clothing, shoes and accessories. Through its web site Net-A-Porter.com, the company offers products from over 300 designers, including Chloe, Stella McCartney, Miu Miu, Christian Louboutin, Burberry, Fendi, Donna Karen, Yves Saint Laurent, Diane von Furstenburg and Oscar de la Renta, among others. The site also includes an online magazine that features style advice, interviews with designers and videos of various runway shows and fashion events, with around 2.5 million views per month. Net-A-Porter delivers its products to 170 countries worldwide, offering next day shipping in the U.S. and Canada. Other online properties include the outlet designer fashion apparel web site TheOutNet.com. Apart from its retail operations, the firm earns revenue from selling advertising space on its web site. Recently the luxury goods retailer Financiere Richemont SA acquired a 93% controlling interest in Net-A-Porter. The firm has continued to operate as separate enterprise alongside Richemont's other luxury goods businesses. In addition, the company has recently launched a web site dedicated to designer menswear. The site, dubbed Mr. Porter, features clothing and accessories from such designers as Lanvin, YSL, Ralph Lauren and Burberry.

FINANCIAL DATA:

Note: Data for latest year may not have been available at press time.

In U.S. $	2012	2011	2010	2009	2008	2007
Revenue						
R&D Expense						
Operating Income						
Operating Margin %						
SGA Expense						
Net Income						
Operating Cash Flow						
Capital Expenditure						
EBITDA						
Return on Assets %						
Return on Equity %						
Debt to Equity						

CONTACT INFORMATION:

Phone: 44-1473-323032 Fax:
Toll-Free: 800-481-1064
Address: 151 Queensway, 4th Fl., London, W24YN United Kingdom

STOCK TICKER/OTHER:

Stock Ticker: Subsidiary Exchange:
Employees: 800 Fiscal Year Ends:
Parent Company: FINANCIERE RICHEMONT SA

SALARIES/BONUSES:

Top Exec. Salary: $ Bonus: $
Second Exec. Salary: $ Bonus: $

OTHER THOUGHTS:

Estimated Female Officers or Directors: 3
Hot Spot for Advancement for Women/Minorities: Y

New Balance Athletic Shoe Inc

www.newbalance.com

NAIC Code: 316210

TYPES OF BUSINESS:

Athletic Footwear & Apparel
Sports Equipment
Leather Boots
Lacrosse Equipment

BRANDS/DIVISIONS/AFFILIATES:

Warrior
Dunham
PF Flyers
Abzorb
Aravon
Brine
Black Box Distribution
Creative Direction

CONTACTS: *Note: Officers with more than one job title may be intentionally listed here more than once.*

Robert DeMartini, CEO
Robert DeMartini, Pres.
John Withee, CFO/Exec. VP
Chris Quinn, Exec. VP-North American Sales & Retail
Carol O'Donnell, VP-Corp. Human Resources
Peter Zappala, VP-Specialty Sales
Stephanie Smith, VP-Retail
Joe Preston, Exec. VP-Global Footwear, Prod. & Mktg.
Duncan Scott, VP-External Prod.
Edith Harmon, VP-R&D Wellness
John Wilson, Exec. VP-Mfg.
Anne Davis, Exec. VP-Admin./Vice Chmn.
Christine Madigan, VP-Responsible Leadership
Jim Connors, VP-Global Design & Dev.
Dave Crosier, Exec. VP-Value Chain
Bill Hayden, VP-Finance
Kerry Kligerman, Exec. VP-Apparel
Paul Heffernan, VP-NBRx Foot Health Solutions
Alan Rosen, VP/Treas.
Edward Haddad, VP-Intellectual Property & Licensed Prod.
James S. Davis, Chmn.
Alan Hed, Exec. VP-Int'l
Kevin Holian, VP-Global Logistics

GROWTH PLANS/SPECIAL FEATURES:

New Balance Athletic Shoe, Inc. is an athletic footwear and apparel company, which began as an arch-support firm in the early 1900s. The firm manufactures men's and women's shoes for running, cross training, basketball, tennis, baseball, cheerleading, hiking and golf. The company also offers fitness apparel and kids' shoes. In addition, New Balance owns Warrior, a cutting-edge global manufacturer of lacrosse equipment; men's footwear maker Dunham; PF-Flyers, a line of active comfort shoes; Aravon and Cobb Hill, both casual footwear lines for women; and Brine, a line of performance products for lacrosse, soccer, hockey and volleyball. New Balance makes use of advanced technology in the production of its shoes, using products like Abzorb foam, a complex blend of rubber and proprietary materials, which provide extra cushioning, and compression properties. The company is also known for its wide selection of shoe widths. It is the only major maker of athletic shoes that still has plants in the U.S.; these plants manufacture more than 28,000 pairs per week. New Balance sells its products in over 120 countries around the world. The firm also maintains numerous licensees, joint ventures and distributors in the U.K., France, Germany, Sweden, Hong Kong, Singapore, Australia, New Zealand, Mexico, Canada, Japan, Brazil and South Africa. New Balance is an active sponsor of athletic and community organizations, including the La Salle Bank Chicago Marathon and the Susan G. Komen Breast Cancer Foundation, and is the official footwear and apparel provider of the U.S. military. In December 2012, the firm announced it would be partnering with companies Black Box Distribution and Creative Direction to begin producing shoes for the skate footwear market.

Employees are offered medical and dental coverage, a 401(k) plan, flexible spending accounts, a flextime work schedule, tuition reimbursement, an employee purchase program, a college saving plan, corporate gift matching, employee discounts and an on-site wellness program.

FINANCIAL DATA: *Note: Data for latest year may not have been available at press time.*

In U.S. $	2012	2011	2010	2009	2008	2007
Revenue	2,175,000,000	2,000,000,000				
R&D Expense						
Operating Income						
Operating Margin %						
SGA Expense						
Net Income						
Operating Cash Flow						
Capital Expenditure						
EBITDA						
Return on Assets %						
Return on Equity %						
Debt to Equity						

CONTACT INFORMATION:

Phone: 617-783-4000 Fax: 617-787-9355
Toll-Free: 800-622-1218
Address: 20 Guest St., Brighton Landing, Boston, MA 02135 United States

STOCK TICKER/OTHER:

Stock Ticker: Private
Employees:
Parent Company:
Exchange:
Fiscal Year Ends: 31-Dec

SALARIES/BONUSES:

Top Exec. Salary: $ Bonus: $
Second Exec. Salary: $ Bonus: $

OTHER THOUGHTS:

Estimated Female Officers or Directors: 5
Hot Spot for Advancement for Women/Minorities: Y

New Look Group plc

NAIC Code: 448120

www.newlook.co.uk

TYPES OF BUSINESS:

Women's & Children's Clothing-Retail
Lingerie Sales
Shoe Sales
Plus-Size Apparel
Men's Apparel

BRANDS/DIVISIONS/AFFILIATES:

Mim
Label Lounge
Idol
Tokyo Laundry
Brave Soul
New Look

CONTACTS: *Note: Officers with more than one job title may be intentionally listed here more than once.*

Anders Kristiansen, CEO
Will Kernan, COO
Alastair Miller, CFO
Tom Singh, Dir.-Commercial
Guy Lister, Dir.-Customer and Multichannel Managing
Alistair McGeorge, Exec. Chmn.

GROWTH PLANS/SPECIAL FEATURES:

New Look Group plc is a British retailer that sells women's, men's teens and kids clothing, shoes and accessories. The firm operates over 1,100 New Look stores in 16 countries, including the U.K., Singapore, Russia and Poland, with a significant market maintained in France, Belgium and the Netherlands. Including its e-commerce capabilities, the company serves over 120 markets worldwide. Annually, NewLook.com receives 355 million customer visits. The largest location is located in Newcastle upon Tyne with 30,000 square feet of merchandise. New Look has an in-house design team of over 30 designers who develop unique and exclusive fashion merchandise, allowing the company to limit its exposure to supply-driven trends. The company's brands include Label Lounge, Idol, Tokyo Laundry and Brave Soul. Additionally, the company supplies other company brands, such as 18 & East, Rocket Dog, Lee and Calvin Klein. The firm's subsidiary Mim operates more than 320 stores located in France and Belgium and caters to a younger clientele than that of New Look stores. Mim offers a slightly lower price point and has a more limited range of sizes.

Employees receive merchandise discounts, competitive bonuses, a company pension scheme and access to learning and development training programs, which often lead to upper management positions hired from within.

FINANCIAL DATA: *Note: Data for latest year may not have been available at press time.*

In U.S. $	2012	2011	2010	2009	2008	2007
Revenue	2,250,000,000	2,332,189,000	2,336,104,000	2,111,493,000	1,866,438,000	
R&D Expense						
Operating Income						
Operating Margin %						
SGA Expense						
Net Income		-4,628,000	31,442,000	-6,548,000	-8,145,000	
Operating Cash Flow						
Capital Expenditure						
EBITDA						
Return on Assets %						
Return on Equity %						
Debt to Equity						

CONTACT INFORMATION:

Phone: 44-1305-765-000 Fax:
Toll-Free:
Address: Mercery Rd., New Look House, Weymouth, DT3 5HJ United Kingdom

STOCK TICKER/OTHER:

Stock Ticker: Private
Employees: 3,500
Parent Company:
Exchange:
Fiscal Year Ends: 31-Mar

SALARIES/BONUSES:

Top Exec. Salary: $ Bonus: $
Second Exec. Salary: $ Bonus: $

OTHER THOUGHTS:

Estimated Female Officers or Directors: 1
Hot Spot for Advancement for Women/Minorities:

New York & Company Inc

www.nyandcompany.com

NAIC Code: 448120

TYPES OF BUSINESS:

Apparel-Women's, Retail

BRANDS/DIVISIONS/AFFILIATES:

City Stretch
City Style
Lerner
Lerner New York
NY&C
New York Style
Limited Brands Inc
New York & Company

CONTACTS: *Note: Officers with more than one job title may be intentionally listed here more than once.*

Gregory Scott, CEO
Laura Weil, COO/Exec. VP
Sheamus G. Toal, CFO/Exec. VP
Eran Cohen, Chief Mktg. Officer/Exec. VP
Suzanne Rosenberg, Dir.-Investor Rel.
Grace Nichols, Chmn.
Kevin L. Finnegan, Exec. VP-Global Sales

GROWTH PLANS/SPECIAL FEATURES:

New York & Company, Inc. is a specialty retailer of women's apparel that targets fashion-conscious women between the ages of 25 and 45. It designs and sources its proprietary branded merchandise sold exclusively through its national network of New York & Company retail stores and online. The company operates about 532 retail stores in 43 states. The stores are typically concentrated in large population centers of the U.S. and are located in shopping malls, lifestyle centers and off-mall locations, including urban street locations. The firm operates a web site, NYandCompany.com, through which customers can place and track orders, apply for the NY&C Rewards credit card and purchase New York & Company gift cards. New York & Company offers a merchandise assortment consisting of casual and wear-to-work apparel and accessories, including pants, jackets, knit tops, blouses, sweaters, denim, t-shirts, activewear, handbags and jewelry. New product lines are introduced in five major deliveries every year: spring, summer, fall, holiday and pre-spring; these deliveries are updated every four weeks to keep merchandise up to date with changing fashions. Limited Brands, Inc. provides warehousing and distribution services to the company. Brand names include New York & Company, Lerner, Lerner New York, New York Style, City Stretch, City Style and NY&C.

New York & Company offers its employees medical, dental, life, disability, vision and prescription drug coverage; medical reimbursement; flexible spending accounts; a 401(k) plan; a retirement plan; adoption assistance; tuition reimbursement; employee discounts; and commuter expense reimbursement.

FINANCIAL DATA: *Note: Data for latest year may not have been available at press time.*

In U.S. $	2012	2011	2010	2009	2008	2007
Revenue	956,456,000	1,021,699,000	1,006,675,000	1,139,853,000	1,194,944,000	
R&D Expense						
Operating Income	-35,570,000	-66,379,000	-23,926,000	-34,255,000	44,880,000	
Operating Margin %	-3.71%	-6.49%	-2.37%	-3.00%	3.75%	
SGA Expense	257,188,000	298,419,000	274,139,000	306,101,000	298,325,000	
Net Income	-38,937,000	-76,461,000	-13,481,000	-19,807,000	-4,857,000	
Operating Cash Flow	-8,554,000	10,803,000	55,300,000	28,311,000	83,556,000	
Capital Expenditure	12,158,000	15,695,000	13,285,000	44,612,000	75,464,000	
EBITDA	2,704,000	-25,238,000	18,570,000	10,710,000	83,614,000	
Return on Assets %	-11.93%	-19.31%	-3.01%	-4.19%	-1.01%	
Return on Equity %	-33.28%	-44.71%	-6.26%	-8.56%	-2.02%	
Debt to Equity			0.03	0.06	0.08	

CONTACT INFORMATION:

Phone: 212 884-2000 Fax: 212 884-2396
Toll-Free: 800-961-9906
Address: 450 W. 33rd St., 5th Fl., New York, NY 10001 United States

STOCK TICKER/OTHER:

Stock Ticker: NWY Exchange: NYS
Employees: 7,949 Fiscal Year Ends: 31-Jan
Parent Company:

SALARIES/BONUSES:

Top Exec. Salary: $775,000 Bonus: $
Second Exec. Salary: $600,000 Bonus: $

OTHER THOUGHTS:

Estimated Female Officers or Directors: 5
Hot Spot for Advancement for Women/Minorities: Y

NEXT plc

NAIC Code: 448140

www.next.co.uk

TYPES OF BUSINESS:

Men's, Women's & Children's Clothing-Retail
Franchising
Catalog & Online Sales
Online Flower & Gift Sales
Call Center Services
Customer Support Services
Housewares

BRANDS/DIVISIONS/AFFILIATES:

Next Clearance
NEXT Directory
NEXT International
NextFlowers.co.uk
Next Photos
NEXT Sourcing
Lipsy Limited
NEXT

CONTACTS: *Note: Officers with more than one job title may be intentionally listed here more than once.*

Simon A. Wolfson, CEO
David W. Keens, Dir.-Finance
Christos E. Angelides, Dir.-Prod.
A.J.R. McKinlay, Corp. Sec.
Andrew J. Varley, Dir.-Property
John O. Barton, Chmn.

GROWTH PLANS/SPECIAL FEATURES:

NEXT plc retails moderately priced clothing for men, women and children as well as housewares and furniture. Its three main retail channels are its 500 NEXT Retail stores in the U.K. and Ireland, targeting 20-30 year-olds looking for trendy but affordable clothes; NEXT International, which operates nearly 200 locations overseas in 30 countries in Europe, Asia and the Middle East; and NEXT Directory, which offers a mail catalog and an e-commerce web site, NextElectric.co.uk, which together serve 2.99 million customers in 35 countries. Its other main business operations include Next Sourcing, which buys, sources and designs products branded NEXT; and Lipsy, which designs and sells young women's clothing through wholesale, retail and website channels. Sourcing operations occur in mainland Hong Kong, Sri Lanka, China, India and the U.K.; the firm also owns garment manufacturing facilities in Sri Lanka. Lesser operations include NEXT Clearance, which sells marked-down merchandise specifically for clearance. Finally, its other business includes the e-commerce web site NextFlowers.co.uk, which retails flowers and gifts; Next Photos, which sells art, photos and camera equipment; and Next Gift List, which organizes wedding gift lists.

NEXT offers its employees a retirement plan and an employee stock ownership plan.

FINANCIAL DATA: *Note: Data for latest year may not have been available at press time.*

In U.S. $	2012	2011	2010	2009	2008	2007
Revenue	5,497,844,000	5,268,733,000	5,442,563,000	5,226,874,000	5,318,901,000	
R&D Expense						
Operating Income	961,495,500	905,576,000	846,461,100	764,179,600	858,124,400	
Operating Margin %	17.48%	17.18%	15.55%	14.62%	16.13%	
SGA Expense						
Net Income	758,747,500	640,837,200	581,722,400	483,144,300	565,745,300	
Operating Cash Flow	802,045,100	687,649,800	861,799,000	637,002,800	762,741,700	
Capital Expenditure	215,849,200	230,548,000	157,533,200	192,682,500	286,467,500	
EBITDA	1,152,740,000	1,096,182,000	1,043,138,000	950,791,000	1,031,315,000	
Return on Assets %	26.04%	23.01%	21.08%	17.83%	22.12%	
Return on Equity %	208.70%	219.30%	265.86%	985.01%	642.64%	
Debt to Equity						

CONTACT INFORMATION:

Phone: 44 1162842429 Fax:
Toll-Free:
Address: Desford Rd., Enderby, Leicester, LE19 4AT United Kingdom

STOCK TICKER/OTHER:

Stock Ticker: NXGPF
Exchange: GREY
Employees: 37,220
Fiscal Year Ends: 31-Jan
Parent Company:

SALARIES/BONUSES:

Top Exec. Salary: $ Bonus: $
Second Exec. Salary: $ Bonus: $

OTHER THOUGHTS:

Estimated Female Officers or Directors: 1
Hot Spot for Advancement for Women/Minorities:

Nike Inc

www.nike.com

NAIC Code: 316210

TYPES OF BUSINESS:

Athletic Shoes/Apparel Manufacturing
Athletic Equipment
Sports Accessories
Retail Stores
Sports Apparel
Plastic Products
Hockey Products
Swimwear

BRANDS/DIVISIONS/AFFILIATES:

All Star
Cole Haan
Converse Inc
Hurley International LLC
Umbro International Limited
Umbro
Chuck Taylor
One Star

CONTACTS: *Note: Officers with more than one job title may be intentionally listed here more than once.*

Mark Parker, CEO
Mark Parker, Pres.
Donald Blair, CFO/VP
John Slusher, VP-Global Sports Mktg.
David Ayre, VP-Global Human Resources
Hans van Alebeek, VP-Tech. & Global Oper.
Eric Sprunk, VP-Merch. & Prod.
Hilary Krane, General Counsel
Gary M. DeStefano, Pres., Global Oper.
Hilary Krane, VP-Corp. Affairs
Bernard Pliska, Corp. Controller/VP
Trevor Edwards, VP-Global Brand & Category Mgmt.
Charlie Denson, Pres., NIKE Brand
Roger Wyett, Pres., NIKE Affiliates
Jeanne Jackson, Pres., Direct to Consumer
Philip H. Knight, Chmn.

GROWTH PLANS/SPECIAL FEATURES:

Nike, Inc. designs, develops and markets footwear, apparel, equipment and accessories. It is one of the largest sellers of athletic footwear and athletic apparel in the world. The company's athletic footwear products are designed primarily for specific athletic use, although a large percentage of its products are worn for casual or leisure purposes. Running, training, basketball, soccer, sport-inspired urban shoes and children's shoes are the firm's top-selling product categories. Nike also markets shoes designed for tennis, golf, baseball, football, lacrosse, walking, outdoor activities, skateboarding, bicycling, volleyball, wrestling, cheerleading, aquatic activities and other athletic and recreational uses. The firm maintains several wholly-owned subsidiaries, including Cole Haan, which designs and distributes a line of dress and casual footwear, apparel and accessories for men and women under the Cole Haan brand name; Converse, Inc., which distributes and licenses footwear, apparel and accessories through brand names Converse, All Star, One Star, Chuck Taylor, Star Chevron and Jack Purcell; Hurley International LLC, which is headquartered in the U.K. and designs/distributes a collection of action sports apparel sold under the Hurley brand; and Umbro International Limited, which markets a line of athletic and casual footwear, apparel and equipment (primarily for soccer) under the trademark name Umbro. Nike sells its products to retail accounts, through Nike-owned retail stores and through a mix of independent distributors and licensees worldwide. Within the U.S. the firm operates 384 Nike Brand and subsidiary retail stores: 184 Nike locations, 109 Cole Haan stores, 62 Converse factory stores and 29 Hurley locations. In the international market, which includes countries within Europe, Asia, South America and Africa, the firm maintains 442 retail stores (373 Nike and 69 Cole Haan stores).

FINANCIAL DATA: *Note: Data for latest year may not have been available at press time.*

In U.S. $	2012	2011	2010	2009	2008	2007
Revenue	24,128,000,000	20,862,000,000	19,014,000,000	19,176,100,000	18,627,000,000	16,325,900,000
R&D Expense						
Operating Income	3,040,000,000	2,815,000,000	2,474,000,000	1,858,500,000	2,433,700,000	2,131,800,000
Operating Margin %	12.59%	13.49%	13.01%	9.69%	13.06%	13.05%
SGA Expense	7,431,000,000	6,693,000,000	6,326,400,000	6,149,600,000	5,953,700,000	5,028,700,000
Net Income	2,223,000,000	2,133,000,000	1,906,700,000	1,486,700,000	1,883,400,000	1,491,500,000
Operating Cash Flow	1,899,000,000	1,812,000,000	3,164,200,000	1,736,100,000	1,936,300,000	1,878,700,000
Capital Expenditure	597,000,000	432,000,000	335,100,000	455,700,000	449,200,000	313,500,000
EBITDA	3,421,000,000	3,206,000,000	2,869,500,000	2,241,800,000	2,899,500,000	2,402,000,000
Return on Assets %	14.59%	14.50%	13.78%	11.57%	16.28%	14.51%
Return on Equity %	21.98%	21.76%	20.67%	18.00%	25.36%	22.41%
Debt to Equity	0.02	0.02	0.04	0.05	0.05	0.05

CONTACT INFORMATION:

Phone: 503 671-6453 Fax: 503 671-6300
Toll-Free: 800-344-6453
Address: 1 Bowerman Dr., Beaverton, OR 97005 United States

STOCK TICKER/OTHER:

Stock Ticker: NKE Exchange: NYS
Employees: 38,000 Fiscal Year Ends: 31-May
Parent Company:

SALARIES/BONUSES:

Top Exec. Salary: $1,550,000 Bonus: $
Second Exec. Salary: $1,320,000 Bonus: $

OTHER THOUGHTS:

Estimated Female Officers or Directors: 2
Hot Spot for Advancement for Women/Minorities: Y

Nisshinbo Holdings Inc

NAIC Code: 313230

www.nisshinbo.co.jp

TYPES OF BUSINESS:

Textiles Manufacturing
Chemicals
Automotive Products
Tissue & Paper Products
Machine Tools
Electronics Manufacturing
Photovoltaic Cell Manufacturing Equipment
Real Estate

BRANDS/DIVISIONS/AFFILIATES:

Nisshinbo Textile Inc
Nisshin Denim Inc
Nisshinbo Yarn Dyed Co Ltd
Nisshinbo Automotive Manufacturing Inc
Nisshinbo Paper Products Inc
Nisshinbo Mechatronics Inc
Japan Radio Co Ltd
TMD Friction Group SA

CONTACTS: *Note: Officers with more than one job title may be intentionally listed here more than once.*

Shizuka Uzawa, Pres.
Yoshihito Onda, Sr. Exec. VP
Masaaki Isobe, Sr. Exec. VP
Masaya Kawata, Sr. Exec. VP
Masahiro Murakami, Exec. Managing Officer
Shizuka Uzawa, Chmn.

GROWTH PLANS/SPECIAL FEATURES:

Nisshinbo Holdings, Inc., formerly Nisshinbo Industries, started out in the early 1900s as a textile manufacturer of cotton yarns and threads. Today, it is a total textile manufacturer, with operations ranging from spinning to finishing, dyeing, marketing and sales. Nisshinbo's textile group spins, weaves, knits, dyes and finishes cotton material for bedding, shirts, denim and polyester and cotton blended fabrics used in uniforms. It also manufactures chemical and synthetic textured yarns, spandex, elastomeric yarns and non-woven all-cotton fabrics. The firm's non-textile activities, which now collectively account for a larger portion of its sales than textiles, encompass automobile brakes and brake assemblies; chemicals, including fuel battery separators and other similar technologies; tissue and paper products, including synthetic paper and toilet paper; electronics, including radio communication equipment and semiconductors; mechatronics, such as precision machinery and photovoltaic module manufacturing; and other businesses. During 2012, electronics accounted for 44.8 % of sales; textiles generated 16.1%; automobile brakes accounted for 12.5%; paper products represented 8%; mechatronics accounted for 6.6%; real estate leasing accounted for 2.4% and chemicals generated 2.2%, while other miscellaneous business activities accounted for 7.4%. Major Nisshinbo subsidiaries include Nisshinbo Textile, Inc.; Nisshin Denim, Inc.; Naigai Shirts Co., Ltd.; Nisshinbo Yarn Dyed Co., Ltd.; Nisshinbo Automotive Manufacturing, Inc.; Nisshinbo Paper Products, Inc.; Nisshinbo Mechatronics, Inc.; Nisshinbo Chemical, Inc.; Japan Radio Co., Ltd.; and Nisshinbo Urban Development Co., Ltd. Recently, the company purchased Japan Radio Co., Ltd. The firm also acquired all the shares of TMD Friction Group SA, a brake friction manufacturer, for approximately $577 million.

FINANCIAL DATA: *Note: Data for latest year may not have been available at press time.*

In U.S. $	2012	2011	2010	2009	2008	2007
Revenue	4,223,288,000	3,624,486,000	2,698,801,000	3,185,970,000	3,589,495,000	3,482,760,000
R&D Expense						
Operating Income	46,425,660	220,906,000	39,734,580	4,542,368	133,977,600	128,600,200
Operating Margin %	1.09%	6.09%	1.47%	.14%	3.73%	3.69%
SGA Expense	641,665,100			466,494,400		485,510,100
Net Income	104,819,600	124,514,300	21,108,650	-14,317,360	136,827,700	168,190,100
Operating Cash Flow	144,431,700	184,021,600	306,576,400	132,919,900	275,870,900	215,450,700
Capital Expenditure	162,322,800	115,685,600	147,515,600	196,713,500	237,483,400	207,401,400
EBITDA	212,856,700	388,327,900	216,397,100	199,029,200	459,157,700	445,152,000
Return on Assets %	1.85%	2.66%	.52%	-.32%	2.73%	3.13%
Return on Equity %	5.10%	6.11%	1.04%	-.63%	5.05%	5.72%
Debt to Equity	0.33	0.13	0.07	0.02	0.01	0.02

CONTACT INFORMATION:

Phone: 81 356958833 Fax:
Toll-Free:
Address: 2-31-11 Ningyo-cho, Nihonbashi, Chuo-ku, Tokyo, 103-8650 Japan

STOCK TICKER/OTHER:

Stock Ticker: NSHBF
Exchange: GREY
Employees: 18,292
Fiscal Year Ends: 31-Mar
Parent Company:

SALARIES/BONUSES:

Top Exec. Salary: $ Bonus: $
Second Exec. Salary: $ Bonus: $

OTHER THOUGHTS:

Estimated Female Officers or Directors:
Hot Spot for Advancement for Women/Minorities:

Nitto Boseki Co Ltd

www.nittobo.co.jp

NAIC Code: 313110

TYPES OF BUSINESS:

Building Materials
Glass Fiber
Silk & Synthetic Textile Manufacturing
Engineering
Chemicals
Real Estate Development
Beverage Bottle Manufacturing
Medical Research

BRANDS/DIVISIONS/AFFILIATES:

Nittobo America Inc

CONTACTS:

Note: Officers with more than one job title may be intentionally listed here more than once.

Katsumi Minamizono, CEO
Yoshitada Shiratori, COO
Yo Usami, Gen. Mgr.-Admin.
Yo Usami, Gen. Mgr.-Planning
Yo Usami, Gen. Mgr.-Acct.

GROWTH PLANS/SPECIAL FEATURES:

Nitto Boseki Co., Ltd. (Nittobo) is a diversified company with lead interests in the production of glass fiber, building materials, environmental services and real estate as well as specialized textiles. The company organizes its business into four segments: Glass Fiber, Environmental and Health, Textiles and Property Management. The Glass Fiber segment provides filament mats, fine-spun glass yarns, glass fabrics, tapes and related products to the industrial and electronics markets. These products are primarily used in the construction and maintenance of airports, wind power generators and printed circuit boards. The Environmental and Health segment is further subdivided into medical and beverage and acoustical engineering. The medical and beverage segment focuses on the development of clinical diagnostic reagents for biotechnology, hematology and immunology applications and the development of soft drinks. Divisions within this segment include chemicals, which produces polymers for uses such as dying and printing, metal chelating and water treatment; and beverage, which manufactures Polyethylene terephthalate (PET) bottles. The acoustical engineering unit provides high-quality acoustic products and services for the design and construction of listening rooms and studios. The Textiles segment offers core spun yarn; woven and knitted stretch fabrics; rayon technical yarn; interlining cloth products; and other products, such as dishcloths and fabric glue. The Property Management segment is responsible for the sale and lease of all constructed properties and buildings, including the Nittobo building, Atago Golf Garden and Renaissance Fukushima. The company maintains more than 20 consolidated subsidiaries, including international operations in Taiwan and China. Subsidiary Nittobo America, Inc., headquartered in California, focuses on the manufacture and sale of medical antiserums.

FINANCIAL DATA:

Note: Data for latest year may not have been available at press time.

In U.S. $	2012	2011	2010	2009	2008	2007
Revenue		1,081,400,000	1,076,110,000	1,273,210,000	1,362,782,000	1,175,568,000
R&D Expense						
Operating Income						
Operating Margin %						
SGA Expense						
Net Income		36,100,000	16,640,000	-100,960,000	59,166,000	39,094,000
Operating Cash Flow						
Capital Expenditure						
EBITDA						
Return on Assets %						
Return on Equity %						
Debt to Equity						

CONTACT INFORMATION:

Phone: 81-3-3514-8670 Fax:
Toll-Free:
Address: 4-1-28, Kudankita, Chiyoda-ku, Tokyo, 102-8489 Japan

STOCK TICKER/OTHER:

Stock Ticker: 3110
Employees: 2,887
Parent Company:
Exchange: Tokyo
Fiscal Year Ends: 31-Mar

SALARIES/BONUSES:

Top Exec. Salary: $ Bonus: $
Second Exec. Salary: $ Bonus: $

OTHER THOUGHTS:

Estimated Female Officers or Directors:
Hot Spot for Advancement for Women/Minorities:

Nordstrom Inc

www.nordstrom.com

NAIC Code: 452111

TYPES OF BUSINESS:

Department Stores
Outlet Stores
Online Retailing
Catalog Sales
Financial Services
Federal Savings Bank

BRANDS/DIVISIONS/AFFILIATES:

Nordstrom fsb
Nordstrom Rack
Last Chance
Nordstrom.com
Treasure & Bond
Hautelook Inc
Nordstrom

CONTACTS: *Note: Officers with more than one job title may be intentionally listed here more than once.*

Blake W. Nordstrom, Pres.
Michael G. Koppel, CFO/Exec. VP
Delena M. Sunday, Exec. VP-Human Resources & Diversity Affairs
R. Michael Richardson, CIO/Exec. VP
Peter E. Nordstrom, Pres., Merch./Exec. VP
Daniel F. Little, Chief Admin. Officer/Exec. VP
Robert B. Sari, General Counsel/Exec. VP/Corp. Sec.
Ken Worzel, Exec. VP-Strategy & Dev.
Robert E. Campbell, VP-Investor Rel./Treas.
James A. Howell, VP-Finance
Erik B. Nordstrom, Pres., Stores/Exec. VP
Geevy S.K. Thomas, Pres., Nordstrom Rack/Exec. VP
James F. Nordstrom, Jr., Pres., Nordstrom Direct/Exec. VP
Mark J. Tritton, Pres., Nordstrom Prod. Group/Exec. VP
Enrique Hernandez, Jr., Chmn.

GROWTH PLANS/SPECIAL FEATURES:

Nordstrom, Inc., founded in 1901, is an upscale fashion apparel and shoe retailers. It sells a wide selection of apparel, shoes and accessories for women, men and children. Nordstrom is comprised of two segments: Retail and Credit. Retail derives its revenues from the sale of designer, luxury and high-end apparel, shoes, cosmetics and accessories. The company operates a total of 225 stores in 30 states, including 116 full-line stores, 105 Nordstrom Rack stores, one Last Chance clearance store, one Treasure & Bond location and two Jeffrey boutiques. The Nordstrom Rack stores serve as outlets for clearance merchandise from the full-line stores. Rack stores also purchase merchandise directly from manufacturers. The Retail segment also generates revenues by serving customers on the Internet at Nordstrom.com and through catalogs. This segment also includes the online private sale subsidiary, Haute Look. The Credit segment includes Nordstrom's wholly-owned savings bank, Nordstrom fsb, through which it offers a private label card, two Nordstrom VISA credit cards and a debit card. It generates income through finance charges and fees on these cards.

Nordstrom offers employees benefits including medical, dental, vision, AD&D, life, and short- and long-term disability insurance; wellness programs; 401(k) plan & profit sharing; employee stock purchase plan; access to the company bank and credit union; employee assistance programs; adoption financial assistance; and merchandise discounts.

FINANCIAL DATA: *Note: Data for latest year may not have been available at press time.*

In U.S. $	2012	2011	2010	2009	2008	2007
Revenue	10,877,000,000	9,700,000,000	8,627,000,000	8,573,000,000	8,828,000,000	
R&D Expense						
Operating Income	1,249,000,000	1,118,000,000	834,000,000	779,000,000	976,000,000	
Operating Margin %	11.48%	11.52%	9.66%	9.08%	11.05%	
SGA Expense	3,036,000,000	2,685,000,000	2,465,000,000	2,386,000,000	2,360,000,000	
Net Income	683,000,000	613,000,000	441,000,000	401,000,000	715,000,000	
Operating Cash Flow	1,177,000,000	1,177,000,000	1,251,000,000	848,000,000	161,000,000	
Capital Expenditure	511,000,000	399,000,000	360,000,000	563,000,000	501,000,000	
EBITDA	1,576,000,000	1,451,000,000	1,115,000,000	1,063,000,000	1,496,000,000	
Return on Assets %	8.56%	8.73%	7.20%	7.12%	13.72%	
Return on Equity %	34.34%	34.12%	31.70%	34.49%	43.55%	
Debt to Equity	1.6	1.37	1.43	1.82	2	

CONTACT INFORMATION:

Phone: 206 628-2111 Fax: 206 628-1795
Toll-Free: 888-282-6060
Address: 1700 7th Ave., Ste. 300, Seattle, WA 98101 United States

STOCK TICKER/OTHER:

Stock Ticker: JWN Exchange: NYS
Employees: 56,500 Fiscal Year Ends: 31-Jan
Parent Company:

SALARIES/BONUSES:

Top Exec. Salary: $709,722 Bonus: $
Second Exec. Salary: $700,972 Bonus: $

OTHER THOUGHTS:

Estimated Female Officers or Directors: 9
Hot Spot for Advancement for Women/Minorities: Y

OMNOVA Solutions Inc

www.omnova.com

NAIC Code: 313320

TYPES OF BUSINESS:

Wall Coverings, Fabrics & Laminates
Roofing Systems
Performance Chemicals
Specialty Chemicals

BRANDS/DIVISIONS/AFFILIATES:

PILOTEC EL45
PILOTEC HDT 12
Diversiwall
Viewnique
Duramax
NovaCryl
Sunbond
StyleCoat

CONTACTS: *Note: Officers with more than one job title may be intentionally listed here more than once.*

Kevin M. McMullen, CEO
Kevin M. McMullen, Pres.
Michael E. Hicks, CFO/Sr. VP
Kris C. Syrvalin, VP-Human Resources Admin.
Douglas E. Wenger, CIO/Sr. VP
James C. LeMay, General Counsel
James C. LeMay, Sr. VP-Corp. Dev.
Chester W. Fox, Treas./VP
David H. Maynard, Pres., Eng. Surfaces
James J. Hohman, Pres., Performance Chemicals/Corp. Sr. VP
Kris C. Syrvalin, Sec./Asst. General Counsel
Kevin M. McMullen, Chmn.

GROWTH PLANS/SPECIAL FEATURES:

OMNOVA Solutions, Inc. designs, develops, produces and markets emulsion polymers, specialty chemicals and decorative and functional surfaces for a variety of commercial, industrial and residential end uses. The firm utilizes about 20 strategically located manufacturing, development and design facilities in North America, Europe and Asia. OMNOVA operates in two business segments: performance chemicals, accounting for approximately 79% of the company's revenues, and engineered surfaces, accounting for 21%. The performance chemicals segment is a North American leader in emulsion polymers and specialty chemicals. Its products include styrene butadiene carpet binders marketed under the NovaGreen and GenFlo brands; in-mold coating primers under the GenGlaze and StyleCoat brands; and paper chemicals such as coating binders and coating chemicals, which are developed by the firm's joint venture with Rohm and Haas, RohmNova. The segment's specialty chemicals include PolyFox fluorosurfactants for high gloss coatings, NovaCryl adhesives, Sunbond nonwoven materials used for disposable hygiene products, AcryGen acrylic top coats used in rubberized running tracks and GenTac latex for tire and rubber goods. The engineered surfaces segment develops, designs, produces and markets a broad line of decorative and functional surfacing products, such as commercial wall coverings; wall services; paper, flat and 3D laminates; digital wall murals; upholstery products; ceiling tile surfaces; pool liner films; and performance films, such as awnings, tents, banners, security shades for cars and movie screens. Brands within this segment include Boltaflex, Boltasoft, Nautolex, Radiance, Duramax, Harmony, Viewnique and Diversiwall. In October 2012, the company launched PILOTEC HDT 12, a new water resistant binder for usage in direct-to-metal anticorrosive paint, and released PILOTEC EL45, a binder for high humidity/low temperature applications.

OMNOVA offers its employees medical, dental and vision coverage; a 401(k) plan; life and AD&D insurance; and adoption assistance.

FINANCIAL DATA: *Note: Data for latest year may not have been available at press time.*

In U.S. $	2012	2011	2010	2009	2008	2007
Revenue	1,125,500,000	1,201,100,000	846,200,000	696,400,000	869,400,000	745,500,000
R&D Expense						
Operating Income	72,000,000	71,800,000	34,400,000	33,700,000	8,700,000	20,100,000
Operating Margin %	6.39%	5.97%	4.06%	4.83%	1.00%	2.69%
SGA Expense	121,200,000	108,600,000	99,600,000	99,900,000	104,800,000	99,100,000
Net Income	27,600,000	-2,800,000	107,900,000	26,200,000	-2,200,000	-6,700,000
Operating Cash Flow	65,300,000	15,700,000	45,000,000	74,200,000	16,400,000	12,700,000
Capital Expenditure	32,800,000	24,100,000	17,300,000	10,400,000	14,800,000	16,200,000
EBITDA	105,400,000	101,600,000	47,800,000	58,900,000	34,900,000	29,700,000
Return on Assets %	3.17%	-.35%	20.28%	7.59%	-.64%	-2.01%
Return on Equity %	21.91%	-2.28%	155.81%	101.35%	-4.31%	-11.77%
Debt to Equity	3.39	3.65	3.15	9.32	4.97	2.21

CONTACT INFORMATION:

Phone: 330 869-4200 Fax: 330 869-4211
Toll-Free:
Address: 175 Ghent Rd., Fairlawn, OH 44333 United States

STOCK TICKER/OTHER:

Stock Ticker: OMN
Exchange: NYS
Employees: 2,300
Fiscal Year Ends: 30-Nov
Parent Company:

SALARIES/BONUSES:

Top Exec. Salary: $776,596 Bonus: $
Second Exec. Salary: $392,500 Bonus: $

OTHER THOUGHTS:

Estimated Female Officers or Directors: 1
Hot Spot for Advancement for Women/Minorities:

Orvis Company Inc (The)

NAIC Code: 451110

www.orvis.com

TYPES OF BUSINESS:

Outdoor Apparel, Retail
Fly-Fishing & Hunting Equipment
Home Furnishings
Catalog & Online Sales
Luggage & Accessories
Guided Trips & Travel Services
Firearms
Log Home Design & Construction

BRANDS/DIVISIONS/AFFILIATES:

Orvis Travel
Rocky Mountain Log Homes Co
C.F. Orvis Company Inc
Orvis Log Homes

CONTACTS: *Note: Officers with more than one job title may be intentionally listed here more than once.*

Leigh Perkins, Jr., CEO
Raymond G. McCready, Pres.
Tom Rosenbauer, Dir.-Mktg., Rod & Tackle
David Perkins, Vice Chmn.
Leigh H. Perkins, Chmn.

GROWTH PLANS/SPECIAL FEATURES:

The Orvis Company, Inc., formerly known as C.F. Orvis Company, Inc., is a retailer of outdoor products. The firm markets fly fishing equipment, outdoor apparel and home products through retailers, outlets and authorized dealers in the U.S. and U.K. The company's outdoor apparel includes men's and women's clothing; outerwear; footwear; and accessories such as insect-repellent clothing, hats, luggage and wallets. Its fly-fishing products include waders/boots, lines, leaders, rods, reels, vests, packs and fly-fishing apparel. The firm's hunting products include shotguns, shotgun accessories and hunting apparel. In addition, Orvis runs fly-fishing and shooting schools and seminars; guided fishing and hunting trips through its series of lodges; and Orvis Travel helps customers plan international vacations with a focus on outdoor activities. The company's home furnishings include bedding, lighting, dining furniture, kitchen furniture, bedroom furniture, tables, chairs, desks, tableware, outdoor furniture and fireplace accessories. It also offers dog and cat products including apparel, beds, crates and toys. The firm operates over 44 retail stores in the U.S. and the U.K. Additionally, the company sells products through 500 authorized Orvis dealers, which primarily sell fly-fishing products and apparel, although there are over 100 full line dealers worldwide. Orvis also markets products through catalogs and an e-commerce web site. Other activities include Orvis Log Homes, a division that offers six log home design packages and complete construction options through Rocky Mountain Log Homes Co.

FINANCIAL DATA: *Note: Data for latest year may not have been available at press time.*

In U.S. $	2012	2011	2010	2009	2008	2007
Revenue	245,000,000	225,000,000				
R&D Expense						
Operating Income						
Operating Margin %						
SGA Expense						
Net Income						
Operating Cash Flow						
Capital Expenditure						
EBITDA						
Return on Assets %						
Return on Equity %						
Debt to Equity						

CONTACT INFORMATION:

Phone: 802-362-1300 Fax: 802-362-0141
Toll-Free: 888-235-9763
Address: 178 Conservation Way, Sunderland, VT 05250 United States

STOCK TICKER/OTHER:

Stock Ticker: Private
Exchange:
Employees:
Fiscal Year Ends: 30-Sep
Parent Company:

SALARIES/BONUSES:

Top Exec. Salary: $ Bonus: $
Second Exec. Salary: $ Bonus: $

OTHER THOUGHTS:

Estimated Female Officers or Directors:
Hot Spot for Advancement for Women/Minorities:

OshKosh B'Gosh Inc

www.oshkoshbgosh.com

NAIC Code: 315240

TYPES OF BUSINESS:

Children's Apparel
Home Furnishings
Retail Stores

BRANDS/DIVISIONS/AFFILIATES:

Carters Inc
OshKosh Baby

CONTACTS:

Note: Officers with more than one job title may be intentionally listed here more than once.

Michael D. Casey, CEO
David Brown, Pres.
Richard F. Westenberger, CFO/Exec. VP
William G. Foglesong, Sr. VP-Mktg.
Jill A. Wilson, Sr. VP-Human Resources & Talent Dev.
Brendan M. Gibbons, General Counsel/Sr. VP-Legal & Corp. Affairs/Sec.
Lisa A. Fitzgerald, Exec. VP/Brand Leader-OshKosh B'gosh
Brian J. Lynch, Exec. VP/Brand Leader-Carter's
James C. Petty, Pres., Retail Stores
Michael D. Casey, Chmn.
Christopher W. Rork, Exec. VP-Supply Chain

GROWTH PLANS/SPECIAL FEATURES:

OshKosh B'Gosh, Inc., a subsidiary of Carter's, Inc., manufactures and sells colorful, seasonal apparel for kids. Founded in 1895, the company was historically a local manufacturer of adult workwear, predominately denim bib overalls for railroad workers and farmers. The children's line was started as a youth version of the adult overalls as a novelty item for proud parents. Presently, the children's apparel line not only includes bib overalls, but also features pants, shorts, footwear and swimwear as well as a layette line for newborns. The OshKosh brand comes in boys and girls sizes 2t to size 12 in addition to OshKosh Baby for infants sizes 0-24 months. The firm's merchandise also includes a variety of licensed products such as eyewear, plush toys, hair accessories, sleepwear and bedding. High-quality home furnishings (including throws, pillows and picture frames), toys and games have also been added to the firm's product line. The company operates approximately 180 outlet and specialty stores. OshKosh also sells its products through third party stores, primarily department stores, in the U.S. and in over 50 countries around the world.

FINANCIAL DATA:

Note: Data for latest year may not have been available at press time.

In U.S. $	2012	2011	2010	2009	2008	2007
Revenue	283,300,000	280,900,000				
R&D Expense						
Operating Income						
Operating Margin %						
SGA Expense						
Net Income						
Operating Cash Flow						
Capital Expenditure						
EBITDA						
Return on Assets %						
Return on Equity %						
Debt to Equity						

CONTACT INFORMATION:

Phone: 920-231-8800 Fax: 920-231-8621
Toll-Free: 800-692-4674
Address: 112 Otter Ave., Oshkosh, WI 54901 United States

STOCK TICKER/OTHER:

Stock Ticker: Subsidiary
Employees:
Parent Company: CARTERS INC
Exchange:
Fiscal Year Ends: 31-Dec

SALARIES/BONUSES:

Top Exec. Salary: $ Bonus: $
Second Exec. Salary: $ Bonus: $

OTHER THOUGHTS:

Estimated Female Officers or Directors: 2
Hot Spot for Advancement for Women/Minorities:

Oxford Industries Inc

www.oxfordinc.com

NAIC Code: 315220

TYPES OF BUSINESS:

Casual & Dress Apparel

BRANDS/DIVISIONS/AFFILIATES:

Tommy Bahama
Ben Sherman
Lilly Pulitzer
Lanier Clothes
Oxford Golf
Lilly Pulitzer Signature Stores
LillyPulitzer.com

CONTACTS: *Note: Officers with more than one job title may be intentionally listed here more than once.*

Thomas C. Chubb, III, CEO
Thomas C. Chubb, III, Pres.
K. Scott Grassmyer, CFO/Sr. VP-Finance/Controller
Christine B. Cole, Sr. VP-Human Resources
Thomas E. Campbell, Sr. VP-Admin.
Thomas E. Campbell, General Counsel/Sr. VP-Law/Sec.
Mark B. Kirby, VP-Oper.
Anne M. Shoemaker, Treas./VP-Capital Markets
Terry R. Pillow, CEO-Tommy Bahama Group
Scott A. Beaumont, CEO-Lilly Pulitzer Group
Wesley Howard, Pres., Lanier Clothes
James B. Bradbeer, Jr., Pres., Lilly Pulitzer Group
J. Hicks Lanier, Chmn.

GROWTH PLANS/SPECIAL FEATURES:

Oxford Industries, Inc. is an international apparel design, sourcing and marketing company that features a diverse portfolio of owned and licensed lifestyle brands, company-owned retail operations and a collection of private label apparel businesses. It operates under four primary brands: Tommy Bahama, Ben Sherman, Lilly Pulitzer and Lanier Clothes. Tommy Bahama designs, sources and markets men's and women's sportswear and related products with an elegant island living theme. The segment's target consumers are affluent 35 and older men and women. Ben Sherman is a London-based designer, marketer and distributor of men's branded sportswear and footwear, primarily targeting fashion conscious men aged 25-40. Lilly Pulitzer manufactures and markets upscale women's and girl's apparel, including dresses and sportswear. The segment's products are offered through Lilly Pulitzer Signature Stores as well as through its e-commerce site LillyPulitzer.com. Lanier Clothes designs and markets branded and private label men's suits, sport coats, suit separates and dress slacks. Its products are sold to national chains, department stores, mass merchants, specialty stores, specialty catalog retailers and discount retailers throughout the U.S. Additionally, through the Oxford Golf brand, the company sells third-party sourced golf apparel and accessories. Besides its own brands, Oxford Industries holds licenses to produce and sell certain categories of apparel and footwear products under various brands, including Kenneth Cole, Geoffrey Beene and Dockers. The firm's direct to consumer operations consist of 113 Tommy Bahama, 19 Lilly Pulitzer and 19 Ben Sherman retail locations, as well as e-commerce sites and 14 Tommy Bahama restaurants.

FINANCIAL DATA: *Note: Data for latest year may not have been available at press time.*

In U.S. $	2012	2011	2010	2009	2008	2007
Revenue	758,913,000	603,947,000	800,658,000	947,516,000	695,798,000	
R&D Expense						
Operating Income	68,807,000	40,662,000	41,154,000	-262,022,000	40,994,000	
Operating Margin %	9.06%	6.73%	5.13%	-27.65%		
SGA Expense	358,582,000	301,002,000	304,330,000	358,071,000	244,033,000	
Net Income	29,380,000	78,658,000	14,624,000	-265,812,000	19,215,000	
Operating Cash Flow	58,380,000	15,761,000	81,569,000	90,363,000	44,137,000	
Capital Expenditure	35,310,000	13,328,000	11,323,000	20,735,000	21,097,000	
EBITDA	86,944,000	59,475,000	61,890,000	-229,399,000	58,727,000	
Return on Assets %	5.50%	15.99%	3.25%	-38.41%		
Return on Equity %	15.29%	55.30%	14.80%	-106.18%		
Debt to Equity	0.5	0.81	1.4	2.08		

CONTACT INFORMATION:

Phone: 404 659-2424   Fax: 404 653-1545
Toll-Free:
Address: 222 Piedmont Ave. NE, Atlanta, GA 30308 United States

STOCK TICKER/OTHER:

Stock Ticker: OXM   Exchange: NYS
Employees: 4,400   Fiscal Year Ends: 31-Jan
Parent Company:

SALARIES/BONUSES:

Top Exec. Salary: $750,000   Bonus: $
Second Exec. Salary: $550,000   Bonus: $

OTHER THOUGHTS:

Estimated Female Officers or Directors: 3
Hot Spot for Advancement for Women/Minorities: Y

Pacific Brands Limited

www.pacificbrands.com.au

NAIC Code: 314999

TYPES OF BUSINESS:

All Other Miscellaneous Textile Product Mills
Hosiery
Footwear
Bed Linens
Carpet Underlay
Uniforms
Apparel
Sporting Goods

BRANDS/DIVISIONS/AFFILIATES:

Berlei
Bonds
Clarks
Dunlop
Everlast
Grosby
Hard Yakka
Hush Puppies

CONTACTS: *Note: Officers with more than one job title may be intentionally listed here more than once.*

John Pollaers, CEO
David Bortolussi, CFO
Peter Bush, Chmn.

GROWTH PLANS/SPECIAL FEATURES:

Pacific Brands Limited manufactures, sources, markets, and sells consumer lifestyle brands across underwear, hosiery, sleepwear, intimate apparel, footwear, bed linen, bedding accessories, bedding, carpet underlay, foams, corporate uniforms, workwear, streetwear, lifestyle apparel and sporting goods. Brand names include Berlei, Bonds, Clarks, Dunlop, Everlast, Grosby, Hard Yakka, Holeproof, Hush Puppies, King Gee, Mooks, Mossimo, Razzamatazz, Sheridan, Slazenger, Tontine, and Volley. The firm produces in excess of 300,000 different products and sells over 200 million units per year. Pacific Brands has operations throughout Australia, New Zealand, the UK, China and Indonesia.

FINANCIAL DATA: *Note: Data for latest year may not have been available at press time.*

In U.S. $	2012	2011	2010	2009	2008	2007
Revenue	1,400,778,000	1,710,616,000	1,847,405,000	2,122,267,000		1,934,321,000
R&D Expense						
Operating Income	136,111,900	196,381,600	188,751,500	213,328,100		202,754,600
Operating Margin %	9.71%	11.48%	10.21%	10.05%		10.48%
SGA Expense						
Net Income	-475,258,800	-139,097,500	55,601,020	-247,283,300		111,745,200
Operating Cash Flow	112,947,400	99,861,840	142,663,100	109,370,200		145,847,000
Capital Expenditure	29,583,850	22,758,430	10,591,420	23,700,190		25,227,270
EBITDA	157,661,700	224,602,900	214,346,900	245,205,700		231,740,500
Return on Assets %	-29.27%	-6.74%	2.44%	-10.53%		4.54%
Return on Equity %	-47.97%	-10.26%	3.97%	-18.44%		8.06%
Debt to Equity	0.5	0.32	0.33	0.45		0.71

CONTACT INFORMATION:

Phone: 61 399474900 Fax: 61 399474951
Toll-Free:
Address: 290 Burwood Rd., Fl. 2, Hawthorn, VIC 3122 Australia

STOCK TICKER/OTHER:

Stock Ticker: PFBDY Exchange: PINX
Employees: 9,022 Fiscal Year Ends: 30-Jun
Parent Company:

SALARIES/BONUSES:

Top Exec. Salary: $ Bonus: $
Second Exec. Salary: $ Bonus: $

OTHER THOUGHTS:

Estimated Female Officers or Directors: 2
Hot Spot for Advancement for Women/Minorities:

Pacific Sunwear of California Inc

www.pacsun.com

NAIC Code: 448140

TYPES OF BUSINESS:

Casual Apparel, Retail
Teen & Young Adult Apparel
Accessories
Footwear
Online Sales

BRANDS/DIVISIONS/AFFILIATES:

PacSun
Pacific Sunwear Outlet
PacSun.com

CONTACTS: *Note: Officers with more than one job title may be intentionally listed here more than once.*

Gary H. Schoenfeld, CEO
Gary H. Schoenfeld, Pres.
Michael Kaplan, CFO/Sr. VP
Craig E. Gosselin, Sr. VP-Human Resources
Charlie Mescher, Sr. VP-Men's Merch.
Craig E. Gosselin, General Counsel
Jon Brewer, Sr. VP-Oper.
Eric Fong, VP-Planning & Allocation
Mondy Beller, VP-e-commerce
Michael L. Henry, Contact-Investor Rel.
Paula M. Lentini, Sr. VP-Retail
Christine Lee, Sr. VP-Women's Merch.
Kristin Gwinner, VP-Human Resources
Wayne Bowman, VP-Construction
Peter Starrett, Chmn.

GROWTH PLANS/SPECIAL FEATURES:

Pacific Sunwear of California, Inc. is a leading specialty retailer of everyday casual apparel, accessories and footwear designed to meet the lifestyle needs of active teens and young adults. The company operates two nationwide primarily mall-based chains of retail stores under the names PacSun and Pacific Sunwear Outlet, with approximately 852 stores in all 50 states and Puerto Rico. PacSun's typical customers are young people who prefer a casual, fun look. Much of the fashion influence at the PacSun stores comes from surfing, skateboarding and snowboarding as well as from brand names associated with these sports. A typical store offers accessories such as sunglasses and hats, footwear and casual apparel, such as shorts, jeans, swimwear and shirts under its heritage brand names including Roxy, Quiksilver, Billabong, O'Neill and DC as well as newer brands like Volcom, Fox Racing and Famous Stars. The company also offers merchandise under its proprietary labels Black Poppie, Bullhead, Kirra, Kirra Girl, On The Byas and Nollie. In addition to its retail operations, Pacific Sunwear operates the e-commerce site PacSun.com. The company also allows many of its heritage brands to set up brand fixtures featuring brand-specific signage and merchandise within its stores, and some have even formed brand shops that create a store within a store experience. Pacific Sunwear operates a 400,000 square foot distribution center in Olathe, Kansas. In November 2012, a new collaboration was announced with Kendall and Kylie Jenner to create a young women's clothing line. This line will feature various tops and bottoms as well as accessories, such as scarves.

The company offers employees medical, dental and vision insurance along with life insurance and a 401(k).

FINANCIAL DATA: *Note: Data for latest year may not have been available at press time.*

In U.S. $	2012	2011	2010	2009	2008	2007
Revenue	833,751,000	929,506,000	1,027,101,000	1,254,886,000	1,454,163,000	
R&D Expense						
Operating Income	-80,376,000	-95,144,000	-81,125,000	-60,901,000	-36,403,000	
Operating Margin %	-9.64%	-10.23%	-7.89%	-4.85%	-2.50%	
SGA Expense	261,169,000	300,530,000	339,728,000	381,008,000	443,578,000	
Net Income	-106,423,000	-96,648,000	-70,302,000	-63,822,000	-30,367,000	
Operating Cash Flow	-47,404,000	-40,885,000	87,451,000	33,915,000	115,641,000	
Capital Expenditure	13,235,000	17,159,000	23,498,000	80,934,000	106,362,000	
EBITDA	-37,871,000	-39,497,000	-10,758,000	15,532,000	43,920,000	
Return on Assets %	-28.13%	-22.00%	-13.43%	-9.65%	-3.98%	
Return on Equity %	-65.00%	-37.10%	-20.71%	-14.92%	-6.15%	
Debt to Equity	0.65	0.13				

CONTACT INFORMATION:

Phone: 714 414-4000 Fax: 714 414-4260
Toll-Free: 800-444-6770
Address: 3450 E. Miraloma Ave., Anaheim, CA 92806 United States

STOCK TICKER/OTHER:

Stock Ticker: PSUN Exchange: NAS
Employees: 11,500 Fiscal Year Ends: 31-Jan
Parent Company:

SALARIES/BONUSES:

Top Exec. Salary: $1,050,000 Bonus: $1,312,500
Second Exec. Salary: $421,919 Bonus: $260,623

OTHER THOUGHTS:

Estimated Female Officers or Directors: 4
Hot Spot for Advancement for Women/Minorities: Y

Paris Blues Inc

www.parisblues.com

NAIC Code: 315240

TYPES OF BUSINESS:

Women's Apparel
Jeans
Junior Apparel

BRANDS/DIVISIONS/AFFILIATES:

CONTACTS:

Note: Officers with more than one job title may be intentionally listed here more than once.

Jose Quant, CEO
Jose Quant, Chmn.

GROWTH PLANS/SPECIAL FEATURES:

Paris Blues, Inc. designs and manufactures low-cost apparel for young women, including denim jeans, shorts and dresses in stonewash, acid-wash and garment-dye styles. The firm also offers shoes and tops. Paris Blues' widest selection stems from its jeans collection, which includes flared leg, low rise and slim fit in a variety of shades. The company reinforces its brand name via creative advertising, sales promotions and solid relationships with retailers. Paris Blues has significant penetration across a wide range of retail channels, including department stores such as Macy's and Kohls, national specialty stores like Nordstrom's, national junior shops such as Alloy and Wet Seal and multiple independent specialty stores and junior catalogs. Paris Blues apparel can also be found at Charlotte Russe, 5-7-9, Kohls, Ross, Marshalls, Bealls and FredMeyer. Many of the firm's brands are advertised with magazines appealing to young women, including Cosmo Girl, Seventeen and Teen People. The company's apparel is also sold through several web sites. Paris Blues also sells a maternity line of blue jeans through a partnership with Mimi Maternity. The firm's in-house design and merchandising team is involved in assisting and supporting licensees with product design and direction for the development of all the company's licensed products. In addition to its corporate headquarters in Rancho Dominquez, California, the firm has a showroom in New York City.

FINANCIAL DATA:

Note: Data for latest year may not have been available at press time.

In U.S. $	2012	2011	2010	2009	2008	2007
Revenue						
R&D Expense						
Operating Income						
Operating Margin %						
SGA Expense						
Net Income						
Operating Cash Flow						
Capital Expenditure						
EBITDA						
Return on Assets %						
Return on Equity %						
Debt to Equity						

CONTACT INFORMATION:

Phone: 310-605-2000 Fax: 310-605-1750
Toll-Free:
Address: 3040 E. Ana St., Rancho Dominguez, CA 90221 United States

STOCK TICKER/OTHER:

Stock Ticker: Private
Exchange:
Employees: 90
Fiscal Year Ends:
Parent Company:

SALARIES/BONUSES:

Top Exec. Salary: $ Bonus: $
Second Exec. Salary: $ Bonus: $

OTHER THOUGHTS:

Estimated Female Officers or Directors:
Hot Spot for Advancement for Women/Minorities:

Parkdale America LLC

www.parkdalemills.com

NAIC Code: 313110

TYPES OF BUSINESS:

Cotton & Polycotton Yarn

BRANDS/DIVISIONS/AFFILIATES:

Parkdale Mills Inc
Unifi
Coolmax
Prematouch Collection
Amsler Yarns
Parkdale

CONTACTS:

Note: Officers with more than one job title may be intentionally listed here more than once.

John Nims, Pres.
Cecilia Mead, VP-Finance
Duke Kimbrell, Chmn.

GROWTH PLANS/SPECIAL FEATURES:

Parkdale America LLC, formed in 1997 as a joint venture between Unifi and Parkdale Mills, is a top North Carolina-based manufacturer of cotton and cotton/polyester blend yarns, which are used by commercial producers of apparel and fabrics. The firm operates 25 plants in the U.S., Mexico and Colombia and uses open-ended, ring spun and air jet-spinning technology to manufacture its products. Parkdale America's products are sold to apparel and textile companies under the Parkdale brand. The firm's products include organic cotton yarn; Coolmax, spun from Invista's Coolmax fibers; premium cotton yarns; Prematouch Collection yarns, made from engineered cotton blends; Amsler Yarns, which are used in denim, novelty jersey knits, linen look twills and upholstery fabrics; course count plied yarns for accent and bath rugs and thermal blankets; and cored yarns, available with yarn cores wrapped with cotton and Lycra cored wrapped with cotton. Parkdale America is 66% owned by Parkdale Mills, Inc. and 34% owned by Unifi.

FINANCIAL DATA:

Note: Data for latest year may not have been available at press time.

In U.S. $	2012	2011	2010	2009	2008	2007
Revenue						
R&D Expense						
Operating Income						
Operating Margin %						
SGA Expense						
Net Income						
Operating Cash Flow						
Capital Expenditure						
EBITDA						
Return on Assets %						
Return on Equity %						
Debt to Equity						

CONTACT INFORMATION:

Phone: 704-874-5000 Fax: 704-874-5175
Toll-Free: 800-331-1843
Address: 531 Cotton Blossom Cir., Gastonia, NC 28054 United States

STOCK TICKER/OTHER:

Stock Ticker: Joint Venture Exchange:
Employees: Fiscal Year Ends: 31-Dec
Parent Company: PARKDALE MILLS INC

SALARIES/BONUSES:

Top Exec. Salary: $ Bonus: $
Second Exec. Salary: $ Bonus: $

OTHER THOUGHTS:

Estimated Female Officers or Directors: 1
Hot Spot for Advancement for Women/Minorities:

Parkdale Mills Inc

www.parkdalemills.com

NAIC Code: 313110

TYPES OF BUSINESS:

Cotton Yarn Manufacturing
Blended Yarns
Cotton Fiber Research
Industrial Yarns
Military Yarns

BRANDS/DIVISIONS/AFFILIATES:

Parkdale America LLC
Coolmax
Permatouch Collection

CONTACTS: *Note: Officers with more than one job title may be intentionally listed here more than once.*

Anderson D. Warlick, CEO
Anderson D. Warlick, Pres.
Gregory S. Sellers, CFO
Charles Dickinson, VP-Mktg.
Freddie Harris, VP-Sales, Hosiery
Duke Kimbrell, Chmn.

GROWTH PLANS/SPECIAL FEATURES:

Parkdale Mills, Inc. is a leading independent producer of spun yarn for use in commercial knitting, weaving and hosiery manufacturing. The company specializes in organic and premium cotton, cotton blend and polyester spun yarns. Parkdale additionally produces industrial yarns, used in transmission belts, sandpaper substrates, book bindings, wallpaper backings and window treatments; home furnishing yarns, used in sheeting, bath rugs, institutional blankets, upholstery, throws and mattress ticking; and yarns with military applications. Some of the company's blended yarns feature advanced fiber technologies, such as Coolmax and Permatouch Collection, to provide apparel manufacturers with desirable performance traits and characteristics. Other specialty products include organic cotton; Egyptian cotton, cotton blends, high tenacity polyester, Amsler yarn, cored yarn, coarse count plied yarn, spandex and nylon. The firm uses air jet, ring spun and rotor spun systems to create its yarns, and manufactures its products through more than 23 domestic plants throughout South Carolina, North Carolina, Georgia, Tennessee, Virginia and Alabama as well as at international facilities in Colombia and Mexico. In an effort to continue to develop new innovative products, the company additionally maintains a dedicated cotton fiber research center, working in partnership with machine and component manufacturers to seek out new applications of technology in spinning and finishing. Parkdale serves markets in the U.S., South America, Central America, Canada, Mexico and the Caribbean. Some of the firm's major client's include Springmaid, Lands End and Jockey International. Additionally, the company holds a 66% interest in the ownership and operation of Parkdale America, a joint venture with the nylon and polyester yarn producer Unifi.

Employees are offered medical, dental and prescription drug coverage; life insurance; AD&D coverage; long-term disability; and a retirement savings plan.

FINANCIAL DATA: *Note: Data for latest year may not have been available at press time.*

In U.S. $	2012	2011	2010	2009	2008	2007
Revenue						
R&D Expense						
Operating Income						
Operating Margin %						
SGA Expense						
Net Income						
Operating Cash Flow						
Capital Expenditure						
EBITDA						
Return on Assets %						
Return on Equity %						
Debt to Equity						

CONTACT INFORMATION:

Phone: 704-874-5000 Fax: 704-874-5175
Toll-Free: 800-331-1843
Address: 531 Cotton Blossom Cir., Gastonia, NC 28054-5245 United States

STOCK TICKER/OTHER:

Stock Ticker: Private
Exchange:
Employees:
Fiscal Year Ends: 30-Sep
Parent Company:

SALARIES/BONUSES:

Top Exec. Salary: $ Bonus: $
Second Exec. Salary: $ Bonus: $

OTHER THOUGHTS:

Estimated Female Officers or Directors:
Hot Spot for Advancement for Women/Minorities:

Passport Brands Inc

passportbrandsinc.com

NAIC Code: 315220

TYPES OF BUSINESS:

Apparel Design
Men's & Women's Apparel
Jeans & Sportswear
Women's Handbags & Accessories

BRANDS/DIVISIONS/AFFILIATES:

I.C. Isaacs and Company Inc
Marithe & Francois Girbaud
Brandina, The Original

CONTACTS: *Note: Officers with more than one job title may be intentionally listed here more than once.*

Robert Stephen Stec, CEO
Robert Arnot, Pres.
Eugene Wielepski, Interim CFO
Marc Goldberg, VP-Sales
Liviu Goldenbergk, VP-Prod.
Liviu Goldenberg, Sr.VP-Oper.
Eugene Wielepski, Treas.
Scott A. Sullens, VP-Mktg. & Brand Dev.

GROWTH PLANS/SPECIAL FEATURES:

Passport Brands, Inc., formerly I.C. Isaacs & Company, Inc., is a designer and marketer of high-end apparel and accessories. Its brand portfolio currently consists of Marithe & Francois Girbaud and Brandina, The Original. Through a license agreement with Girbaud Design, Inc., the firm offers a men's collection under the Marithe & Francois Girbaud brand in the U.S., Puerto Rico and the U.S. Virgin Islands. The Girbaud collection includes full lines of bottoms consisting of jeans and casual pants in a variety of fabrications, including denim, stretch denim, cotton twill and nylon; tops such as cotton t-shirts, polo shirts, knit tops, woven tops and sweaters; and outerwear and leather sportswear. These collections target men ages 16-40. Brandina, The Original, based in Italy, produces women's bags, shoes and accessories influenced by the Italian Rivera. The company has an exclusive licensing agreement to distribute the Brandina brand within the U.S. and two U.S. Brandina stores in Charleston, South Carolina and La Jolla, California.

FINANCIAL DATA: *Note: Data for latest year may not have been available at press time.*

In U.S. $	2012	2011	2010	2009	2008	2007
Revenue						37,213,730
R&D Expense						
Operating Income						-15,393,770
Operating Margin %						-41.36%
SGA Expense						23,380,050
Net Income						-15,643,440
Operating Cash Flow						4,387,540
Capital Expenditure						424,152
EBITDA						-14,573,950
Return on Assets %						-78.48%
Return on Equity %						-146.39%
Debt to Equity						0.65

CONTACT INFORMATION:

Phone: 410 342-8200 Fax:
Toll-Free:
Address: 252 W. 37th St., 6th Fl., Baltimore, MD 10018 United States

STOCK TICKER/OTHER:

Stock Ticker: PBIX Exchange: PINX
Employees: 69 Fiscal Year Ends: 31-Dec
Parent Company:

SALARIES/BONUSES:

Top Exec. Salary: $ Bonus: $
Second Exec. Salary: $ Bonus: $

OTHER THOUGHTS:

Estimated Female Officers or Directors:
Hot Spot for Advancement for Women/Minorities:

Patagonia Inc

www.patagonia.com

NAIC Code: 315210

TYPES OF BUSINESS:

Outdoor Apparel, Manufacturing
Outdoor Equipment
Catalog Sales
Surfing Apparel
Surfing Equipment

BRANDS/DIVISIONS/AFFILIATES:

Lotus Designs
Point Blanks
Water Girl USA Inc
Lost Arrow Corporation
Beneficial T's
PCR Clothing
Patagonia Music
Regulator

CONTACTS: *Note: Officers with more than one job title may be intentionally listed here more than once.*

Casey Sheahan, CEO
Rose Marcario, COO
Casey Sheahan, Pres.
Rose Marcario, CFO
Rob BonDurant, VP-Mktg. & Comm.
Rob BonDurant, VP-Merch.
Dmitri Siegel, VP-Global e-commerce
Jenn Rapp, Dir.-Comm. & Public Rel.
John Collins, VP-Sales, Americas & Asia Pacific

GROWTH PLANS/SPECIAL FEATURES:

Patagonia, Inc., the primary subsidiary and brand name of the Lost Arrow Corporation, markets and sells outdoor clothing and equipment developed and designed for a broad range of extreme sports, such as rock climbing, mountaineering, white-water kayaking, mountain biking, surfing and skiing. Patagonia sells its products through 53 company-owned retail outlets worldwide: 26 in the U.S., 16 in Japan, five in Europe, two in Australia, two in Chile, one in Argentina and one in Canada, as well as through hundreds of other outdoor gear and clothing stores. Patagonia has pioneered many materials now considered essential for outdoor enthusiasts, such as polypropylene, Capilene and Synchilla, which are designed to retain warmth even when the material becomes moist from precipitation or sweat. The firm's Regulator system combines soft shells, insulation and moisture transition layers into jackets and pants to create an integrated clothing system designed to keep people warm and dry even in the most adverse weather conditions. The company also introduced Beneficial T's, made from organically grown cotton, and PCR Clothing, a fleece product line made from recycled plastic soda bottles. Patagonia also markets gear specifically for whitewater paddling under the Lotus Designs brand name and gear for surfing under the Point Blanks name. Water Girl USA, Inc. is the company's subsidiary dedicated to making surfing, paddling, swimming and kayaking clothing for women. The firm is committed to promoting and helping to conserve nature and contributes its time, services and at least 1% of its sales to hundreds of global environmental initiatives. Through its Patagonia Music initiative, the firm features exclusive songs from world-class musicians on its website, which when purchased on iTunes go towards supporting non-profit environmental groups. In August 2011, the firm announced the construction of its second Chicago-based retail store.

FINANCIAL DATA: *Note: Data for latest year may not have been available at press time.*

In U.S. $	2012	2011	2010	2009	2008	2007
Revenue	360,000,000	340,000,000	330,000,000	315,000,000	300,000,000	290,000,000
R&D Expense						
Operating Income						
Operating Margin %						
SGA Expense						
Net Income						
Operating Cash Flow						
Capital Expenditure						
EBITDA						
Return on Assets %						
Return on Equity %						
Debt to Equity						

CONTACT INFORMATION:

Phone: 805-643-8616 Fax: 800-543-5522
Toll-Free: 800-638-6464
Address: 259 W. Santa Clara St., Ventura, CA 93001 United States

STOCK TICKER/OTHER:

Stock Ticker: Subsidiary Exchange:
Employees: 1,350 Fiscal Year Ends: 30-Apr
Parent Company: LOST ARROW CORPORATION

SALARIES/BONUSES:

Top Exec. Salary: $ Bonus: $
Second Exec. Salary: $ Bonus: $

OTHER THOUGHTS:

Estimated Female Officers or Directors: 2
Hot Spot for Advancement for Women/Minorities:

Paul Reinhart AG

www.reinhart.com

NAIC Code: 313110

TYPES OF BUSINESS:

Cotton Processing

BRANDS/DIVISIONS/AFFILIATES:

CONTACTS:

Note: Officers with more than one job title may be intentionally listed here more than once.

Paul Jurg Reinhart, Pres.
Thomas Paul Reinhart, VP

GROWTH PLANS/SPECIAL FEATURES:

Paul Reinhart AG is a leading global cotton merchant based in Switzerland. The firm offers organic, fair-trade and extra long staple (ELS) cotton. In order to control for quality and provide its customers with logistics and documentation services, the firm acts as an intermediary between cotton growers and spinners. The cotton is certified as organic by the Institute for Marketecology (IMO), ECOCERT and Control Union World Group. The company sources its organic cotton from Uganda, India, the U.S., Turkey, Syria, Egypt, Kyrgyzstan and various West African countries. Reinhart's fair trade cotton is sourced from Cameroon, Mali and India. Reinhart's organic and fair trade cotton is certified by Fairtrade Labeling Organizations (FLO) International and is produced in concert with organizations such as Helvetas, a private development organization that supports farmer's associations and provides guarantees for production; Cotton Made in Africa (CMIA), which supports farmer training programs and seeks to establish a retail demand for African-grown cotton; and SEKEM, an Egyptian organic and fair trade organization involved in processing cotton to semi-manufactured and finished goods sold under SEKEM labels. The organic cotton and fair trade divisions are managed by the same team of specialists, and much of the organic cotton traded by the firm is also fair trade certified. Reinhart's ELS cotton is grown in Egypt, The U.S., Central Asia, Sudan Barakat, China and Israel. ELS cotton is desirable for its softness and durability and is mainly used in luxury items with high thread counts. Paul Reinhart has subsidiaries and representatives engaged in buying and selling operations in Brazil, China, Egypt, Greece, Latvia, Russia, Switzerland, Tajikistan, Turkmenistan, the UAE and Uzbekistan as well as agents around the world. It also has investments in ginning operations in the Ivory Coast and Burkina Faso.

FINANCIAL DATA:

Note: Data for latest year may not have been available at press time.

In U.S. $	2012	2011	2010	2009	2008	2007
Revenue						
R&D Expense						
Operating Income						
Operating Margin %						
SGA Expense						
Net Income						
Operating Cash Flow						
Capital Expenditure						
EBITDA						
Return on Assets %						
Return on Equity %						
Debt to Equity						

CONTACT INFORMATION:

Phone: 41-52-264-81-81 Fax: 41-52-212-00-55
Toll-Free:
Address: Technikumstrasse 82, Winterthur, CH-8401 Switzerland

STOCK TICKER/OTHER:

Stock Ticker: Private
Exchange:
Employees:
Fiscal Year Ends:
Parent Company:

SALARIES/BONUSES:

Top Exec. Salary: $ Bonus: $
Second Exec. Salary: $ Bonus: $

OTHER THOUGHTS:

Estimated Female Officers or Directors:
Hot Spot for Advancement for Women/Minorities:

Payless ShoeSource Inc

www.payless.com

NAIC Code: 448210

TYPES OF BUSINESS:

Discount Shoe Stores
Online Retail
Shoe Dyeing
Beauty Products

BRANDS/DIVISIONS/AFFILIATES:

Payless Holdings
Airwalk
Payless Kids
Rugged Outback
Dexter
Lela Rose
Smartfit
Brash

CONTACTS: *Note: Officers with more than one job title may be intentionally listed here more than once.*

LuAnn Via, CEO
LuAnn Via, Pres.
Vincent DeSantis, Chief Mktg. Officer/Sr. VP
Michele A. Bergerac, Sr. VP-Merch.
Robert Mingione, Head-Design Team
Christopher Kittle, Dir.-Men's Footwear Design
Bernard Figueroa, Head-Women's Footwear

GROWTH PLANS/SPECIAL FEATURES:

Payless ShoeSource, Inc. is owned by Payless Holdings and is one of the world's largest family footwear retailers. The firm's core business, the Payless ShoeSource chain, features fashionable, quality footwear for women, men and children sold at affordable prices in a self-service shopping format. The Payless ShoeSource chain operates approximately 4,500 stores throughout all 50 states and in Puerto Rico, Guam, Saipan, the U.S. Virgin Islands, Canada, Central America, the Caribbean and South America. Locations for Payless ShoeSource stores include regional malls, shopping centers, central business districts and freestanding buildings. Payless stores sell a broad assortment of quality footwear, including athletic, casual, dress, sandals, work and fashion boots, slippers and accessories, such as handbags, bath and beauty products and hosiery. The firm's brand portfolio includes Airwalk, Brash, American Eagle, Champion, Smartfit, Rugged Outback, Christian Siriano and Dexter. Designer collections have been created through the firm's relationship with New York-based fashion designers and include such brands as Alice + Olivia and Lela Rose. In addition to its stores, the firm operates Payless.com, which allows customers to shop online. Payless Holdings was formed following the sale and breakup of the company's former parent company, Collective Brands, Inc., in October 2012.

FINANCIAL DATA: *Note: Data for latest year may not have been available at press time.*

In U.S. $	2012	2011	2010	2009	2008	2007
Revenue	11,250,000,000	11,000,000,000				
R&D Expense						
Operating Income						
Operating Margin %						
SGA Expense						
Net Income						
Operating Cash Flow						
Capital Expenditure						
EBITDA						
Return on Assets %						
Return on Equity %						
Debt to Equity						

CONTACT INFORMATION:

Phone: 785-233-5171 Fax: 785-368-7519
Toll-Free: 877-474-6379
Address: 3231 SE 6th Ave., Topeka, KS 66607 United States

STOCK TICKER/OTHER:

Stock Ticker: Subsidiary Exchange:
Employees: Fiscal Year Ends: 31-Jan
Parent Company: Payless Holdings

SALARIES/BONUSES:

Top Exec. Salary: $ Bonus: $
Second Exec. Salary: $ Bonus: $

OTHER THOUGHTS:

Estimated Female Officers or Directors: 2
Hot Spot for Advancement for Women/Minorities:

Pendleton Woolen Mills Inc

www.pendleton-usa.com

NAIC Code: 315240

TYPES OF BUSINESS:

Men's & Women's Casualwear
Wool Textiles & Apparel
Woolen Blankets
Sweater Manufacturing
Accessories & Luggage
Retail Stores
Online & Catalog Sales

BRANDS/DIVISIONS/AFFILIATES:

Heritage Collection
49er Jacket
Portland Collection
Pendleton-usa.com

CONTACTS: *Note: Officers with more than one job title may be intentionally listed here more than once.*

Mark Korros, CEO
C. M. Bishop, III, Pres.
Dennis Simmonds, CFO
John Bishop, VP
Charles Bishop, VP
Peter Bishop, VP
Broughton Bishop, Vice Chmn.
John Bishop, Chmn.

GROWTH PLANS/SPECIAL FEATURES:

Pendleton Woolen Mills, Inc. is a long-standing family business based in Oregon that traces its roots back to the late 19th century. The company manufactures wool products including men's and women's apparel, blankets and home accessories through seven owned and operated facilities. In the early 1900s, Pendleton's original line of goods consisted of woven woolen blankets featuring traditional Native American patterns, with its blankets becoming prized objects among Native American nations including the Nez Perce, Navajo, Hopi and Zuni and subsequently among the broader American public. Company products now include wool heavyweight and lightweight men's shirts, pants, sweaters, loungewear, western wear and women's casual skirts (including Pendleton's signature reversible wool skirt), dresses, loungewear and accessories. Pendleton also offers furniture, luggage, bedding products, rugs and window treatments as well as an extensive line of wool blankets. The firm uses various types of wool, such as boiled, double-faced, Jacquards, Luxe, seasonless, tricotine, virgin and whisperwool. Pendleton products are available through department stores and select outdoors and western-oriented boutiques as well as around 70 Pendleton shops, nationally distributed catalogs and online at Pendleton-usa.com. Products are also available in targeted markets in Japan, Europe and Canada. The company actively seeks to promote and reinforce the heritage of its products, maintaining items like the 49er Jacket (which launched Pendleton's women's wear line in 1949), creating special series of blankets and apparel themed around various aspects of Americana (such as the National Parks system) and launching the Heritage Collection. Pendleton is a sixth generation family-owned business. In recent years, in order to reach a higher-end market, the firm has introduced collaborative collections with fashion designers, including the Portland Collection, which features designs inspired by Pendleton's archives.

Pendleton offers its employees medical, dental and vision coverage; life and disability insurance; profit sharing; a 401(k) plan; tuition assistance; and merchandise discounts.

FINANCIAL DATA: *Note: Data for latest year may not have been available at press time.*

In U.S. $	2012	2011	2010	2009	2008	2007
Revenue						
R&D Expense						
Operating Income						
Operating Margin %						
SGA Expense						
Net Income						
Operating Cash Flow						
Capital Expenditure						
EBITDA						
Return on Assets %						
Return on Equity %						
Debt to Equity						

CONTACT INFORMATION:

Phone: 503-226-4801 Fax: 503-535-5777
Toll-Free: 877-996-6599
Address: 220 NW Broadway, Portland, OR 97208 United States

STOCK TICKER/OTHER:

Stock Ticker: Private Exchange:
Employees: 1,000 Fiscal Year Ends: 31-Dec
Parent Company:

SALARIES/BONUSES:

Top Exec. Salary: $ Bonus: $
Second Exec. Salary: $ Bonus: $

OTHER THOUGHTS:

Estimated Female Officers or Directors:
Hot Spot for Advancement for Women/Minorities:

People's Garment PCL

www.pg.co.th

NAIC Code: 315190

TYPES OF BUSINESS:

Apparel Manufacturing
Textile Manufacturing

GROWTH PLANS/SPECIAL FEATURES:

People's Garment PCL manufactures apparel and textiles. Its apparel is comprised of women's wear, menswear, swimwear and children's wear, largely manufactured under third-party brands, including Elle, Arrow, bsc, 360, Lacoste, Streamline, Primo Linea and Proud. People's Garment also produces uniforms, including flame retardant suits for firefighters and factory workers. The firm's textiles cover various styles of knitted fabrics, including jersey, rib, pique and interlock. Its research and development team has developed new textiles such as Cool Mode, an environmentally friendly fabric, and Smart Hybrid, a textile designed to absorb water and sweat and dry faster. Some of these new textiles were developed in conjunction with the Thai Textile Institute. The company and the Institute are currently conducting textile research in the fields of biotechnology and nanotechnology. The firm distributes its products through trading firms, directly to consumers and through ICC International, of which People's Garment maintains a minority interest. People's Garment was founded in 1980 as part of the Saha Group, a Thai distributor of consumer products and services provided by affiliated companies.

BRANDS/DIVISIONS/AFFILIATES:

Cool Mode
Smart Hybrid
Saha Group

CONTACTS: *Note: Officers with more than one job title may be intentionally listed here more than once.*

Somporn Tiyaviboonsiri, Managing Dir.
Boonkiet Chokwatana, Pres.
Sunan Niyomnaitham, CFO
Suvimol Charoensrichai, Sec.
Somporn Tiyaviboonsiri, Chmn.

FINANCIAL DATA: *Note: Data for latest year may not have been available at press time.*

In U.S. $	2012	2011	2010	2009	2008	2007
Revenue		41,200,000	33,000,000	32,440,000	34,110,000	
R&D Expense						
Operating Income						
Operating Margin %						
SGA Expense						
Net Income		3,600,000	4,010,000	4,230,000	3,860,000	
Operating Cash Flow						
Capital Expenditure						
EBITDA						
Return on Assets %						
Return on Equity %						
Debt to Equity						

CONTACT INFORMATION:

Phone: 66-2-685-6500 Fax: 66-2-294-5159
Toll-Free:
Address: 666 Rama 3 Rd., Bangpongpang, Yannawa, Bangkok, 10120 Thailand

STOCK TICKER/OTHER:

Stock Ticker: PG
Employees: 1,330
Parent Company:
Exchange: Bangkok
Fiscal Year Ends: 31-Dec

SALARIES/BONUSES:

Top Exec. Salary: $ Bonus: $
Second Exec. Salary: $ Bonus: $

OTHER THOUGHTS:

Estimated Female Officers or Directors:
Hot Spot for Advancement for Women/Minorities: Y

Performance + Lifestyle Group (PLG) www.wolverineworldwide.com

NAIC Code: 316210

TYPES OF BUSINESS:

Children's Footwear
Adult Footwear
Athletic Footwear
Retail Stores & Outlets
Online Sales

BRANDS/DIVISIONS/AFFILIATES:

Wolverine Worldwide Inc
Stride Rite Corp.
Keds
Sperry Top-Sider
Stride Rite
Saucony
Pro-Keds
Collective Brands Performance + Lifestyle Group

CONTACTS: *Note: Officers with more than one job title may be intentionally listed here more than once.*

Sharon John, Pres., Stride Rite Children's Group
Rick Blackshaw, Pres., Keds
Richie Woodworth, Pres., Saucony
Craig Reingold, Pres., Sperry Top-Spider

GROWTH PLANS/SPECIAL FEATURES:

Performance + Lifestyle Group (PLG), formerly Collective Brands Performance + Lifestyle Group, markets performance and lifestyle footwear for children and adults under the well-known brands Stride Rite, Sperry Top-Sider, Saucony and Keds. The company designs and markets children's footwear for consumers 0-10 years old through its Stride Rite Children's Group, including dress and casual footwear, boots, sandals and athletic shoes. Through the Sperry and Sperry Top-Sider labels, PLG designs and markets nautical performance, outdoor recreational, dress-casual and casual footwear for men and women. Under the Saucony and Saucony Originals brands, it designs and markets technical running, athletic lifestyle, outdoor trail shoes and fashion athletic shoes as well as athletic apparel. The Keds brand specializes in fashion/athletic and casual footwear for adults and children under the Keds, Pro-Keds and Grasshoppers labels. PLG primarily functions as a wholesaler, selling its footwear through department stores, specialty and boutique stores and independent shoe stores. Products are also marketed outside the U.S. and Canada through distributors and licensees. The company's retail includes 345 stores, owned or licensed primarily in North America and 25 consumer-direct retail websites. In October 2012, PLG's former parent company. Collective Brands. Inc., was acquired jointly by Wolverine Worldwide (WWW), Golden Gate Capital and Blum Capital Partners with WWW gaining sole ownership over PLG. The acquisition of PLG's brands expanded WWW's portfolio to a total of 16 brands marketed in approximately 200 countries and territories. The company will continue to operate from its headquarters in Lexington, Massachusetts.

Parent company Wolverine Worldwide offers its employees medical, dental, life and disability insurance; flexible spending accounts; tuition assistance; 401(k) plan and wellness programs.

FINANCIAL DATA: *Note: Data for latest year may not have been available at press time.*

In U.S. $	2012	2011	2010	2009	2008	2007
Revenue						
R&D Expense						
Operating Income						
Operating Margin %						
SGA Expense						
Net Income						
Operating Cash Flow						
Capital Expenditure						
EBITDA						
Return on Assets %						
Return on Equity %						
Debt to Equity						

CONTACT INFORMATION:

Phone: 617-824-6000 Fax:
Toll-Free:
Address: 191 Spring St., Lexington, MA 02420-1919 United States

STOCK TICKER/OTHER:

Stock Ticker: Subsidiary Exchange:
Employees: 3,100 Fiscal Year Ends: 30-Nov
Parent Company: Wolverine Worldwide Inc

SALARIES/BONUSES:

Top Exec. Salary: $ Bonus: $
Second Exec. Salary: $ Bonus: $

OTHER THOUGHTS:

Estimated Female Officers or Directors: 1
Hot Spot for Advancement for Women/Minorities: Y

Perry Ellis International Inc

www.pery.com

NAIC Code: 315220

TYPES OF BUSINESS:

Men's & Women's Apparel
Retail Stores
Golf Apparel
Swimwear
Accessories & Fragrances

BRANDS/DIVISIONS/AFFILIATES:

Perry Ellis America
Axis
Original Penguin
Cubavera
Havanera Co
Perry Ellis
Jantzen
Natural Issue

CONTACTS: *Note: Officers with more than one job title may be intentionally listed here more than once.*

George Feldenkreis, CEO
Oscar Feldenkreis, COO
Oscar Feldenkreis, Pres.
Anita Britt, CFO
Luis Paez, CIO
Cory Shade, General Counsel/Sr. VP
Fanny Hanono, Treas./Corp. Sec.
Steve Harriman, Pres., Perry Ellis Bottoms Div.
Jack Voith, Pres., Golf & Sportswear Div.
George Feldenkreis, Chmn.

GROWTH PLANS/SPECIAL FEATURES:

Perry Ellis International, Inc. is a leading designer, distributor and licensor of a broad line of men's and women's apparel, accessories and fragrances. Its products include dress and casual shirts, golf sportswear, sweaters, dress and casual pants and shorts, jeans wear, active wear and men's and women's swimwear for all major levels of retail distribution. The company maintains a diversified portfolio of more than 30 brands, including Perry Ellis, Perry Ellis America, Axis, Jantzen, Cubavera, Havanera Co., Natural Issue, Grand Slam, C&C California, Laundry, Original Penguin and Manhattan. The company also licenses the Nike and JAG brands for swimwear and accessories and the PGA Tour brand for golf apparel. The firm's wholesale business accounts for 97% of the company's total revenues, and the licensing business accounts for 3%. Perry Ellis sources its products from Asia (82%), the Middle East (12%) and the Americas (6%). The company distributes apparel primarily to department stores, national and regional chains, mass merchants, specialty stores, sporting goods stores, the corporate wear market and independent retailers in the U.S., Canada, and the U.K. In addition, the firm operates 40 Perry Ellis retail outlet stores, 22 Original Penguin retail stores, two Perry Ellis full-price retail stores and one Cubavera retail location. Perry Ellis is currently the licensor in 131 license agreements (three worldwide, 44 domestic and 84 international) for various products.

Employees are offered health insurance, short- and long-term disability coverage, life insurance, a 401(k) plan, flexible spending accounts, an employee assistance program and tuition reimbursement.

FINANCIAL DATA: *Note: Data for latest year may not have been available at press time.*

In U.S. $	2012	2011	2010	2009	2008	2007
Revenue	980,592,000	790,288,000	754,202,000	851,297,000	863,866,000	
R&D Expense						
Operating Income	55,385,000	49,838,000	34,863,000	4,328,000	62,483,000	
Operating Margin %	5.64%	6.30%	4.62%	.50%	7.23%	
SGA Expense	248,618,000	220,018,000	200,356,000	236,840,000	215,873,000	
Net Income	25,517,000	24,112,000	13,167,000	-12,890,000	28,173,000	
Operating Cash Flow	712,000	21,004,000	88,795,000	-4,982,000	91,292,000	
Capital Expenditure	13,667,000	6,237,000	2,734,000	10,577,000	18,955,000	
EBITDA	67,520,000	61,152,000	47,899,000	16,656,000	76,276,000	
Return on Assets %	3.61%	3.86%	2.26%	-2.17%	4.77%	
Return on Equity %	7.62%	8.46%	5.11%	-4.93%	10.83%	
Debt to Equity	0.53	0.75	0.53	0.91	0.64	

CONTACT INFORMATION:

Phone: 305 592-2830 Fax: 305 594-2307
Toll-Free: 800-994-0073
Address: 3000 N.W. 107th Ave., Miami, FL 33172 United States

STOCK TICKER/OTHER:

Stock Ticker: PERY Exchange: NAS
Employees: 2,400 Fiscal Year Ends: 31-Jan
Parent Company:

SALARIES/BONUSES:

Top Exec. Salary: $1,000,000 Bonus: $
Second Exec. Salary: $1,000,000 Bonus: $

OTHER THOUGHTS:

Estimated Female Officers or Directors: 3
Hot Spot for Advancement for Women/Minorities: Y

Phat Fashions LLC

www.phatfarm.com

NAIC Code: 315220

TYPES OF BUSINESS:

Apparel Manufacturing-Hip Hop Styles
Urban Apparel
Retail Stores

BRANDS/DIVISIONS/AFFILIATES:

Kellwood Co
Phat Farm
Baby Phat
Phat Premium
KLS
Fabulosity

CONTACTS: *Note: Officers with more than one job title may be intentionally listed here more than once.*

Jill Granoff, CEO-Kellwood Co.

GROWTH PLANS/SPECIAL FEATURES:

Phat Fashions, LLC, a subsidiary of Kellwood Co., is an apparel company founded by hip-hop mogul Russell Simmons. Begun as a small niche operation, Phat Fashions has become a widely recognized cultural symbol. Its founding ideology was to create clothing for a forward-thinking, multi-ethnic youth market that combined high quality with street credibility, and it has set itself up as not just a clothing company but also the bearer of a message intended to break stereotypes and ethnic boundaries. Phat Fashions designs, manufactures and sells mainly urban-oriented apparel targeted at male and female hip-hop fans of all ages and sizes, including infants, toddlers, children and plus-sizes. It has two brands of contemporary men's clothing: Phat Farm and Phat Premium, which aim to integrate preppy Ivy League culture with urban street aesthetics; and it also has three lines of women's clothing, Baby Phat, Fabulosity and KLS, run by Kimora Lee Simmons, Russell Simmons' former wife. The firm's product categories include shorts, hats, sport jackets, shirts, jeans, outerwear, loungewear, swimwear, lingerie, footwear, pants, handbags, beauty products, fragrances, home decor and accessories. Its products are sold at department and specialty stores across the world and at company stores in Montreal and New York. In order to keep its inventory seasonal, the firm does not restock most online items; once an item is sold out, it often remains so indefinitely.

FINANCIAL DATA: *Note: Data for latest year may not have been available at press time.*

In U.S. $	2012	2011	2010	2009	2008	2007
Revenue						
R&D Expense						
Operating Income						
Operating Margin %						
SGA Expense						
Net Income						
Operating Cash Flow						
Capital Expenditure						
EBITDA						
Return on Assets %						
Return on Equity %						
Debt to Equity						

CONTACT INFORMATION:

Phone: 212-391-9443 Fax: 201-617-8686
Toll-Free: 866-444-3323
Address: 512 7th Ave., 43rd Fl., New York, NY 10018 United States

STOCK TICKER/OTHER:

Stock Ticker: Subsidiary
Exchange:
Employees: 27
Fiscal Year Ends: 31-Dec
Parent Company: KELLWOOD CO

SALARIES/BONUSES:

Top Exec. Salary: $ Bonus: $
Second Exec. Salary: $ Bonus: $

OTHER THOUGHTS:

Estimated Female Officers or Directors: 2
Hot Spot for Advancement for Women/Minorities: Y

Phoenix Footwear Group, Inc.

www.phoenixfootwear.com

NAIC Code: 316210

TYPES OF BUSINESS:

Womens Footwear (except Athletic) Manufacturing
Men's Footwear

GROWTH PLANS/SPECIAL FEATURES:

Phoenix Footwear Group, Inc. specializes in quality comfort women's and men's footwear. It designs, develops, markets and sells footwear in a wide range of sizes and widths under the brands Trotters and SoftWalk. These two core brands make up its women's footwear business. In addition to these brands, Phoenix Footwear owns H.S. Trask, a men's brand positioned to appeal to consumers with tastes for styles suitable to the American Old West. H.S. Trask utilizes original leathers, including bison and elk from the U.S. The firm maintains an open stock system, which provides customers with the ability to replenish their inventory at retail, and supports special consumer orders as well.

BRANDS/DIVISIONS/AFFILIATES:

Trotters
SoftWalk
H.S. Trask

CONTACTS: *Note: Officers with more than one job title may be intentionally listed here more than once.*

James R. Riedman, CEO
Greg Slack, CFO
Greg Slack, Sec.
Greg Slack, Treas.
James R. Riedman, Chmn.

FINANCIAL DATA: *Note: Data for latest year may not have been available at press time.*

In U.S. $	2012	2011	2010	2009	2008	2007
Revenue			17,262,000	19,939,000	75,070,000	82,871,000
R&D Expense						
Operating Income			-3,607,000	-6,708,000	-17,021,000	-15,750,000
Operating Margin %			-20.89%	-33.64%	-22.67%	-19.00%
SGA Expense			8,500,000	11,550,000	32,267,000	34,921,000
Net Income			-1,705,000	-6,991,000	-19,460,000	-1,344,000
Operating Cash Flow			-3,916,000	5,736,000	-3,005,000	-3,798,000
Capital Expenditure			37,000	60,000	689,000	350,000
EBITDA			-3,327,000	-6,378,000	-15,828,000	-14,253,000
Return on Assets %			-13.40%	-29.21%	-38.60%	-1.53%
Return on Equity %			-46.44%	-88.14%	-91.68%	-4.27%
Debt to Equity			0.59			

CONTACT INFORMATION:

Phone: 706 602-9688 Fax: 760 602-9684
Toll-Free: 888-218-7275
Address: 5937 Darwin Ct., Ste. 109, Carlsbad, CA 92008 United States

STOCK TICKER/OTHER:

Stock Ticker: PXFG Exchange: PINX
Employees: 52 Fiscal Year Ends: 31-Dec
Parent Company:

SALARIES/BONUSES:

Top Exec. Salary: $353,077 Bonus: $10,000
Second Exec. Salary: $321,154 Bonus: $

OTHER THOUGHTS:

Estimated Female Officers or Directors:
Hot Spot for Advancement for Women/Minorities:

Pittards plc

NAIC Code: 316110

www.pittards.com

TYPES OF BUSINESS:

Leather Manufacturing
Leather Treatments & Technologies

BRANDS/DIVISIONS/AFFILIATES:

Armortan
Parados
Firebloc
Microspike
Ethiopia Tannery Share Company
Daines & Hathaway
Gripster

CONTACTS: *Note: Officers with more than one job title may be intentionally listed here more than once.*

Reg H. Hankey, CEO
Jill Williams, Dir.-Finance
Jill Williams, Company Sec.
Stephen D. Boyd, Chmn.

GROWTH PLANS/SPECIAL FEATURES:

Pittards plc produces technically advanced cow, sheep and goat leathers from its facilities in Yeovil, England; Ethiopia; and Taiwan. Various manufacturers use the company's leather in such brand-name products as Dents and Kenzo dress gloves; Bennett safety gloves, often used by the military, police, firefighters and industrial workers; Rawlings sporting goods; Clarks shoes; Nike football and batting gloves; Head racquetball gloves; Jeffries Saddlery saddles; and Ona Polo gloves. Pittards supplies many major footwear makers, including athletic shoe manufacturers such as New Balance and Merrell. Additionally, the company produces a line of leather garments and accessories under its own Pittards name as well as its Daines & Hathaway subsidiary. The firm operates in three principal divisions. The glove leather division produces leather largely for sport gloves for baseball, golfing, football, cycling and hiking, as well as dress gloves, military gloves and service gloves. The shoe and leather goods division produces leather largely for footwear and sports equipment, leather goods and saddlery. Lastly, the raw materials division procures and part-processes hides and wool sheepskins of U.K. origin that it sells worldwide. Pittards has also introduced several proprietary innovations in leather treatment. Its Armortan technology enhances the abrasion resistance of glove leather. Firebloc is a fire-resistant coating that resists deterioration from washing. Microspike, an anti-microbial treatment, prevents microbe buildup that causes odors, discoloration and deterioration. The company's Custom Image Generation (CIG) technology enables the permanent application of designs on leather. Gripster technology improves the fit and performance of sports and functional leather gloves. The Parados chemical- and acid-resistant leather was specially developed for the military, riot police, fire fighters and high-tech industrial sectors. Wholly-owned subsidiaries include the Ethiopia Tannery Share Company, which Pittards managed for several years on behalf of the Ethiopian government before acquiring.

FINANCIAL DATA: *Note: Data for latest year may not have been available at press time.*

In U.S. $	2012	2011	2010	2009	2008	2007
Revenue		60,800,000	59,100,000	40,270,000	37,490,000	41,000,000
R&D Expense						
Operating Income						
Operating Margin %						
SGA Expense						
Net Income		5,800,000	6,060,000	2,620,000	-540,000	-3,100,000
Operating Cash Flow						
Capital Expenditure						
EBITDA						
Return on Assets %						
Return on Equity %						
Debt to Equity						

CONTACT INFORMATION:

Phone: 44-1935-474-321 Fax: 44-1935-427-145
Toll-Free:
Address: Sherborne Rd., Yeovil, BA21 5BA United Kingdom

STOCK TICKER/OTHER:

Stock Ticker: PTD
Exchange: London
Employees: 871
Fiscal Year Ends: 31-Dec
Parent Company:

SALARIES/BONUSES:

Top Exec. Salary: $ Bonus: $
Second Exec. Salary: $ Bonus: $

OTHER THOUGHTS:

Estimated Female Officers or Directors: 1
Hot Spot for Advancement for Women/Minorities:

Plains Cotton Cooperative Association

www.pcca.com

NAIC Code: 313110

TYPES OF BUSINESS:

Cotton Production
Marketing Cooperative
Denim Manufacturing
Online Cotton Trading System

BRANDS/DIVISIONS/AFFILIATES:

ACG Denim
TELCOT
Telmark LP
DENIMATRIX

CONTACTS:

Note: Officers with more than one job title may be intentionally listed here more than once.

Wally Darneille, CEO
Wally Darneille, Pres.
Lonnie Winters, VP-Mktg.
Jim Taylor, VP-Human Resources
Joe Tubb, VP-Info. Systems
Jack Matthews, VP-Prod. Dev. & Fabric Sales
Bryan Gregory, VP-Textile Mfg.
John Johnson, Corp. Sec./Dir.-Legislative Affairs
Jay Cowart, VP-Warehouse Oper.
John Johnson, Dir.-Public Rel.
Sam Hill, VP-Finance/Treas.
Greg Bell, VP-Grower Svcs. & Gin Bookkeeping
Stan Kirby, VP-Telmark LP
Eddie Smith, Chmn.
Marco Cotton, Pres., DENIMATRIX

GROWTH PLANS/SPECIAL FEATURES:

Plains Cotton Cooperative Association (PCCA) is a farmer-owned cotton marketing cooperative headquartered in Lubbock, Texas. It is one of the largest handlers of U.S. cotton, marketing millions of bales each year. In addition to cotton marketing, PCCA also owns a denim mill in Littlefield, Texas as well as cotton warehouse operations in Sweetwater, Rule, Big Spring and Memphis Texas; Altus, Oklahoma; and Liberal, Kansas. The company also operates an electronic marketing division. This division provides the company's members with access to an online trading system, known as TELCOT, to market their cotton, allowing online buyers to bid on a blind basis for cotton offered for sale on the online system, much like the floor trade on a stock exchange. The firm's textiles division includes ACG Denim and DENIMATRIX. ACG Denim operates two denim manufacturing facilities that can produce as much as 36 million combined linear yards of denim annually. DENIMATRIX is a Guatemalan denim jean manufacturer; its clients include American Eagle Outfitters, Abercrombie & Fitch, Anthropologie, Urban Outfitters, Lacose, Replay and Quiksilver. PCCA members can share in textile earnings by signing into the company's mill option program. The company has a subsidiary, Telmark LP, which offers cotton ginning services to members in Texas and Oklahoma.

FINANCIAL DATA:

Note: Data for latest year may not have been available at press time.

In U.S. $	2012	2011	2010	2009	2008	2007
Revenue						
R&D Expense						
Operating Income						
Operating Margin %						
SGA Expense						
Net Income						
Operating Cash Flow						
Capital Expenditure						
EBITDA						
Return on Assets %						
Return on Equity %						
Debt to Equity						

CONTACT INFORMATION:

Phone: 806-763-8011 Fax: 806-762-7400
Toll-Free: 800-333-8011
Address: 3301 E. 50th St., Lubbock, TX 79404 United States

STOCK TICKER/OTHER:

Stock Ticker: Cooperative
Exchange:
Employees: 800
Fiscal Year Ends: 30-Jun
Parent Company:

SALARIES/BONUSES:

Top Exec. Salary: $ Bonus: $
Second Exec. Salary: $ Bonus: $

OTHER THOUGHTS:

Estimated Female Officers or Directors:
Hot Spot for Advancement for Women/Minorities:

Point Blank Enterprises Inc

www.pointblanksolutionsinc.com

NAIC Code: 339113A

TYPES OF BUSINESS:

Protective Apparel
Bullet Resistant & Fragmentation Vests

BRANDS/DIVISIONS/AFFILIATES:

Protective Apparel Corporation of America
Point Blank Body Armor Inc
Protective Products Enterprises
Paraclete
Sun Capital Partners

CONTACTS:

Note: Officers with more than one job title may be intentionally listed here more than once.

Daniel Gaston, CEO

GROWTH PLANS/SPECIAL FEATURES:

Point Blank Enterprises, Inc. (PBEI), owned by Sun Capital Partners, designs, manufactures and distributes bullet, fragmentation and stab-resistant apparel and related ballistic accessories through subsidiaries Point Blank Body Armor, Inc.; Protective Apparel Corporation of America; Protective Products Enterprises; and Paraclete. In addition to body armor products that protect individuals from bullets, knives and other sharp instruments, the company designs advanced systems that protect users from specific threats. The firm's products are used domestically and internationally by military, law enforcement, security and corrections personnel as well as governmental agencies. The company's vests are constructed with special blended fabrics, stainless steel and flexible woven fabrics. Additional body armor products offered by the firm include tactical police jackets, unique vests for special agents, corrections vests and K-9 protection. All of the company's products are sold through direct sale contracts, a corporate sales force, sales agents and a network of distributors. Sales to the U.S. military comprise the largest portion of the firm's business, followed by sales to federal, state and local law enforcement agencies, including correctional facilities. In May 2012, Point Blank Enterprises became the parent company of Protective Products Enterprises LLC through a completed merger.

FINANCIAL DATA:

Note: Data for latest year may not have been available at press time.

In U.S. $	2012	2011	2010	2009	2008	2007
Revenue						
R&D Expense						
Operating Income						
Operating Margin %						
SGA Expense						
Net Income						
Operating Cash Flow						
Capital Expenditure						
EBITDA						
Return on Assets %						
Return on Equity %						
Debt to Equity						

CONTACT INFORMATION:

Phone: 954-630-0900 Fax: 954-630-9225
Toll-Free: 800-413-5155
Address: 2102 SW 2nd St., Pompano Beach, FL 33069 United States

STOCK TICKER/OTHER:

Stock Ticker: Private Exchange:
Employees: 1,300 Fiscal Year Ends: 31-Dec
Parent Company: SUN CAPITAL PARTNERS

SALARIES/BONUSES:

Top Exec. Salary: $ Bonus: $
Second Exec. Salary: $ Bonus: $

OTHER THOUGHTS:

Estimated Female Officers or Directors:
Hot Spot for Advancement for Women/Minorities:

Polartec LLC

www.polartec.com

NAIC Code: 313210

TYPES OF BUSINESS:

Textiles Manufacturing
Polyester Fleece Manufacturing
Technical Fabrics

BRANDS/DIVISIONS/AFFILIATES:

Hardface
Versa Capital Management Inc
Polartec Power Dry
Polartec Power Stretch
Polartec Classic
Polartec Thermal Pro
Polartec Wind Pro
Polartec Power Shield

CONTACTS: *Note: Officers with more than one job title may be intentionally listed here more than once.*

Gary Smith, CEO
Joe Robinson, CFO
Jonathan Adelman, Exec. VP-Global Sales
Kathy Skala, Dir.-Human Resources
Moshe Rock, Sr. VP-R&D
David Alvers, CIO
Robert Stubbs, Dir.-Tech.
Roman Mikolajewicz, Dir.-Oper.
Peter Sobha, Dir.-Corp. Credit & Collections
Allon Cohne, Dir.-Global Mktg.
Greg Segall, Chmn.-Versa Capital Mgmt., Inc.
Alan Savidge, Dir.-Bus. Dev.-Asia Pacific

GROWTH PLANS/SPECIAL FEATURES:

Polartec LLC, owned by Versa Capital Management, Inc., designs and manufactures high-tech fabrics. The company produces more than 300 styles of Polartec fabric for brand partners all over the world. The firm is also a significant supplier for the U.S. Military. The original inventor of fleece, Polartec has designed and currently manufactures a variety of performance fabrics for next-to-skin, insulation and weather protection applications. Next-to-skin fabrics are designed to provide moisture wicking capabilities, fast drying capabilities, breathability and a comfortable texture, and consist of the Polartec Power Dry and Polartec Power Stretch brand fabrics. Polartec's insulation fabrics feature air pockets which trap warm air next to the body, and consist of the Polartec Classic and Polartec Thermal Pro brand fabrics. The company's weather protection fabrics provide wind and water resistance, and consist of the Polartec Wind Pro, Polartec Windbloc, Polartec NeoShell and Polartec Power Shield brand fabrics. Polartec's fabric technologies can be woven into existing fabrics. Its Hardface Technology provides durability and abrasion resistance in a soft, breathable and water repellant fabric. The company's Polygiene Odor Resistant Technology provides permanent odor protection by inhibiting the growth of odor causing bacteria in the fabric. Additional fabric technologies include Durable Water Repellency, Sun Protection Technology, Flame Resistant Technology and Recycled Fiber Technology.

FINANCIAL DATA: *Note: Data for latest year may not have been available at press time.*

In U.S. $	2012	2011	2010	2009	2008	2007
Revenue						
R&D Expense						
Operating Income						
Operating Margin %						
SGA Expense						
Net Income						
Operating Cash Flow						
Capital Expenditure						
EBITDA						
Return on Assets %						
Return on Equity %						
Debt to Equity						

CONTACT INFORMATION:

Phone: 978-691-4369 Fax:
Toll-Free: 800-252-6688
Address: 46 Stafford St., Lawrence, MA 01841 United States

STOCK TICKER/OTHER:

Stock Ticker: Private Exchange:
Employees: 800 Fiscal Year Ends: 31-Jan
Parent Company: VERSA CAPITAL MANAGEMENT INC

SALARIES/BONUSES:

Top Exec. Salary: $ Bonus: $
Second Exec. Salary: $ Bonus: $

OTHER THOUGHTS:

Estimated Female Officers or Directors: 1
Hot Spot for Advancement for Women/Minorities:

Polymer Group Inc

NAIC Code: 313110

www.polymergroupinc.com

TYPES OF BUSINESS:

Synthetic Fiber Manufacturing
Nonwoven Materials

BRANDS/DIVISIONS/AFFILIATES:

Blackstone Group LP (The)
Arium

CONTACTS: *Note: Officers with more than one job title may be intentionally listed here more than once.*

Veronica M. Hagen, CEO
Veronica M. Hagen, Pres.
Dennis Norman, CFO/Exec. VP
John Heironimus, Chief Mktg. Officer/VP
Mary Tomasello, VP-Human Resources & Employee Comm.
Dan Rikard, General Counsel/Sr. VP/Sec.
Michael Modak, Sr. VP-Global Growth & Innovation Officer
Bob Dale, VP/Gen. Mgr.-Asia
Scott Tracey, Sr. VP/Gen. Mgr.-Americas
Mike S. Zafirovski, Chmn.
Jean-Marc, Sr. VP/Gen. Mgr.-Europe
Mike Hale, Sr. VP-Global Supply Chain

GROWTH PLANS/SPECIAL FEATURES:

Polymer Group, Inc. (PGI) is a global, technology-driven developer, manufacturer and marketer of engineered materials. It is one of the world's largest producers of technologies for nonwovens. Nonwovens provide certain qualities similar to those of textiles, but at a lower cost. The company supplies engineered materials to a number of the largest consumer and industrial product manufacturers in the world. Its product offerings are sold principally to converters that manufacture a wide range of end-use products used in the hygiene, health care, fabrics, wiping and industrial industries. For example, the firm produces baby wipes, feminine napkins and baby diapers for the hygiene industry; surgical gowns and face masks for the health care industry; wet wipes for the wipes industry; and automotive insulation and fabric softener sheets for the industrial industry. PGI has invested in advanced technology in order to increase capacity, improve quality and develop new high-value fabric structures. Working as a developmental partner with its major customers, the company utilizes its technological capabilities and depth of research and development resources to develop and manufacture new products to specifically meet their needs. The firm operates 15 manufacturing facilities located in nine countries, including its joint venture operations in Argentina and China. The company is owned by The Blackstone Group. Recently, PGI released Arium, a series of submicron fibers that can deliver improved performance properties for industrial, health care, filtration and other industrial applications.

FINANCIAL DATA: *Note: Data for latest year may not have been available at press time.*

In U.S. $	2012	2011	2010	2009	2008	2007
Revenue	910,000,000	900,000,000	882,652,000	1,073,272,000	1,145,637,000	1,059,663,000
R&D Expense						
Operating Income						
Operating Margin %						
SGA Expense						
Net Income			16,195,000	-1,404,000	5,353,000	-41,146,000
Operating Cash Flow						
Capital Expenditure						
EBITDA						
Return on Assets %						
Return on Equity %						
Debt to Equity						

CONTACT INFORMATION:

Phone: 704-697-5100 Fax: 704-697-5116
Toll-Free:
Address: 9335 Harris Corners Pkwy., Ste. 300, Charlotte, NC 28269 United States

STOCK TICKER/OTHER:

Stock Ticker: Private Exchange:
Employees: 3,100 Fiscal Year Ends: 31-Dec
Parent Company: BLACKSTONE GROUP LP (THE)

SALARIES/BONUSES:

Top Exec. Salary: $ Bonus: $
Second Exec. Salary: $ Bonus: $

OTHER THOUGHTS:

Estimated Female Officers or Directors: 2
Hot Spot for Advancement for Women/Minorities: Y

PPR S.A. (KERING)

www.kering.com

NAIC Code: 315240

TYPES OF BUSINESS:

Catalog & Online Sales
Luxury Goods & Fashions Manufacturing
Book, Software & Consumer Electronics Stores
Appliance & Furniture Stores
Watches & Jewelry
Lingerie, Shoes & Leather Goods

BRANDS/DIVISIONS/AFFILIATES:

Bottega Veneta
Gucci
Redcats
Fnac
Brioni
Volcom Inc
Yves Saint Laurent
Kering

CONTACTS: *Note: Officers with more than one job title may be intentionally listed here more than once.*

Francois-Henri Pinault, CEO
Jean-Marc Duplaix, CFO
Belen Essioux Trujillo, Sr. VP-Human Resources
Charles de Fleurieu, Dir.-Mergers & Acquisitions
Federico Barbieri, Sr. VP-e-Bus.
Louise Beveridge, Sr. VP-Comm.
Jean-Michel Noir, Chmn./CEO-Redcats
Patrizio di Marco, CEO/Pres., Gucci
Alexis Babeau, Managing Dir.-Luxury Goods
Jean-Francois Palus, Group Managing Dir.
Francois-Henri Pinault, Chmn.
Franz Koch, CEO-Puma SE

GROWTH PLANS/SPECIAL FEATURES:

PPR SA is a leading European retailer and manufacturer specializing in apparel and accessories. The company operates in four segments: Luxury Goods, Sport & Lifestyle, Redcats and Fnac. The Luxury Goods sector designs and manufactures premium leather goods, shoes, watches, jewelry, eyewear, perfume, scarves and ready-to-wear clothing under a variety of brand names, such as Gucci, Bottega Veneta, Yves Saint Laurent (YSL), Balenciaga, Boucheron, Sergio Rossi, Girard-Perregaux, JeanRichard, Alexander McQueen and Stella McCartney. This segment operates about 680 stores across its brand portfolio worldwide, including 376 directly operated Gucci stores, 170 Bottega Veneta and 83 YSL. Subsidiary Luxury Goods International, based in Switzerland, operates PPR's international distribution and logistics business for its luxury brands. The Sport & Lifestyle segment, which includes the Puma, Volcom, Cobra, Electric and Tretorn brands, manufactures and retails footwear, sports apparel and accessories. Redcats is one of the largest catalog shopping companies in the world. It has numerous multi-line brands in Europe, including La Redoute, Ellos, Empire, La Maison de Valerie, Daxon Cyrillus and Daxon, and in the U.S., with brands such as the Woman Within, La Redoute, Jessica London, Roaman's, King Size, Avenue Brylane Home and BrylaneHome Kitchen catalogues. It also has 70 different e-commerce sites. Fnac is a leading retailer of cultural and technology products, with 154 stores located in France, Belgium, Brazil, Spain, Italy, Portugal and Switzerland. In 2011, PPR acquired Volcom, Inc., a producer of skateboard culture-inspired clothing and accessories, and luxury Italian fashion brand Brioni. Also in 2011, the company sold its Conforama furniture business to Steinhoff International Holdings. In November 2012, Redcats agreed to sell The Golf Warehouse and The Sportsman's Guide catalogues to Northern Tool + Equipment. In 2013, the firm announced plans to change its name to Kering in June of that year.

FINANCIAL DATA: *Note: Data for latest year may not have been available at press time.*

In U.S. $	2012	2011	2010	2009	2008	2007
Revenue		16,176,600,000	20,625,600,000	20,371,500,000	21,212,800,000	23,544,300,000
R&D Expense						
Operating Income						
Operating Margin %						
SGA Expense						
Net Income		1,304,900,000	1,362,090,000	1,213,810,000	1,135,530,000	1,137,010,000
Operating Cash Flow						
Capital Expenditure						
EBITDA						
Return on Assets %						
Return on Equity %						
Debt to Equity						

CONTACT INFORMATION:

Phone: 33-1-45-64-61-00 Fax: 33-1-45-64-60-00
Toll-Free:
Address: 10 Ave. Hoche, Paris, 75381 France

STOCK TICKER/OTHER:

Stock Ticker: PPRUY Exchange: PINX
Employees: 52,107 Fiscal Year Ends: 31-Dec
Parent Company:

SALARIES/BONUSES:

Top Exec. Salary: $ Bonus: $
Second Exec. Salary: $ Bonus: $

OTHER THOUGHTS:

Estimated Female Officers or Directors: 3
Hot Spot for Advancement for Women/Minorities: Y

Prada SpA Group

www.pradagroup.com

NAIC Code: 315240

TYPES OF BUSINESS:

Designer Apparel
Leather Goods
Retail Stores

BRANDS/DIVISIONS/AFFILIATES:

Church's
Church & Co
Prada Donna
Prada Uomo
Granello
Miu Miu
I Pellettieri d'Italia
Car Shoe

CONTACTS: *Note: Officers with more than one job title may be intentionally listed here more than once.*

Patrizio Bertelli, CEO
Donatello Galli, CFO
Miuccia Prada, Chmn.

GROWTH PLANS/SPECIAL FEATURES:

Prada SpA Group (formerly I Pellettieri d'Italia) manufactures and sells upscale apparel, shoes, accessories and fragrances for men and women. The company was originally created through the merger of leather goods businesses run by Miuccia Prada and her husband Patrizio Bertelli, who together continue to own approximately 95% of the business. Prada started out as a small family luggage business, based in a shop in the Galleria Emmanuele Vittorio in Italy. It is now a multi-label operation with approximately 388 directly operated stores around the world in over 70 countries, 26 franchise stores and a network of department stores. Its original brand lines include Prada, Prada Donna, Prada Uomo, Granello and Miu Miu (a less expensive apparel line). The firm has also acquired luxury brands such as English shoe manufacturer Church & Co. The Prada store in New York was the first of a string of epicenters, or flagship retail stores. The company currently produces work under the brands Prada, Miu Miu, Church's and Car Shoe. Production takes place exclusively in Italy, with the exception of Church & Co. shoes for men which are made in the U.K. In 2012, the firm opened 15 new Prada brand stores and five Miu Miu stores internationally.

FINANCIAL DATA: *Note: Data for latest year may not have been available at press time.*

In U.S. $	2012	2011	2010	2009	2008	2007
Revenue	4,320,000,000	3,352,320,000	2,684,690,000	1,600,000,000	1,700,000,000	
R&D Expense						
Operating Income						
Operating Margin %						
SGA Expense						
Net Income	809,300,000	605,000,000				
Operating Cash Flow						
Capital Expenditure						
EBITDA						
Return on Assets %						
Return on Equity %						
Debt to Equity						

CONTACT INFORMATION:

Phone: 39-02-550281 Fax: 39-02-55028859
Toll-Free:
Address: via Antonio Fogazzaro 28, Milano, 20135 Italy

STOCK TICKER/OTHER:

Stock Ticker: 1913 Exchange: Hong Kong
Employees: 7,855 Fiscal Year Ends: 31-Jan
Parent Company:

SALARIES/BONUSES:

Top Exec. Salary: $ Bonus: $
Second Exec. Salary: $ Bonus: $

OTHER THOUGHTS:

Estimated Female Officers or Directors: 1
Hot Spot for Advancement for Women/Minorities:

Propex Operating Company LLC

www.propexglobal.com

NAIC Code: 313210

TYPES OF BUSINESS:

Textiles Manufacturing
Carpet Backing & Industrial Fabrics
Geotextiles
Civil Engineering Fabrics

BRANDS/DIVISIONS/AFFILIATES:

Wayzata Investment Partners LLC
Polybac
Matrix
Actionbac
Petromat
Fibermesh
Novocon
Geotex

CONTACTS:

Note: Officers with more than one job title may be intentionally listed here more than once.

Michael K. Gorey, CEO
Michael K. Gorey, Pres.
Mark J. Thomas, CFO
Ralph T. Bruno, Exec. VP-Global Sales & Mktg.
Randal D. Powell, Exec. VP-Oper.
Dave Gartshor, VP-Sales, Furnishing Solutions
John Maynie, Dir.-Mktg.
John W. Goers, VP-Sales, Geotextile Systems

GROWTH PLANS/SPECIAL FEATURES:

Propex Operating Company LLC is a leading independent manufacturer of primary and secondary carpet backing and woven, non-woven and stitch-bonded fabrics and fibers used in furniture, bedding, automotive applications, geosynthetics, industrial applications and concrete reinforcement. Propex operates globally in North America, Europe, Mexico, Germany, Hungary, the U.K and Brazil, from which it markets its products to various countries. The company's Polybac and Actionbac primary and secondary carpet backing are highly durable synthetic woven products with moisture and mildew resistance not present in natural fiber alternatives. Propex's primary carpet backing, Matrix, is a patented combination of woven and non-woven fabrics for critical applications such as high-end commercial and artificial turf products. The company's synthetic and steel fibers provide secondary reinforcement and help to minimize concrete cracking in various applications, while its geosynthetic fabrics improve the service life of roadways and drainage systems, minimize erosion and provide protection from storm water sediment run-off. Its Fibermesh brand concrete fibers are used as secondary reinforcement for residential and commercial concrete slabs, pavements, mines, tunnels and pre-cast concrete. Additionally, Propex offers steel fibers for more demanding applications, such as composite metal decks, industrial slabs and airport runways through its Novocon brand. Other products include Fibercast, with microsynthetic fibers; Enduro, with macrosynthetic fibers; and Novomesh with blended microsynthetic fibers combined with macrosynthetic or steel fibers. Its polypropylene geotextile products are marketed under the brand names Geotex, Petromat, Landlok, Pyramat, Petrotac and ArmorMax. The company additionally engineers fabrics that are used for liquid filtration, protective membranes and recreational and agricultural applications. Aside from these interests, the firm also manufactures Curv, a fabric used to create lightweight and durable luggage. Propex is owned by private equity firm Wayzata Investment Partners LLC.

FINANCIAL DATA:

Note: Data for latest year may not have been available at press time.

In U.S. $	2012	2011	2010	2009	2008	2007
Revenue						
R&D Expense						
Operating Income						
Operating Margin %						
SGA Expense						
Net Income						
Operating Cash Flow						
Capital Expenditure						
EBITDA						
Return on Assets %						
Return on Equity %						
Debt to Equity						

CONTACT INFORMATION:

Phone: Fax: 423-899-5005
Toll-Free: 800-621-1273
Address: 1110 Market St., Ste. 300, Chattanooga, TN 37402 United States

STOCK TICKER/OTHER:

Stock Ticker: Private Exchange:
Employees: Fiscal Year Ends: 31-Dec
Parent Company: WAYZATA INVESTMENT PARTNERS LLC

SALARIES/BONUSES:

Top Exec. Salary: $ Bonus: $
Second Exec. Salary: $ Bonus: $

OTHER THOUGHTS:

Estimated Female Officers or Directors:
Hot Spot for Advancement for Women/Minorities:

Puig Beauty & Fashion Group SL

www.puig.com

NAIC Code: 315240

TYPES OF BUSINESS:

Women's Apparel
Luxury Perfumes
Beauty & Personal Care Products

BRANDS/DIVISIONS/AFFILIATES:

Nina Ricci
Paco Rabanne
Carolina Herrera
Mango
Comme des Garcons
Pacha
Jean Paul Gaultier
Vitesse

CONTACTS: *Note: Officers with more than one job title may be intentionally listed here more than once.*

Marc Puig Guasch, CEO
Javier Bach, COO
Joan Albiol, CFO
Eulalia Alfonso, Chief Human Resources Officer
Marc Puig Guasch, Chmn.

GROWTH PLANS/SPECIAL FEATURES:

Puig Beauty & Fashion Group, SL, based in Spain, is a fragrance, cosmetics and fashion multinational company. The family-owned and -operated company maintains operations in 21 countries. The firm is divided into three core business areas: fashion, fragrance and cosmetics. Puig's fashion division oversees a portfolio composed of couture lines Carolina Herrera, Jean Paul Gaultier, Paco Rabanne and Nina Ricci. The fragrance and cosmetic divisions include perfume and prestige cosmetics products, represented by five major brands: Carolina Herrera, Nina Ricci, Paco Rabanne, Prada and Comme des Garcons. The fragrance and cosmetic divisions also include mass distribution products that fall into the prestige, an esteemed yet low priced, category. Brand names in this category include Mango, Pacha, Vitesse and Payot, among many others. The company also conceives and designs product lines for celebrities such as Shakira, Antonio Banderas, Rafael Marquez and Rosario Flores. Company products are distributed in 130 countries worldwide.

FINANCIAL DATA: *Note: Data for latest year may not have been available at press time.*

In U.S. $	2012	2011	2010	2009	2008	2007
Revenue						
R&D Expense						
Operating Income						
Operating Margin %						
SGA Expense						
Net Income						
Operating Cash Flow						
Capital Expenditure						
EBITDA						
Return on Assets %						
Return on Equity %						
Debt to Equity						

CONTACT INFORMATION:

Phone: 34-93-400-70-00 Fax: 34-93-202-16-01
Toll-Free:
Address: Travessera de Gracia, 9, Barcelona, 08021 Spain

STOCK TICKER/OTHER:

Stock Ticker: Private Exchange:
Employees: 5,000 Fiscal Year Ends: 31-Dec
Parent Company:

SALARIES/BONUSES:

Top Exec. Salary: $ Bonus: $
Second Exec. Salary: $ Bonus: $

OTHER THOUGHTS:

Estimated Female Officers or Directors: 2
Hot Spot for Advancement for Women/Minorities: Y

PUMA SE

www.puma.com

NAIC Code: 316210

TYPES OF BUSINESS:

Athletic Shoes Manufacturing
Athletic Apparel
Sports Equipment
Retail Stores

BRANDS/DIVISIONS/AFFILIATES:

PUMA
Tretorn
Kering Group
Cobra Golf

CONTACTS: *Note: Officers with more than one job title may be intentionally listed here more than once.*

Klaus Bauer, COO
Michael Lammermann, CFO
Michael Lammermann, General Manager-Legal
Ulf Santjer, Dir.-Corp. Comm.
Michael Lammermann, General Manager-Finance
Stefano Caroti, Chief Commercial Officer
Jean-Francois Palus, Chmn.

GROWTH PLANS/SPECIAL FEATURES:

PUMA SE (PUMA) is an international footwear, apparel and accessories company based in Germany. Luxury fashion conglomerate Kering Group, formerly PPR, owns an 82.4% controlling stake in the company. PUMA develops and markets footwear, t-shirts, shorts, pants, jackets, caps, swimwear, socks and head- and wristbands, as well as a variety of athletic equipment under the PUMA, Tretorn and Cobra Golf brands. The firm designs products for multiple sports, including running, soccer, football, tennis, baseball, cricket and motocross and relies on celebrity endorsement as well as high-profile advertising campaigns to fulfill its marketing strategies. PUMA has provided apparel to the national soccer team of Cameroon, Belgium's Club Brugge, tennis star Serena Williams, baseball player Johnny Damon, motocross medalist Travis Pastrana and a variety of teams playing in the World Athletics Championships for the Jamaican Federation. The company has also partnered with fashion designers such as Rudolf Dassler, Alexander McQueen, Yasuhiro Mihara and Sergio Rossi. Products are distributed in more than 120 countries through sporting goods stores, department stores, online stores and PUMA sports boutiques located around the world. Approximately 48.8% of its revenues are derived from footwear, 35.2% from apparel and 16% from accessories. In August 2012, the company opened its first sustainable PUMA store, located in Bangalore, India, and in October 2012, it launched its first series of biodegradable and recyclable products.

FINANCIAL DATA: *Note: Data for latest year may not have been available at press time.*

In U.S. $	2012	2011	2010	2009	2008	2007
Revenue	4,373,763,000	4,023,803,000	3,619,150,000	3,290,586,000	3,375,501,000	3,173,977,000
R&D Expense		77,000,000	63,600,000	58,100,000	55,100,000	57,500,000
Operating Income	151,377,400	445,573,700	410,270,100	257,288,000	435,143,100	497,459,200
Operating Margin %	3.46%	11.07%	11.33%	7.81%	12.89%	15.67%
SGA Expense		1,515,111,000	1,376,036,000	1,380,851,000	1,273,068,000	
Net Income	93,875,370	308,103,800	270,393,200	168,360,500	309,708,500	359,721,900
Operating Cash Flow	209,548,000	169,564,000	226,531,200	415,351,700	292,992,800	579,967,900
Capital Expenditure	108,585,200	95,078,900	73,816,530	72,880,450	159,400,900	
EBITDA	253,142,600	528,082,400	506,686,300	352,768,100	510,296,900	497,459,200
Return on Assets %	2.74%	9.30%	9.23%	6.55%	12.37%	15.03%
Return on Equity %	4.38%	15.38%	15.39%	10.60%	20.03%	24.58%
Debt to Equity						0.09

CONTACT INFORMATION:

Phone: 49 09132810 Fax: 49 09132812246
Toll-Free:
Address: Puma-Way 1, Herzogenaurach, 91074 Germany

STOCK TICKER/OTHER:

Stock Ticker: PMMAF Exchange: GREY
Employees: 10,836 Fiscal Year Ends: 31-Dec
Parent Company: Kering

SALARIES/BONUSES:

Top Exec. Salary: $ Bonus: $
Second Exec. Salary: $ Bonus: $

OTHER THOUGHTS:

Estimated Female Officers or Directors:
Hot Spot for Advancement for Women/Minorities:

PVH Corp

NAIC Code: 315240

www.pvh.com

TYPES OF BUSINESS:

Designer Apparel
Sportswear
Footwear
Outlet Stores
Neckwear
Jeans
Dress Shirts
Apparel Wholesale

BRANDS/DIVISIONS/AFFILIATES:

Calvin Klein
Van Heusen
IZOD
Bass
Wamaco
ARROW
Phillips-Van Heusen Corporation
Tommy Hilfiger Corp.

CONTACTS: *Note: Officers with more than one job title may be intentionally listed here more than once.*

Emanuel Chirico, CEO
Michael Shaffer, COO
Allen E. Sirkin, Pres.
Michael A. Shaffer, CFO/Exec. VP
Michael Kelly, Exec. VP-PVH Mktg. Group
David F. Kozel, Sr. VP-Human Resources
Jon D. Peters, CIO
Jon D. Peters, Exec. VP-Tech. & Logistics
Mark D. Fischer, General Counsel/Sec./Sr. VP
Dana Perlman, Sr. VP-Bus. Dev./Treas.
Dana Perlman, Sr. VP-Investor Rel.
Bruce Goldstein, Controller/Sr. VP
Paul T. Murry, CEO/Pres., Calvin Klein
Melanie Steiner, Chief Risk Officer
Steve Shiffman, Pres./COO-Retail Oper.
Margaret P. Lachance, Pres., Van Heusen Retail
Emanuel Chirico, Chmn.
Fred Gehring, CEO-PVH & Tommy Hilfiger Int'l Oper.
Bill McRaith, Chief Supply Chain Officer

GROWTH PLANS/SPECIAL FEATURES:

PVH Corp., formerly Phillips-Van Heusen Corporation, manufactures, distributes and sells clothing and shoes for men, women and children. The firm operates under several brand names, including owned brands such as Calvin Klein, Van Heusen, Tommy Hilfiger, ARROW, IZOD and Bass as well as licensed brands such as CHAPS, Trump, MICHAEL Michael Kors, Kenneth Cole and DKNY. PVH distributes to department stores and other wholesale clients as well as selling through outlet stores operating under the firm's brand names. It currently maintains over 1,000 retail locations across its brand portfolio in the U.S., Canada, Europe and Japan. The company operates in three primary segments: Tommy Hilfiger, which consists of Tommy Hilfiger North America and Tommy Hilfiger International; Calvin Klein, which includes Calvin Klein Licensing and Other Calvin Klein Apparel, which is comprised of the Calvin Klein dress furnishings, sportswear and outlet retail divisions; and Heritage Brands, which consists of the Wholesale Dress Furnishings, Wholesale Sportswear and Retail units. In October 2012, PVH bought the Wamaco group for approximately $2.9 billion. This has given PVH full control over the Calvin Klein brand, as Wamaco held licensing rights for Calvin Klein jeans and underwear.

PVH offers employees medical, dental and life insurance; an employee assistance program; flexible spending accounts; domestic partner benefits; a 401(k); a pension plan; a 529 college savings program; group auto and home owner's insurance; discounts on company merchandise; education assistance; and college scholarships for children of employees.

FINANCIAL DATA: *Note: Data for latest year may not have been available at press time.*

In U.S. $	2012	2011	2010	2009	2008	2007
Revenue	5,890,624,000	4,636,848,000	2,398,731,000	2,491,935,000	2,425,175,000	
R&D Expense						
Operating Income	559,653,000	203,395,000	243,812,000	173,748,000	308,495,000	
Operating Margin %	9.50%	4.38%	10.16%	6.97%	12.72%	
SGA Expense	2,481,370,000	2,071,416,000	938,791,000	1,028,784,000	882,492,000	
Net Income	317,881,000	53,805,000	161,910,000	91,771,000	183,319,000	
Operating Cash Flow	490,721,000	352,227,000	214,452,000	238,747,000	221,040,000	
Capital Expenditure	169,841,000	100,995,000	23,856,000	88,141,000	94,749,000	
EBITDA	692,930,000	352,271,000	294,996,000	235,309,000	375,164,000	
Return on Assets %	4.71%	1.18%	7.13%	4.19%	8.79%	
Return on Equity %	13.29%	3.14%	14.94%	9.38%	19.31%	
Debt to Equity	0.72	1.04	0.34	0.4	0.41	

CONTACT INFORMATION:

Phone: 212 381-3500 Fax: 212 247-5309
Toll-Free: 866-214-6694
Address: 200 Madison Ave., New York, NY 10016 United States

STOCK TICKER/OTHER:

Stock Ticker: PVH
Exchange: NYS
Employees: 10,500
Fiscal Year Ends: 31-Jan
Parent Company:

SALARIES/BONUSES:

Top Exec. Salary: $900,000 Bonus: $500,000
Second Exec. Salary: $1,250,000 Bonus: $

OTHER THOUGHTS:

Estimated Female Officers or Directors: 11
Hot Spot for Advancement for Women/Minorities: Y

QST Industries Inc

www.qst.com

NAIC Code: 313240

TYPES OF BUSINESS:

Apparel Components
Linings
Elastic & Waistbands
Home Furnishing Products

GROWTH PLANS/SPECIAL FEATURES:

QST Industries, Inc., founded in 1880, is one of the world's largest industrial apparel components suppliers. The family-owned and -operated company, with offices across the U.S. and over 30 countries around the globe, provides products and services such as linings, interlinings, under-collar cloth, elastics, insulations, waistbands, pocketing, printing and die and table cutting as well as home furnishing products. Its brands include Q-Loop, a product designed to eliminate belt-loop fraying, and Ban-Rol, one of the world's leading waistband inserts. The components provided by the company cut such apparel production costs as warehousing, marker making, cut planning, table and die-cutting equipment, unnecessary inventory and interest expense and production losses related to cutting errors. Pocketing and lining products come in a variety of fabrics and weaves, including poly/rayon and poly/cotton blends as well as schreinered, calendered, textured, fancy, plain weave and soft and medium finishes. Elastics designed for use in men's, women's and children's apparel and swimwear include knitted insert, exposed, woven and innovative elastics such as Quik/cord and single unit band cord. The firm's DuPont Dacron Insulation Fiberfill is used in domestic and Asian production of insulated garments. John Solomon, Inc., a division of the company, produces non-woven stabilizers for embroidery.

BRANDS/DIVISIONS/AFFILIATES:

John Solomon Inc
Q-Loop
Ban-Rol
Quik/cord
DuPont Dacron Insulation Fiberfill

CONTACTS: *Note: Officers with more than one job title may be intentionally listed here more than once.*

Thomas Danch, CEO
Thomas Danch, Pres.
Jeffrey Carlevato, CFO
Hector Ballon, VP-Sales
Don Mungovan, VP-Global IT
Jeffrey Carlevato, Corp. Sec.
Jeffrey Carlevato, Treas.
Alexander Danch, Corp. VP
Michael Danch, Exec. VP
Ely Lionheart, Chmn.

FINANCIAL DATA: *Note: Data for latest year may not have been available at press time.*

In U.S. $	2012	2011	2010	2009	2008	2007
Revenue						
R&D Expense						
Operating Income						
Operating Margin %						
SGA Expense						
Net Income						
Operating Cash Flow						
Capital Expenditure						
EBITDA						
Return on Assets %						
Return on Equity %						
Debt to Equity						

CONTACT INFORMATION:

Phone: 312-930-9400 Fax: 312-648-0312
Toll-Free:
Address: 550 W. Adams St., Ste. 200, Chicago, IL 60661 United States

STOCK TICKER/OTHER:

Stock Ticker: Private
Employees: 462
Parent Company:
Exchange:
Fiscal Year Ends: 31-Dec

SALARIES/BONUSES:

Top Exec. Salary: $ Bonus: $
Second Exec. Salary: $ Bonus: $

OTHER THOUGHTS:

Estimated Female Officers or Directors:
Hot Spot for Advancement for Women/Minorities:

Quiksilver Inc

NAIC Code: 315210

www.quiksilverinc.com

TYPES OF BUSINESS:

Sports Apparel & Equipment
Snow & Surf Apparel & Equipment
Accessories
Swimwear
Retail Stores

BRANDS/DIVISIONS/AFFILIATES:

Quicksilver
Roxy
Roxy Girl
DC
Teenie Wahine

CONTACTS: *Note: Officers with more than one job title may be intentionally listed here more than once.*

Andrew P Mooney, CEO
Craig Stevenson, COO
Andrew P Mooney, Pres.
Richard Shields, CFO
Carol Scherman, Exec. VP-Global Human Resources
Charles S. Exon, Chief Admin. Officer
Charles S. Exon, General Counsel/Sec.
Craig Stevenson, Pres., Global Brand
Rob Colby, Pres., Quiksilver Americas
Greg Healy, Pres., Quiksilver Asia Pacific
Robert B. McKnight, Jr., Chmn.
Pierre Agnes, Pres., Quicksilver Europe
Kasey Mazzone, Global Head-Supply Chain

GROWTH PLANS/SPECIAL FEATURES:

Quiksilver, Inc. is an internationally diversified firm that designs, produces, retails and distributes branded apparel, footwear, accessories and related products. It operates in three segments: the Americas, which includes the U.S., Latin America and Canada; Europe, which includes countries located in Europe, Africa and the Middle East; and Asia/Pacific, which includes Australia, Japan, New Zealand and Indonesia. The company's brands are focused on different sports within the outdoor market. The Quiksilver and Roxy brands are rooted in the sport of surfing and are leading brands representing the boardriding lifestyle, which includes surfing, skateboarding and snowboarding. Quiksilver has grown to include shirts, shorts, t-shirts, pants, jackets, fleece, snowboard wear, footwear, hats, backpacks, wetsuits, watches, eyewear and other accessories. In addition, the brand has expanded its target market to include boys, toddlers and infants. The Roxy brand includes sportswear, footwear, backpacks, snowboard wear, swimwear, backpacks, snowboard boots, skis, fragrance, beauty care, bedroom furnishings and other accessories for young women. The brand also includes the Teenie Wahine and Roxy Girl brands for girls and infants. The firm's DC brand specializes in technical shoes made for skateboarding and snowboarding as well as sandals and general branded apparel. The company's products are sold in over 90 countries in its 825 owned or licensed stores as well as a wide range of other distribution channels, including surf shops, snowboard shops, skate shops and other specialty stores and select department stores. In 2012, Quiksilver accounted for 39% of the firm's revenues; DC, 30%; Roxy, 26%; and 5% for other brands.

FINANCIAL DATA: *Note: Data for latest year may not have been available at press time.*

In U.S. $	2012	2011	2010	2009	2008	2007
Revenue	2,013,239,000	1,953,061,000	1,837,620,000	1,977,526,000	2,264,636,000	2,426,035,000
R&D Expense						
Operating Income	56,968,000	41,512,000	123,525,000	68,548,000	138,856,000	-15,522,000
Operating Margin %	2.82%	2.12%	6.72%	3.46%	6.13%	-.63%
SGA Expense	916,144,000	895,949,000	832,066,000	851,746,000	915,933,000	971,411,000
Net Income	-10,756,000	-21,258,000	-9,684,000	-192,042,000	-226,265,000	-121,119,000
Operating Cash Flow	-13,539,000	54,149,000				
Capital Expenditure	66,081,000	89,590,000				
EBITDA	111,869,000	96,771,000	183,303,000	112,380,000	201,129,000	49,474,000
Return on Assets %	-.61%	-1.22%	-.54%	-9.54%	-9.40%	-4.76%
Return on Equity %	-1.80%	-3.48%	-1.81%	-36.35%	-30.44%	-13.70%
Debt to Equity	1.23	1.18	1.14	1.99	1.31	0.87

CONTACT INFORMATION:

Phone: 714 889-2200 Fax: 714 645-0313
Toll-Free:
Address: 15202 Graham St., Huntington Beach, CA 92649 United States

STOCK TICKER/OTHER:

Stock Ticker: ZQK Exchange: NYS
Employees: 6,600 Fiscal Year Ends: 31-Oct
Parent Company:

SALARIES/BONUSES:

Top Exec. Salary: $992,300 Bonus: $62,500
Second Exec. Salary: $617,500 Bonus: $38,700

OTHER THOUGHTS:

Estimated Female Officers or Directors: 2
Hot Spot for Advancement for Women/Minorities:

Ralph Lauren Corporation

www.polo.com

NAIC Code: 315240

TYPES OF BUSINESS:

Men's & Women's Branded Fashions
Apparel Design & Marketing
Accessories
Fragrances
Home Furnishings
Cosmetics
Retail Stores

BRANDS/DIVISIONS/AFFILIATES:

Polo
Ralph by Ralph Lauren
Club Monaco
Black Label
Purple Label
Blue Label
Rugby
Polo.com

CONTACTS: *Note: Officers with more than one job title may be intentionally listed here more than once.*

Ralph Lauren, CEO
Roger N. Farah, COO
Roger N. Farah, Pres.
Robert Madore, Interim CFO
Mitchell A. Kosh, Sr. VP-Human Resources
Jackwyn L. Nemerov, Exec. VP
Ralph Lauren, Chmn.
Daniel Lalonde, Pres., Ralph Lauren Int'l

GROWTH PLANS/SPECIAL FEATURES:

Ralph Lauren Corporation (Polo) designs, markets and distributes premium lifestyle products. Polo licenses the manufacturing of its products to companies worldwide. Capitalizing on the creative force of its founder, Ralph Lauren, the firm's various brand names have become recognizable cultural symbols across the globe. The firm offers four categories of lifestyle products: apparel products (including extensive collections of men's, women's and children's clothing); home products (including bedding and bath products, furniture, fabric, wallpaper, paints, tabletop and giftware); accessories (including footwear, eyewear, jewelry, leather goods, handbags and luggage); and fragrance products (consisting of fragrances and skin care products). The company markets its products through department, specialty, golf and Ralph Lauren stores, as well as via the Internet and mail order catalogs. Polo also offers wholesale products to upscale (and certain mid-tier) department stores, specialty stores and golf/pro shops worldwide. Its brands include Ralph by Ralph Lauren, Romance, Polo, Lauren for Men, Purple Label, Blue Label, Black Label, RRL, RLX Ralph Lauren, Polo Jeans Co., Big Pony, Pink Pony, Polo Sport and Club Monaco. It operates company-owned stores that inclucde 103 Ralph Lauren, 59 Club Monaco, 16 Rugby stores and 201 factory retail stores in North America, Japan, Europe and Latin America.

FINANCIAL DATA: *Note: Data for latest year may not have been available at press time.*

In U.S. $	2012	2011	2010	2009	2008	2007
Revenue	6,859,500,000	5,660,300,000	4,978,900,000	5,018,900,000	4,880,100,000	4,295,400,000
R&D Expense						
Operating Income	1,039,400,000	845,100,000	706,900,000	595,500,000	653,400,000	652,600,000
Operating Margin %	15.15%	14.93%	14.19%	11.86%	13.38%	15.19%
SGA Expense	2,915,200,000	2,442,700,000	2,157,000,000	2,036,000,000	1,932,500,000	1,663,400,000
Net Income	681,000,000	567,600,000	479,500,000	406,000,000	419,800,000	400,900,000
Operating Cash Flow	885,300,000	688,700,000	906,500,000	774,200,000	695,400,000	796,100,000
Capital Expenditure	272,200,000	255,000,000	201,300,000	185,000,000	217,100,000	184,000,000
EBITDA	1,068,500,000	869,100,000	892,700,000	798,500,000	869,100,000	809,600,000
Return on Assets %	13.09%	11.78%	10.64%	9.30%	10.33%	11.71%
Return on Equity %	19.57%	17.67%	16.38%	15.84%	17.77%	18.28%
Debt to Equity	0.07	0.1	0.1	0.14	0.22	0.19

CONTACT INFORMATION:

Phone: 212 318-7000 Fax:
Toll-Free: 800-377-7656
Address: 650 Madison Ave., New York, NY 10022 United States

STOCK TICKER/OTHER:

Stock Ticker: RL Exchange: NYS
Employees: 24,000 Fiscal Year Ends: 31-Mar
Parent Company:

SALARIES/BONUSES:

Top Exec. Salary: $1,250,000 Bonus: $
Second Exec. Salary: $900,000 Bonus: $

OTHER THOUGHTS:

Estimated Female Officers or Directors: 3
Hot Spot for Advancement for Women/Minorities: Y

Ratti SpA

NAIC Code: 313210

www.ratti.it

TYPES OF BUSINESS:

Silk Fabrics, Manufacturing
Ties, Scarves & Accessories
Weaving
Cashmere, Linen & Microfiber Fabrics
Upholstery Fabrics

BRANDS/DIVISIONS/AFFILIATES:

Braghenti
Campi
Ratti Donna
Ratti Accessori
Ratti Uomo
Ratti Casa

CONTACTS: *Note: Officers with more than one job title may be intentionally listed here more than once.*

Sergio Tamborini, CEO
Roberto Brachetti, Corp. Sec.
Donatella Ratti, Chmn.

GROWTH PLANS/SPECIAL FEATURES:

Ratti SpA is a leading Italian textile manufacturer and producer of fabrics, accessories and home furnishings, specializing in silk and other natural fibers. It creates over 6.5 million yards of fabric every year. The company operates a mill facility and distributes products worldwide, with offices in Italy, Japan and Germany as well as a U.S. office in New York. Ratti's printworks facility in Guanzate has a complete production cycle, with photoengraving, preparation, dyeing, printing and finishing all at one facility. Braghenti, a subsidiary of Ratti, is a weaving company that produces fabric for clothing and accessories in silk, linen, wool, cotton and other skillfully blended fibers, while Campi, another subsidiary, produces yarn-dyed silk fabrics for ties and scarves. The firm produces several lines including Ratti Donna, a collection of printed, plain and yarn-dyed fabrics for high-fashion women's apparel; Ratti Uomo, the high-end men's clothing brand consisting of printed, plain and jacquard fabrics for ties, scarves and shirts; Ratti Casa, a fabric line for furniture, window coverings and other home accessories; and Ratti Accessori scarves, shawls, beachwear and underwear. Licensed brands include Leonard Foulards, Leonard Lingerie, Leonard Soleil, Valentino Foulards, Valentino Cravatte, Chantal Thomass Pleasure and Chantal Thomass Beachwear.

FINANCIAL DATA: *Note: Data for latest year may not have been available at press time.*

In U.S. $	2012	2011	2010	2009	2008	2007
Revenue			93,000,000	91,000,000	129,600,000	147,300,000
R&D Expense						
Operating Income						
Operating Margin %						
SGA Expense						
Net Income			-1,600,000	-5,900,000	-17,500,000	-7,700,000
Operating Cash Flow						
Capital Expenditure						
EBITDA						
Return on Assets %						
Return on Equity %						
Debt to Equity						

CONTACT INFORMATION:

Phone: 0039-031-35351 Fax: 0039-031-3535249
Toll-Free:
Address: Via Madonna 30, Guanzate, 22070 Italy

STOCK TICKER/OTHER:

Stock Ticker: RAT Exchange: Milan
Employees: 548 Fiscal Year Ends: 31-Dec
Parent Company:

SALARIES/BONUSES:

Top Exec. Salary: $ Bonus: $
Second Exec. Salary: $ Bonus: $

OTHER THOUGHTS:

Estimated Female Officers or Directors: 2
Hot Spot for Advancement for Women/Minorities: Y

Raymond Ltd

www.raymondindia.com

NAIC Code: 313210

TYPES OF BUSINESS:

Fabric & Textiles Manufacturing
Wool & Cashmere
Silks
Denim
Apparel-Suits
Automotive & Engineering Tools

BRANDS/DIVISIONS/AFFILIATES:

Raymond Apparel Ltd
Everblue Apparel Ltd
Raymond UCO Denim
Raymond Shop (The)
Raymond Zambaiti Ltd
J.K. Helene Curtis Ltd
ColorPlus Fashions Pvt Ltd
Ring Plus Aqua Ltd

CONTACTS:
Note: Officers with more than one job title may be intentionally listed here more than once.

Gautam H. Singhania, Managing Dir.
Aniruddha Deshmukh, Pres.
Harshall Sunder, CFO
K. A. Narayan, Pres., Human Resources
Harshall Jayavant, Pres., Eng.
Thomas Fernandes, Corp. Sec.
Robert Lobo, Pres., Oper.-Apparel Group
Rakesh Pandey, Pres., Retail & Bus. Dev.
Harshall Sunder, Pres., Finance
S.L. Pokharna, Pres., Commercial
Shreyas Joshi, Pres., Apparel Group
Gautam H. Singhania, Chmn.

GROWTH PLANS/SPECIAL FEATURES:

Raymond, Ltd. is India's largest exporter of worsted fabrics and ready-made garments to over 55 countries, including Australia, Canada, the U.S. and Japan. The company has two core business segments: textiles; and files and tools. The textile segment is comprised of three divisions: worsted, denim and shirting. The worsted suiting fabrics division manufactures pure wool and wool-blended fabrics utilizing fibers such as bamboo, polyester, linen, Alpaca, cashmere, mohair and angora wool. The division has a capacity of 41.5 million yards of worsted fabrics annually, producing nearly 20,000 different designs and colors sold at approximately 3,000 stores. The denim division, operated by joint-venture company Raymond UCO Denim, is capable of producing approximately 43.7 million yards of denim at its facilities in India and Romania, which is manufactured into garments at the firm's jeans wear facility, EverBlue Apparel Ltd. The shirting division, operated by joint-venture company Raymond Zambaiti Ltd., produces cotton and linen shirting, manufactured in India by Celebrations Apparel Ltd. The files and tools segment manufactures and markets steel files, high speed steel (HSS) cutting tools (primarily drills) and hand tools. In addition to its core businesses, the company has operations in furnishings, personal care, auto parts and engineering through its many subsidiaries. A few of these subsidiaries include Raymond Apparel Ltd., ColorPlus Fashions Pvt. Ltd., Ring Plus Aqua Ltd. and J.K. Helene Curtis Ltd. Raymond suits are available in India through 650 domestic retailers, as well as 600 exclusive international retail shops, called The Raymond Shop.

Raymond offers employees educational, housing and transportation benefits as well as discounted tuition for employees' children enrolled in specific schools.

FINANCIAL DATA:
Note: Data for latest year may not have been available at press time.

In U.S. $	2012	2011	2010	2009	2008	2007
Revenue	670,201,400	558,604,700	461,439,000	470,944,100	440,890,600	375,483,300
R&D Expense						
Operating Income	58,635,540	59,424,930	51,062,120	-713,274	-10,603,630	102,316,800
Operating Margin %	8.74%	10.63%	11.06%	-.15%	-2.40%	27.24%
SGA Expense	86,696,220	113,894,900	98,396,110	136,347,700	101,364,100	2,720,229
Net Income	28,663,350	7,000,074	-9,227,350	-42,276,390	1,955,803	25,532,290
Operating Cash Flow	42,402,830	41,595,040	51,963,330	59,997,740	-168,654	8,559,027
Capital Expenditure	45,278,920				26,307,330	127,058,100
EBITDA	98,490,770	51,252,960	54,104,200	9,039,523	56,144,150	71,492,820
Return on Assets %	4.09%	1.06%	-1.27%	-6.25%	.30%	4.87%
Return on Equity %	11.96%	3.15%	-3.81%	-17.16%	.73%	10.13%
Debt to Equity	0.64	1.33	1.43	1.49	1.05	0.91

CONTACT INFORMATION:

Phone: 91 2352232514   Fax: 91 2352232514
Toll-Free:
Address: Plot No. 156/H, No. 2, Village Zadgaon, Ratanagiri, Maharashtra 415 612 India

STOCK TICKER/OTHER:

Stock Ticker: RYWKF   Exchange: GREY
Employees: Fiscal Year Ends: 31-Mar
Parent Company:

SALARIES/BONUSES:

Top Exec. Salary: $ Bonus: $
Second Exec. Salary: $ Bonus: $

OTHER THOUGHTS:

Estimated Female Officers or Directors:
Hot Spot for Advancement for Women/Minorities:

Recreational Equipment Inc (REI)

www.rei.com

NAIC Code: 451110

TYPES OF BUSINESS:

Outdoor Gear & Clothing, Retail
Sporting Equipment Retail & Rental
Adventure Travel Services
Catalog & Online Sales

BRANDS/DIVISIONS/AFFILIATES:

REI
REI Adventures
REI Gear and Apparel
Novara

CONTACTS:

Note: Officers with more than one job title may be intentionally listed here more than once.

Brian Unmacht, Interim CEO
Eric Artz, Interim COO
Eric Artz, CFO/Treas.
Angela Owen, Sr. VP-Mktg.
Michelle Clements, Sr. VP-Human Resources
Bill Baumann, Sr. VP-IT
Lee Fromson, Sr. VP-Merch.
Catherine Walker, General Counsel/Sr. VP/Corp. Sec.
Brad Brown, Sr. VP-e-Commerce & Direct Sales
Michael Collins, VP-Public Affairs
Sue Sallee, VP-Finance & Acct.
Tim Spangler, Sr. VP-Retail
Kathleen Peterson, VP-REI Private Brands
John Hamlin, Chmn.
Rick Bingle, VP-Supply Chain

GROWTH PLANS/SPECIAL FEATURES:

Recreational Equipment, Inc. (REI) is one of the largest consumer cooperatives in the U.S. The firm offers quality outdoor gear, clothing and footwear selected for performance and durability in outdoor recreation, including hiking, climbing, camping, bicycling, paddling and winter sports. Today, REI has more than 5.1 million active members served by approximately 129 retail stores in 32 states. Stores include a variety of facilities for testing equipment, including bike test trails, climbing pinnacles and camp stove demonstration tables. While anyone may shop at the stores, customers who pay a small fee to become members receive special discounts and a share in the company's profits through an annual patronage refund based on their purchases. REI's e-commerce site is one of the largest outdoor online stores, offering a comprehensive library of product information, expert gear advice and outdoor recreation information. In addition to nationally recognized brands, REI sells private label apparel and accessories under the REI Gear and Apparel brand. It also offers mountain, road and touring bikes under the private Novara name. Through REI Adventures, the company has been operating small group tours throughout the world for more than 20 years, avoiding standard tourist routes and emphasizing outdoor activities. Each year, REI Adventures plans domestic and international bicycling, trekking, kayaking, hiking, camping and mountaineering adventures. The firm invests millions of dollars on an annual basis to build trails, clean up the environment and teach children outdoor ethics.

Employees are offered health, life and disability plans; tuition reimbursement; adoption and relocation assistance; an employee discount program; and a public transit subsidy.

FINANCIAL DATA:

Note: Data for latest year may not have been available at press time.

In U.S. $	2012	2011	2010	2009	2008	2007
Revenue	1,930,000,000	1,798,009,000	1,658,751,000	1,455,351,000	1,434,569,000	1,342,024,000
R&D Expense						
Operating Income						
Operating Margin %						
SGA Expense						
Net Income		30,168,000	30,230,000	29,807,000	14,465,000	41,427,000
Operating Cash Flow						
Capital Expenditure						
EBITDA						
Return on Assets %						
Return on Equity %						
Debt to Equity						

CONTACT INFORMATION:

Phone: 253-891-2500 Fax: 253-891-2523
Toll-Free: 800-999-4734
Address: 6750 S. 228th St., Kent, WA 98032 United States

STOCK TICKER/OTHER:

Stock Ticker: Private Exchange:
Employees: 9,100 Fiscal Year Ends: 31-Dec
Parent Company:

SALARIES/BONUSES:

Top Exec. Salary: $ Bonus: $
Second Exec. Salary: $ Bonus: $

OTHER THOUGHTS:

Estimated Female Officers or Directors: 8
Hot Spot for Advancement for Women/Minorities: Y

Red Wing Shoe Company Inc

www.redwingshoe.com

NAIC Code: 316210

TYPES OF BUSINESS:

Footwear Manufacturing
Retail-Mobile Truck Stores
Leather Tanning Operations

BRANDS/DIVISIONS/AFFILIATES:

Vasque
Irish Setter
Worx
Carhartt Footwear
Red Wing Europe
Red Wing International
S.B. Foot Tanning Company

CONTACTS: *Note: Officers with more than one job title may be intentionally listed here more than once.*

David D. Murphy, COO
David D. Murphy, Pres.
Rick Bawek, CFO/Exec. VP
William J. Sweasy, Chmn.

GROWTH PLANS/SPECIAL FEATURES:

Red Wing Shoe Company, Inc. has been hand-making work boots since 1905. Red Wing shoes are used predominately by workmen and laborers in a wide variety of industries, including the military. The company offers shoe styles for different lines of work, including waterproof, insulated, puncture-resistant, static-dissipative, comfort force, steel-toe and tech-toe (plastic) featured footwear. Distributed through retailers all over the U.S., Canada and in several other countries, Red Wing markets shoes under the Red Wing, Vasque, Irish Setter, Worx and Carhartt Footwear brands. The Red Wing brand represents the firm's premium boot line and shoes while Worx is made as a more affordable, basic work boot. Vasque specializes in technical boots and shoes for outdoor activities such as hiking and mountaineering. Irish Setter focuses on boots and casual shoes for hunting, fishing/boating and snow activities. Subsidiaries of the firm include Red Wing Europe; Red Wing International; and S.B. Foot Tanning Company. The company operates a manufacturing facility and leather tanning site in Red Wing, Minnesota; footwear factories in Kentucky and Missouri; and other warehouses worldwide. The firm operates over 430 retail stores. Red Wing maintains a North American headquarters in Houston, Texas. The firm has a flagship store in Red Wing, Minnesota. The 3-story location houses a shoe museum, retail store and factory outlet center.

FINANCIAL DATA: *Note: Data for latest year may not have been available at press time.*

In U.S. $	2012	2011	2010	2009	2008	2007
Revenue						
R&D Expense						
Operating Income						
Operating Margin %						
SGA Expense						
Net Income						
Operating Cash Flow						
Capital Expenditure						
EBITDA						
Return on Assets %						
Return on Equity %						
Debt to Equity						

CONTACT INFORMATION:

Phone: 651-388-8211 Fax: 651-388-7415
Toll-Free: 800-733-9464
Address: 314 Main St., Red Wing, MN 55066 United States

STOCK TICKER/OTHER:

Stock Ticker: Private Exchange:
Employees: Fiscal Year Ends: 30-Nov
Parent Company:

SALARIES/BONUSES:

Top Exec. Salary: $ Bonus: $
Second Exec. Salary: $ Bonus: $

OTHER THOUGHTS:

Estimated Female Officers or Directors:
Hot Spot for Advancement for Women/Minorities:

Redcats USA

NAIC Code: 454113

www.redcatsusa.com

TYPES OF BUSINESS:

Apparel, Catalogs
Online Sales
Plus-Size Women's Apparel
Kitchen & Home Products
Big & Tall Men's Apparel
Home Furnishings
Children's Apparel

BRANDS/DIVISIONS/AFFILIATES:

PPR SA
BrylaneHome
Women Within
Jessica London
KingSize
Roaman's
Stella McCartney Kids USA
OneStopPlus.com

CONTACTS: *Note: Officers with more than one job title may be intentionally listed here more than once.*

Jean-Michel Noir, Chmn./CEO-Redcats Group

GROWTH PLANS/SPECIAL FEATURES:

Redcats USA, part of the Redcats Group, is a leading direct retailer operating specialty catalogs and e-commerce sites. It functions as the U.S. home shopping division of PPR SA, a luxury goods corporation based in France. The firm markets men's, women's and children's apparel, home and lifestyle products through six primary brands: Jessica London, Roaman's and Woman Within, which include women's wear and accessories; KingSize, which offers casual collections for men; Stella McCartney Kids USA, a children's clothing line from luxury fashion designer Stella McCartney; and BrylaneHome, a home goods brand. The Jessica London, Roaman's and Woman Within brands include plus-sized women's traditional apparel, including casual clothes, work clothes, special occasion looks, outerwear, swimwear, intimates and shoes. The e-commerce site OneStopPlus.com brings together all these brands in an online fashion mall setting. The KingSize brand sells men's shirts, underwear, sleepwear, active wear, swimwear, business clothing and accessories such as belts and shoes, all designed specifically for big and tall men. Stella McCartney Kids USA sells clothing, accessories and gifts for boys and girls in sizes 2 to 14, as well as newborn and baby apparel in sizes up to 24 months. BrylaneHome provides home and lifestyle products such as bedding, towels, slipcovers, curtains and linens; seasonal items; dining accessories; home accents; and artwork, lamps and other home accessories. The company also operates The Bargain Catalog Outlet, which offers surplus and discontinued items from all Redcats brands at a 30% to 80% discount. In November 2012, Redcats agreed to sell the operations of The Golf Warehouse and The Sportsman's Guide to Northern Tool + Equipment.

FINANCIAL DATA: *Note: Data for latest year may not have been available at press time.*

In U.S. $	2012	2011	2010	2009	2008	2007
Revenue						
R&D Expense						
Operating Income						
Operating Margin %						
SGA Expense						
Net Income						
Operating Cash Flow						
Capital Expenditure						
EBITDA						
Return on Assets %						
Return on Equity %						
Debt to Equity						

CONTACT INFORMATION:

Phone: 212-613-9500　　Fax: 212-613-9690
Toll-Free:
Address: 463 7th Ave., Fl. 22, New York, NY 10018 United States

STOCK TICKER/OTHER:

Stock Ticker: Subsidiary　　Exchange:
Employees:　　Fiscal Year Ends: 31-Dec
Parent Company: Kering

SALARIES/BONUSES:

Top Exec. Salary: $　　Bonus: $
Second Exec. Salary: $　　Bonus: $

OTHER THOUGHTS:

Estimated Female Officers or Directors:
Hot Spot for Advancement for Women/Minorities: Y

Reebok CCM Hockey

www.thehockeycompany.com

NAIC Code: 339920

TYPES OF BUSINESS:

Hockey Equipment & Apparel
NHL Replica Jerseys

BRANDS/DIVISIONS/AFFILIATES:

Adidas AG
Reebok International Ltd
JOFA
CCM
Reebok
KOHO
Rbk
Hockey Company (The)

CONTACTS: *Note: Officers with more than one job title may be intentionally listed here more than once.*

Phil Dube, Gen. Mgr.
James Coughlin, VP-Sales, North America
Igor Landau, Chmn.-Adidas AG
Herbert Hainer, CEO-Adidas AG
Uli Becker, Pres., Reebok

GROWTH PLANS/SPECIAL FEATURES:

Reebok-CCM Hockey, formerly known as The Hockey Company, is a division of Reebok International Ltd., which is itself a subsidiary of Adidas AG. It is headquartered in Montreal, Canada with operations in Europe and North America. The company is a designer, manufacturer and marketer of hockey equipment such as skates, sticks, helmets, protective equipment products and related apparel. Reebok-CCM Hockey brings together five global hockey brand names (Reebok, CCM, Rbk, JOFA and KOHO), and it holds the exclusive license to produce and sell replica jerseys for the National Hockey League (NHL). The firm also serves as the sole supplier of on-ice game jerseys to NHL teams. In addition to its NHL license, the firm maintains exclusive agreements with the American Hockey League, the Canadian Hockey League and the ECHL (formerly the East Coast Hockey League). The firm's brands are endorsed by professional hockey players such as Alexander Ovechkin, Sidney Crosby, Robert Luongo, Jean-Sebastien Giguere, Marty Turco, Matt Duchene, Jonathan Bemier and Corey Crawford. Reebok-CCM Hockey distributes its products to countries around the world through diverse retail channels, including specialty shops and sporting goods retailers. The company has recently extended its exclusive partnership with the American Hockey League for two more years (through the 2015-16 season).

FINANCIAL DATA: *Note: Data for latest year may not have been available at press time.*

In U.S. $	2012	2011	2010	2009	2008	2007
Revenue						
R&D Expense						
Operating Income						
Operating Margin %						
SGA Expense						
Net Income						
Operating Cash Flow						
Capital Expenditure						
EBITDA						
Return on Assets %						
Return on Equity %						
Debt to Equity						

CONTACT INFORMATION:

Phone: 514-461-8000 Fax:
Toll-Free:
Address: 3400 Raymond Lasnier, Montreal, QC H4R 3L3 Canada

STOCK TICKER/OTHER:

Stock Ticker: Subsidiary
Exchange:
Employees:
Fiscal Year Ends: 31-Dec
Parent Company: ADIDAS AG

SALARIES/BONUSES:

Top Exec. Salary: $ Bonus: $
Second Exec. Salary: $ Bonus: $

OTHER THOUGHTS:

Estimated Female Officers or Directors:
Hot Spot for Advancement for Women/Minorities:

Reebok International Ltd

NAIC Code: 316210

www.reebok.com

TYPES OF BUSINESS:

Athletic Shoes
Apparel
Accessories
Athletic Equipment
Online Sales
Retail Stores

BRANDS/DIVISIONS/AFFILIATES:

Adidas AG
Reebok-CCM Hockey
Sports Licensed Division
Reebok Step
CCM
Rbk
JOFA
KOHO

CONTACTS: *Note: Officers with more than one job title may be intentionally listed here more than once.*

Uli Becker, Pres.
Matt O'Toole, Chief Mktg. Officer
Bill Holmes, Head-Human Resources
David Mischler, Head-Brand Oper.
David Baxter, Pres., Sports Licensed Div.
John Warren, Gen. Mgr.-Sports Licensed Div.
Charlie Maurath, Head-Latin America

GROWTH PLANS/SPECIAL FEATURES:

Reebok International, Ltd., a subsidiary of Adidas AG, operates under the brand names Reebok, Reebok-CCM Hockey and the Sports Licensed Division. Under the Reebok label, the firm offers a varied mix of athletic apparel, footwear and special collections for men, women and children, accessories such as socks and caps as well as exercise equipment such as the Reebok Step. The Reebok-CCM Hockey label specializes in the design and production of professional hockey athletic gear, including skates, apparel, pads and equipment. It brings together a number of individual brands, including CCM, Rbk, JOFA and KOHO. The Sports Licensed Division manufactures performance apparel, uniforms and accessories for professional athletic teams in the National Football League (NFL), Major League Baseball (MLB), National Hockey League (NHL) and the National Lacrosse League (NLL). Additionally, Reebok markets and distributes athletic equipment, including treadmills, heart rate monitors, recumbent bikes, upright bikes and elliptical machines. The company operates retail stores across the U.S. as well as an online store. The firm's online store features a shopping guide, which offers sizing and product information, as well as a comprehensive portfolio of products available for direct consumer purchase. Reebok has regional offices in Amsterdam, Montreal, Hong Kong and Panama.

Employees are offered medical, dental and prescription drug coverage; a 401(k) plan; development and training courses; company cafes; sporting event tickets; and fitness facilities.

FINANCIAL DATA: *Note: Data for latest year may not have been available at press time.*

In U.S. $	2012	2011	2010	2009	2008	2007
Revenue		2,821,175,000	2,661,486,000	2,230,660,000		
R&D Expense						
Operating Income						
Operating Margin %						
SGA Expense						
Net Income						
Operating Cash Flow						
Capital Expenditure						
EBITDA						
Return on Assets %						
Return on Equity %						
Debt to Equity						

CONTACT INFORMATION:

Phone: 781-401-5000 Fax: 781-401-7402
Toll-Free:
Address: 1895 J.W. Foster Blvd., Canton, MA 02021 United States

STOCK TICKER/OTHER:

Stock Ticker: Subsidiary
Employees:
Parent Company: ADIDAS AG
Exchange:
Fiscal Year Ends: 31-Dec

SALARIES/BONUSES:

Top Exec. Salary: $ Bonus: $
Second Exec. Salary: $ Bonus: $

OTHER THOUGHTS:

Estimated Female Officers or Directors:
Hot Spot for Advancement for Women/Minorities:

Reliance Industries Ltd (Reliance Group)

www.ril.com

NAIC Code: 325110

TYPES OF BUSINESS:

Petrochemicals Refining & Manufacturing
Textiles & Fabrics
Oil & Gas Exploration
Power Generation
Supermarkets
Mall Development

BRANDS/DIVISIONS/AFFILIATES:

Reliance Super
Vimal
Reliance Fresh
Reliance Mart
Reliance Trends
Reliance Digital
Reliance Wellness
Reliance Footprint

CONTACTS: *Note: Officers with more than one job title may be intentionally listed here more than once.*

Mukesh D. Ambani, Managing Dir.
Hital R. Meswani, Exec. Dir.
Nikhil R. Meswani, Exec. Dir.
P.K. Kapil, Exec. Dir.
Mukesh D. Ambani, Chmn.

GROWTH PLANS/SPECIAL FEATURES:

Reliance Industries, Ltd. (Reliance Group or RIL) is a conglomerate that operates in several business sectors, including energy, textiles, retail and petrochemicals. RIL produces and distributes plastics and intermediates, polyester filament yarn, polymer intermediates, chemicals, textiles, oil and natural gas. It has one of the largest marketing networks in India, with textile products sold under labels such as flagship brand Vimal. Its latest textile products include anti-microbial and anti-bacterial work-wear fabrics, heavy-duty textiles for home furnishing and automobiles, silk-amino suiting materials, fire-retardant and water-repellent tent fabrics and insect and mosquito repellent nets at World Health Organization standards. RIL's petroleum exploration and production segment provides crude oil and natural gas to refining, power, fertilizer, petrochemical and other industries, and the company also conducts its own refining and marketing under the Reliance Gas brand name. The firm's petrochemicals, chemicals, acrylics and polyesters division, through brands such as Repol and Relene, supplies a wide variety of products and materials to technology partners. RIL's retail segment includes the grocery store Reliance Fresh, hypermarkets Reliance Super and Reliance Mart, Reliance Trends apparel retailers, Reliance Digital electronics stores, Reliance Wellness pharmacies, Reliance Footprint shoe stores and iStore, an Apple products store.

FINANCIAL DATA: *Note: Data for latest year may not have been available at press time.*

In U.S. $	2012	2011	2010	2009	2008	2007
Revenue	65,963,970,000	48,908,990,000	37,487,990,000	27,825,130,000	25,234,910,000	20,933,630,000
R&D Expense						
Operating Income	4,126,923,000	4,839,234,000	5,426,796,000	3,014,190,000	4,433,896,000	2,685,530,000
Operating Margin %	6.25%	9.89%	14.47%	10.83%	17.57%	12.82%
SGA Expense	1,701,810,000	1,894,314,000	1,451,097,000	1,294,257,000	1,377,727,000	1,077,364,000
Net Income	3,627,916,000	3,545,948,000	4,493,924,000	2,750,848,000	3,592,265,000	2,221,745,000
Operating Cash Flow	4,504,858,000	6,134,179,000	3,770,934,000	2,996,881,000	2,986,881,000	3,081,804,000
Capital Expenditure	3,014,094,000	6,231,074,000	4,283,157,000	5,125,629,000	4,921,130,000	5,148,362,000
EBITDA	7,935,526,000	7,934,074,000	8,002,902,000	4,433,300,000	5,682,316,000	3,954,475,000
Return on Assets %	6.21%	6.80%	9.69%	7.11%	12.69%	10.52%
Return on Equity %	12.13%	13.01%	18.63%	14.18%	24.21%	19.61%
Debt to Equity	0.38	0.54	0.45	0.62	0.56	0.46

CONTACT INFORMATION:

Phone: 91 2222785000 Fax:
Toll-Free:
Address: Makers Chambers IV, 222, Fl. 3, Nariman Point, Mumbai, Maharashtra 400 021 India

STOCK TICKER/OTHER:

Stock Ticker: RLNIY
Employees: 23,365
Parent Company:
Exchange: GREY
Fiscal Year Ends: 31-Mar

SALARIES/BONUSES:

Top Exec. Salary: $ Bonus: $
Second Exec. Salary: $ Bonus: $

OTHER THOUGHTS:

Estimated Female Officers or Directors:
Hot Spot for Advancement for Women/Minorities:

RG Barry Corp

NAIC Code: 316210

www.rgbarry.com

TYPES OF BUSINESS:

Slippers & Comfort Footwear
Handbags & Accessories

BRANDS/DIVISIONS/AFFILIATES:

Terrasoles
Dearfoams
Utopia by Dearfoams
DF by Dearfoams
Foot Petals
baggallini
Glamour Toez
KIVA

CONTACTS: *Note: Officers with more than one job title may be intentionally listed here more than once.*

Greg A. Tunney, CEO
Greg A. Tunney, Pres.
Jose G. Ibarra, CFO
Yvonne E. Kalucis, Sr. VP-Human Resources
Glenn Evans, Sr. VP-Global Oper.
Jose G. Ibarra, Sr. VP-Finance
Mark Zobel, Pres., Foot Petals
Dennis Eckols, Pres., baggallini
Nancy N. Coons, Pres., Dearfoams
Lee F. Smith, Sr. VP-Creative Svcs.
Gordon Zacks, Chmn.

GROWTH PLANS/SPECIAL FEATURES:

R.G. Barry Corp. is a designer, purchaser, marketer and distributer of retail accessory products. Its product portfolio encompasses footwear, such as sandals and slippers; foot and shoe care products, such as cushioned insoles; and handbags, tote bags and travel accoutrements. The firm operates in two segments: footwear and accessories. Through the firm's footwear segment the company has sold over 1 billion pairs of shoes since its inception in 1947. The firm's footwear brands include Dearfoams, DF by Dearfoams, Utopia by Dearfoams, Angel Treads and Terrasoles. The accessories segment includes the firm's subsidiaries baggallini and Foot Petals. Baggallini offers handbags, tote bags and travel products under the Le Bagg and baggallini trademarks. Foot Petals offers around 30 foot and shoe care products under the Fab Feet, Glamour Toez, Heavenly Heelz, Killer Kushionz and Foot Petals brand names. Product lines are sold through department stores, national chain stores, discount stores and warehouse clubs as well as drug stores, supermarkets, specialty catalogs and Internet retailers. The company also supplies many retailer customers with footwear under their own private labels. Most of its products are made by third-party manufacturers in Asia. The company's largest customer, Wal-Mart, accounted for approximately 29% of its total 2012 revenues. In January 2013, the firm acquired the outdoor travel and accessory brand KIVA from KIVA Designs, Inc.

Employees are offered medical, dental, vision and life insurance; educational assistance; flex-time and telecommuting; product discounts; and a flexible spending account.

FINANCIAL DATA: *Note: Data for latest year may not have been available at press time.*

In U.S. $	2012	2011	2010	2009	2008	2007
Revenue	155,938,000	129,568,000	123,787,000	113,817,000	109,499,000	105,332,000
R&D Expense						
Operating Income	23,373,000	11,680,000	14,736,000	10,496,000	14,215,000	12,103,000
Operating Margin %	14.98%	9.01%	11.90%	9.22%	12.98%	11.49%
SGA Expense	43,795,000	36,483,000	36,623,000	32,971,000	32,126,000	30,367,000
Net Income	14,549,000	7,510,000	9,400,000	6,992,000	9,785,000	25,145,000
Operating Cash Flow	25,317,000	2,538,000	6,813,000	17,255,000	8,686,000	16,065,000
Capital Expenditure	1,834,000	676,000	1,181,000	1,365,000	1,571,000	633,000
EBITDA	27,200,000	14,099,000	15,944,000	12,065,000	15,614,000	13,282,000
Return on Assets %	11.91%	7.53%	11.86%	9.77%	14.88%	39.58%
Return on Equity %	21.25%	12.82%	18.70%	15.21%	23.79%	69.43%
Debt to Equity	0.27	0.38				

CONTACT INFORMATION:

Phone: 614 864-6400 Fax: 614 866-9787
Toll-Free: 800-848-7560
Address: 13405 Yarmouth Rd. NW, Pickerington, OH 43147 United States

STOCK TICKER/OTHER:

Stock Ticker: DFZ
Employees: 163
Parent Company:
Exchange: NAS
Fiscal Year Ends: 30-Jun

SALARIES/BONUSES:

Top Exec. Salary: $514,154 Bonus: $
Second Exec. Salary: $271,615 Bonus: $

OTHER THOUGHTS:

Estimated Female Officers or Directors: 3
Hot Spot for Advancement for Women/Minorities: Y

Rocky Brands Inc

www.rockyboots.com

NAIC Code: 316210

TYPES OF BUSINESS:

Footwear
Outdoor & Work Apparel, Footwear & Accessories
Duty & Western Apparel, Footwear & Accessories
Retail Stores

BRANDS/DIVISIONS/AFFILIATES:

Rocky
Durango
Georgia Boot
Lehigh
Rocky Outdoor Gear

CONTACTS: *Note: Officers with more than one job title may be intentionally listed here more than once.*

David N. Sharp, CEO
David Sharp, Pres.
James E. McDonald, CFO/Exec. VP
Richard Simms, Pres., Retail Sales
James E. McDonald, Treas.
Jason S. Brooks, Pres., U.S. Wholesale Sales
Mike Brooks, Chmn.
Gary Adam, Pres., Int'l Sales

GROWTH PLANS/SPECIAL FEATURES:

Rocky Brands, Inc. is a designer, manufacturer and marketer of footwear. The firm's portfolio of brand names includes Rocky Outdoor Gear, Georgia Boot, Durango and Lehigh. The company also licenses the Michelin and Mossy Oak brands. Its products are organized around five target markets: outdoor, work, duty, commercial military and western. The outdoor product lines consist of footwear, apparel and accessory items, which are typically waterproof and insulated. The work product lines consist of footwear and apparel marketed to industrial and construction workers as well as workers in the hospitality industry. The duty product line consists of footwear products marketed to law enforcement, security personnel and postal employees who are required to spend a majority of time at work on their feet. The commercial military product line consists of durable, lightweight products marketed as alternatives to government issued military boots. The western product line consists of authentic footwear products marketed to farmers and ranchers who generally live in rural communities in North America. The firm's products are distributed through three distinct business segments: wholesale, retail and military. The wholesale segment distributes products through a wide range of distribution channels representing over 10,000 retail store locations in the U.S. and Canada. Wholesale channels vary by product line and include sporting goods stores, outdoor retailers, independent shoe retailers, hardware stores, catalogs, mass merchants, uniform stores, farm store chains and other specialty retailers. The retail business includes direct sales of products to consumers through the Lehigh Safety Shoes mobile and retail stores (including a fleet of trucks supported by small warehouses that include retail stores); the Rocky outlet store; and the company's web site. The military division sells footwear under the Rocky label to the U.S. military.

The firm offers employees benefits including medical, dental, prescription, vision, AD&D, disability and life insurance; an employee assistance program; product discounts; flexible spending accounts; and 401(k).

FINANCIAL DATA: *Note: Data for latest year may not have been available at press time.*

In U.S. $	2012	2011	2010	2009	2008	2007
Revenue	228,317,700	239,599,100	252,792,300	229,485,600	259,538,100	275,266,800
R&D Expense						
Operating Income	13,606,830	12,797,060	17,069,460	8,773,979	-9,884,646	-13,289,760
Operating Margin %	5.95%	5.34%	6.75%	3.82%	-3.80%	-4.82%
SGA Expense	66,679,760	75,133,700	72,303,260	75,072,210	87,496,050	96,409,460
Net Income	8,854,765	8,306,895	7,683,732	1,174,807	1,167,139	-23,104,530
Operating Cash Flow	17,993,550	6,529,558	14,055,200	35,917,000	18,333,840	16,458,630
Capital Expenditure	6,201,003	7,605,944	4,769,146	4,998,274	4,849,860	5,910,402
EBITDA	19,635,390	18,672,980	23,360,440	15,689,780	16,288,840	-7,138,264
Return on Assets %	5.07%	4.84%	4.62%	.65%	.56%	-9.97%
Return on Equity %	7.30%	7.49%	8.19%	1.43%	1.43%	-24.86%
Debt to Equity	0.18	0.3	0.32	0.66	1.07	1.26

CONTACT INFORMATION:

Phone: 614 753-1951 Fax:
Toll-Free: 877-795-2410
Address: 39 E. Canal St., Nelsonville, OH 45764 United States

STOCK TICKER/OTHER:

Stock Ticker: RCKY
Employees: 2,450
Parent Company:
Exchange: NAS
Fiscal Year Ends: 31-Dec

SALARIES/BONUSES:

Top Exec. Salary: $475,000 Bonus: $
Second Exec. Salary: $320,000 Bonus: $

OTHER THOUGHTS:

Estimated Female Officers or Directors:
Hot Spot for Advancement for Women/Minorities:

Ross Stores Inc

NAIC Code: 448140

www.rossstores.com

TYPES OF BUSINESS:

Discount Apparel Stores
Home Furnishings

BRANDS/DIVISIONS/AFFILIATES:

dd's DISCOUNTS
Ross Dress for Less

CONTACTS: *Note: Officers with more than one job title may be intentionally listed here more than once.*

Michael A. Balmuth, CEO
Michael B. O'Sullivan, COO
Michael B. O'Sullivan, Co-Pres.
John G. Call, CFO/Sr. VP
Ken Caruana, Exec. VP-Mktg.
Ken Caruana, Exec. VP-Human Resources
Michael K. Kobayashi, CIO/Exec. VP
Barbara Rentler, Chief Merch. Officer/Pres.
Mark LeHocky, General Counsel/Corp. Sec./Sr. VP
James S. Fassio, Chief Dev. Officer/Co-Pres.
Douglas Baker, Gen. Merch. Officer/Sr. VP
Ken Caruana, Exec. VP-Strategy
Gary L. Cribb, Exec. VP-Stores & Loss Prevention
Vanessa Baqu-Stanton, Exec. VP-Planning
Norman A. Ferber, Chmn.
Michael K. Kobayashi, Exec. VP-Supply Chain, Allocation, CIO

GROWTH PLANS/SPECIAL FEATURES:

Ross Stores, Inc. operates 1,125 off-price retail apparel and home accessories stores in 29 states, Washington, D.C. and Guam, 1,037 of which operate under the Ross Dress for Less brand. The company also operates 88 dd's DISCOUNTS locations in California, Florida, Texas and Arizona. Most of the stores are located in community and neighborhood strip shopping centers in heavily populated urban and suburban areas. The company's chains target value-conscious women and men ages 18-54. Ross offers new, in-season, name-brand and designer apparel, accessories, footwear and home merchandise at savings of 20%-60% off department and specialty store regular prices, while dd's DISCOUNTS, targeting lower-income customers, offers similar merchandise, but at savings of up to 70% off department and specialty store prices. The company's stores are supplied by four distribution processing facilities. Ross has combined a network of approximately 7,800 vendors and manufacturers, purchasing the vast majority of its merchandise directly from the manufacturer. By purchasing later in the merchandise buying cycle than department and specialty stores, Ross takes advantage of imbalances between retailers' demand for products and manufacturers' supply of those products. In addition, the company typically does not require that manufacturers provide promotional and markdown allowances, return privileges, split shipments, drop shipments to stores or delayed deliveries of merchandise, further enabling Ross to provide significant discounts on in-season merchandise. Sales of women's products account for approximately 29% of the firm's revenues; home accents/bed and bath, 25%; men's products, 13%; accessories, lingerie, jewelry and fragrances,13%; shoes, 12%; and children's products, 8%. In October of 2011 Ross opened 12 Ross Dress for Less locations in the greater Chicago area.

Ross offers its employees medical, dental, vision and life insurance; sick pay; health care spending accounts; holiday and personal days; a commuter reimbursement account; a 401(k) plan; and an employee stock purchasing plan.

FINANCIAL DATA: *Note: Data for latest year may not have been available at press time.*

In U.S. $	2012	2011	2010	2009	2008	2007
Revenue	8,608,291,000	7,866,100,000	7,184,213,000	6,486,139,000	5,975,212,000	
R&D Expense						
Operating Income	1,053,144,000	906,590,000	726,122,000	495,206,000	421,091,000	
Operating Margin %	12.23%	11.52%	10.10%	7.63%	7.04%	
SGA Expense	1,304,065,000	1,229,775,000	1,130,813,000	1,034,357,000	935,901,000	
Net Income	657,170,000	554,797,000	442,757,000	305,441,000	261,051,000	
Operating Cash Flow	820,105,000	673,066,000	888,384,000	583,439,000	353,559,000	
Capital Expenditure	416,271,000	198,651,000	158,487,000	224,418,000	236,121,000	
EBITDA	1,224,036,000	1,068,414,000	886,972,000	645,465,000	555,619,000	
Return on Assets %	20.48%	18.85%	17.28%	12.92%	11.03%	
Return on Equity %	46.51%	44.56%	41.11%	31.05%	27.76%	
Debt to Equity	0.1	0.11	0.12	0.15	0.15	

CONTACT INFORMATION:

Phone: 925 965-4400 Fax:
Toll-Free:
Address: 4440 Rosewood Dr., Pleasanton, CA 94588 United States

STOCK TICKER/OTHER:

Stock Ticker: ROST Exchange: NAS
Employees: 49,500 Fiscal Year Ends: 31-Jan
Parent Company:

SALARIES/BONUSES:

Top Exec. Salary: $1,238,024 Bonus: $
Second Exec. Salary: $1,017,163 Bonus: $

OTHER THOUGHTS:

Estimated Female Officers or Directors: 4
Hot Spot for Advancement for Women/Minorities: Y

Royal Ten Cate NV

www.tencate.com

NAIC Code: 313110

TYPES OF BUSINESS:

Technical Fibers
Synthetic Fabrics
Aerospace Composites
Bullet-Resistant Materials
Printer Components
Tentcloth & Awning Fabric
Artificial Grass Fibers
Fluid Systems

BRANDS/DIVISIONS/AFFILIATES:

TenCate Protective Fabrics
TenCate Outdoor Fabrics
TenCate Grass
TenCate Advanced Armor
TenCate Geosynthetics
Amber Composites Ltd
TenCate Advanced Composites
Performance Materials Corporation Baycomp

CONTACTS: *Note: Officers with more than one job title may be intentionally listed here more than once.*

Loek de Vries, CEO
Loek de Vries, Pres.
Bert Cornelese, CFO
J.C.M. Hovers, Chmn.-Supervisory Board

GROWTH PLANS/SPECIAL FEATURES:

Royal Ten Cate NV is a leading producer of textiles and synthetic materials used in a variety of industrial applications. Ten Cate has three main business sectors: advanced textiles and composites; geosynthetics & grass; and other activities. The advanced textiles and composites division operates via TenCate Protective Fabrics, TenCate Outdoor Fabrics and TenCate Advanced Armor. The division's products include chemically-treated textiles used in fire-fighting clothing, military uniforms, aircraft materials, tentcloths and awnings. It also produces components such as fiber reinforced thermoplastic and thermoset materials, which are primarily used in aircraft parts. Its armor technologies are used in land-, sea- and air-based vehicles; individual personnel body armor; ballistic fabrics; and civilian vehicles. The geosynthetics & grass sector operates through TenCate Geosynthetics and TenCate Grass. The division produces geosynthetic materials, such as erosion protection fabrics, dike reinforcement materials and other construction, agriculture and civil engineering materials. Its synthetic grass offerings are used in a variety of sports, including soccer, football, field hockey, tennis and golf as well as general landscaping. Lastly, the other operations division consists of its technical components business operating through TenCate Enbi, which manufactures foam and rubber rollers for paper transport and image transfer in printers, fax machines and copiers. Another company within this division, Xennia Technology, Ltd., specializes in inkjet printers and related technologies. The firm is a commercially decentralized organization, with its head office in the Netherlands and operating companies in many European countries, North America, Asia and Australia. In 2012, subsidiary TenCate Advanced Composites acquired Performance Materials Corporation Baycomp. In 2013, the company acquired thermoset composites company Amber Composites Ltd.

FINANCIAL DATA: *Note: Data for latest year may not have been available at press time.*

In U.S. $	2012	2011	2010	2009	2008	2007
Revenue		1,522,867,000	1,316,528,000	1,126,103,000	1,380,851,000	1,184,809,000
R&D Expense						
Operating Income		119,818,100	99,759,300	37,175,720	112,062,000	92,805,560
Operating Margin %		7.86%	7.57%	3.30%	8.11%	7.83%
SGA Expense			98,823,220	231,345,300		
Net Income		77,962,020	60,711,420	30,890,610	68,333,780	62,182,400
Operating Cash Flow		65,926,720	38,914,150	193,634,700	65,124,360	37,175,720
Capital Expenditure		34,367,480	28,483,550	23,134,530	64,188,280	84,113,400
EBITDA		184,140,100	160,604,400	95,881,250	168,628,000	136,533,800
Return on Assets %		6.19%	5.61%	2.91%	6.34%	7.66%
Return on Equity %		12.96%	11.21%	6.35%	15.09%	16.90%
Debt to Equity		0.58	0.44	0.49		0.71

CONTACT INFORMATION:

Phone: 31 546544911 Fax: 31 546814145
Toll-Free:
Address: Stationsstraat 11, 13th Fl., Almelo, 7607 GX Netherlands

STOCK TICKER/OTHER:

Stock Ticker: KNKCF Exchange: GREY
Employees: 4,271 Fiscal Year Ends: 31-Dec
Parent Company:

SALARIES/BONUSES:

Top Exec. Salary: $ Bonus: $
Second Exec. Salary: $ Bonus: $

OTHER THOUGHTS:

Estimated Female Officers or Directors:
Hot Spot for Advancement for Women/Minorities:

Russell Brands LLC

www.russell-brands.com

NAIC Code: 315210

TYPES OF BUSINESS:

Athletic Apparel & Equipment
Athletic Uniforms

BRANDS/DIVISIONS/AFFILIATES:

Spalding
Russell Athletic
Berkshire Hathaway Inc
JERZEES
Russell Outdoors
Bike Athletic
AAI
Dudley Sports

CONTACTS:

Note: Officers with more than one job title may be intentionally listed here more than once.

John F. Ward, CEO
Robert D. Koney, Jr., CFO/Sr. VP
Victoria W. Beck, Principal Acct. Officer/VP/Corp. Controller
Calvin Johnston, VP/Pres., Brooks Sports
John F. Ward, Chmn.

GROWTH PLANS/SPECIAL FEATURES:

Russell Brands, LLC, a Berkshire Hathaway company, is a leading branded athletic, active wear and outdoor gear company that markets athletic uniforms, apparel, footwear, sporting goods, athletic equipment and accessories for a wide variety of sports, outdoor and fitness activities. The firm's well-known brands include Russell Athletic, Russell Outdoors, JERZEES, Bike Athletic, Dudley Sports, AAI and Spalding. The company markets and manufactures a variety of apparel products, including fleece, t-shirts, casual shirts, jackets, athletic shorts, socks and camouflage attire for men, women, boys and girls. Russell is also a leading supplier of uniforms and related apparel to college, high school and other organized sports teams. The company is one of the largest providers of basketball equipment in the world, due in part to its eight-year global partnership with the National Basketball Association. Russell markets its products in the U.S., Canada and approximately 100 additional countries through mass merchandisers, sporting goods dealers, department and sports specialty stores, college stores, online retailers, mail-order houses, art wear distributors, screen printers and embroiderers.

FINANCIAL DATA:

Note: Data for latest year may not have been available at press time.

In U.S. $	2012	2011	2010	2009	2008	2007
Revenue						
R&D Expense						
Operating Income						
Operating Margin %						
SGA Expense						
Net Income						
Operating Cash Flow						
Capital Expenditure						
EBITDA						
Return on Assets %						
Return on Equity %						
Debt to Equity						

CONTACT INFORMATION:

Phone: 270-781-6400 Fax: 270-781-6588
Toll-Free:
Address: 1 Fruit of the Loom Dr., Bowling Green, KY 42103 United States

STOCK TICKER/OTHER:

Stock Ticker: Subsidiary Exchange:
Employees: 13,000 Fiscal Year Ends: 31-Dec
Parent Company: BERKSHIRE HATHAWAY INC

SALARIES/BONUSES:

Top Exec. Salary: $ Bonus: $
Second Exec. Salary: $ Bonus: $

OTHER THOUGHTS:

Estimated Female Officers or Directors: 1
Hot Spot for Advancement for Women/Minorities:

Ryohin Keikaku Co Ltd (Muji)

http://ryohin-keikaku.jp/

NAIC Code: 442110

TYPES OF BUSINESS:

Home Furnishings Stores
Cafes, Restaurants & Foods
Apparel & Household Goods
Campsites
Trucking
Online Sales
Trucking

BRANDS/DIVISIONS/AFFILIATES:

MUJI
IDEE Co Ltd
Cafe Muji/Meal Muji
MUJI Campsite
MUJI.net Co Ltd
RK Trucks Co Ltd

CONTACTS: *Note: Officers with more than one job title may be intentionally listed here more than once.*

Masaaki Kanai, Pres.
Takashi Kato, Managing Dir.
Junichi Tokue, Sr. Exec. Officer
Wai Lan Dora Hung, Exec. Officer
Hiroto Oki, Exec. Officer
Tadamitsu Matsui, Chmn.

GROWTH PLANS/SPECIAL FEATURES:

Ryohin Keikaku Co., Ltd., commonly known as MUJI, is a Japanese company operating in several divisions, including MUJI Products; Cafe MUJI/Meal MUJI; MUJI Campsite; MUJI.net Co., Ltd., a housing business; and IDEE Co., Ltd., a home furnishings business. MUJI offers over 7,000 items, including clothing, household products, foods and other items. The company's strategy is to offer quality goods at low prices through careful materials selection, streamlined production processes and simple packaging. MUJI products are primarily sold in Japan and internationally. In Japan, the company directly operates approximately 256 stores and also supplies its products to 116 other retail outlets. In the U.S. and internationally, MUJI products are sold in 163 stores. These stores are primarily concentrated in Europe and Southeast Asia. The Cafe MUJI/Meal MUJI business operates cafes and restaurants with the aim of offering simple foods at reasonable prices. The MUJI Campsite division operates three campgrounds in Japan. The company's housing business, MUJI.net Co., Ltd., is focused on developing adaptable residential housing which uses various movable and removable partitions to optimize structures for different family needs over time. This design concept is intended to provide maximum flexibility at affordable housing prices. The IDEE division plans, manufactures and sells home furnishings, interiors, antiques, curtains and rugs and other products. The company distributes its products through wholly-owned subsidiary R.K. Trucks Co., Ltd. MUJI's U.S. presence consists of four shops around Manhattan.

FINANCIAL DATA: *Note: Data for latest year may not have been available at press time.*

In U.S. $	2012	2011	2010	2009	2008	2007
Revenue	1,983,779,000	1,883,045,000	1,822,881,000	1,812,650,000		
R&D Expense						
Operating Income	171,875,200	154,752,200	157,357,400	191,748,000		
Operating Margin %	8.66%	8.21%	8.63%	10.57%		
SGA Expense	736,553,800	705,525,400	672,793,700	641,754,100		
Net Income	98,529,300	87,496,240	83,566,200	77,220,250		
Operating Cash Flow	108,315,400	79,658,420	128,544,600	126,039,600		
Capital Expenditure	32,297,570	26,920,210	35,359,210	51,313,160		
EBITDA	200,342,900	176,005,600	176,640,200	165,551,500		
Return on Assets %	8.86%	7.98%	7.84%	7.53%		
Return on Equity %	11.11%	10.28%	10.29%	9.88%		
Debt to Equity						

CONTACT INFORMATION:

Phone: 81 339894403 Fax: 81
Toll-Free:
Address: 4-26-3, Higashi-Ikebukuro, Toshima-ku, Tokyo, 170-8424 Japan

STOCK TICKER/OTHER:

Stock Ticker: RYKKF
Employees: 2,734
Parent Company:
Exchange: GREY
Fiscal Year Ends: 28-Feb

SALARIES/BONUSES:

Top Exec. Salary: $ Bonus: $
Second Exec. Salary: $ Bonus: $

OTHER THOUGHTS:

Estimated Female Officers or Directors: 1
Hot Spot for Advancement for Women/Minorities:

Saks Inc

NAIC Code: 452111

www.saksincorporated.com

TYPES OF BUSINESS:

Department Stores, Luxury
Outlet Stores
Online Sales

BRANDS/DIVISIONS/AFFILIATES:

Saks Fifth Avenue
OFF 5TH
Saks.com

CONTACTS: *Note: Officers with more than one job title may be intentionally listed here more than once.*

Stephen I. Sadove, CEO
Michael Rodgers, COO/Exec. VP
Ronald L. Frasch, Pres.
Kevin Wills, CFO/Exec. VP
Denise Incandela, Chief Mktg. Officer/Exec. VP
Christine A. Morena, Exec. VP-Human Resources
Michael Rodgers, CIO
Ronald L. Frasch, Chief Merch. Officer
Michael Brizel, General Counsel/Exec. VP
Michael Burgess, Pres., Saks Direct
Julia Bentley, Dir.-Investor Rel.
Jennifer de Winter, Exec. VP-Stores
Robert T. Wallstrom, Pres., OFF 5TH/Exec. VP
Stephen I. Sadove, Chmn.

GROWTH PLANS/SPECIAL FEATURES:

Saks, Inc. is a luxury fashion retailer operating 108 locations across the U.S., including Saks Fifth Avenue (SFA) and OFF 5TH. The company's 45 SFA stores are principally free-standing stores in exclusive shopping destinations or anchor stores in upscale regional malls. They typically offer a wide assortment of distinctive luxury fashion apparel, shoes, accessories, jewelry, cosmetics and gifts. SFA maintains key relationships with many of the leading American and European fashion houses, including Giorgio Armani, Chanel, Gucci, Prada, David Yurman, Louis Vuitton and Burberry, among others. OFF 5TH is intended to be the premier luxury off-price retailer in the U.S. The 65 OFF 5TH stores are primarily located in upscale mixed-use and off-price centers and offer luxury apparel, shoes, accessories, cosmetics and decorative home furnishings, targeting the value-conscious customer. Additionally, Saks maintains an e-commerce site, Saks.com. Through an agreement with HSBC, the company maintains 621,000 active proprietary credit card accounts, which represented approximately 36.3% of Saks' 2011 sales. During 2011, women's apparel accounted for roughly 34.8% of the firm's sales; accessories, 19.3%; men's apparel, 16.2%; women's shoes, 12.2%; cosmetics and fragrances, 12.2%; and other, 5.3%. In February 2012, Saks disclosed plans to add an additional distribution and fulfillment facility in LaVergne, Tennessee. In September 2012, it announced plans to close two SFA locations in the U.S. (Austin, Texas and Highland Park, Illinois) and opened a licensed store in Almaty, Kazakhstan.

The firm offers its employees medical, dental and vision insurance; a 401(k); life insurance; disability coverage; a 529 college savings plan; and associate discounts at company stores and online.

FINANCIAL DATA: *Note: Data for latest year may not have been available at press time.*

In U.S. $	2012	2011	2010	2009	2008	2007
Revenue	3,013,593,000	2,785,745,000	2,631,532,000	3,029,743,000	3,282,640,000	
R&D Expense						
Operating Income	148,344,000	90,098,000	-54,485,000	-135,388,000	98,020,000	
Operating Margin %	4.92%	3.23%	-2.07%	-4.46%	2.98%	
SGA Expense	868,417,000	814,452,000	778,139,000	874,206,000	959,164,000	
Net Income	74,790,000	47,846,000	-57,919,000	-154,941,000	47,473,000	
Operating Cash Flow	272,697,000	123,881,000	192,205,000	-1,497,000	71,479,000	
Capital Expenditure	82,097,000	55,721,000	74,577,000	127,305,000	140,985,000	
EBITDA	268,539,000	208,907,000	82,084,000	4,906,000	251,929,000	
Return on Assets %	3.50%	2.23%	-2.69%	-6.83%	1.93%	
Return on Equity %	6.31%	4.28%	-5.68%	-14.47%	4.17%	
Debt to Equity	0.3	0.3	0.46	0.65	0.21	

CONTACT INFORMATION:

Phone: 212 940-5305 Fax:
Toll-Free:
Address: 12 E. 49th St., New York, NY 10017 United States

STOCK TICKER/OTHER:

Stock Ticker: SKS Exchange: NYS
Employees: 15,000 Fiscal Year Ends: 31-Jan
Parent Company:

SALARIES/BONUSES:

Top Exec. Salary: $1,037,500 Bonus: $
Second Exec. Salary: $1,022,500 Bonus: $

OTHER THOUGHTS:

Estimated Female Officers or Directors: 7
Hot Spot for Advancement for Women/Minorities: Y

Salvatore Ferragamo SpA

www.ferragamo.com

NAIC Code: 316210

TYPES OF BUSINESS:

Shoes & Accessories Manufacturing
Retail Stores
Leather Goods
Leather Apparel
Watches
Ladies' Shoes
Men's Shoes
Fragrance

BRANDS/DIVISIONS/AFFILIATES:

Ferragamo Parfums SpA
Trinity Luxury Brands Holdings Ltd
Ferrinch (L) Ltd

CONTACTS: *Note: Officers with more than one job title may be intentionally listed here more than once.*

Michele Norsa, CEO
Ferruccio Ferragamo, Pres.
Ernesto Greco, CFO
Alessandro Corsi, Dir.-Investor Rel.
Giovanna Gentile Ferragamo, Vice Chmn.
Ferruccio Ferragamo, Chmn.

GROWTH PLANS/SPECIAL FEATURES:

Salvatore Ferragamo SpA is a leading international designer of luxury shoes, leather products, accessories, jewelry, fragrances and apparel for men and women. Founded in 1927, the firm is still controlled by the Ferragamo family and headquartered in Florence, Italy. The company's products are sold in over 338 Ferragamo directly operated stores, 268 third-party operated Ferragamo stores and department stores and boutiques in 90 countries. Ferragamo shoes, which are aimed at fashion-conscious consumers, have a reputation for high quality, conservative styling and comfort. The company has outfitted such celebrities as Judy Garland (supplying the ruby red slippers worn in The Wizard of Oz), Marilyn Monroe (metal-reinforced stilettos) and Madonna. Though the firm initially made its name producing high-end shoes, Ferragamo now designs and creates a vast product line, including men's and women's ready-to-wear clothing collections, footwear, handbags, watches, fragrances and eyewear. Nearly all of the company's merchandise is manufactured in Italy, as the use of fine Italian leather is of critical importance across its entire product line. During 2012, shoes, leather goods and handbags accounted for approximately 75.1 % of its revenue. Ferragamo owns and operates Ferragamo Parfums SpA and partners with Luxottica to produce eyewear sold under the Ferragamo name. The group has regional headquarters in New York and Miami, and also has store locations across Europe and in the U.S., Central and South America, Japan, Hong Kong, China, Taiwan, Korea, Singapore, Thailand, Malaysia, Indonesia and Australia. In 2012, the Asia Pacific region generated 36.5% of the group's revenue; Europe, 25.1%; North America, 22.3%; Japan, 11.6%; and Latin America, 4.5 %. In December 2012, Ferragamo purchased a 30% stake in Trinity Luxury Brands Holdings Ltd and Ferrich (L) Ltd., both wholly-owned subsidiaries of Trinity Ltd., for $25.5 million.

FINANCIAL DATA: *Note: Data for latest year may not have been available at press time.*

In U.S. $	2012	2011	2010	2009	2008	2007
Revenue		1,309,170,000	1,045,201,000	828,561,200		
R&D Expense		34,197,000	27,690,000	24,826,000		
Operating Income		209,479,800	115,588,400	48,777,750		
Operating Margin %		16.00%	11.05%	5.88%		
SGA Expense		630,966,800	499,177,600	412,320,100		
Net Income		138,083,700	65,361,060	-27,958,010		
Operating Cash Flow		155,069,500	162,594,300	30,885,260		
Capital Expenditure		56,599,360	29,096,020	27,975,390		
EBITDA		270,672,600	173,764,400	104,717,800		
Return on Assets %		12.48%	8.44%	-3.93%		
Return on Equity %		32.74%	22.53%	-10.81%		
Debt to Equity						

CONTACT INFORMATION:

Phone: 39 5533601 Fax: 39 553360444
Toll-Free: 866-337-724-266
Address: Palazzo Feroni, Via del Tornabuoni, 2, Florence, 50123 Italy

STOCK TICKER/OTHER:

Stock Ticker: SFRGF Exchange: GREY
Employees: 2,827 Fiscal Year Ends:
Parent Company:

SALARIES/BONUSES:

Top Exec. Salary: $ Bonus: $
Second Exec. Salary: $ Bonus: $

OTHER THOUGHTS:

Estimated Female Officers or Directors: 3
Hot Spot for Advancement for Women/Minorities: Y

Samsonite International SA

NAIC Code: 316998

www.samsonite.com

TYPES OF BUSINESS:

Luggage Manufacturing
Clothing & Accessories
Retail Stores
Online Sales

BRANDS/DIVISIONS/AFFILIATES:

Cosmolite
Samsonite Black Label
American Tourister
Samsonite OutLab
Samsonite
High Sierra Sport Company

CONTACTS: *Note: Officers with more than one job title may be intentionally listed here more than once.*

Timothy C. Parker, CEO
Jean-Remy Roussel, COO
Kyle F. Gendreau, CFO
William Yue, Dir.-Investor Rel.
Thomas Korbas, Pres., Americas
Timothy C. Parker, Chmn.
Ramesh Dungarmal Tainwala, Pres., Asia-Pacific & Middle East

GROWTH PLANS/SPECIAL FEATURES:

Samsonite International SA is one of the most recognized designers, manufacturers and distributors of luggage and travel-related consumer products in the world. It makes softside and hardside suitcases, garment bags, business cases and casual bags in the mid-tier and luxury markets. The firm's family of brands includes Samsonite, a luggage and travel accessories line; Samsonite Black Label, a top-level Italian luggage brand; Samsonite OutLab, a line of outdoor travel items such as daypacks, outdoor backpacks, laptop backpacks, duffels, and messenger bags; American Tourister, a moderately priced luggage line; and the Cosmolite luggage collection for the Middle East market. Through a licensing agreement, the firm markets a collection of outdoor items that are shock absorbent and weather resistant under the Timberland brand name. The company's products are available in more than 100 countries at numerous retail locations, including 200 company-owned stores, department and specialty stores, warehouse clubs, electronic stores, office superstores and bookstores. Recently, the firm sold its Lambertson Truex brand, which produces luxury handbags, shoes, gloves, belts and accessories constructed of fine leathers and exotic skins, to Tiffany & Co. In July 2012, Samsonite International acquired High Sierra Sport Company, a U.S.-based manufacturer and distributor of causal bags and luggage, for $4 billion.

FINANCIAL DATA: *Note: Data for latest year may not have been available at press time.*

In U.S. $	2012	2011	2010	2009	2008	2007
Revenue		1,565,147,000	1,215,307,000			
R&D Expense						
Operating Income		209,930,000	543,602,000			
Operating Margin %		13.41%	44.72%			
SGA Expense		647,324,000	519,170,000			
Net Income		103,618,000	366,814,000			
Operating Cash Flow		64,502,000	34,441,000			
Capital Expenditure		37,172,000	29,575,000			
EBITDA		249,668,000	565,993,000			
Return on Assets %		5.40%	21.32%			
Return on Equity %		10.15%	46.53%			
Debt to Equity			0.32			

CONTACT INFORMATION:

Phone: 508-851-1400 Fax: 508-851-8715
Toll-Free:
Address: 575 West St., Ste. 110, Mansfield, MA 02048 United States

STOCK TICKER/OTHER:

Stock Ticker: SMSEY Exchange: GREY
Employees: 6,640 Fiscal Year Ends: 31-Dec
Parent Company:

SALARIES/BONUSES:

Top Exec. Salary: $ Bonus: $
Second Exec. Salary: $ Bonus: $

OTHER THOUGHTS:

Estimated Female Officers or Directors:
Hot Spot for Advancement for Women/Minorities:

Samsung Group

www.samsung.com

NAIC Code: 334310

TYPES OF BUSINESS:

Electronics
Appliances
Photography
Chemical
Textiles
Apparel
Mobile
Television

BRANDS/DIVISIONS/AFFILIATES:

Samsung Electronics Co Ltd
Samsung Electro-Mechanics Co Ltd
Samsung Life Insurance Co Ltd
Samsung Semiconductor Inc
Samsung SDI Co Ltd
Samsung Fine Chemicals Co Ltd
Cheil Industries Inc
Samsung Petrochemical Co Ltd

CONTACTS: *Note: Officers with more than one job title may be intentionally listed here more than once.*

Oh-Hyun Kwon, CEO
Juhwa Yoon, Pres.
Juhwa Yoon, CFO
Gee Sung Choi, Head-Corp. Strategy Group
Woosung Han, Pres., Samsung Austin Semiconductor
Dale Sohn, Pres., Samsung Telecommunications America
Charlie Bae, CEO/Pres., Samsung Semiconductor, Inc.
Yangkyu Kim, CEO/Pres., Samsung Electronics North America

GROWTH PLANS/SPECIAL FEATURES:

Samsung Group is a conglomerate based in South Korea, designing, producing and selling a vast array of products across multiple industries worldwide. The firm is composed of several businesses, including Samsung Electronics. The company's main focus is consumer electronics such as semiconductors, cell phones, laptop computers and digital displays. Samsung's strategic research and development alliances with major players like Sony, IBM, Sanyo, Dell, Toshiba, HP and Disney allow it to remain on the forefront of technological innovation. It is the world's largest producer of flash memory, DRAM and SRAM as well as the largest seller of projection televisions in the U.S. Major subsidiaries include Samsung Corporation; Samsung C&T America, Inc.; Samsung America Venture Capital Company; Samsung Economic Research Institute; Samsung Electronics America, Inc.; Samsung Electronics Co., Ltd.; Samsung Heavy Industries Co., Ltd.; Samsung Life Insurance Co., Ltd.; Samsung Semiconductor, Inc.; and Samsung Telecommunications America, LLP. Through Samsung C&T America, Inc., the firm's textile and apparel brands include FUBU, Coogi, BB Dakota, Girbaud, True Religion footwear, Flow Society, Levi's and .Mountain Gear Samsung currently has production bases in Korea and China as well as production and sales bases spread across Indonesia, Malaysia, Thailand, Vietnam, India, the Philippines, Russia, Australia, Singapore and Japan. Moreover, the firm has facilities in multiple European countries, including the Ukraine, the Netherlands, the U.K., Germany, Portugal, France, Italy, Sweden, Poland, Spain and Russia as well as in North and South American countries such as the U.S., Canada, Mexico, Brazil, Panama and Colombia. In June 2012, the firm acquired SEHF Korea from its subsidiary SEHF (Samsung Electronics Hainan Fiber Optics); SEHF Korea was later (September 2012) merged with Samsung Electronics. In January 2013, Samsung Electronics America acquired the medical imaging company NeuroLogica.

FINANCIAL DATA: *Note: Data for latest year may not have been available at press time.*

In U.S. $	2012	2011	2010	2009	2008	2007
Revenue		111,279,349,760	103,389,036,544	82,686,599,168	67,194,429,440	58,189,164,544
R&D Expense						
Operating Income						
Operating Margin %						
SGA Expense						
Net Income		9,237,498,880	12,191,638,528	8,887,802,880	5,089,715,712	6,838,920,704
Operating Cash Flow						
Capital Expenditure						
EBITDA						
Return on Assets %						
Return on Equity %						
Debt to Equity						

CONTACT INFORMATION:

Phone: 82 222550114 Fax: 82 222550117
Toll-Free: 866-726-4249
Address: 1320-10 Seocho-2-dong, Seocho-gu, Seoul, 137 857 South Korea

STOCK TICKER/OTHER:

Stock Ticker: Private Exchange:
Employees: 344,000 Fiscal Year Ends: 31-Dec
Parent Company:

SALARIES/BONUSES:

Top Exec. Salary: $ Bonus: $
Second Exec. Salary: $ Bonus: $

OTHER THOUGHTS:

Estimated Female Officers or Directors:
Hot Spot for Advancement for Women/Minorities:

Saucony Inc

NAIC Code: 316210

www.saucony.com

TYPES OF BUSINESS:

Athletic Footwear
Outlet Stores
Athletic Apparel

BRANDS/DIVISIONS/AFFILIATES:

Saucony
Stride Rite Corp
Saucony Originals
Hind

CONTACTS:

Note: Officers with more than one job title may be intentionally listed here more than once.

John H Fisher, CEO
Michael Umana, COO
Richie Woodworth, Pres.
Chris Linder, Chief Mktg. Officer/Sr. VP
Samuel S. Ward, Sr. VP-Tech.
Sharon Barbano, VP-Public Rel. & Events
Brian Enge, Sr. VP-Apparel Div./Gen. Mgr.-Apparel Div.

GROWTH PLANS/SPECIAL FEATURES:

Saucony, Inc., a subsidiary of Stride Rite Corp., designs, develops and markets athletic footwear, athletic apparel and casual leather footwear. The company's principal products include running, walking and outdoor trail shoes as well as athletic apparel, all sold under the Saucony brand name. The firm also sells technical running shoe models from the early 1980s, which were reintroduced in 1998 as Saucony Originals, the firm's classic footwear line. Saucony sells athletic apparel primarily under the Hind brand name. The firm's Spot-bilt brand is the oldest division of Saucony, providing shoes and cleats for football and soccer coaches and officials as well as casual leather walking and workplace footwear. The company targets products to athletes who participate regularly in their sport of choice through the firm's Loyal to the Sport advertising campaign. It believes that consumers in this market are more brand-loyal than those who buy athletic footwear for casual use. Saucony's products are sold in the U.S. through retail locations and factory outlet stores. Outside the U.S., its products are sold in more than 50 countries through over 20 distributors and through subsidiaries located in Canada, the Netherlands and the U.K.

Saucony offers its employees group medical, life, disability, travel accident and dental insurance; flexible spending accounts; 401(k) and stock purchase option plans; wellness programs; and product discounts.

FINANCIAL DATA:

Note: Data for latest year may not have been available at press time.

In U.S. $	2012	2011	2010	2009	2008	2007
Revenue						
R&D Expense						
Operating Income						
Operating Margin %						
SGA Expense						
Net Income						
Operating Cash Flow						
Capital Expenditure						
EBITDA						
Return on Assets %						
Return on Equity %						
Debt to Equity						

CONTACT INFORMATION:

Phone: 617-824-6000 Fax: 617-824-6549
Toll-Free: 800-282-6575
Address: 191 Spring St., Lexington, MA 02420 United States

STOCK TICKER/OTHER:

Stock Ticker: Subsidiary Exchange:
Employees: Fiscal Year Ends: 31-Dec
Parent Company: STRIDE RITE CORP

SALARIES/BONUSES:

Top Exec. Salary: $ Bonus: $
Second Exec. Salary: $ Bonus: $

OTHER THOUGHTS:

Estimated Female Officers or Directors: 1
Hot Spot for Advancement for Women/Minorities:

Sculptz Inc

NAIC Code: 454113

www.silkies.com

TYPES OF BUSINESS:

Hosiery Products, Retail
Direct Marketing
Online Sales

BRANDS/DIVISIONS/AFFILIATES:

PainVanish
Silkies.com
EnchantressHosiery.com
Silkies.co.uk
Silkies
Silkies Hosiery
Silkies Enriche
HCI Direct Inc

CONTACTS: *Note: Officers with more than one job title may be intentionally listed here more than once.*

Jean P. Vernor, CEO
Jean P. Vernor, Pres.
Sue Dudek, Media Contact

GROWTH PLANS/SPECIAL FEATURES:

Sculptz, Inc., formerly HCI Direct, Inc., markets, manufactures and distributes women's hosiery and beauty products as Silkies Hosiery. It markets to consumers throughout the U.S. via mailings and its e-commerce site, Silkies.com; throughout Canada through EnchantressHosiery.com; and the U.K. through Silkies.co.uk. The company's main product line includes hosiery, knee-highs, socks and tights for women and girls. Other products include the Silkies Enriche brand of skincare products and the PainVanish brand pain relief cream. Silkies.com retails slimming shapewear such as waist cinchers and control underwear; bras; panties and briefs; therapeutic products, such as Sheer Renu firm-support therapeutic knee-highs, pantyhose and thigh-highs; slips and camisoles; socks; sleepwear; casualwear; gifts; skincare products; and accessories. The web site also features a sizing guide and offers gift certificate purchasing. The firm's Silkies brand-name hosiery (of which Sculptz has sold over 1.5 million) is marketed through a continuous product shipment program. The program involves mailing to customers a specially priced introductory hosiery offer, the acceptance of which enrolls customers in the program and results in additional shipments of hose being mailed on a regular basis upon payment of a prior shipment. Sculptz relies on sophisticated statistical, regression, segmentation and other financial analyses to accurately target, test and acquire customers through direct-mail solicitation.

FINANCIAL DATA: *Note: Data for latest year may not have been available at press time.*

In U.S. $	2012	2011	2010	2009	2008	2007
Revenue						
R&D Expense						
Operating Income						
Operating Margin %						
SGA Expense						
Net Income						
Operating Cash Flow						
Capital Expenditure						
EBITDA						
Return on Assets %						
Return on Equity %						
Debt to Equity						

CONTACT INFORMATION:

Phone: Fax:
Toll-Free: 877-745-5437
Address: 1150 Northbrook Dr., Ste. 300, Trevose, PA 19053 United States

STOCK TICKER/OTHER:

Stock Ticker: Private
Employees:
Parent Company:
Exchange:
Fiscal Year Ends: 31-Dec

SALARIES/BONUSES:

Top Exec. Salary: $ Bonus: $
Second Exec. Salary: $ Bonus: $

OTHER THOUGHTS:

Estimated Female Officers or Directors: 2
Hot Spot for Advancement for Women/Minorities:

Sears Canada Inc

NAIC Code: 452111

www.searstravel.ca

TYPES OF BUSINESS:

Department Stores
Travel Agencies
Parts and Service Network
Portrait Studios
Floor Coverings
Appliance Stores

BRANDS/DIVISIONS/AFFILIATES:

Sears Travel
Flowers by Sears
Sears Appliances and Mattresses
Corbeil Electrique Inc
S.L.H. Transport
Sears Home
Sears Home Services
Sears.ca

CONTACTS: *Note: Officers with more than one job title may be intentionally listed here more than once.*

Calvin McDonald, CEO
Calvin McDonald, Pres.
Sharon R. Driscoll, CFO/Sr. VP
Chris Brockbank, Sr. VP-Mktg.
Ismat Mirza, Sr. VP-Human Resources
Steven Goldsmith, Exec. VP-Merch. Apparel & Accessories
Thomas J. Fitzgerald, Chief Admin. Officer/Exec. VP
Klaudio Leshnjani, General Counsel/Sr. VP
Peter Kalen, Sr. VP-Finance & Home Svcs.
Dennis Singh, Sr. VP-Retail Stores
Doug Campbell, Sr. VP-Major Appliances
Lora Tisi, Sr. VP-Merch. Home & Hardlines
William C. Crowley, Chmn.
Tim Flemming, Sr. VP-Corp. Procurement & Supply Chain

GROWTH PLANS/SPECIAL FEATURES:

Sears Canada, Inc. is a multi-channel retail network in Canada. The company operates through two business segments. The Merchandising Operations segment includes the sales of goods and services through its Full-Line, Sears Home, Appliances and Mattresses retail units; its wholly-owned subsidiary Corbeil Electrique, Inc.; and its catalogue and Internet operations. The Real Estates Joint Venture Operations segment includes income from the company's joint venture interests in shopping centers across Canada. The firm operates 196 department stores, which offer a broad range of merchandise including apparel, home fashions and appliances; 285 locally-owned dealer stores, offering major appliances, home electronics, lawn and garden equipment and home furnishings; 48 Sears Home stores; 30 Sears Home Services showrooms; 20 Floor Covering Centers; 4 Appliances and Mattresses stores; 11 outlet stores; 105 Sears Travel offices; 24 Corbeil appliance stores; 1,734 catalogue merchandise pick-up locations; and a nationwide home maintenance, repair and installation network. In addition, the company publishes one of the largest general merchandise catalogue in Canada, issued to more than 3 million households, and offers shopping online at Sears.ca. Sears offers many specialty services, including a parts and service network; travel offices; hair care centers; optical services; health food shops; portrait studios; income tax services; carpet; upholstery and air duct cleaning; Flowers by Sears; and home installed products and services. Sears Travel offers online travel service at Searstravel.ca as well as through the toll-free number 1-866-Fly-Sears. Sears also owns S.L.H. Transport, a trucking fleet operation comprising more than 620 trucks. In March 2012, Sears was one of the first retailers in Canada to launch an online points redemption feature, which allows Sears credit card holders to earn points as they shop. These points can be redeemed for discounts. In April 2012, Sears announced the sale of its subsidiary Cantrex Group to Nationwide Markenting, Group LLC.

FINANCIAL DATA: *Note: Data for latest year may not have been available at press time.*

In U.S. $	2012	2011	2010	2009	2008	2007
Revenue	4,685,839,000	5,029,215,000	5,275,512,000	5,815,784,000	5,815,784,000	
R&D Expense						
Operating Income	-65,023,330	220,227,200	350,679,600	395,719,200	395,719,200	
Operating Margin %	-1.38%	4.37%	6.64%	6.80%	6.80%	
SGA Expense	1,776,324,000					
Net Income	-60,965,710	151,957,800	238,080,800	292,757,200	292,757,200	
Operating Cash Flow	86,224,380	99,817,410	503,246,100	168,289,700	168,289,700	
Capital Expenditure	85,514,300	63,298,840	66,646,380	97,991,480	97,991,480	
EBITDA	61,675,800	342,767,300	497,261,100	583,282,600	563,806,000	
Return on Assets %	-2.24%	5.06%	7.03%	9.20%	8.83%	
Return on Equity %	-4.88%	11.27%	14.82%	22.18%	19.13%	
Debt to Equity	0.08	0.13	0.02	0.22	0.22	

CONTACT INFORMATION:

Phone: 416 362-1711 Fax: 416 941-2321
Toll-Free: 866-359-7327
Address: 290 Yonge St., Ste. 700, Toronto, ON M5B 2C3 Canada

STOCK TICKER/OTHER:

Stock Ticker: SCC Exchange: TSE
Employees: 11,240 Fiscal Year Ends: 31-Jan
Parent Company:

SALARIES/BONUSES:

Top Exec. Salary: $ Bonus: $
Second Exec. Salary: $ Bonus: $

OTHER THOUGHTS:

Estimated Female Officers or Directors: 4
Hot Spot for Advancement for Women/Minorities: Y

Sears Holdings Corporation

www.sears.com

NAIC Code: 452111

TYPES OF BUSINESS:

Department Stores
Catalog Sales
Online Sales
Automotive Parts & Services
Appliances & Electronics
Health & Beauty Products

BRANDS/DIVISIONS/AFFILIATES:

Sears, Roebuck & Co
Kmart Holding Corporation
Kenmore
Craftsman
Sears Auto Centers
Lands' End
Sears Canada Inc
DieHard

CONTACTS: *Note: Officers with more than one job title may be intentionally listed here more than once.*

Edward S. Lampert, CEO
Robert A. Schriesheim, CFO/Exec. VP
Robert A. Riecker, Chief Acct. Officer/Controller/VP
Edward S. Lampert, Chmn.

GROWTH PLANS/SPECIAL FEATURES:

Sears Holdings Corporation is the parent firm of Kmart Holding Corporation and Sears, Roebuck & Co. As a broadline retailer, the company owns 2,019 full-line and 54 specialty retail stores in the U.S. operating through Sears and Kmart; and 475 full-line and specialty retail stores in Canada operating through Sears Canada, Inc., a 51%-owned subsidiary. The firm operates in three segments: Kmart, Sears domestic and Sears Canada. Kmart operates 1,221 stores in 49 states, Puerto Rico, Guam and the U.S. Virgin Islands, including both discount stores and super centers. Most Kmart stores are one-floor, free-standing units that carry a wide assortment of general merchandise and proprietary Sears brand products, including Kenmore, Craftsman and DieHard. Of these stores, 896 operate in-store pharmacies. The super centers generally operate 24-hours-a-day and combine a full-service grocery along with the general merchandise selection of a discount store. There are also 21 Sears Auto Centers operating in Kmart stores. Sears Auto Centers offer a variety of professional automotive repair and maintenance services as well as a full assortment of automotive accessories. Sears domestic operates 798 broadline stores in the U.S. and Puerto Rico that offer products such as home appliances; consumer electronics; tools; fitness, lawn and garden equipment; certain automotive services and products; home fashion products; apparel; footwear; and accessories, including the proprietary Lands' End brand merchandise. The segment also operates 10 Sears essentials/grand stores, offering health and beauty products, pantry goods, household products and toys. Sears' specialty stores include 36 free standing Auto Centers and 14 Lands' End retail stores. Sears Canada operates 118 full-line stores, 357 specialty stores, 1,512 pick-up locations and 101 travel offices. In 2012, eight Sears Essentials/Grad stores were converted to Kmart stores. In October 2012, the Sears Hometown and Outlet Stores officially separated from Sears Holdings and began trading separately on Nasdaq under the symbol SHOS.

Employees are offered medical and dental care; a flex spending account; and disability, survivor and retirement benefits.

FINANCIAL DATA: *Note: Data for latest year may not have been available at press time.*

In U.S. $	2012	2011	2010	2009	2008	2007
Revenue	41,567,000,000	43,326,000,000	44,043,000,000	46,770,000,000	50,703,000,000	
R&D Expense						
Operating Income	-1,501,000,000	474,000,000	713,000,000	302,000,000	1,586,000,000	
Operating Margin %	-3.61%	1.09%	1.61%	.64%	3.12%	
SGA Expense	10,664,000,000	10,571,000,000	10,654,000,000	11,060,000,000	11,468,000,000	
Net Income	-3,140,000,000	133,000,000	235,000,000	53,000,000	826,000,000	
Operating Cash Flow	-275,000,000	130,000,000	1,507,000,000	992,000,000	1,547,000,000	
Capital Expenditure	432,000,000	441,000,000	361,000,000	497,000,000	570,000,000	
EBITDA	-609,000,000	1,396,000,000	1,611,000,000	1,437,000,000	2,787,000,000	
Return on Assets %	-13.75%	.54%	.93%	.20%	2.87%	
Return on Equity %	-49.09%	1.51%	2.54%	.52%	7.06%	
Debt to Equity	0.48	0.31	0.18	0.22	0.24	

CONTACT INFORMATION:

Phone: 847 286-2500 Fax:
Toll-Free:
Address: 3333 Beverly Rd., Hoffman Estates, IL 60179 United States

STOCK TICKER/OTHER:

Stock Ticker: SHLD Exchange: NAS
Employees: 312,000 Fiscal Year Ends: 31-Jan
Parent Company:

SALARIES/BONUSES:

Top Exec. Salary: $815,385 Bonus: $850,000
Second Exec. Salary: $815,385 Bonus: $600,000

OTHER THOUGHTS:

Estimated Female Officers or Directors: 2
Hot Spot for Advancement for Women/Minorities:

Seven & I Holdings Co Ltd

NAIC Code: 445120

www.7andi.com

TYPES OF BUSINESS:

Retail Operations
Insurance-Health, Property & Specialty
IT/Information Services
Convenience Stores
Restaurants
Superstores & Supermarkets
Department Stores
Banking Services & Electronic Cash Cards

BRANDS/DIVISIONS/AFFILIATES:

Seven-Eleven Japan Co Ltd
7-Eleven Inc
Ito Yokado Co Ltd
York Mart Co
Sogo & Seibu Co Ltd
K.K. York Insurance
York-Benimaru Co Ltd
Seven & I Food Systems Co Ltd

CONTACTS: *Note: Officers with more than one job title may be intentionally listed here more than once.*

Toshifumi Suzuki, CEO
Noritoshi Murata, COO
Noritoshi Murata, Pres.
Kunio Takahashi, CFO
Katsuhiro Goto, Chief Admin. Officer
Masao Eguchi, Exec. Officer
Yoshihiro Tanaka, Exec. Officer
Akihiko Shimizu, Exec. Officer
Shinobu Matsumoto, Exec. Officer
Toshifumi Suzuki, Chmn.

GROWTH PLANS/SPECIAL FEATURES:

Seven & I Holdings Co., Ltd. is a holding company that performs the planning, management and operations of a large network of companies based in Japan. The firm operates in eight primary business areas: convenience stores; general merchandise stores; department stores; grocery stores; food-services; financial services, including insurance; IT/service businesses; and other. The company owns and operates approximately 41,900 stores in 16 countries. The company's convenience store subsidiaries include Seven-Eleven Japan Co., Ltd.; 7-Eleven, Inc.; Seven-Eleven (Hawaii), Inc.; Seven-Eleven (China) Co., Ltd.; Seven-Eleven Beijing Co., Ltd.; and Seven-Eleven (Changdu) Co., Ltd., a joint venture owned by Seven-Eleven Japan Co., Ltd., and Ito-Yokado Co., Ltd. The general merchandise segment consists of a number of retail chains, including Japanese fashion giant Ito-Yokado Co., Ltd. Seven & I's department stores segment, through subsidiary Sogo & Seibu Co., Ltd., operates 27 stores under brands Sogo, Seibu and Robinson. The grocery stores segment includes York-Benimaru Co., Ltd. and York Mart Co. The food services segment includes Seven & I Food Systems Co., Ltd., a holding company established to integrate the former Denny's Japan Co., Ltd.; Famil Co., Ltd.; and York Bussan K.K. The financial services segment includes subsidiaries Seven Financial Service Co., Ltd.; Seven Card Service Co., Ltd.; Seven CS Card Service Co., Ltd.; Seven Bank, Ltd.; and K.K. York Insurance. The IT/services segment's operations primarily consist of online marketplace and auction web sites under subsidiary Seven & i Netmedia Co., Ltd. The other segment of business encompasses real estate, security and manufacturing and processing businesses. In early 2012, 7-Eleven acquired 55 U.S. convenience stores from Sam's Mart LLC, and 23 stores from Strasburger Enterprises, Inc. In August 2012, 7-Eleven, Inc. acquired 74 U.S. convenience stores from Prima Marketing LLC, and 176 U.S. convenience stores and the wholesale fuel business from Tetco, Inc.

FINANCIAL DATA: *Note: Data for latest year may not have been available at press time.*

In U.S. $	2012	2011	2010	2009	2008	2007
Revenue	53,287,580,000	50,441,250,000	50,654,830,000	62,902,320,000	58,158,260,000	53,879,980,000
R&D Expense						
Operating Income	3,251,578,000	2,709,233,000	2,523,530,000	3,138,086,000	3,129,424,000	3,193,440,000
Operating Margin %	6.10%	5.37%	4.98%	4.98%	5.38%	5.92%
SGA Expense	15,761,440,000			17,573,660,000	17,896,230,000	16,506,530,000
Net Income	1,445,508,000	1,246,490,000	499,604,800	1,028,011,000	1,454,649,000	1,485,388,000
Operating Cash Flow	5,150,711,000	3,457,176,000	3,587,157,000	3,451,387,000	5,181,194,000	1,750,248,000
Capital Expenditure	2,333,574,000	3,424,399,000	1,863,128,000	1,777,046,000	2,162,078,000	2,665,178,000
EBITDA	4,349,751,000	4,203,649,000	3,780,819,000	4,074,303,000	4,261,264,000	4,307,757,000
Return on Assets %	3.40%	3.02%	1.21%	2.42%	3.39%	3.68%
Return on Equity %	7.48%	6.53%	2.55%	4.89%	6.71%	7.60%
Debt to Equity	0.25	0.25	0.25	0.25	0.24	0.29

CONTACT INFORMATION:

Phone: 81 362383000 Fax:
Toll-Free:
Address: 8-8, Nibancho, Chiyoda-ku, Tokyo, 102-8452 Japan

STOCK TICKER/OTHER:

Stock Ticker: SVNDF
Exchange: PINX
Employees: 51,888
Fiscal Year Ends: 28-Feb
Parent Company:

SALARIES/BONUSES:

Top Exec. Salary: $ Bonus: $
Second Exec. Salary: $ Bonus: $

OTHER THOUGHTS:

Estimated Female Officers or Directors: 2
Hot Spot for Advancement for Women/Minorities:

Shaw Industries

www.shawfloors.com

NAIC Code: 314110

TYPES OF BUSINESS:

Carpet and Rug Mills
Flooring

BRANDS/DIVISIONS/AFFILIATES:

Berkshire Hathaway Inc
Kathy Ireland
Tuftex
Inside/Out
ShawMark
Queen Commercial
Anso
Philadelphia Commercial

CONTACTS:

Note: Officers with more than one job title may be intentionally listed here more than once.

Vance D. Bell, CEO
Randy Merritt, Pres.
Kenneth G. Jackson, CFO/Exec. VP
Reggie Newton, VP-Mktg.
Paul Richard, VP-Human Resources
David Morgan, VP-Manufacturing
Steve Decarlo, VP-Bus. Dev.
Paul Murray, VP-Sustainability & Environmental Affairs

GROWTH PLANS/SPECIAL FEATURES:

Shaw Industries, a Berkshire Hathaway company, is one of the world's largest global manufacturers and distributors of carpeting, rugs, hardwood, laminate and ceramic tile for residential and commercial applications worldwide. Its customers include architects, designers, corporate and hospitality end-users, institutions, government agencies, contractors and management companies. Shaw designs and manufactures tufted and woven carpet, rugs, laminate and wood flooring in a broad range of prices, patterns, colors and textures for residential and commercial use. The firm also provides installation services and sells ceramic tile. Products are marketed under about 30 brand and trade names and under certain private labels. Brands include Kathy Ireland, Tuftex, Inside/Out, Anso, ShawMark, Philadelphia Commercial and Queen Commercial. Shaw sells its wholesale products to over 24,000 retailers, distributors and commercial users throughout the U.S., Canada and Mexico and, to a lesser degree, exports to additional overseas markets. Domestically, its products are marketed through its own residential and commercial sales personnel to various retailers and distributors. The firm maintains 12 carpet full-service distribution facilities, 3 hard surface and 4 rug full-service distribution facilities and 30 redistribution centers.

Shaw offers its employees tuition reimbursement; adoption assistance; employee discounts; relocation flexible spending accounts; in-depth training programs; and medical, dental, vision, prescription drug, disability and life insurance.

FINANCIAL DATA:

Note: Data for latest year may not have been available at press time.

In U.S. $	2012	2011	2010	2009	2008	2007
Revenue						
R&D Expense						
Operating Income						
Operating Margin %						
SGA Expense						
Net Income						
Operating Cash Flow						
Capital Expenditure						
EBITDA						
Return on Assets %						
Return on Equity %						
Debt to Equity						

CONTACT INFORMATION:

Phone: 706-278-3812 Fax:
Toll-Free: 800-441-7429
Address: 616 E. Walnut Ave., Dalton, GA 30722 United States

STOCK TICKER/OTHER:

Stock Ticker: Subsidiary Exchange:
Employees: Fiscal Year Ends: 31-Dec
Parent Company: BERKSHIRE HATHAWAY INC

SALARIES/BONUSES:

Top Exec. Salary: $ Bonus: $
Second Exec. Salary: $ Bonus: $

OTHER THOUGHTS:

Estimated Female Officers or Directors:
Hot Spot for Advancement for Women/Minorities:

Shoe Carnival Inc

NAIC Code: 448210

www.shoecarnival.com

TYPES OF BUSINESS:

Shoes, Retail
Accessories
Online Sales

BRANDS/DIVISIONS/AFFILIATES:

UNR8ED
Solanz
93 Octance
Donna Lawrence
Oak Meadow
Via Nova
Victoria Spenser
Y-NOT?

CONTACTS: *Note: Officers with more than one job title may be intentionally listed here more than once.*

Clifton E. Sifford, CEO
W. Kerry Jackson, COO/Exec. VP
Clifton E. Sifford, Pres.
W. Kerry Jackson, CFO/Treas.
Todd A. Beurman, Sr. VP-Mktg.
Sean M. Georges, Sr. VP-Human Resources/In-House Counsel
Terry L. Clements, CIO/Sr. VP
Clifton E. Sifford, Chief Merch. Officer
David M. Groff, VP-Admin.
David A. Kapp, Corp. Sec./Sr. VP-Planning & Allocation
Timothy T. Baker, Exec. VP-Store Oper. & Real Estate
David M. Groff, VP- Bus. Dev.
Jeffrey D. Hanson, VP-e-commerce
Kathy A. Yearwood, Chief Acct. Officer/Sr. VP/Controller
Mitchell A. Chandler, Sr. VP/Mgr.-Athletics & Children's Merch.
Steven D. Meyer, Sr. VP-Store Oper.
Michael C. Smith, VP-Real Estate
Lisa A. Rosenbaum, VP/Mgr.-Women's Non-Athletics & Accessories Merch.
J. Wayne Weaver, Chmn.
W. Dwight Burton, Jr., VP-Dist.

GROWTH PLANS/SPECIAL FEATURES:

Shoe Carnival, Inc. is one of the largest family footwear retailers in the U.S. The company operates 352 stores in 32 states and Puerto Rico. These stores are primarily located in strip malls and offer an assortment of both name brand and private label dress, casual and athletic footwear for men, women and children as well as boots, sandals and accessories. On average, each location carries approximately 28,500 pairs of shoes. Athletic merchandise is the highest earning product group, accounting for 37% of sales, with women's merchandise accounting for 26%. Besides footwear from nationally-recognized brands, the company sells merchandise under a variety of private labels, including Shoe Carnival, The Carnival, Carnival Lites, Donna Lawrence, Oak Meadow, Via Nova, Victoria Spenser, Innocence, Y-NOT?, UNR8ED, Solanz and 93 Octance. The company emphasizes the creation of an exciting, fun retail environment through the use of marketing tactics, such as utilizing bold colors and graphics as well as setting up a stage and microphone at each store front's entry and announcing various games and contests throughout the day, offering prizes to customers, such as on-the-spot discounts on purchases. In conjunction with its extensive in-store promotions, the firm uses television, radio and newspaper ads to announce special sale products. In 2011, Shoe Carnival, Inc. launched ShoeCarnival.com, allowing customers the option of shopping online for more than 1,500 styles of merchandise.

Shoe Carnival employees receive health benefits, paid time off, a 401(k), a discount stock purchase plan, flexible spending accounts and professional training programs.

FINANCIAL DATA: *Note: Data for latest year may not have been available at press time.*

In U.S. $	2012	2011	2010	2009	2008	2007
Revenue	762,534,000	739,189,000	682,422,000	647,572,000	658,680,000	
R&D Expense						
Operating Income	42,137,000	42,385,000	25,130,000	8,375,000	19,132,000	
Operating Margin %	5.52%	5.73%	3.68%	1.29%	2.90%	
SGA Expense	182,716,000	179,154,000	168,476,000	165,953,000	166,717,000	
Net Income	26,382,000	26,821,000	15,166,000	5,319,000	12,807,000	
Operating Cash Flow	30,875,000	29,423,000	27,897,000	32,078,000	19,879,000	
Capital Expenditure	21,260,000	14,412,000	9,794,000	18,204,000	18,434,000	
EBITDA	56,666,000	56,285,000	40,151,000	25,368,000	35,628,000	
Return on Assets %	7.21%	8.16%	5.01%	1.81%	4.24%	
Return on Equity %	9.80%	11.26%	7.11%	2.65%	6.30%	
Debt to Equity						

CONTACT INFORMATION:

Phone: 812 867-6471 Fax: 812 867-4055
Toll-Free: 800-430-7463
Address: 7500 E. Columbia St., Evansville, IN 47715 United States

STOCK TICKER/OTHER:

Stock Ticker: SCVL Exchange: NAS
Employees: 4,300 Fiscal Year Ends: 31-Jan
Parent Company:

SALARIES/BONUSES:

Top Exec. Salary: $775,000 Bonus: $
Second Exec. Salary: $460,000 Bonus: $

OTHER THOUGHTS:

Estimated Female Officers or Directors: 2
Hot Spot for Advancement for Women/Minorities: Y

Simply Fashion Stores Ltd

www.simplyfashions.com

NAIC Code: 448120

TYPES OF BUSINESS:

Apparel-Women's, Retail
Women's Shoes & Accessories, Retail
Super Plus Sized Women's Apparel, Retail

BRANDS/DIVISIONS/AFFILIATES:

Simply Fashions
Fashion Trend

CONTACTS:

Note: Officers with more than one job title may be intentionally listed here more than once.

Rodney Barstein, CEO
Rodney Barstein, Pres.
Mark Barstein, Mgr.-Merch./Exec. VP
Kathy Hymer, Divisional Mgr.-Merch., Juniors & Accessories

GROWTH PLANS/SPECIAL FEATURES:

Simply Fashion Stores, Ltd. owns and operates over 300 Simply Fashions and Fashion Trend stores in 21 U.S. states. The firm offers low-priced, ready-to-wear ladies' clothing, lingerie and accessories. Its apparel line is available in a variety of sizes ranging from juniors, misses and plus sizes up to size 36. The company's stores are located in strip centers, urban downtown store fronts, free-standing buildings and second generation urban malls in major cities including Detroit, Miami, New Orleans, St. Louis, Chicago, Atlanta, Baltimore and Dallas. Simply Fashion stores stock ladies' casual wear, accessories and lingerie in junior, misses, plus and super plus sizes, with most prices under $30. The firm's criteria for stores include a population of at least 5,000 African Americans within a one-mile radius and 25,000 within a three-mile radius as well as a space of approximately 3,000 square feet with a minimum of 30 feet of frontage. It is preferred that the retail location share space with grocery, discount, shoe, beauty supply, variety, check cashing, dollar store, nail salon and rent-to-own businesses.

Employees are offered a 401(k); discounts on merchandise; paid vacations, holidays and sick days; and health, dental and life insurance.

FINANCIAL DATA:

Note: Data for latest year may not have been available at press time.

In U.S. $	2012	2011	2010	2009	2008	2007
Revenue						
R&D Expense						
Operating Income						
Operating Margin %						
SGA Expense						
Net Income						
Operating Cash Flow						
Capital Expenditure						
EBITDA						
Return on Assets %						
Return on Equity %						
Debt to Equity						

CONTACT INFORMATION:

Phone: 205-951-1700 Fax: 205-951-1717
Toll-Free: 800-746-7596
Address: 2500 Crestwood Blvd., Birmingham, AL 35210 United States

STOCK TICKER/OTHER:

Stock Ticker: Private
Employees:
Parent Company:
Exchange:
Fiscal Year Ends: 31-Dec

SALARIES/BONUSES:

Top Exec. Salary: $ Bonus: $
Second Exec. Salary: $ Bonus: $

OTHER THOUGHTS:

Estimated Female Officers or Directors: 1
Hot Spot for Advancement for Women/Minorities:

Skechers USA Inc

NAIC Code: 316210

www.skx.com

TYPES OF BUSINESS:

Shoes, Manufacturing
Retail Stores
E-commerce Sites
Factory Outlet Stores
Concept Stores

BRANDS/DIVISIONS/AFFILIATES:

Skechers USA
Skechers Sport
Skechers Active
Skechers Goperform
Skechers Kids
Mark Nason
Zoo York
Unltd By Mark Ecko

CONTACTS: *Note: Officers with more than one job title may be intentionally listed here more than once.*

Robert Greenberg, CEO
David Weinberg, COO/Exec. VP
Michael Greenberg, Pres.
David Weinberg, CFO
Mark Nason, Exec. VP-Product Dev.
Philip G. Paccione, General Counsel/Corp. Sec.
Philip G. Paccione, Exec. VP-Bus. Affairs
Robert Greenberg, Chmn.

GROWTH PLANS/SPECIAL FEATURES:

Skechers U.S.A., Inc., is a manufacturer and marketer of footwear operating 281 domestic and 49 international Skechers retail concept stores, factory outlets and warehouse outlets. The company targets fashion-conscious consumers ages 12-24. Skechers' brand lines include Skechers USA, Skechers Sport, Skechers Active, BOBS from Sketchers, Skechers Work and Skechers Kids. The firm also offers the specialty fitness brands Shape-ups by Skechers, Shape-ups Liv by Skechers, Tone-ups by Skechers and Skechers GOperform. The company also offers a designer line for women, a high-end men's line, a men's urban-focused specialty line and a licensed urban-targeted men's, women's and children's line. Fashion lines include Mark Nason, Zoo York, Unltd. By Mark Ecko, Rhino Red and Kitson. Skechers sells its products to numerous department and specialty stores, such as Bloomingdale's, Nordstrom, Macy's, Dillard's, JCPenney, Footlocker, Famous Footwear, Journeys and Jarman. Skechers products are also sold by top wholesale retailers in more than 120 countries and territories throughout the world; and on its e-commerce site Skechers.com. The firm also operates MarkNason.com to promote that brand. In Europe, the company's direct distribution operations are handled through a warehouse in Liege, Belgium; and a Swiss subsidiary that oversees the company's international sales, marketing and distribution.

FINANCIAL DATA: *Note: Data for latest year may not have been available at press time.*

In U.S. $	2012	2011	2010	2009	2008	2007
Revenue	1,567,425,000	1,613,574,000	2,011,436,000	1,438,095,000	1,440,743,000	1,394,181,000
R&D Expense						
Operating Income	22,319,000	-133,793,000	196,740,000	72,582,000	57,892,000	112,930,000
Operating Margin %	1.42%	-8.29%	9.78%	5.04%	4.01%	8.10%
SGA Expense	668,111,000	765,099,000	719,734,000	550,083,000	540,491,000	491,238,000
Net Income	9,512,000	-67,484,000	136,148,000	54,699,000	55,396,000	75,686,000
Operating Cash Flow	-3,447,000	164,919,000	-47,379,000	115,109,000	-21,829,000	101,355,000
Capital Expenditure	52,452,000	122,248,000	82,310,000	39,841,000	72,461,000	31,175,000
EBITDA	66,245,000	-87,962,000	226,015,000	94,784,000	83,092,000	140,820,000
Return on Assets %	.72%	-5.21%	11.83%	5.84%	6.50%	9.67%
Return on Equity %	1.10%	-7.66%	16.46%	7.73%	8.55%	14.07%
Debt to Equity	0.14	0.08	0.05	0.02	0.02	0.02

CONTACT INFORMATION:

Phone: 310 318-3100 Fax:
Toll-Free:
Address: 228 Manhattan Beach Blvd., Manhattan Beach, CA 90266 United States

STOCK TICKER/OTHER:

Stock Ticker: SKX
Employees: 2,434
Parent Company:
Exchange: NYS
Fiscal Year Ends:

SALARIES/BONUSES:

Top Exec. Salary: $1,000,000 Bonus: $
Second Exec. Salary: $1,000,000 Bonus: $

OTHER THOUGHTS:

Estimated Female Officers or Directors:
Hot Spot for Advancement for Women/Minorities:

Spanx Inc

NAIC Code: 315240

www.spanx.com

TYPES OF BUSINESS:

Women's Lingerie
Body Shaping Products

BRANDS/DIVISIONS/AFFILIATES:

Haute Contour
ASSETS
SPANK for Men
Skinny Britches
On Top and In Control
Slim Cognito
MAMA Spanx
Hide & Sleek

CONTACTS: *Note: Officers with more than one job title may be intentionally listed here more than once.*

Laurie Ann Goldman, CEO
Sara Blakely, Pres.

GROWTH PLANS/SPECIAL FEATURES:

Spanx, Inc., a privately held company established by independent entrepreneur and inventor Sara Blakely in 1998 and headquartered in Atlanta, Georgia, designs and manufacturers a variety of over 200 body shaping apparel products for men and women. The company's products, which can be purchased in a variety of sizes and slimming levels online or in department stores, includes intimate apparel, bras, panties, legwear, swimwear and activewear. Spanx offers its products through many different trademarked collections: On Top and In Control, which manufactures silhouette slimming shirts; In-Power, made for tummy-taming, with mega compression zones and cling free fabric; Skinny Britches, which includes slimming capris, shorts and skorts; Simplicity, which includes bodysuits and slips; SPANX for Men, which includes figure-enhancing undershirts and underwear exclusively for men; Tight-End Tights, which includes patterned and micro fishnets tights; Swim; Slim Cognito, which includes seamless bodysuits; Hide & Sleek, which includes Hi-Rise shapers; Mama Spanx; for expecting mothers; and the two best selling lines Haute Contour and ASSETS. The Haute Contour line designs fashionable hosiery and undergarments for women, including thigh and tummy-taming shapewear, slips, panties and camisoles from lace, microfibers and mesh. The ASSETS product line manufactures an array of affordable figure-flattering, intimate apparel and shaping undergarments. The ASSET'S high-waist shaper with built in panty, also the company's best selling product, is a slimming hosiery product designed to give women an instant hourglass figure using a brief-style bottom that allows the piece to be worn in place of other undergarments. SPANX International ships to international retailers and wholesale distributors throughout 50 countries.

FINANCIAL DATA: *Note: Data for latest year may not have been available at press time.*

In U.S. $	2012	2011	2010	2009	2008	2007
Revenue	260,000,000	225,000,000				
R&D Expense						
Operating Income						
Operating Margin %						
SGA Expense						
Net Income						
Operating Cash Flow						
Capital Expenditure						
EBITDA						
Return on Assets %						
Return on Equity %						
Debt to Equity						

CONTACT INFORMATION:

Phone: Fax:
Toll-Free: 888-806-7311
Address: 3344 Peachtree Rd., Ste. 1700, Atlanta, GA 30326 United States

STOCK TICKER/OTHER:

Stock Ticker: Private
Employees: 125
Parent Company:
Exchange:
Fiscal Year Ends:

SALARIES/BONUSES:

Top Exec. Salary: $ Bonus: $
Second Exec. Salary: $ Bonus: $

OTHER THOUGHTS:

Estimated Female Officers or Directors: 2
Hot Spot for Advancement for Women/Minorities:

Sport Chalet Inc

NAIC Code: 451110

www.sportchalet.com

TYPES OF BUSINESS:

Sporting Goods Stores
Equipment Repair
Scuba Training
Equipment Rental
Sports-Related Services
Online Sales

BRANDS/DIVISIONS/AFFILIATES:

CONTACTS:

Note: Officers with more than one job title may be intentionally listed here more than once.

Craig L. Levra, CEO
Craig L. Levra, Pres.
Howard K. Kaminsky, CFO
Brad Morton, Sr. VP-Sales
Ted Jackson, CIO/VP-IT
Tom Tennyson, Chief Merch. Officer/Exec. VP
Howard K. Kaminsky, Sec.
Tim A. Anderson, Exec. VP-Retail Oper. & Loss Prevention
Dennis D. Trausch, Exec. VP-Growth & Dev.
Howard K. Kaminsky, Exec. VP-Finance
Tim A. Anderson, Exec. VP-Specialty Svcs.
Craig L. Levra, Chmn.

GROWTH PLANS/SPECIAL FEATURES:

Sport Chalet, Inc. is a leading operator of full-service sporting goods superstores in California, Arizona, Utah and Nevada. The company's 55 stores feature a number of specialty sporting goods divisions, offering a large assortment of brand-name merchandise at competitive prices. The stores offer traditional sporting goods merchandise, including footwear, apparel and equipment, as well as specialty products, such as snowboarding, skateboarding, mountaineering and Scuba merchandise. The company purchases its merchandise from over 1,000 vendors, with its largest vendor, Nike, Inc., accounting for 10% of its total inventory purchases in 2011. A typical Sport Chalet store features multiple specialty shops for beginners to experts in several sports, consisting of products such as traditional and nontraditional sports equipment and athletic footwear for a variety of sports. Additionally, the stores feature custom fitting services, including custom wet suits, scuba masks and boot mountings. Sport Chalet stores also generally offer more than 50 services for customers, such as backpacking, canyoneering, and kayaking instruction; custom golf club fitting; snowboard and ski rental and repair; Scuba training and certification; Scuba boat charters; team sales; racquet stringing; and bicycle tune-ups and repair. The stores are designed to feature a natural and outdoor-feeling color scheme; 30-foot clear ceilings; a pool for water sports instruction and demonstrations; and a 100-foot shoe wall. To keep its stores stocked, the firm operates a 326,000 square foot distribution center in Ontario, California. The firm's Team Sales division is a full-service team dealer primarily offering in-house silk screen, embroidery, embellishment and custom artwork on footwear, uniforms and equipment. The division's target market includes high schools, universities, youth sport leagues, athletic teams, booster clubs and recreational organizations.

Employees receive discounted scuba lessons, scuba scholarships, discounts from vendors, free or discounted ski lift tickets and free equipment rentals.

FINANCIAL DATA:

Note: Data for latest year may not have been available at press time.

In U.S. $	2012	2011	2010	2009	2008	2007
Revenue	349,883,000	362,483,000	353,695,000	372,652,000	402,534,000	388,209,200
R&D Expense						
Operating Income	-3,280,000	-646,000	-14,651,000	-44,229,000	-4,120,000	12,245,240
Operating Margin %	-.93%	-.17%	-4.14%	-11.86%	-1.02%	3.15%
SGA Expense	89,203,000	92,647,000	85,894,000	107,651,000	105,697,000	107,775,700
Net Income	-5,072,000	-3,015,000	-8,274,000	-52,247,000	-3,362,000	7,098,714
Operating Cash Flow	5,297,000	2,774,000	-2,938,000	-10,739,000	16,374,000	10,614,760
Capital Expenditure	3,023,000	1,205,000	738,000	16,245,000	22,433,000	22,855,770
EBITDA	6,170,000	9,705,000	-2,007,000	-29,986,000	8,778,000	23,664,170
Return on Assets %	-3.98%	-2.26%	-5.71%	-32.41%	-1.96%	4.67%
Return on Equity %	-24.90%	-12.84%	-29.25%	-90.03%	-3.94%	8.66%
Debt to Equity						

CONTACT INFORMATION:

Phone: 818 949-5300 Fax: 818 949-5301
Toll-Free:
Address: 1 Sport Chalet Dr., La Canada, CA 91011 United States

STOCK TICKER/OTHER:

Stock Ticker: SPCHA
Exchange: NAS
Employees: 3,100
Fiscal Year Ends: 30-Apr
Parent Company:

SALARIES/BONUSES:

Top Exec. Salary: $380,000 Bonus: $150,000
Second Exec. Salary: $295,000 Bonus: $

OTHER THOUGHTS:

Estimated Female Officers or Directors:
Hot Spot for Advancement for Women/Minorities:

Sport Supply Group Inc

www.sportsupplygroup.com

NAIC Code: 454113

TYPES OF BUSINESS:

Sporting Goods, Direct Sales
Online Sales
Catalogs
Telemarketing
Physical Education Products

BRANDS/DIVISIONS/AFFILIATES:

Onex Corporation
BSN Sports
Tomark Sports
New England Camp
Dixie Sporting Goods
Kessler Team Sports
Salkeld Sports
ONCAP Investment Partners II LP

CONTACTS: *Note: Officers with more than one job title may be intentionally listed here more than once.*

Adam Blumenfeld, CEO
Terrence M. Babilla, COO
Terrence M. Babilla, Pres.
John E. Pitts, CFO
Kurt Hagen, Exec. VP-Sales & Mktg.
Terrence M. Babilla, General Counsel/Sec.

GROWTH PLANS/SPECIAL FEATURES:

Sport Supply Group, Inc. (SSG) manufactures, distributes and markets sporting related goods. The company's primary products include sporting goods equipment, soft good athletic apparel, recreational and leisure products, physical education products and footwear. The firm primarily sells its products through direct catalogs, e-commerce sites, telemarketers and a direct sales team through BSN (Best Supply Network) Brand. SSG offers brand name and private-label products, including Nike, Wilson, Rawlings and Under Armour, among others. Catalogs include BSN Sports, League Direct, Tomark Sports and New England Camp. The company's web site, BSNSports.com, enables consumers to place orders, access account information, track orders and perform routine customer service inquiries. In addition, the company has licenses and marketing alliances with national organizations such as YMCA, YWCA, Little League Baseball and American Youth Football. Through its dealer division, SSG offers sporting goods and equipment to the institutional market. The division consists of four regional dealers: Dixie Sporting Goods, based in Richmond, Virginia, which supplies customers in Virginia, North Carolina, South Carolina, Ohio, Maryland and West Virginia; Orlando Team Sports, which serves the Florida market; Kessler Team Sports, which covers the Midwest and South Central regions; and Salkeld Sports, which serves Illinois. The company markets over 50,000 products to the U.S. institutional market, which comprises schools, colleges, universities, government agencies, military facilities, athletic clubs, athletic teams and dealers, youth sports leagues and recreational organizations. The firm is owned by private equity firm ONCAP Investment Partners II L.P., an investment platform of Onex Corporation.

FINANCIAL DATA: *Note: Data for latest year may not have been available at press time.*

In U.S. $	2012	2011	2010	2009	2008	2007
Revenue						
R&D Expense						
Operating Income						
Operating Margin %						
SGA Expense						
Net Income						
Operating Cash Flow						
Capital Expenditure						
EBITDA						
Return on Assets %						
Return on Equity %						
Debt to Equity						

CONTACT INFORMATION:

Phone: 972-484-9484 Fax:
Toll-Free:
Address: 1901 Diplomat Dr., Farmers Branch, TX 75234 United States

STOCK TICKER/OTHER:

Stock Ticker: Private Exchange:
Employees: Fiscal Year Ends: 31-Mar
Parent Company: ONEX CORPORATION

SALARIES/BONUSES:

Top Exec. Salary: $ Bonus: $
Second Exec. Salary: $ Bonus: $

OTHER THOUGHTS:

Estimated Female Officers or Directors:
Hot Spot for Advancement for Women/Minorities:

Sports Authority Inc (The)

www.sportsauthority.com

NAIC Code: 451110

TYPES OF BUSINESS:

Sporting Goods Stores
Athletic Apparel & Footwear
Team Sports Equipment

BRANDS/DIVISIONS/AFFILIATES:

SportsAuthority.com
Sports Authority Field at Mile High
Leonard Green & Partners LP
Gart Sports Company
Sims
SA Gear
Alpine Design

CONTACTS: *Note: Officers with more than one job title may be intentionally listed here more than once.*

Darrell Webb, CEO
Thomas T. Hendrickson, CFO/Exec. VP
Greg Waters, Exec. VP-Merch.
Paul Gaudet, Exec. VP-Store Oper.
Darrell Webb, Chmn.

GROWTH PLANS/SPECIAL FEATURES:

The Sports Authority, Inc., owned by private equity firm Leonard Green & Partners LP, is a full-line sporting goods dealer. Formerly known as Gart Sports Company, the firm operates more than 450 stores in 45 states. These stores offer an extensive product selection for sports and leisure activities, such as golf, tennis, skiing, cycling, hunting, fishing and bowling. Merchandise includes men's, women's and children's athletic apparel; athletic footwear; and team sports equipment. Sports Authority stores are located primarily in regional strip or power centers, although some stores are located in malls and stand-alone locations. Sports Authority emphasizes features such as specialty stores within stores, interactive kiosks, in-store customer clinics, demonstrations, small batting cages or basketball courts and a wide array of audiovisual entertainment. The firm's merchandise assortment includes over 700 brand names, such as Adidas, Champion, Mongoose, Nike, Reebok and Wilson. Additionally, the company markets its own private label brands such as Sims, which offers snowboards and seasonal clothing; SA Gear, which sells fitness equipment and men's sportswear; and Alpine Design, a line of outdoor goods and seasonal clothing. The company tailors merchandise to reflect customer preferences at each store location, which may mean the elimination of items such as stationary bicycles or swimming goods at certain stores. Sports Authority also maintains an e-commerce site, SportsAuthority.com, operated by GSI Commerce, Inc. Through its joint venture with Aeon Co., Ltd. the company has several Sports Authority locations in Japan through a licensing agreement. The company also owns the corporate naming rights to Sports Authority Field at Mile High, home to the NFL's Denver Broncos. In August 2011, the firm announced the opening of six new stores nationwide in Miami, Los Angeles, Chicago, Milwaukee and Salt Lake City.

FINANCIAL DATA: *Note: Data for latest year may not have been available at press time.*

In U.S. $	2012	2011	2010	2009	2008	2007
Revenue	3,400,000,000	3,200,000,000	3,200,000,000	3,050,000,000	3,030,000,000	2,980,000,000
R&D Expense						
Operating Income						
Operating Margin %						
SGA Expense						
Net Income						
Operating Cash Flow						
Capital Expenditure						
EBITDA						
Return on Assets %						
Return on Equity %						
Debt to Equity						

CONTACT INFORMATION:

Phone: 303-200-5050 Fax: 303-832-4738
Toll-Free:
Address: 1050 W. Hampden Ave., Englewood, CO 80110 United States

STOCK TICKER/OTHER:

Stock Ticker: Private Exchange:
Employees: 14,240 Fiscal Year Ends: 31-Jan
Parent Company: LEONARD GREEN & PARTNERS LP

SALARIES/BONUSES:

Top Exec. Salary: $ Bonus: $
Second Exec. Salary: $ Bonus: $

OTHER THOUGHTS:

Estimated Female Officers or Directors: 1
Hot Spot for Advancement for Women/Minorities:

Sportsman's Guide Inc (The)

www.sportsmansguide.com

NAIC Code: 451110

TYPES OF BUSINESS:

Outdoor & Hunting Products
Online & Catalog Sales
Outdoor Apparel & Footwear
Boating Supplies

BRANDS/DIVISIONS/AFFILIATES:

Redcats USA Inc
Redcats Group
SportsmansGuide.com
BargainOutfitters.com
Buyers' Club (The)
Buyer's Advantage Catalog
WorkwearSavings.com
BoatingSavings.com

CONTACTS: *Note: Officers with more than one job title may be intentionally listed here more than once.*

Jay Berlin, CEO
Matt Sullivan, VP-Mktg.
John M. Casler, Exec. VP-Merch., Mktg. & Creative Svcs.
Dale D. Monson, Sr. VP-Oper.
Tim Arland, Sr. VP-e-commerce

GROWTH PLANS/SPECIAL FEATURES:

The Sportsman's Guide, Inc., a subsidiary of Redcats USA, Inc. (which is owned by Redcats Group), is a multi-channel direct marketer of value-priced outdoor gear and clothing, golf equipment and general merchandise. The company sells its products through main and specialty catalogs and several e-commerce sites, including SportsmansGuide.com and BargainOutfitters.com, a liquidation outlet. The firm's main catalog is mailed monthly and offers merchandise across a broad range of categories. Additional specialized catalogs are also offered during the course of each year and focus on individual categories such as camping, government surplus, ammunition and shooting supplies, gifts and hunting. The Buyers' Club, which customers can join for an annual fee, offers 10% discounts on most merchandise (5% on ammunition), a monthly Buyer's Advantage Catalog and exclusive special offers, discounts and coupons. The Sportsman's Guide e-commerce site offers a selection similar to its print catalog in addition to certain specialty categories available exclusively online. More than 65% of the company's sales come through its web site. An online resource center features a variety of articles, advice columns and information about outdoor lifestyles and pursuits, including a section specifically aimed at female outdoor enthusiasts. In addition, customers can find maps, fish and game forecasts, local guide and outfitter listings, ballistics charts, useful links and other information on the web site. Merchandise orders can be placed 24-hours-a-day by phone or online, and the company ships all orders directly from its 330,000-square-foot warehouse. Recently, the firm launched the two new e-commerce sites: WorkwearSavings.com and BoatingSavings.com, each featuring discounted, name-brand accessories, clothing and gear in the work wear and boating & marine categories, respectively.

FINANCIAL DATA: *Note: Data for latest year may not have been available at press time.*

In U.S. $	2012	2011	2010	2009	2008	2007
Revenue						
R&D Expense						
Operating Income						
Operating Margin %						
SGA Expense						
Net Income						
Operating Cash Flow						
Capital Expenditure						
EBITDA						
Return on Assets %						
Return on Equity %						
Debt to Equity						

CONTACT INFORMATION:

Phone: 651-451-3030 Fax: 651-450-6130
Toll-Free: 888-844-0667
Address: 411 Farwell Ave. S., St. Paul, MN 55075 United States

STOCK TICKER/OTHER:

Stock Ticker: Subsidiary
Exchange:
Employees:
Fiscal Year Ends: 31-Dec
Parent Company: REDCATS USA

SALARIES/BONUSES:

Top Exec. Salary: $ Bonus: $
Second Exec. Salary: $ Bonus: $

OTHER THOUGHTS:

Estimated Female Officers or Directors:
Hot Spot for Advancement for Women/Minorities:

Springs Global USA Inc

www.springs.com

NAIC Code: 313240

TYPES OF BUSINESS:

Bedding Products
Baby Apparel
Bath Products
Window Products
Sewing Fabrics

BRANDS/DIVISIONS/AFFILIATES:

Santista
Springmaid
Wabasso
Artex
Dundee
Springs Global Participacoes SA
Court of Versailles

CONTACTS:

Note: Officers with more than one job title may be intentionally listed here more than once.

Josue Christiano Gomes da Silva, CEO
Tom O'Connor, Pres.
Gustavo Kawassaki, Head-Investor Rel.
Josue Christiano Gomes da Silva, Chmn.

GROWTH PLANS/SPECIAL FEATURES:

Springs Global USA, Inc., a subsidiary of Brazil-based company Springs Global Participacoes SA, supplies leading retailers with a complete line of coordinated home furnishings designed to simplify home decorating for every consumer. The firm also produces and markets bed and bath products for institutional and hospitality customers, home sewing fabrics and baby bedding and apparel products. Bedding products include sheets, comforters, bedspreads, blankets, bed skirts, quilts, pillow shams, pillows and mattress pads. Bath products include towels, bath and accent rugs, shower curtains, ceramics and other bath accessories. Window products include window hardware and decorative rods; horizontal blinds in a range of widths and materials; motorized blinds; pleated and cellular shades; and soft window treatments such as drapes, valances and balloon shades. Through licensing agreements, the company has extended its product line to include kitchen and table linen items, decorative napkin rings, flannel and knit sheets, toilet seat covers, blankets, throws and fabric-covered lampshades. The firm's major brands are Springmaid, Wabasso, Dundee, Artex, Santista and Court of Versailles. Springs Global has operations in Argentina, Brazil, Mexico and the U.S., with its North American headquarters located in Fort Mill, South Carolina. It also has additional supply offices in Turkey, Pakistan, Thailand, China, Japan, India, Bangladesh and Vietnam.

Employees are offered medical, vision, and dental plans; a 401(k) plan; life insurance and disability coverage; flexible spending accounts; and an employee assistance program.

FINANCIAL DATA:

Note: Data for latest year may not have been available at press time.

In U.S. $	2012	2011	2010	2009	2008	2007
Revenue						
R&D Expense						
Operating Income						
Operating Margin %						
SGA Expense						
Net Income						
Operating Cash Flow						
Capital Expenditure						
EBITDA						
Return on Assets %						
Return on Equity %						
Debt to Equity						

CONTACT INFORMATION:

Phone: 803-547-1500 Fax: 803-547-1636
Toll-Free: 888-926-7888
Address: 205 N. White St., Fort Mill, SC 29715 United States

STOCK TICKER/OTHER:

Stock Ticker: Subsidiary Exchange:
Employees: 20,000 Fiscal Year Ends: 31-Dec
Parent Company: SPRINGS GLOBAL PARTICIPACOES SA

SALARIES/BONUSES:

Top Exec. Salary: $ Bonus: $
Second Exec. Salary: $ Bonus: $

OTHER THOUGHTS:

Estimated Female Officers or Directors:
Hot Spot for Advancement for Women/Minorities:

St John Knits International Inc

www.sjk.com

NAIC Code: 315240

TYPES OF BUSINESS:

Women's Apparel
Women's Knits & Accessories
Boutiques & Outlet Stores
Yarn & Fabric Manufacturing
Footwear

GROWTH PLANS/SPECIAL FEATURES:

St. John Knits International, Inc., founded in 1962 by fashion model Marie Gray and her husband Robert, is an international designer, manufacturer and distributor of fine women's clothing; accessories, such as eyewear, shoes, belts, handbags, scarves and jewelry; and fragrances. The company offers several product lines including evening wear, jackets, tops, dresses, bottoms, jewelry, belts and a line of caviar colored clothing. The firm markets its products to wholesale clients, which include high-end department stores in the U.S. such as Neiman Marcus and Saks Fifth Avenue, as well as retailers in more than 25 countries throughout Europe, the Middle East, South America and Asia. Additionally, the company maintains company-operated boutiques and outlets throughout the U.S. and Canada. The company is dedicated to maintaining consistently high product quality, and to that end controls almost every stage in the design and manufacturing process. St. John has developed a vertically integrated manufacturing process, including twisting and dyeing its own brand of wool and rayon yarns and knitting (including its trademark Santana knit), constructing, pressing and finishing garments at company-owned facilities. Vestar Capital Partners Inc. owns 78% of the firm.

BRANDS/DIVISIONS/AFFILIATES:

Vestar Capital Partners Inc
Santana knit

CONTACTS: *Note: Officers with more than one job title may be intentionally listed here more than once.*

Bruce A. Fetter, COO
Bruce A. Fetter, Pres.
Roger G. Ruppert, CFO
Robert Green, Exec. VP-Sales & Mktg.
Roger G. Ruppert, Exec. VP-Finance

FINANCIAL DATA: *Note: Data for latest year may not have been available at press time.*

In U.S. $	2012	2011	2010	2009	2008	2007
Revenue	253,900,000	269,500,000				
R&D Expense						
Operating Income						
Operating Margin %						
SGA Expense						
Net Income						
Operating Cash Flow						
Capital Expenditure						
EBITDA						
Return on Assets %						
Return on Equity %						
Debt to Equity						

CONTACT INFORMATION:

Phone: 949 863-1171 Fax: 949 223-3390
Toll-Free:
Address: 17622 Armstrong Ave., Irvine, CA 92614 United States

STOCK TICKER/OTHER:

Stock Ticker: Private Exchange:
Employees: 4,925 Fiscal Year Ends: 31-Oct
Parent Company: VESTAR CAPITAL PARTNERS INC

SALARIES/BONUSES:

Top Exec. Salary: $ Bonus: $
Second Exec. Salary: $ Bonus: $

OTHER THOUGHTS:

Estimated Female Officers or Directors:
Hot Spot for Advancement for Women/Minorities:

Stage Stores Inc

NAIC Code: 452111

www.stagestoresinc.com

TYPES OF BUSINESS:

Department Stores
Online Sales

BRANDS/DIVISIONS/AFFILIATES:

Stage
Bealls
Palais Royal
Peebles
Goody's
Steele's

CONTACTS:

Note: Officers with more than one job title may be intentionally listed here more than once.

Michael Glazer, CEO
Edward Record, COO
Michael Glazer, Pres.
Oded Shein, CFO/Exec. VP
Bill Gentner, Sr. VP-Mktg.
Ron Lucas, Exec. VP-Human Resources
Steven Hunter, CIO/Exec. VP
Steven Lawrence, Chief Merch. Officer
Richard Stasyszen, Sr. VP-Finance/Controller
William Montgoris, Chmn.

GROWTH PLANS/SPECIAL FEATURES:

Stage Stores, Inc. is a specialty department store retailer offering moderately priced, nationally recognized brand-name and private-label apparel, accessories, cosmetics and footwear. The company operates 810 department stores located in 40 states under the Stage, Bealls, Goody's, Peebles and Palais Royal names. Additionally, the firm's has recently introduced a new off-price concept store under the name Steele's, which currently has three locations. Approximately 90% of sales consist of nationally recognized brands, such as Liz Claiborne, Levi Strauss, Calvin Klein, IZOD, Chaps, Clinique, Nike and New Balance, and the remaining 10% consists of the company's private labels. In targeting small and midsize markets, the company has developed a store format that is smaller than typical department stores yet large enough to offer a carefully chosen but broad selection of merchandise. The firm attempts to locate its stores by, or in the vicinity of, other tenants that it believes will help attract additional foot traffic to the area, such as grocery stores, drug stores or major discount stores. Approximately 65% of the company's stores operate in towns with populations under 50,000 people. It currently distributes all merchandise to its stores through three distribution centers located in Jacksonville, Texas; South Hill, Virginia; and Jeffersonville, Ohio.

Employees are offered medical and dental insurance, life insurance, flexible spending accounts, product discounts, a 401(k) plan and tuition reimbursements.

FINANCIAL DATA:

Note: Data for latest year may not have been available at press time.

In U.S. $	2012	2011	2010	2009	2008	2007
Revenue	1,511,919,000	1,470,590,000	1,431,927,000	1,515,820,000	1,545,606,000	
R&D Expense						
Operating Income	51,096,000	62,767,000	45,827,000	51,859,000	89,788,000	
Operating Margin %	3.37%	4.26%	3.20%	3.42%	5.80%	
SGA Expense	359,504,000	354,057,000	341,592,000	357,725,000	354,926,000	
Net Income	30,960,000	37,640,000	28,721,000	-65,535,000	53,080,000	
Operating Cash Flow	78,055,000	77,875,000	120,936,000	162,783,000	124,535,000	
Capital Expenditure	45,731,000	36,990,000	42,707,000	99,841,000	95,311,000	
EBITDA	112,776,000	62,767,000	50,831,000	15,493,000	139,726,000	
Return on Assets %	4.04%	4.71%	3.66%	-7.99%	6.25%	
Return on Equity %	6.86%	7.79%	6.20%	-13.50%	9.71%	
Debt to Equity	0.08	0.05	0.08	0.1	0.18	

CONTACT INFORMATION:

Phone: 800 579-2302 Fax: 713 660-3330
Toll-Free: 800-324-3244
Address: 10201 Main St., Houston, TX 77025 United States

STOCK TICKER/OTHER:

Stock Ticker: SSI Exchange: NYS
Employees: 13,500 Fiscal Year Ends: 31-Jan
Parent Company:

SALARIES/BONUSES:

Top Exec. Salary: $709,423 Bonus: $
Second Exec. Salary: $582,750 Bonus: $

OTHER THOUGHTS:

Estimated Female Officers or Directors: 3
Hot Spot for Advancement for Women/Minorities: Y

Stefanel SpA

www.stefanel.com

NAIC Code: 315190

TYPES OF BUSINESS:

Knitwear & Casual wear Design & Manufacturing
Casual Apparel
Retail Apparel Stores
Online Sales
Apparel Distribution

BRANDS/DIVISIONS/AFFILIATES:

Marithe
Francoise Girbaud
Stefanel
Interfashion SpA
Girbaud.com

CONTACTS: *Note: Officers with more than one job title may be intentionally listed here more than once.*

Luciano Santel, CEO
Giuseppe Stefanel, Pres.
Federico Girotto, Dir.-Admin.
Federico Girotto, Dir.-Investor Rel.
Federico Girotto, Dir.-Finance & Control
Giuseppe Stefanel, Chmn.

GROWTH PLANS/SPECIAL FEATURES:

Stefanel SpA designs, manufactures and markets brand and licensed knitwear and casualwear for women and men. The firm's knitwear and casual clothes are retailed under the Stefanel brand. Stefanel's knitwear is produced almost entirely in a single factory in Salgareda, Italy. Stefanel also offers a complete line of jeanswear and casualwear for men and women under the Marithe and Francoise Girbaud brand. The company distributes Marithe and Francois Girbaud through subsidiary Interfashion SpA. The firm also retails through several e-commerce sites, including Girbaud.com.

FINANCIAL DATA: *Note: Data for latest year may not have been available at press time.*

In U.S. $	2012	2011	2010	2009	2008	2007
Revenue		254,720,000	251,100,000	248,500,000	363,670,000	417,950,000
R&D Expense						
Operating Income						
Operating Margin %						
SGA Expense						
Net Income		38,633,000	-48,600,000	-68,780,000	-37,470,000	-35,690,000
Operating Cash Flow						
Capital Expenditure						
EBITDA						
Return on Assets %						
Return on Equity %						
Debt to Equity						

CONTACT INFORMATION:

Phone: 39-0422-8191 Fax: 39-0422-819342
Toll-Free:
Address: Via Postumia 85, Ponte di Piave, 31047 Italy

STOCK TICKER/OTHER:

Stock Ticker: STEF
Exchange: Milan
Employees: 1,250
Fiscal Year Ends: 31-Dec
Parent Company:

SALARIES/BONUSES:

Top Exec. Salary: $ Bonus: $
Second Exec. Salary: $ Bonus: $

OTHER THOUGHTS:

Estimated Female Officers or Directors:
Hot Spot for Advancement for Women/Minorities:

Stein Mart Inc

NAIC Code: 452111

www.steinmart.com

TYPES OF BUSINESS:

Discount Department Stores
Apparel
Linens
Accessories
Gifts
Shoes

BRANDS/DIVISIONS/AFFILIATES:

Boutique Ladies
Stein Mart Boutique

CONTACTS: *Note: Officers with more than one job title may be intentionally listed here more than once.*

Jay Stein, Interim CEO
D. Hunt Hawkins, COO/Exec. VP
Gregory W. Kleffner, CFO
Glori Katz, Sr. VP-Mktg. & Advertising
Brian R. Morrow, Chief Merch. Officer/Exec. VP
Linda L. Tasseff, Dir.-Investor Rel.
Gregory W. Kleffner, Exec. VP-Finance
Gary L. Pierce, Sr. VP/Dir-Stores
Joseph Martinolich, Sr. VP-Internal Audit, Safety & Security
Jay Stein, Chmn.

GROWTH PLANS/SPECIAL FEATURES:

Stein Mart, Inc. is a retail chain that offers current-season, fashionable and primarily branded merchandise. The company currently operates 263 stores in 29 states and Washington, D.C., typically in neighborhood shopping centers in metropolitan areas. Although comparable in quality and presentation, it typically charges up to 60% less than traditional department and specialty stores. Target customers are fashion-conscious, value-seeking women ages 35-55 with above-average annual household incomes. The company's focused assortment of merchandise features moderate and designer brand-name apparel for women and men as well as accessories, home decor, gifts, linens and shoes. Stein Mart purchases its products from roughly 1,300 vendors. Ladies' and boutique apparel accounts for 45% of sales; men's, 20%; ladies' accessories, 13%; home, 11%; shoes, 8%; and other, 3%. The company strives to keep prices low through its vendor relationships, tight control over corporate and store expenses and efficient management of inventory. To this end, Stein Mart's shoe department inventory is exclusively supplied and owned by DSW, Inc. A unique store feature is its Stein Mart Boutique, which is designed to be a store within a store. Store employees known as Boutique Ladies provide customers with specific fashion advice and personal service as wells as acting as personal shoppers. Other services include the Stein Mart Preferred Customer rewards program, a private label credit card and a co-branded Stein Mart Platinum MasterCard.

Stein Mart offers its employees medical, dental, vision and life insurance; flexible spending accounts; a 401(k) plan; a discount stock purchase plan; and an employee assistance program.

FINANCIAL DATA: *Note: Data for latest year may not have been available at press time.*

In U.S. $	2012	2011	2010	2009	2008	2007
Revenue	1,160,367,000	1,181,510,000	1,219,109,000	1,326,469,000	1,457,645,000	1,501,296,000
R&D Expense						
Operating Income	32,257,000	53,146,000	34,647,000	-80,129,000	-5,786,000	57,924,000
Operating Margin %	2.77%	4.49%	2.84%	-6.04%	-.39%	3.85%
SGA Expense	289,114,000	288,592,000	314,115,000	394,767,000	388,572,000	376,611,000
Net Income	19,828,000	48,753,000	23,553,000	-71,301,000	-4,530,000	37,176,000
Operating Cash Flow	59,568,000	49,632,000	98,329,000	19,356,000	17,843,000	30,319,000
Capital Expenditure	33,449,000	29,550,000	7,585,000	19,281,000	25,898,000	48,759,000
EBITDA	51,194,000	70,474,000	53,870,000	-54,377,000	20,282,000	81,916,000
Return on Assets %	4.35%	11.59%	5.51%	-15.66%	-.96%	7.43%
Return on Equity %	7.82%	21.06%	11.71%	-32.18%	-1.66%	12.15%
Debt to Equity				0.53	0.1	

CONTACT INFORMATION:

Phone: 904 346-1500   Fax: 904 398-4341
Toll-Free: 888-783-4662
Address: 1200 Riverplace Blvd., Jacksonville, FL 32207 United States

STOCK TICKER/OTHER:

Stock Ticker: SMRT   Exchange: NAS
Employees: 11,500   Fiscal Year Ends: 31-Jan
Parent Company:

SALARIES/BONUSES:

Top Exec. Salary: $539,000   Bonus: $
Second Exec. Salary: $526,850   Bonus: $

OTHER THOUGHTS:

Estimated Female Officers or Directors: 3
Hot Spot for Advancement for Women/Minorities: Y

Stella International Holdings Ltd

www.stella.com.hk

NAIC Code: 316210

TYPES OF BUSINESS:

Rubber and Plastics Footwear Manufacturing

GROWTH PLANS/SPECIAL FEATURES:

Stella International Holdings, Ltd. is a developer and manufacturer of footwear products. It operates in two segments: Manufacturing and Retail. The Manufacturing segment operates production facilities in China and Vietnam. In addition to providing footwear manufacturing services for other companies, Stella International produces its own lines of footwear: Stella Luna, What For and JKJY. Clients include Clarks, Deckers, ECCO, Rockport, Timberland, Wolverine, Cole Haan, Guess, Bally and Via Spiga. The Retail segment focuses on marketing the firm's brands in stores located throughout China, Taiwan, Thailand, the Philippines, Lebanon, the United Arab Emirates and Kuwait. In 2012, Stella International opened a Stella Luna store in Paris.

BRANDS/DIVISIONS/AFFILIATES:

Stella Luna
What For
JKJY

CONTACTS: *Note: Officers with more than one job title may be intentionally listed here more than once.*

Lawrence Li-Ming Chen, CEO
Daniel Takuen Shih, Chmn.

FINANCIAL DATA: *Note: Data for latest year may not have been available at press time.*

In U.S. $	2012	2011	2010	2009	2008	2007
Revenue	1,550,003,000	1,494,531,000	1,293,521,000	1,008,598,000	1,102,116,000	937,164,000
R&D Expense	49,504,000	44,753,000	39,621,000	35,660,000	38,491,000	33,025,000
Operating Income	166,404,000	156,389,000	129,460,000	108,333,000	131,563,000	103,691,000
Operating Margin %	10.73%	10.46%	10.00%	10.74%	11.93%	11.06%
SGA Expense	191,435,000	161,896,000	125,264,000	102,254,000	104,691,000	84,299,000
Net Income	152,588,000	142,784,000	121,328,000	102,081,000	124,978,000	114,623,000
Operating Cash Flow	143,993,000	59,469,000	49,818,000	144,640,000	87,115,000	49,115,000
Capital Expenditure	53,261,000	41,163,000	36,912,000	12,070,000	15,571,000	23,772,000
EBITDA	194,510,000	182,240,000	154,302,000	129,752,000	150,337,000	134,797,000
Return on Assets %	13.73%	13.84%	12.67%	11.26%	14.32%	16.15%
Return on Equity %	17.24%	17.18%	15.45%	13.43%	16.82%	19.56%
Debt to Equity						

CONTACT INFORMATION:

Phone: 852-2956-1339 Fax: 852-2956-1383
Toll-Free:
Address: The Gateway, 25 Canton Rd., Tower 2, Stes. 3003-04, Kowloon, Hong Kong

STOCK TICKER/OTHER:

Stock Ticker: SLNLF
Exchange: PINX
Employees: 21,000
Fiscal Year Ends: 31-Dec
Parent Company:

SALARIES/BONUSES:

Top Exec. Salary: $ Bonus: $
Second Exec. Salary: $ Bonus: $

OTHER THOUGHTS:

Estimated Female Officers or Directors:
Hot Spot for Advancement for Women/Minorities:

Steven Madden Ltd

NAIC Code: 316210

www.stevemadden.com

TYPES OF BUSINESS:

Shoes, Manufacturing
Retail Stores
Online Sales
Women's Jewelry & Accessories
Brand Licensing

BRANDS/DIVISIONS/AFFILIATES:

Steve Madden
Steven
Superga
Betsey Johnson
SM Canada
Big Buddha
Betseyville
Madden Girl

CONTACTS: *Note: Officers with more than one job title may be intentionally listed here more than once.*

Edward R. Rosenfeld, CEO
Awadesh Sinha, COO
Arvind Dharia, CFO
Robert Schmertz, Dir.-Brand
Arvind Dharia, Corp. Sec.
Amelia N. Varela, Exec. VP-Wholesale & Retail
Edward R. Rosenfeld, Chmn.

GROWTH PLANS/SPECIAL FEATURES:

Steven Madden, Ltd. designs, sources, markets and sells shoes for men, women and children. Headquartered in New York, it distributes products through its retail stores, its e-commerce web site and department and specialty stores throughout the U.S. and Canada as well as Europe, Latin America and the Asia-Pacific. The company operates in five segments: wholesale footwear, wholesale accessories, retail, first cost and licensing. The wholesale footwear division is comprised of the Women's, Men's, Kid's, Madden Girl, Steven, Elizabeth and James and other licensed brands. Madden Women's wholesale, the company's largest division, designs, sources and markets the Steve Madden brand to a variety of stores and boutiques. Madden men's wholesale designs and markets men's casual and athletic shoes. The Kid's division designs shoes for young girls below the age of 13. Madden Girl markets young women's shoes geared for females ages 13-20. Steven designs, sources and markets footwear under the Steven brand. Through a license agreement with Dualstar Entertainment Group, LLC, the firm produces luxury footwear under the Elizabeth and James brand. The wholesale accessories segment designs, sources and markets name brand and private label fashion handbags and accessories. Its brands include Betsey Johnson, Big Buddha and Betseyville in addition to Steve Madden brands. The retail segment, operating through Steven Madden Retail, Inc., manages over 109 retail stores. The first cost division acts as a buying agent for footwear products sold under private labels for mass merchandisers. The licensing segment is responsible for the licensing of the Steve Madden and Betsey Johnson brands. In 2012, the company opened twenty new stores, acquired seven stores as part of the acquisition of the privately held SM Canada and established the Superga and Betsey Johnson online stores.

FINANCIAL DATA: *Note: Data for latest year may not have been available at press time.*

In U.S. $	2012	2011	2010	2009	2008	2007
Revenue	1,227,072,000	987,264,000	635,418,000	503,550,000	457,046,000	431,050,000
R&D Expense						
Operating Income	178,976,000	153,770,000	121,624,000	78,968,000	30,612,000	52,914,000
Operating Margin %	14.58%	15.57%	19.14%	15.68%	6.69%	12.27%
SGA Expense	283,689,000	226,893,000	176,859,000	157,149,000	156,212,000	138,841,000
Net Income	119,626,000	97,319,000	75,725,000	50,107,000	27,976,000	35,690,000
Operating Cash Flow	143,341,000	74,968,000	86,873,000	64,342,000	41,779,000	59,907,000
Capital Expenditure	20,102,000	15,477,000	3,424,000	3,399,000	8,314,000	12,965,000
EBITDA	196,878,000	169,903,000	135,860,000	90,442,000	55,614,000	64,636,000
Return on Assets %	16.57%	17.89%	19.55%	16.38%	10.15%	13.78%
Return on Equity %	21.72%	23.38%	24.22%	21.14%	13.27%	16.70%
Debt to Equity						

CONTACT INFORMATION:

Phone: 718 446-1800 Fax: 718 446-5599
Toll-Free:
Address: 52-16 Barnett Ave., Long Island City, NY 11104 United States

STOCK TICKER/OTHER:

Stock Ticker: SHOO Exchange: NAS
Employees: 1,620 Fiscal Year Ends: 31-Dec
Parent Company:

SALARIES/BONUSES:

Top Exec. Salary: $660,000 Bonus: $
Second Exec. Salary: $575,000 Bonus: $

OTHER THOUGHTS:

Estimated Female Officers or Directors: 1
Hot Spot for Advancement for Women/Minorities:

Stowe-Pharr Mills Inc

www.pharryarns.com

NAIC Code: 313110

TYPES OF BUSINESS:

Yarn & Fiber Manufacturing
Fire-Resistant Materials
Dyeing
Carpeting & Upholstery Fabrics
Apparel & Hosiery Fabrics

BRANDS/DIVISIONS/AFFILIATES:

Pharr Yarns

GROWTH PLANS/SPECIAL FEATURES:

Stowe-Pharr Mills, Inc., which operates under the name Pharr Yarns, is a leading diversified sales yarn spinner. Based in McArdenville, North Carolina, the company serves a wide variety of customers from offices and operation centers across the U.S. and Europe. U.S. locations include Georgia, South Carolina, North Carolina and California. The company produces carpet yarn and high performance yarn. Carpet yarn is sold to woven carpet and rug manufacturers. Types of yarns include natural yarns, solution dyed bulked continuous filament (BCF) yarns and spaced dyed yarns in long or short spacing. The division utilizes five yarn production plants in the U.S. The high performance yarns division produces yarn that is created from fibers that are flame or heat resistant, chemical resistant, cut resistant or has significantly high strength. The yarns are marketed toward technical, commercial, protective, industrial and apparel markets.

CONTACTS: *Note: Officers with more than one job title may be intentionally listed here more than once.*

J. M. Carstarphen, III, CEO
William P. Carstarphen, COO
William P. Carstarphen, Pres.
Frederick Erickson, CFO
Frederick Erickson, Sec.
Frederick Erickson, Treas.
J. M. Carstarphen, III, Chmn.

FINANCIAL DATA: *Note: Data for latest year may not have been available at press time.*

In U.S. $	2012	2011	2010	2009	2008	2007
Revenue						
R&D Expense						
Operating Income						
Operating Margin %						
SGA Expense						
Net Income						
Operating Cash Flow						
Capital Expenditure						
EBITDA						
Return on Assets %						
Return on Equity %						
Debt to Equity						

CONTACT INFORMATION:

Phone: 704-824-3551 Fax: 704-824-0072
Toll-Free:
Address: 100 Main St., McAdenville, NC 28101 United States

STOCK TICKER/OTHER:

Stock Ticker: Private
Exchange:
Employees:
Fiscal Year Ends: 31-Mar
Parent Company:

SALARIES/BONUSES:

Top Exec. Salary: $ Bonus: $
Second Exec. Salary: $ Bonus: $

OTHER THOUGHTS:

Estimated Female Officers or Directors:
Hot Spot for Advancement for Women/Minorities:

Sunrise Brands LLC

NAIC Code: 315240

www.sunrisebrands.com

TYPES OF BUSINESS:

Casual Clothing
Casual Apparel
Jeans
Private Label Apparel
e-Commerce

BRANDS/DIVISIONS/AFFILIATES:

Dylan George
American Rag CIE
Isaac Mizrahi Jeans
Rerock
G.E.T
Sketchers
Superdry
Seven7 Jeans

CONTACTS: *Note: Officers with more than one job title may be intentionally listed here more than once.*

Gerard Guez, CEO
Donald Waldman, COO
Peter Akaragian, CFO
Todd Kay, Vice Chmn.
Gerard Guez, Chmn.

GROWTH PLANS/SPECIAL FEATURES:

Sunrise Brands LLC is a brand development, design and sourcing company for branded and private label apparel, serving specialty retailers, mass merchants and national department stores. Products are manufactured in woven and knit fabrications and include jeans wear, casual pants, shorts, skirts, dresses, t-shirts, shirts and other tops and jackets. Its company-owned and licensed brands include a mix of high-end fashion apparel such as American Rag CIE, Isaac Mizrahi Jeans, Dylan George, G.E.T., Rerock, Sketchers, Superdry and Seven7 Jeans. Its private label business provides merchandise design and sourcing services for many of the leading fashion and specialty retailers in the U.S. Its clients include American Eagle, Chico's, Express, Forever 21, Sears, Wal-Mart and New York & Company. The company's design services include fashion trend identification, product design, branding, product sample and test item development, fabric choice coordination and procurement, retail floor and fixture planning, production monitoring and final product delivery. The firm provides sourcing solutions through an international network of independent contractors, suppliers and manufacturers across Asia, the Middle East, Africa and Central and South America. In addition to its Los Angeles headquarters, Sunrise Brands has a showroom in New York City and an office in Hong Kong to support its private label sourcing business.

FINANCIAL DATA: *Note: Data for latest year may not have been available at press time.*

In U.S. $	2012	2011	2010	2009	2008	2007
Revenue						
R&D Expense						
Operating Income						
Operating Margin %						
SGA Expense						
Net Income						
Operating Cash Flow						
Capital Expenditure						
EBITDA						
Return on Assets %						
Return on Equity %						
Debt to Equity						

CONTACT INFORMATION:

Phone: 323-780-8250 Fax: 323-881-0369
Toll-Free:
Address: 801 S. Figuera St., Ste. 2500, Los Angeles, CA 90017 United States

STOCK TICKER/OTHER:

Stock Ticker: Private
Exchange:
Employees: 250
Fiscal Year Ends: 31-Dec
Parent Company:

SALARIES/BONUSES:

Top Exec. Salary: $ Bonus: $
Second Exec. Salary: $ Bonus: $

OTHER THOUGHTS:

Estimated Female Officers or Directors:
Hot Spot for Advancement for Women/Minorities:

Superior Uniform Group Inc

www.superioruniformgroup.com

NAIC Code: 315280

TYPES OF BUSINESS:

Uniform Design & Manufacture
Controlled Environment Wear
Online Sales
Uniform Accessories

BRANDS/DIVISIONS/AFFILIATES:

Fashion Seal
Martin's Uniforms
UniVogue
Fashion Seal Healthcare
Worklon
Office Gurus (The)
Blade Sportswear Inc

CONTACTS: *Note: Officers with more than one job title may be intentionally listed here more than once.*

Michael Benstock, CEO
Alan D. Schwartz, Pres.
Andrew D. Demott, Jr., CFO/Exec. VP
Jordan M. Alpert, General Counsel/Sec.
Andrew D. Demott, Jr., Treas.
Peter Benstock, Exec. VP
Sidney Kirschner, Chmn.

GROWTH PLANS/SPECIAL FEATURES:

Superior Uniform Group, Inc. manufactures and sells a range of uniforms, career apparel and accessories. The firm operates in two segments: Uniforms and Related Products; and Remote Staffing solutions. Through its signature marketing brands, Fashion Seal, Fashion Seal Healthcare, Martin's Uniforms, UniVogue, Worklon and Blade, the company provides uniform and clothing needs in the following markets: cleanroom, commercial, entertainment, government and public safety, health care, hospitality, laundry services, textile rental, promotional products, restaurant and food services, retail and transportation. The firm's primary brands, Fashion Seal and Fashion Seal Healthcare, cater to lodging/hospitality employees as well as restaurant employees and health care providers. Martin's Uniforms provides uniforms primarily to the public safety, transportation, governmental services, private security and commercial markets. UniVogue specializes in the lodging, hospitality and foodservice industries. The Worklon brand manufactures cleanroom, electrostatic-sensitive and controlled environment apparel and accessories for the automotive, semiconductor, food processing and pharmaceutical markets. Blade Sportswear, Inc. is a manufacturer and distributor of uniforms that cater to major U.S. restaurants and foodservice chains. Superior services its Remote Staffing Solutions segment through multiple The Office Gurus entities, including its direct and indirect subsidiaries in El Salvador, Costa Rica, and the U.S., and an affiliated entity in Belize, collectively known as TOG. TOG is a near-shore premium provider of cost effective bilingual telemarketing and total office support solutions.

The firm offers employees disability, medical and dental insurance; health care spending and dependent care accounts; life insurance; a 401(k) plan and pension plan; access to an on-site cafeteria; and an employee discount program.

FINANCIAL DATA: *Note: Data for latest year may not have been available at press time.*

In U.S. $	2012	2011	2010	2009	2008	2007
Revenue	119,486,000	112,373,000	105,877,900	102,801,900	123,745,200	120,457,900
R&D Expense						
Operating Income	4,621,000	5,613,000	5,747,000	2,696,882	4,140,333	5,835,505
Operating Margin %	3.86%	4.99%	5.42%	2.62%	3.34%	4.84%
SGA Expense	33,886,000	34,646,000	31,696,430	30,402,390	34,263,750	33,784,790
Net Income	3,031,000	4,136,000	3,807,000	1,966,882	2,133,773	2,549,328
Operating Cash Flow	9,158,000	-855,000	6,455,157	17,455,960	10,426,790	839,201
Capital Expenditure	1,647,000	2,974,000	771,315	1,091,284	2,271,090	1,162,745
EBITDA	6,951,000	8,595,000	8,324,178	5,764,400	7,697,439	9,194,151
Return on Assets %	3.79%	5.33%	5.15%	2.56%	2.54%	2.94%
Return on Equity %	5.10%	6.77%	6.28%	3.25%	3.20%	3.52%
Debt to Equity		0.01			0.05	0.03

CONTACT INFORMATION:

Phone: 727 397-9611 Fax: 727 391-5401
Toll-Free: 866-687-7118
Address: 10055 Seminole Blvd., Seminole, FL 33772 United States

STOCK TICKER/OTHER:

Stock Ticker: SGC Exchange: NAS
Employees: 647 Fiscal Year Ends: 31-Dec
Parent Company:

SALARIES/BONUSES:

Top Exec. Salary: $441,324 Bonus: $
Second Exec. Salary: $354,911 Bonus: $

OTHER THOUGHTS:

Estimated Female Officers or Directors:
Hot Spot for Advancement for Women/Minorities: Y

Swank Inc

NAIC Code: 315990

www.swankinc.com

TYPES OF BUSINESS:

Leather Accessories-Importing & Distribution
Men's Jewelry

BRANDS/DIVISIONS/AFFILIATES:

Swank

CONTACTS:
Note: Officers with more than one job title may be intentionally listed here more than once.

John A. Tulin, CEO
Eric P. Luft, Pres.
Jerold R. Kassner, CFO/Exec. VP
William F. Rubin, Sr. VP-Regional Sales
James E. Tulin, Sr. VP-Merch.
Arthur T. Gately, III, VP-Admin.
Jerold R. Kassner, Sec.
Paul Duckett, Sr. VP-Retail Store Oper.
Jerold R. Kassner, Treas.
Melvin Goldfeder, Sr. VP-Special Markets
Christine Myerson, VP-Merch.
John A. Tulin, Chmn.
Paul Duckett, Sr. VP-Dist.

GROWTH PLANS/SPECIAL FEATURES:

Swank, Inc. is engaged in the import, sale and distribution of men's belts, leather accessories, suspenders and men's jewelry. The company's products include belts, wallets and other small leather goods, including billfolds, key cases, card holders and other items; and men's jewelry, including cuff links, tie clips, chains and tacs, bracelets, neck chains, vest chains, collar pins, key rings and money clips. The firm distributes these products under its own brand name, Swank, as well as licensed brands Geoffrey Beene, Kenneth Cole, Pierre Cardin, Chaps, Donald Trump, U.S. Polo Association, Tommy Hilfiger, Guess, Buffalo David Bitton, Nautica and Tumi. Additionally, the company distributes men's costume jewelry under the brands Geoffrey Beene, Claiborne, Kenneth Cole, Chaps, Donald Trump, U.S. Polo Association, Guess, Buffalo David Bitton, Nautica and Tumi; men's suspenders under the Geoffrey Beene and Pierre Cardin brands; and women's leather accessories under the Pierre Cardin and Buffalo David Bitton brands. Swank no longer manufactures the products it sells, which are instead sourced from third-party vendors; it purchases most of its small leather goods, chiefly wallets, from a single supplier in India. Swank sells its products both domestically and internationally, principally through department stores, but also through national chain stores, specialty stores, mass merchandisers, catalog retailers and U.S. military retail exchanges. The company's three largest customers include Macy's, Inc.; Kohl's Department Stores; and The TJX Companies, Inc.

FINANCIAL DATA:
Note: Data for latest year may not have been available at press time.

In U.S. $	2012	2011	2010	2009	2008	2007
Revenue		138,620,000	132,702,000	114,798,000	113,967,000	128,615,000
R&D Expense						
Operating Income						
Operating Margin %						
SGA Expense						
Net Income		5,938,000	4,174,000	1,778,000	2,090,000	4,947,000
Operating Cash Flow						
Capital Expenditure						
EBITDA						
Return on Assets %						
Return on Equity %						
Debt to Equity						

CONTACT INFORMATION:

Phone: 212-867-2600 Fax:
Toll-Free:
Address: 90 Park Ave., 13th Fl., New York, NY 10016 United States

STOCK TICKER/OTHER:

Stock Ticker: SNKI Exchange: PINX
Employees: 256 Fiscal Year Ends: 31-Dec
Parent Company:

SALARIES/BONUSES:

Top Exec. Salary: $ Bonus: $
Second Exec. Salary: $ Bonus: $

OTHER THOUGHTS:

Estimated Female Officers or Directors: 1
Hot Spot for Advancement for Women/Minorities:

Swift Galey

www.swiftgaley.com

NAIC Code: 313210

TYPES OF BUSINESS:

Textiles & Fabrics Manufacturing
Sportswear Fabrics
Denim, Corduroy, Poplin & Twill
Uniform Fabrics

BRANDS/DIVISIONS/AFFILIATES:

Patriarch Partners LLC
Swift Denim
Galey & Lord
Swift Galey Denim North America

CONTACTS:

Note: Officers with more than one job title may be intentionally listed here more than once.

John Heldrich, CEO
Robert McCormack, Pres., Apparel Mktg. Group, Woven Div./Exec. VP
Larry Galbraith, Pres., Swift Galey D.N.A.
Rick Waide, Pres., Int'l Mktg. & Jeanswear

GROWTH PLANS/SPECIAL FEATURES:

Swift Galey is a leading manufacturer and marketer of denim, khaki, twill, corduroy and poplins. The firm is owned by Patriarch Partners, LLC, a private equity firm. Shortly after Patriarch Partners acquired Galey & Lord, a leading producer of premium twill and corduroy, and its subsidiary Swift Denim, one of the world's largest denim fabric brands, it merged the two into Swift Galey. The fusion of these two companies' areas of expertise allowed the firm to serve several divisions of the fabrics industry, namely jeanswear, sportswear, uniforms, workwear and interior fabrics. The company maintains manufacturing locations in the U.S. and Cambodia and utilizes such technologies as Zeyneth XDR, FlameX & ArcBloc, Purista and H2Tech in its production techniques. Swift Galey operates a joint venture with Denim North America, called Swift Galey Denim North America (D.N.A.), which produces denim fabrics for international markets. In the uniforms market, the firm supplies cotton twills, poplins, denim and corduroy; synthetic fabrics; and yarn dyes. It supplies apparel for law enforcement, medical, culinary, industrial laundry, military, retail and other customers. Some major clients of the firm include Abercrombie & Fitch and The Gap.

FINANCIAL DATA:

Note: Data for latest year may not have been available at press time.

In U.S. $	2012	2011	2010	2009	2008	2007
Revenue						
R&D Expense						
Operating Income						
Operating Margin %						
SGA Expense						
Net Income						
Operating Cash Flow						
Capital Expenditure						
EBITDA						
Return on Assets %						
Return on Equity %						
Debt to Equity						

CONTACT INFORMATION:

Phone: 843-378-2211 Fax:
Toll-Free:
Address: 670 N. Main Steet, Society Hill, NC 29593-0157 United States

STOCK TICKER/OTHER:

Stock Ticker: Private Exchange:
Employees: 4,790 Fiscal Year Ends: 30-Sep
Parent Company: PATRIARCH PARTNERS LLC

SALARIES/BONUSES:

Top Exec. Salary: $ Bonus: $
Second Exec. Salary: $ Bonus: $

OTHER THOUGHTS:

Estimated Female Officers or Directors:
Hot Spot for Advancement for Women/Minorities:

TAL Apparel Group

www.talgroup.com

NAIC Code: 315220

TYPES OF BUSINESS:

Apparel Manufacturing
Men's Dress Shirts
Knitted Clothing
Tailored Suits
Women's Clothing
Outerwear & Rainwear
Nanotextiles

BRANDS/DIVISIONS/AFFILIATES:

SofTAL
WOR

CONTACTS:
Note: Officers with more than one job title may be intentionally listed here more than once.

Harry Lee, Managing Dir.

GROWTH PLANS/SPECIAL FEATURES:

TAL Apparel Group, a Hong Kong-based company, has manufactured and distributed garments since 1947. The company makes outerwear, shirts, polo shirts, blouses, wool pants and cotton pants, with products for both men and women. TAL produces one of every six dress shirts sold in the U.S. From its factories across Southeast Asia and in the U.S., the company produces its clothing for labels such as Eddie Bauer, Tommy Hilfiger, Banana Republic, Talbots, Callaway, Giordano and Givenchy. The firm manufactures its tailored suits with traditional hand stitching. Machinists sew most of the garments TAL sends out, but the company has invested in computerized cutting machines to expedite production. The firm's continual investment in research and development has resulted in innovations that include patents for non-iron, non-wrinkle shirts; deodorant technology; and its pucker free seam innovation, for which it holds patents worldwide. Its proprietary SofTAL fabric is made from natural fiber designed to be as low maintenance as possible, and the WOR nano-technology combines wrinkle free and water and oil repellent finishes in one treatment. The company often ships directly to its customers' individual stores. Daily, customers in the U.S. inform TAL's sales team electronically what they have sold. The company manages the inventory of the products that it manufactures, as well as forecasting sales in some cases, and replenishes a store's stock when a product runs low. Traditionally, 80% of the company's production has gone to the U.S., although TAL's current goal is to increase its Asian market share.

FINANCIAL DATA:
Note: Data for latest year may not have been available at press time.

In U.S. $	2012	2011	2010	2009	2008	2007
Revenue	550,000,000	500,000,000				
R&D Expense						
Operating Income						
Operating Margin %						
SGA Expense						
Net Income						
Operating Cash Flow						
Capital Expenditure						
EBITDA						
Return on Assets %						
Return on Equity %						
Debt to Equity						

CONTACT INFORMATION:

Phone: 852-2738-6211 Fax: 852-2738-6228
Toll-Free:
Address: 49 Austin Rd., TAL Bldg., 4th Fl., Hong Kong, Hong Kong

STOCK TICKER/OTHER:

Stock Ticker: Private
Employees: 23,000
Parent Company:
Exchange:
Fiscal Year Ends:

SALARIES/BONUSES:

Top Exec. Salary: $ Bonus: $
Second Exec. Salary: $ Bonus: $

OTHER THOUGHTS:

Estimated Female Officers or Directors:
Hot Spot for Advancement for Women/Minorities:

Talbots Inc (The)

www.talbotsinc.com

NAIC Code: 448120

TYPES OF BUSINESS:

Women's Apparel, Retail
Online & Catalog Sales
Footwear
Accessories

BRANDS/DIVISIONS/AFFILIATES:

Talbots Woman
Talbots Misses
Talbots Petites
Talbots Classics National Bank
Talbots Classics Finance Company
Talbots.com
Sycamore Partners

CONTACTS: *Note: Officers with more than one job title may be intentionally listed here more than once.*

Trudy F. Sullivan, CEO
Michael Scarpa, COO
Trudy F. Sullivan, Pres.
Michael Scarpa, CFO/Treas.
John Kovac, CIO/Sr. VP
Richard T. O'Connell, Jr., Corp. Sec./Exec. VP-Legal
Katherine Goodman, VP-e-commerce
Julie Lorigan, Sr. VP-Media Rel.
Julie Lorigan, Sr. VP-Investor Rel.
Benedetta Casamento, Exec. VP-Finance
Lesli Gilbert, Sr. VP-Stores
Richard T. O'Connell, Jr., Exec. VP-Real Estate, Store Planning & Design
Gary M. Pfeiffer, Chmn.
Gregory Poole, Chief Supply Chain Officer/Exec. VP

GROWTH PLANS/SPECIAL FEATURES:

The Talbots, Inc. is a leading international specialty retailer of women's apparel, shoes and accessories. The firm currently operates 517 stores in the U.S. and Canada. The company offers classic sportswear, casual wear, dresses, coats, sweaters, accessories and shoes, consisting almost exclusively of privately branded merchandise in misses, petites, woman and woman petite sizes. The firm's target demographic is women 32 years and older. In addition to brick-and-mortar retail operations, Talbots maintains a significant direct marketing sales division that includes catalogs, red-line phone sales and an e-commerce site, Talbots.com. During 2011, it circulated approximately 41 million catalogs worldwide. Talbots' 2011 retail operation accounts for about 80% of its overall business, with catalog & internet sales accounting for the rest. Through wholly-owned subsidiaries Talbots Classics National Bank and Talbots Classics Finance Company, the firm manages and administers a Talbots credit card program. Recently, the firm began opening new upscale design concept stores that offer an exclusive selection of merchandise only available at these new locations. As part of a store rationalization initiative, Talbots plans to shutter around 110 stores by the end of 2013. In mid-2012, the company was acquired by the private equity company Sycamore Partners.

The company offers its employees an associate merchandise discount, health and dental insurance, paid time off, life and disability insurance and tuition assistance.

FINANCIAL DATA: *Note: Data for latest year may not have been available at press time.*

In U.S. $	2012	2011	2010	2009	2008	2007
Revenue	1,140,000,000	1,213,060,000	1,235,632,000	1,495,170,000	1,708,115,000	1,772,306,000
R&D Expense						
Operating Income						
Operating Margin %						
SGA Expense						
Net Income	-111,900,000	10,815,000	-29,412,000	-555,659,000	-188,841,000	31,576,000
Operating Cash Flow						
Capital Expenditure						
EBITDA						
Return on Assets %						
Return on Equity %						
Debt to Equity						

CONTACT INFORMATION:

Phone: 781-749-7600 Fax: 781-741-4369
Toll-Free: 800-825-2687
Address: 1 Talbots Dr., Hingham, MA 02043 United States

STOCK TICKER/OTHER:

Stock Ticker: Private Exchange:
Employees: 9,096 Fiscal Year Ends: 31-Jan
Parent Company: Sycamore Partners

SALARIES/BONUSES:

Top Exec. Salary: $ Bonus: $
Second Exec. Salary: $ Bonus: $

OTHER THOUGHTS:

Estimated Female Officers or Directors: 6
Hot Spot for Advancement for Women/Minorities: Y

Talon International Inc

NAIC Code: 339993

www.talonzippers.com

TYPES OF BUSINESS:

Apparel Trim
Packaging Materials
Zippers & Fasteners
Labels

BRANDS/DIVISIONS/AFFILIATES:

Talon
Talon KidZip
Tag-It
TekFit

CONTACTS:

Note: Officers with more than one job title may be intentionally listed here more than once.

Lonnie D. Schnell, CEO
Larry Dyne, Pres.
Lonnie D. Schnell, CFO
James E. Reeder, Corp. Controller/VP
Mark Dyne, Chmn.

GROWTH PLANS/SPECIAL FEATURES:

Talon International, Inc. manufactures and distributes a range of zippers and trim items to manufacturers of fashion apparel, specialty retailers and mass merchandisers under its trademarked Talon, TekFit and Tag-It brand names. The company provides full-service outsourced trim design, sourcing and management services and supplies specified trim items for 956 active customers and manufacturers, including Abercrombie & Fitch, Fat Face, Wal-Mart, Kohl's, JC Penney, Polo Ralph Lauren, Guess and Express. Talon operates in three product groups: Talon, Trim and TekFit. The Talon segment focuses on the design, engineering, testing and distribution of zippers under the Talon brand and trade names to apparel brands and manufacturers worldwide, in countries such as China, Europe, Hong Kong, Taiwan, India, Mexico, Indonesia and Bangladesh. The company also designs the Talon KidZip, a specialty zipper for children's apparel. The Trim products group operates as a supplier, designer and sourcing agent of trim products, including woven, leather, synthetic, embroidered and novelty labels and tapes. The TekFit group distributes a proprietary stretch waistband using its patented technology. Headquartered in California, Talon has facilities around the world, including such locations as Hong Kong, London, Taiwan, Bangladesh, New York and India. In 2012, Talon zippers accounted for 49.4% of revenue; Trim, 50.5%; and Tekfit, 0.1%.

FINANCIAL DATA:

Note: Data for latest year may not have been available at press time.

In U.S. $	2012	2011	2010	2009	2008	2007
Revenue	44,600,870	41,668,510	41,459,750	38,675,790	48,170,980	40,529,560
R&D Expense						
Operating Income	994,768	1,339,248	1,471,408	289,076	-5,961,883	-3,170,679
Operating Margin %	2.23%	3.21%	3.54%	.74%	-12.37%	-7.82%
SGA Expense	13,465,630	11,864,520	10,988,980	11,023,500	15,109,500	14,189,760
Net Income	679,347	729,133	-1,466,820	-2,692,977	-8,358,786	-4,921,707
Operating Cash Flow	3,852,671	3,041,604	939,964	185,890	363,208	2,137,218
Capital Expenditure	374,796	173,401	91,285	543,117	819,876	725,498
EBITDA	1,520,207	1,945,235	1,724,937	1,022,574	-4,818,117	-1,266,203
Return on Assets %	-14.87%	-14.06%	-25.18%	-18.29%	-44.34%	-20.59%
Return on Equity %						-1015.95%
Debt to Equity						

CONTACT INFORMATION:

Phone: 818 444-4100 Fax: 818 444-4105
Toll-Free:
Address: 21900 Burbank Blvd., Ste. 270, Woodland Hills, CA 91367 United States

STOCK TICKER/OTHER:

Stock Ticker: TALN
Exchange: PINX
Employees: 171
Fiscal Year Ends: 31-Dec
Parent Company:

SALARIES/BONUSES:

Top Exec. Salary: $310,573 Bonus: $
Second Exec. Salary: $284,612 Bonus: $

OTHER THOUGHTS:

Estimated Female Officers or Directors:
Hot Spot for Advancement for Women/Minorities:

Tandy Brands Accessories Inc

www.tandybrands.com

NAIC Code: 315280

TYPES OF BUSINESS:

Belt Manufacturing
Handbags & Accessories
Small Leather Goods
Neckwear
Hair Accessories

BRANDS/DIVISIONS/AFFILIATES:

GROWTH PLANS/SPECIAL FEATURES:

Tandy Brands Accessories, Inc. is a U.S. designer, manufacturer and marketer of fashion accessories for men, women and children. The company's primary product line is belts (60.2% of total sales), with an extended portfolio of small leather goods and accessories including wallets and handbags as well as socks, cold weather goods, neckwear, eyewear, sporting goods accessories and gift accessories. The company's merchandise is marketed under a broad portfolio of licensed and proprietary brand names, including Wolverine, Eddie Bauer, Bone Collector, Totes, Rolfs, Canterbury, Prince Gardner, Princess Gardner, Amity, Surplus, Eileen West and private brands for major retail customers. Private brands account for approximately 61.2% of Tandy Brands' sales; licensed brands, 27%; and proprietary brands, 11.8%. Tandy Brands sells its products through all major retail distribution channels throughout the U.S. and Canada, including mass merchants, national chain stores, men's and women's specialty stores, department stores, sporting goods stores, catalog retailers, grocery stores, drug stores and the retail exchange operations of the U.S. military. The firm has manufacturing facilities in Texas, Mexico and Canada, although the majority of its products are outsourced to contract manufacturers in lower-cost environments, including India, China, Italy, the Dominican Republic and Taiwan.

CONTACTS: *Note: Officers with more than one job title may be intentionally listed here more than once.*

N. Roderick (Rod) McGeachy, III, CEO
N. Roderick (Rod) McGeachy, III, Pres.
Robert D. Martin, Interim CFO
Robert J. McCarten, Exec. VP-Sales & Merch., Gifts
Sue Elliott, Chief Performance Officer
David Crabtree, Dir.-IT
Hilda McDuff, Exec. VP-Merch.
Franklin Dominguez, Dir.-Oper., Tandy Brands California
Susan E. Dunn-Shea, VP-Planning & Inventory Control
Chuck Talley, Chief Acct. Officer/VP
Dan Weaver, VP-Design
Dan Pearson, Gen. Mgr./VP-Gifts & Outdoor
Jennifer Snellgrove, Gen. Mgr./VP-Belts
N. Roderick (Rod) McGeachy, III, Chmn.
Sean A. Aherne, VP-Int'l Bus./Pres., H.A. Sheldon Canada Ltd.
Vaneet Grover, Sr. VP-Supply Chain

FINANCIAL DATA: *Note: Data for latest year may not have been available at press time.*

In U.S. $	2012	2011	2010	2009	2008	2007
Revenue	117,601,000	123,767,000	141,887,000	129,017,000	149,257,000	195,809,000
R&D Expense						
Operating Income	-2,064,000	-12,165,000	-4,018,000	-15,272,000	-44,920,000	3,935,000
Operating Margin %	-1.75%	-9.82%	-2.83%	-11.83%	-30.09%	2.00%
SGA Expense	37,130,000	44,693,000	50,520,000	53,031,000	56,604,000	61,204,000
Net Income	-3,652,000	-13,476,000	1,214,000	-15,111,000	-49,288,000	1,934,000
Operating Cash Flow	4,913,000	-10,344,000	2,000	3,719,000	5,642,000	9,101,000
Capital Expenditure	664,000	958,000	5,083,000	542,000	594,000	2,204,000
EBITDA	347,000	-9,224,000	803,000	-13,100,000	-41,131,000	9,534,000
Return on Assets %	-5.99%	-19.12%	1.76%	-21.38%	-44.79%	1.38%
Return on Equity %	-12.62%	-36.05%	2.83%	-30.34%	-59.46%	1.81%
Debt to Equity						

CONTACT INFORMATION:

Phone: 214 549-5200 Fax:
Toll-Free:
Address: 3631 W. Davis St., Ste. A, Dallas, TX 75211 United States

STOCK TICKER/OTHER:

Stock Ticker: TBAC
Exchange: NAS
Employees: 516
Fiscal Year Ends: 30-Jun
Parent Company:

SALARIES/BONUSES:

Top Exec. Salary: $ Bonus: $
Second Exec. Salary: $ Bonus: $

OTHER THOUGHTS:

Estimated Female Officers or Directors: 5
Hot Spot for Advancement for Women/Minorities: Y

Target Corporation

NAIC Code: 452112

www.target.com

TYPES OF BUSINESS:

Discount Department Stores
Online Sales
Catalog Sales
Groceries
Credit Cards

BRANDS/DIVISIONS/AFFILIATES:

SuperTarget
Target.com
Market Pantry
Archer Farms
Merona
REDcards
Xhilaration
Choxie

CONTACTS: *Note: Officers with more than one job title may be intentionally listed here more than once.*

Gregg W. Steinhafel, CEO
Gregg W. Steinhafel, Pres.
Jeffrey J. Jones, II, Chief Mktg. Officer/Exec. VP
Jodeen A. Kozlak, Exec. VP-Human Resources
Beth M. Jacob, CIO
Beth M. Jacob, Exec. VP-Tech. Svcs.
Kathryn A. Tesija, Exec. VP-Merch.
Timothy R. Baer, General Counsel/Exec. VP/Corp. Sec.
John D. Griffith, Exec. VP-Property Dev.
Laysha L. Ward, Pres., Community Rel. & Target Foundation
Terrence J. Scully, Pres., Target Financial & Retail Svcs.
Tina M. Schiel, Exec. VP-Stores
Terrence J. Scully, Pres., Target Financial & Retail Svcs.
Gregg W. Steinhafel, Chmn.
Anthony S. Fisher, Pres., Target Canada
Kathryn A. Tesija, Exec. VP-Supply Chain

GROWTH PLANS/SPECIAL FEATURES:

Target Corporation operates large-format general merchandise and food discount stores in the U.S., which include Target and SuperTarget stores. The firm operates in three segments: U.S. retail operations; credit cards; and Canadian. Target operates approximately 1,763 stores in 49 states and Washington, D.C. Approximately 251 of these locations are SuperTarget stores, which combine grocery and general merchandise in a single format. SuperTarget stores feature coffee bars, bakeries, banking areas, pharmacies and photo services. Target carries brands such as Room Essentials, Durabuilt, Garden Place and Sutton and Dodge, among many others. In addition, Target sells merchandise under its own private-label brands, including Market Pantry, Archer Farms, Merona, Xhilaration, Choxie, Trutech, Target and Target Limited Edition. Target stores derive the largest percentage of sales from household essentials (approximately 25%); electronics, music, movies, sporting goods and toys for 19%; apparel and accessories for 19%; food and pet supplies for 19%; andhome furniture and decor for 18%. The company's proprietary credit card products, called REDcards, provide discounts in an attempt to gain customer loyalty. REDcards are sold under the Target Visa and Target Card names. Target also sells merchandise via its e-commerce site, Target.com. Target has 37 distribution centers, and operates three stores in India. In 2011 and 2012, Target was experimenting with everyday reduced pricing, instead of ongoing sale events. The strategy was not particularly successful. Meanwhile, the company plans its first international expansion by opening 60 stores in Canada. In addition, it is testing smaller â€˜city' stores in dense urban areas. Stores averaged about $290 in sales per square foot of store space in 2011.

Target offers employees benefits including medical, dental, vision, prescription drug, life and disability insurance; wellness programs; flexible spending accounts; childcare discounts; 401(k) plan; tuition reimbursement; and 10% team member discount on store merchandise.

FINANCIAL DATA: *Note: Data for latest year may not have been available at press time.*

In U.S. $	2012	2011	2010	2009	2008	2007
Revenue	69,865,000,000	67,390,000,000	65,357,000,000	64,948,000,000	63,367,000,000	
R&D Expense						
Operating Income	5,322,000,000	5,252,000,000	4,673,000,000	4,402,000,000	5,272,000,000	
Operating Margin %	7.61%	7.79%	7.15%	6.77%	8.31%	
SGA Expense	14,106,000,000	13,469,000,000	13,078,000,000	14,563,000,000	13,704,000,000	
Net Income	2,929,000,000	2,920,000,000	2,488,000,000	2,214,000,000	2,849,000,000	
Operating Cash Flow	5,434,000,000	5,271,000,000	5,881,000,000	4,430,000,000	4,125,000,000	
Capital Expenditure	4,368,000,000	2,129,000,000	1,729,000,000	3,547,000,000	4,369,000,000	
EBITDA	7,456,000,000	7,339,000,000	6,699,000,000	6,256,000,000	6,931,000,000	
Return on Assets %	6.48%	6.61%	5.61%	4.99%	6.95%	
Return on Equity %	18.71%	18.94%	17.12%	15.25%	18.41%	
Debt to Equity	0.84	1	0.98	1.27	0.98	

CONTACT INFORMATION:

Phone: 612 304-6073 Fax: 612 370-5502
Toll-Free:
Address: 1000 Nicollet Mall, Minneapolis, MN 55403 United States

STOCK TICKER/OTHER:

Stock Ticker: TGT Exchange: NYS
Employees: 365,000 Fiscal Year Ends: 31-Jan
Parent Company:

SALARIES/BONUSES:

Top Exec. Salary: $1,500,000 Bonus: $1,250,000
Second Exec. Salary: $1,025,000 Bonus: $423,325

OTHER THOUGHTS:

Estimated Female Officers or Directors: 9
Hot Spot for Advancement for Women/Minorities: Y

Ted Baker plc

www.tedbaker-london.com

NAIC Code: 315220

TYPES OF BUSINESS:

Men's, Women's & Children's Apparel
Housewares
Watches
Footwear
Retail Stores
Online Sales

BRANDS/DIVISIONS/AFFILIATES:

Ted Baker Pashion
Ted Baker Endurance
Global
Archive
Teddy Boy
Teddy Girl

CONTACTS: *Note: Officers with more than one job title may be intentionally listed here more than once.*

Raymond Stuart Kelvin, CEO
Lindsay Dennis Page, Dir.-Finance
Charles F. Anderson, Sec.
David Bernstein, Chmn.

GROWTH PLANS/SPECIAL FEATURES:

Ted Baker plc began as a British designer men's clothing brand, but has since extended beyond its roots, repositioning itself as a lifestyle brand. The firm's collection now includes general womenswear, menswear, beachwear, formal clothing, handbags, watches, eyewear, footwear, fragrance and accessories, encompassing collections such as Ted Baker Pashion, Ted Baker Endurance, Global, Archive, Teddy Girl and Teddy Boy. The primary collections, which maintain the other collections, are the Ted Baker Menswear and Ted Baker Womenswear lines. The Ted Baker men's clothing line includes innovations such as the wrinkle-proof suit and anti-bacterial t-shirts. The company's retail division consists of 33 stores, 10 outlets and 165 concessions throughout the U.K. and Europe; 13 stores and two outlets in the U.S.; and 23 stores throughout Middle East, Asia and Australia. Ted Baker shop locations in the U.S. include Dallas, Texas; Orlando and Miami, Florida; Chicago, Illinois; Scottsdale, Arizona; Las Vegas, Nevada; Los Angeles, Santa Monica and Costa Mesa, California; and New York, New York. Many independent boutique stores and leading department stores (such as Nordstrom) also stock Ted Baker products. Ted Baker also sells apparel through its e-commerce web site. The brand was created and named after a trademarked character called Ted Baker, a stylish, charming and well-dressed British gentleman. The company is noted for avoiding the use of traditional advertising, preferring to rely primarily on word of mouth to grow the brand.

FINANCIAL DATA: *Note: Data for latest year may not have been available at press time.*

In U.S. $	2012	2011	2010	2009	2008	2007
Revenue		343,600,000	291,600,000	231,240,000	208,078,000	183,820,000
R&D Expense						
Operating Income						
Operating Margin %						
SGA Expense						
Net Income		28,000,000	26,800,000	19,040,000	22,300,000	21,090,000
Operating Cash Flow						
Capital Expenditure						
EBITDA						
Return on Assets %						
Return on Equity %						
Debt to Equity						

CONTACT INFORMATION:

Phone: 44-20-7255-4800 Fax: 44-20-7255-4961
Toll-Free:
Address: 6A St. Pancras Way, Ugly Brown Bldg., London, NW1 OTB United Kingdom

STOCK TICKER/OTHER:

Stock Ticker: TBK
Employees: 1,700
Parent Company:
Exchange: London
Fiscal Year Ends: 31-Jan

SALARIES/BONUSES:

Top Exec. Salary: $ Bonus: $
Second Exec. Salary: $ Bonus: $

OTHER THOUGHTS:

Estimated Female Officers or Directors: 1
Hot Spot for Advancement for Women/Minorities:

Tefron Ltd

NAIC Code: 315240

www.tefron.com

TYPES OF BUSINESS:

Underwear & Intimate Apparel
Swimwear
Active Wear

BRANDS/DIVISIONS/AFFILIATES:

Lea by Lea Gottlieb
Macro Clothing Ltd

CONTACTS: *Note: Officers with more than one job title may be intentionally listed here more than once.*

Amit Meridor, CEO
Eliezer Parnafes, CFO
Arnon Tiberg, Chmn.

GROWTH PLANS/SPECIAL FEATURES:

Tefron, Ltd. designs and manufactures men's underwear and women's intimate apparel, active-wear and swimwear for well-known brands and designer labels marketed worldwide. The company's customers include Victoria's Secret, Calvin Klein, JC Penny and adidas as well as other retailers and designer labels. The firm's product lines include knitted briefs, bras, tank tops, boxers, leggings, t-shirts, nightwear, bodysuits, swimwear and active-wear. The company operates in two segments: Seamless (also called Hi-Tex) and Cut & Sew. Tefron's Seamless division manufactures intimate apparel and active-wear products. Its operations are based on the Hi-Tex manufacturing process, which uses a single machine to transform yarn directly into a nearly complete garment, replacing the knitting, cutting and significant sewing functions performed in traditional manufacturing. The process allows the company to produce a wider range of fabrics, styles and product lines. The manufacturing for the Seamless division takes place mainly in Israel. The Cut & Sew division manufactures intimate apparel, active-wear and swimwear products. The manufacturing process in this division involves a vertically integrated production process, with Tefron involved in all steps of the process, from the design of the product to the knitting, dyeing, cutting and sewing of the product. Manufacturing and production in the Cut & Sew division is concentrated in Jordan and the Far East. Macro Clothing, Ltd., the company's swimwear manufacturing subsidiary, produces third-party branded swimwear as well as its own brand, Lea by Lea Gottlieb.

FINANCIAL DATA: *Note: Data for latest year may not have been available at press time.*

In U.S. $	2012	2011	2010	2009	2008	2007
Revenue			86,044,000	115,538,000	173,829,000	158,614,000
R&D Expense						
Operating Income			-22,806,000	-20,926,000	-19,228,000	1,752,000
Operating Margin %			-26.50%	-18.11%	-11.06%	1.10%
SGA Expense			15,900,000	17,621,000	23,365,000	17,715,000
Net Income			-22,686,000	-17,393,000	-17,579,000	483,000
Operating Cash Flow			-2,420,000	-1,476,000	-7,718,000	3,034,000
Capital Expenditure			113,000	686,000	3,374,000	6,376,000
EBITDA			-13,273,000	-11,208,000	-9,984,000	11,315,000
Return on Assets %			-24.93%	-15.06%	-11.94%	.29%
Return on Equity %			-54.72%	-31.48%	-23.12%	.56%
Debt to Equity			0.55			0.15

CONTACT INFORMATION:

Phone: 972 49900881 Fax: 972 49900054
Toll-Free:
Address: Industrial Center Teradyon, Misgav, 20179 Israel

STOCK TICKER/OTHER:

Stock Ticker: TFRFF Exchange: PINX
Employees: 1,149 Fiscal Year Ends: 31-Dec
Parent Company:

SALARIES/BONUSES:

Top Exec. Salary: $ 225 Bonus: $
Second Exec. Salary: $ 180 Bonus: $ 29

OTHER THOUGHTS:

Estimated Female Officers or Directors: 2
Hot Spot for Advancement for Women/Minorities:

Teijin Ltd

www.teijin.com

NAIC Code: 313110

TYPES OF BUSINESS:

Textiles Manufacturing
Polyester Fibers
Plastics & Films
Pharmaceuticals
Software & IT Solutions
Nanotextiles
Polyester Recycling Technology
Trading & Retail

BRANDS/DIVISIONS/AFFILIATES:

GRANDIT
Toho Tenax Co Ltd
Pyromex

CONTACTS: *Note: Officers with more than one job title may be intentionally listed here more than once.*

Shigeo Ohyagi, CEO
Shigeo Ohyagi, Pres.
Yoshihisa Sonobe, CFO/Exec. VP
Kentaro Arao, Chief Mktg. Officer
Yasuhiro Hayakawa, Chief Human Resources Officer
Takashi Takahashi, CTO/Exec. VP
Yo Goto, Chief Eng. Officer

GROWTH PLANS/SPECIAL FEATURES:

Teijin, Ltd., through its roughly 150 group companies, serves the fibers, plastics, pharmaceuticals and films industries. It operates five main groups: IT business; advanced fibers & composites business; health care business; electric materials & performance polymer products business; and products converting business. The IT Business provides IT services to the health care industry as well as e-commerce, digital content management and the proprietary web-based enterprise resource planning solution, GRANDIT. Its advanced fibers & composite materials products include high performance fibers, which includes Aramid Fibers; and carbon fibers & composites including flame resistant Pyromex. The high-performance unit markets Aramid fibers, used in protective clothing, automotive parts and sporting goods and artificial leather. The carbon fibers & composites unit markets carbon and flame-resistant fibers, used in aircraft components, golf clubs and industrial applications. The health care business includes pharmaceuticals for treatment of bone, joint, cardiovascular and respiratory conditions; medical equipment; and oxygen delivery equipment. The electronic materials & performance polymer business produces Polyethylene Terephthalate (PET) film, used in LCDs and food and beverage packaging; Polyethylene Naphthalate (PEN) film, used in digital components; and processed film, used in semiconductor materials and flexible panel displays. The products converting business consists of the sales and trading of fibrous raw material, apparel, industrial materials and performance polymer products; polyester/recycled polyester fibers/textiles; and closed-loop recycling of polyester products.

FINANCIAL DATA: *Note: Data for latest year may not have been available at press time.*

In U.S. $	2012	2011	2010	2009	2008	2007
Revenue	9,511,919,000	9,080,894,000	8,526,291,000	10,503,220,000	11,541,000,000	11,239,980,000
R&D Expense				37,630,000,000	36,282,000,000	
Operating Income	379,020,500	540,630,800	149,575,300	200,020,000	725,465,000	835,673,200
Operating Margin %	3.98%	5.95%	1.75%	1.90%	6.28%	7.43%
SGA Expense	2,128,389,000			1,815,166,000	1,725,487,000	2,044,900,000
Net Income	133,365,200	280,357,600	-397,267,900	-478,318,000	140,423,700	379,911,100
Operating Cash Flow	597,499,500	858,730,100	895,469,800	449,694,400	598,301,100	1,073,858,000
Capital Expenditure	307,734,300	283,397,000	379,855,500	844,401,600	877,534,100	779,272,100
EBITDA	946,549,200	1,172,554,000	345,899,000	489,351,000	955,177,500	835,673,200
Return on Assets %	1.57%	3.17%	-4.20%	-4.54%	1.25%	3.51%
Return on Equity %	4.15%	9.05%	-12.35%	-12.67%	3.59%	10.19%
Debt to Equity	0.45	0.59	0.69	0.57	0.31	0.3

CONTACT INFORMATION:

Phone: 81 662682132 Fax:
Toll-Free:
Address: 6-7, Minami-Hommachi 1-chome, Chuo-ku, Osaka, 541-8587 Japan

STOCK TICKER/OTHER:

Stock Ticker: TINLY
Exchange: PINX
Employees: 17,452
Fiscal Year Ends: 31-Mar
Parent Company:

SALARIES/BONUSES:

Top Exec. Salary: $ Bonus: $
Second Exec. Salary: $ Bonus: $

OTHER THOUGHTS:

Estimated Female Officers or Directors:
Hot Spot for Advancement for Women/Minorities: Y

Texwinca Holdings Limited

www.texwinca.com

NAIC Code: 313110

TYPES OF BUSINESS:

Yarn Spinning Mills
Apparel Sales and Manufacturing
Vehicle Sales
Vehicle Repair
Generator Sales
Generator Service

BRANDS/DIVISIONS/AFFILIATES:

Dongguan Texwinca Textile & Garment Ltd.
Megawell Group
Baleno Group
Winca (Dongguan) Motor Service, Ltd.
Winca Trading, Ltd.

GROWTH PLANS/SPECIAL FEATURES:

Texwinca Holdings Limited is principally engaged in the production and sales of dyed yarns and knitted fabrics. The firm's sales network covers the U.S., Europe and Asia. Texwinca has a seven million square foot factory, with the capacity to produce 9 million pounds of dyed wool every month, 25 million pounds knitted fabric and 25 million lbs dyed and finished fabric. Products include single knit, double knit, velour and polar fleece, jacquard and fabrics blended with spandex (lycra), in addition to flat knit collars and cuffs and dyed yarns. The firm has offices in Hong Kong, China and Macau. Texwinca has several subsidiaries including Dongguan Texwinca Textile & Garment Ltd. (knit fabric production); Megawell Group (garment sales and manufacturing); Baleno Group (apparel retailing); Winca (Dongguan) Motor Service, Ltd. (vehicle sales and repair); and Winca Trading, Ltd. (generator service and sales).

CONTACTS: *Note: Officers with more than one job title may be intentionally listed here more than once.*

Poon Bun Chak, CEO
Poon Bun Chak, Chmn.

FINANCIAL DATA: *Note: Data for latest year may not have been available at press time.*

In U.S. $	2012	2011	2010	2009	2008	2007
Revenue	1,775,709,000	1,530,098,000	1,359,334,000	1,289,648,000	1,247,269,000	1,070,463,000
R&D Expense						
Operating Income	142,140,300	184,430,000	152,919,100	128,984,700	136,432,000	78,117,580
Operating Margin %	8.00%	12.05%	11.24%	10.00%	10.93%	7.29%
SGA Expense	425,705,400	354,835,300	322,052,700	319,833,900	280,491,900	262,909,300
Net Income	123,681,100	162,216,900	141,989,300	114,949,600	123,722,200	80,727,140
Operating Cash Flow	122,345,900	162,029,700	102,799,700	135,644,900	145,937,500	77,199,670
Capital Expenditure	29,870,360	22,131,750	18,434,140	21,078,520	24,038,800	66,097,540
EBITDA	194,918,000	232,611,700	199,863,000	179,212,000	203,527,700	138,747,000
Return on Assets %	7.24%	11.04%	14.63%	12.57%	14.55%	11.25%
Return on Equity %	16.34%	21.99%	24.65%	21.62%	26.21%	20.43%
Debt to Equity	0.07	0.1	0.1	0.12	0.21	0.17

CONTACT INFORMATION:

Phone: +852 24818018 Fax: +852 22331111
Toll-Free:
Address: Metroplaza Tower II, 223 Hing Fong Rd., Fl. 16, Hong Kong, Hong Kong

STOCK TICKER/OTHER:

Stock Ticker: TXWHF Exchange: GREY
Employees: 23,853 Fiscal Year Ends: 31-Mar
Parent Company:

SALARIES/BONUSES:

Top Exec. Salary: $ Bonus: $
Second Exec. Salary: $ Bonus: $

OTHER THOUGHTS:

Estimated Female Officers or Directors:
Hot Spot for Advancement for Women/Minorities:

Tilly's Inc

www.tillys.com

NAIC Code: 448140

TYPES OF BUSINESS:

Retail Surf and Board Wear Stores
E-commerce

BRANDS/DIVISIONS/AFFILIATES:

RSQ
Full Tilt
Blue Crown
Infamous
Tillys.com

CONTACTS: *Note: Officers with more than one job title may be intentionally listed here more than once.*

Daniel Griesemer, CEO
Craig DeMerit, COO/VP
Daniel Griesemer, Pres.
Bill Langsdorf, CFO/Sr. VP
Cheryl Rudich, VP-Mktg.
Barb Arneklev-Krol, VP-Human Resources
Craig DeMerit, CIO
Debbie Anker-Morris, VP/Gen. Merch. Mgr.
Christopher M. Lal, General Counsel/VP/Sec.
Hezy Shaked, Chief Strategy Officer
Rochelle Myers, VP-Finance/Controller
Shelly Bueno-Johnson, VP-Stores
Tilly Levine, VP-Vendor Rel.
Carolyn S. McNamara, VP-Merch. Planning & Allocation
Jeff Sinkey, VP-Real Estate
Hezy Shaked, Chmn.

GROWTH PLANS/SPECIAL FEATURES:

Tilly's, Inc. is a specialty retailer of surf-inspired clothing and accessories based in Irvine, California. Targeting teens and young adults, the company offers apparel, footwear and lifestyle accessories inspired by the West Coast surf, action sport and music culture. The firm currently operates 161 retail stores in 25 states, with an average store size of approximately 7,800 square feet. These stores offer numerous popular and emerging brands centered on the youth culture of its target customers, including Billabong, Element, Hurley, Levi's, LRG, Neff, RVCA, Uggs and Volcom. Tilly's also sells extensive lines of proprietary merchandise under the RSQ, Full Tilt, Blue Crown and Infamous labels. In addition to its bricks-and-mortar stores, the company operates an e-commerce site at Tillys.com, which accounts for over 10% of its annual revenue. The firm handles merchandise distribution and fulfillment from a 126,000-square-foot distribution facility co-located at its headquarters in Irvine. Tilly's uses its own fleet of trucks to ship merchandise to local (Southern California) stores and third party distributors to ship merchandise to stores outside this area. In May 2012, the company completed an initial public offering of common stock.

FINANCIAL DATA: *Note: Data for latest year may not have been available at press time.*

In U.S. $	2012	2011	2010	2009	2008	2007
Revenue	400,624,000	332,604,000	282,764,000			
R&D Expense						
Operating Income	34,925,000	24,947,000	21,422,000			
Operating Margin %	8.71%	7.50%	7.57%			
SGA Expense	94,217,000	77,668,000	65,912,000			
Net Income	34,340,000	24,416,000	20,863,000			
Operating Cash Flow	52,584,000	41,702,000	35,256,000			
Capital Expenditure	20,223,000	15,674,000	17,514,000			
EBITDA	50,054,000	39,239,000	35,337,000			
Return on Assets %	24.82%	19.81%	18.07%			
Return on Equity %	161.82%	40.03%	34.83%			
Debt to Equity		0.07	0.08			

CONTACT INFORMATION:

Phone: 949 609-5599 Fax:
Toll-Free:
Address: 10 Whatney, Irvine, CA 92618 United States

STOCK TICKER/OTHER:

Stock Ticker: TLYS
Exchange: NYS
Employees: 3,100
Fiscal Year Ends: 31-Jan
Parent Company:

SALARIES/BONUSES:

Top Exec. Salary: $659,615 Bonus: $
Second Exec. Salary: $640,000 Bonus: $

OTHER THOUGHTS:

Estimated Female Officers or Directors: 7
Hot Spot for Advancement for Women/Minorities: Y

Timberland Co

NAIC Code: 316210

www.timberland.com

TYPES OF BUSINESS:

Shoes, Manufacturing
Apparel & Accessories
Retail Stores
Online Sales

BRANDS/DIVISIONS/AFFILIATES:

Timberland PRO
Timberland Boot Company
SmartWool
IPATH
VF Corp

CONTACTS: *Note: Officers with more than one job title may be intentionally listed here more than once.*

Eric C. Wiseman, CEO-VF Corp.
Eric C. Wiseman, Pres., VF Corp.
Steve Rendle, Pres., Outdoor & Action Sports America-VF Corp.
Eric C. Wiseman, Chmn.-VF Corp.

GROWTH PLANS/SPECIAL FEATURES:

Timberland Co. designs, develops, engineers, markets and distributes footwear, apparel and accessories products for men, women and children under the Timberland, Timberland PRO, SmartWool, Timberland Boot Company, IPATH and howies brands. The firm's products are sold primarily through independent retailers, better-grade department stores, athletic stores and other national retailers as well as through Timberland specialty stores, Timberland factory outlet stores, Timberland footwear stores and the Internet. Timberland products are sold throughout the U.S., Canada, Europe, Asia, Latin America, Africa and the Middle East. Timberland PRO is dedicated to providing footwear for on-the-job use. This line features higher-grade leather combined with steel or soft toes and waterproofing options. The line also features customized fit through an Ergomorphic foam footbed. The SmartWool brand includes apparel and accessories such as performance socks, lifestyle socks, cycling jerseys and shorts, sweaters, hats, scarves and technical ski gloves. IPATH includes skateboard-related casual footwear, apparel and accessories. Roughly 79% of revenue is derived from footwear, the rest being generated by apparel and accessories. Sales to North America generated 40%, while the rest of the world generated the remaining 60%. Not too long ago, Timberland announced a licensing agreement partnership with Parigi Group which will design, source and market Timberland's kids' apparel in the U.S. Recently, the firm was acquired by VF Corp., a leading brand-name apparel producer, for approximately $2 billion.

Timberland Co offers employees health insurance, an employee assistance plan, disability coverage and retirement savings plans.

FINANCIAL DATA: *Note: Data for latest year may not have been available at press time.*

In U.S. $	2012	2011	2010	2009	2008	2007
Revenue	1,440,000,000	1,500,000,000	1,429,484,000	1,285,876,000	1,364,550,000	1,436,451,000
R&D Expense						
Operating Income						
Operating Margin %						
SGA Expense						
Net Income			96,622,000	56,644,000	42,906,000	39,999,000
Operating Cash Flow						
Capital Expenditure						
EBITDA						
Return on Assets %						
Return on Equity %						
Debt to Equity						

CONTACT INFORMATION:

Phone: 603-772-9500 Fax: 603-926-9239
Toll-Free: 888-802-9947
Address: 200 Domain Dr., Stratham, NH 03885 United States

STOCK TICKER/OTHER:

Stock Ticker: Subsidiary
Employees: 5,600
Parent Company: VF CORP
Exchange:
Fiscal Year Ends: 31-Dec

SALARIES/BONUSES:

Top Exec. Salary: $ Bonus: $
Second Exec. Salary: $ Bonus: $

OTHER THOUGHTS:

Estimated Female Officers or Directors:
Hot Spot for Advancement for Women/Minorities: Y

TJX Companies Inc (The)

www.tjx.com

NAIC Code: 448140

TYPES OF BUSINESS:

Discount Apparel Stores
Domestics
Footwear
Jewelry
Home Furnishings
Accessories

BRANDS/DIVISIONS/AFFILIATES:

T.J. Maxx
Marshalls
HomeGoods
Winners
T.K. Maxx
HomeSense

CONTACTS: *Note: Officers with more than one job title may be intentionally listed here more than once.*

Carol Meyrowitz, CEO
Ernie Herrman, Pres.
Scott Goldenberg, CFO/Exec. VP
Jeffrey G. Naylor, Chief Admin. Officer
Jerome Rossi, Sr. VP-e-commerce
Richard Sherr, Sr. Exec. VP/Group Pres., Marmaxx
Jerome R. Rossi, Sr. Exec. VP/Group Pres.
Nan Stutz, Sr. Exec. VP/Group Pres., HomeGoods
Bernard Cammarata, Chmn.
Michael MacMillan, Group Pres., Europe/Sr. Exec. VP
Jermone Rossi, Sr. VP-Logistics & Procurement

GROWTH PLANS/SPECIAL FEATURES:

The TJX Companies, Inc. is a low-price apparel and home fashions retailer, operating over 2,900 stores. The firm operates through six chain names: T.J. Maxx, Marshalls and HomeGoods in the U.S. and Canada; Winners and HomeSense in Canada; and HomeSense and T.K. Maxx in Europe. TJX's 1,880 T.J. Maxx stores and Marshalls stores offer brand-name family apparel, including footwear and accessories, and home fashions, including home basics, accent furniture and giftware, in the U.S., Washington, D.C., Puerto Rico and Canada. The chains are similar, although Marshalls features a full-line shoe department and larger men's and juniors' departments while T.J. Maxx carries an extended line of jewelry and accessories. TJX's stores offer merchandise sold at 20% to 60% below department and specialty store regular prices. TJX also operates 374 U.S. HomeGoods stores, which sell low-priced home fashions. The chain offers a broad array of home basics, giftware, accent furniture, lamps, rugs, wall decor, decorative accessories, children's furniture and seasonal merchandise. Its 110 international HomeSense stores resemble the HomeGoods chain. Moreover, the firm operates the 216-store Winners chain, patterned after T.J. Maxx, in Canada. The company's 332 T.K. Maxx stores in the U.K., Ireland and Germany resemble its domestic stores, although with a slight name change. The company purchases its inventory from over 15,000 vendors worldwide. In 2012, TJX completed the consolidation of 90 of its A.J. Wright locations to the T.J. Maxx, Marshalls and HomeGoods brands; while it closed the remaining 72 A.J. Wright stores. Also in 2012, the firm opened six additional Marshalls stores in Canada as a continuation of its 2011 introduction of the chain.

TJX offers its employees medical, dental, vision, disability and life insurance; a 401(k) plan; a profit sharing plan; group auto and home insurance; a college savings program; store discounts; a mortgage discount program; a scholarship award program; credit union access; paid vacation; and flexible spending accounts. Corporate employees have access to basketball courts, fitness classes and indoor golf driving ranges.

FINANCIAL DATA: *Note: Data for latest year may not have been available at press time.*

In U.S. $	2012	2011	2010	2009	2008	2007
Revenue	23,191,460,000	21,942,190,000	20,288,440,000	18,999,500,000	18,647,130,000	
R&D Expense						
Operating Income	2,411,414,000	2,203,229,000	1,951,562,000	1,465,231,000	1,241,091,000	
Operating Margin %	10.39%	10.04%	9.61%	7.71%	6.65%	
SGA Expense	3,890,144,000	3,710,053,000	3,328,944,000	3,139,518,000	3,323,587,000	
Net Income	1,496,090,000	1,343,141,000	1,213,572,000	880,617,000	771,750,000	
Operating Cash Flow	1,916,034,000	1,976,481,000	2,271,926,000	1,154,572,000	1,361,107,000	
Capital Expenditure	803,330,000	707,134,000	429,282,000	582,932,000	526,987,000	
EBITDA	2,943,798,000	2,671,158,000	2,436,058,000	1,918,167,000	1,708,555,000	
Return on Assets %	18.40%	17.40%	17.79%	13.78%	12.16%	
Return on Equity %	47.42%	44.85%	48.31%	41.28%	34.90%	
Debt to Equity	0.24	0.25	0.27	0.17	0.4	

CONTACT INFORMATION:

Phone: 508 390-1000 Fax: 508 390-2091
Toll-Free:
Address: 770 Cochituate Rd., Framingham, MA 01701 United States

STOCK TICKER/OTHER:

Stock Ticker: TJX Exchange: NYS
Employees: 166,000 Fiscal Year Ends: 31-Jan
Parent Company:

SALARIES/BONUSES:

Top Exec. Salary: $1,426,924 Bonus: $
Second Exec. Salary: $1,205,770 Bonus: $

OTHER THOUGHTS:

Estimated Female Officers or Directors: 3
Hot Spot for Advancement for Women/Minorities: Y

Tod's SpA

NAIC Code: 316210

www.todsgroup.com

TYPES OF BUSINESS:

Shoes & Leather Goods
Fashion Apparel

BRANDS/DIVISIONS/AFFILIATES:

Formapura Srl
Hogan
Fay
Driving Shoe
D-Bag
Edmond Srl

CONTACTS:

Note: Officers with more than one job title may be intentionally listed here more than once.

Diego Della Valle, CEO
Cinzia Oglio, Mgr.-Investor Relations
Andrea Della Valle, Vice Chmn.
Stefano Sincini, Co-CEO/Exec. Dir./Gen. Mgr.
Diego Della Valle, Chmn.

GROWTH PLANS/SPECIAL FEATURES:

Tod's SpA is an Italian designer of luxury leather goods. The company is best known for its finely crafted shoes, which it has been producing since its foundation in the late 1920s. Besides shoes, which generate over 70% of annual revenue, Tod's produces high-end leather accessories such as handbags and apparel. Its most recognized products include its Driving Shoe and D-Bag. Substantially all of Tod's goods are produced in six company-owned manufacturing facilities in Italy. Additionally, Tod's produces merchandise under the Hogan and Fay brands and owns licenses for the French shoe designer Roger Vivier. Hogan is a brand of luxury sportswear and accessories with an international appeal. Fay produces a line of high-quality apparel that is set apart by the technical treatment of fabrics, a close attention to detail and mix of style and functionality. The firm maintains subsidiaries across the globe to oversee the distribution of its products, with operations throughout Europe, the U.S. and Asia. Its merchandise is sold through 176 company-owned boutiques and 70 franchised stores as well as select independent multi-brand stores. During 2012, the company purchased Formapura Srl, an Italian creative agency, and Edmond Srl, an Italian pharmaceutical and chemical plant.

FINANCIAL DATA:

Note: Data for latest year may not have been available at press time.

In U.S. $	2012	2011	2010	2009	2008	2007
Revenue		1,185,300,000	1,166,200,000	1,053,720,000		
R&D Expense						
Operating Income						
Operating Margin %						
SGA Expense						
Net Income		179,100,000	157,750,000	123,900,000		
Operating Cash Flow						
Capital Expenditure						
EBITDA						
Return on Assets %						
Return on Equity %						
Debt to Equity						

CONTACT INFORMATION:

Phone: 39-0734-8661 Fax:
Toll-Free:
Address: 1 Via Filippo Della Valle, Sant'Elpidio a Mare, 63019 Italy

STOCK TICKER/OTHER:

Stock Ticker: TODGF Exchange: GREY
Employees: 3,194 Fiscal Year Ends: 31-Dec
Parent Company:

SALARIES/BONUSES:

Top Exec. Salary: $ Bonus: $
Second Exec. Salary: $ Bonus: $

OTHER THOUGHTS:

Estimated Female Officers or Directors: 2
Hot Spot for Advancement for Women/Minorities:

Toho Tenax Co Ltd

www.tohotenax.com

NAIC Code: 313110

TYPES OF BUSINESS:

Textiles Manufacturing
Carbon Fiber
Heat-resistant Fabrics
Composite Materials

BRANDS/DIVISIONS/AFFILIATES:

Teijin Ltd
Toho Telax America Inc
Tenax Tenax Europe GmbH
TENAX
Pyromex
Toho Chemical Engineering & Construction Co Ltd
Toho Machinery Co Ltd
Diversified Structural Composites

CONTACTS: *Note: Officers with more than one job title may be intentionally listed here more than once.*

Jun Suzuki, Pres.

GROWTH PLANS/SPECIAL FEATURES:

Toho Tenax Co., Ltd., a wholly-owned subsidiary of Teijin, Ltd., is primarily engaged in the production and sale of carbon fiber, oxidized PAN fibers and advanced composite materials. The company's primary product is carbon fiber, marketed under the TENAX brand, which has high tensile strength, good fatigue and abrasion characteristics, high dimensional stability and good electromagnetic shielding performance. Carbon fiber products include filament, chopped fiber, pre-impregnated composite fibers, woven fabrics and metal-coated fibers. These products have applications in a broad range of industrial, aeronautics and aerospace applications, with Toho Tenax notably supplying materials for the next-generation Airbus A380. The materials are also used in golf clubs, fishing rods, tennis racquets and yacht bodies, among others. The firm is aggressively developing new applications for its carbon fiber in civil engineering, construction and other industries. The company also produces Pyromex, an oxidized PAN fiber with high flame, heat and chemical resistance and electrical insulation capabilities, as well as processability, draping properties and a soft texture. Pyromex is used in anti-spattering sheets for welding, protective garments for furnaces, flameproof sheets, heat insulators and gland packing. Finally, the company markets advanced composite materials, which have applications in robotics and high-speed rotation technologies. Internationally, carbon fiber manufacturing and marketing is handled by subsidiaries Toho Tenax America, Inc., based in the U.S.; and Toho Tenax Europe GmbH in Germany. Other subsidiaries and affiliates include Toho Machinery Co., Ltd.; Toho Chemical Engineering & Construction Co., Ltd.; Toho Textile Co., Ltd.; and Diversified Structural Composites.

FINANCIAL DATA: *Note: Data for latest year may not have been available at press time.*

In U.S. $	2012	2011	2010	2009	2008	2007
Revenue						
R&D Expense						
Operating Income						
Operating Margin %						
SGA Expense						
Net Income						
Operating Cash Flow						
Capital Expenditure						
EBITDA						
Return on Assets %						
Return on Equity %						
Debt to Equity						

CONTACT INFORMATION:

Phone: 81-3-3506-6800 Fax: 81-3-3506-2261
Toll-Free:
Address: 3-2-1 Kasumigaseki, Chiyoda-ku, Tokyo, 100-8585 Japan

STOCK TICKER/OTHER:

Stock Ticker: Subsidiary
Exchange:
Employees: 1,328
Fiscal Year Ends: 31-Mar
Parent Company: TEIJIN LTD

SALARIES/BONUSES:

Top Exec. Salary: $ Bonus: $
Second Exec. Salary: $ Bonus: $

OTHER THOUGHTS:

Estimated Female Officers or Directors:
Hot Spot for Advancement for Women/Minorities:

Tommy Hilfiger Corp

NAIC Code: 315240

www.tommy.com

TYPES OF BUSINESS:

Fashion Apparel
Jeans & Sportswear
Children's Apparel
Footwear
Fragrances
Home Goods
Intimate Apparel
Swimwear

BRANDS/DIVISIONS/AFFILIATES:

Hilfiger Sport
Hilfiger Denim
Tommy Hilfiger Children's Wear
Tommy Hilfiger Corporate Foundation
PVH Corp

CONTACTS: *Note: Officers with more than one job title may be intentionally listed here more than once.*

Fred Gehring, CEO
Ludo Onnink, COO
Ludo Onnink, CFO
Avery Baker, Chief Marketing Officer
Avery Baker, Exec. VP-Global Comm.
Tommy Hilfiger, Principle Designer
Daniel Grieder, CEO-Tommy Hilfiger Europe
Gary Sheinbaum, CEO-Tommy Hilfiger North America

GROWTH PLANS/SPECIAL FEATURES:

Tommy Hilfiger Corp. designs, sources and markets men's and women's sportswear, jeans and children's wear under the Tommy Hilfiger trademarks. It is wholly-owned by PVH Corp. (formerly Phillips-Van Heusen Corporation). Tommy Hilfiger operates over 1,000 retail stores in a distribution network of more than 90 countries. The firm offers a broad array of apparel, accessories, footwear, fragrance and home furnishings. Tommy Hilfiger's products are purchased by department and specialty stores throughout the world and operate through a number of clothing divisions. The Tommy Hilfiger division offers casual sportswear and accessories targeting men and women ages 24-45. The Hilfiger Denim division supplies denim separates for men and women and targets those between the ages of 18-28. In addition, the Tommy Hilfiger Children's wear division offers casual sportswear and denim lines. Though the firm was taken private several years ago by Apax Partners, original founder Tommy Hilfiger remains the firm's principal designer and overseer of the design process. The Tommy Hilfiger Corporate Foundation, established in 1995, serves as the charitable outreach arm of the company, regularly sponsoring a variety of educational and cultural programs throughout the U.S. In 2012, the firm established a joint venture with Inbrands SA to distribute their brand in Brazil, giving the firm direct control over its brand image in Brazil.

FINANCIAL DATA: *Note: Data for latest year may not have been available at press time.*

In U.S. $	2012	2011	2010	2009	2008	2007
Revenue	3,216,990,000	3,050,000,000	2,250,000,000			
R&D Expense						
Operating Income						
Operating Margin %						
SGA Expense						
Net Income	420,933,000	281,839,000				
Operating Cash Flow						
Capital Expenditure						
EBITDA						
Return on Assets %						
Return on Equity %						
Debt to Equity						

CONTACT INFORMATION:

Phone: 852-2216-0668 Fax: 852-2312-1368
Toll-Free:
Address: 850-870 Lai Chi Kok Rd., Novel Industrial Bldg., Hong Kong, Hong Kong

STOCK TICKER/OTHER:

Stock Ticker: Subsidiary
Exchange:
Employees: 6,662
Fiscal Year Ends: 31-Mar
Parent Company: PVH CORP

SALARIES/BONUSES:

Top Exec. Salary: $ Bonus: $
Second Exec. Salary: $ Bonus: $

OTHER THOUGHTS:

Estimated Female Officers or Directors: 1
Hot Spot for Advancement for Women/Minorities: Y

TOMS Shoes Inc

www.toms.com

NAIC Code: 316210

TYPES OF BUSINESS:

Shoe Manufacturing

BRANDS/DIVISIONS/AFFILIATES:

One for One
Friends of TOMS
TOMS Classics
TOMS Cordones
Stitchouts
Tiny TOMS

CONTACTS:

Note: Officers with more than one job title may be intentionally listed here more than once.

Blake Mycoskie, Chief Shoe Giver (CEO)
Doug Piwinski, Sr. VP-Global Mktg. & Comm.
Christine Simeoni, Dir.-Human Resources
Anna Hordov, Head-Affiliate Online Mktg. & Paid Advertising
Keith Eshelman, Dir.-Alternative Retail
Pete Ferrer, VP-Sales & Mktg.

GROWTH PLANS/SPECIAL FEATURES:

TOMS Shoes, Inc. is a footwear manufacturer that pledges to donate a pair of shoes to an underprivileged child for every pair sold. Through its One for One program, the company has donated over a million pairs of shoes to needy children in more than 50 countries around the world. The firm also partners with other brands, such as Ralph Lauren and Element Skateboards, to offers variations of the One for One idea. The company's designs are based primarily on the traditional Argentinean alpargata, a light slipper-like shoe made of cotton or canvas, sold as the TOMS Classics. Additionally, the company offers TOMS Cordones, a shoe that can be worn without laces; Stitchouts, designed specifically for men; the Wedge and the Wrapboot, designed for women; Youth and Tiny TOMS children's shoes; and vegan-friendly TOMS shoe options. TOMS shoes are manufactured in China, Argentina, Ethiopia and Russia. The firm markets its products in venues such as department retailers, specialty grocery stores and other similar markets. Friends of TOMS, an affiliated non-profit organization, allows TOMS employees and volunteers to participate in Shoe Drop operations in which they hand out shoes to children in person. Beyond shoes, the company also sells shirts, hoodies and eyewear accessories.

FINANCIAL DATA:

Note: Data for latest year may not have been available at press time.

In U.S. $	2012	2011	2010	2009	2008	2007
Revenue						
R&D Expense						
Operating Income						
Operating Margin %						
SGA Expense						
Net Income						
Operating Cash Flow						
Capital Expenditure						
EBITDA						
Return on Assets %						
Return on Equity %						
Debt to Equity						

CONTACT INFORMATION:

Phone: 310-566-3170 Fax: 310-463-4872
Toll-Free: 800-975-8667
Address: 3025 Olympic Blvd., Ste. C, Santa Monica, CA 90404 United States

STOCK TICKER/OTHER:

Stock Ticker: Private
Employees:
Parent Company:
Exchange:
Fiscal Year Ends:

SALARIES/BONUSES:

Top Exec. Salary: $ Bonus: $
Second Exec. Salary: $ Bonus: $

OTHER THOUGHTS:

Estimated Female Officers or Directors: 2
Hot Spot for Advancement for Women/Minorities:

TOPSHOP

NAIC Code: 448120

www.topshop.com

TYPES OF BUSINESS:

Retail, Women's Clothing & Accessories
Online Retail
Retail, Men's Clothing and Accessories
Infant Apparel

BRANDS/DIVISIONS/AFFILIATES:

TopShop.com
Boutique
Unique
Topman.com
TOPMAN
Topshop To Go
Topshop Express
Style Advisor Membership

CONTACTS: *Note: Officers with more than one job title may be intentionally listed here more than once.*

Mary Homer, Managing Dir.
Justin Cooke, Chief Mktg. Officer
Kelly Clarke, Mgr.-Area
Tim Whitmore, Dir.-Creative
Craig McGregor, Dir.-Retail
Pilar Garcia, Dir.-Trading

GROWTH PLANS/SPECIAL FEATURES:

TOPSHOP, owned by The Arcadia Group, operates a chain of affordable, fashion-forward retail stores for women, children and infants. It has over 313 U.K.-based and 600 international outlet locations. TOPSHOP's flagship store is a four-story location of nearly 90,000 square feet, located in London's high-traffic Oxford Circus area. The store draws 28,000 to 30,000 shoppers on an average day. The company's apparel is organized into several lines, including Boutique, Mainline, Premium and Unique. The Boutique items include wardrobe basics in cashmere blends and silk-based fabrics. The Mainline items include affordable seasonal designs. The Premium items include special fabrics and limited-edition designs. Some of the materials used are silk, chiffon, virgin wool and cashmere. Unique items feature products from the runway to the store. Additionally, TOPSHOP offers Tall and Petite specialty sized pieces cut to fit women over 5 feet 7 inches or under 5 feet 3 inches; Maternity, for women who are expecting but still want to be stylish and trend conscious; and a line of makeup products. The company offers unique shopper services including the Style Advisor personal shopping service available in 18 stores in the U.K., and Style Advisor Membership, a VIP-shopper card that earns loyalty rewards. Other services include Topshop To Go, in which store associates bring merchandise to shopper's homes to try on; and Topshop Express, in which shoppers text apparel needs to the flagship store in London, and the requested merchandise is then couriered anywhere in the London area on the same day. The firm also retails men's clothing and accessories through TOPMAN stores and the Topman.com web site.

FINANCIAL DATA: *Note: Data for latest year may not have been available at press time.*

In U.S. $	2012	2011	2010	2009	2008	2007
Revenue	2,000,000,000	1,850,000,000				
R&D Expense						
Operating Income						
Operating Margin %						
SGA Expense						
Net Income						
Operating Cash Flow						
Capital Expenditure						
EBITDA						
Return on Assets %						
Return on Equity %						
Debt to Equity						

CONTACT INFORMATION:

Phone: 845-121-4519 Fax: 44-20-7927-0577
Toll-Free: 866-853-8559
Address: 70 Berners St., Colegrave House, London, W1T 3NL United Kingdom

STOCK TICKER/OTHER:

Stock Ticker: Private Exchange:
Employees: 18,000 Fiscal Year Ends:
Parent Company: ARCADIA GROUP (THE)

SALARIES/BONUSES:

Top Exec. Salary: $ Bonus: $
Second Exec. Salary: $ Bonus: $

OTHER THOUGHTS:

Estimated Female Officers or Directors: 3
Hot Spot for Advancement for Women/Minorities: Y

Toray Industries Inc

www.toray.com

NAIC Code: 313110

TYPES OF BUSINESS:

Synthetic Fibers & Textiles
Chemicals
Films, Resins & Plastics
IT & Telecommunications Products
Construction Materials
Carbon Fiber Composite Materials
Pharmaceuticals & Medical Products
Engineering Services

BRANDS/DIVISIONS/AFFILIATES:

CONTACTS: *Note: Officers with more than one job title may be intentionally listed here more than once.*

Sadayuki Sakakibara, CEO
Akihiro Nikkaku, COO
Akihiro Nikkaku, Pres.
Chiaki Tanaka, Exec. VP-Plastics & Chemicals Div.
Junichi Fujikawa, Exec. VP-Electronic & Info Materials Div.
Sadayuki Sakakibara, Chmn.
Shinichi Koizumi, Exec. VP-Int'l Oper.

GROWTH PLANS/SPECIAL FEATURES:

Toray Industries, Inc. produces synthetic fibers and textiles as well as films, resins, chemicals, composites and other products. It operates in six segments: Fibers & Textiles, Plastics & Chemicals, IT-related Products, Carbon Fiber Composite Materials, Environment & Engineering and Life Science & Other Businesses. The Fibers & Textiles segment manufactures nylon, acrylic and polyester knitted fabrics, staple fibers, filament yarns, man-made suede, cellulose fibers and apparel products. Besides apparel, its products are used for car airbags, seat belts and bag filters. The Plastics & Chemicals segment manufactures numerous resins, films and chemicals used in office automation equipment, automobile parts, prepaid cards, labels and food packaging materials. The IT-related Products segment produces plastic products and films used in flat panel displays, plasma display panel (PDP) paste materials, small and medium-sized liquid crystal color filters, circuit materials, semiconductor materials, IT-related equipment and data storage materials such as video tapes. The Carbon Fiber Composite Materials segment produces carbon fibers, composites and molded products used in aircraft structures, automobiles, fishing rods, golf shafts, pressure vessels and windmills. The Environment & Engineering segment provides construction materials, water treatment membranes and related equipment and industrial machinery and equipment. Lastly, the Life Science & Other Businesses segment offers medical products; pharmaceuticals; and DNA chips and other bio-tools.

FINANCIAL DATA: *Note: Data for latest year may not have been available at press time.*

In U.S. $	2012	2011	2010	2009	2008	2007
Revenue	17,686,330,000	17,141,790,000	15,137,120,000	16,383,260,000	18,366,190,000	17,217,140,000
R&D Expense						
Operating Income	1,199,285,000	1,114,294,000	446,521,400	400,863,900	1,151,501,000	1,140,301,000
Operating Margin %	6.78%	6.50%	2.94%	2.44%	6.26%	6.62%
SGA Expense	2,446,744,000			2,532,804,000	2,586,355,000	2,478,563,000
Net Income	714,955,300	644,893,700	-157,624,600	-181,761,500	535,164,400	652,152,600
Operating Cash Flow	1,162,423,000	1,438,572,000	1,850,514,000	428,040,200	1,228,744,000	863,261,400
Capital Expenditure	967,624,400	600,338,400	671,680,300	1,128,389,000	1,613,565,000	1,373,888,000
EBITDA	1,939,647,000	1,777,958,000	905,534,300	844,368,300	1,964,652,000	1,828,359,000
Return on Assets %	4.07%	3.70%	-.91%	-1.01%	2.85%	3.64%
Return on Equity %	10.52%	10.88%	-3.01%	-3.07%	8.12%	10.39%
Debt to Equity	0.47	0.55	0.91	0.97	0.69	0.62

CONTACT INFORMATION:

Phone: 81 332455111 Fax: 81 332455054
Toll-Free:
Address: 1-1, Nihonbashi-Muromachi 2-chome, Chuo-ku, Tokyo, 103-8666 Japan

STOCK TICKER/OTHER:

Stock Ticker: TRYIF
Employees: 38,740
Parent Company:
Exchange: PINX
Fiscal Year Ends: 31-Mar

SALARIES/BONUSES:

Top Exec. Salary: $ Bonus: $
Second Exec. Salary: $ Bonus: $

OTHER THOUGHTS:

Estimated Female Officers or Directors:
Hot Spot for Advancement for Women/Minorities:

Tory Burch LLC

www.toryburch.com

NAIC Code: 315240

TYPES OF BUSINESS:

Women's Apparel
Women's Shoes
Women's Accessories
Retail Stores
Eyewear
Children's Shoes

BRANDS/DIVISIONS/AFFILIATES:

Reva
Luxottica Group
The Tory Burch Foundation
ToryBurch.com
Robinson

CONTACTS: *Note: Officers with more than one job title may be intentionally listed here more than once.*

Tory Burch, CEO

GROWTH PLANS/SPECIAL FEATURES:

Tory Burch LLC, headquartered in New York and established in 2004, is a lifestyle brand that manufactures and designs apparel and accessories for women. The firm focuses the design of its products on an American classic sportswear concept. The Tory Burch collection consists of a variety of products, such as ready-to-wear clothing, shoes, jewelry and handbags. The Tory Burch Clothing brand designs products such as tunics, tops, dresses, skirts, sweaters, jackets and outerwear, denim, swimwear and pants and shorts for women and children. The Tory Burch shoe line manufactures a variety of heels, wedges, sandals, flip flops, and flats through the Reva brand. Handbag's are produced under the Robinson brand and include totes, clutches, top handles and shoulder bags. Additional accessories include belts, hats, scarves, gloves, eye wear, wristlets, wallets, cosmetic cases, dog wear and phone cases. Since its establishment, the company continues to expand its product portfolio through partnerships with other brands, such as the Luxottica Group, which designs and manufacturers an exclusive collection of Tory Burch eyewear. Products can be purchased at 83 Tory Burch boutiques, located across the U.S., Middle East, Europe and Asia; over 800 department store chains; and online at ToryBurch.com. Founder and designer, Tory Burch also maintains the Tory Burch Foundation, which provides economic support through microfinancing and grants to small business owning women.

FINANCIAL DATA: *Note: Data for latest year may not have been available at press time.*

In U.S. $	2012	2011	2010	2009	2008	2007
Revenue	800,000,000	505,000,000	330,000,000			
R&D Expense						
Operating Income						
Operating Margin %						
SGA Expense						
Net Income						
Operating Cash Flow						
Capital Expenditure						
EBITDA						
Return on Assets %						
Return on Equity %						
Debt to Equity						

CONTACT INFORMATION:

Phone: 212-683-2323 Fax:
Toll-Free:
Address: 11 West 19th St. 7th Fl., New York, NY 10011 United States

STOCK TICKER/OTHER:

Stock Ticker: Private Exchange:
Employees: Fiscal Year Ends:
Parent Company:

SALARIES/BONUSES:

Top Exec. Salary: $ Bonus: $
Second Exec. Salary: $ Bonus: $

OTHER THOUGHTS:

Estimated Female Officers or Directors:
Hot Spot for Advancement for Women/Minorities:

Toyobo Co Ltd

www.toyobo-global.com

NAIC Code: 313210

TYPES OF BUSINESS:

Textile & Fiber Manufacturing
Advanced Materials
Biomedical Products
Industrial Textiles
Plastics & Films
Engineering
Pharmaceuticals
Water Treatment Membranes

BRANDS/DIVISIONS/AFFILIATES:

Munsingwear
Toyobo Biologics Inc
Zylon
Dyneema
Toyobo Europe GmbH
Toyobo China Co Ltd
Toyobo U.S.A. Inc

CONTACTS: *Note: Officers with more than one job title may be intentionally listed here more than once.*

Ryuzo Sakamoto, COO
Ryuzo Sakamoto, Pres.
Kunio Yano, Controlling Supervisor-Personnel & Labor
Toshiyuki Matsui, Head-Bioscience & Medical Div.
Shinichi Onizuka, Gen. Mgr.-Production Tech. Operations
Seiji Narahara, Global Head-Bus. Planning Div.
Hiroshi Takahashi, Controlling Supervisor-Finance, Acct. & Control
Hiroshi Takabayashi, Head-Plastic Div./Head-Films Div.
Shigeki Sano, Head-Fiber & Textiles
Shigeo Takenaka, Head-Specialty Chemical Div.
Toshitake Suzuki, Gen. Mgr.-Industrial Films Operating Dept.
Ryuzo Sakamoto, Chmn.
Masakazu Saito, Gen. Mgr.-China Bus. Operating Dept.
Jiro Suwa, Controlling Supervisor-Procurement Operations

GROWTH PLANS/SPECIAL FEATURES:

Toyobo Co. Ltd., founded as a textile business in 1882, has expanded into domains that utilize its core technologies of polymerization, modification, processing and biotechnology. Currently, the company operates in four segments: textiles; life science; films and functional polymers; and industrial materials. The textiles segment produces a range of fabric materials; functional textiles for sports clothing, underwear and uniforms; synthetic fibers and acrylic and acrylate fibers. The firm also operates trading companies in textiles and manufactures Munsingwear brand apparel. The division has subsidiaries in Japan, Malaysia, Indonesia, El Salvador and Brazil. Toyobo's life science segment, operating primarily through Toyobo Biologics, Inc., produces pharmaceuticals through contract manufacturing, including injections, pharmaceutical intermediaries and raw pharmaceuticals; bioproducts, including enzymes for diagnostics, diagnostic systems and reagents for research; medical membranes, equipment and devices, such as artificial kidney hollow fiber and anti-clotting materials; and water treatment membranes for seawater desalination. The films and functional polymers segment produces industrial films for LCD and optical use; synthetic paper; packaging films for food packaging, including PET (Polyethylene terephthalate), polyolefin and nylon; heat-shrink PET films; and functional polymers, including engineering plastics, industrial adhesives and coatings, photosensitive printing plates, acrylate polymers and electronic materials. The segment also conducts rubber and coat processing. The industrial materials segment produces fibers for airbags and tire cords and a variety of air filter products as well as high-performance fibers. These include Zylon, a fiber with high thermal resistance, and Dyneema, a super-strong fiber. Other miscellaneous activities include real estate development, logistics, linen supply, construction and engineering. Recently, Toyobo established Toyobo Europe GmbH and Toyobo China Co. Ltd. to strengthen its market presence in Europe and China. In January 2013, the firm announced that it merged two of its U.S. based subsidiaries, Toyobo Specialties Inc. and Toyobo America, Inc., to create Toyobo U.S.A., Inc.

FINANCIAL DATA: *Note: Data for latest year may not have been available at press time.*

In U.S. $	2012	2011	2010	2009	2008	2007
Revenue	3,891,128,000	3,791,686,000	3,548,981,000	4,088,921,000	4,803,075,000	4,750,181,000
R&D Expense						
Operating Income	203,794,200	232,573,700	127,687,300	125,015,300	301,432,800	338,840,600
Operating Margin %	5.23%	6.13%	3.59%	3.05%	6.27%	7.13%
SGA Expense	627,871,000				688,435,900	677,592,100
Net Income	51,068,230	46,258,670	23,313,030	-139,221,300	52,304,020	149,987,200
Operating Cash Flow	158,003,100	375,346,500	323,131,600	-4,097,037	259,204,400	301,310,400
Capital Expenditure	185,468,900	156,466,700	160,641,700	243,183,600	335,211,100	164,182,100
EBITDA	360,717,400	296,400,600	255,764,200	39,678,920	391,868,300	438,249,400
Return on Assets %	1.04%	.94%	.47%	-2.66%	.93%	2.62%
Return on Equity %	3.57%	3.56%	2.03%	-10.97%	3.57%	10.41%
Debt to Equity	0.56	0.54	0.73	0.97	0.44	0.56

CONTACT INFORMATION:

Phone: 81-6-6348-3111 Fax: 81-6-6348-3206
Toll-Free:
Address: 2-8, Dojima Hama 2-chome, Kita-ku, Osaka, 530-8230 Japan

STOCK TICKER/OTHER:

Stock Ticker: TYOBY Exchange: PINX
Employees: 10,178 Fiscal Year Ends: 31-Mar
Parent Company:

SALARIES/BONUSES:

Top Exec. Salary: $ Bonus: $
Second Exec. Salary: $ Bonus: $

OTHER THOUGHTS:

Estimated Female Officers or Directors:
Hot Spot for Advancement for Women/Minorities:

True Religion Apparel Inc

NAIC Code: 448140

www.truereligionbrandjeans.com

TYPES OF BUSINESS:

Jeans & Apparel Retail

BRANDS/DIVISIONS/AFFILIATES:

True Religion Brand Jeans
True Religion Japan KK
TrueReligionBrandJeans.com
True Religion Brand Jeans Germany GmbH
TrueReligionBrandJeans.com

CONTACTS: *Note: Officers with more than one job title may be intentionally listed here more than once.*

Jeffrey Lubell, CEO
Lynne Koplin, Pres.
Peter Collins, CFO
Kelly Gvildys, Sr. VP-Oper.
Jeffrey Lubell, Chmn.

GROWTH PLANS/SPECIAL FEATURES:

True Religion Apparel, Inc. is a designer, manufacturer and marketer of the True Religion Brand Jeans brand. This brand includes high fashion designer jeans and related sportswear apparel. True Religion products fall into three categories: knits, including hoodies, t-shirts and sweats; denim fabrics, which includes its line of jeans; and non-denim fabrics, such as corduroy and twill. The firm's jeans usually sell for $168-398 per pair. True Religion operates in four segments: U.S. Wholesale, International, U.S. Consumer Direct and Core Services. The U.S. Wholesale division, which accounts for approximately 20.5% of the company's net sales, markets the firm's products to approximately 570 boutique and specialty stores; and major retail chains such as Saks Fifth Avenue, Neiman Marcus, Nordstrom and Bloomingdale's. The International segment (18.8% of sales) distributes products through subsidiary True Religion Japan, K.K. and joint venture True Religion Brand Jeans Germany GmbH (40%-owned by UNIFA Premium GmbH), and distributors/sales agents who sell to upscale boutiques in their respective geographic areas. It also operates 10 retail and six outlet locations in Japan, Germany, the U.K., the Netherlands and Canada. The U.S. Consumer Direct segment (59.9% of sales) includes True Religion Apparel's e-commerce web site (TrueReligionBrandJeans.com) and its 109 self-branded retail locations (including 30 outlets). The company's Core Services division handles its licensing business, including the licensing of the True Religion brand and logo to other companies. True Religion's average branded retail store measures approximately 1,700 square feet, while its outlet stores measure roughly 2,500 square feet. Roughly 44% of the firm's sales are derived from women's attire; 53% are derived from men's products; and 3% from children's items.

FINANCIAL DATA: *Note: Data for latest year may not have been available at press time.*

In U.S. $	2012	2011	2010	2009	2008	2007
Revenue	467,285,000	419,798,000	363,714,000	311,001,000	270,000,000	173,256,000
R&D Expense						
Operating Income	78,117,000	74,611,000	69,922,000	77,597,000	68,876,000	47,142,000
Operating Margin %	16.71%	17.77%	19.22%	24.95%	25.50%	27.20%
SGA Expense	221,406,000	197,218,000	160,057,000	117,965,000	88,125,000	51,685,000
Net Income	46,015,000	44,967,000	43,496,000	47,332,000	44,371,000	27,845,000
Operating Cash Flow	64,095,000	71,768,000	67,761,000	66,490,000	49,062,000	9,810,000
Capital Expenditure	21,994,000	20,769,000	17,622,000	20,210,000	18,187,000	8,993,000
EBITDA	91,490,000	87,361,000	80,264,000	84,089,000	72,303,000	49,186,000
Return on Assets %	12.01%	13.70%	16.54%	23.88%	31.72%	28.79%
Return on Equity %	14.54%	16.38%	19.46%	27.83%	37.36%	34.22%
Debt to Equity						

CONTACT INFORMATION:

Phone: 323 266-3072 Fax: 323 266-8060
Toll-Free:
Address: 2263 E. Vernon Ave., Vernon, CA 90058 United States

STOCK TICKER/OTHER:

Stock Ticker: TRLG Exchange: NAS
Employees: 1,663 Fiscal Year Ends: 31-Dec
Parent Company:

SALARIES/BONUSES:

Top Exec. Salary: $1,034,000 Bonus: $
Second Exec. Salary: $750,000 Bonus: $

OTHER THOUGHTS:

Estimated Female Officers or Directors: 2
Hot Spot for Advancement for Women/Minorities:

Tween Brands Inc

www.tweenbrands.com

NAIC Code: 448130

TYPES OF BUSINESS:

Apparel-Children's, Retail
Footwear
Sportswear
Jewelry
Accessories
Online & Catalog Sales

BRANDS/DIVISIONS/AFFILIATES:

Too Inc
Ascena Retail Group Inc
Justice
ShopJustice.com
Brothers
Shopbrothers.com

CONTACTS:

Note: Officers with more than one job title may be intentionally listed here more than once.

Michael W. Rayden, CEO
Rolando de Aguiar, CFO/Exec. VP
Michael Keane, Sr. VP-Human Resources
Lece Lohr, Chief Merch. Officer/Exec. VP
Chris Kaighn, Sr. VP-Stores & Store Oper.
Alan J. Hochman, Sr. VP-Real Estate & Store Planning
Scott M. Bracale, Pres., The Agency
Michael W. Rayden, Chmn.
Ronnie Robinson, Exec. VP-Supply Chain

GROWTH PLANS/SPECIAL FEATURES:

Tween Brands, Inc., formerly Too, Inc. is a clothing and accessories retailer that primarily works through its subsidiary Justice, which markets to girls aged 7-14 years old. Justice sells apparel, underwear, sleepwear, swimwear, lifestyle and personal care products as well as non-apparel merchandise, such as candy, electronic toys and games, through 942 retail locations and an e-commerce site at ShopJustice.com. Justice operates in both shopping mall locations and non-mall locations such as power centers, which attract customers intent on apparel shopping. Tween Brands' clothing and other products are developed internally. The company has three direct sourcing offices located in Shanghai, Hong Kong and Seoul. Tween Brands stores feature furniture, fixtures, lighting and music to create a shopping experience matching the energetic lifestyle of young girls. To keep store atmosphere fresh, the firm continually reassesses the layout of its stores. In March 2011, the firm launched the Brothers brand and website, selling clothes online geared to boys in the same age bracket as the Justice demographic. The company is a wholly-owned subsidiary of Ascena Retail Group, Inc.

The firm offers its employees medical, prescription, dental and vision coverage; life and disability insurance; health savings accounts and flexible savings accounts; adoption assistance; an employee stock purchase plan; paid time off; a 401(k); and an employee assistance plan.

FINANCIAL DATA:

Note: Data for latest year may not have been available at press time.

In U.S. $	2012	2011	2010	2009	2008	2007
Revenue	1,306,700,000	1,150,000,000	1,050,000,000	995,062,000	1,013,971,000	883,683,000
R&D Expense						
Operating Income						
Operating Margin %						
SGA Expense						
Net Income	172,500,000	129,300,000		-17,145,000	52,574,000	64,821,000
Operating Cash Flow						
Capital Expenditure						
EBITDA						
Return on Assets %						
Return on Equity %						
Debt to Equity						

CONTACT INFORMATION:

Phone: 614-775-3500 Fax:
Toll-Free:
Address: 8323 Walton Pkwy., New Albany, OH 43054 United States

STOCK TICKER/OTHER:

Stock Ticker: Subsidiary
Exchange:
Employees: 3,350
Fiscal Year Ends: 31-Jul
Parent Company: ASCENA RETAIL GROUP INC

SALARIES/BONUSES:

Top Exec. Salary: $ Bonus: $
Second Exec. Salary: $ Bonus: $

OTHER THOUGHTS:

Estimated Female Officers or Directors: 2
Hot Spot for Advancement for Women/Minorities: Y

Under Armour Inc

NAIC Code: 315210

www.underarmour.com

TYPES OF BUSINESS:

Compression Performance Apparel
Outdoor and Sports Apparel
Shirts
Footwear
Gloves

BRANDS/DIVISIONS/AFFILIATES:

HEATGEAR
COLDGEAR
ALLSEASONGEAR

CONTACTS: *Note: Officers with more than one job title may be intentionally listed here more than once.*

Kevin A. Plank, CEO
Kip J. Fulks, COO
Kevin A. Plank, Pres.
Brad Dickerson, CFO
Adam Peake, Sr. VP-U.S. Sales
Byron K. Adams, Jr., Chief Performance Officer
Matthew Mirchin, Sr. VP-Global Brand & Sports Mktg.
Henry B. Stafford, Sr. VP-Apparel
Leanne Fremar, Exec. Dir.-Womens/Sr. VP
Kevin A. Plank, Chmn.
Karl-Heinz (Charlie) Maurath, Pres., Int'l

GROWTH PLANS/SPECIAL FEATURES:

Under Armour, Inc. (UA) develops, markets and distributes branded performance apparel, footwear and accessories for men, women and youth. It offers several lines of apparel and accessories that utilize a variety of synthetic microfiber fabrications engineered to replace traditional cotton products in the world of athletics and fitness. UA's active wear and sports apparel accessories are designed to wick perspiration away from the skin, help regulate body temperature, enhance comfort and mobility and improve performance regardless of weather conditions. Its products are designed and merchandised along three gear lines: HEATGEAR, for hot weather; COLDGEAR, for cold weather; and ALLSEASONGEAR, for weather between the extremes. Within each product line, UA's garments come in three fit types: compression (tight fitting), fitted (athletic cut) and loose (relaxed). Annually, apparel accounts for 76% of the firm's revenues, while footwear and accessories (such as gloves for baseball batting, football, golf and running) account for 13% and 9% respectively. Licensing arrangements for the sale of UA products represent the remaining 2% of revenue. The majority of sales are generated through wholesale channels, including leading sporting goods retailers such as Academy Sports and Outdoors, Dick's Sporting Goods, Hibbett Sporting Goods, The Sports Authority and Cabela's. Unaffiliated manufacturers operating in 14 countries manufacture virtually all of the company's products. International professional football, baseball, basketball, hockey, rugby and soccer players as well as athletes in major collegiate sports and junior athletes of all levels use the firm's products. The company is the official supplier of footwear to the NFL (National Football League).

FINANCIAL DATA: *Note: Data for latest year may not have been available at press time.*

In U.S. $	2012	2011	2010	2009	2008	2007
Revenue	1,834,921,000	1,472,684,000	1,063,927,000	856,411,000	725,244,000	606,561,000
R&D Expense						
Operating Income	208,695,000	162,767,000	112,355,000	85,273,000	76,925,000	86,265,000
Operating Margin %	11.37%	11.05%	10.56%	9.95%	10.60%	14.22%
SGA Expense	670,602,000	550,069,000	418,152,000	327,752,000	278,023,000	218,779,000
Net Income	128,778,000	96,919,000	68,477,000	46,785,000	38,229,000	52,558,000
Operating Cash Flow	199,761,000	15,218,000	50,114,000	119,041,000	69,516,000	-14,628,000
Capital Expenditure	50,650,000	79,392,000	30,182,000	19,845,000	39,194,000	34,084,000
EBITDA	251,777,000	197,004,000	142,498,000	113,011,000	98,272,000	100,887,000
Return on Assets %	12.40%	12.15%	11.21%	9.05%	8.70%	15.45%
Return on Equity %	17.72%	17.10%	15.26%	12.79%	12.50%	21.24%
Debt to Equity	0.06	0.11	0.01	0.02	0.03	0.03

CONTACT INFORMATION:

Phone: 410 454-6428 Fax: 410 367-2400
Toll-Free: 888-727-6687
Address: 1020 Hull St., Fl. 3, Baltimore, MD 21230 United States

STOCK TICKER/OTHER:

Stock Ticker: UA Exchange: NYS
Employees: 1,800 Fiscal Year Ends: 31-Dec
Parent Company:

SALARIES/BONUSES:

Top Exec. Salary: $116,666 Bonus: $900,000
Second Exec. Salary: $540,000 Bonus: $

OTHER THOUGHTS:

Estimated Female Officers or Directors: 2
Hot Spot for Advancement for Women/Minorities:

Unifi Inc

www.unifi.com

NAIC Code: 313110

TYPES OF BUSINESS:

Fiber Manufacturing
Polyester & Nylon Yarn

BRANDS/DIVISIONS/AFFILIATES:

aio
Sorbtek
A.M.Y.
Unifi Central America Ltda De CV
Reflexx
Repreve
Unifi Textiles Suzhou Co Ltd

CONTACTS: *Note: Officers with more than one job title may be intentionally listed here more than once.*

William L. Jasper, CEO
R. Roger Berrier, Jr., COO
R. Roger Berrier, Jr., Pres.
Ronald L. Smith, CFO/VP
Thomas H. Caudle, Jr., VP-Mfg.
Charles F. McCoy, General Counsel/VP/Sec.
William L. Jasper, Chmn.

GROWTH PLANS/SPECIAL FEATURES:

Unifi, Inc. is a producer and processor of multi-filament polyester and nylon yarns, including specialty and value-added yarns with enhanced performance characteristics. The company manufactures partially oriented, textured, dyed, twisted and beamed polyester yarns as well as textured nylon and nylon covered spandex products. Unifi sells its products to other yarn manufacturers, knitters and weavers that produce fabric for the furnishings, apparel, automotive, hosiery, industrial and other end-use markets. Nylon products include textured nylon and covered spandex products, which the company sells to other yarn manufacturers, knitters and weavers that produce fabric for the apparel, hosiery and sock markets, among others. The texturing process, which is common to both polyester and nylon, involves the processing of partially oriented yarn (POY), which is either natural- or solution-dyed raw polyester or natural nylon filament fiber. POY is made from petroleum-based chemicals that react to form a polymer, which is melted and extruded through microscopic holes to form filaments, which are then combined to form a yarn. The company currently has approximately 20 PVA yarns in its portfolio, sold under brand names such as A.M.Y., Reflexx, aio, Sorbtek and Repreve. Unifi has manufacturing operations in North and South America and participates in joint ventures in Israel and the U.S. It also sells specialty and PVA products in Asia (primarily China) through subsidiary Unifi Textiles Suzhou Co., Ltd.

FINANCIAL DATA: *Note: Data for latest year may not have been available at press time.*

In U.S. $	2012	2011	2010	2009	2008	2007
Revenue	705,086,000	712,812,000	616,753,000	553,663,000	713,346,000	690,308,000
R&D Expense						
Operating Income	8,632,000	28,692,000	25,388,000	-26,560,000	-1,231,000	-14,338,000
Operating Margin %	1.22%	4.02%	4.11%	-4.79%	-.17%	-2.07%
SGA Expense	43,482,000	44,659,000	46,183,000	39,122,000	47,786,000	52,060,000
Net Income	11,491,000	25,089,000	10,685,000	-48,996,000	-16,151,000	-116,305,000
Operating Cash Flow	43,309,000	11,880,000	20,581,000	16,619,000	13,087,000	10,897,000
Capital Expenditure	6,354,000	20,539,000	13,112,000	15,259,000	12,809,000	7,840,000
EBITDA	52,057,000	77,589,000	67,676,000	10,865,000	37,304,000	-69,482,000
Return on Assets %	2.25%	4.81%	2.17%	-9.17%	-2.57%	-16.69%
Return on Equity %	3.90%	8.96%	4.23%	-17.79%	-5.33%	-34.06%
Debt to Equity	0.39	0.56	0.63	0.74	0.66	0.78

CONTACT INFORMATION:

Phone: 336 294-4410 Fax: 336 316-5422
Toll-Free:
Address: 7201 W. Friendly Ave., Greensboro, NC 27410 United States

STOCK TICKER/OTHER:

Stock Ticker: UFI Exchange: NYS
Employees: 2,700 Fiscal Year Ends: 30-Jun
Parent Company:

SALARIES/BONUSES:

Top Exec. Salary: $685,231 Bonus: $257,625
Second Exec. Salary: $460,231 Bonus: $173,250

OTHER THOUGHTS:

Estimated Female Officers or Directors:
Hot Spot for Advancement for Women/Minorities:

Uniqlo Co Ltd

NAIC Code: 448140

www.uniqlo.com

TYPES OF BUSINESS:

Apparel, Retail
Men's, Women's & Children's Apparel
Fast Fashion

BRANDS/DIVISIONS/AFFILIATES:

Fast Retailing Co Ltd
UNIQLO KIDS
BODY by UNIQLO

CONTACTS: *Note: Officers with more than one job title may be intentionally listed here more than once.*

Tadashi Yanai, CEO
Tadashi Yanai, Pres.
Tadashi Yanai, Chmn.
Nubuo Domae, CEO/Pres., UNIQLO USA, Inc.

GROWTH PLANS/SPECIAL FEATURES:

Uniqlo Co. Ltd., a subsidiary of Fast Retailing Co., Ltd., is a trendy clothing store operator based in Japan. Its name stands for Unique Clothing Warehouse and reflects the fact that much of each store's stock is kept on the floor for the purpose of customer interaction. The company specializes in fast fashion; that is, clothing that is fashion-forward, moderately priced and constantly changing in style and color. In total, the company operates over 1,137 stores, with 845 locations in Japan. The company also retails through BODY by UNIQLO, a sports centered store, and UNIQLO KIDS, a children's store. Roughly 46% of Uniqlo's sales are generated by women's categories. Uniqlo maintains its ability to sell affordable, quality clothing by controlling every step of the process: R&D; manufacturing; distribution and retail. The company works with around 70 factories throughout Asia to create its three main clothing sections: Sophisticated Minimalism, Sophisticated Fashion and The Preppy Student. The company's approach allows its customers to create their own style with basic pieces, rather than being branded by the company's sense of fashion. Uniqlo operates global flagship stores in New York City, London, Paris and Shanghai. Recently, the firm announced the opening of its largest global flagship store, UNIQLO New York Fifth Avenue, and its second largest global flagship store, UNIQLO New York 34th Street. In October 2012, Uniqlo opened its first U.S. West Coast store in San Francisco.

FINANCIAL DATA: *Note: Data for latest year may not have been available at press time.*

In U.S. $	2012	2011	2010	2009	2008	2007
Revenue	10,339,110,000	9,133,153,000	9,071,497,000	7,626,758,000	6,529,107,000	5,847,218,000
R&D Expense						
Operating Income	1,407,800,000	1,295,521,000	1,473,798,000	1,209,506,000	974,081,800	723,249,500
Operating Margin %	13.61%	14.18%	16.24%	15.85%	14.91%	12.36%
SGA Expense	3,885,684,000	3,444,640,000	3,211,977,000	2,592,801,000	2,295,555,000	2,042,195,000
Net Income	797,742,100	605,136,900	686,710,200	554,402,600	484,619,400	353,759,100
Operating Cash Flow	1,421,082,000	636,354,500	986,662,300	659,244,500	972,333,800	209,828,400
Capital Expenditure	266,975,400	210,440,800	190,935,300	110,330,500	175,727,300	199,441,100
EBITDA	1,649,892,000	1,333,363,000	1,526,703,000	1,253,816,000	1,085,125,000	838,434,200
Return on Assets %	12.69%	10.44%	12.71%	11.47%	11.38%	8.59%
Return on Equity %	20.39%	18.09%	22.63%	19.09%	17.33%	13.56%
Debt to Equity	0.02	0.04	0.02	0.06	0.06	0.08

CONTACT INFORMATION:

Phone: 81 362720050 Fax: 81 362720076
Toll-Free:
Address: 717-1 Sayama, Yamaguchi City, Yamaguchi 754-0894 Japan

STOCK TICKER/OTHER:

Stock Ticker: FRCOF Exchange: PINX
Employees: 18,798 Fiscal Year Ends: 31-Aug
Parent Company: FAST RETAILING CO LTD

SALARIES/BONUSES:

Top Exec. Salary: $ Bonus: $
Second Exec. Salary: $ Bonus: $

OTHER THOUGHTS:

Estimated Female Officers or Directors: 1
Hot Spot for Advancement for Women/Minorities:

United Retail Group Inc

www.unitedretail.com

NAIC Code: 448120

TYPES OF BUSINESS:

Plus-Sized Apparel-Women's, Retail
Accessories
Footwear
Lingerie
Fragrances
Catalog & Online Sales

BRANDS/DIVISIONS/AFFILIATES:

Avenue
Avenue.com
Avenue Body
Redcats USA

CONTACTS: *Note: Officers with more than one job title may be intentionally listed here more than once.*

Gary White, CEO
Raphael Benaroya, Pres.
George R. Remeta, CFO
George R. Remeta, Chief Admin. Officer
Julie L. Daly, Pres., Shop @ Home
Kenneth P. Carroll, Sr. VP
Raphael Benaroya, Chmn.

GROWTH PLANS/SPECIAL FEATURES:

United Retail Group, Inc., a subsidiary of Redcats USA, is a leading nationwide specialty retailer of plus-size women's apparel and accessories, offering private-label merchandise under the Avenue trademark and Avenue Body plus-size lingerie, hosiery, accessories and gifts. The firm currently operates more than 500 Avenue retail stores across the U.S. The majority of stores are located in strip shopping centers, with the company's national distribution facility located in Troy, Ohio. The firm targets fashion-conscious women between the ages of 25-55 who wear size 14 and larger. Each store offers a selection of casual wear, career apparel, specialty items and accessories. Casual wear, which accounts for the majority of the company's sales, includes jeans, slacks, tee shirts, skirts, active wear and sweaters. Career apparel consists of skirts, soft blouses and dresses. Specialty items include sleepwear and lingerie, while accessories consist of earrings, pins, scarves, socks, hosiery and a selection of gift items. The company also markets its own Avenue brand jeans. Most Avenue products are custom designed by the company, allowing it to update its merchandise frequently to reflect customer demand and fashion trends. United Retail develops new Avenue brand apparel assortments approximately 4-6 times per year. The company also sells merchandise through its e-commerce site Avenue.com.

FINANCIAL DATA: *Note: Data for latest year may not have been available at press time.*

In U.S. $	2012	2011	2010	2009	2008	2007
Revenue						
R&D Expense						
Operating Income						
Operating Margin %						
SGA Expense						
Net Income						
Operating Cash Flow						
Capital Expenditure						
EBITDA						
Return on Assets %						
Return on Equity %						
Debt to Equity						

CONTACT INFORMATION:

Phone: 201-845-0880 Fax: 201-909-2210
Toll-Free: 800-441-1362
Address: 365 W. Passaic St., Rochelle Park, NJ 07662 United States

STOCK TICKER/OTHER:

Stock Ticker: Subsidiary
Employees:
Parent Company: REDCATS USA
Exchange:
Fiscal Year Ends: 31-Jan

SALARIES/BONUSES:

Top Exec. Salary: $ Bonus: $
Second Exec. Salary: $ Bonus: $

OTHER THOUGHTS:

Estimated Female Officers or Directors: 1
Hot Spot for Advancement for Women/Minorities: Y

Unitika Ltd

NAIC Code: 313110

www.unitika.co.jp

TYPES OF BUSINESS:

Textiles Manufacturing
Advanced Materials
Polymers
Yarn & Fabrics
Environmental Facilities & Systems
Health Care Products
Recreation Centers
Nanotextiles

BRANDS/DIVISIONS/AFFILIATES:

Unitika Fibers Ltd
Unitika Textiles Ltd
Nippon Ester
Unitika Spinning Co Ltd
Osaka Dyeing Co Ltd
Unitika Mate Co Ltd
Unitika Berkshire Co Ltd
Unitika America Corp

CONTACTS: *Note: Officers with more than one job title may be intentionally listed here more than once.*

Kenji Yasue, CEO
Kenji Yasue, Pres.
Hiroyuki Shime, Dir.-Oper. Div.
Takuro Matsunaga, CEO/Pres., Unitika Trading Co., Ltd.

GROWTH PLANS/SPECIAL FEATURES:

Unitika, Ltd. is comprised of an extensive network of Japanese companies and group business units, with core operations in fibers and textiles as well as advanced materials, polymers, environmental businesses and health and amenity ventures. A number of related group support companies and overseas offices complement Unitika's portfolio of manufacturing, research and management companies. Within the firm's well-established fibers and textiles divisions (operating as Unitika Fibers Ltd. and Unitika Textiles Ltd., respectively), the company manufactures and markets a comprehensive line of yarn and fabrics for use in apparel, commercial and industrial applications. Representing both natural and synthetic materials, Unitika products can be found in everything from fishing line and construction barricades to kimonos and stockings. The firm manages all aspects of production, from fiber processing through its Nippon Ester unit and Okazaki and Tokiwa plants, to spinning through Unitika Spinning Co. Ltd. and dyeing through I-Tex Co. Ltd. and Osaka Dyeing Co. Ltd. Through Unitika Mate Co. Ltd., the firm manufactures men's and women's sportswear and sporting goods, while Unitika Berkshire Co. Ltd. produces pantyhose, socks and other garments. Unitika is additionally engaged in advanced materials manufacturing (including carbon fibers, metal fibers and phenol thermosetting resins) and production of polymers (including nylon and polyester films and resins, spunbonded synthetic materials and biodegradable plastics). The firm's environmental businesses include water treatment facilities, incinerators and air pollution prevention systems. Health and amenity operations range from medical products such as catheters, enzymes and diagnostic reagents, functional food ingredients and dietary supplements. Products produced by the company include Terramac film, a biodegradable material made from plants; Castlon, a 100% plant-derived nylon fiber made from castor beans; and Uniecolo, a polyester fabric made from recycled PET (Polyethylene terephthalate) bottles. Unitika America Corp. handles the firm's U.S. import and export business activities.

FINANCIAL DATA: *Note: Data for latest year may not have been available at press time.*

In U.S. $	2012	2011	2010	2009	2008	2007
Revenue		2,323,900,000	2,323,500,000	2,672,200,000	2,236,220,000	2,205,700,000
R&D Expense						
Operating Income						
Operating Margin %						
SGA Expense						
Net Income		31,400,000	38,700,000	-178,300,000	25,970,000	25,600,000
Operating Cash Flow						
Capital Expenditure						
EBITDA						
Return on Assets %						
Return on Equity %						
Debt to Equity						

CONTACT INFORMATION:

Phone: 81-6-6281-5695 Fax: 81-6-6281-5697
Toll-Free:
Address: 4-1-3, Kyutaro-machi, Chuo-ku, Osaka, 541-8566 Japan

STOCK TICKER/OTHER:

Stock Ticker: 3103 Exchange: Tokyo
Employees: 4,845 Fiscal Year Ends: 31-Mar
Parent Company:

SALARIES/BONUSES:

Top Exec. Salary: $ Bonus: $
Second Exec. Salary: $ Bonus: $

OTHER THOUGHTS:

Estimated Female Officers or Directors:
Hot Spot for Advancement for Women/Minorities:

Urban Outfitters Inc

www.urbanoutfittersinc.com

NAIC Code: 448140

TYPES OF BUSINESS:

Casual Apparel Stores
Household & Gift Merchandise
Accessories
Wholesale Distribution
Online Sales
Footwear
Catalog Sales
Furniture

BRANDS/DIVISIONS/AFFILIATES:

Urban Outfitters
Anthropologie
Free People
Terrain
BHLDN
Leifsdottir

CONTACTS: *Note: Officers with more than one job title may be intentionally listed here more than once.*

Richard A. Hayne, CEO
Richard A. Hayne, Pres.
Frank J. Conforti, CFO
Glen A. Bodzy, General Counsel/Sec.
Frank J. Conforti, Chief Acct. Officer/Controller
Margaret Hayne, Pres., Free People
Wendy McDevitt, Pres., Terrain Brand
David W. McCreight, CEO-Anthroplogie Group
Wendy Wurtzburger, Chief Merch. & Design Officer-Anthropologie Group
Richard A. Hayne, Chmn.
Tedford Marlow, CEO-Urban Outfitters Group

GROWTH PLANS/SPECIAL FEATURES:

Urban Outfitters, Inc. is a clothing and accessories company that operates specialty retail stores under the Urban Outfitters, Anthropologie, Terrain, Free People and BHLDN brands, and operates a wholesale division under the Free People brand. Urban Outfitters targets ages 18-28 through a style-conscious merchandise mix and store environment. Urban Outfitters operate 209 stores in North America and Europe, and these stores have 55,000 to 60,000 stock keeping units (SKUs). This includes women's and men's apparel, footwear, accessories, apartment wares and gifts. Anthropologie targets upscale female customers from 28-45 years old. Anthropologie operates approximately 177 stores offering women's casual apparel, accessories, home furnishings, gifts and decorative items and carrying 45,000 to 50,000 SKUs. Free People, targeting contemporary women aged 25-30, designs, develops and markets young women's casual apparel. Free People retail stores average 1,400 square feet and carry up to 7,000 SKUs, while operating 77 stores. Terrain is targeted to men and women interested in outdoor living and gardening. Currently, there are two Terrain stores. Products include lifestyle home and garden items combined with antiques, live plants and flowers. Additionally, Terrain offers landscape and design service solutions. BHLDN operates two locations, and is focused on creating unique and memorable weddings. Products include wedding dresses, bridesmaid frocks, assorted jewelry, footwear, decorations and lingerie.

Urban Outfitters offers employees medical, dental, and vision coverage; wellness programs (discounted gym memberships, smoking cessation, nutrition counseling, and alternative health); flexible spending accounts; a 401(k); short-term and long-term disability, AD&D and life insurance; employee discounts; and paid holidays.

FINANCIAL DATA: *Note: Data for latest year may not have been available at press time.*

In U.S. $	2012	2011	2010	2009	2008	2007
Revenue	2,473,801,000	2,274,102,000	1,937,815,000	1,834,618,000	1,507,724,000	
R&D Expense						
Operating Income	284,725,000	414,203,000	338,984,000	299,435,000	224,945,000	
Operating Margin %	11.50%	18.21%	17.49%	16.32%	14.91%	
SGA Expense	575,811,000	522,417,000	447,161,000	414,043,000	351,827,000	
Net Income	185,251,000	272,958,000	219,893,000	199,364,000	160,231,000	
Operating Cash Flow	282,702,000	385,113,000	325,394,000	251,570,000	254,353,000	
Capital Expenditure	190,010,000	143,642,000	109,260,000	112,553,000	115,370,000	
EBITDA	392,837,000	515,308,000	431,334,000	381,384,000	294,962,000	
Return on Assets %	11.30%	15.91%	14.83%	16.13%	15.69%	
Return on Equity %	14.95%	20.15%	18.70%	20.90%	20.96%	
Debt to Equity						

CONTACT INFORMATION:

Phone: 215 454-5500 Fax: 215 568-1549
Toll-Free: 800-282-2200
Address: 5000 S. Broad St., Philadelphia, PA 19112 United States

STOCK TICKER/OTHER:

Stock Ticker: URBN Exchange: NAS
Employees: 16,000 Fiscal Year Ends: 31-Jan
Parent Company:

SALARIES/BONUSES:

Top Exec. Salary: $856,538 Bonus: $5,000
Second Exec. Salary: $597,692 Bonus: $6,088

OTHER THOUGHTS:

Estimated Female Officers or Directors: 3
Hot Spot for Advancement for Women/Minorities: Y

Valentino Fashion Group SpA

www.valentinofashiongroup.com

NAIC Code: 315240

TYPES OF BUSINESS:

Apparel Manufacturing
Fashion and Luxury Items
Couture Fashion
Perfume
Eyewear
Wedding Gowns
Watches

BRANDS/DIVISIONS/AFFILIATES:

Valentino
Valentino Garavani
Valentino Roma
Mayhoola for Investments
R.E.D. Valentino
Proenza Schouler

CONTACTS: *Note: Officers with more than one job title may be intentionally listed here more than once.*

Stefano Sassi, CEO
Maria Grazia Chiuri, Designer/Creative Dir.
Pier Paolo Piccioli, Designer/Creative Dir.
Alessandra Facchinetti, Dir- Women's Collections
Stefano Sassi, Chmn.
Graziano De Boni, CEO-Valentino USA/Pres., Valentino USA

GROWTH PLANS/SPECIAL FEATURES:

Valentino Fashion Group SpA is a fashion and luxury company headquartered in Milan, Italy. Maintaining an international presence, the firm specializes in clothing, accessories and footwear for men and women. Valentino's activities are divided between the following two business units: Valentino, which includes the brands Valentino, Valentino Garavani, Valentino Roma and R.E.D. Valentino; and licensed brands, which include MCS Marlboro Classics and M Missoni. Valentino offers exclusive products in the couture, ready-to-wear, designer, diffusion and bridge segments. Valentino Garavani is a provider of prestigious handbags, footwear, small leather goods, belts and other accessories for men and women. It was created when Marzotto SpA, a firm dedicated to the textile and fashion business, vested its clothing sector to the newly formed Valentino Fashion Group. Marzotto formed a separate firm to operate its textile business, allowing Valentino to be exclusively concerned with fashion and luxury items. Since the fashion group's inception, Valentino has begun an expansion process to include several new product lines, most notably perfume, eyewear and wedding gowns. The firm currently operates in over 90 countries, with more than 150 single-brand boutiques and 98 directly managed shops. Additionally, the company owns interest in U.S.-based fashion label Proenza Schouler. In July 2012, Valentino's was sold to Mayhoola for Investments, a Qatari investment group, for $856 million.

FINANCIAL DATA: *Note: Data for latest year may not have been available at press time.*

In U.S. $	2012	2011	2010	2009	2008	2007
Revenue	385,000,000	375,000,000	356,493,700	301,205,560		
R&D Expense						
Operating Income						
Operating Margin %						
SGA Expense						
Net Income						
Operating Cash Flow						
Capital Expenditure						
EBITDA						
Return on Assets %						
Return on Equity %						
Debt to Equity						

CONTACT INFORMATION:

Phone: 39-02-624921 Fax: 39-02-62492584
Toll-Free:
Address: Via Turati 16/18, Milan, 20121 Italy

STOCK TICKER/OTHER:

Stock Ticker: Private Exchange:
Employees: Fiscal Year Ends: 31-Dec
Parent Company: Mayhoola for Investments

SALARIES/BONUSES:

Top Exec. Salary: $ Bonus: $
Second Exec. Salary: $ Bonus: $

OTHER THOUGHTS:

Estimated Female Officers or Directors: 2
Hot Spot for Advancement for Women/Minorities:

Vans Inc

www.vans.com

NAIC Code: 448210

TYPES OF BUSINESS:

Shoes, Retail
Sports Apparel & Accessories
Online Sales
Helmets
Snowboard Bindings
Concert Tours

BRANDS/DIVISIONS/AFFILIATES:

Vans Triple Crown Series
VF Corp
PRO-TEC
Vans Warped Tour
SWITCH

CONTACTS: *Note: Officers with more than one job title may be intentionally listed here more than once.*

Scott J. Blechman, CFO/VP
Eric Wiseman, CEO/Pres./Chmn.-VF Corp.

GROWTH PLANS/SPECIAL FEATURES:

Vans, Inc., a subsidiary of VF Corp., is a sports retailing company that merchandises, designs, sources and distributes Vans-branded footwear, apparel and accessories primarily for skateboarding, snowboarding, surfing, BMX bike riding and motor cross. Vans' products are also sold under the SWITCH brand of step-in snowboard boot bindings and the PRO-TEC line of protective helmets and pads sold through Pro-Tec, Inc. The company distributes its products through more than 350 Vans retail stores in the U.S., Mexico and Europe as well as on the Internet. Vans was founded in 1966 in Southern California as a domestic manufacturer of vulcanized canvas shoes, many of which appealed to skateboard enthusiasts and the Southern California skate and surf culture. Its primary strategy is to continue to build and reinforce the authenticity and credibility of the Vans brand for a core customer base of 10-24 year-old sports enthusiasts. The company sponsors a skateboarding team, the Vans Triple Crown of Surfing as well as many skateboarding events. Many of these events are available on videos on its web site and through cell phone downloads. The brand is also a driving force behind the Vans Warped Tour, a music festival that features more than 40 bands traveling to over 40 cities in North America. Vans' sister brands owned by VF Corp. include Timberland, Kipling, lucy, Napapijri, Reef, 7 For All Mankind, Splendid and Ella Moss.

Vans offers its employees medical, dental and vision benefits; a 401(k) program; retirement contributions; flexible spending accounts; company-paid flu shots; an employee assistance program; life and AD&D insurance; short- and long-term disability; on-site gym and lunch vendors; a tobacco cessation program; a scholarship program for employees' children; and education assistance.

FINANCIAL DATA: *Note: Data for latest year may not have been available at press time.*

In U.S. $	2012	2011	2010	2009	2008	2007
Revenue						
R&D Expense						
Operating Income						
Operating Margin %						
SGA Expense						
Net Income						
Operating Cash Flow						
Capital Expenditure						
EBITDA						
Return on Assets %						
Return on Equity %						
Debt to Equity						

CONTACT INFORMATION:

Phone: 714-889-6100 Fax:
Toll-Free: 888-691-8889
Address: 6550 Katella Ave., Cypress, CA 90630 United States

STOCK TICKER/OTHER:

Stock Ticker: Subsidiary
Exchange:
Employees:
Fiscal Year Ends: 31-May
Parent Company: VF CORP

SALARIES/BONUSES:

Top Exec. Salary: $ Bonus: $
Second Exec. Salary: $ Bonus: $

OTHER THOUGHTS:

Estimated Female Officers or Directors:
Hot Spot for Advancement for Women/Minorities:

Varsity Brands Inc

NAIC Code: 315210

www.varsity.com

TYPES OF BUSINESS:

Cheerleader & Dance Team Uniforms
Camps, Events & Competitions

BRANDS/DIVISIONS/AFFILIATES:

Herff Jones Inc
American Cheerleader
Varsity/Intropa
Varsity.com
Varsity TV
Universal Dance Association

CONTACTS:

Note: Officers with more than one job title may be intentionally listed here more than once.

Jeffrey G. Webb, CEO
Jeffrey G. Webb, Pres.
John M. Nichols, CFO
Nicole Lauchaire, VP-Corp. Mktg.
Marlene Cota, VP-Bus. Dev. & Corp. Alliances
Jackie Martin, Mgr.-Content, Varsity.com
Nicole Lauchaire, VP-Comm.
Gregory C. Webb, Sr. VP-Camps & Events
Kimberly Williams, VP-Design
J. Kristyn Shepherd, Sr. VP-Special Events
Sheila Noone, VP-Public Rel.
Jeffrey G. Webb, Chmn.

GROWTH PLANS/SPECIAL FEATURES:

Varsity Brands, Inc. is a top designer, marketer and supplier of cheerleader and dance team uniforms and accessories. Thousands of catalogs featuring the firm's uniforms and accessories are mailed annually to schools, school spirit advisors and coaches. The company website, Varsity.com, allows customers to design their own uniforms with numerous colors and styles, and also features Varsity TV, a library for online videos from competitions and events. Varsity promotes its products through active association with championships and specials, including the National High School Cheerleading Championship, the International All-Levels Championship, the National All Star Cheerleading Championship and the All Star Challenge. With an annual attendance of over 325,000 participants, Varsity is also a leading operator of cheerleading and dance team training camps, competitions and clinics. Most of its dance team camps and competitions are operated through division Universal Dance Association. Besides its training camps, the firm also offers various educational materials through its web site, such as online articles featuring information on topics including becoming a cheerleader, being a male cheerleader, the history of cheerleading, being the parent of a cheerleader, coaching cheerleaders, transitioning from high school to college cheerleading squads, being on a dance team, practical tips on developing cheerleading skills, cheerleading safety and other issues. Additionally, tour company subsidiary Varsity/Intropa specializes in organizing trips for cheerleaders, bands, choirs and orchestras, church groups, garden clubs, dance and theater groups and other school-affiliated groups. The company is owned by Heff Jones, Inc. Recently, the company acquired American Cheerleader magazine from McFadden Performing Arts Media.

FINANCIAL DATA:

Note: Data for latest year may not have been available at press time.

In U.S. $	2012	2011	2010	2009	2008	2007
Revenue						
R&D Expense						
Operating Income						
Operating Margin %						
SGA Expense						
Net Income						
Operating Cash Flow						
Capital Expenditure						
EBITDA						
Return on Assets %						
Return on Equity %						
Debt to Equity						

CONTACT INFORMATION:

Phone: 901-387-4300   Fax: 800-218-0818
Toll-Free: 800-533-8022
Address: 6745 Lenox Ctr. Ct., Ste. 300, Memphis, TN 38115 United States

STOCK TICKER/OTHER:

Stock Ticker: Private   Exchange:
Employees: Fiscal Year Ends: 31-Dec
Parent Company: HERFF JONES INC

SALARIES/BONUSES:

Top Exec. Salary: $ Bonus: $
Second Exec. Salary: $ Bonus: $

OTHER THOUGHTS:

Estimated Female Officers or Directors: 6
Hot Spot for Advancement for Women/Minorities: Y

Velcro Industries NV

www.velcro.com

NAIC Code: 339993

TYPES OF BUSINESS:

Hook-and-Loop Fastener Manufacturing
Fastening Technology
Adhesive Products
Twist Ties

BRANDS/DIVISIONS/AFFILIATES:

Velcro
Cohere Ltd
Bundle Ties

CONTACTS:

Note: Officers with more than one job title may be intentionally listed here more than once.

Wil De Hollander, CEO
Wil De Hollander, Pres.

GROWTH PLANS/SPECIAL FEATURES:

Velcro Industries NV, a subsidiary of Cohere, Ltd., owns, manufactures and markets the trademarked hook-and-loop fastening technology that bears its name. The technology can hold up to nine pounds per square inch and is capable of over 10,000 separations and refastenings. The firm owns over 300 active patents. The company and its subsidiaries offer hundreds of products for apparel, construction, medical, military, packaging, personal care, transportation, agriculture, electronics, displays and industrial cleaning. Consumer applications include a wide variety of sew-on, iron-on and adhesive backed products for crafts and other uses, both indoor and outdoor. Industrial applications include Bundle Ties for food such as lettuce or herbs; flame and adhesive laminates for baby seats, children's costumes, medical braces and other products; and metal and plastic hardware for outdoor applications, such as backpacks, camping equipment, hiking gear and other products. Velcro also hosts a blog on its web site which offers interviews, commercials and do-it-yourself ideas involving Velcro products. The firm has subsidiaries in Canada, the U.S., Mexico, Australia, China, Hong Kong, Italy, Spain, France, Germany and the U.K. In June 2012, the company released a new catheter securement system and antibacterial fasteners for the medical and nursing market.

Velcro's U.S. affiliate offers its employees a 401(k) savings plan; medical, dental and life insurance; flexible spending accounts; nutritional counseling; acupuncture service; chiropractic service; and massage therapy.

FINANCIAL DATA:

Note: Data for latest year may not have been available at press time.

In U.S. $	2012	2011	2010	2009	2008	2007
Revenue						
R&D Expense						
Operating Income						
Operating Margin %						
SGA Expense						
Net Income						
Operating Cash Flow						
Capital Expenditure						
EBITDA						
Return on Assets %						
Return on Equity %						
Debt to Equity						

CONTACT INFORMATION:

Phone: Fax:
Toll-Free: 800-564-6253
Address: 15 Pietermaai, Curacao, Curacao

STOCK TICKER/OTHER:

Stock Ticker: Subsidiary
Exchange:
Employees: 2,923
Fiscal Year Ends: 30-Sep
Parent Company: COHERE LTD

SALARIES/BONUSES:

Top Exec. Salary: $ Bonus: $
Second Exec. Salary: $ Bonus: $

OTHER THOUGHTS:

Estimated Female Officers or Directors:
Hot Spot for Advancement for Women/Minorities:

Vera Bradley Inc

www.verabradley.com

NAIC Code: 448120

TYPES OF BUSINESS:

Women's Fashion Accessories

BRANDS/DIVISIONS/AFFILIATES:

VeraBradley.com
Vera Bradley Baby

CONTACTS:

Note: Officers with more than one job title may be intentionally listed here more than once.

Michael C. Ray, CEO
Kevin J. Sierks, Interim CFO
Matthew C. Wojewuczki, Exec. VP-Oper.
C. Roddy Mann, Exec. VP-Strategy & Bus. Dev.
Kevin J. Sierks, VP-Corp. Controller
Barbara Bradley Baekgaard, Chief Creative Officer/Dir.
Kimberly F. Colby, Exec. VP-Design
David O. Thompson, VP-Direct Sales

GROWTH PLANS/SPECIAL FEATURES:

Vera Bradley, Inc. is a designer, producer, marketer and retailer of accessories for women. Its products include a variety of handbags, accessories and travel and leisure items, with its most established products being printed paisley handbags. In addition to its handbags, the firm sells sunglasses, luggage, laptop cases, home accessories and stationery. Its target customers are young girls, teens, college students, young professionals, mothers and grandmothers. The company sells products in both direct and indirect sales channels. The direct business consists of 65 full-price Vera Bradley stores, 11 outlet stores and its e-commerce site VeraBradley.com as well as an annual outlet sale in Fort Wayne, Indiana and 15 department store locations in Japan. Indirect sales consist of sales to approximately 3,400 independent retailers, several national retailers and third-party e-commerce sites. In 2012, 40.4% its net revenues were derived from the sale of handbags; 31.7% from accessories; 14.7% from travel and leisure items; and the remaining 13.2% from other items such as promotional products. In March 2013, the company launched Vera Bradley Baby, a new line of apparel items, bags and gifts focused on newborns and infants.

Employees of Vera Bradley receive health, dental, life and disability insurance; flexible spending accounts; a 401(k); a matching gifts program; tuition reimbursement; an employee assistance program; and product discounts.

FINANCIAL DATA:

Note: Data for latest year may not have been available at press time.

In U.S. $	2012	2011	2010	2009	2008	2007
Revenue	460,843,000	366,057,000	288,940,000	238,577,000		
R&D Expense						
Operating Income	96,171,000	53,319,000	45,712,000	27,191,000		
Operating Margin %	20.86%	14.56%	15.82%	11.39%		
SGA Expense	169,427,000	163,053,000	116,168,000	109,195,000		
Net Income	57,921,000	46,198,000	43,219,000	23,671,000		
Operating Cash Flow	51,516,000	58,466,000	66,006,000	50,187,000		
Capital Expenditure	18,836,000	11,365,000	5,844,000	14,949,000		
EBITDA	105,628,000	61,730,000	56,378,000	34,538,000		
Return on Assets %	27.22%	25.68%	28.46%	15.78%		
Return on Equity %	61.51%	64.96%	63.63%	40.84%		
Debt to Equity	0.2	1.04	0.32	0.91		

CONTACT INFORMATION:

Phone: 877 708-8372 Fax:
Toll-Free: 877-708-8372
Address: 2208 Production Rd., Fort Wayne, IN 46808 United States

STOCK TICKER/OTHER:

Stock Ticker: VRA
Employees: 1,427
Parent Company:
Exchange: NAS
Fiscal Year Ends: 31-Jan

SALARIES/BONUSES:

Top Exec. Salary: $608,430 Bonus: $
Second Exec. Salary: $474,721 Bonus: $

OTHER THOUGHTS:

Estimated Female Officers or Directors: 2
Hot Spot for Advancement for Women/Minorities: Y

VF Corp

www.vfc.com

NAIC Code: 315240

TYPES OF BUSINESS:

Jeans-Manufacturing
Swimsuits
Outdoor Gear & Apparel
Image Wear
Outlet Stores
Footwear

BRANDS/DIVISIONS/AFFILIATES:

Timberland
lucy
Earthkeepers
SmartWool
Wrangler
Rock and Republic Enterprises Inc
Timberland Company
Kipling

CONTACTS:

Note: Officers with more than one job title may be intentionally listed here more than once.

Eric C. Wiseman, CEO
Eric C. Wiseman, Pres.
Robert K. Shearer, CFO/Sr. VP
Susan Larson Williams, VP-Human Resources
Martin Schneider, CIO/VP
Candace S. Cummings, VP-Admin.
Candace S. Cummings, General Counsel/Sec.
Stephen Dull, VP-Strategy & Innovation
Bradley Batten, Chief Acct. Officer/VP/Controller
Michael T. Gannaway, VP-VF Direct/Customer Teams
Franklin L. Terkelsen, VP-Mergers & Acquisitions
Steve Rendle, Pres., Outdoor & Action Sports Americas/VP
Scott Baxter, Pres., Jeanswear Americas & Imagewear/VP
Eric C. Wiseman, Chmn.
Karl Heinz Salzburger, Pres., Int'l/VP

GROWTH PLANS/SPECIAL FEATURES:

VF Corp., organized in 1899, is one of the world's largest brand-name apparel manufacturers and a leading producer of jeanswear, outerwear, packs, footwear, sportswear and occupational apparel. VF products are sold globally throughout the US, Canada, Europe, Asia and Latin America. The company divides its brands into five business groups, called coalitions: outdoor and action sports, which includes outerwear, sportswear, footwear, equipment, backpacks, daypacks, luggage and accessories; jeanswear, which consists of jeans as well as shorts, casual pants, knit and woven tops and outerwear; imagewear, which includes occupational apparel, uniforms and owned and licensed sports and lifestyle apparel; sportswear, which includes outerwear, underwear, swimwear, sleepwear, luggage and accessories; and contemporary brands which focuses on lifestyle brands. The outdoor and action sports coalition includes the firm's largest brand, the North Face, as well as Timberland, Earthkeepers, SmartWool, Vans, JanSport, Eastpak, Kipling, Napapijri, Reef and lucy. In Jeanswear, Lee and Wrangler are its largest brand names. The company's imagewear coalition includes its licensed sports and image businesses. The licensed sports business consists of VF's owned and licensed high profile athletic apparel, marketed under the Majestic and Harley-Davidson brand names. The image business includes the firm's uniforms and career occupational clothing. These brands include Red Cap work clothes; Chef Designs apparel for restaurant and food service staff; Bulwark flame resistant and protective clothing; and Horace Small apparel for law enforcement and public safety officials. Sportswear brands include Nautica and Kipling. The company's contemporary brands includes its premium upscale lifestyle brands such as John Varvatos, Splendid, Ella Moss and 7 For All Mankind. In 2011, the company acquired Rock and Republic Enterprises, Inc. and Timberland Company.

FINANCIAL DATA:

Note: Data for latest year may not have been available at press time.

In U.S. $	2012	2011	2010	2009	2008	2007
Revenue	10,879,850,000	9,459,232,000	7,702,589,000	7,220,286,000	7,642,600,000	7,219,359,000
R&D Expense						
Operating Income	1,465,267,000	1,244,791,000	820,860,000	736,817,000	938,995,000	965,441,000
Operating Margin %	13.46%	13.15%	10.65%	10.20%	12.28%	13.37%
SGA Expense	3,596,708,000	3,085,839,000	2,574,790,000	2,336,394,000	2,419,925,000	2,173,896,000
Net Income	1,085,999,000	888,089,000	571,362,000	461,271,000	602,748,000	591,621,000
Operating Cash Flow	1,275,000,000	1,081,371,000	1,001,282,000	973,485,000	678,401,000	820,576,000
Capital Expenditure	282,830,000	249,128,000	125,250,000	95,594,000	134,808,000	120,230,000
EBITDA	1,753,436,000	1,441,056,000	1,001,346,000	911,027,000	1,108,178,000	1,118,919,000
Return on Assets %	11.46%	11.26%	8.83%	7.14%	9.35%	9.93%
Return on Equity %	22.50%	21.17%	14.88%	12.51%	16.90%	17.29%
Debt to Equity	0.27	0.4	0.24	0.24	0.32	0.32

CONTACT INFORMATION:

Phone: 336 424-6000 Fax:
Toll-Free:
Address: 105 Corporate Ctr. Blvd., Greensboro, NC 27408 United States

STOCK TICKER/OTHER:

Stock Ticker: VFC
Exchange: NYS
Employees: 58,000
Fiscal Year Ends: 31-Dec
Parent Company:

SALARIES/BONUSES:

Top Exec. Salary: $1,200,000 Bonus: $
Second Exec. Salary: $816,547 Bonus: $

OTHER THOUGHTS:

Estimated Female Officers or Directors: 5
Hot Spot for Advancement for Women/Minorities: Y

Victoria plc

NAIC Code: 314110

www.victoriaplc.com

TYPES OF BUSINESS:

Carpets & Carpet Yarns
Carpet Tiles
Vinyl Flooring

BRANDS/DIVISIONS/AFFILIATES:

Victoria Carpets Ltd
Munster Carpets Ltd
Westwood Yarns Ltd
Victoria Carpet Company Pty Ltd
Colin Campbell & Sons Limited
Navan Carpets
Victoria Luxury Flooring

CONTACTS: *Note: Officers with more than one job title may be intentionally listed here more than once.*

Terry A. Danks, Company Sec./Dir.-Finance, Victoria Carpets Ltd.
Neil Glover, Dir.-Oper., Victoria Carpets Ltd.
Michael Oakley, Chmn.-Victoria Carpet Company Pty. Ltd.
Trevor Chippendale, Managing Dir.-Westwood Yarns
Geoffrey Wilding, Exec. Chmn.
Chris Dragan, Pres., Colin Campbell & Sons Ltd.

GROWTH PLANS/SPECIAL FEATURES:

Victoria plc, in operation since 1895, manufactures and sells carpets, carpet yarns and other floor coverings. The company is organized into three regional divisions: the U.K. and Ireland, through Victoria Carpets Ltd., Munster Carpets Ltd. and Navan Carpets; Australia, operating as Victoria Carpet Company Pty. Ltd.; and Canada, through a 50% interest in Canadian carpet distributor Colin Campbell & Sons Ltd. In the U.K., Victoria Carpets Ltd. manufactures and markets tufted carpets in facilities that also support the introduction of new product offerings and custom contract work. The export segment targets 40 markets globally, with contracts largely for the leisure and hospitality sectors. The U.K. carpet unit is supported by Westwood Yarns Ltd., which produces high-quality dry woolen spun carpet yarns for Victoria Carpets and other select business partners. Westwood produces about 70 tons of yarn each week, most of which is utilized by Victoria group companies. Additionally, through its Victoria Luxury Flooring division, Victoria distributes luxury vinyl tiles primarily to the contract market in the U.K. In Ireland, Munster Carpets specializes in Wilton woven carpet for office and commercial environments, while Navan Carpets provides carpets to the Irish residential market as well as hotels, golf clubs, casinos and ocean liners. In Australia, the Victoria Carpet Company, headquartered outside of Melbourne, manages the manufacture and sale of tufted carpeting both within Australia and worldwide. Colin Campbell & Sons, jointly owned with Ontario retailer Elte Carpets, specializes in high-end wool and natural fiber carpets and floor coverings for the architectural and design community, with showrooms in Vancouver, Calgary and Edmonton.

FINANCIAL DATA: *Note: Data for latest year may not have been available at press time.*

In U.S. $	2012	2011	2010	2009	2008	2007
Revenue			109,500,000	100,180,000	90,270,000	109,700,000
R&D Expense						
Operating Income						
Operating Margin %						
SGA Expense						
Net Income			1,900,000	640,000	3,710,000	4,000,000
Operating Cash Flow						
Capital Expenditure						
EBITDA						
Return on Assets %						
Return on Equity %						
Debt to Equity						

CONTACT INFORMATION:

Phone: 44-1562-749-300 Fax: 44-1562-749-649
Toll-Free:
Address: Worcester Rd., Kidderminster, DY10 1JR United Kingdom

STOCK TICKER/OTHER:

Stock Ticker: VCP Exchange: London
Employees: 647 Fiscal Year Ends: 31-Mar
Parent Company:

SALARIES/BONUSES:

Top Exec. Salary: $ Bonus: $
Second Exec. Salary: $ Bonus: $

OTHER THOUGHTS:

Estimated Female Officers or Directors:
Hot Spot for Advancement for Women/Minorities: Y

Victorias Secret

www.victoriassecret.com

NAIC Code: 448120

TYPES OF BUSINESS:

Intimate Apparel-Women's, Retail
Cosmetics
Fragrances
Personal Care Products
Online Sales
Catalogs
Women's Shoes
General Women's Apparel

BRANDS/DIVISIONS/AFFILIATES:

PINK
Victoria's Secret Direct
Victoria's Secret Beauty
Angels
IPEX
Body by Victoria
La Senza
Limited Brands Inc

CONTACTS: *Note: Officers with more than one job title may be intentionally listed here more than once.*

Lori Greeley, CEO
Larry Fultz, Exec. VP-Human Resources
Brian VanOoyen, VP-Merch. Planning
Bridget Ryan-Berman, CEO-Victoria's Secret Direct
Sharen Jester Turney, CEO/Pres.,Victoria's Secret Megabrand & Intimate

GROWTH PLANS/SPECIAL FEATURES:

Victoria's Secret, a wholly-owned subsidiary of Limited Brands, Inc., purchases, distributes and sells lingerie, personal care products and women's apparel through over 1,000 retail stores, the Internet and direct mail channels. The stores offer branded merchandise such as IPEX, PINK, Very Sexy, Body by Victoria, VS Cotton, BioFit, Dream Angels, Beauty Rush and Angels. The firm also owns several standalone PINK stores in the U.S. and Canada, which sell intimate apparel, denim, casual apparel and body products targeted to young women ages 13-25, and 108 standalone Victoria's Secret Beauty and Accessories Stores. Victoria's Secret Beauty offers a complete line of fragrance, cosmetics and body products for skin and hair. In addition to its retail stores, the company operates Victoria's Secret Direct, which consists of the famous Victoria's Secret Catalog and an e-commerce site, VictoriasSecret.com. Street apparel such as dresses, pants, skirts, shorts, tops and shoes are available from the catalog and online, but not in stores. Once each year, Victoria's Secret conducts a televised fashion show featuring some of the world's top models and performances by acclaimed musicians such as Maroon 5 or Kanye West. The Victoria's Secret group also includes the operations of La Senza, a Canadian lingerie store chain owned by Limited Brands. For 2013, sales per square foot per year were in the $800 range, with average sales per store of about $5 million. Approximately 25% of the firm's sales are made via the catalogs and online as of 2013.

Employees are offered medical, dental, vision and prescription drug coverage; a 401(k) plan; a discount stock purchase plan; life insurance; discounts on products; tuition reimbursement; a commuter discount program; and an employee assistance program.

FINANCIAL DATA: *Note: Data for latest year may not have been available at press time.*

In U.S. $	2012	2011	2010	2009	2008	2007
Revenue	6,574,000,000	6,121,000,000	5,520,000,000	5,604,000,000	5,601,046,000	5,606,993,000
R&D Expense						
Operating Income						
Operating Margin %						
SGA Expense						
Net Income	1,188,000,000	1,081,000,000	889,000,000	579,000,000	405,000,000	
Operating Cash Flow						
Capital Expenditure						
EBITDA						
Return on Assets %						
Return on Equity %						
Debt to Equity						

CONTACT INFORMATION:

Phone: 614-577-7111 Fax:
Toll-Free: 800-411-5116
Address: 4 Limited Pkwy. E, Reynoldsburg, OH 43068 United States

STOCK TICKER/OTHER:

Stock Ticker: Subsidiary Exchange:
Employees: 45,000 Fiscal Year Ends: 31-Jan
Parent Company: LIMITED BRANDS INC

SALARIES/BONUSES:

Top Exec. Salary: $ Bonus: $
Second Exec. Salary: $ Bonus: $

OTHER THOUGHTS:

Estimated Female Officers or Directors: 3
Hot Spot for Advancement for Women/Minorities: Y

Vincenzo Zucchi SpA

NAIC Code: 313210

www.gruppozucchi.com

TYPES OF BUSINESS:

Textiles Manufacturing
Linens & Home Furnishings
Retail Stores

BRANDS/DIVISIONS/AFFILIATES:

Bassetti
Descamps
Jalla
Sartoria Notturna
Bassetti Schweiz AG
Welspun Zucchi Textiles Private Ltd
Masconi USA Ltd
Bassetti Deutschland GmbH

CONTACTS: *Note: Officers with more than one job title may be intentionally listed here more than once.*

Matteo Zucchi, Managing Dir.
Stefano Crespi, CFO
Stefano Crespi, Chief Admin. Officer
Matteo Zucchi, Chmn.

GROWTH PLANS/SPECIAL FEATURES:

Vincenzo Zucchi SpA manufactures, markets and distributes a variety of household linens, including sheets, table cloths, towels, bathrobes, carpets, bedspreads, kitchen articles, accessories, curtains and various other decorative items. Headquartered in Milan, Italy, the company markets its products under brand names such as Zucchi, Bassetti, Descamps, Jalla, Ecoemotion, Sartoria Notturna, Solotuo and Clic Clac. The firm is also a licensed manufacturer for prominent brands such as Diesel, Laura Ashley and Lacoste. The firm operates in three divisions: Zucchi & Affiliates, which is responsible for producing items under the company's own brands as well as licensed brands; Mascioni, which is engaged in the finishing of fabrics for group companies and third parties and the packaging and marketing of textiles; and Hospitality, which markets textiles to the hospitality industry. The company's affiliates and subsidiaries include Welspun Zucchi Textiles Private, Ltd.; Intesa Srl; Bassetti Espanola SA; Bassetti Deutschland GmbH; Bassetti Schweiz AG; Ta Tria Epsilon-Bianca SA; Basitalia Leasing SpA.; Descamps SAS; Mascioni SpA; Mascioni USA Ltd.; and Mascioni Hospitality, Inc., among others. Additionally, the firm holds the Zucchi Collection, an antique collection of handblocks used for the production of printed fabrics. The collection includes over 12,000 designs and some 56,000 printing blocks dating from 1785 to 1935.

FINANCIAL DATA: *Note: Data for latest year may not have been available at press time.*

In U.S. $	2012	2011	2010	2009	2008	2007
Revenue			246,400,000	320,920,000	374,150,000	438,900,000
R&D Expense						
Operating Income						
Operating Margin %						
SGA Expense						
Net Income			-26,900,000	-68,120,000	-54,810,000	-3,700,000
Operating Cash Flow						
Capital Expenditure						
EBITDA						
Return on Assets %						
Return on Equity %						
Debt to Equity						

CONTACT INFORMATION:

Phone: 39-02-439-221 Fax: 39-02-481-8854
Toll-Free:
Address: Via Tiziano, 9/a, Milan, 20145 Italy

STOCK TICKER/OTHER:

Stock Ticker: ZUCR Exchange: Milan
Employees: 1,344 Fiscal Year Ends: 31-Dec
Parent Company:

SALARIES/BONUSES:

Top Exec. Salary: $ Bonus: $
Second Exec. Salary: $ Bonus: $

OTHER THOUGHTS:

Estimated Female Officers or Directors:
Hot Spot for Advancement for Women/Minorities:

Volcom Inc

www.volcom.com

NAIC Code: 315210

TYPES OF BUSINESS:

Activewear
Sports-Specific Apparel
Music & Film Production
Sporting Events

BRANDS/DIVISIONS/AFFILIATES:

Volcom
Electric Visual Evolution LLC
Laguna Surf & Sport
Volcom Entertainment
Veeco Productions
Volcom Pipe House
PPR SA

CONTACTS:

Note: Officers with more than one job title may be intentionally listed here more than once.

Richard Woolcott, CEO
Jason W. Steris, COO
Jason W. Steris, Pres.
Tom Ruiz, Exec. VP-Sales
David A. Unter, VP-Finance
Ryan N. Immegart, VP-Mktg.
Ethan F. Anderson, Exec. VP/Creative Dir.
John W. Fearnley, Sr.VP-Production & Design
Richard R. Woolcott, Chmn.

GROWTH PLANS/SPECIAL FEATURES:

Volcom, Inc., a subsidiary of PPR SA, is a designer, marketer and distributor of lifestyle clothing, footwear and accessories for young men and women. Its products are sold throughout the U.S. and in over 40 countries worldwide through traditional and online retailers, as well as 20 company-operated Volcom branded retail stores and additional licensed stores worldwide. Volcom products, marketed under the Volcom brand name, are targeted at participants of boardsports, including skateboarding, snowboarding and surfing. Products are organized into six primary categories: mens, girls, boys, footwear, girls swim and snow. Principal products include t-shirts, fleece, swimwear, denim, boardshorts, bottoms, tops, jackets, outerwear, slip-on footwear, music and film. The firm's executive, warehousing and distribution offices are located in Costa Mesa, California. Volcom employs a multi-faceted advertising and promotion strategy through internal marketing and art departments that create its advertisements and manage its various grassroots programs. These include an athlete sponsorship program; Volcom sponsored sporting events; Volcom Entertainment, the firm's own music label; Film, through Veeco Productions; and the Volcom Pipe House located in front of the renowned Pipeline surf break on the North Shore of Oahu Hawaii, which serves as a headquarters for marketing, research and development for top Volcom-sponsored riders. Additionally, the company owns the retail operations of Laguna Surf & Sport and Electric Visual Evolution LLC, an action sports lifestyle brand.

FINANCIAL DATA:

Note: Data for latest year may not have been available at press time.

In U.S. $	2012	2011	2010	2009	2008	2007
Revenue	354,000,000	349,000,000	323,181,000	280,639,000	334,304,000	268,613,000
R&D Expense						
Operating Income						
Operating Margin %						
SGA Expense						
Net Income			22,290,000	21,744,000	21,709,000	33,335,000
Operating Cash Flow						
Capital Expenditure						
EBITDA						
Return on Assets %						
Return on Equity %						
Debt to Equity						

CONTACT INFORMATION:

Phone: 949-646-2175 Fax: 949-646-5247
Toll-Free:
Address: 1740 Monrovia Ave., Costa Mesa, CA 92627 United States

STOCK TICKER/OTHER:

Stock Ticker: Subsidiary
Exchange:
Employees: 543
Fiscal Year Ends: 31-Dec
Parent Company: Kering

SALARIES/BONUSES:

Top Exec. Salary: $ Bonus: $
Second Exec. Salary: $ Bonus: $

OTHER THOUGHTS:

Estimated Female Officers or Directors:
Hot Spot for Advancement for Women/Minorities:

W L Gore & Associates Inc

NAIC Code: 313320

www.gore.com

TYPES OF BUSINESS:

Polymer Materials
Waterproof/Windproof Clothing
Cable Assemblies
Health Care Products
Wire & Cable Manufacturing
Electrochemical Materials
Filtration & Sealant Materials
Fabrics & Fibers

BRANDS/DIVISIONS/AFFILIATES:

GORE PROCEL
ePTFE
GORE-TEX
WINDSTOPPER
ELIXER
GORE PROTECTIVE WEAR
CLEANSTREAM

CONTACTS: *Note: Officers with more than one job title may be intentionally listed here more than once.*

Terri Kelly, CEO
Terri Kelly, Pres.
David Abeyta, Head-Aortic Segment

GROWTH PLANS/SPECIAL FEATURES:

W. L. Gore & Associates, Inc. (Gore) designs, produces and markets a variety of products based on high-performance fluorocarbon polymer materials, primarily polytetrafluoroethylene (PTFE). Rapidly expanded PTFE, the basis for a number of Gore's solutions, is known as GORE expanded PTFE (ePTFE). It exhibits a high strength-to-weight ratio, chemical inertness, biocompatibility, a low coefficient of friction, low water absorption and good weathering properties. Gore's products cover a wide range of categories, including consumer products, cables and cable assemblies, electronic materials, fabrics, medical and health care, sealants and venting. Consumer products include its famous GORE-TEX outerwear and footwear as well as WINDSTOPPER outerwear, ELIXER guitar cables and strings, GORE PROCEL cast liner, CLEANSTREAM vacuum cleaner filters and GORE PROTECTIVE WEAR. Additional Gore products include fuel cell components, battery and capacitor technologies, electromagnetic interference (EMI) shielding, cleanroom garments, fire and safety service fabrics, transparent screen fabric, disk drive filtration products, industrial liquid filtration products, geochemical imaging capabilities, surgical meshes, aircraft sealants and automotive vents. Gore operates facilities located in approximately 25 countries in the Asia Pacific, Australia, Europe, North America and South America. It maintains manufacturing centers in Germany, the U.K., the U.S., China and Japan.

Employees are offered tuition assistance, a 401(k), a profit sharing plan, an employee stock purchase plan, comprehensive medical coverage and domestic partner benefits.

FINANCIAL DATA: *Note: Data for latest year may not have been available at press time.*

In U.S. $	2012	2011	2010	2009	2008	2007
Revenue		3,000,000,000	2,800,000,000	2,500,000,000	2,500,000,000	
R&D Expense						
Operating Income						
Operating Margin %						
SGA Expense						
Net Income						
Operating Cash Flow						
Capital Expenditure						
EBITDA						
Return on Assets %						
Return on Equity %						
Debt to Equity						

CONTACT INFORMATION:

Phone: 410-506-7787 Fax:
Toll-Free: 888-914-4673
Address: 555 Papermill Rd., Newark, DE 19711 United States

STOCK TICKER/OTHER:

Stock Ticker: Private Exchange:
Employees: 9,500 Fiscal Year Ends: 31-Mar
Parent Company:

SALARIES/BONUSES:

Top Exec. Salary: $ Bonus: $
Second Exec. Salary: $ Bonus: $

OTHER THOUGHTS:

Estimated Female Officers or Directors: 1
Hot Spot for Advancement for Women/Minorities: Y

Wacoal Holdings Corp

www.wacoalholdings.jp

NAIC Code: 315240

TYPES OF BUSINESS:

Intimate Apparel
Lingerie
Sportswear
Online Sales
Catalog Sales

BRANDS/DIVISIONS/AFFILIATES:

Wacoal
Wing
Wacoal France SA
Wacoal UK Ltd
Wacoal America Inc
Wacoal Sports Science Corp
Eveden Group Limited
Peach John Co Ltd

CONTACTS:

Note: Officers with more than one job title may be intentionally listed here more than once.

Yoshikata Tsukamoto, Pres.
Ikuo Otani, Gen Mgr.-Corp. Planning
Hideo Kawanaka, Exec. VP

GROWTH PLANS/SPECIAL FEATURES:

Wacoal Holdings Corp. is one of Japan's leading makers of lingerie under such brands as Wacoal and Wing. The company is composed of 62 subsidiaries and nine affiliated companies. Wacoal manufactures an extensive range of bra sizes and styles. It also designs and manufactures nightwear; children's underwear, outerwear and sportswear; hosiery; and other apparel and textile products. Additionally, the company engages in business operations related to cultural projects; the restaurant industry; and the interior design of commercial buildings, such as the manufacturing and rental of fixtures and mannequins. Although the bulk of the company's sales revenue is generated in Japan, the company has extended its sales distribution to throughout China and Southeast Asia, the U.K., France and the U.S. The firm's primary customers include department and specialty stores throughout Japan as well as general merchandise stores; however, Wacoal also has its own chain of approximately 155 company-owned stores in Japan, including 18 outlet stores. Wacoal also sells online and through catalogs such as LOVE BODY, and provides online tools to help customers find correctly fitting lingerie. A few of the firm's many subsidiaries include Wacoal Corp.; Lecien Corporation; House of Rose Co., Ltd.; Wacoal China Co., Ltd.; Shanghai Yadie Fashions Co., Ltd.; Wacoal France SA; and Wacoal UK, Ltd. Subsidiary Peach John Co., Ltd. specializes in mail-order and e-commerce sales of women's apparel and other related products in Japan, and maintains numerous Peach John stores. The firm's Nanasai Co., Ltd. subsidiary is engaged in the manufacture and rental of mannequins and in-store fixtures as well as providing commercial interior design services. The company also maintains several U.S. subsidiaries, including Wacoal International Corp., Wacoal America, Inc. and Wacoal Sports Science Corp. In April 2012, the firm acquired Eveden Group Limited.

FINANCIAL DATA:

Note: Data for latest year may not have been available at press time.

In U.S. $	2012	2011	2010	2009	2008	2007
Revenue	1,913,773,000	1,845,070,000	1,818,027,000	1,917,992,000	1,842,887,000	1,856,787,000
R&D Expense						
Operating Income	115,529,800	47,372,000	42,417,700	112,768,700	150,534,800	143,892,100
Operating Margin %	6.03%	2.56%	2.33%	5.87%	8.16%	7.74%
SGA Expense	886,529,900	863,851,500	872,758,000	861,702,700	769,847,400	768,010,200
Net Income	76,964,180	29,113,460	28,100,330	58,226,920	55,211,130	100,744,300
Operating Cash Flow	112,000,500	111,933,700	105,198,100	90,936,420	158,150,200	104,203,300
Capital Expenditure	39,567,580	37,107,130	64,049,610	46,848,730	32,118,700	28,296,500
EBITDA	166,553,500	94,688,320	89,377,760	146,302,100	203,888,700	197,807,900
Return on Assets %	3.16%	1.19%	1.15%	2.30%	2.01%	3.66%
Return on Equity %	4.08%	1.54%	1.49%	2.98%	2.62%	4.75%
Debt to Equity						

CONTACT INFORMATION:

Phone: 81 756821028 Fax: 81 756615603
Toll-Free:
Address: 29 Nakajima-cho, Kisshoin, Minami-ku, Kyoto, 601-8530 Japan

STOCK TICKER/OTHER:

Stock Ticker: WACLY Exchange: PINX
Employees: 15,983 Fiscal Year Ends: 31-Mar
Parent Company:

SALARIES/BONUSES:

Top Exec. Salary: $ Bonus: $
Second Exec. Salary: $ Bonus: $

OTHER THOUGHTS:

Estimated Female Officers or Directors:
Hot Spot for Advancement for Women/Minorities:

Walking Company Holdings Inc (The) www.thewalkingcompany.com

NAIC Code: 448140

TYPES OF BUSINESS:

Casual Apparel Stores
Accessories & Gifts
Catalog & Online Sales

BRANDS/DIVISIONS/AFFILIATES:

Walking Company (The)
Big Dog Holdings Inc
Big Dog USA Inc
BigDogs.com

GROWTH PLANS/SPECIAL FEATURES:

The Walking Company Holdings, Inc., formerly Big Dog Holdings, Inc., is the parent company to two retail entities: Big Dog USA, Inc. and The Walking Company. The Walking Company, a retailer of comfort footwear and accessories, operates over 200 specialty stores in malls across the country. The chain sells walk-wear and walking accessories from around the world under the Ecco, Dansko, Mephisto, Ugg, Pikolinos, MBT, Keen, Beautifeel, Sofft, Naot, Earth, Born, Ariat, Raffini, Olukai, Josef Seibel, Suzzato and Sierra West brands. Big Dogs USA develops, markets and retails active-wear, casual sportswear, accessories and gifts online at BigDogs.com.

The firm offers its employees a 401(k) savings plan and medical, dental and vision insurance.

CONTACTS: *Note: Officers with more than one job title may be intentionally listed here more than once.*

Andrew D. Feshbach, CEO
Andrew D. Feshbach, Pres.
Roberta J. Morris, CFO
Michael Grenley, Sr. VP-Merch.
Anthony J. Wall, General Counsel/Sec./Exec. VP
Lee M. Cox, Sr. VP-Retail Oper.
Fred Kayne, Chmn.

FINANCIAL DATA: *Note: Data for latest year may not have been available at press time.*

In U.S. $	2012	2011	2010	2009	2008	2007
Revenue						
R&D Expense						
Operating Income						
Operating Margin %						
SGA Expense						
Net Income						
Operating Cash Flow						
Capital Expenditure						
EBITDA						
Return on Assets %						
Return on Equity %						
Debt to Equity						

CONTACT INFORMATION:

Phone: 805-963-8727 Fax: 805-962-9460
Toll-Free: 800-642-9265
Address: 121 Gray Ave., Ste. 300, Santa Barbara, CA 93101 United States

STOCK TICKER/OTHER:

Stock Ticker: Private
Exchange:
Employees:
Fiscal Year Ends: 31-Dec
Parent Company:

SALARIES/BONUSES:

Top Exec. Salary: $ Bonus: $
Second Exec. Salary: $ Bonus: $

OTHER THOUGHTS:

Estimated Female Officers or Directors: 1
Hot Spot for Advancement for Women/Minorities:

Sales, profits and employees may be estimates. Financial information, benefits and other data can change quickly and may vary from those stated here.

Wal-Mart Stores Inc

www.walmartstores.com

NAIC Code: 452112

TYPES OF BUSINESS:

Discount Department Stores
Supermarkets
Warehouse Membership Clubs
Online Sales
Pharmacies
Vision Centers
Auto Repair Centers

BRANDS/DIVISIONS/AFFILIATES:

Sam's Club
Wal-Mart Supercenter
Marketside
Best Price
Massmart Holdings Limited
Yihaodian
Neighborhood Market

CONTACTS: *Note: Officers with more than one job title may be intentionally listed here more than once.*

Michael T. Duke, CEO
Michael T. Duke, Pres.
Charles Holley, Jr., CFO/Exec. VP
Stephen Quinn, Chief Mktg. Officer/Exec. VP-Wal-Mart U.S.
M. Susan Chambers, Exec. VP-People
Rollin L. Ford, CIO/Exec. VP
Jeffrey J. Gearhart, General Counsel/Exec. VP/Sec.
Leslie A. Dach, Exec. VP-Corp. Affairs
Carol Schumacher, VP-Investor Rel.
Steven P. Whaley, Controller/Sr. VP
Ignacio Perez Lizuar, Exec. VP-Oper., Sam's Club
Eduardo Solorzano, Pres./CEO-Wal-Mart Latin America
Rosalind Brewer, CEO/Pres., Sam's Club
William S. Simon, CEO/Pres., Wal-Mart U.S.
S. Robson Walton, Chmn.
C. Douglas McMillon, CEO/Pres., Wal-Mart Int'l
Ed Kolodzieski, Exec. VP-Global Sourcing

GROWTH PLANS/SPECIAL FEATURES:

Wal-Mart Stores, Inc., one of the world's largest retailers, operates through a massive base of Wal-Mart stores, Wal-Mart Supercenters, Sam's Clubs, Marketside, Neighborhood Markets and WalMart.com. The company operates in three business segments: Wal-Mart Stores, representing roughly 62.1% of sales; Sam's Club, 11.8%; and International, 26.1%. The company serves over 200 million weekly customers in over 9,600 stores in 28 countries, with 69 different banners. Wal-Mart offers a wide variety of discount merchandise in six merchandise units, including family apparel; grocery, consisting of meat, produce, deli, bakery, dairy, frozen foods, beauty aids, household chemicals, paper goods and pet supplies; entertainment, such as electronics, toys and photo-processing services; hardlines, which consists of fabrics, books, hardware, sporting goods and seasonal merchandise; health and wellness, including pharmacy and optical services; and home, including home furnishings, housewares and small appliances. Wal-Mart Supercenters, located in 48 states, are larger stores that combine a full-line supermarket with a discount department store. Sam's Club is a members-only warehouse club that sells merchandise at warehouse prices to consumers and small businesses. The International segment includes wholly-owned subsidiaries in Argentina, Brazil, Canada, Japan, Puerto Rico and the U.K.; majority-owned subsidiaries in Central America, Chile, South Africa and Mexico; joint ventures in India and China; and minority-owned subsidiaries in China. The company is currently shifting its focus away from its supercenters. New stores will focus more on small formats, including the Marketside concept. Wal-Mart is expanding into India with a joint venture opening several Best Price wholesale distribution centers. In September 2010, the company partnered with T-Mobile to offer the Walmart Family Mobile post-paid cell phone service. In May 2011, Wal-Mart agreed to acquire minority interest in Chinese e-commerce firm Yihaodian. In June 2011, the firm acquired 51% of South African company Massmart Holdings Limited.

FINANCIAL DATA: *Note: Data for latest year may not have been available at press time.*

In U.S. $	2012	2011	2010	2009	2008	2007
Revenue	446,950,000,000	421,849,000,000	408,214,000,000	405,607,000,000	378,799,000,000	
R&D Expense						
Operating Income	26,558,000,000	25,542,000,000	23,950,000,000	22,798,000,000	21,996,000,000	
Operating Margin %	5.94%	6.05%	5.86%	5.62%	5.80%	
SGA Expense	85,265,000,000	81,020,000,000	79,607,000,000	76,651,000,000	70,288,000,000	
Net Income	15,699,000,000	16,389,000,000	14,335,000,000	13,400,000,000	12,731,000,000	
Operating Cash Flow	24,255,000,000	23,643,000,000	26,249,000,000	23,147,000,000	20,354,000,000	
Capital Expenditure	13,510,000,000	12,699,000,000	12,184,000,000	11,499,000,000	14,937,000,000	
EBITDA	34,850,000,000	33,384,000,000	31,288,000,000	29,821,000,000	28,618,000,000	
Return on Assets %	8.39%	9.32%	8.58%	8.19%	8.09%	
Return on Equity %	22.45%	23.53%	21.07%	20.63%	20.17%	
Debt to Equity	0.66	0.63	0.51	0.52	0.51	

CONTACT INFORMATION:

Phone: 479 273-4000 Fax: 479 273-1986
Toll-Free: 800-925-6278
Address: 702 SW 8th St., Bentonville, AR 72716 United States

STOCK TICKER/OTHER:

Stock Ticker: WMT Exchange: NYS
Employees: 2,000,000 Fiscal Year Ends: 31-Jan
Parent Company:

SALARIES/BONUSES:

Top Exec. Salary: $1,315,731 Bonus: $
Second Exec. Salary: $929,748 Bonus: $

OTHER THOUGHTS:

Estimated Female Officers or Directors: 7
Hot Spot for Advancement for Women/Minorities: Y

Warnaco Group Inc (The)

www.warnaco.com

NAIC Code: 315240

TYPES OF BUSINESS:

Lingerie Manufacturing
Swimwear
Sportswear
Retail Stores
Online Sales

BRANDS/DIVISIONS/AFFILIATES:

Warnaco Retail Inc
Warnaco Swimwear Inc
Warnaco Fashion SAS
Speedo
Warner's
PVH Corp
Warnaco Logistics BV
Olga

CONTACTS: *Note: Officers with more than one job title may be intentionally listed here more than once.*

Emanuel Chirico, Pres.
Emanuel Chirico, Chmn./CEO-PVH Corp.

GROWTH PLANS/SPECIAL FEATURES:

The Warnaco Group, Inc. designs, sources, markets, licenses and distributes a broad line of intimate apparel, sportswear and swimwear worldwide. The company sells its products under several highly recognized brand names, including Warner's, Olga, Calvin Klein, Speedo and Chaps. Warnaco operates in three business segments: intimate apparel, sportswear and swimwear. The firm relies on its highly recognizable brands to sell at a broad range of price points and across consumer age groups. The company's products are distributed domestically and internationally in over 100 countries. Warnaco sells its products through major department stores, independent retailers, chain stores, membership clubs, specialty, off-price and other stores, mass merchandisers and the Internet. Its customers include such leading retailers as Macy's, JC Penney, Kohl's, Sears, Target and Costco. Warnaco operates approximately 1,851 Calvin Klein retail stores worldwide, including 409 free-standing stores, one online store dedicated to Calvin Klein Underwear and 1,441 shop-in-shop/concession stores. The firm also operates SpeedoUSA.com, an online Speedo store marketing swimwear, accessories and aquatic exercise equipment. The company focuses its advertising on brand and/or product-specific advertising, primarily through point-of-sale (POS) displays, visuals and in-store promotions. Subsidiaries of the firm include Warnaco Swimwear, Inc.; Warnaco Retail, Inc.; Warnaco Logistics BV; and Warnaco Fashion SAS. In February 2013, Warnaco was acquired by PVH Corp.

FINANCIAL DATA: *Note: Data for latest year may not have been available at press time.*

In U.S. $	2012	2011	2010	2009	2008	2007
Revenue		2,513,388,032	2,295,750,912	2,019,624,960	2,065,022,976	1,860,120,064
R&D Expense						
Operating Income						
Operating Margin %						
SGA Expense						
Net Income		127,450,000	138,580,992	95,998,000	47,254,000	79,107,000
Operating Cash Flow						
Capital Expenditure						
EBITDA						
Return on Assets %						
Return on Equity %						
Debt to Equity						

CONTACT INFORMATION:

Phone: 212 287-8000 Fax: 212 687-6771
Toll-Free:
Address: 501 7th Ave., New York, NY 10018 United States

STOCK TICKER/OTHER:

Stock Ticker: Subsidiary
Exchange:
Employees: 7,136
Fiscal Year Ends: 31-Dec
Parent Company: PVH CORP

SALARIES/BONUSES:

Top Exec. Salary: $ Bonus: $
Second Exec. Salary: $ Bonus: $

OTHER THOUGHTS:

Estimated Female Officers or Directors:
Hot Spot for Advancement for Women/Minorities: Y

Warnaco Swimwear Inc

www.speedousa.com

NAIC Code: 315210

TYPES OF BUSINESS:

Swimwear Apparel
Online Sales
Aquatic Fitness Equipment
Swim Accessories

BRANDS/DIVISIONS/AFFILIATES:

Warnaco Group Inc (The)
Speedo
SpeedoUSA.com
FastSkin FS-Pro
FastSkin LZR Racer
Aqualab
Calvin Klein

CONTACTS: *Note: Officers with more than one job title may be intentionally listed here more than once.*

Martha Olson, Pres.
James B. Gerson, Pres.
James Christl, CFO

GROWTH PLANS/SPECIAL FEATURES:

Warnaco Swimwear, Inc., the swimwear marketing arm of its parent company, the Warnaco Group, Inc., designs, manufactures and markets swimwear, swim accessories and active fitness apparel. Its products are primarily sold under the Speedo brand name, which the firm has licensed in North America and the Caribbean, from British firm the Pentland Group, whose Speedo International affiliate provides the company with various apparel accessories. Throughout the U.S., Mexico, Canada and the Caribbean, Warnaco distributes Speedo fitness and fashion swimwear, Speedo swimwear for kids, Speedo footwear, Speedo active apparel and aquatic fitness equipment. Speedo is sold in over 11,000 stores in North America and is available in approximately 170 countries worldwide. Other distribution channels include department stores, independent retailers, chain stores, sporting goods stores, team dealers, membership clubs and other retailers as well as the company's SpeedoUSA.com web site. The swimwear accessories line includes a diverse range of products such as swim goggles; swimming fins; swim caps; kickboards; nose clips; earplugs; duffle bags and backpacks; watches; waterproof digital cameras and MP3 players; swimming vests; masks and snorkels; and aquatic exercise gear such as leg trainers, barbells and swimming gloves. The company often forms long-term promotional contracts with Olympic medalists and world-champion swimmers to better market its products. Warnaco products also include FastSkin FS-Pro, a line of high-quality skin-like bodysuits for swim racing; and FastSkin LZR Racer, which was co-developed by Speedo's research and development facility, Aqualab and NASA. Warnaco also markets Calvin Klein-brand swimwear, which is distributed in the U.S., Europe, Canada, Mexico and Asia through department stores, high-end specialty stores and boutiques, including Saks, Bloomingdale's, Nordstrom, Macy's and Dillard's.

FINANCIAL DATA: *Note: Data for latest year may not have been available at press time.*

In U.S. $	2012	2011	2010	2009	2008	2007
Revenue						
R&D Expense						
Operating Income						
Operating Margin %						
SGA Expense						
Net Income						
Operating Cash Flow						
Capital Expenditure						
EBITDA						
Return on Assets %						
Return on Equity %						
Debt to Equity						

CONTACT INFORMATION:

Phone: 213-482-7622 Fax:
Toll-Free: 888-477-3336
Address: 1201 W. 5th St., #1100, Los Angeles, CA 90017 United States

STOCK TICKER/OTHER:

Stock Ticker: Subsidiary Exchange:
Employees: Fiscal Year Ends: 31-Dec
Parent Company: WARNACO GROUP INC (THE)

SALARIES/BONUSES:

Top Exec. Salary: $ Bonus: $
Second Exec. Salary: $ Bonus: $

OTHER THOUGHTS:

Estimated Female Officers or Directors: 1
Hot Spot for Advancement for Women/Minorities:

Weiqiao Textile Company Limited

NAIC Code: 313110

www.wqfz.com

TYPES OF BUSINESS:

Cotton Spun Yarn
Denim Fabric

BRANDS/DIVISIONS/AFFILIATES:

Shandong Luteng Textile Co., Ltd.
Binzhou Weiqiao Technology Industrial Park Co.
Weihai Weiqiao Textile Co., Ltd.
Shandong Weiqiao Cotton Industry Co., Ltd.
Weihai Weiqiao Technology Industrial Park Co., Ltd
Shandong Binteng Textile Co., Ltd.

GROWTH PLANS/SPECIAL FEATURES:

Weiqiao Textile Company Limited is a Chinese cotton textile manufacturer. The firm is primarily engaged in the production, sale and distribution of cotton yarn, grey fabric and denim. The company is one of the largest textile manufacturers in the world. Its annual production is approximately 1.3 billion yards of grey fabric, 111 million yards of denim and 618,000 tons of cotton yarn. There are a total of four production areas, including Weiqiao Production Base, Binzhou Production Base, Weihai Production Base and Zouping Production Base. Subsidiaries include Shandong Luteng Textile Co., Ltd.; Binzhou Weiqiao Technology Industrial Park Co., Ltd.; Weihai Weiqiao Textile Co., Ltd.; Shandong Weiqiao Cotton Industry Co., Ltd.; Weihai Weiqiao Technology Industrial Park Co., Ltd.; and Shandong Binteng Textile Co., Ltd.

CONTACTS: *Note: Officers with more than one job title may be intentionally listed here more than once.*

Zhang Hongxia, CEO
Zhang Hongxia, Chmn.

FINANCIAL DATA: *Note: Data for latest year may not have been available at press time.*

In U.S. $	2012	2011	2010	2009	2008	2007
Revenue	2,452,110,000	2,449,550,000	2,876,579,000	2,304,985,000	2,645,914,000	2,989,496,000
R&D Expense						
Operating Income	126,015,500	132,155,100	442,365,300	275,489,800	244,647,900	407,140,000
Operating Margin %	5.13%	5.39%	15.37%	11.95%	9.24%	13.61%
SGA Expense	73,751,500	66,106,170	78,739,050	70,592,440	86,726,120	97,646,140
Net Income	77,028,930	38,487,200	262,992,300	143,173,200	95,854,980	302,028,700
Operating Cash Flow	713,607,600	-250,887,100	-17,693,100	182,496,700	287,400,600	344,680,400
Capital Expenditure	36,305,740	83,164,700	115,541,700	22,785,330	413,528,000	205,444,900
EBITDA	524,829,800	354,815,800	654,173,200	479,225,300	418,466,800	407,140,000
Return on Assets %	1.66%	.85%	5.90%	3.30%	2.20%	7.44%
Return on Equity %	3.12%	1.59%	10.96%	6.45%	4.44%	14.89%
Debt to Equity	0.29	0.3	0.33	0.33	0.37	0.37

CONTACT INFORMATION:

Phone: 0543-4161171 Fax:
Toll-Free:
Address: No. 1 Weifang Rd., Zouping Economic Dev. Zone, Shandong, 256200 China

STOCK TICKER/OTHER:

Stock Ticker: WQTEF
Exchange: GREY
Employees: 102,000
Fiscal Year Ends: 31-Dec
Parent Company:

SALARIES/BONUSES:

Top Exec. Salary: $ Bonus: $
Second Exec. Salary: $ Bonus: $

OTHER THOUGHTS:

Estimated Female Officers or Directors: 1
Hot Spot for Advancement for Women/Minorities:

Wet Seal Inc (The)

www.wetsealinc.com

NAIC Code: 448120

TYPES OF BUSINESS:

Casual Apparel-Young Women's, Retail
Accessories
Footwear
Online Retailing

BRANDS/DIVISIONS/AFFILIATES:

Wet Seal
Arden B
ArdenB.com
WetSeal.com
Blink by Wet Seal
iRunway

CONTACTS: *Note: Officers with more than one job title may be intentionally listed here more than once.*

Kenneth Seipel, COO
Kenneth Seipel, Pres.
Steve Benrubi, CFO/Exec. VP
Dianne Ippolito, VP-Human Resources
Barbara Cook, VP-Store Oper.
David Towers, VP-e-commerce
Jennifer Ehrhardt, Corp. Controller/VP
Sharon Huges, Pres./Chief Merch. Officer-Arden B
Nick Dialynas, VP-Real Estate
Charles Torok, VP-Logistics

GROWTH PLANS/SPECIAL FEATURES:

The Wet Seal, Inc. is a specialty clothing retailer offering moderately priced, fashionable and contemporary apparel and accessories designed for female consumers aged 15-39. The company operates 554 stores in 47 states and Puerto Rico, under the names Wet Seal and Arden B. All of the company's retail concepts offer a wide range of women's apparel, such as tops, sweaters, bottoms, dresses, outerwear, shoes and accessories. Most of the company's stores are located in regional, high-traffic shopping malls, which contain at least one anchor department store. The company also operates two e-commerce sites, WetSeal.com and ArdenB.com. The firm's 472 Wet Seal locations target trendy young women, age 15-23, by providing an ever-changing array of new apparel and accessories. Wet Seal offers a 10% discount through its frequent buyer program, which has a $20 one-year membership fee. The firm's 81 Arden B stores offer a unique mix of high quality European and custom in-house designs. The chain provides everyday, wear-to-work, special occasion and casual apparel to fashion-conscious and sophisticated women, ages 25-35. Arden B offers a loyalty program based on a point system, offering awards redeemable for merchandise. The company's iPhone application, iRunway, offers customers the ability to view, purchase and share outfits created using Wet Seal products. The company does not design any of its apparel in-house, instead it relies on buying teams at each of its divisions to identify emerging fashion trends and work with vendors to develop consistent fashion themes. Recently, the firm has introduced a new store concept, Blink by Wet Seal, focused on denim for young women.

FINANCIAL DATA: *Note: Data for latest year may not have been available at press time.*

In U.S. $	2012	2011	2010	2009	2008	2007
Revenue	620,097,000	581,194,000	560,918,000	592,960,000	611,163,000	
R&D Expense						
Operating Income	25,000,000	25,247,000	22,852,000	32,157,000	19,257,000	
Operating Margin %	4.03%	4.34%	4.07%	5.42%	3.15%	
SGA Expense	165,933,000	150,432,000	141,633,000	154,671,000	177,468,000	
Net Income	15,082,000	12,570,000	93,430,000	30,154,000	23,232,000	
Operating Cash Flow	61,900,000	50,080,000	40,456,000	58,340,000	51,470,000	
Capital Expenditure	26,486,000	30,727,000	21,304,000	23,023,000	38,523,000	
EBITDA	44,612,000	42,397,000	38,440,000	48,857,000	39,294,000	
Return on Assets %	4.31%	3.47%	30.58%	12.54%	10.74%	
Return on Equity %	5.83%	4.63%	42.67%	20.16%	18.99%	
Debt to Equity			0.01	0.01	0.02	

CONTACT INFORMATION:

Phone: 949 699-3900 Fax: 949 583-0715
Toll-Free:
Address: 26972 Burbank, Foothill Ranch, CA United States

STOCK TICKER/OTHER:

Stock Ticker: WTSL
Exchange: NAS
Employees: 6,982
Fiscal Year Ends: 31-Jan
Parent Company:

SALARIES/BONUSES:

Top Exec. Salary: $603,365 Bonus: $
Second Exec. Salary: $494,327 Bonus: $

OTHER THOUGHTS:

Estimated Female Officers or Directors: 7
Hot Spot for Advancement for Women/Minorities: Y

Weyco Group Inc

NAIC Code: 316210

www.weycogroup.com

TYPES OF BUSINESS:

Footwear, Wholesale
Men's Footwear, Retail

BRANDS/DIVISIONS/AFFILIATES:

Florsheim
Nunn Bush
Brass Boot
Stacy Adams
Umi
Rafters
Bogs
Combs Company (The)

CONTACTS: *Note: Officers with more than one job title may be intentionally listed here more than once.*

Thomas W. Florsheim, Jr., CEO
John W. Florsheim, COO
John W. Florsheim, Pres.
John F. Wittkowske, CFO/Sr. VP
Matthew J. Engerman, VP-Sales, Nunn Bush Brand
George Sotiros, VP-IT
Keven Ringgold, VP-Design
John F. Wittkowske, Corp. Sec.
Judy Anderson, VP-Finance/Treas.
William Combs, VP/Pres., BOGS & Rafters Brands
Brian Flannery, VP/Pres., Stacy Adams Brand
Peter S. Grossman, Sr. VP/Pres., Nunn Bush Brand & Retail Div.
Kevin Schiff, VP/Pres., Florsheim Brand
Thomas W. Florsheim, Jr., Chmn.
Allison Woss, VP-Purchasing

GROWTH PLANS/SPECIAL FEATURES:

Weyco Group, Inc. is a distributor of men's footwear. The firm sells mid-priced formal and casual men's footwear under the brands Stacy Adams, Florsheim, Nunn Bush, Nunn Bush NXXT, SAO by Stacy Adams, Brass Boot, BOGS, Umi and Rafters. Products consist of both mid-priced leather dress shoes and leather, casual footwear in addition to all weather boots. All of Weyco Group's shoes are purchased from outside suppliers. The firm operates in two segments: North American wholesale and North American retail. Wholesale distribution sales, which include both wholesale sales and licensing revenues, account for approximately 74% of Weyco Group's total sales. At the wholesale level, shoes are marketed nationwide through over 10,000 shoe, clothing and department stores. The North American retail operation, accounting for 9% of total sales, consists of 26 stores, primarily under the Florsheim banner, and an Internet business. In addition to the sale of the company's brands in these retail stores, other branded footwear and accessories are sold in order to enhance selection. Moreover, the firm provides distribution and retail services in Europe, South Africa, Australia and Asia, accounting for 17% of total sales. Weyco Group has licensing agreements with third parties who sell its branded shoes overseas and specialty shoe, apparel and accessory manufacturers in the U.S. It maintains a 950,000-square-foot office and distribution center in Glendale, Wisconsin, as well as office and distribution properties in Canada, Italy, Australia, South Africa, Australia and China. In 2011, the firm acquired The Combs Company, the owner of the BOGS and Rafters footwear brands, for a total consideration of $33.1 million.

FINANCIAL DATA: *Note: Data for latest year may not have been available at press time.*

In U.S. $	2012	2011	2010	2009	2008	2007
Revenue	293,471,000	271,100,000	229,231,000	225,305,000	221,432,000	232,616,700
R&D Expense						
Operating Income	29,797,000	23,197,000	18,781,000	16,780,000	24,499,000	34,132,010
Operating Margin %	10.15%	8.55%	8.19%	7.44%	11.06%	14.67%
SGA Expense	85,090,000	83,525,000	71,516,000	67,696,000	56,639,000	55,285,340
Net Income	18,957,000	15,251,000	13,668,000	12,821,000	17,025,000	22,900,980
Operating Cash Flow	17,987,000	17,143,000	98,000	37,902,000	15,744,000	24,151,670
Capital Expenditure	9,695,000	8,344,000	1,665,000	1,318,000	2,178,000	2,727,362
EBITDA	35,136,000	28,477,000	24,233,000	23,077,000	29,239,000	38,890,310
Return on Assets %	6.78%	6.13%	6.34%	6.44%	8.94%	12.06%
Return on Equity %	11.14%	9.01%	8.11%	7.93%	10.66%	14.82%
Debt to Equity						

CONTACT INFORMATION:

Phone: 414 908-1600 Fax: 414 908-1601
Toll-Free:
Address: 333 W. Estabrook Blvd., Milwaukee, WI 53212 United States

STOCK TICKER/OTHER:

Stock Ticker: WEYS Exchange: NAS
Employees: 621 Fiscal Year Ends: 31-Dec
Parent Company:

SALARIES/BONUSES:

Top Exec. Salary: $563,500 Bonus: $
Second Exec. Salary: $538,000 Bonus: $

OTHER THOUGHTS:

Estimated Female Officers or Directors: 4
Hot Spot for Advancement for Women/Minorities: Y

Wolford AG

www.wolford.com

NAIC Code: 315240

TYPES OF BUSINESS:

Apparel Manufacturing
Retail Stores
Intimate Apparel
Men's Socks
Swimwear
Knit Clothing

BRANDS/DIVISIONS/AFFILIATES:

Wolford

CONTACTS:

Note: Officers with more than one job title may be intentionally listed here more than once.

Holger Dahmen, CEO
Thomas Melzer, CFO
Axel Dreher, Chief Production Officer
Sabine Labonte, Head-Corp. Comm.
Theresa Jordis, Chmn.

GROWTH PLANS/SPECIAL FEATURES:

Wolford AG manufactures and distributes women's lingerie, swimwear, legwear and ready-to-wear collections, which evolved from its bodywear line. The company also offers a men's collection, which comprises socks and knee-highs with a soft band. Wolford sells its products through its own specialty stores, boutiques and department stores under the Wolford brand name. While the company also sells its products in department stores and factory outlets, a significant portion of its revenues come from the company's 260 boutiques situated throughout the world. Around 111 of these boutiques are owned by Wolford, while the remaining 95 are owned and operated by various partners. Wolford maintains 15 international subsidiaries and distributes its products in 68 countries. Germany remains the largest market for Wolford's fashions, with a growing presence in markets in Asia. Its core focus remains its existing products, but the firm also invests in research and development of new products and fabrics. In recent years, the firm has produced co-branded, collaborative collections with leading fashion houses such as Vivienne Westwood, Karl Lagerfeld, Missoni, Zac Posen and Valentino. Wolford plans to expand its presence through additional boutiques and shop-in-shops in the U.S., Spain, the Netherlands, Central and Eastern Europe and Asia/Oceania.

FINANCIAL DATA:

Note: Data for latest year may not have been available at press time.

In U.S. $	2012	2011	2010	2009	2008	2007
Revenue	206,023,000	203,464,800	192,618,400	197,035,300	225,593,800	189,466,400
R&D Expense						
Operating Income	9,355,443	9,798,074	6,020,327	2,920,567	15,163,150	10,278,150
Operating Margin %	4.54%	4.81%	3.12%	1.48%	6.72%	5.42%
SGA Expense			91,006,950	94,334,050		
Net Income	1,820,005	6,753,143	3,424,713	-1,602,033	9,592,137	8,614,603
Operating Cash Flow	9,720,514	18,967,640	29,608,180	7,247,928	5,097,620	10,143,090
Capital Expenditure	11,053,760	8,113,132	11,162,080	19,731,210	10,504,150	
EBITDA	20,728,800	20,227,330	15,952,130	11,937,680	24,562,720	10,278,150
Return on Assets %	.93%	3.48%	1.70%	-.77%	4.85%	4.79%
Return on Equity %	1.62%	6.18%	3.27%	-1.53%	9.34%	9.30%
Debt to Equity	0.21	0.12	0.2	0.24	0.2	0.34

CONTACT INFORMATION:

Phone: 43 55746900 Fax: 43 557479544
Toll-Free:
Address: Wolfordstrasse 1, Bregenz, A-6901 Austria

STOCK TICKER/OTHER:

Stock Ticker: WLFDY
Exchange: PINX
Employees: 1,649
Fiscal Year Ends: 30-Apr
Parent Company:

SALARIES/BONUSES:

Top Exec. Salary: $ Bonus: $
Second Exec. Salary: $ Bonus: $

OTHER THOUGHTS:

Estimated Female Officers or Directors: 2
Hot Spot for Advancement for Women/Minorities:

Wolverine World Wide Inc

NAIC Code: 316210

www.wolverineworldwide.com

TYPES OF BUSINESS:

Shoe Manufacturing
Retail Stores
Work & Industrial Footwear
Slippers & Moccasins
Tanning Operations

BRANDS/DIVISIONS/AFFILIATES:

Hush Puppies
Sebago
Wolverine
HyTest
Merrell
Chaco
Performance + Lifestyle Group
Keds

CONTACTS: *Note: Officers with more than one job title may be intentionally listed here more than once.*

Blake W. Krueger, CEO
Blake W. Krueger, Pres.
Donald T. Grimes, CFO/Sr. VP/Treas.
Pamela L. Linton, Sr. VP-Global Human Resources
R. Paul Guerre, General Counsel/VP/Sec.
Michael Jeppesen, Pres., Global Oper. Group
Christi Cowdin, Dir.-Corp. Comm.
Christi Cowdin, Dir.-Investor Rel.
Michael D. Stornant, VP-Corp. Finance
James D. Zwiers, Sr. VP/Pres., Performance Group
Mark Neal, Pres., Lifestyle Group
Ted Gedra, Pres., Heritage Group
Douglas M. Jones, Corp. Controller
Blake W. Krueger, Chmn.
Daniel Le Vesconte, Managing Dir.-EMEA

GROWTH PLANS/SPECIAL FEATURES:

Wolverine World Wide, Inc. designs, manufactures and markets a broad line of casual shoes, rugged outdoor and work footwear. Its products are marketed under recognized brand names, including Bates, Cat Footwear, Harley-Davidson Footwear, Hush Puppies, HyTest, Merrell, Patagonia Footwear, Chaco, Cushe, Sebago and Wolverine. The company's footwear products are organized under three operating units. The Performance Group produces merchandise under the Patagonia, Chaco, Cushe, Merrell and Saucony branded product lines. The Heritage Group consists of Wolverine boots, shoes and branded apparel; Sebago footwear and apparel; and footwear under the Cat, Bates, Harley-Davidson and HyTest brands. The Lifestyle Group includes Hush Puppies footwear and apparel, as well as Sperry Top-Sider, Stride Rite and Keds branded merchandise; and Soft Style footwear. The firm also licenses its brands for use on non-footwear products including apparel, eyewear, watches, socks, gloves, handbags and other accessories. Products are marketed worldwide in roughly 190 countries through company-owned wholesale operations, licensees, distributors and retailers, including department stores, footwear chains, catalogs, specialty retailers, mass merchants and Internet retailers. Wolverine's consumer-direct operations consist of Wolverine Retail's 99 legacy company-owned retail stores in North America and the U.K. and 38 e-commerce sites; and 345 owned or licensed stores that it acquired in the Collective Brands transaction. Additionally, through its Wolverine Leathers division, the company operates a tannery that produces pigskin leather for the shoe and leather goods industries. In October 2012, the firm acquired Collective Brands' Performance + Lifestyle Group operations, which consists of the Sperry Top-Sider, Saucony, Stride Rite and Keds brands, for approximately $1.24 billion.

Employees are offered disability, health, dental and life insurance; a pension plan; a 401(k); a discount stock purchase plan; tuition reimbursement; wellness programs; and product discounts.

FINANCIAL DATA: *Note: Data for latest year may not have been available at press time.*

In U.S. $	2012	2011	2010	2009	2008	2007
Revenue	1,640,838,000	1,409,068,000	1,248,517,000	1,101,056,000	1,220,568,000	1,198,972,000
R&D Expense						
Operating Income	113,724,000	170,218,000	142,247,000	85,621,000	140,838,000	138,780,000
Operating Margin %	6.93%	12.08%	11.39%	7.77%	11.53%	11.57%
SGA Expense	514,436,000	386,534,000	347,499,000	316,378,000	345,183,000	333,151,000
Net Income	80,686,000	123,287,000	104,470,000	61,912,000	95,821,000	92,886,000
Operating Cash Flow	91,640,000	78,814,000	67,866,000	168,609,000	93,470,000	123,288,000
Capital Expenditure	14,942,000	19,397,000	16,370,000	11,670,000	24,126,000	17,879,000
EBITDA	136,426,000	186,212,000	159,998,000	103,807,000	164,130,000	163,832,000
Return on Assets %	4.65%	15.05%	13.98%	9.02%	14.70%	14.18%
Return on Equity %	13.21%	21.96%	20.36%	13.57%	21.08%	18.89%
Debt to Equity	1.89					

CONTACT INFORMATION:

Phone: 616 866-5500 Fax:
Toll-Free: 800-789-8586
Address: 9431 Courtland Dr. NE, Rockford, MI 49351 United States

STOCK TICKER/OTHER:

Stock Ticker: WWW
Exchange: NYS
Employees: 4,435
Fiscal Year Ends: 31-Dec
Parent Company:

SALARIES/BONUSES:

Top Exec. Salary: $853,906 Bonus: $
Second Exec. Salary: $427,827 Bonus: $50,000

OTHER THOUGHTS:

Estimated Female Officers or Directors: 4
Hot Spot for Advancement for Women/Minorities: Y

Worldtex Inc

www.abbasigroup.ae/worldtex.html

NAIC Code: 313210

TYPES OF BUSINESS:

Textiles
Logistics

BRANDS/DIVISIONS/AFFILIATES:

Abbasi Group

CONTACTS:
Note: Officers with more than one job title may be intentionally listed here more than once.

Maxwell Joseph, Gen. Mgr.
Ismail Akil Abbasi, Chmn.

GROWTH PLANS/SPECIAL FEATURES:

Worldtex, Inc., headquartered in Dubai, is a global producer and supplier of an extensive range of textiles and fabrics. Working from offices in Dubai, Korea, China and Poland, Worldtex offers a multitude of manmade and natural fiber textiles using a variety of fibers such as silk, wool, cotton, Merino, cashmere, mohair, linen, ramie, hemp, polynosic, polyester, polypropylene, acrylic and cupra, as well as many others, including other vegetable and natural fibers and a variety of blends. The firm sources its textiles from a variety of internationally recognized manufactures to ensure that its clients are receiving a quality product. Fabrics are sourced from Japan, China, Korea and Poland for redistribution to the United Arab Emirates, Iran, Turkey, India, Pakistan, the U.S. and other countries in North America, Europe and the Middle East. Additionally, Worldtex offers several support services such as procurement of fabrics and administrative support with emphasis on quality control, inspection and testing, careful handling, packaging, storage, up-to-date record control, shipment, processing of legal documents and timely delivery. Worldtex is part of the Abbasi Group of companies, which has interests in real estate, trading, textiles, coffee chains, watches, diamond jewelry, leather, stained glass and retail.

FINANCIAL DATA:
Note: Data for latest year may not have been available at press time.

In U.S. $	2012	2011	2010	2009	2008	2007
Revenue						
R&D Expense						
Operating Income						
Operating Margin %						
SGA Expense						
Net Income						
Operating Cash Flow						
Capital Expenditure						
EBITDA						
Return on Assets %						
Return on Equity %						
Debt to Equity						

CONTACT INFORMATION:

Phone: 9714-3539500 Fax: 9714-3530900
Toll-Free:
Address: P.O. Box 6637, Dubai, United Arab Emirates

STOCK TICKER/OTHER:

Stock Ticker: Subsidiary
Exchange:
Employees: 2,072
Fiscal Year Ends: 31-Dec
Parent Company: ABBASI GROUP

SALARIES/BONUSES:

Top Exec. Salary: $ Bonus: $
Second Exec. Salary: $ Bonus: $

OTHER THOUGHTS:

Estimated Female Officers or Directors:
Hot Spot for Advancement for Women/Minorities:

Xerium Technologies Inc

NAIC Code: 313230

www.xerium.com

TYPES OF BUSINESS:

Machine Clothing
Roll Covers
Nanotextiles

BRANDS/DIVISIONS/AFFILIATES:

Huyck.Wangner
Mount Hope
Stowe Woodward
Weavexx
Robec
Xibe

CONTACTS: *Note: Officers with more than one job title may be intentionally listed here more than once.*

Harold Bevis, CEO
Harold Bevis, Pres.
Clifford Pietrafitta, CFO
John B. Ardevol, CTO/Pres., Xerium Europe Rolls
Kevin McDougall, General Counsel/Exec. VP
David Pretty, Pres., Xerium Europe PMC
David Pretty, Pres., Xerium North America
Thomas Johnson, Pres., Xerium Asia
Harold Bevis, Chmn.
Eduardo Fracasso, Pres., Xerium South America

GROWTH PLANS/SPECIAL FEATURES:

Xerium Technologies, Inc. is a global manufacturer of clothing and roll covers used in the production of paper. It markets its products through brands such as Huyck Wangner, Weavexx, Stowe Woodward, Mount Hope, Robec and Xibe. The company has a network of 30 manufacturing facilities in 13 countries in North America, Europe, South America and the Asia Pacific region. Machine clothing and roll covers are essential to the formation/processing of paper. Clothing products are highly engineered synthetic textile belts used during the forming, pressing and drying stages of paper production. Roll cover products provide a surface with the mechanical properties to process paper cost-effectively and produce the sheet qualities desired by the paper manufacturer. The firm's clothing products include forming fabrics, press felts and dryer fabrics. Because paper-making machines vary widely in size and design, the company customizes clothing and covers to each individual paper-making machine. Xerium also provides refurbishment services for previously installed roll covers, as well as mechanical maintenance and repair services for roll and manufacture spreader roll mechanisms. The company has introduced a number of roll cover and spreader roll innovations in recent years, including continuous pressure sensing paper machine rolls to maximize performance; composite calendar roll covers that use nanoparticle technology to improve roll cover durability and paper gloss; and covers that increase abrasion and moisture resistance. It also manufactures clothing products used in other industrial applications, including steel, plastics, leather and textiles manufacturing. The clothing segment represents about 66% of the firm's total revenue, while the roll cover segment accounts for the remaining 34%. Xerium has roughly 270 issued patents covering 60 different inventions.

FINANCIAL DATA: *Note: Data for latest year may not have been available at press time.*

In U.S. $	2012	2011	2010	2009	2008	2007
Revenue	538,740,000	586,960,000	548,334,000	500,091,000	638,139,000	615,426,000
R&D Expense	11,681,000	12,097,000	11,427,000	11,309,000	11,740,000	10,189,000
Operating Income	16,396,000	61,101,000	45,264,000	-31,471,000	82,645,000	-99,084,000
Operating Margin %	3.04%	10.40%	8.25%	-6.29%	12.95%	-16.10%
SGA Expense	139,784,000	141,419,000	147,681,000	122,977,000	172,287,000	149,375,000
Net Income	-18,035,000	8,190,000	-73,086,000	-111,993,000	26,596,000	-150,212,000
Operating Cash Flow	39,322,000	45,208,000	20,734,000	16,131,000	77,068,000	89,020,000
Capital Expenditure	21,705,000	30,154,000	27,928,000	19,532,000	39,028,000	47,859,000
EBITDA	57,119,000	102,086,000	44,373,000	10,671,000	136,753,000	-48,732,000
Return on Assets %	-2.80%	1.19%	-10.48%	-14.88%	3.12%	-15.96%
Return on Equity %		99.69%				-260.04%
Debt to Equity			25			

CONTACT INFORMATION:

Phone: 508 616-9468 Fax: 508 616-9487
Toll-Free:
Address: 8537 6 Forks Rd., Ste. 300, Raleigh, NC 27615 United States

STOCK TICKER/OTHER:

Stock Ticker: XRM
Exchange: NYS
Employees: 3,448
Fiscal Year Ends: 31-Dec
Parent Company:

SALARIES/BONUSES:

Top Exec. Salary: $795,000 Bonus: $
Second Exec. Salary: $440,374 Bonus: $

OTHER THOUGHTS:

Estimated Female Officers or Directors: 1
Hot Spot for Advancement for Women/Minorities:

YKK Corp

www.ykk.com

NAIC Code: 339993

TYPES OF BUSINESS:

Zippers & Fasteners
Machinery
Aluminum Products
Architectural Products
Precision Machinery
Real Estate

BRANDS/DIVISIONS/AFFILIATES:

YKK Fastening Product Sales Inc
YKK Corporation of America
YKK AP Inc
YKK AP America Inc
YKK AP Facade Pte Ltd

CONTACTS: *Note: Officers with more than one job title may be intentionally listed here more than once.*

Tadahiro Yoshida, CEO
Masayuki Sarumaru, Pres.
Yasuharu Terada, VP-Human Resources Department
Ichiro Narasaki, VP-IT Strategy
Wataru Otani, Exec. VP-Eng. & Machinery Group
Yusei Sassa, Exec. VP-Korube Mfg. Center
Akira Yoshida, Sr. VP-Admin.
Katsuya Yumoto, VP-Legal
Atsushi Endo, VP-Bus. Dev.
Akira Yoshida, Sr. VP-Finance
Masanori Honda, Exec. VP-Fastening Prod. Group
Masahiro Nakagawa, Pres., YKK Do Brasil Ltda.
Ken Nakano, Pres., YKK Holding Asia Pte. Ltd.
Masaru Kawakami, Pres., YKK Fastening Products Sales, Inc.
Tadahiro Yoshida, Chmn.
Alex Gregory, Pres./Chmn.-YKK Corporation of America

GROWTH PLANS/SPECIAL FEATURES:

YKK Corp. is a Japanese conglomerate of more than 100 affiliated companies in 70 countries, engaged primarily in the manufacture of fastening products, architectural products and machinery and engineering products. Founded in 1934 as a fastening products company, YKK is most famous for its zippers, which enjoy an extremely high degree of market saturation. Its operations are divided into three primary segments: Fastening Products, Architectural Products and Machinery & Engineering. The Fastening Products group operates through three product business divisions: the slide fastener division, the textile and plastic products division and the snap fastener and button division. Products include zippers and sliders, snaps and buttons, plastic notions, synthetic webbing and hook and loop fasteners. YKK's Architectural Products segment, operating under the name YKK AP, Inc., provides aluminum profile products, windows, doors, metal exterior products and roofing materials as well as facade design and engineering services through three divisions: the residential architectural products division; the building architectural products division and the aluminum profile products sales division. Subsidiaries in this division include YKK AP America, Inc. and YKK AP Facade Pte. Ltd. The company's Machinery & Engineering group develops and produces machines, equipment and dies for internal use. YKK's Research & Development Center focuses on new materials research, with recent developments including a nano-crystalline aluminum alloy providing high strength and corrosion resistance as well as excellent transcriptability; and eco-friendly recycling technologies. Other business activities include domestic real estate operations; and aluminum smelting. The company's U.S. subsidiary, YKK Corporation of America, produces fastening products and residential and commercial architectural products through operating companies across Canada, the U.S., Mexico, the Caribbean, Central America and South America.

FINANCIAL DATA: *Note: Data for latest year may not have been available at press time.*

In U.S. $	2012	2011	2010	2009	2008	2007
Revenue	5,358,060,000	5,363,490,000	5,475,600,000			
R&D Expense						
Operating Income						
Operating Margin %						
SGA Expense						
Net Income	160,734,000	99,747,600	37,663,000			
Operating Cash Flow						
Capital Expenditure						
EBITDA						
Return on Assets %						
Return on Equity %						
Debt to Equity						

CONTACT INFORMATION:

Phone: 81-3-3864-2000 Fax: 81-3-3866-5500
Toll-Free:
Address: 1 Kanda Izumi-cho, Chiyoda-ku, Tokyo, 101-8642 Japan

STOCK TICKER/OTHER:

Stock Ticker: Private Exchange:
Employees: 39,000 Fiscal Year Ends: 31-Mar
Parent Company:

SALARIES/BONUSES:

Top Exec. Salary: $ Bonus: $
Second Exec. Salary: $ Bonus: $

OTHER THOUGHTS:

Estimated Female Officers or Directors:
Hot Spot for Advancement for Women/Minorities:

YOOX SpA

www.yoox.com

NAIC Code: 454111

TYPES OF BUSINESS:

Online Apparel Sales

BRANDS/DIVISIONS/AFFILIATES:

Yoox.com
TheCorner.com

GROWTH PLANS/SPECIAL FEATURES:

YOOX SpA, based in Italy, is an online fashion apparel retailer. With operations in the U.S., Europe, Japan and China, the firm delivers men's, women's and children's fashions and accessories to 100 countries across the globe. Its top designers include Jil Sander, Dolce & Gabbana, Prada, Armani, Roberto Cavalli and Missoni, among others. Besides clothing and accessories, YOOX sells art, fashion, lifestyle and design books; and home furnishings and decor. In addition to its multi-brand web stores, YOOX.com and TheCorner.com, the company operates several sites dedicated to single designers. It powers e-commerce sites for such major fashion retailers as Emporio Armani, Diesel, Valentino, DSquared, Jil Sander, Moschino, Y-3, Alberta Ferretti and others. In 2012, the firm's multibrand sites averaged 5.7 million unique visitors monthly.

CONTACTS: *Note: Officers with more than one job title may be intentionally listed here more than once.*

Federico Marchetti, CEO
Giuseppe Guillot, COO
Francesco Guidotti, CFO
Massimiliano Benedetti, Chief Mktg. Officer
Alessandro Montanari, Dir.-Human Resources & Organization
Gabriele Tazzari, Dir.-R&D
Gabriele Tazzari, Dir.-Strategy, Governance & Innovation
Paolo Mascio, Dir.-Online Stores
Silvia Scagnelli, Mgr.-Investor Rel.
Alessandra Rossi, Dir.-Commercial
Alberto Grignolo, Gen. Mgr.
Davide di Dario, Dir.-Demand Planning
Luca Martines, Dir.-Asia Pacific
Federico Marchetti, Chmn.
Clement Kwan, Dir.-North America

FINANCIAL DATA: *Note: Data for latest year may not have been available at press time.*

In U.S. $	2012	2011	2010	2009	2008	2007
Revenue	502,706,600	389,392,900	286,557,900	203,553,100	135,664,600	
R&D Expense						
Operating Income	25,288,850	21,961,750	20,069,540	11,821,340	6,404,119	
Operating Margin %	5.03%	5.64%	7.00%	5.80%	4.72%	
SGA Expense	107,488,600	79,767,320	62,678,520	40,940,090	29,458,410	
Net Income	13,617,280	13,372,560	12,191,760	5,480,075	3,212,089	
Operating Cash Flow	36,077,830	19,910,400	-660,604			
Capital Expenditure	25,684,680	14,154,850	15,296,870	3,864,670	4,831,506	
EBITDA	42,905,860	33,856,650	26,394,760	11,821,340	6,404,119	
Return on Assets %	4.69%	6.17%	7.39%	4.88%	3.70%	
Return on Equity %	11.04%	13.22%	14.85%	11.62%	14.61%	
Debt to Equity				0.01	1.13	

CONTACT INFORMATION:

Phone: 39 516184211 Fax:
Toll-Free:
Address: Via Nannetti 1, Zola Predosa, 40069 Italy

STOCK TICKER/OTHER:

Stock Ticker: YXOXF
Exchange: GREY
Employees: 469
Fiscal Year Ends: 31-Dec
Parent Company:

SALARIES/BONUSES:

Top Exec. Salary: $ Bonus: $
Second Exec. Salary: $ Bonus: $

OTHER THOUGHTS:

Estimated Female Officers or Directors: 3
Hot Spot for Advancement for Women/Minorities: Y

Yue Yuen Industrial (Holdings) Limited

www.yueyuen.com

NAIC Code: 316210

TYPES OF BUSINESS:

Shoe Manufacturing-General
Sports & Casual Apparel
Accessories
Retail Stores

BRANDS/DIVISIONS/AFFILIATES:

CONTACTS:

Note: Officers with more than one job title may be intentionally listed here more than once.

David N. F. Tsai, Managing Dir.
Jerry Shum, Dir.-Investor Rel.
Tsai Chi Neng, Chmn.

GROWTH PLANS/SPECIAL FEATURES:

Yue Yuen Industrial (Holdings) Limited is a manufacturer of branded athletic and casual apparel and footwear. The company is one of the largest manufacturers of its kind, with production facilities in China, Vietnam and Indonesia. Yue Yuen's fully integrated supply chain infrastructure design allows it to offer its clients services such as dedicated and flexible production facilities and research and development capabilities. The company operates two primary business divisions: footwear and shoe component manufacturing, and wholesale and retail sportswear sales. The footwear division manufactures over 500 lines of footwear for major international brands including Nike, adidas, Reebok, Asics, New Balance, Puma, Timberland and Rockport. During 2012, the company produced over 300 million pairs of shoes. Its wholesale and retail sportswear sales division operates a network of more than 3,650 directly operated retail stores/counters and sells to more than 2,200 sub-distributors across the Greater China region. Yue Yuen operates in three primary geographic regions: Asia, which accounted for 42.5% of 2012 sales; the U.S., which made up 28.1% of sales; and Europe, which contributed 20.2% of sales. Other areas, including South America and Canada, contributed the remaining 9.2%.

FINANCIAL DATA:

Note: Data for latest year may not have been available at press time.

In U.S. $	2012	2011	2010	2009	2008	2007
Revenue	9,193,226,000	7,045,373,000	5,788,208,000	5,016,902,000	4,919,937,000	4,114,090,000
R&D Expense						
Operating Income	568,919,000	506,949,000	470,938,000	614,202,000	518,813,000	317,698,000
Operating Margin %		7.19%	8.13%	12.24%	10.54%	7.72%
SGA Expense	1,466,279,000	1,055,679,000	894,875,000	817,720,000	749,036,000	533,720,000
Net Income	611,322,000	483,525,000	495,838,000	461,962,000	490,744,000	368,932,000
Operating Cash Flow	649,688,000	558,014,000	466,560,000	980,403,000	247,061,000	378,880,000
Capital Expenditure	302,286,000	482,047,000	348,818,000	260,638,000	355,925,000	324,612,000
EBITDA	988,663,000	721,244,000	654,583,000	789,714,000	672,214,000	575,658,000
Return on Assets %		7.37%	8.35%	8.66%	10.09%	9.33%
Return on Equity %		11.41%	13.32%	14.23%	16.95%	15.93%
Debt to Equity		0.1	0.19	0.19	0.17	0.33

CONTACT INFORMATION:

Phone: 852 31830888 Fax: 852 31830818
Toll-Free:
Address: 9 Canton Rd., The Gateway, Tower 6, 33/F, Hong Kong, Hong Kong

STOCK TICKER/OTHER:

Stock Ticker: YUEIF
Exchange: PINX
Employees: 460,000
Fiscal Year Ends: 31-Dec
Parent Company:

SALARIES/BONUSES:

Top Exec. Salary: $ Bonus: $
Second Exec. Salary: $ Bonus: $

OTHER THOUGHTS:

Estimated Female Officers or Directors:
Hot Spot for Advancement for Women/Minorities: Y

Zappos.com Inc

NAIC Code: 454111

www.zappos.com

TYPES OF BUSINESS:

Online Shoe Retail
Outlet Sales
Shoes
Housewares
Clothing
Bags & Handbags
Watches

BRANDS/DIVISIONS/AFFILIATES:

Zappos.com
Zappos IP Inc
Zappos Retail Inc
6pm.com
Zappos Gift Cards Inc
Zappos Insights Inc
6pm.com LLC
Zappos Merchandising Inc

CONTACTS: *Note: Officers with more than one job title may be intentionally listed here more than once.*

Tony Hsieh, CEO
Chris Nielsen, COO
Chris Nielsen, CFO

GROWTH PLANS/SPECIAL FEATURES:

Zappos.com, Inc. is an online retailer of shoes, handbags and other accessories, offering nearly 3 million products for women, men and children from over 1,136 brands. Its inventory includes shoes, bags and handbags, clothing, beauty products, eyewear, sporting goods, housewares and watches. Due to massive growth, the firm was restructured into seven separate companies under the Zappos family umbrella. Zappos.com, Inc., the management company for the group, provides finance, treasury, accounting, help desk, human resources, information technology, legal and training services to the Zappos family of companies. Zappos Gift Cards, Inc. sells gift cards and gift certificates for use on the company's web sites. Zappos Insights, Inc. provides business consulting and corporate culture advising services to third-party companies and individuals. Zappos IP, Inc. provides web site software development, project management, marketing, and content services to the Zappos family of companies and follows customer experience trends. Zappos Merchandising, Inc. focuses on product purchasing, brand acquisition and vendor relationship management. Zappos Retail, Inc. is in charge of the actual sale of products on Zappos.com. Finally, 6pm.com LLC operates e-commerce site 6pm.com, which sells discount branded shoes, clothing, bags and accessories for men, women and children.

Employee benefits include medical, dental and vision coverage; life insurance; flexible spending accounts; an employee assistance program; and on-site wellness services.

FINANCIAL DATA: *Note: Data for latest year may not have been available at press time.*

In U.S. $	2012	2011	2010	2009	2008	2007
Revenue	1,300,000,000	1,200,000,000	1,100,000,000	1,000,000,000	960,000,000	700,000,000
R&D Expense						
Operating Income						
Operating Margin %						
SGA Expense						
Net Income						
Operating Cash Flow						
Capital Expenditure						
EBITDA						
Return on Assets %						
Return on Equity %						
Debt to Equity						

CONTACT INFORMATION:

Phone: 702-943-7777 Fax: 702-943-7778
Toll-Free: 800-927-7671
Address: 2280 Corporate Cir., Ste. 100, Henderson, NV 89074 United States

STOCK TICKER/OTHER:

Stock Ticker: Subsidiary Exchange:
Employees: 1,400 Fiscal Year Ends: 31-Dec
Parent Company: AMAZON.COM INC

SALARIES/BONUSES:

Top Exec. Salary: $ Bonus: $
Second Exec. Salary: $ Bonus: $

OTHER THOUGHTS:

Estimated Female Officers or Directors:
Hot Spot for Advancement for Women/Minorities:

Zara International Inc

www.zara.com

NAIC Code: 448140

TYPES OF BUSINESS:

Apparel Manufacturing
Apparel Stores-Men, Women & Children
Household Linens
Teen Women's Clothing
Maternity Clothing
E-commerce

BRANDS/DIVISIONS/AFFILIATES:

Industria De Diseno Textil SA (INDITEX)
Zara Home
Zara
Zara for Mum
Kiddy's Class

CONTACTS: *Note: Officers with more than one job title may be intentionally listed here more than once.*

Pablo Isla Alvarez de Tejera, CEO

GROWTH PLANS/SPECIAL FEATURES:

Zara International, Inc. is the primary subsidiary of Industria de Diseno Textile (Inditex), one of the largest fashion retail groups in the world. Zara is considered Inditex's flagship brand, with around 1,830 Zara brand stores in 82 markets. Zara sells women's, men's and children's apparel, offering chic, trendy looks for those on a budget. Zara also offers the maternity collection, Zara for Mum. Zara Kids, carries children's appeal designed to suit children from toddlers to the pre-teens. The company also has a juniors focused Zara store, Kiddy's Class, in Spain. Additionally, the company operates Zara Home, an independent store concept at approximately 300 locations primarily in Europe. Zara Home offers an exclusive line of textile products for the bed, table and bathroom as well as silverware, household items, small decorative objects and cosmetics. The company also makes and markets special edition seasonal fragrances for men and women. Services offered at Zara include alterations, gift cards and the Affinity Card, which is Zara's own credit card and is currently only available in Spain, Greece, Italy, Portugal and Mexico. In 2012, Zara opened its largest store in the U.S. with the New Global Concept Store in New York City.

FINANCIAL DATA: *Note: Data for latest year may not have been available at press time.*

In U.S. $	2012	2011	2010	2009	2008	2007
Revenue						
R&D Expense						
Operating Income						
Operating Margin %						
SGA Expense						
Net Income						
Operating Cash Flow						
Capital Expenditure						
EBITDA						
Return on Assets %						
Return on Equity %						
Debt to Equity						

CONTACT INFORMATION:

Phone: 34-981-185-400 Fax: 34-981-185-544
Toll-Free:
Address: Ave. de la Diputacion s/n, Edificio Inditex, Arteixo, 15142 Spain

STOCK TICKER/OTHER:

Stock Ticker: Subsidiary Exchange:
Employees: Fiscal Year Ends: 31-Jan
Parent Company: INDITEX (INDUSTRIA DE DISENO TEXTIL SA)

SALARIES/BONUSES:

Top Exec. Salary: $ Bonus: $
Second Exec. Salary: $ Bonus: $

OTHER THOUGHTS:

Estimated Female Officers or Directors:
Hot Spot for Advancement for Women/Minorities:

Zegna (Ermenegildo Zegna Holditalia SpA)

www.zegna.com

NAIC Code: 315240

TYPES OF BUSINESS:

High-End Menswear Manufacturing
Fabric Manufacturing
Accessories
Womenswear

BRANDS/DIVISIONS/AFFILIATES:

Ermenegildo Zegna
Zegna Sport
Z Zegna
Gucci
Agnona

CONTACTS:

Note: Officers with more than one job title may be intentionally listed here more than once.

Ermenegildo Zegna, CEO
Paolo Zegna, Pres.
Benedetta Zegna, Mgr.-Talent
Anna Zegna, Dir.-Image
Paolo Zegna, Chmn.

GROWTH PLANS/SPECIAL FEATURES:

Zegna (Ermenegildo Zegna Holditalia SpA) is a world leader in high-end menswear and sportswear, accessories and fabrics. Founded as a textile mill, the company produces its own high-quality fabrics and supplies them to top designers and tailors, including a line of clothing for Gucci. Zegna fabrics are woven from extra-fine materials like inner Mongolian cashmere, South African kid mohair and superfine fibers from Australian merino wool. In addition to this exclusive production process, the company attaches high importance to design. The company also carries a line of prescription eyewear, a sunglass collection and, through a partnership with Perofil, a luxury underwear line. Over four generations of family members have owned and operated Zegna. With production plants in Spain, Switzerland, Mexico and Turkey, Zegna is present in 560 mono-brand stores (311 of which are under direct group ownership) in 86 countries. The firm generates 46% of its revenues from its Asian exports and 44% from exports to Europe and America. These stores include three men's product lines: Ermenegildo Zegna, the contemporary style with its precious and exclusive materials; Zegna Sport, a balance between a luxurious and a sportive look; and Z Zegna, the younger, more urban brand. The women's clothing line, Agnona, focuses on contemporary styles. Recently, the firm announced a licensing agreement with Estee Lauder Companies, Inc., which allows the exclusive rights for said company to market fragrances under the firm's brands. The company planned 50 new stores for 2012.

FINANCIAL DATA:

Note: Data for latest year may not have been available at press time.

In U.S. $	2012	2011	2010	2009	2008	2007
Revenue	1,670,000,000	1,150,000,000	1,000,000,000	1,067,990,000	1,013,500,000	1,072,840,000
R&D Expense						
Operating Income						
Operating Margin %						
SGA Expense						
Net Income						
Operating Cash Flow						
Capital Expenditure						
EBITDA						
Return on Assets %						
Return on Equity %						
Debt to Equity						

CONTACT INFORMATION:

Phone: 39-02-422091 Fax: 39-02-42209101
Toll-Free:
Address: Via Savona 56/a, Milan, 20144 Italy

STOCK TICKER/OTHER:

Stock Ticker: Private
Exchange:
Employees: 7,000
Fiscal Year Ends: 31-Dec
Parent Company:

SALARIES/BONUSES:

Top Exec. Salary: $ Bonus: $
Second Exec. Salary: $ Bonus: $

OTHER THOUGHTS:

Estimated Female Officers or Directors: 2
Hot Spot for Advancement for Women/Minorities:

Zumiez Inc

www.zumiez.com

NAIC Code: 448140

TYPES OF BUSINESS:

Action Sports Apparel
Action Sports Footwear, Equipment & Accessories

BRANDS/DIVISIONS/AFFILIATES:

Zumiez
Blue Tomato

CONTACTS:

Note: Officers with more than one job title may be intentionally listed here more than once.

Richard M. Brooks, CEO
Lynn K. Kilbourne, Pres.
Christopher Work, CFO
Lynn K. Kilbourne, Gen. Mgr.-Merch.
Chris K. Visser, General Counsel/Corp. Sec
Brian Leith, Dir.-Investor Rel.
Brian Leith, Dir.-Finance
Ford K. Wright, Exec. VP-Stores
Thomas D. Campion, Chmn.

GROWTH PLANS/SPECIAL FEATURES:

Zumiez, Inc. is a mall-based specialty retailer of action sports-related apparel, footwear, equipment and accessories operating under the Zumiez brand name. The company operates more than 500 stores (472 in the U.S., 20 in Canada and 8 in Europe) which cater to young men and women between the ages of 12 and 24 who seek popular brands representing a lifestyle centered on action sports such as skateboarding, surfing, snowboarding, BMX (bicycle motocross) and motocross. In addition to apparel and footwear, the stores offer sports equipment, including skateboards, snowboards, boots and bindings; and accessories, including wallets, jewelry and other miscellaneous novelties and DVDs. Zumiez promotes its brand through a multi-faceted marketing approach featuring extensive grassroots marketing events as well as through a differentiated merchandising strategy, store design and comprehensive training programs aimed at providing an experience for its customers that it believes is consistent with their attitudes, fashion tastes and identities. The company's stores, which average 2,900 square feet, feature couches and action sports oriented video game stations that are intended to encourage customers to stay in the store longer and to interact with each other and store associates. To increase customer traffic, Zumiez generally locates its stores near busy areas of the mall such as food courts, movie theaters, music or game stores and other popular teen retailers. Besides store operations, Zumiez manages an e-commerce site that provides current information regarding upcoming events and promotions, store locations and merchandise selection. The company purchased the Austrian sports action retailer Blue Tomato in June 2012 and now operates in Europe under the new brand name.

The firm's employees receive a 401(k) retirement plan and an employee stock purchase plan, among other benefits.

FINANCIAL DATA:

Note: Data for latest year may not have been available at press time.

In U.S. $	2012	2011	2010	2009	2008	2007
Revenue	555,874,000	478,849,000	407,603,000	408,669,000	381,416,000	
R&D Expense						
Operating Income	60,232,000	37,367,000	12,735,000	24,608,000	38,945,000	
Operating Margin %	10.83%	7.80%	3.12%	6.02%	10.21%	
SGA Expense	141,444,000	133,030,000	122,003,000	109,927,000	98,042,000	
Net Income	37,351,000	24,203,000	9,131,000	17,204,000	25,326,000	
Operating Cash Flow	68,065,000	48,692,000	45,116,000	38,337,000	34,286,000	
Capital Expenditure	25,508,000	29,361,000	16,548,000	28,349,000	30,722,000	
EBITDA	60,232,000	37,367,000	12,735,000	44,078,000	53,707,000	
Return on Assets %	11.25%	8.61%	3.69%	7.65%	13.21%	
Return on Equity %	14.97%	11.54%	4.92%	10.34%	19.52%	
Debt to Equity						

CONTACT INFORMATION:

Phone: 425 551-1500 Fax: 425 551-1555
Toll-Free: 877-828-6929
Address: 6300 Merrill Creek Pkwy., Ste. B, Lynnwood, WA 98203 United States

STOCK TICKER/OTHER:

Stock Ticker: ZUMZ
Employees: 4,840
Parent Company:
Exchange: NAS
Fiscal Year Ends: 31-Jan

SALARIES/BONUSES:

Top Exec. Salary: $631,324 Bonus: $
Second Exec. Salary: $489,113 Bonus: $

OTHER THOUGHTS:

Estimated Female Officers or Directors: 2
Hot Spot for Advancement for Women/Minorities:

ADDITIONAL INDEXES

Contents:

INDEX OF FIRMS NOTED AS HOT SPOTS FOR ADVANCEMENT FOR WOMEN & MINORITIES

Puig Beauty & Fashion Group SL
PVH Corp
Ralph Lauren Corporation
Ratti SpA
Recreational Equipment Inc (REI)
Redcats USA
RG Barry Corp
Ross Stores Inc
Saks Inc
Salvatore Ferragamo SpA
Sears Canada Inc
Shoe Carnival Inc
Stage Stores Inc
Stein Mart Inc
Superior Uniform Group Inc
Talbots Inc (The)
Tandy Brands Accessories Inc
Target Corporation
Teijin Ltd
Tilly's Inc
Timberland Co
TJX Companies Inc (The)
Tommy Hilfiger Corp
TOPSHOP
Tween Brands Inc
United Retail Group Inc
Urban Outfitters Inc
Varsity Brands Inc
Vera Bradley Inc
VF Corp
Victoria plc
Victorias Secret
W L Gore & Associates Inc
Wal-Mart Stores Inc
Warnaco Group Inc (The)
Wet Seal Inc (The)
Weyco Group Inc
Wolverine World Wide Inc
YOOX SpA
Yue Yuen Industrial (Holdings) Limited

INDEX OF SUBSIDIARIES, BRAND NAMES AND AFFILIATIONS

Brand or subsidiary, followed by the name of the related corporation

INDEX OF SUBSIDIARIES, BRAND NAMES AND AFFILIATIONS, CONT.

INDEX OF SUBSIDIARIES, BRAND NAMES AND AFFILIATIONS, CONT.

INDEX OF SUBSIDIARIES, BRAND NAMES AND AFFILIATIONS, CONT.

INDEX OF SUBSIDIARIES, BRAND NAMES AND AFFILIATIONS, CONT.

INDEX OF SUBSIDIARIES, BRAND NAMES AND AFFILIATIONS, CONT.

INDEX OF SUBSIDIARIES, BRAND NAMES AND AFFILIATIONS, CONT.

INDEX OF SUBSIDIARIES, BRAND NAMES AND AFFILIATIONS, CONT.

INDEX OF SUBSIDIARIES, BRAND NAMES AND AFFILIATIONS, CONT.

INDEX OF SUBSIDIARIES, BRAND NAMES AND AFFILIATIONS, CONT.

INDEX OF SUBSIDIARIES, BRAND NAMES AND AFFILIATIONS, CONT.

INDEX OF SUBSIDIARIES, BRAND NAMES AND AFFILIATIONS, CONT.

INDEX OF SUBSIDIARIES, BRAND NAMES AND AFFILIATIONS, CONT.

INDEX OF SUBSIDIARIES, BRAND NAMES AND AFFILIATIONS, CONT.

INDEX OF SUBSIDIARIES, BRAND NAMES AND AFFILIATIONS, CONT.

INDEX OF SUBSIDIARIES, BRAND NAMES AND AFFILIATIONS, CONT.

INDEX OF SUBSIDIARIES, BRAND NAMES AND AFFILIATIONS, CONT.

INDEX OF SUBSIDIARIES, BRAND NAMES AND AFFILIATIONS, CONT.

INDEX OF SUBSIDIARIES, BRAND NAMES AND AFFILIATIONS, CONT.

A Short Apparel & Textiles Industry Glossary

3PL: See "Third-Party Logistics (3PL)."

Acetate Fiber: A fast-drying, wrinkle-resistant fiber made from cellulose acetate, used in apparel, draperies, bed linen, carpets and cigarette filters.

Acrylic Fiber: Any manufactured fiber that contains at least 85% acrylonitrile by weight, used in apparel, faux fur, carpets, draperies and upholstery.

Activewear: Any garment made to provide comfort, warmth or support during vigorous exercise.

Ageing: The degrading of textiles due to oxidation and light exposure, or a similar process used on printed fabric to achieve a desired effect such as a particular color.

Air Jet Loom: A shuttle-less loom that uses an air jet to propel the yarn through the shed at very high speeds.

Alpaca: The hair of the Alpaca sheep or a cotton blend thereof, used in suits, dresses and coat pilings.

Angora: The hair of the Angora goat or Angora rabbit. Rabbit hair must be described by law as Angora rabbit hair.

Anidex Fiber: A manufactured fiber in which the fiber-forming substance is any long-chain synthetic polymer composed at least 50% by weight of one or more esters of a monohydric alcohol and acrylic acid.

APAC: Asia Pacific Advisory Committee. A multi-country committee representing the Asia and Pacific region.

Aramid Fiber: A manufactured fiber in which the fiber-forming substance is a long-chain synthetic polyamide in which at least 85% of the amide (-CO-NH-) linkages are attached directly between two aromatic rings.

Art Linen: A plain-weave linen often used for needle work.

ASEAN: Association of Southeast Asian Nations. A regional economic development association established in 1967 by five original member countries: Indonesia, Malaysia, Philippines, Singapore, and Thailand. Brunei joined on 8 January 1984, Vietnam on 28 July 1995, Laos and Myanmar on 23 July 1997, and Cambodia on 30 April 1999.

Azlon Fiber: A manufactured fiber in which the fiber-forming substance is composed of any regenerated, naturally occurring protein.

Baby Boomer: Generally refers to people born from 1946 to 1964. In the U.S., the initial number of Baby Boomers totaled about 78 million. The term evolved to describe the children of soldiers and war industry workers who were involved in World War II and who began forming families after the war's end. In 2011, the oldest Baby Boomers began reaching the traditional retirement age of 65.

Backed Fabric: A fabric that has an extra warp or filling for weight or warmth. The most common backing fabrics are satin and twill.

Bale: A tightly wrapped bundle of unfinished material, usually cotton. The industry standard for cotton bales is 480 lb.

Basketweave: A plain weave with two or more warp and filling threads interwoven in a basket-like pattern.

Bespoke: Refers to an item that is custom-made for the customer. Traditionally, bespoke meant crafted completely under the control of and to the desire of the customer. In clothing, a high level of service such as this is available only from firms that specialize in custom tailoring. A truly bespoke man's suit can cost from $5,000 to $10,000 from a tailor on London's famous Saville Row. Elsewhere, what many apparel consumers are buying is actually made-to-measure, which many firms now call bespoke. In a man's suit, for example, that means that a factory-built suit is altered during production to fit a customer's specific measurements.

Biocouture: The creation of textiles through the growth of cellular matter, such as bacterial cellulose, and the use of those textiles for manufacture of garments.

BPO: See "Business Process Outsourcing (BPO)."

Brand: A combination of several factors that separates a product, group of products, company or organization from competing products or companies. Brand can include identification, such as name, logo, label design or color scheme. More importantly, brand includes "brand equity" factors such as reputation, perceived value of products or services, customer satisfaction levels, communications activities, and corporate values in such areas as environmental matters, workplace and sensitivity to factors such as human rights.

Brand Extension: Using the strength of an existing product (or line of products) to establish a new product. For example, a maker of laundry detergent might "extend its brand" by introducing household cleaners and dish detergents.

Brand Name: See "Brand."

Branding: A marketing strategy that places a focus on the brand name of a product, service or firm in order to increase the brand's market share, increase sales, establish credibility, improve satisfaction, raise the profile of the firm and increase profits. Also, see "Brand."

Broadwoven Gray Fabrics: A broad classification of fabrics including: denim and other cotton fabrics; blanketing fabrics; other woolen or worsted fabrics with 36% or more wool in them; manmade fiber or silk broadwoven fabrics; tire cord; and tire cord fabrics.

B-to-B, or B2B: See "Business-to-Business."

B-to-C, or B2C: See "Business-to-Consumer."

Business Process Outsourcing (BPO): The process of hiring another company to handle business activities. BPO is one of the fastest-growing segments in the offshoring sector. Services include human resources management, billing and purchasing and call centers, as well as many types of customer service or marketing activities, depending on the industry involved. Also, see "Knowledge Process Outsourcing (KPO)."

Business-to-Business: An organization focused on selling products, services or data to commercial customers rather than individual consumers. Also known as B2B.

Business-to-Consumer: An organization focused on selling products, services or data to individual consumers rather than commercial customers. Also known as B2C.

CAFTA-DR: See "Central American-Dominican Republic Free Trade Agreement (CAFTA-DR)."

Calico: A fabric of closely woven inexpensive fibers, usually cotton.

Captive Offshoring: Used to describe a company-owned offshore operation. For example, Microsoft owns and operates significant captive offshore research and development centers in China and elsewhere that are offshore from Microsoft's U.S. home base. Also see "Offshoring."

Cellulose: The primary material of plant cell walls such as wood, cotton, linen and hemp. It is also used in the production of rayon, acetate and triacetate fibers.

CEM: Contract electronic manufacturing. See "Contract Manufacturing."

Central American-Dominican Republic Free Trade Agreement (CAFTA-DR): A trade agreement signed into law in 2005 that aimed to open up the Central American and Dominican Republic markets to American goods. Member nations include Guatemala, Nicaragua, Costa Rica, El Salvador, Honduras and the Dominican Republic. Before the law was signed, products from those countries could enter the U.S. almost tariff-free, while American goods heading into those countries faced stiff tariffs. The goal of this agreement was to create U.S. jobs while at the same time offering the non-U.S. member citizens a chance for a better quality of life through access to U.S.-made goods.

Chambray: A plain woven fabric with a colored wrap and white filling thread, made of cotton, silk or synthetic fiber, used mostly in dresses, shirts, aprons and active wear.

Chasing: A method of ordering apparel from a manufacturer in which a small, short-term order is placed and fulfilled, followed by another rapid order if the first delivery sold well for the retailer.

Chav: Generally used by marketers to describe a young, lower-class consumer who acts in a brash manner and wears designer-type clothing. This segment was first described in the U.K.

Cheese Cloth: A thin, loose-woven cotton cloth that derives its name from its use in cheese making.

Chevron: A broad term used to describe zig-zag or herringbone weaves.

Chino: A cotton twill fabric composed of combed two-ply yarn, often used in military uniforms.

Chintz: A plain-weave, glazed fabric with a polished look, usually made of cotton and used for draperies, slip covers, dresses and sportswear.

Collaborative Planning, Forecasting and Replenishment (CPFR): An inventory management practice that evaluates and fills inventory needs throughout the supply chain. It is a cooperative effort between suppliers and retailers.

Comfort Stretch: A term denoting fabric that has a 10% to 15% stretch to allow comfortable wearing.

Commerce Chain Management (CCM): Refers to Internet-based tools to facilitate sales, distribution, inventory management and content personalization in the e-commerce industry. Also see "Supply Chain."

Contract Manufacturing: A business arrangement whereby a company manufactures products that will be sold under the brand names of its client companies. For example, a large number of consumer electronics, such as laptop computers, are manufactured by contract manufacturers for leading brand-name computer companies such as Dell and Apple. Many other types of products, such as shoes and apparel, are made under contract manufacturing. Also see "Original Equipment Manufacturer (OEM)" and "Original Design Manufacturer (ODM)."

Converter: A person or organization that buys unfinished fabric and arranges for finishing processes according to a buyer's specifications.

Copyright: A form of protection provided to the authors of "original works of authorship," including literary, dramatic, musical, artistic, and certain other intellectual works, both published and unpublished. In the U.S., the 1976 Copyright Act generally gives the owner of copyright the exclusive right to reproduce the copyrighted work, to prepare derivative works, to distribute copies or recordings of the copyrighted work, to perform the copyrighted work publicly, or to display the copyrighted work publicly. Other nations have similar laws. In the United States, copyrights are registered by the Copyright Office of the Library of Congress. There are also cooperative, international copyright agreements and agencies.

Corduroy: A cut-pile fabric with vertical ridges, usually made of cotton.

CPFR: See "Collaborative Planning, Forecasting and Replenishment (CPFR)."

Crepe: A fabric with a crimped or grainy surface, usually a plain weave made of cotton, wool, rayon or synthetic fibers.

CRM: See "Customer Relationship Management (CRM)."

Customer Relationship Management (CRM): Refers to the automation, via sophisticated software, of business processes involving existing and prospective customers. CRM may cover aspects such as sales (contact management and contact history), marketing (campaign management and telemarketing) and customer service (call center history and field service history). Well known providers of CRM software include Salesforce, which delivers via a Software as a Service model (see "Software as a Service (Saas)"), Microsoft and Oracle.

Customs and Shipping Broker: A firm specializing in the preparation of shipping and customs documents for international shipments. Brokers often have offices at major freight gateways, including border crossings, seaports and airports.

DBA: Doing business as.

Demand Chain: A similar concept to a supply chain, but with an emphasis on the end user.

Demographics: The breakdown of the population into statistical categories such as age, income, education and sex.

Denier: A measure of filament weight in fiber. The standard weight is usually expressed as the number of grams per 9,000 meters of filament.

Denim: True denim is a cotton twill weave made of colored yarn, usually blue, used for jeans, jackets, overalls and upholstery.

Design Patent: A patent that may be granted by the U.S. Patent and Trademark Office to anyone who invents a new, original, and ornamental design for an article of manufacture.

Direct Marketing: A form of non-store retailing in which customers are exposed to merchandise through catalogs, direct-mail brochures, telemarketing or television. Direct marketing may be used to generate direct-response purchases, store traffic, sales leads or a combination thereof.

Direct Selling: A form of marketing which involves manufacturing and then selling merchandise or services through direct mail or through salespeople who contact consumers directly or by telephone at home or place of work (e.g., Mary Kay cosmetics).

Direct Yarn Number: A yarn sizing system in which higher numbers indicate coarser yarn.

Direct-Mail Merchandising: The process of offering merchandise or services through catalogs or brochures mailed directly to customers.

Direct-Response Marketing: The use of direct marketing with the intent of soliciting an order from the customer, rather than generating store traffic. Direct-response methods may include mail-order catalogs or brochures, infomercials, telemarketing and online banner ads. See "Direct Marketing."

Distributor: An individual or business involved in marketing, warehousing and/or shipping of products manufactured by others to a specific group of end users. Distributors do not sell to the general public. In order to develop a competitive advantage, distributors often focus on serving one industry or one set of niche clients. For example, within the medical industry, there are major distributors that focus on providing pharmaceuticals, surgical supplies or dental supplies to clinics and hospitals.

Dobby: A loom attachment that allows for the weaving of geometric figures.

Donegal: Originally a rough wool form Donegal, Ireland, now broadly used to indicate a cloth with colored slubs woven in for a sporty look. Used mostly in coats and sports jackets.

Double-Cloth Construction: A fabric of two cloths woven together, often of different weaves.

Drill: A cotton twill weave fabric, often called khaki when the appropriate tan color.

Dry Spinning: The process of evaporating the solvent in the fiber-forming solution by air jet rather than perception. See "Wet Spinning."

Duck: A heavy, plain-weave cotton fabric. There are many weights of duck. The heaviest-weight duck is canvas. Lighter-weight duck is used in clothing such as aprons, while heavier weights are used for tents and sails.

Echo Boomers: See "Generation Y."

ECM: Electronic Contract Manufacturing. See "Contract Manufacturing."

E-Commerce: The use of online, Internet-based sales methods. The phrase is used to describe both business-to-consumer and business-to-business sales.

Elastoester: A manufactured fiber in which the fiber-forming substance is a long-chain synthetic polymer composed at least 50% by weight of aliphatic polyester and at least 35% by weight of polyester.

Electronic Data Interchange (EDI): An accepted standard format for the exchange of data between various companies' networks. EDI allows for the transfer of e-mail as well as orders, invoices and other files from one company to another.

EMEA: The region comprised of Europe, the Middle East and Africa.

EMS: Electronics Manufacturing Services. See "Contract Manufacturing."

Enterprise Resource Planning (ERP): An integrated information system that helps manage all aspects of a business, including accounting, ordering and human resources, typically across all locations of a major corporation or organization. ERP is considered to be a critical tool for management of large organizations. Suppliers of ERP tools include SAP and Oracle.

Enzyme Finishing: A process that uses enzymes to produce a desired effect. This process is largely used to fade or weather denim.

ERP: See "Enterprise Resource Planning (ERP)."

EU: See "European Union (EU)."

EU Competence: The jurisdiction in which the European Union (EU) can take legal action.

European Community (EC): See "European Union (EU)."

European Union (EU): A consolidation of European countries (member states) functioning as one body to facilitate trade. Previously known as the European Community (EC). The EU has a unified currency, the Euro. See europa.eu.int.

Factory Outlet Stores: Off-price retail stores owned by manufacturers.

FASB: See "Financial Accounting Standards Board (FASB)."

Fast Fashion: A retail trend followed by apparel stores that rapidly change their merchandise offerings to include the latest, trendiest fashions possible.

Filament Yarn: A yarn with several continuous filaments twisted together.

Filling: In woven fabric, the thread running at angles to the warp.

Financial Accounting Standards Board (FASB): An independent organization that establishes the Generally Accepted Accounting Principles (GAAP).

Flannel: A soft, medium-weight, plain or twill fabric, usually made of wool or cotton.

Flash Sale: Online sales events, generally advertised to by email to people who have asked to be notified. Flash sales are very limited in time. They tend to offer apparel, accessories and travel.

Flax: Fiber of the flax plant that has been spun into linen.

Fluoropolymer: A manufactured fiber containing at least 95% of a long-chain polymer synthesized from aliphatic fluorocarbon monomers.

Fourth-Party Logistics (4PL): A service that integrates a company's third-party logistics providers into a single entity for ease of use. Often formed by a telecommunications company, a 4PL is also called a lead logistics provider. A 4PL service provider provides a top layer of business processes, generally technology-driven, to the client's supply chain. Also see "Third-Party Logistics (3PL)."

Franchise: 1) A contractual agreement between a franchisor (for example, a company or organization owning all rights to a brand, type of business, retail operation, restaurant concept or sports league) and a franchisee (person or organization desiring to license the use of those rights for a specific purpose within a specific region) that allows the franchisee to operate a retail outlet or other type of business using a brand, trade secrets, formulas and format developed and supported by the franchisor. Typically, a franchisee pays an upfront fee and then continuing fees to the franchisor. 2) A generic term used to describe a very well established business or brand.

Franchisee: See "Franchise."

Franchisor: See "Franchise."

Free Alongside Ship (FAS): In shipping, a term that means that the exporter's obligations are fulfilled when the goods are placed on the dock alongside the ship that will carry them. The price includes delivery to the dock, but does not include the cost of further shipping. The buyer must clear the goods for export and pay further costs of shipment.

GAAP: See "Generally Accepted Accounting Principles (GAAP)."

Gabardine: A tightly woven, twill worsted fabric that is usually made of wool, used primarily in year-round suits.

GDP: See "Gross Domestic Product (GDP)."

Gel Spinning: A process for high-tensile-strength fibers in which a gel is extruded through the spinneret instead of a liquid, resulting in stronger bonds. See "Melt Spinning."

Generally Accepted Accounting Principles (GAAP): A set of accounting standards administered by the Financial Accounting Standards Board (FASB) and enforced by the U.S. Security and Exchange Commission (SEC). GAAP is primarily used in the U.S.

Generation M: A very loosely defined term that is sometimes used to refer to young people who have grown up in the digital age. "M" may refer to any or all of media-saturated, mobile or multi-tasking. The term was most notably used in a Kaiser Family Foundation report published in 2005, "Generation M: Media in the Lives of 8-18 year olds." Also, see "Generation Y" and "Generation Z."

Generation X: A loosely-defined and variously-used term that describes people born between approximately 1965 and 1980, but other time frames are recited. Generation X is often referred to as a group influential in defining tastes in consumer goods, entertainment and/or political and social matters.

Generation Y: Refers to people born between approximately 1982 and 2002. In the U.S., they number more than 90 million, making them the largest generation segment in the nation's history. They are also known as Echo Boomers, Millenials or the Millenial Generation. These are children of the Baby Boom generation who will be filling the work force as Baby Boomers retire.

Generation Z: Some people refer to Generation Z as people born after 1991. Others use the beginning date of 2001, or refer to the era of 1994 to 2004. Members of Generation Z are considered to be natural and rapid adopters of the latest technologies.

Gross Domestic Product (GDP): The total value of a nation's output, income and expenditures produced with a nation's physical borders.

Gross Margin: The difference between the retail price the customer pays for merchandise and the cost of the merchandise (the price the retailer paid at wholesale).

Gross National Product (GNP): A country's total output of goods and services from all forms of economic activity measured at market prices for one calendar year. It differs from Gross Domestic Product (GDP) in that GNP includes income from investments made in foreign nations.

Group of Eight (G8): The eight major industrial countries, including the United States, Japan, Germany, France, the United Kingdom, Italy, Russia and Canada, whose leaders meet at annual economic summits to coordinate economic policies. The group was formerly known as the G7, before Russia was admitted in June 2002.

Haute Couture: Hand-sewn to order clothes using fine materials. Couture designation is awarded by the French government to those fashion houses that employ at least 20 people, create at least 25 designs for twice yearly shows in Paris and have had collections for at least two years.

Head End: The beginning of a new piece of fabric as it comes off the loom. Also, a sample of fabric.

Herringbone: A twill weave design in which the twill is reversed to produce a chevron weave.

Higg Index: The Higg Index 1.0 is an indicator-based tool for apparel that enables companies to evaluate material types, products, facilities and processes based on a range of environmental and product design choices. The Index asks practice-based, qualitative questions to gauge environmental sustainability performance and drive behavior for improvement. It is based largely on the Eco Index and Nike's Apparel Environmental Design Tool, however it has been significantly enhanced through a pilot testing period.

Houndstooth: A cloth pattern with small checks in a diagonal pattern.

Hybrid Fabric: A fabric incorporating two or more different kinds of yarn.

IFRS: See "International Financials Reporting Standards (IFRS)."

Industrial Fabric: A fabric used for industrial purposes such as paper manufacturing of air filtration.

Industry Code: A descriptive code assigned to any company in order to group it with firms that operate in similar businesses. Common industry codes include the NAICS (North American Industrial

Classification System) and the SIC (Standard Industrial Classification), both of which are standards widely used in America, as well as the International Standard Industrial Classification of all Economic Activities (ISIC), the Standard International Trade Classification established by the United Nations (SITC) and the General Industrial Classification of Economic Activities within the European Communities (NACE).

Initial Public Offering (IPO): A company's first effort to sell its stock to investors (the public). Investors in an up-trending market eagerly seek stocks offered in many IPOs because the stocks of newly public companies that seem to have great promise may appreciate very rapidly in price, reaping great profits for those who were able to get the stock at the first offering. In the United States, IPOs are regulated by the SEC (U.S. Securities Exchange Commission) and by the state-level regulatory agencies of the states in which the IPO shares are offered.

Intellectual Property (IP): The exclusive ownership of original concepts, ideas, designs, engineering plans or other assets that are protected by law. Examples include items covered by trademarks, copyrights and patents. Items such as software, engineering plans, fashion designs and architectural designs, as well as games, books, songs and other entertainment items are among the many things that may be considered to be intellectual property. (Also, see "Patent.")

International Financials Reporting Standards (IFRS): A set of accounting standards established by the International Accounting Standards Board (IASB) for the preparation of public financial statements. IFRS has been adopted by much of the world, including the European Union, Russia and Singapore.

Intimate Apparel: Underclothing such as bras, panties, briefs and lingerie.

Inventory Management: The process of acquiring and maintaining a proper assortment of merchandise while keeping ordering, shipping, handling and other related costs in check. Since it is expensive to own and stock inventory, proper management is vital. Dillard's and Wal-Mart are famous for advanced inventory management.

IP: See "Intellectual Property (IP)."

IT-Enabled Services (ITES): The portion of the Information Technology industry focused on providing business services, such as call centers, insurance claims processing and medical records transcription, by utilizing the power of IT, especially the Internet. Most ITES functions are considered to be back-office procedures. Also, see "Business Process Outsourcing (BPO)."

Jacquard: A fabric with intricate, decorative weaving.

Just-in-Time (JIT) Delivery: Refers to a supply chain practice whereby manufacturers receive components on or just before the time that they are needed on the assembly line, rather than bearing the cost of maintaining several days' or weeks' supply in a warehouse. This adds greatly to the cost-effectiveness of a manufacturing plant and puts the burden of warehousing and timely delivery on the supplier of the components.

Keystone Markup: A markup that is equal to the cost of the merchandise, essentially taking the cost of an item and doubling it to determine the retail price.

Knowledge Process Outsourcing (KPO): The use of outsourced and/or offshore workers to perform business tasks that require judgment and analysis. Examples include such professional tasks as patent research, legal research, architecture, design, engineering, market research, scientific research, accounting and tax return preparation. Also, see "Business Process Outsourcing (BPO)."

LAC: An acronym for Latin America and the Caribbean.

Lawn: A light, crisp cloth made of linen or cotton.

LDCs: See "Least Developed Countries (LDCs)."

Leased Department: A department in a retail store operated by an outside party. The outside party either pays fixed rent or a percentage of sales to the retailer for the space.

Least Developed Countries (LDCs): Nations determined by the U.N. Economic and Social Council to be the poorest and weakest members of the international community. There are currently 50 LDCs, of which 34 are in Africa, 15 are in Asia Pacific and the remaining one (Haiti) is in Latin

America. The top 10 on the LDC list, in descending order from top to 10th, are Afghanistan, Angola, Bangladesh, Benin, Bhutan, Burkina Faso, Burundi, Cambodia, Cape Verde and the Central African Republic. Sixteen of the LDCs are also Landlocked Least Developed Countries (LLDCs) which present them with additional difficulties often due to the high cost of transporting trade goods. Eleven of the LDCs are Small Island Developing States (SIDS), which are often at risk of extreme weather phenomenon (hurricanes, typhoons, Tsunami); have fragile ecosystems; are often dependent on foreign energy sources; can have high disease rates for HIV/AIDS and malaria; and can have poor market access and trade terms.

Leno Weave: A weave in which the wrap yarns are twisted between filling yarns to produce a firm, strong cloth.

Linen: A fiber made from the flax plant.

Logistics Services: Ancillary activities that support or assist logistical operations.

LOHAS: Lifestyles of Health and Sustainability. A marketing term that refers to consumers who choose to purchase and/or live with items that are natural, organic, less polluting, etc. Such consumers may also prefer products powered by alternative energy, such as hybrid cars.

Loom: A textile machine for weaving yarn into cloth.

Loom Finished: Fabric that is sold exactly as it came off the loom.

Loss Leaders: The use of very low prices to attract customers where merchandise may be sold at or below cost.

Lycra: A spandex fiber trademarked by Dupont.

Make-to-Order (MTO): A term applied to products that are manufactured only when an order is confirmed. MTO differs from ETO in that no additional engineering or design is required to make the product.

Make-to-Stock (MTS): Refers to products that are manufactured in anticipation of sales, with inventories of finished goods stocked in warehouses awaiting customer orders.

Manufactured Fiber: Fiber that is manufactured by polymers, minerals, or modified natural fibers.

Manufacturing Resource Planning (MRP II): A methodology that supports effective planning with regard to all resources of a manufacturing company, linking MRP with sales and operations planning, production planning and master production scheduling.

Market Segmentation: The division of a consumer market into specific groups of buyers based on demographic factors.

Marketing: Includes all planning and management activities and expenses associated with the promotion of a product or service. Marketing can encompass advertising, customer surveys, public relations and many other disciplines. Marketing is distinct from selling, which is the process of sell-through to the end user.

Material Requirements Planning (MRP): An integrated management technique that uses bill of material data, inventory data and master production schedules to calculate requirements for material. MRP generates recommendations to release or reschedule replenishment orders for materials.

Melamine: A manufactured fiber in which the fiber-forming substance is a synthetic polymer composed at least 50% by weight of a cross-linked melamine polymer.

Melt Spinning: A process of spinning in which the fiber-forming substance is melted and then extruded through the spinneret.

Mercerization: A treatment that increases the luster of cotton yarn.

Metalized Fabric: Cloth that contains metal.

Metric Count: The number of kilometers per kilogram of yarn.

Metric Ton (MT): A measurement of weight equivalent to 1,000 kilograms.

MFA: See "Multifiber Agreement (MFA)."

Microfiber: An ultra-fine filament that is usually 0.1 tex per filament or 10 microns in diameter.

Millenials: See "Generation Y."

Modacrylic: A manufactured fiber in which the fiber-forming substance is any long-chain synthetic polymer composed of less than 85% but at least 35% by weight of acrylonitrile units.

Mote: A small piece of seed in cotton fabric, usually removed by sodium hydroxide. If not removed, the mote will damage the fabric.

Multifiber Agreement (MFA): The apparel and textiles industry has long been ruled by complex import and export agreements that limit the amounts of particular garments (such as t-shirts, pants and sweaters) and textiles (such as yarns and fabrics) that may be exported to specific markets around the world. In an effort to safeguard domestic production, the United States, Canada and several additional countries now part of the European Union established the Multifiber Agreement (MFA) in 1973. Under the provisions of the agreement, a quota system was put in place that established the maximum numbers of products produced in developing countries that could be legally exported to MFA member countries. These amounts differed from country to country and were based on historic purchasing patterns

Muslin: A wide variety of inexpensive, medium-weight, plain-weave cotton or cotton blend fabric, usually with less than a 180 thread count.

NAFTA: See "North American Free Trade Agreement (NAFTA)."

NAICS: North American Industrial Classification System. See "Industry Code."

Nainsook: A fine, lightweight, plain-weave cotton fabric used in lingerie and blouses.

Nano Textile: See "Smart Fabric."

Natural Fiber: Fibers made of plant, animal or mineral fibers.

Nonwoven Fabrics: Fabrics that are bonded together by entangling fibers mechanically, thermally or chemically. They are flat, porous sheets that are made directly from separate fibers or from molten plastic or plastic film. They are not made by weaving or knitting and do not require converting the fibers to yarn. Typical uses of nonwovens include disposable diapers, carpeting and upholstery fabrics, sterile surgical gowns and drapings, filters and automotive headliners.

North American Free Trade Agreement (NAFTA): A trade agreement signed in December 1992 by U.S. President George H. W. Bush, Canadian Prime Minister Brian Mulroney and Mexican President Carlos Salinas de Gortari. The agreement eliminates tariffs on most goods originating in and traveling between the three member countries. It was approved by the legislatures of the three countries and had entered into force by January 1994. When it was created, NAFTA formed one of the largest free-trade areas of its kind in the world.

Novelty Weave: A variation of plain, satin or twill weaves.

Nylon Fiber: The first synthetic fiber, known for its reliance and strength.

Nytril: A manufactured fiber containing at least 85% of a long-chain polymer of vinylidene dinitrile (—CH2—C(CN)2—) where the vinylidene dinitrile content is no less than every other unit in the polymer chain.

ODM: See "Original Design Manufacturer (ODM)."

OECD: See "Organisation for Economic Co-operation and Development (OECD)."

OEM: See "Original Equipment Manufacturer (OEM)."

Offshoring: The rapidly growing tendency among U.S., Japanese and Western European firms to send knowledge-based and manufacturing work overseas. The intent is to take advantage of lower wages and operating costs in such nations as China, India, Hungary and Russia. The choice of a nation for offshore work may be influenced by such factors as language and education of the local workforce, transportation systems or natural resources. For example, China and India are graduating high numbers of skilled engineers and scientists from their universities. Also, some nations are noted for large numbers of workers skilled in the English language, such as the Philippines and India. Also see "Captive Offshoring" and "Outsourcing."

Onshoring: The opposite of "offshoring." Providing or maintaining manufacturing or services within or nearby a company's domestic location. Sometimes referred to as reshoring.

Organdy: A very thin, stiff, wiry muslin fabric usually used in ties and dresses.

Organisation for Economic Co-operation and Development (OECD): A group of more than 30 nations that are strongly committed to the market economy and democracy. Some of the OECD members include Japan, the U.S., Spain, Germany, Australia, Korea, the U.K., Canada and Mexico. Although not members, Estonia, Israel and Russia are invited to member talks; and Brazil, China, India, Indonesia and South Africa have enhanced engagement policies with the OECD. The Organisation provides statistics, as well as social and economic data; and researches social changes, including patterns in evolving fiscal policy, agriculture, technology, trade, the environment and other areas. It publishes over 250 titles annually; publishes a corporate magazine, the OECD Observer; has radio and TV studios; and has centers in Tokyo, Washington, D.C., Berlin and Mexico City that distributed the Organisation's work and organizes events.

Original Design Manufacturer (ODM): A contract manufacturer that offers complete, end-to-end design, engineering and manufacturing services. ODMs design and build products, such as consumer electronics, that client companies can then brand and sell as their own. For example, a large percentage of laptop computers, cell phones and PDAs are made by ODMs. Also see "Original Equipment Manufacturer (OEM)" and "Contract Manufacturing."

Original Equipment Manufacturer (OEM): A company that manufactures a product or component for sale to a customer that will integrate the component into a final product or assembly. The OEM's customer will distribute the end product or resell it to an end user. For example, a personal computer made under a brand name by a given company may contain various components, such as hard drives, graphics cards or speakers, manufactured by several different OEM "vendors," but the firm doing the final assembly/manufacturing process is the final manufacturer. Also see "Original Design Manufacturer (ODM)" and "Contract Manufacturing."

Ottoman: A heavy cloth with large filling yarns made of cotton or wool, used mainly for coats.

Outbound Logistics: The process related to the movement and storage of products from the end of the production line to the end user.

Outsourcing: The hiring of an outside company to perform a task otherwise performed internally by the company, generally with the goal of lowering costs and/or streamlining work flow. Outsourcing contracts are generally several years in length. Companies that hire outsourced services providers often prefer to focus on their core strengths while sending more routine tasks outside for others to perform. Typical outsourced services include the running of human resources departments, telephone call centers and computer departments. When outsourcing is performed overseas, it may be referred to as offshoring. Also see "Offshoring."

Oxford Cloth: A soft, stout fabric made with a basketweave, most commonly used in shirts.

Patent: An intellectual property right granted by a national government to an inventor to exclude others from making, using, offering for sale, or selling the invention throughout that nation or importing the invention into the nation for a limited time in exchange for public disclosure of the invention when the patent is granted. In addition to national patenting agencies, such as the United States Patent and Trademark Office, and regional organizations such as the European Patent Office, there is a cooperative international patent organization, the World Intellectual Property Organization, or WIPO, established by the United Nations.

PBI: A manufactured fiber in which the fiber-forming substance is a long-chain aromatic polymer having reoccurring imidazole groups as an integral part of the polymer chain.

PBT: See "Polybutylene Terephthalate Fiber (PBT)."

Pebble-Weave Fabric: A fabric with a rough, irregular texture caused by a special weave or yarn.

Percale: A closely woven fabric that generally has an 80x80 thread count, used mostly for sheeting.

Pile: The surface of a textile that has loops or tufts of yarn, used most often in reference to carpets.

Piqué: A medium-weight fabric with raised cords in the direction of the warp.

Plain Weave: Either plain, satin or twill weave.

Plated: Fabric that is woven with thread of two different colors.

Plissé: A fabric with a crinkled or pleated effect made from cotton, rayon or acetate.

Point-of-Sale (POS): The time when a purchase is made.

Poly Lactic Acid: A manufactured fiber, also known as PLA, in which the fiber-forming substance is composed at least 85% by weight of lactic acid ester units derived from naturally occurring sugars.

Polyamide: A synthetic polymer of compounds linked by amides.

Polybutylene Terephthalate Fiber (PBT): A polyester fiber that is noted for its elastic properties, often used in comfort stretch clothing.

Polyester: A category of polymers that contain the ester functional group in their main chain. The most important polyester is polyethylene terephthalate (PET), an extremely common thermoplastic. Polyester fibers are used in a wide variety of fabrics and textiles.

Polyester Fiber: A manufactured fiber in which the fiber-forming substance is any long-chain synthetic polymer composed at least 85% by weight of an ester of a substituted aromatic carboxylic acid, including but not restricted to substituted terephthalic units.

Polyester Stable Fibers: Material used in synthetic fibers, textiles, apparel and other industrial products such as diapers, non-wovens and insulation.

Polytrimethylene Terephthalate Fiber: A synthetic fiber noted for its dyeability and colorfastness, used mostly in carpets.

Pongee: A fabric with a fine weave and a high luster used mostly in underwear.

Poplin: A plain-weave cotton fabric characterized by a rib effect.

POS: See "Point of Sale (POS)."

Power Stretch: Fabric with a 30% to 50% stretch, used mostly in figure-controlling garments.

Pre-Boomer: A term occasionally used to describe people who were born between 1935 and 1945. They are somewhat older than Baby Boomers (born between 1946 and 1962). Also see "Baby Boomer."

Printcloth: A medium-weight cotton plain weave with a thread count between 20 and 40.

Product Lifecycle (Product Life Cycle): The prediction of the life of a product or brand. Stages are described as Introduction, Growth, Maturity and finally Sales Decline. These stages track a product from its initial introduction to the market through to the end of its usefulness as a commercially viable product. The goal of Product Lifecycle Management is to maximize production efficiency, consumer acceptance and profits. Consequently, critical processes around the product need to be adjusted during its lifecycle, including pricing, advertising, promotion, distribution and packaging.

Product Lifecycle Management (PLM): See "Product Lifecycle (Product Life Cycle)."

Projectile Loom: A shuttle-less loom that uses a small projective to pass the yarn through the shed.

Radio Frequency Identification (RFID): A technology that applies a special microchip-enabled tag to an individual item or piece of merchandise or inventory. RFID technology enables wireless, computerized tracking of that inventory item as it moves through the supply chain from factory to transport to warehouse to retail store or end user. Also known as radio tags.

Rapier Loom: A loom that uses a thin metal shaft to pass the thread through the shed.

Ratiné: A loosely constructed cotton fabric with a spongy texture.

Rayon: A manufactured fiber composed of regenerated cellulose, as well as manufactured fibers composed of regenerated cellulose in which substituents have replaced not more than 15% of the hydrogens of the hydroxyl groups. Where the fiber is composed of cellulose precipitated from an organic solution in which no substitution of the hydroxyl groups takes place and no chemical intermediates are formed, the term lyocell may be used as a generic description of the fiber.

Ready-To-Wear: Designer clothes that are mass produced and sold off the rack in retail stores.

Reed: A comb-like device in a loom that beats threads against the woven fabric.

Regional Center: A shopping center or mall which includes up to three department stores plus shopping or specialty stores rather than convenience stores. Super-regionals are similar but have at least four department stores and at least 1 million square feet of retail space.

Reshore: With regard to manufacturing, the return of some of the business to plants based in the country where sales are typically made. An example would be if Nike should shift some of its production from Asia to the U.S.

Retail Chain: A firm that consists of multiple retail units under common ownership and usually has some centralization of decision-making in defining and implementing its strategy.

Reverse Logistics: A specialized segment of logistics focusing on the movement and management of products and resources after sale and delivery to the customer. Includes product returns and repair for credit.

RFID: See “Radio Frequency Identification (RFID).”

Ribbing: A corded effect on the cloth, either vertical or horizontal.

Rubber: (1) A manufactured fiber in which the fiber-forming substance is a hydrocarbon such as natural rubber, polyisoprene, polybutadiene, copolymers of dienes and hydrocarbons, or amorphous (noncrystalline) polyolefins; (2) a manufactured fiber in which the fiber-forming substance is a copolymer of acrylonitrile and a diene (such as butadiene) composed of not more than 50% but at least 10% by weight of acrylonitrile units; or (3) a manufactured fiber in which the fiber-forming substance is a polychloroprene or a copolymer of chloroprene in which at least 35% by weight of the fiber-forming substance is composed of chloroprene units.

Running Bale (RB): One RB equals approximately 1.056 Bales or a 507-lb. Bale. See “Bale.”

SaaS: See “Software as a Service (SaaS).”

Sailcloth: A heavy canvas-like fabric made of cotton.

Sales Per Square Foot of Selling Space: Net sales divided by square feet of selling space.

Same-Store Sales: Sales dollars generated only by those stores that have been open more than a year and have historical data to compare this year's sales to the same time-frame last year.

Saran: A manufactured fiber in which the fiber-forming substance is any long-chain synthetic polymer composed at least 80% by weight of vinylidene chloride units.

Sateen: A strong, satin woven fabric with a high sheen.

Satin Weave: A basic weave that has long floats of yarn on the fabric face. A float is created when yarn is interlaced over the weft producing a smooth shiny surface.

SCM: See “Supply Chain Management.”

Seersucker: A lightweight fabric with crinkled stripes, usually made from cotton or manufactured fibers.

Selvage: The edge of a woven fabric that runs parallel to the warp and prevents unraveling.

Semi-Couture: Machine made clothes made with fine materials in limited quantities.

Serge: A woolen fabric with a twill weave.

Shed: In a loom, an opening between warps that permits the shuttle to be passed through.

Shuttle: In a loom, a bobbin that carries thread between the warp.

SIC: Standard Industrial Classification. See “Industry Code.”

Six Sigma: A quality enhancement strategy designed to reduce the number of products coming from a manufacturing plant that do not conform to specifications. Six Sigma states that no more than 3.4 defects per million parts is the goal of high-quality output. Motorola invented the system in the 1980s in order to enhance its competitive position against Japanese electronics manufacturers.

Slub Yarn: A yarn of irregular diameter or color.

Smart Fabric: A fabric that has integral, active functionality. Smart fabrics can include those that can generate or store electricity, have a human interface, include wireless devices such as sensors or RFID, or incorporate some sort of assistive technology. Smart fabrics may be able to sense and react to the environment. By some definitions, smart fabrics may be those that incorporate nanotechnology ("nanotextiles"), incorporate certain elements to resist bacteria, shed dirt or stains, or resist odors.

Soft Goods: Apparel and linens.

Software as a Service (SaaS): Refers to the practice of providing users with software applications that are hosted on remote servers and accessed via the Internet. Excellent examples include the CRM (Customer Relationship Management) software provided in SaaS format by Salesforce. An earlier technology that operated in a similar, but less sophisticated, manner was called ASP or Application Service Provider.

Spandex: A manufactured, lightweight, highly elastic fabric used mostly in sporting goods for its ability to stretch and its resistance to water.

Specialty Department Store: A store with a department store format that focuses on apparel and soft home goods (e.g., Neiman Marcus).

Specialty Store: A store specializing in one category of merchandising, frequently fashion-related.

Spindle Hours: Spindle hours are the number of machine hours fibers were spun into yarn/thread.

Sportswear: Apparel worn for sport or casual purposes.

Standard Moisture Regain: The standard for acceptable moisture gain expressed as a percentage of the cloth's dry weight.

Subsidiary, Wholly-Owned: A company that is wholly controlled by another company through stock ownership.

Sulfar: A manufactured fiber in which the fiber-forming substance is a long-chain synthetic polysulfide in which at least 85% of the sulfide (—S—) linkages are attached directly to two aromatic rings.

Supply Chain: The complete set of suppliers of goods and services required for a company to operate its business. For example, a manufacturer's supply chain may include providers of raw materials, components, custom-made parts and packaging materials.

Supply Chain Management: The coordination of all supply chain activities. Also referred to as "SCM."

Swatch: A sample of fabric used to judge color or quality.

Technical Textiles: Textiles designed for special uses such as industrial, aerospace, military, marine, medical, construction, geotextile, transportation and high-technology applications.

Tex: A measurement of yarn density expressed in grams per kilometer.

Textronics: Textile electronics; new kinds of fabric that are woven with nano-engineered electronics. Also know as "smart fabrics," these materials would have the ability to react to their environment by changing color, shape or electronic properties.

Third-Party Logistics (3PL): A specialist firm in logistics, which may provide a variety of transportation, warehousing and logistics-related services to buyers or sellers. These tasks were previously performed in-house by the customer. When 3PL services are provided within the client's own facilities, it can also be referred to as insourcing. Also see "Fourth-Party Logistics (4PL)."

Thread Count: The number of threads per square inch in a fabric. The higher the thread count, the finer the fabric.

Ticking: A tightly woven, durable fabric used to cover mattresses, pillow and work clothes, usually made of cotton.

Tire Cord: A textile that includes high tenacity yarn of nylon, polyamides, polyester and viscose rayon.

To Order: Handmade clothes made by designers who do not have the haute couture designation.

Trade Secrets: Proprietary formulas, research, manuals, methods of operation and concepts owned by a business. Generally trade secrets are carefully guarded by the owner and are shared only on a need-to-know basis.

Twill Weave: A basic weave constructed by interlacing warp and filling yarns, which produces a diagonal effect.

Twist: A measurement of the number and the direction of turns in yarn manufacturing.

Unbalanced Plain Weave: A plain weave that uses heavier yarn in one direction than the other.

Union Cloth: A textile woven of two or more cloths.

Value Added Tax (VAT): A tax that imposes a levy on businesses at every stage of manufacturing based on the value it adds to a product. Each business in the supply chain pays its own VAT and is subsequently repaid by the next link down the chain; hence, a VAT is ultimately paid by the consumer, being the last link in the supply chain, making it comparable to a sales tax. Generally, VAT only applies to goods bought for consumption within a given country; export goods are exempt from VAT, and purchasers from other countries taking goods back home may apply for a VAT refund.

Velour: A medium-weight, closely woven fabric with a thick pile. It closely resembles velvet.

Vendor: Any firm, such as a manufacturer or distributor, from which a retailer obtains merchandise.

Vinyl: A manufactured fiber in which the fiber-forming substance is any long-chain synthetic polymer composed at least 50% by weight of vinyl alcohol units, and in which the total of the vinyl alcohol units and any one or more of the various acetal units is at least 85% by weight of the fiber.

Warehouse Store: A discount retailer offering merchandise in a no-frills environment. These stores often cut costs by reducing services and product variety.

Warp Knit: A type of knitted fabric construction in which the yarns are formed into stitches in a lengthwise manner. Warp knits are generally less elastic than weft knits.

Weft Knit: A type of knitted fabric in which yarns are formed into stitches in a widthwise manner.

Wet Spinning: A fiber-forming process in which a spinneret is immersed in a chemical bath in which the fiber-forming chemicals are dissolved. As the spinneret is withdrawn from the bath, the fibers are drawn out through the holes in the head of the spinneret and solidify.

Wholesale Club: A retail store that sells a limited assortment of general merchandise to customers who are members of the club. Memberships are generally fee-based, margins are small and there is little customer service provided to the members. Groceries are often sold in bulk sizes. The largest wholesale club chain is Sam's, owned by Wal-Mart. Costco is Sam's biggest competitor in the U.S.

World Trade Organization (WTO): One of the only globally active international organizations dealing with the trade rules between nations. Its goal is to assist the free flow of trade goods, ensuring a smooth, predictable supply of goods to help raise the quality of life of member citizens. Members form consensus decisions that are then ratified by their respective parliaments. The WTO's conflict resolution process generally emphasizes interpreting existing commitments and agreements, and discovers how to ensure trade policies to conform to those agreements, with the ultimate aim of avoiding military or political conflict.

Worsted Fabric: A tightly woven wool or wool-blend fabric such as gabardine.

Woven Fabrics: Fabrics that are created by weaving or knitting. In some cases this requires converting basic fibers into yarn prior to weaving.

WTO: See "World Trade Organization (WTO)."

CPSIA information can be obtained at www.ICGtesting.com
Printed in the USA
LVOW021513050713

341612LV00016B/838/P

9 781608 797028